Springer Texts in Statistics

Advisors:
George Casella   Stephen Fienberg   Ingram Olkin

Springer Science+Business Media, LLC

Springer Texts in Statistics

Alfred: Elements of Statistics for the Life and Social Sciences
Berger: An Introduction to Probability and Stochastic Processes
Bilodeau and Brenner: Theory of Multivariate Statistics
Blom: Probability and Statistics: Theory and Applications
Brockwell and Davis: Introduction to Times Series and Forecasting, Second Edition
Chow and Teicher: Probability Theory: Independence, Interchangeability, Martingales, Third Edition
Christensen: Advanced Linear Modeling: Multivariate, Time Series, and Spatial Data: Nonparametric Regression and Response Surface Maximization, Second Edition
Christensen: Log-Linear Models and Logistic Regression, Second Edition
Christensen: Plane Answers to Complex Questions: The Theory of Linear Models, Third Edition
Creighton: A First Course in Probability Models and Statistical Inference
Davis: Statistical Methods for the Analysis of Repeated Measurements
Dean and Voss: Design and Analysis of Experiments
du Toit, Steyn, and Stumpf: Graphical Exploratory Data Analysis
Durrett: Essentials of Stochastic Processes
Edwards: Introduction to Graphical Modelling, Second Edition
Finkelstein and Levin: Statistics for Lawyers
Flury: A First Course in Multivariate Statistics
Jobson: Applied Multivariate Data Analysis, Volume I: Regression and Experimental Design
Jobson: Applied Multivariate Data Analysis, Volume II: Categorical and Multivariate Methods
Kalbfleisch: Probability and Statistical Inference, Volume I: Probability, Second Edition
Kalbfleisch: Probability and Statistical Inference, Volume II: Statistical Inference, Second Edition
Karr: Probability
Keyfitz: Applied Mathematical Demography, Second Edition
Kiefer: Introduction to Statistical Inference
Kokoska and Nevison: Statistical Tables and Formulae
Kulkarni: Modeling, Analysis, Design, and Control of Stochastic Systems
Lange: Applied Probability
Lehmann: Elements of Large-Sample Theory
Lehmann: Testing Statistical Hypotheses, Second Edition
Lehmann and Casella: Theory of Point Estimation, Second Edition
Lindman: Analysis of Variance in Experimental Design
Lindsey: Applying Generalized Linear Models

(continued after index)

Angela Dean   Daniel Voss

Design and Analysis of Experiments

With 83 Illustrations

Springer

Angela Dean
Departments of Statistics
Ohio State University
Columbus, OH 43210
USA
amd@stat.ohio-state.edu

Daniel Voss
Department of Mathematics and Statistics
Wright State University
Dayton, OH 45435
USA
dvoss@math.wright.edu

Editoral Board

George Casella
Department of Statistics
University of Florida
Gainesville, FL 32611
USA

Stephen Fienberg
Department of Statistics
Carnegie Mellon University
Pittsburgh, PA 15213
USA

Ingram Olkin
Department of Statistics
Stanford University
Stanford, CA 94305
USA

SAS®, SAS/STAT®, and JMP® are registered trademarks of the SAS Institute, Inc.
Minitab® is a registered trademark of Minitab Inc.
Robust Design® is a registered trademark of the American Supplier Institue.

Library of Congress Cataloging-in-Publication Data
Dean, A.M.
   Design and analysis of experiments / A.M. Dean, D.T. Voss.
       p.   cm. — (Springer texts in statistics)
   Includes bibliographical references (p. –) and index.
   ISBN 978-1-4757-7292-0   ISBN 978-0-387-22634-7 (eBook)
   DOI 10.1007/978-0-387-22634-7

   1. Experimental design.   I. Voss, D.T.   II. Title.   III. Series.
QA279.D43   1999
001.4'34—dc21   98-20302

ISBN 978-1-4757-7292-0   Printed on acid-free paper.

© 1999 Springer Science+Business Media New York
Originally published by Springer-Verlag New York, Inc. in 1999
Softcover reprint of the hardcover 1st edition 1999

All rights reserved. This work may not be translated or copied in whole or in part without the written permission of the publisher Springer Science+Business Media, LLC ,
except for brief excerpts in connection with reviews or scholarly analysis. Use in
connection with any form of information storage and retrieval, electronic adaptation, computer software, or by similar or dissimilar methodology now known or hereafter developed is forbidden.
The use of general descriptive names, trade names, trademarks, etc., in this publication, even if the former are not especially identified, is not to be taken as a sign that such names, as understood by the Trade Marks and Merchandise Marks Act, may accordingly be used freely by anyone.

9 8 7 6 5 4   SPIN 11013419

springeronline.com

Preface

Our initial motivation for writing this book was the observation from various students that the subject of design and analysis of experiments can seem like "a bunch of miscellaneous topics." We believe that the identification of the objectives of the experiment and the practical considerations governing the design form the heart of the subject matter and serve as the link between the various analytical techniques. We also believe that learning about design and analysis of experiments is best achieved by the planning, running, and analyzing of a simple experiment.

With these considerations in mind, we have included throughout the book the details of the planning stage of several experiments that were run in the course of teaching our classes. The experiments were run by students in statistics and the applied sciences and are sufficiently simple that it is possible to discuss the planning of the entire experiment in a few pages, and the procedures can be reproduced by readers of the book. In each of these experiments, we had access to the investigators' actual report, including the difficulties they came across and how they decided on the treatment factors, the needed number of observations, and the layout of the design. In the later chapters, we have included details of a number of published experiments. The outlines of many other student and published experiments appear as exercises at the ends of the chapters.

Complementing the practical aspects of the design are the statistical aspects of the analysis. We have developed the theory of estimable functions and analysis of variance with some care, but at a low mathematical level. Formulae are provided for almost all analyses so that the statistical methods can be well understood, related design issues can be discussed, and computations can be done by hand in order to check computer output.

We recommend the use of a sophisticated statistical package in conjunction with the book. Use of software helps to focus attention on the statistical issues rather than on the calculation. Our particular preference is for the SAS software, and we have included the elementary use of this package at the end of most chapters. Many of the SAS program files and data sets used in the book can be found at www.springer-ny.com. However, the book can equally well be used with any other statistical package. Availability of statistical software has also helped shape the book in that we can discuss more complicated analyses—the analysis of unbalanced designs, for example.

The level of presentation of material is intended to make the book accessible to a wide audience. Standard linear models under normality are used for all analyses. We have avoided using calculus, except in a few optional sections where least squares estimators are obtained. We have also avoided using linear algebra, except in an optional section on the canonical analysis of second-order response surface designs. Contrast coefficients are listed in the form of a vector, but these are interpreted merely as a list of coefficients.

This book reflects a number of personal preferences. First and foremost, we have not put side conditions on the parameters in our models. The reason for this is threefold. Firstly, when side conditions are added to the model, all the parameters appear to be estimable. Consequently, one loses the perspective that in factorial experiments, main effects can be interpreted only as averages over any interactions that happen to be present. Secondly, the side conditions that are the most useful for hand calculation do not coincide with those used by the SAS software. Thirdly, if one feeds a nonestimable parametric function into a computer program such as PROC GLM in SAS, the program will declare the function to be "nonestimable," and the user needs to be able to interpret this statement. A consequence is that the traditional solutions to the normal equations do not arise naturally. Since the traditional solutions are for nonestimable parameters, we have tried to avoid giving these, and instead have focused on the estimation of functions of $E[Y]$, all of which are estimable.

We have concentrated on the use of prespecified models and preplanned analyses rather than exploratory data analysis. We have emphasized the experimentwise control of error rates and confidence levels rather than individual error rates and confidence levels.

We rely upon residual plots rather than formal tests to assess model assumptions. This is because of the additional information provided by residual plots. For example, plots to check homogeneity of variance also indicate when a variance-stabilizing transformation should be effective. Likewise, nonlinear patterns in a normal probability plot may indicate whether inferences under normality are likely to be liberal or conservative. Except for some tests for lack of fit, we have, in fact, omitted all details of formal testing for model assumptions, even though they are readily available in many computer packages.

The book starts with basic principles and techniques of experimental design and analysis of experiments. It provides a checklist for the planning of experiments, and covers analysis of variance, inferences for treatment contrasts, regression, and analysis of covariance. These basics are then applied in a wide variety of settings. Designs covered include completely randomized designs, complete and incomplete block designs, row-column designs, single replicate designs with confounding, fractional factorial designs, response surface designs, and designs involving nested factors and factors with random effects, including split-plot designs.

In the last few years, "Taguchi methods" have become very popular for industrial experimentation, and we have incorporated some of these ideas. Rather than separating Taguchi methods as special topics, we have interspersed them throughout the chapters via the notion of including "noise factors" in an experiment and analyzing the variability of the response as the noise factors vary.

We have introduced factorial experiments as early as Chapter 3, but analyzed them as one-way layouts (i.e., using a cell means model). The purpose is to avoid introducing factorial experiments halfway through the book as a totally new topic, and to emphasize that many factorial experiments are run as completely randomized designs. We have analyzed contrasts in a two-factor experiment both via the usual two-way analysis of variance model (where the contrasts are in terms of the main effect and interaction parameters) and also via a cell-means model (where the contrasts are in terms of the treatment combination parameters). The purpose of this is to lay the groundwork for Chapters 13–15, where these contrasts are used in confounding and fractions. It is also the traditional notation used in conjunction with Taguchi methods.

The book is not all-inclusive. For example, we do not cover recovery of interblock information for incomplete block designs with random block effects. We do not provide extensive tables of incomplete block designs. Also, careful coverage of unbalanced models involving random effects is beyond our scope. Finally, inclusion of SAS graphics is limited to low-resolution plots.

The book has been classroom tested successfully over the past five years at The Ohio State University, Wright State University, and Kenyon College, for junior and senior undergraduate students majoring in a variety of fields, first-year graduate students in statistics, and senior graduate students in the applied sciences. These three institutions are somewhat different. The Ohio State University is a large land-grant university offering degrees through the Ph.D., Wright State University is a mid-sized university with few Ph.D. programs, and Kenyon College is a liberal arts undergraduate college. Below we describe typical syllabi that have been used.

At OSU, classes meet for five hours per week for ten weeks. A typical class is composed of 35 students, about a third of whom are graduate students in the applied statistics master's program. The remaining students are undergraduates in the mathematical sciences or graduate students in industrial engineering, biomedical engineering, and various applied sciences. The somewhat ambitious syllabus covers Chapters 1–7 and 10, Sections 11.1–11.4, and Chapters 13, 15, and 17. Students taking these classes plan, run, and analyze their own experiments, usually in a team of four or five students from several different departments. This project serves the function of giving statisticians the opportunity of working with scientists and seeing the experimental procedure firsthand, and gives the scientists access to colleagues with a broader statistical training. The experience is usually highly rated by the student participants.

Classes at WSU meet four hours per week for ten weeks. A typical class involves about 10 students who are either in the applied statistics master's degree program or who are undergraduates majoring in mathematics with a statistics concentration. Originally, two quarters (20 weeks) of probability and statistics formed the prerequisite, and the course covered much of Chapters 1–4, 6, 7, 10, 11, and 13, with Chapters 3 and 4 being primarily review material. Currently, students enter with two additional quarters in applied linear models, including regression, analysis of variance, and methods of multiple comparisons, and the course covers Chapters 1 and 2, Sections 3.2, 6.7, and 7.5, Chapters 10, 11, and 13, Sections 15.1–15.2, and perhaps Chapter 16. As at OSU, both of these syllabi are ambitious. During the second half of the course, the students plan, run, and analyze their own experiments, working in groups of one to three. The students provide written and oral reports on the projects, and the discussions during the oral reports are of mutual enjoyment and benefit. A leisurely topics course has also been offered as a sequel, covering the rest of Chapters 14–17.

At Kenyon College, classes meet for three hours a week for 15 weeks. A typical class is composed of about 10 junior and senior undergraduates majoring in various fields. The syllabus covers Chapters 1–7, 10, and 17.

For some areas of application, random effects, nested models, and split-plot designs, which are covered in Chapters 17–19, are important topics. It is possible to design a syllabus that reaches these chapters fairly rapidly by covering Chapters 1–4, 6, 7, 17, 18, 10, 19.

We owe a debt of gratitude to many. For reading of, and comments on, prior drafts, we thank Bradley Hartlaub, Jeffrey Nunemacher, Mark Irwin, an anonymous reviewer, and the many students who suffered through the early drafts. We thank Baoshe An, James Clark, Amy Ferketich, and Dionne Pratt for checking a large number of exercises, and Lisa Abrams, Paul Burte, Kathryn Collins, Yuming Deng, Joseph Mesaros, Dionne Pratt, Kari Rabe, Joseph Whitmore, and many others for catching numerous typing errors. We are grateful to Peg Steigerwald, Terry England, Dolores Wills, Jill McClane, and Brian J. Williams for supplying hours of typing skills. We extend our thanks to all the many students in classes at The Ohio State University, Wright State University, and the University of Wisconsin at Madison whose imagination and diligence produced so many wonderful experiments; also to Brian H. Williams and Bob Wardrop for supplying data sets; to Nathan Buurma, Colleen Brensinger, and James Colton for library searches; and to the publishers and journal editors who gave us permission to use data and descriptions of experiments. We are especially grateful to the SAS Institute for permission to reproduce portions of SAS programs and corresponding output, and to John Kimmel for his patience and encouragement throughout this endeavor.

This book has been ten years in the making. In the view of the authors, it is "a work in progress temporarily cast in stone"—or in print, as it were. We are wholly responsible for any errors and omissions, and we would be most grateful for comments, corrections, and suggestions from readers so that we can improve any future editions.

Finally, we extend our love and gratitude to Jeff, Nancy, Tommy, and Jimmy, often neglected during this endeavor, for their enduring patience, love, and support.

Angela Dean
Columbus, Ohio

Daniel Voss
Dayton, Ohio

Contents

Preface v

1. Principles and Techniques 1

    1.1. Design: Basic Principles and Techniques 1
        1.1.1. The Art of Experimentation 1
        1.1.2. Replication 2
        1.1.3. Blocking 3
        1.1.4. Randomization 3
    1.2. Analysis: Basic Principles and Techniques 5

2. Planning Experiments 7

    2.1. Introduction 7
    2.2. A Checklist for Planning Experiments 7
    2.3. A Real Experiment—Cotton-Spinning Experiment 14
    2.4. Some Standard Experimental Designs 17
        2.4.1. Completely Randomized Designs 18
        2.4.2. Block Designs 18
        2.4.3. Designs with Two or More Blocking Factors 19
        2.4.4. Split-Plot Designs 21
    2.5. More Real Experiments 22
        2.5.1. Soap Experiment 22
        2.5.2. Battery Experiment 26
        2.5.3. Cake-Baking Experiment 29
        Exercises 31

3. Designs with One Source of Variation 33

    3.1. Introduction 33
    3.2. Randomization 34
    3.3. Model for a Completely Randomized Design 35
    3.4. Estimation of Parameters 37

		3.4.1.	Estimable Functions of Parameters	37
		3.4.2.	Notation	37
		3.4.3.	Obtaining Least Squares Estimates	38
		3.4.4.	Properties of Least Squares Estimators	40
		3.4.5.	Estimation of σ^2	42
		3.4.6.	Confidence Bound for σ^2	43
	3.5.	One-Way Analysis of Variance	44
		3.5.1.	Testing Equality of Treatment Effects	44
		3.5.2.	Use of p-Values	48
	3.6.	Sample Sizes	49
		3.6.1.	Expected Mean Squares for Treatments	50
		3.6.2.	Sample Sizes Using Power of a Test	51
	3.7.	A Real Experiment—Soap Experiment, Continued	53
		3.7.1.	Checklist, Continued	53
		3.7.2.	Data Collection and Analysis	54
		3.7.3.	Discussion by the Experimenter	56
		3.7.4.	Further Observations by the Experimenter	56
	3.8.	Using SAS Software	57
		3.8.1.	Randomization	57
		3.8.2.	Analysis of Variance	58
		Exercises	61

4. Inferences for Contrasts and Treatment Means   67

4.1.	Introduction		67
4.2.	Contrasts		68
	4.2.1.	Pairwise Comparisons	69
	4.2.2.	Treatment Versus Control	70
	4.2.3.	Difference of Averages	70
	4.2.4.	Trends	71
4.3.	Individual Contrasts and Treatment Means		73
	4.3.1.	Confidence Interval for a Single Contrast	73
	4.3.2.	Confidence Interval for a Single Treatment Mean	75
	4.3.3.	Hypothesis Test for a Single Contrast or Treatment Mean . .	75
4.4.	Methods of Multiple Comparisons		78
	4.4.1.	Multiple Confidence Intervals	78
	4.4.2.	Bonferroni Method for Preplanned Comparisons	80
	4.4.3.	Scheffé Method of Multiple Comparisons	83
	4.4.4.	Tukey Method for All Pairwise Comparisons	85
	4.4.5.	Dunnett Method for Treatment-Versus-Control Comparisons .	87
	4.4.6.	Hsu Method for Multiple Comparisons with the Best Treatment	89
	4.4.7.	Combination of Methods	91
	4.4.8.	Methods Not Controlling Experimentwise Error Rate	92

	4.5.	Sample Sizes	92
	4.6.	Using SAS Software	94
		4.6.1. Inferences on Individual Contrasts	94
		4.6.2. Multiple Comparisons	96
		Exercises	97

5. Checking Model Assumptions   103

	5.1.	Introduction	103
	5.2.	Strategy for Checking Model Assumptions	104
		5.2.1. Residuals	104
		5.2.2. Residual Plots	105
	5.3.	Checking the Fit of the Model	107
	5.4.	Checking for Outliers	107
	5.5.	Checking Independence of the Error Terms	109
	5.6.	Checking the Equal Variance Assumption	111
		5.6.1. Detection of Unequal Variances	112
		5.6.2. Data Transformations to Equalize Variances	113
		5.6.3. Analysis with Unequal Error Variances	116
	5.7.	Checking the Normality Assumption	119
	5.8.	Using SAS Software	122
		5.8.1. Using SAS to Generate Residual Plots	122
		5.8.2. Transforming the Data	126
		Exercises	127

6. Experiments with Two Crossed Treatment Factors   135

	6.1.	Introduction	135
	6.2.	Models and Factorial Effects	136
		6.2.1. The Meaning of Interaction	136
		6.2.2. Models for Two Treatment Factors	138
		6.2.3. Checking the Assumptions on the Model	140
	6.3.	Contrasts	141
		6.3.1. Contrasts for Main Effects and Interactions	141
		6.3.2. Writing Contrasts as Coefficient Lists	143
	6.4.	Analysis of the Two-Way Complete Model	145
		6.4.1. Least Squares Estimators for the Two-Way Complete Model	146
		6.4.2. Estimation of σ^2 for the Two-Way Complete Model	147
		6.4.3. Multiple Comparisons for the Complete Model	149
		6.4.4. Analysis of Variance for the Complete Model	152
	6.5.	Analysis of the Two-Way Main-Effects Model	158
		6.5.1. Least Squares Estimators for the Main-Effects Model	158
		6.5.2. Estimation of σ^2 in the Main-Effects Model	162
		6.5.3. Multiple Comparisons for the Main-Effects Model	163
		6.5.4. Unequal Variances	165

		6.5.5.	Analysis of Variance for Equal Sample Sizes	165
		6.5.6.	Model Building	168
	6.6.		Calculating Sample Sizes	168
	6.7.		Small Experiments	169
		6.7.1.	One Observation per Cell	169
		6.7.2.	Analysis Based on Orthogonal Contrasts	169
		6.7.3.	Tukey's Test for Additivity	172
		6.7.4.	A Real Experiment—Air Velocity Experiment	173
	6.8.		Using SAS Software	175
		6.8.1.	Contrasts and Multiple Comparisons	177
		6.8.2.	Plots	181
		6.8.3.	One Observation per Cell	182
		Exercises		183

7. Several Crossed Treatment Factors — 193

7.1.	Introduction			193
7.2.	Models and Factorial Effects			194
	7.2.1.	Models		194
	7.2.2.	The Meaning of Interaction		195
	7.2.3.	Separability of Factorial Effects		197
	7.2.4.	Estimation of Factorial Contrasts		199
7.3.	Analysis—Equal Sample Sizes			201
7.4.	A Real Experiment—Popcorn–Microwave Experiment			205
7.5.	One Observation per Cell			211
	7.5.1.	Analysis Assuming That Certain Interaction Effects Are Negligible		211
	7.5.2.	Analysis Using Normal Probability Plot of Effect Estimates		213
	7.5.3.	Analysis Using Confidence Intervals		215
7.6.	Design for the Control of Noise Variability			217
	7.6.1.	Analysis of Design-by-Noise Interactions		218
	7.6.2.	Analyzing the Effects of Design Factors on Variability		221
7.7.	Using SAS Software			223
	7.7.1.	Normal Probability Plots of Contrast Estimates		224
	7.7.2.	Voss–Wang Confidence Interval Method		224
	7.7.3.	Identification of Robust Factor Settings		226
	7.7.4.	Experiments with Empty Cells		227
	Exercises			231

8. Polynomial Regression — 243

8.1.	Introduction		243
8.2.	Models		244
8.3.	Least Squares Estimation (Optional)		248
	8.3.1.	Normal Equations	248

		8.3.2. Least Squares Estimates for Simple Linear Regression	248
	8.4.	Test for Lack of Fit	249
	8.5.	Analysis of the Simple Linear Regression Model	251
	8.6.	Analysis of Polynomial Regression Models	255
		8.6.1. Analysis of Variance	255
		8.6.2. Confidence Intervals	257
	8.7.	Orthogonal Polynomials and Trend Contrasts (Optional)	258
		8.7.1. Simple Linear Regression	258
		8.7.2. Quadratic Regression	260
		8.7.3. Comments	261
	8.8.	A Real Experiment—Bean-Soaking Experiment	262
		8.8.1. Checklist	262
		8.8.2. One-Way Analysis of Variance and Multiple Comparisons	264
		8.8.3. Regression Analysis	267
	8.9.	Using SAS Software	268
		Exercises	273

9. Analysis of Covariance   277

	9.1.	Introduction	277
	9.2.	Models	278
		9.2.1. Checking Model Assumptions and Equality of Slopes	279
		9.2.2. Model Extensions	279
	9.3.	Least Squares Estimates	280
		9.3.1. Normal Equations (Optional)	280
		9.3.2. Least Squares Estimates and Adjusted Treatment Means	281
	9.4.	Analysis of Covariance	282
	9.5.	Treatment Contrasts and Confidence Intervals	286
		9.5.1. Individual Confidence Intervals	286
		9.5.2. Multiple Comparisons	287
	9.6.	Using SAS Software	288
		Exercises	292

10. Complete Block Designs   295

	10.1.	Introduction	295
	10.2.	Blocks, Noise Factors or Covariates?	296
	10.3.	Design Issues	297
		10.3.1. Block Sizes	297
		10.3.2. Complete Block Design Definitions	298
		10.3.3. The Randomized Complete Block Design	299
		10.3.4. The General Complete Block Design	300
		10.3.5. How Many Observations?	301
	10.4.	Analysis of Randomized Complete Block Designs	301
		10.4.1. Model and Analysis of Variance	301

		10.4.2. Multiple Comparisons	305
	10.5.	A Real Experiment—Cotton-Spinning Experiment	306
		10.5.1. Design Details	306
		10.5.2. Sample-Size Calculation	307
		10.5.3. Analysis of the Cotton-Spinning Experiment	307
	10.6.	Analysis of General Complete Block Designs	309
		10.6.1. Model and Analysis of Variance	309
		10.6.2. Multiple Comparisons for the General Complete Block Design	312
		10.6.3. Sample-Size Calculations	315
	10.7.	Checking Model Assumptions	316
	10.8.	Factorial Experiments	317
	10.9.	Using SAS Software	320
		Exercises	324

11. Incomplete Block Designs — 339

	11.1.	Introduction	339
	11.2.	Design Issues	340
		11.2.1. Block Sizes	340
		11.2.2. Design Plans and Randomization	340
		11.2.3. Estimation of Contrasts (Optional)	342
		11.2.4. Balanced Incomplete Block Designs	343
		11.2.5. Group Divisible Designs	345
		11.2.6. Cyclic Designs	346
	11.3.	Analysis of General Incomplete Block Designs	348
		11.3.1. Contrast Estimators and Multiple Comparisons	348
		11.3.2. Least Squares Estimation (Optional)	351
	11.4.	Analysis of Balanced Incomplete Block Designs	354
		11.4.1. Multiple Comparisons and Analysis of Variance	354
		11.4.2. A Real Experiment—Detergent Experiment	355
	11.5.	Analysis of Group Divisible Designs	360
		11.5.1. Multiple Comparisons and Analysis of Variance	360
	11.6.	Analysis of Cyclic Designs	362
	11.7.	A Real Experiment—Plasma Experiment	362
	11.8.	Sample Sizes	368
	11.9.	Factorial Experiments	369
		11.9.1. Factorial Structure	369
	11.10.	Using SAS Software	372
		11.10.1. Analysis of Variance and Estimation of Contrasts	372
		11.10.2. Plots	377
		Exercises	378

12. Designs with Two Blocking Factors — 387

- 12.1. Introduction — 387
- 12.2. Design Issues — 388
  - 12.2.1. Selection and Randomization of Row–Column Designs — 388
  - 12.2.2. Latin Square Designs — 389
  - 12.2.3. Youden Designs — 391
  - 12.2.4. Cyclic and Other Row–Column Designs — 392
- 12.3. Model for a Row–Column Design — 394
- 12.4. Analysis of Row–Column Designs (Optional) — 395
  - 12.4.1. Least Squares Estimation (Optional) — 395
  - 12.4.2. Solution for Complete Column Blocks (Optional) — 397
  - 12.4.3. Formula for ssE (Optional) — 398
  - 12.4.4. Analysis of Variance for a Row–Column Design (Optional) — 399
  - 12.4.5. Confidence Intervals and Multiple Comparisons — 401
- 12.5. Analysis of Latin Square Designs — 401
  - 12.5.1. Analysis of Variance for Latin Square Designs — 401
  - 12.5.2. Confidence Intervals for Latin Square Designs — 403
  - 12.5.3. How Many Observations? — 405
- 12.6. Analysis of Youden Designs — 406
  - 12.6.1. Analysis of Variance for Youden Designs — 406
  - 12.6.2. Confidence Intervals for Youden Designs — 407
  - 12.6.3. How Many Observations? — 407
- 12.7. Analysis of Cyclic and Other Row–Column Designs — 408
- 12.8. Checking the Assumptions on the Model — 409
- 12.9. Factorial Experiments in Row–Column Designs — 410
- 12.10. Using SAS Software — 410
  - 12.10.1. Factorial Model — 413
  - 12.10.2. Plots — 415
  - Exercises — 415

13. Confounded Two-Level Factorial Experiments — 421

- 13.1. Introduction — 421
- 13.2. Single replicate factorial experiments — 422
  - 13.2.1. Coding and notation — 422
  - 13.2.2. Confounding — 422
  - 13.2.3. Analysis — 423
- 13.3. Confounding Using Contrasts — 424
  - 13.3.1. Contrasts — 424
  - 13.3.2. Experiments in Two Blocks — 425
  - 13.3.3. Experiments in Four Blocks — 430
  - 13.3.4. Experiments in Eight Blocks — 432
  - 13.3.5. Experiments in More Than Eight Blocks — 433

13.4.	Confounding Using Equations	433
	13.4.1. Experiments in Two Blocks	433
	13.4.2. Experiments in More Than Two Blocks	435
13.5.	A Real Experiment—Mangold Experiment	437
13.6.	Plans for Confounded 2^p Experiments	441
13.7.	Multireplicate Designs	441
13.8.	Complete Confounding: Repeated Single-Replicate Designs	442
	13.8.1. A Real Experiment—Decontamination Experiment	442
13.9.	Partial Confounding	446
13.10.	Comparing the Multireplicate Designs	449
13.11.	Using SAS Software	452
	Exercises	454

14. Confounding in General Factorial Experiments 461

14.1.	Introduction	461
14.2.	Confounding with Factors at Three Levels	462
	14.2.1. Contrasts	462
	14.2.2. Confounding Using Contrasts	463
	14.2.3. Confounding Using Equations	464
	14.2.4. A Real Experiment—Dye Experiment	467
	14.2.5. Plans for Confounded 3^p Experiments	470
14.3.	Designing Using Pseudofactors	471
	14.3.1. Confounding in 4^p Experiments	471
	14.3.2. Confounding in $2^p \times 4^q$ Experiments	472
14.4.	Designing Confounded Asymmetrical Experiments	472
14.5.	Using SAS Software	475
	Exercises	477

15. Fractional Factorial Experiments 483

15.1.	Introduction	483
15.2.	Fractions from Block Designs; Factors with 2 Levels	484
	15.2.1. Half-Fractions of 2^p Experiments; 2^{p-1} Experiments	484
	15.2.2. Resolution and Notation	487
	15.2.3. A Real Experiment—Soup Experiment	487
	15.2.4. Quarter-Fractions of 2^p Experiments; 2^{p-2} Experiments	490
	15.2.5. Smaller Fractions of 2^p Experiments	494
15.3.	Fractions from Block Designs; Factors with 3 Levels	496
	15.3.1. One-Third Fractions of 3^p Experiments; 3^{p-1} Experiments	496
	15.3.2. One-Ninth Fractions of 3^p Experiments; 3^{p-2} Experiments	501
15.4.	Fractions from Block Designs; Other Experiments	501
	15.4.1. $2^p \times 4^q$ Experiments	501
	15.4.2. $2^p \times 3^q$ Experiments	502
15.5.	Blocked Fractional Factorial Experiments	503

15.6.	Fractions from Orthogonal Arrays	506
	15.6.1. 2^p Orthogonal Arrays	506
	15.6.2. Saturated Designs	512
	15.6.3. $2^p \times 4^q$ Orthogonal Arrays	513
	15.6.4. 3^p Orthogonal Arrays	514
15.7.	Design for the Control of Noise Variability	515
	15.7.1. A Real Experiment—Inclinometer Experiment	516
15.8.	Using SAS Software	521
	15.8.1. Fractional Factorials	521
	15.8.2. Design for the Control of Noise Variability	524
	Exercises	529

16. Response Surface Methodology — 547

16.1.	Introduction	547
16.2.	First-Order Designs and Analysis	549
	16.2.1. Models	549
	16.2.2. Standard First-Order Designs	551
	16.2.3. Least Squares Estimation	552
	16.2.4. Checking Model Assumptions	553
	16.2.5. Analysis of Variance	553
	16.2.6. Tests for Lack of Fit	554
	16.2.7. Path of Steepest Ascent	559
16.3.	Second-Order Designs and Analysis	561
	16.3.1. Models and Designs	561
	16.3.2. Central Composite Designs	562
	16.3.3. Generic Test for Lack of Fit of the Second-Order Model	564
	16.3.4. Analysis of Variance for a Second-Order Model	564
	16.3.5. Canonical Analysis of a Second-Order Model	566
16.4.	Properties of Second-Order Designs: CCDs	569
	16.4.1. Rotatability	569
	16.4.2. Orthogonality	570
	16.4.3. Orthogonal Blocking	571
16.5.	A Real Experiment: Flour Production Experiment, Continued	573
16.6.	Box–Behnken Designs	576
16.7.	Using SAS Software	579
	16.7.1. Analysis of a Standard First-Order Design	579
	16.7.2. Analysis of a Second-Order Design	582
	Exercises	586

17. Random Effects and Variance Components — 593

17.1.	Introduction	593
17.2.	Some Examples	594
17.3.	One Random Effect	596

17.3.1. The Random-Effects One-Way Model 596
17.3.2. Estimation of σ^2 597
17.3.3. Estimation of σ_T^2 598
17.3.4. Testing Equality of Treatment Effects 601
17.3.5. Confidence Intervals for Variance Components 603
17.4. Sample Sizes for an Experiment with One Random Effect 607
17.5. Checking Assumptions on the Model 610
17.6. Two or More Random Effects 610
17.6.1. Models and Examples 610
17.6.2. Checking Model Assumptions 613
17.6.3. Estimation of σ^2 613
17.6.4. Estimation of Variance Components 614
17.6.5. Confidence Intervals for Variance Components 616
17.6.6. Hypothesis Tests for Variance Components 620
17.6.7. Sample Sizes 622
17.7. Mixed Models 622
17.7.1. Expected Mean Squares and Hypothesis Tests 622
17.7.2. Confidence Intervals in Mixed Models 625
17.8. Rules for Analysis of Random and Mixed Models 627
17.8.1. Rules—Equal Sample Sizes 627
17.8.2. Controversy (Optional) 628
17.9. Block Designs and Random Blocking Factors 630
17.10. Using SAS Software 632
17.10.1. Checking Assumptions on the Model 632
17.10.2. Estimation and Hypothesis Testing 635
Exercises 639

18. Nested Models   645

18.1. Introduction 645
18.2. Examples and Models 646
18.3. Analysis of Nested Fixed Effects 648
18.3.1. Least Squares Estimates 648
18.3.2. Estimation of σ^2 649
18.3.3. Confidence Intervals 650
18.3.4. Hypothesis Testing 650
18.4. Analysis of Nested Random Effects 654
18.4.1. Expected Mean Squares 654
18.4.2. Estimation of Variance Components 656
18.4.3. Hypothesis Testing 657
18.4.4. Some Examples 658
18.5. Using SAS Software 662
18.5.1. Voltage Experiment 662
Exercises 667

19. Split-Plot Designs — 675

19.1. Introduction — 675
19.2. Designs and Models — 676
19.3. Analysis of a Split-Plot Design with Complete Blocks — 678
    19.3.1. Split-Plot Analysis — 678
    19.3.2. Whole-Plot Analysis — 680
    19.3.3. Contrasts Within and Between Whole Plots — 681
    19.3.4. A Real Experiment—Oats Experiment — 681
19.4. Split-Split-Plot Designs — 684
19.5. Split-Plot Confounding — 686
19.6. Using SAS Software — 687
Exercises — 691

A. Tables — 695

Bibliography — 725

Index of Authors — 731

Index of Experiments — 733

Index of Subjects — 735

1 Principles and Techniques

1.1 Design: Basic Principles and Techniques
1.2 Analysis: Basic Principles and Techniques

1.1 Design: Basic Principles and Techniques
1.1.1 The Art of Experimentation

One of the first questions facing an experimenter is, "How many observations do I need to take?" or alternatively, "Given my limited budget, how can I gain as much information as possible?" These are not questions that can be answered in a couple of sentences. They are, however, questions that are central to the material in this book. As a first step towards obtaining an answer, the experimenter must ask further questions, such as, "What is the main purpose of running this experiment?" and "What do I hope to be able to show?"

Typically, an experiment may be run for one or more of the following reasons:

(i) to determine the principal causes of variation in a measured response,

(ii) to find the conditions that give rise to a maximum or minimum response,

(iii) to compare the responses achieved at different settings of controllable variables,

(iv) to obtain a mathematical model in order to predict future responses.

Observations can be collected from *observational studies* as well as from *experiments*, but only an experiment allows conclusions to be drawn about cause and effect. For example, consider the following situation:

The output from each machine on a factory floor is constantly monitored by any successful manufacturing company. Suppose that in a particular factory, the output from a particular machine is consistently of low quality. What should the managers do? They could conclude that the machine needs replacing and pay out a large sum of money for a new one. They could decide that the machine operator is at fault and dismiss him or her. They could conclude that the humidity in that part of the factory is too high and install a new air conditioning system. In other words, the machine output has been observed under the current operating conditions (an observational study), and although it has been very effective in showing the management that a problem exists, it has given them very little idea about the *cause* of the poor quality.

It would actually be a simple matter to determine or rule out some of the potential causes. For example, the question about the operator could be answered by moving all the operators from machine to machine over several days. If the poor output follows the operator, then it is safe to conclude that the operator is the cause. If the poor output remains with the original machine, then the operator is blameless, and the machine itself or the factory humidity is the most likely cause of the poor quality. This is an "experiment." The experimenter has control over a possible cause in the difference in output quality between machines. If this particular cause is ruled out, then the experimenter can begin to vary other factors such as humidity or machine settings.

It is more efficient to examine all possible causes of variation simultaneously rather than one at a time. Fewer observations are usually needed, and one gains more information about the system being investigated. This simultaneous study is known as a "factorial experiment." In the early stages of a project, a list of all factors that conceivably could have an important effect on the response of interest is drawn up. This may yield a large number of factors to be studied, in which case special techniques are needed to gain as much information as possible from examining only a subset of possible factor settings.

The art of designing an experiment and the art of analyzing an experiment are closely intertwined and need to be studied side by side. In designing an experiment, one must take into account the analysis that will be performed. In turn, the efficiency of the analysis will depend upon the particular experimental design that is used to collect the data. Without these considerations, it is possible to invest much time, effort, and expense in the collection of data which seem relevant to the purpose at hand but which, in fact, contribute little to the research questions being asked. A guiding principle of experimental design is to "keep it simple." Interpretation and presentation of the results of experiments are generally clearer for simpler experiments.

Three basic techniques fundamental to experimental design are replication, blocking, and randomization. The first two help to increase precision in the experiment; the last is used to decrease bias. These techniques are discussed briefly below and in more detail throughout the book.

1.1.2 Replication

There is a difference between "replication" and "repeated measurements." For example, suppose four subjects are each assigned to a drug and a measurement is taken on each subject. The result is four independent observations on the drug. This is "replication." On the other hand, if one subject is assigned to a drug and then measured four times, the measurements are not independent. We call them "repeated measurements." The variation recorded in repeated measurements taken at the same time reflects the variation in the measurement process, while the variation recorded in repeated measurements taken over a time interval reflects the variation in the single subject's response to the drug over time. Neither reflects the variation in independent subjects' responses to the drug. We need to know about the latter variation in order to generalize any conclusion about the drug so that it is relevant to all similar subjects.

1.1.3 Blocking

The experimental conditions under which an experiment is run should be representative of those to which the conclusions of the experiment are to be applied. For inferences to be broad in scope, the experimental conditions should be rather varied. However, an unfortunate consequence of increasing the scope of the experiment is an increase in the variability of the response. Blocking is a technique that can often be used to help deal with this problem.

To block an experiment is to divide, or partition, the observations into groups called blocks in such a way that the observations in each block are collected under relatively similar experimental conditions. If blocking is done well, then comparisons of two or more treatments are made more precisely than similar comparisons from an unblocked design.

1.1.4 Randomization

The purpose of randomization is to prevent systematic and personal biases from being introduced into the experiment by the experimenter. A random assignment of subjects or experimental material to treatments prior to the start of the experiment ensures that observations that are favored or adversely affected by unknown sources of variation are observations "selected in the luck of the draw" and not systematically selected.

Lack of a random assignment of experimental material or subjects leaves the experimental procedure open to *experimenter bias*. For example, a horticulturist may assign his or her favorite variety of experimental crop to the parts of the field that look the most fertile, or a medical practitioner may assign his or her preferred drug to the patients most likely to respond well. The preferred variety or drug may then appear to give better results no matter how good or bad it actually is.

Lack of random assignment can also leave the procedure open to *systematic bias*. Consider, for example, an experiment involving drying time of three paints applied to sections of a wooden board, where each paint is to be observed four times. If no random assignment of order of observation is made, many experimenters would take the four observations on paint 1, followed by those on paint 2, followed by those on paint 3. This order might be perfectly satisfactory, but it could equally well prove to be disastrous. Observations taken over time could be affected by differences in atmospheric conditions, fatigue of the experimenter, systematic differences in the wooden board sections, etc. These could all conspire to ensure that any measurements taken during the last part of the experiment are, say, underrecorded, with the result that paint 3 appears to dry faster than the other paints when, in fact, it may be less good. The order 1, 2, 3, 1, 2, 3, 1, 2, 3, 1, 2, 3 helps to solve the problem, but it does not remove it completely (especially if the experimenter takes a break after every three observations).

There are also analytical reasons to support the use of a random assignment. It will be seen in Chapters 3 and 4 that common forms of analysis of the data depend on the F and t distributions. It can be shown that a random assignment ensures that these distributions are the correct ones to use. The interested reader is referred to Kempthorne (1977).

To understand the meaning of randomization, consider an experiment to compare the effects on blood pressure of three exercise programs, where each program is observed four times, giving a total of 12 observations. Now, given 12 subjects, imagine making a list of all possible assignments of the 12 subjects to the three exercise programs so that 4 subjects are assigned to each program. (There are $12!/(4!4!4!)$, or 34,650 ways to do this.) If the assignment of subjects to programs is done in such a way that every possible assignment has the same chance of occurring, then the assignment is said to be a *completely random assignment*. Completely randomized designs, discussed in Chapters 3–7 of this book, are randomized in this way. It is, of course, possible that a random assignment itself could lead to the order 1, 1, 1, 1, 2, 2, 2, 2, 3, 3, 3, 3. If the experimenter expressly wishes to avoid certain assignments, then a different type of design should be used. An experimenter should not look at the resulting assignment, decide that it does not look very random, and change it.

Without the aid of an objective randomizing device, it is not possible for an experimenter to make a random assignment. In fact, it is not even possible to select a single number at random. This is borne out by a study run at the University of Delaware and reported by Professor Hoerl in the *Royal Statistical Society News and Notes* (January 1988). The study, which was run over several years, asked students to pick a number at random between 0 and 9. The numbers 3 and 7 were selected by about 40% of the students. This is twice as many as would be expected if the numbers were truly selected at random.

The most frequently used objective mechanism for achieving a random assignment in experimental design is a random number generator. A random number generator is a computer program that gives as output a very long string of digits that are integers between 0 and 9 inclusive and that have the following properties. All integers between 0 and 9 occur approximately the same number of times, as do all pairs of integers, all triples, and so on. Furthermore, there is no discernible pattern in the string of digits, and hence the name "random" numbers.

The random numbers in Appendix Table A.1 are part of a string of digits produced by a random number generator (in SAS® version 6.09 on a DEC Model 4000 MODEL 610 computer at Wright State University). Many experimenters and statistical consultants will have direct access to their own random number generator on a computer or calculator and will not need to use the table. The table is divided into six sections (pages), each section containing six groups of six rows and six groups of six columns. The grouping is merely a device to aid in reading the table. To use the table, a random starting place must be found. An experimenter who always starts reading the table at the same place always has the same set of digits, and these could not be regarded as random. The grouping of the digits by six rows and columns allows a random starting place to be obtained using five rolls of a fair die. For example, the five rolls giving 3, 1, 3, 5, 2 tells the experimenter to find the digit that is in section 3 of the table, row group 1, column group 3, row 5, column 2. Then the digits can be read singly, or in pairs, or triples, etc. from the starting point across the rows.

The most common random number generators on computers or calculators generate n-digit real numbers between zero and one. Single digit random numbers can be obtained from an n-digit real number by reading the first digit after the decimal point. Pairs of digits can be obtained by reading the first two digits after the decimal point, and so on. The use of random numbers for randomization is shown in Sections 3.2 and 3.8.

1.2 Analysis: Basic Principles and Techniques

In the analysis of data, it is desirable to provide both graphical and statistical analyses. Plots that illustrate the relative responses of the factor settings under study allow the experimenter to gain a feel for the practical implications of the statistical results and to communicate effectively the results of the experiment to others. In addition, data plots allow the proposed model to be checked and aid in the identification of unusual observations, as discussed in Chapter 5. Statistical analysis quantifies the relative responses of the factors, thus clarifying conclusions that might be misleading or not at all apparent in plots of the data.

The purpose of an experiment can range from exploratory (discovering new important sources of variability) to confirmatory (confirming that previously discovered sources of variability are sufficiently major to warrant further study), and the philosophy of the analysis depends on the purpose of the experiment. In the early stages of experimentation the analysis may be exploratory, and one would plot and analyze the data in any way that assists in the identification of important sources of variation. In later stages of experimentation, analysis is usually confirmatory in nature. A mathematical model of the response is postulated and hypotheses are tested and confidence intervals are calculated.

In this book, we use *linear models* to model our response and the *method of least squares* for obtaining estimates of the parameters in the model. These are described in Chapter 3. Our models include random "error variables" that encompass all the sources of variability not explicity present in the model. We operate under the assumption that the error terms are normally distributed. However, most of the procedures in this book are generally fairly robust to nonnormality, provided that there are no extreme observations among the data.

It is rare nowadays for experimental data to be analyzed by hand. Most experimenters and statisticians have access to a computer package that is capable of producing, at the very least, a basic analysis of data for the simplest experiments. To the extent possible, for each design discussed, we shall present useful plots and methods of analysis that can be obtained from most statistical software packages. We will also develop many of the mathematical formulas that lie behind the computer analysis. This will enable the reader more easily to appreciate and interpret statistical computer package output and the associated manuals. Computer packages vary in sophistication, flexibility, and the statistical knowledge required of the user. The SAS software is one of the better packages for analyzing experimental data. It can handle every model discussed in this book, and although it requires some knowledge of experimental design on the part of the user, it is easy to learn. We provide some basic SAS statements and output at the end of most chapters to illustrate data analysis. A reader who wishes to use a different computer package can run the equivalent analyses on his or her own package and compare the output with those shown. It is important that every user know exactly the capabilities of his or her own package and also the likely size of rounding errors. It is not our intent to teach the best use of the SAS software, and those readers who have access to the SAS package may find better ways of achieving the same analyses.

2 Planning Experiments

2.1 Introduction
2.2 A Checklist for Planning Experiments
2.3 A Real Experiment—Cotton-Spinning Experiment
2.4 Some Standard Experimental Designs
2.5 More Real Experiments
Exercises

2.1 Introduction

Although planning an experiment is an exciting process, it is extremely time-consuming. This creates a temptation to begin collecting data without giving the experimental design sufficient thought. Rarely will this approach yield data that have been collected in exactly the right way and in sufficient quantity to allow a good analysis with the required precision. This chapter gives a step by step guide to the experimental planning process. The steps are discussed in Section 2.2 and illustrated via real experiments in Sections 2.3 and 2.5. Some standard experimental designs are described briefly in Section 2.4.

2.2 A Checklist for Planning Experiments

The steps in the following checklist summarize a very large number of decisions that need to be made at each stage of the experimental planning process. The steps are not independent, and at any stage, it may be necessary to go back and revise some of the decisions made at an earlier stage.

CHECKLIST

(a) Define the objectives of the experiment.

(b) Identify all sources of variation, including:

    (i) treatment factors and their levels,

    (ii) experimental units,

    (iii) blocking factors, noise factors, and covariates.

(c) Choose a rule for assigning the experimental units to the treatments.

(d) Specify the measurements to be made, the experimental procedure, and the anticipated difficulties.

(e) Run a pilot experiment.

(f) Specify the model.

(g) Outline the analysis.

(h) Calculate the number of observations that need to be taken.

(i) Review the above decisions. Revise, if necessary.

A short description of the decisions that need to be made at each stage of the checklist is given below. Only after all of these decisions have been made should the data be collected.

(a) **Define the objectives of the experiment.**

A list should be made of the precise questions that are to be addressed by the experiment. It is this list that helps to determine the decisions required at the subsequent stages of the checklist. It is advisable to list only the essential questions, since side issues will unnecessarily complicate the experiment, increasing both the cost and the likelihood of mistakes. On the other hand, questions that are inadvertently omitted may be unanswerable from the data. In compiling the list of objectives, it can often be helpful to outline the conclusions expected from the analysis of the data. The objectives may need to be refined as the remaining steps of the checklist are completed.

(b) **Identify all sources of variation.** A source of variation is *anything* that could cause an observation to have a different numerical value from another observation. Some sources of variation are minor, producing only small differences in the data. Others are major and need to be planned for in the experiment. It is good practice to make a list of every conceivable source of variation and then label each as either major or minor. Major sources of variation can be divided into two types: those that are of particular interest to the experimenter, called "treatment factors," and those that are not of interest, called "nuisance factors."

(i) Treatment factors and their levels.

Although the term *treatment factor* might suggest a drug in a medical experiment, it is used to mean any substance or item whose effect on the data is to be studied. At this stage in the checklist, the treatment factors and their *levels* should be selected. The levels are the specific types or amounts of the treatment factor that will actually be used in the experiment. For example, a treatment factor might be a drug or a chemical additive or temperature or teaching method, etc. The levels of such treatment factors might be the different amounts of the drug to be studied, different types of chemical additives to be considered, selected temperature settings in the range of interest, different teaching methods to be compared, etc. Few experiments involve more than four levels per treatment factor.

If the levels of a treatment factor are quantitative (i.e., can be measured), then they are usually chosen to be equally spaced. For convenience, treatment factor levels can be coded. For example, temperature levels 60°, 70°, 80°, ... might be coded as 1, 2, 3, ... in the plan of the experiment, or as 0, 1, 2, ... With the latter coding, level 0 does not necessarily signify the absence of the treatment factor. It is merely a label. Provided that the experimenter keeps a clear record of the original choice of levels, no information is lost by working with the codes.

When an experiment involves more than one treatment factor, every observation is a measurement on some combination of levels of the various treatment factors. For example, if there are two treatment factors, temperature and pressure, whenever an observation is taken at a certain pressure, it must necessarily be taken at some temperature, and vice versa. Suppose there are four levels of temperature coded 1, 2, 3, 4 and three levels of pressure coded 1, 2, 3. Then there are twelve combinations of levels coded 11, 12, ..., 43, where the first digit of each pair refers to the level of temperature and the second digit to the level of pressure. Treatment factors are often labeled $F_1, F_2, F_3, \ldots$ or $A, B, C, \ldots$. The combinations of their levels are called *treatment combinations* and an experiment involving two or more treatment factors is called a *factorial experiment*. We will use the term *treatment* to mean a level of a treatment factor in a single factor experiment, or to mean a treatment combination in a factorial experiment.

(ii) Experimental units.

Experimental units are the "material" to which the levels of the treatment factor(s) are applied. For example, in agriculture these would be individual plots of land, in medicine they would be human or animal subjects, in industry they might be batches of raw material, factory workers, etc. If an experiment has to be run over a period of time, with the observations being collected sequentially, then the times of day can also be regarded as experimental units.

Experimental units should be representative of the material and conditions to which the conclusions of the experiment will be applied. For example, the conclusions of an experiment that uses university students as experimental units may not apply to all adults in the country. The results of a chemical experiment run in an 80° laboratory may not apply in a 60° factory. Thus it is important to consider carefully the scope of the experiment in listing the objectives in step (a).

(iii) Blocking factors, noise factors, and covariates.

An important part of designing an experiment is to enable the effects of the nuisance factors to be distinguished from those of the treatment factors. There are several ways of dealing with nuisance factors, depending on their nature.

It may be desirable to limit the scope of the experiment and to fix the level of the nuisance factor. This action may necessitate a revision of the objectives listed in step (a) since the conclusions of the experiment will not be so widely applicable. Alternatively, it may be possible to hold the level of a nuisance factor constant for one group of experimental units, change it to a different fixed value for a second group, change it again for a third, and so on. Such a nuisance factor is called a *blocking factor*, and experimental units measured under the same level of the blocking factor are said to be in the same *block* (see Chapter 10). For example, suppose that temperature was expected to have an effect on the observations in an experiment, but it was not itself a factor of interest. The entire experiment could be run at a single temperature, thus limiting the conclusions to that particular temperature. Alternatively, the experimental units could be divided into blocks with each block of units being measured at a different fixed temperature.

Even when the nuisance variation is not measured, it is still often possible to divide the experimental units into blocks of like units. For example, plots of land or times of day that are close together are more likely to be similar than those far apart. Subjects with similar characteristics are more likely to respond in similar ways to a drug than subjects with different characteristics. Observations made in the same factory are more likely to be similar than observations made in different factories.

Sometimes nuisance variation is a property of the experimental units and can be measured before the experiment takes place, (e.g., the blood pressure of a patient in a medical experiment, the I.Q. of a pupil in an educational experiment, the acidity of a plot of land in an agricultural experiment). Such a measurement is called a *covariate* and can play a major role in the analysis (see Chapter 9). Alternatively, the experimental units can be grouped into blocks, each block having a similar value of the covariate. The covariate would then be regarded as a blocking factor.

If the experimenter is interested in the variability of the response as the experimental conditions are varied, then nuisance factors are deliberately included in the experiment and not removed via blocking. Such nuisance factors are called *noise factors*, and experiments involving noise factors form the subject of *robust design*, discussed in Chapters 7 and 15.

(c) **Choose a rule by which to assign the experimental units to the levels of the treatment factors.**

The assignment rule, or the *experimental design*, specifies which experimental units are to be observed under which treatments. The choice of design, which may or may not involve blocking factors, depends upon all the decisions made so far in the checklist. There are several standard designs that are used often in practice, and these are introduced in Section 2.4. Further details and more complicated designs are discussed later in the book.

The actual assignment of experimental units to treatments should be done at random, subject to restrictions imposed by the chosen design. The importance of a random assignment was discussed in Section 1.1.4. Methods of randomization are given in Section 3.1.

There are some studies in which it appears to be impossible to assign the experimental units to the treatments either at random or indeed by any method. For example, if the study is to investigate the effects of smoking on cancer with human subjects as the experimental units, it is neither ethical nor possible to assign a person to smoke a given number of cigarettes per day. Such a study would therefore need to be done by observing people who have themselves chosen to be light, heavy, or nonsmokers throughout their lives. This type of study is an *observational study* and not an experiment. Although many of the analysis techniques discussed in this book could be used for observational studies, cause and effect conclusions are not valid, and such studies will not be discussed further.

(d) **Specify the measurements to be made, the experimental procedure, and the anticipated difficulties.**

The data (or observations) collected from an experiment are measurements of a response variable (e.g., the yield of a crop, the time taken for the occurrence of a chemical reaction, the output of a machine). The units in which the measurements are to be made should be specified, and these should reflect the objectives of the experiment. For example, if the experimenter is interested in detecting a difference of 0.5 gram in the response variable arising from two different treatments, it would not be sensible to take measurements to the nearest gram. On the other hand, it would be unnecessary to take measurements to the nearest 0.01 gram. Measurements to the nearest 0.1 gram would be sufficiently sensitive to detect the required difference, if it exists.

There are usually unforeseen difficulties in collecting data, but these can often be identified by taking a few practice measurements or by running a pilot experiment (see step (e)). Listing the anticipated difficulties helps to identify sources of variation required by step (b) of the checklist, and also gives the opportunity of simplifying the experimental procedure before the experiment begins.

Precise directions should be listed as to how the measurements are to be made. This might include details of the measuring instruments to be used, the time at which the measurements are to be made, the way in which the measurements are to be recorded. It is important that everyone involved in running the experiment follow these directions exactly. It is advisable to draw up a data collection sheet that shows the order in which the observations are to be made and also the units of measurement.

(e) **Run a pilot experiment.**

A pilot experiment is a mini experiment involving only a few observations. No conclusions are necessarily expected from such an experiment. It is run to aid in the completion of the checklist. It provides an opportunity to practice the experimental technique and to identify unsuspected problems in the data collection. If the pilot experiment is large enough, it can also help in the selection of a suitable model for the main experiment. The observed experimental error in the pilot experiment can help in the calculation of the number of observations required by the main experiment (step (h)).

At this stage, steps (a)–(d) of the checklist should be reevaluated and changes made as necessary.

(f) **Specify the model.**

The model must indicate explicitly the relationship that is believed to exist between the response variable and the major sources of variation that were identified at step (b). The techniques used in the analysis of the experimental data will depend upon the form of the model. It is important, therefore, that the model represent the true relationship reasonably accurately.

The most common type of model is the linear model, which shows the response variable set equal to a linear combination of terms representing the major sources of variation plus an error term representing all the minor sources of variation taken together. A pilot experiment (step (e)) can help to show whether or not the data are reasonably well described by the model.

There are two different types of treatment or block factors that need to be distinguished, since they lead to somewhat different analyses. The effect of a factor is said to be a *fixed effect* if the factor levels have been specifically selected by the experimenter and if the experimenter is interested in comparing the effects on the response variable of these specific levels. This is the most common type of factor and is the type considered in the early chapters. A model containing only fixed-effect factors (apart from the response and error random variables) is called a *fixed-effects model*.

Occasionally, however, a factor has an extremely large number of possible levels, and the levels included in the experiment are a random sample from the population of all possible levels. The effect of such a factor is said to be a *random effect*. Since the levels are not specifically chosen, the experimenter has little interest in comparing the effects on the response variable of the particular levels used in the experiment. Instead, it is the variability of the response due to the entire population of levels that is of interest. Models for which all factors are random effects are called *random-effects models*. Models for which some factors are random effects and others are fixed effects are called *mixed models*. Experiments involving random effects will be considered in Chapters 17–18.

(g) **Outline the analysis.**

The type of analysis that will be performed on the experimental data depends on the objectives determined at step (a), the design selected at step (c), and its associated model specified in step (f). The entire analysis should be outlined (including hypotheses to be tested and confidence intervals to be calculated). The analysis not only determines the calculations at step (h), but also verifies that the design is suitable for achieving the objectives of the experiment.

(h) **Calculate the number of observations needed.**

At this stage in the checklist, a calculation should be done for the number of observations that are needed in order to achieve the objectives of the experiment. If too few observations are taken, then the experiment may be inconclusive. If too many are taken, then time, energy, and money are needlessly expended.

Formulae for calculating the number of observations are discussed in Sections 3.6 and 4.5 for the completely randomized design, and in later chapters for more complex designs. The formulae require a knowledge of the size of the experimental variability. This is the amount of variability in the data caused by the sources of variation designated as minor in step (b) (plus those sources that were forgotten!). Estimating the size of the experimental error prior to the experiment is not easy, and it is advisable to err on the large side. Methods of estimation include the calculation of the experimental error in a pilot experiment (step (e)) and previous experience of working with similar experiments.

(i) **Review the above decisions. Revise if necessary.**

Revision is necessary when the number of observations calculated at step (h) exceeds the number that can reasonably be taken within the time or budget available. Revision must begin at step (a), since the scope of the experiment usually has to be narrowed. If revisions are not necessary, then the data collection may commence.

It should now be obvious that a considerable amount of thought needs to precede the running of an experiment. The data collection is usually the most costly and the most time-consuming part of the experimental procedure. Spending a little extra time in planning helps to ensure that the data can be used to maximum advantage. No method of analysis can save a badly designed experiment.

Although an experimental scientist welltrained in the principles of design and analysis of experiments may not need to consult a statistician, it usually helps to talk over the checklist with someone not connected with the experiment. Step (a) in the checklist is often the most difficult to complete. A consulting statistician's first question to a client is usually, "Tell me *exactly* why you are running the experiment. *Exactly* what do you want to show?" If these questions cannot be answered, it is not sensible for the experimenter to go away, collect some data, and worry about it later. Similarly, it is essential that a consulting statistician understand reasonably well not only the purpose of the experiment but also the experimental technique. It is not helpful to tell an experimenter to run a pilot experiment that eats up most of the budget.

The experimenter needs to give clear directions concerning the experimental procedure to all persons involved in running the experiment and in collecting the data. It is also necessary to check that these directions are being followed exactly as prescribed. An amusing anecdote told by M. Salvadori (1980) in his book *Why Buildings Stand Up* illustrates this point. The story concerns a quality control study of concrete. Concrete consists of cement, sand, pebbles, and water and is mixed in strictly controlled proportions in a concrete plant. It is then carried to a building site in a revolving drum on a large truck. A sample of concrete is taken from each truckload and, after seven days, is tested for compressive strength. Its strength depends partly upon the ratio of water to cement, and decreases as the proportion of water increases. The story continues:

> During the construction of a terminal at J. F. Kennedy Airport in New York, the supervising engineer noticed that all the concrete reaching the site before noon showed good seven-day strength, but some of the concrete batches arriving shortly after noon did not measure up. Puzzled by this phenomenon, he investigated all its most plausible causes until he decided, in desperation, not only to be at the plant during the mixing, but also to follow the trucks as they went from the plant to the site. By doing so unobtrusively, he was able to catch a truck driver regularly stopping for beer and

a sandwich at noon, and before entering the restaurant, hosing extra water into the drums so that the concrete would not harden before reaching the site. The prudent engineer must not only be cautious about material properties, but be aware, most of all, of human behavior.

This applies to prudent experimenters, too! In the chapters that follow, most of the emphasis falls on the statistical analysis of well-designed experiments. It is crucial to keep in mind the ideas in these first sections while reading the rest of the book. Unfortunately, there are no nice formulae to summarize everything. Both the experimenter and the statistical consultant should use the checklist and lots of common sense!

2.3 A Real Experiment—Cotton-Spinning Experiment

The experiment to be described was reported in the November 1953 issue of the *Journal of Applied Statistics* by Robert Peake, of the British Cotton Industry Research Association. Although the experiment was run many years ago, the types of decisions involved in planning experiments have changed very little. The original report was not written in checklist form, but all of the relevant details were provided by the author in the article.

CHECKLIST

(a) **Define the objectives of the experiment.**

At an intermediate stage of the cotton-spinning process, a strand of cotton (known as "roving") thicker than the final thread is produced. Roving is twisted just before it is wound onto a bobbin. As the degree of twist increases, so does the strength of the cotton, but unfortunately, so does the production time and hence, the cost. The twist is introduced by means of a rotary guide called a "flyer." The purpose of the experiment was twofold; first, to investigate the way in which different degrees of twist (measured in turns per inch) affected the breakage rate of the roving, and secondly, to compare the ordinary flyer with the newly devised special flyer.

(b) **Identify all sources of variation.**

(i) Treatment factors and their levels.

There are two treatment factors, namely "type of flyer" and "degree of twist." The first treatment factor, flyer, has two levels, "ordinary" and "special." We code these as 1 and 2, respectively. The levels of the second treatment factor, twist, had to be chosen within a feasible range. A pilot experiment was run to determine this range, and four nonequally spaced levels were selected, 1.63, 1.69, 1.78, and 1.90 turns per inch. Coding these levels as 1, 2, 3, and 4, there are eight possible treatment combinations, as shown in Table 2.1.

The two treatment combinations 11 and 24 were omitted from the experiment, since the pilot experiment showed that these did not produce satisfactory roving. The experiment was run with the six treatment combinations 12, 13, 14, 21, 22, 23.

(ii) Experimental units.

Table 2.1 Treatment combinations for the cotton-spinning experiment

	Twist			
Flyer	1.63	1.69	1.78	1.90
Ordinary	(11)	12	13	14
Special	21	22	23	(24)

An experimental unit consisted of the thread on the set of full bobbins in a machine on a given day. It was not possible to assign different bobbins in a machine to different treatment combinations. The bobbins needed to be fully wound, since the tension, and therefore the breakage rate, changed as the bobbin filled. It took nearly one day to wind each set of bobbins completely.

(iii) Blocking factors, noise factors, and covariates.

Apart from the treatment factors, the following sources of variation were identified: the different machines, the different operators, the experimental material (cotton), and the atmospheric conditions.

There was some discussion among the experimenters over the designation of the blocking factors. Although similar material was fed to the machines and the humidity in the factory was controlled as far as possible, it was still thought that the experimental conditions might change over time. A blocking factor representing the day of the experiment was contemplated. However, the experimenters finally decided to ignore the day-to-day variability and to include just one blocking factor, each of whose levels represented a machine with a single operator. The number of experimental units per block was limited to six to keep the experimental conditions fairly similar within a block.

(c) **Choose a rule by which to assign the experimental units to the treatments.**

A randomized complete block design, which is discussed in detail in Chapter 10, was selected. The six experimental units in each block were randomly assigned to the six treatment combinations. The design of the final experiment was similar to that shown in Table 2.2.

(d) **Specify the measurements to be made, the experimental procedure, and the anticipated difficulties.**

Table 2.2 Part of the design for the cotton-spinning experiment

	Time Order					
Block	1	2	3	4	5	6
I	22	12	14	21	13	23
II	21	14	12	13	22	23
III	23	21	14	12	13	22
IV	23	21	12	...	...	...
⋮	⋮	⋮	⋮	⋮	⋮	⋮

It was decided that a suitable measurement for comparing the effects of the treatment combinations was the number of breaks per hundred pounds of material. Since the job of machine operator included mending every break in the roving, it was easy for the operator to keep a record of every break that occurred.

The experiment was to take place in the factory during the normal routine. The major difficulties were the length of time involved for each observation, the loss of production time caused by changing the flyers, and the fact that it was not known in advance how many machines would be available for the experiment.

(e) **Run a pilot experiment.**

The experimental procedure was already well known. However, a pilot experiment was run in order to identify suitable levels of the treatment factor "degree of twist" for each of the flyers; see step (b).

(f) **Specify the model.**

The model was of the form

Breakage rate = constant + effect of treatment combination
+ effect of block + error .

Models of this form and the associated analyses are discussed in Chapter 10.

(g) **Outline the analysis.**

The analysis was planned to compare differences in the breakage rates caused by the six flyer/twist combinations. Further, the trend in breakage rates as the degree of twist was increased was of interest for each flyer separately.

(h) **Calculate the number of observations that need to be taken.**

The experimental variability was estimated from a previous experiment of a somewhat different nature. This allowed a calculation of the required number of blocks to be done (see Section 10.5.2). The calculation was based on the fact that the experimenters wished to detect a true difference in breakage rates of at least 2 breaks per 100 pounds with high probability. The calculation suggested that 56 blocks should be observed (a total of 336 observations!).

(i) **Review the above decisions. Revise, if necessary.**

Since each block would take about a week to observe, it was decided that 56 blocks would not be possible. The experimenters decided to analyze the data after the first 13 blocks had been run. The effect of decreasing the number of observations from the number calculated is that the requirements stated in step (h) would not be met. The probability of detecting differences of 2 breaks per 100 lbs was substantially reduced.

The results from the 13 blocks are shown in Table 2.3, and the data from five of these are plotted in Figure 2.1. The data show that there are certainly differences in blocks. For example, results in block 5 are consistently above those for block 1. The breakage rate appears to be somewhat higher for treatment combinations 12 and 13 than for 23. However, the observed differences may not be any larger than the inherent variability in the data.

2.4 Some Standard Experimental Designs

Table 2.3 Data from the cotton-spinning experiment

Treatment Combination	Block Number					
	1	2	3	4	5	6
12	6.0	9.7	7.4	11.5	17.9	11.9
13	6.4	8.3	7.9	8.8	10.1	11.5
14	2.3	3.3	7.3	10.6	7.9	5.5
21	3.3	6.4	4.1	6.9	6.0	7.4
22	3.7	6.4	8.3	3.3	7.8	5.9
23	4.2	4.6	5.0	4.1	5.5	3.2
Treatment Combination	Block Number					
	7	8	9	10	11	12	13
12	10.2	7.8	10.6	17.5	10.6	10.6	8.7
13	8.7	9.7	8.3	9.2	9.2	10.1	12.4
14	7.8	5.0	7.8	6.4	8.3	9.2	12.0
21	6.0	7.3	7.8	7.4	7.3	10.1	7.8
22	8.3	5.1	6.0	3.7	11.5	13.8	8.3
23	10.1	4.2	5.1	4.6	11.5	5.0	6.4

Source: Peake, R.E. (1953). Copyright © 1953 Royal Statistical Society. Reprinted with permission.

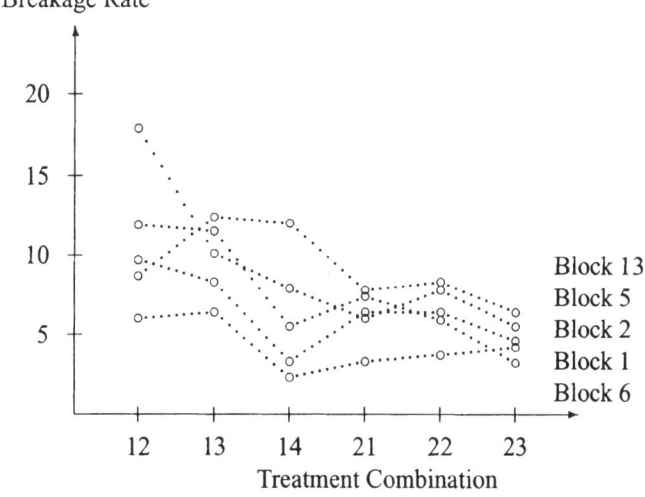

Figure 2.1 A subset of the data for the cotton-spinning experiment

Therefore, it is important to subject these data to a careful statistical analysis. This will be done in Section 10.5.

2.4 Some Standard Experimental Designs

An experimental design is a rule that determines the assignment of the experimental units to the treatments. Although experiments differ from each other greatly in most respects, there are some standard designs that are used frequently. These are described briefly in this section.

2.4.1 Completely Randomized Designs

A *completely randomized design* is the name given to a design in which the experimenter assigns the experimental units to the treatments completely at random, subject only to the number of observations to be taken on each treatment. Completely randomized designs are used for experiments that involve no blocking factors. They are discussed in depth in Chapters 3–9 and again in some of the later chapters. The mechanics of the randomization procedure are illustrated in Section 3.1. The statistical properties of the design are completely determined by specification of $r_1, r_2, \ldots, r_v$, where r_i denotes the number of observations on the ith treatment, $i = 1, \ldots, v$.

The model is of the form

Response = constant + effect of treatment + error .

Factorial experiments often have a large number of treatments. This number can even exceed the number of available experimental units, so that only a subset of the treatment combinations can be observed. Special methods of design and analysis are needed for such experiments, and these are discussed in Chapter 15.

2.4.2 Block Designs

A *block design* is a design in which the experimenter partitions the experimental units into blocks, determines the allocation of treatments to blocks, and assigns the experimental units within each block to the treatments completely at random. Block designs are discussed in depth in Chapters 10–14.

In the analysis of a block design, the blocks are treated as the levels of a single blocking factor even though they may be defined by a combination of levels of more than one nuisance factor. For example, the cotton-spinning experiment of Section 2.3 is a block design with each block corresponding to a combination of a machine and an operator. The model is of the form

Response = constant + effect of block
+ effect of treatment + error .

The simplest block design is the *complete block design*, in which each treatment is observed the same number of times in each block. Complete block designs are easy to analyze. A complete block design whose blocks contain a single observation on each treatment is called a *randomized complete block design* or, simply, a *randomized block design*.

When the block size is smaller than the number of treatments, so that it is not possible to observe every treatment in every block, a block design is called an *incomplete block design*. The precision with which treatment effects can be compared and the methods of analysis that are applicable depend on the choice of the design. Some standard design choices, and

Table 2.4 Schematic plans of experiments with two blocking factors

		(i) Crossed Blocking Factors					(ii) Nested Blocking Factors		
		Block Factor 1					Block Factor 1		
		1	2	3			1	2	3
Block	1	*	*	*		1	*		
Factor	2	*	*	*		2	*		
2	3	*	*	*		3	*		
					Block	4		*	
					Factor	5		*	
					2	6		*	
						7			*
						8			*
						9			*

appropriate methods of randomization, are covered in Chapter 11. Incomplete block designs for factorial experiments are discussed in Chapter 13.

2.4.3 Designs with Two or More Blocking Factors

When an experiment involves two major sources of variation that have each been designated as blocking factors, these blocking factors are said to be either *crossed* or *nested*. The difference between these is illustrated in Table 2.4. Each experimental unit occurs at some combination of levels of the two blocking factors, and an asterisk denotes experimental units that are to be assigned to treatment factors. It can be seen that when the block factors are crossed, experimental units are used from all possible combinations of levels of the blocking factors. When the block factors are nested, a particular level of one of the blocking factors occurs at only one level of the other blocking factor.

Crossed blocking factors A design involving two crossed blocking factors is sometimes called a "row–column" design. This is due to the pictorial representation of the design, in which the levels of one blocking factor are represented by rows and the levels of the second are represented by columns (see Table 2.4(i)). An intersection of a row and a column is called a "cell." Experimental units in the same cell should be similar. The model is of the form

Response = constant + effect of row block + effect of column block
+ effect of treatment + error.

Some standard choices of row–column designs with one experimental unit per cell are discussed in Chapter 12, and an example is given in Section 2.5.3 (page 29) of a row–column design with six experimental units per cell.

Table 2.5 A Latin square for the cotton-spinning experiment

Machine with Operator	\multicolumn{6}{c}{Days}					
	1	2	3	4	5	6
1	12	13	14	21	22	23
2	13	14	21	22	23	12
3	14	21	22	23	12	13
4	22	23	12	13	14	21
5	23	12	13	14	21	22
6	21	22	23	12	13	14

The example shown in Table 2.5 is a basic design (prior to randomization) that was considered for the cotton-spinning experiment. The two blocking factors were "machine with operator" and "day." Notice that if the column headings are ignored, the design looks like a randomized complete block design. Similarly, if the row headings are ignored, the design with columns as blocks looks like a randomized complete block design. Such designs are called Latin squares and are discussed in Chapter 12. For the cotton-spinning experiment, which was run in the factory itself, the experimenters could not guarantee that the same six machines would be available for the same six days, and this led them to select a randomized complete block design. Had the experiment been run in a laboratory, so that every machine was available on every day, the Latin square design would have been used, and the day-to-day variability could have been removed from the analysis of treatments.

Nested (or hierarchical) blocking factors. Two blocking factors are said to be nested when observations taken at two different levels of one blocking factor are automatically at two different levels of the second blocking factor (see Table 2.4(ii)). As an example, consider an experiment to compare the effects of a number of diets (the treatments) on the weight (the response variable) of piglets (the experimental units). Piglets vary in their metabolism, as do human beings. Therefore, the experimental units are extremely variable. However, some of this variability can be controlled by noting that piglets from the same litter are more likely to be similar than piglets from different litters. Also, litters from the same sow are more likely to be similar than litters from different sows. The different sows can be regarded as blocks, the litters regarded as subblocks, and the piglets as the experimental units within the subblocks. A piglet belongs only to one litter (piglets are nested within litters), and a litter belongs only to one sow (litters are nested within sows). The random assignment of piglets to diets would be done separately litter by litter in exactly the same way as for any block design.

In the industrial setting, the experimental units may be samples of some experimental material (e.g., cotton) taken from several different batches that have been obtained from several different suppliers. The samples, which are to be assigned to the treatments, are "nested within batches," and the batches are "nested within suppliers." The random assignment of samples to treatment factor levels is done separately batch by batch.

In an ordinary block design, the experimental units can be thought of as being nested within blocks. In the above two examples, an extra "layer" of nesting is apparent. Experimental units are nested within subblocks, subblocks are nested within blocks. The subblocks can be assigned at random to the levels of a further treatment factor. When this is done, the design is often known as a *split-plot design* (see Section 2.4.4).

2.4.4 Split-Plot Designs

A *split-plot design* is a design with at least one blocking factor where the experimental units within each block are assigned to the treatment factor levels as usual, and *in addition,* the blocks are assigned at random to the levels of a further treatment factor. This type of design is used when the levels of one (or more) treatment factors are easy to change, while the alteration of levels of other treatment factors are costly, or time-consuming. For example, this type of situation occurred in the cotton-spinning experiment of Section 2.3. Setting the degree of twist involved little more than a turn of a dial, but changing the flyers involved stripping down the machines. The experiment was, in fact, run as a randomized complete block design, as shown in Table 2.2. However, it could have been run as a split-plot design, as shown in Table 2.6. The time slots have been grouped into blocks, which have been assigned at random to the two flyers. The three experimental units within each cell have been assigned at random to degrees of twist.

Split-plot designs also occur in medical and psychological experiments. For example, suppose that several subjects are assigned at random to the levels of a drug. In each time-slot each subject is asked to perform one of a number of tasks, and some response variable is measured. The subjects can be regarded as blocks, and the time-slots for each subject can be regarded as experimental units within the blocks. The blocks and the experimental units are each assigned to the levels of the treatment factors—the subject to drugs and the time-slots to tasks. Split-plot designs are discussed in detail in Chapter 19.

Table 2.6 A split-plot design for the cotton-spinning experiment

	Time Order					
	1	2	3	4	5	6
	Block I			Block II		
Machine I		Flyer 2			Flyer 1	
	Twist 2	Twist 1	Twist 3	Twist 2	Twist 4	Twist 3
Machine II		Flyer 2			Flyer 1	
	Twist 1	Twist 2	Twist 3	Twist 4	Twist 2	Twist 3
Machine III		Flyer 1			Flyer 2	
	Twist 4	Twist 2	Twist 3	Twist 3	Twist 1	Twist 2
⋮	⋮	⋮	⋮	⋮	⋮	⋮

In a split-plot design, the effect of a treatment factor whose levels are assigned to the experimental units is generally estimated more precisely than a treatment factor whose levels are assigned to the blocks. It was this reason that led the experimenters of the cotton-spinning experiment to select the randomized complete block design in Table 2.2 rather than the split-plot design of Table 2.6. They preferred to take the extra time in running the experiment rather than risk losing precision in the comparison of the flyers.

2.5 More Real Experiments

Three experiments are described in this section. The first, called the "soap experiment," was run as a class project by Suyapa Silvia in 1985. The second, called the "battery experiment," was run by one of the authors. Both of these experiments are designed as completely randomized designs. The first has one treatment factor at three levels while the second has two treatment factors, each at two levels. The soap and battery experiments are included here to illustrate the large number of decisions that need to be made in running even the simplest investigations. Their data are used in Chapters 3–5 to illustrate methods of analysis. The third experiment, called the "cake-baking experiment," includes some of the more complicated features of the designs discussed in Section 2.4.

2.5.1 Soap Experiment

The checklist for this experiment has been obtained from the experimenter's report. Our comments are in parentheses. The reader is invited to critically appraise the decisions made by this experimenter and to devise alternative ways of running her experiment.

CHECKLIST (Suyapa Silvia, 1985)

(a) **Define the objectives of the experiment.**
The purpose of this experiment is to compare the extent to which three particular types of soap dissolve in water. It is expected that the experiment will answer the following questions: Are there any differences in weight loss due to dissolution among the three soaps when allowed to soak in water for the same length of time? What are these differences?
Generalizations to other soaps advertised to be of the same type as the three used for this experiment cannot be made, as each soap differs in terms of composition, i.e., has different mixtures of ingredients. Also, because of limited laboratory equipment, the experimental conditions imposed upon these soaps cannot be expected to mimic the usual treatment of soaps, i.e., use of friction, running water, etc. Conclusions drawn can only be discussed in terms of the conditions posed in this experiment, although they could give indications of what the results might be under more normal conditions.
(We have deleted the details of the actual soaps used).

(b) **Identify all sources of variation.**

(i) Treatment factors and their levels

The treatment factor, soap, has been chosen to have three levels: regular, deodorant, and moisturizing brands, all from the same manufacturer. The particular brands used in the experiment are of special interest to this experimenter.

The soap will be purchased at local stores and cut into cubes of similar weight and size—about 1″ cubes. The cubes will be cut out of each bar of soap using a sharp hacksaw so that all sides of the cube will be smooth. They will then be weighed on a digital laboratory scale showing a precision of 10 mg. The weight of each cube will be made approximately equal to the weight of the smallest cube by carefully shaving thin slices from it. A record of the preexperimental weight of each cube will be made.

(Note that the experimenter has no control over the age of the soap used in the experiment. She is assuming that the bars of soap purchased will be typical of the population of soap bars available in the stores. If this assumption is not true, then the results of the experiment will not be applicable in general. Each cube should be cut from a different bar of soap purchased from a random sample of stores in order for the experiment to be as representative as possible of the populations of soap bars.)

(ii) Experimental units

The experiment will be carried out using identical metal muffin pans. Water will be heated to 100°F (approximate hot bath temperature), and each section will be quickly filled with 1/4 cup of water. A pilot study indicated that this amount of water is enough to cover the tops of the soaps. The water-filled sections of the muffin pans are the experimental units, and these will be assigned to the different soaps as described in step (c).

(iii) Blocking factors, noise factors, and covariates

(Apart from the differences in the composition of the soaps themselves, the initial sizes of the cubes were not identical, and the sections of the muffin pan were not necessarily all exposed to the same amount of heat. The initial sizes of the cubes were measured by weight. These could have been used as covariates, but the experimenter chose instead to measure the weight changes, that is, "final weight minus initial weight." The sections of the muffin pan could have been grouped into blocks with levels such as "outside sections," "inside sections," or such as "center of heating vent" and "off-center of heating vent." However, the experimenter did not feel that the experimental units would be sufficiently variable to warrant blocking. Other sources of variation include inaccuracies of measuring initial weights, final weights, amounts and temperature of water. All of these were designated as minor. No noise factors were incorporated into the experiment.)

(c) **Choose a rule by which to assign the experimental units to the levels of the treatment factors.**

An equal number of observations will be made on each of the three treatment factor levels. Therefore, r cubes of each type of soap will be prepared. These cubes will be randomly matched to the experimental units (muffin pan sections) using a random-number table.

Table 2.7 Weight loss for soaps in the soap experiment

Soap (Level)	Cube	Pre-weight (grams)	Post-weight (grams)	Weightloss (grams)
Regular (1)	1	13.14	13.44	−0.30
	2	13.17	13.27	−0.10
	3	13.17	13.31	−0.14
	4	13.17	12.77	0.40
Deodorant (2)	5	13.03	10.40	2.63
	6	13.18	10.57	2.61
	7	13.12	10.71	2.41
	8	13.19	10.04	3.15
Moisturizing (3)	9	13.14	11.28	1.86
	10	13.19	11.16	2.03
	11	13.06	10.80	2.26
	12	13.00	11.18	1.82

(This assignment rule defines a completely randomized design with r observations on each treatment factor level, see Chapter 3).

(d) **Specify the measurements to be made, the experimental procedure, and the anticipated difficulties.**

The cubes will be carefully placed in the water according to the assignment rule described in paragraph (c). The pans will be immediately sealed with aluminum foil in order to prevent excessive moisture loss. The pans will be positioned over a heating vent to keep the water at room temperature. Since the sections will be assigned randomly to the cubes, it is hoped that if water temperature differences do exist, these will be randomly distributed among the three treatment factor levels. After 24 hours, the contents of the pans will be inverted onto a screen and left to drain and dry for a period of 4 days in order to ensure that the water that was absorbed by each cube has been removed thoroughly. The screen will be labeled with the appropriate soap numbers to keep track of the individual soap cubes.

After the cubes have dried, each will be carefully weighed. These weights will be recorded next to the corresponding preexperimental weights to study the changes, if any, that may have occurred. The analysis will be carried out on the differences between the post- and preexperimental weights.

Expected Difficulties (i) The length of time required for a cube of soap to dissolve noticeably may be longer than is practical or assumed. Therefore, the data may not show any differences in weights.

(ii) Measuring the partially dissolved cubes may be difficult with the softer soaps (e.g., moisturizing soap), since they are likely to lose their shape.

(iii) The drying time required may be longer than assumed and may vary with the soaps, making it difficult to know when they are completely dry.

(iv) The heating vent may cause the pan sections to dry out prematurely.

2.5 More Real Experiments

(After the experiment was run, Suyapa made a list of the actual difficulties encountered. They are reproduced below. Although she had run a pilot experiment, it failed to alert her to these difficulties ahead of time, since not all levels of the treatment factor had been observed.)

Difficulties Encountered (i) When the cubes were placed in the warm water, it became apparent that some soaps absorbed water very quickly compared to others, causing the tops of these cubes to become exposed eventually. Since this had not been anticipated, no additional water was added to these chambers in order to keep the experiment as designed. This created a problem, since the cubes of soap were not all completely covered with water for the 24-hour period.

(ii) The drying time required was also different for the regular soap compared with the other two. The regular soap was still moist, and even looked bigger, when the other two were beginning to crack and separate. This posed a real dilemma, since the loss of weight due to dissolution could not be judged unless all the water was removed from the cubes. The soaps were observed for two more days after the data was collected and the regular soap did lose part of the water it had retained.

(iii) When the contents of the pans were deposited on the screen, it became apparent that the dissolved portion of the soap had become a semisolid gel, and a decision had to be made to regard this as "nonusable" and not allow it to solidify along with the cubes (which did not lose their shape).

(The remainder of the checklist together with the analysis is given in Sections 3.6.2 and 3.7. The calculations at step (h) showed that four observations should be taken on each soap type. The data were collected and are shown in Table 2.7. A plot of the data is shown in Figure 2.2.)

The weightloss for each cube of soap measured in grams to the nearest 0.01 gm is the difference between the initial weight of the cube (pre-weight) and the weight of the same

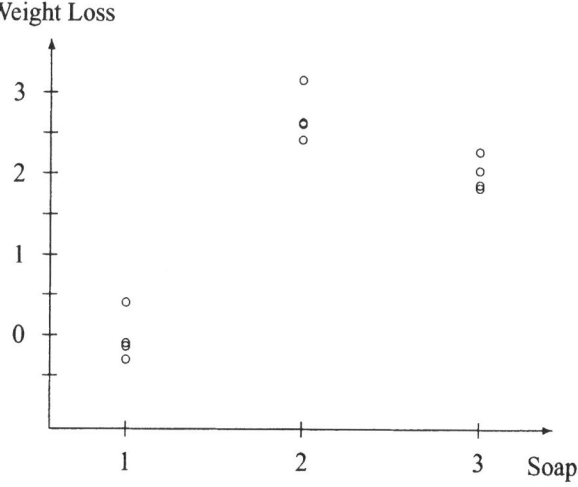

Figure 2.2
Weight loss (in grams) for the soap experiment

cube at the end of the experiment (post-weight). Negative values indicate a weight gain, while positive values indicate a weight loss (a large value being a greater loss). As can be seen, the regular soap cubes experienced the smallest changes in weight, and in fact, appear to have retained some of the water. Possible reasons for this will be examined in the discussion section (see Section 3.6.2). The data show a clear difference in the weight loss of the different soap types. This will be verified by a statistical hypothesis test (Section 3.6.2).

2.5.2 Battery Experiment

CHECKLIST

(a) **Define the objectives of the experiment.**

Due to the frequency with which his family needed to purchase flashlight batteries, one of the authors (Dan Voss) was interested in finding out which type of nonrechargeable battery was the most economical. In particular, Dan was interested in comparing the lifetime per unit cost of the particular name brand that he most often purchased with the store brand where he usually shopped. He also wanted to know whether it was worthwhile paying the extra money for alkaline batteries over heavy duty batteries.

A further objective was to compare the lifetimes of the different types of battery regardless of cost. This was due to the fact that whenever there was a power cut, all the available flashlights appeared to have dead batteries! (Only the first objective will be discussed in Chapters 3 and 4. The second objective will be addressed in Chapter 5.)

(b) **Identify all sources of variation.**

There are several sources of variation that are easy to identify in this experiment. Clearly, different duty batteries such as alkaline and heavy duty could well be an important factor in the lifetime per unit cost, as could the brand of the battery. These two sources of variation are the ones of most interest in the experiment and form the levels of the two treatment factors "duty" and "brand." Dan decided not to include regular duty batteries in the experiment.

Other possible sources of variation include the date of manufacture of the purchased battery, and whether the lifetime was monitored under continuous running conditions or under the more usual setting with the flashlight being turned on and off, the temperature of the environment, the age and variability of the flashlight bulbs.

The first of these could not be controlled in the experiment. The batteries used in the experiment were purchased at different times and in different locations in order to give a wide representation of dates of manufacture. The variability caused by this factor would be measured as part of the natural variability (error variability) in the experiment along with measurement error. Had the dates been marked on the packets, they could have been included in the analysis of the experiment as covariates. However, the dates were not available.

The second of these possible sources of variation (running conditions) was fixed. All the measurements were to be made under constant running conditions. Although this did not mimic the usual operating conditions of flashlight batteries, Dan thought that the relative ordering of the different battery types in terms of life per unit cost would be the same. The continuous running setting was much easier to handle in an experiment since each observation was expected to take several hours and no sophisticated equipment was available.

The third source of variation (temperature) was also fixed. Since the family living quarters are kept at a temperature of about 68 degrees in the winter, Dan decided to run his experiment at this usual temperature. Small fluctuations in temperature were not expected to be important.

The variability due to the age of the flashlight bulb was more difficult to handle. A decision had to be made whether to use a new bulb for each observation and risk muddling the effect of the battery with that of the bulb, or whether to use the same bulb throughout the experiment and risk an effect of the bulb age from biasing the data. A third possibility was to divide the observations into blocks and to use a single bulb throughout a block, but to change bulbs between blocks. Since the lifetime of a bulb is considerably longer than that of a battery, Dan decided to use the same bulb throughout the experiment.

(i) Treatment factors and their levels

There are two treatment factors each having two levels. These are battery "duty" (level 1 = alkaline, level 2 = heavy duty) and "brand" (level 1 = name brand, level 2 = store brand). This gives four treatment combinations coded 11, 12, 21, 22. In Chapters 3–5, we will recode these treatment combinations as 1, 2, 3, 4, and we will often refer to them as the four different treatments or the four different levels of the factor "battery type." Thus, the levels of battery type are:

Level	Treatment Combination
1	alkaline, name brand (11)
2	alkaline, store brand (12)
3	heavy duty, name brand (21)
4	heavy duty, store brand (22)

(ii) Experimental units

The experimental units in this experiment are the time slots. These were assigned at random to the battery types so as to determine the order in which the batteries were to be observed. Any fluctuations in temperature during the experiment form part of the variability between the time slots and are included in the error variability.

(iii) Blocking factors, noise factors, and covariates

As mentioned above, it was decided not to include a blocking factor representing different flashlight bulbs. Also, the date of manufacture of each battery was not available, and small fluctuations in room temperature were not thought to be important. Consequently, there were no covariates in the experiment, and no noise factors were incorporated.

(c) **Choose a rule by which to assign the experimental units to the levels of the treatment factor.**

Since there were to be no blocking factors, a completely randomized design was selected, and the time slots were assigned at random to the four different battery types.

Table 2.8 Data for the battery experiment

Battery Type	Life (min)	Unit Cost ($)	Life per Unit Cost	Time Order
1	602	0.985	611	1
2	863	0.935	923	2
1	529	0.985	537	3
4	235	0.495	476	4
1	534	0.985	542	5
1	585	0.985	593	6
2	743	0.935	794	7
3	232	0.520	445	8
4	282	0.495	569	9
2	773	0.935	827	10
2	840	0.935	898	11
3	255	0.520	490	12
4	238	0.495	480	13
3	200	0.520	384	14
4	228	0.495	460	15
3	215	0.520	413	16

(d) **Specify the measurements to be made, the experimental procedure, and the anticipated difficulties.**

The first difficulty was in deciding exactly how to measure lifetime of a flashlight battery. First, a flashlight requires two batteries. In order to keep the cost of the experiment low, Dan decided to wire a circuit linking just one battery to a flashlight bulb. Although this does not mimic the actual use of a flashlight, Dan thought that as with the constant running conditions, the relative lifetimes per unit cost of the four battery types would be preserved. Secondly, there was the difficulty in determining when the battery had run down. Each observation took several hours, and it was not possible to monitor the experiment constantly. Also, a bulb dims slowly as the battery runs down, and it is a judgment call as to when the battery is flat. Dan decided to deal with both of these problems by including a small clock in the circuit. The clock stopped before the bulb had completely dimmed, and the elapsed time on the clock was taken as a measurement of the battery life. The cost of a battery was computed as half of the cost of a two-pack, and the lifetime per unit cost was measured in minutes per dollar (min/$).

(e) **Run a pilot experiment.**

A few observations were run as a pilot experiment. This ensured that the circuit did indeed work properly. It was discovered that the clock and the bulb had to be wired in parallel and not in series, as Dan had first thought! The pilot experiment also gave a rough idea of the length of time each observation would take (at least four hours), and provided a very rough estimate of the error variability that was used at step (h) to calculate that four observations were needed on each treatment combination.

Difficulties encountered   The only difficulty encountered in running the main experiment was that during the fourth observation, it was discovered that the clock was running

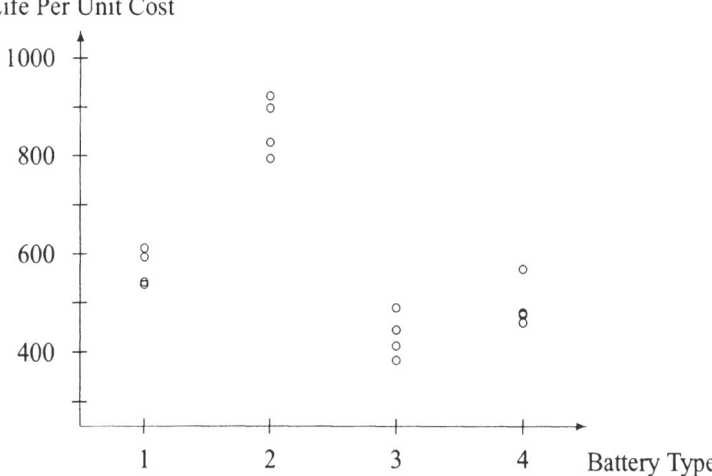

Figure 2.3
Battery life per unit cost versus battery

but the bulb was out. This was due to a loose connection. The connection was repaired, a new battery inserted into the circuit, and the clock reset.

Data The data collected in the main experiment are shown in Table 2.8 and plotted in Figure 2.3. The experiment was run in 1993.

2.5.3 Cake-Baking Experiment

The following factorial experiment was run in 1979 by the baking company Spillers Ltd. (in the U.K.) and was reported in the *Bulletin in Applied Statistics* in 1980 by S. M. Lewis and A. M. Dean.

CHECKLIST

(a) **Define the objectives of the experiment.**

The experimenters at Spillers, Ltd. wanted to know how "cake quality" was affected by adding different amounts of glycerine and tartaric acid to the cake mix.

(b) **Identify all sources of variation.**

(i) Treatment factors and their levels

The two treatment factors of interest were glycerine and tartaric acid. Glycerine was called the "first treatment factor" and labeled F_1, while tartaric acid was called the "second treatment factor" and labeled F_2. The experimenters were very familiar with the problems of cake baking and determinations of cake quality. They knew exactly which amounts of the two treatment factors they wanted to compare. They selected four equally spaced amounts of glycerine and three equally spaced amounts of tartaric acid. These were coded as 1, 2, 3, 4 for glycerine and 1, 2, 3 for tartaric acid. Therefore, the twelve coded treatment combinations were 11, 12, 13, 21, 22, 23, 31, 32, 33, 41, 42, 43.

(ii) Identify the experimental units

Table 2.9 Basic design for the baking experiment

Oven Codes	Time of Day Codes											
	1						2					
1	11	13	22	24	32	34	12	14	21	23	31	33
2	12	14	21	23	32	34	11	13	22	24	31	33
3	12	14	22	24	31	33	11	13	21	23	32	34

Before the experimental units can be identified, it is necessary to think about the experimental procedure. One batch of cake-mix was divided into portions. One of the twelve treatment combinations (i.e., a certain amount of glycerine and a certain amount of tartaric acid) was added to each portion. Each portion was then thoroughly mixed and put into a container for baking. The containers were placed on a tray in an oven at a given temperature for the required length of time. The experimenters required an entire tray of cakes to make one measurement of cake quality. Only one tray would fit on any one shelf of an oven. An experimental unit was, therefore, "an oven shelf with a tray of containers of cake-mix," and these were assigned at random to the twelve treatment combinations.

(iii) Blocking factors, noise factors, and covariates

There were two crossed blocking factors. The first was time of day with two levels (morning and afternoon). The second was oven, which had three levels, one level for each of the three ovens that were available on the day of the experiment. Each cell (defined by oven and time of day) contained six experimental units, since an oven contained six shelves (see Table 2.9). Each set of six experimental units was assigned at random to six of the twelve treatment combinations, and it was decided in advance which six treatment combinations should be observed together in a cell (see step (c) of the checklist).

Although the experimenters expected differences in the ovens and in different runs of the same oven, their experience showed that differences between the shelves of their industrial ovens were very minor. Otherwise, a third blocking factor representing oven shelf would have been needed.

It was possible to control carefully the amount of cake mix put into each container, and the experimenters did not think it was necessary to monitor the precooked weight of each cake. Small differences in these weights would not affect the measurement of the quality. Therefore, no covariates were used in the analysis.

(c) **Choose a rule by which to assign the experimental units to the levels of the treatment factors.**

Since there were two crossed blocking factors, a row–column design with six experimental units per cell was required. It was not possible to observe every treatment combination in every cell. However, it was thought advisable to observe all twelve treatment combinations in each oven, either in the morning or the afternoon. This precaution was taken so that if one of the ovens failed on the day of the experiment, the treatment combinations could still all be observed twice each. The basic design (before

Table 2.10 Some simple experiments

1.	Compare the growth rate of bean seeds under different watering and lighting schedules.
2.	Does the boiling point of water differ with different concentrations of salt?
3.	Compare the strengths of different brands of paper towel.
4.	Do different makes of popcorn give different proportions of unpopped kernels? What about cooking methods?
5.	Compare the effects of different locations of an observer on the speed at which subjects locate the occurrences of the letter "e" in a written passage.
6.	Do different colored candles burn at different speeds?
7.	Compare the proportions of words remembered from lists of related or unrelated words, and under various conditions such as silence and distraction.
8.	Compare the effects of different colors of exam paper on students' performance in an examination.

randomization) that was used by Spillers is shown in Table 2.9. The experimental units (the trays of containers on the six oven shelves) need to be assigned at random to the 6 treatment combinations cell by cell. The oven codes need to be assigned to the actual ovens at random, and the time of day codes 1 and 2 to morning and afternoon.

Exercises

Exercises 1–7 refer to the list of experiments in Table 2.10.

1. Table 2.10 gives a list of experiments that can be run as class projects. Select a simple experiment of interest to you, but preferably not on the list. Complete steps (a)–(c) of the checklist with the intention of actually running the experiment when the checklist is complete.

2. For experiments 1 and 7 in Table 2.10, complete steps (a) and (b) of the checklist. There may be more than one treatment factor. Give precise definitions of their levels.

3. For experiment 2, complete steps (a)–(c) of the checklist.

4. For experiment 3, complete steps (a)–(c) of the checklist.

5. For experiment 4, list sources of variation. Decide which sources can be controlled by limiting the scope of the experiment or by specifying the exact experimental procedure to be followed. Of the remaining sources of variation, decide which are minor and which are major. Are there any blocking factors in this experiment?

6. For experiment 6, specify what measurements should be made, how they should be made, and list any difficulties that might be expected.

7. For experiment 8, write down all the possible sources of variation. In your opinion, should this experiment be run as a completely randomized design, a block design, or a design with more than one blocking factor? Justify your answer.

8. Read critically through the checklists in Section 2.5. Would you suggest any changes? Would you have done anything differently? If you had to criticize these experiments, which points would you address?

9. The following description was given by Clifford Pugh in the 1953 volume of Applied Statistics.
"The widespread use of detergents for domestic dish washing makes it desirable for manufacturers to carry out tests to evaluate the performance of their products.... Since foaming is regarded as the main criterion of performance, the measure adopted is the number of plates washed before the foam is reduced to a thin surface layer. The five main factors which may affect the number of plates washed by a given product are (i) the concentration of detergent, (ii) the temperature of the water, (iii) the hardness of the water, (iv) the type of "soil" on the plates, and (v) the method of washing used by the operator.... The difficulty of standardizing the soil is overcome by using the plates from a works canteen (cafeteria) for the test and adopting a randomized complete block technique in which plates from any one course form a block.... One practical limitation is the number of plates available in any one block. This permits only four ... tests to be completed (in a block)."
Draw up steps (a)–(d) of a checklist for an experiment of the above type and give an example of a design that fits the requirements of your checklist.

3. Designs with One Source of Variation

3.1 Introduction
3.2 Randomization
3.3 Model for a Completely Randomized Design
3.4 Estimation of Parameters
3.5 One-Way Analysis of Variance
3.6 Sample Sizes
3.7 A Real Experiment—Soap Experiment, Continued
3.8 Using SAS Software
Exercises

3.1 Introduction

In working through the checklist in Chapter 2, the experimenter must choose an experimental design at step (c). A design is the rule that determines the assignment of the experimental units to treatments. The simplest possible design is the *completely randomized design*, where the experimental units are assigned to the treatments completely at random, subject to the number of observations to be taken on each treatment. Completely randomized designs involve no blocking factors.

Two ways of calculating the required number of observations (sample sizes) on each treatment are presented in Sections 3.6 and 4.5. The first method chooses sample sizes to obtain desired powers of hypothesis tests, and the second chooses sample sizes to achieve desired lengths of confidence intervals. We sometimes refer to the list of treatments and the corresponding sample sizes as the design, with the understanding that the assignment of experimental units to treatments is to be done completely at random.

In this chapter, we discuss the random assignment procedure for the completely randomized design, we introduce the method of least squares for estimating model parameters, and we develop a procedure for testing equality of the treatment parameters. The SAS computer analysis is described at the end of the chapter.

3.2 Randomization

In this section we provide a procedure for randomization that is very easily applied using a computer, but can equally well be done by hand. On a computer, the procedure requires the availability of software that stores data in rows and columns (like spreadsheet software, a SAS data set, or a Minitab worksheet), that includes a function that randomly generates real numbers between zero and one, and that includes the capacity to sort rows by the values in one column.

We use r_i to denote the number of observations to be taken on the ith treatment, and $n = \Sigma r_i$ to denote the total number of observations (and hence the required number of experimental units). We code the treatments from 1 to v and label the experimental units 1 to n.

Step 1: Enter into one column r_1 1's, then r_2 2's, ..., and finally r_v v's, giving a total of $n = \Sigma r_i$ entries. These represent the treatment labels.

Step 2: Enter into another column $n = \Sigma r_i$ random numbers, including enough digits to avoid ties. (The random numbers can be generated by a computer program or read from Table A.1).

Step 3: Reorder both columns so that the random numbers are put in ascending order. This arranges the treatment labels into a random order.

Step 4: Assign experimental unit t to the treatment whose label is in row t.

If the number n of experimental units is a k-digit integer, then the list in step 2 should be a list of k-digit random numbers. To obtain k-digit random numbers from Table A.1, a random starting place is found as described in Section 1.1.4, page 3. The digits are then read across the rows in groups of k (ignoring spaces).

Table 3.1 Randomization

Unsorted Treatments	Unsorted Random Numbers	Sorted Treatments	Sorted Random Numbers	Experimental Unit
1	0.533	3	0.139	1
1	0.683	2	0.379	2
2	0.702	3	0.411	3
2	0.379	1	0.533	4
3	0.411	1	0.683	5
3	0.962	2	0.702	6
3	0.139	3	0.962	7

We illustrate the randomization procedure using the SAS computer package in Section 3.8.1, page 57. The procedure can equally well be done using the random digits in Table A.1 and sorting by hand.

Example 3.2.1

Consider a completely randomized design for three treatments and sample sizes $r_1 = r_2 = 2, r_3 = 3$. The unrandomized design (step 1 of the randomization procedure) is 1 1 2 2 3 3 3, and is listed in column 1 of Table 3.1. Suppose step 2 generates the random numbers in column 2 of Table 3.1. In step 3, columns 1 and 2 are sorted so that the entries in column 2 are in ascending order. This gives columns 3 and 4. In step 4, the entries in column 3 are matched with experimental units 1–7 in order, so that column 3 contains the design after randomization. Treatment 1 is in rows 4 and 5, so experimental units 4 and 5 are assigned to treatment 1. Likewise, units 2 and 6 are assigned to treatment 2, and units 1, 3 and 7 are assigned to treatment 3. The randomly ordered treatments are then 3 2 3 1 1 2 3, and the experimental units 1–7 are assigned to the treatments in this order. □

3.3 Model for a Completely Randomized Design

A model is an equation that shows the dependence of the response variable upon the levels of the treatment factors. (Models involving block effects or covariates are considered in later chapters.)

Let Y_{it} be a random variable that represents the response obtained on the tth observation of the ith treatment. Let the parameter μ_i denote the "true response" of the ith treatment, that is, the response that would always be obtained from the ith treatment if it could be observed under *identical* experimental conditions and measured without error. Of course, this ideal situation can never happen—there is always some variability in the experimental procedure even if only caused by inaccuracies in reading measuring instruments. Sources of variation that are deemed to be minor and ignored during the planning of the experiment also contribute to variation in the response variable. These sources of nuisance variation are usually represented by a single variable ϵ_{it}, called an *error variable*, which is a random variable with zero mean. The model is then

$$Y_{it} = \mu_i + \epsilon_{it}, \quad t = 1, \ldots, r_i, \quad i = 1, \ldots, v,$$

where v is the number of treatments and r_i is the number of observations to be taken on the ith treatment. An alternative way of writing this model is to replace the parameter μ_i by $\mu + \tau_i$, so that the model becomes

$$Y_{it} = \mu + \tau_i + \epsilon_{it}, \quad t = 1, \ldots, r_i, \quad i = 1, \ldots, v.$$

In this model, $\mu + \tau_i$ denotes the true mean response for the ith treatment, and examination of differences between the parameters μ_i in the first model is equivalent to examination of differences between the parameters τ_i in the second model.

It will be seen in Section 3.4 that unique estimates of the parameters in the second formulation of the model cannot be obtained. Nevertheless, many experimenters prefer this model. The parameter μ is a constant, and the parameter τ_i represents the positive or negative deviation of the response from this constant when the ith treatment is observed. This deviation is called the "effect" on the response of the ith treatment.

The above models are *linear models*, that is, the response variable is written as a linear function of the parameters. Any model that is not, or cannot, be transformed into a linear model cannot be treated by the methods in this book. Linear models often provide reasonably good approximations to more complicated models, and they are used extensively in practice.

The specific forms of the distributions of the random variables in a model need to be identified before any statistical analyses can be done. The error variables represent all the minor sources of variation taken together, including all the measurement errors. In many experiments, it is reasonable to assume that the error variables are independent and that they have a normal distribution with zero mean and unknown variance σ^2, which must be estimated. We call these assumptions the *error assumptions*. It will be shown in Chapter 5 that plots of the experimental data give good indications of whether or not the error assumptions are likely to be true. Proceeding with the analysis when the constant variance, normality, or independence assumptions are violated can result in a totally incorrect analysis.

A complete statement of the model for any experiment should include the list of error assumptions. Thus, for a completely randomized design with v specifically selected treatments (fixed effects), the model is

$$Y_{it} = \mu + \tau_i + \epsilon_{it}, \qquad (3.3.1)$$

$$\epsilon_{it} \sim N(0, \sigma^2),$$

ϵ_{it}'s are mutually independent,

$$t = 1, \ldots, r_i, \quad i = 1, \ldots, v,$$

where "$\sim N(0, \sigma^2)$" denotes "has a normal distribution with mean 0 and variance σ^2." This is sometimes called a *one-way analysis of variance model*, since the model includes only one major source of variation, namely the treatment effect, and because the standard analysis of data using this model involves a comparison of measures of variation.

Notice that it is unnecessary to specify the distribution of Y_{it} in the model, as it is possible to deduce this from the stated information. Since Y_{it} is modeled as the sum of a treatment mean $\mu + \tau_i$ and a normally distributed random variable ϵ_{it}, it follows that

$$Y_{it} \sim N(\mu + \tau_i, \sigma^2).$$

Also, since the ϵ_{it}'s are mutually independent, the Y_{it}'s must also be mutually independent. Therefore, if the model is a true representation of the behavior of the response variable, then the data values y_{it} for the ith treatment form a random sample from a $N(\mu + \tau_i, \sigma^2)$ distribution.

3.4 Estimation of Parameters

3.4.1 Estimable Functions of Parameters

A function of the parameters of any model is said to be *estimable* if and only if it can be written as the expected value of a linear combination of the response variables. Only estimable functions of the parameters have unique linear unbiased estimates. Since it makes no sense to work with functions that have an infinite possible number of values, it is important that the analysis of the experiment involve only the estimable functions. For the one-way analysis of variance model (3.3.1), every estimable function is of the form

$$E\left[\sum_i \sum_t a_{it} Y_{it}\right] = \sum_i \sum_t a_{it} E[Y_{it}]$$
$$= \sum_i \sum_t a_{it}(\mu + \tau_i) = \sum_i b_i(\mu + \tau_i),$$

where $b_i = \sum_t a_{it}$ and the a_{it}'s are real numbers. Any function not of this form is *nonestimable*.

Clearly, $\mu + \tau_1$ is estimable, since it can be obtained by setting $b_1 = 1$ and $b_2 = b_3 = \cdots = b_v = 0$. Similarly, each $\mu + \tau_i$ is estimable. If we choose $b_i = c_i$ where $\sum c_i = 0$, we see that $\sum c_i \tau_i$ is estimable. Any such function $\sum c_i \tau_i$ for which $\sum_i c_i = 0$ is called a *contrast*, so all contrasts are estimable in the one-way analysis of variance model. For example, setting $b_1 = 1, b_2 = -1, b_3 = \cdots = b_v = 0$ shows that $\tau_1 - \tau_2$ is estimable. Similarly, each $\tau_i - \tau_s, i \neq s$, is estimable. Notice that there are no values of b_i that give $\mu, \tau_1, \tau_2, \ldots$, or τ_v separately as the expected value. Therefore, these parameters are not individually estimable.

3.4.2 Notation

We write the ith treatment sample mean as

$$\overline{Y}_{i.} = \frac{1}{r_i}\left(\sum_{t=1}^{r_i} Y_{it}\right)$$

and the corresponding observed sample mean as $\overline{y}_{i.}$. The "dot" notation means "add over all values of the subscript replaced with a dot," and the "bar" means "divide by the number of terms that have been added up." This notation will be extremely useful throughout this book. For example, in the next subsection we write

$$\frac{1}{n}\sum_{i=1}^{v}\sum_{t=1}^{r_i} y_{it} = \frac{1}{n}\sum_{i=1}^{v} y_{i.} = \frac{1}{n} y_{..} = \overline{y}_{..}, \text{ where } n = \sum_{i=1}^{v} r_i = r_.,$$

so that $\overline{y}_{..}$ is the average of all of the observations. Note that if the summation applies to a subscript on two variables, the dot notation cannot be used. For example, $\sum r_i \hat{\tau}_i$ cannot be written as $r_. \hat{\tau}_.$, since $r_. \hat{\tau}_.$ denotes $(\sum r_i)(\sum \hat{\tau}_i)$.

3.4.3 Obtaining Least Squares Estimates

The *method of least squares* is used to obtain estimates and estimators for estimable functions of parameters in linear models. We shall show that the ith treatment sample mean $\overline{Y}_{i.}$ and its observed value $\overline{y}_{i.}$ are the "least squares estimator" and "least squares estimate," respectively, of $\mu + \tau_i$. Least squares solutions for the parameters $\mu, \tau_1, \ldots, \tau_v$ are any set of corresponding values $\hat{\mu}, \hat{\tau}_1, \ldots, \hat{\tau}_v$ that minimize the sum of squared errors

$$\sum_{i=1}^{v}\sum_{t=1}^{r_i} e_{it}^2 = \sum_{i=1}^{v}\sum_{t=1}^{r_i}(y_{it} - \mu - \tau_i)^2. \qquad (3.4.2)$$

The estimated model $\hat{y}_{it} = \hat{\mu} + \hat{\tau}_i$ is the model that best fits the data in the sense of minimizing (3.4.2).

Finding least squares solutions is a standard problem in calculus.* The sum of squared errors (3.4.2) is differentiated with respect to each of the parameters $\mu, \tau_1, \ldots, \tau_v$ in turn. Then each of the $v + 1$ resulting derivatives is set equal to zero, yielding a set of $v + 1$ equations. These $v + 1$ equations are called the *normal equations*. Any solution to the normal equations gives a minimum value of the sum of squared errors (3.4.2) and provides a set of least squares solutions for the parameters.

The reader is asked to verify in Exercise 6 that the normal equations for the one-way analysis of variance model (3.3.1) are those shown in (3.4.3). The first equation in (3.4.3) is obtained by setting the derivative of the sum of squared errors of (3.4.2) with respect to μ equal to zero, and the other v equations are obtained by setting the derivatives with respect to each τ_i in turn equal to zero. We put "hats" on the parameters at this stage to denote solutions. The $v + 1$ normal equations are

$$y_{..} - n\hat{\mu} - \sum_{i} r_i \hat{\tau}_i = 0, \qquad (3.4.3)$$
$$y_{i.} - r_i \hat{\mu} - r_i \hat{\tau}_i = 0, \quad i = 1, \ldots, v,$$

and include $v + 1$ unknown parameters. From the last v equations, we obtain

$$\hat{\mu} + \hat{\tau}_i = \overline{y}_{i.}, \quad i = 1, \ldots, v,$$

so the least squares solution for the ith treatment mean $\mu + \tau_i$ is the corresponding sample mean $\overline{y}_{i.}$.

There is a problem in solving the normal equations to obtain least squares solutions for each parameter $\mu, \tau_1, \ldots, \tau_v$ individually. If the last v normal equations (3.4.3) are added together, the first equation results. This means that the $v + 1$ equations are not distinct (not linearly independent). The last v normal equations *are* distinct, since they each contain a different τ_i. Thus, there are exactly v distinct normal equations in $v + 1$ unknown parameters, and there is no unique solution for the parameters. This is not surprising, in view of the fact that we have already seen in Section 3.4.1 that these parameters are not individually estimable. For practical purposes, any one of the infinite number of solutions

---

*Readers without a background in calculus may note that the least squares solutions for the parameters, individually, are not unique and then may skip forward to Section 3.4.4.

3.4 Estimation of Parameters

will be satisfactory, since they lead to identical solutions for the estimable parameters. To obtain any one of these solutions, it is necessary to add a further equation to the set of normal equations. *Any* extra equation can be added, provided that it is not a linear combination of the equations already present. The trick is to add whichever equation will aid most in solving the entire set of equations.

One obvious possibility is to add the equation $\hat{\mu} = 0$, in which case the normal equations become

$$\hat{\mu} = 0,$$
$$y_{..} - \sum_i r_i \hat{\tau}_i = 0,$$
$$y_{i.} - r_i \hat{\tau}_i = 0, \quad i = 1, \ldots, v.$$

It is then a simple matter to solve the last v equations for the $\hat{\tau}_i$'s, yielding $\hat{\tau}_i = y_{i.}/r_i = \bar{y}_{i.}$. Thus, one solution to the normal equations is

$$\hat{\mu} = 0,$$
$$\hat{\tau}_i = \bar{y}_{i.}, \quad i = 1, \ldots, v.$$

A more common solution is obtained by adding the extra equation $\sum_i r_i \hat{\tau}_i = 0$ to (3.4.3). In this case, the normal equations become

$$\sum_i r_i \hat{\tau}_i = 0,$$
$$y_{..} - n\hat{\mu} = 0,$$
$$y_{i.} - r_i \hat{\mu} - r_i \hat{\tau}_i = 0, \quad i = 1, \ldots, v,$$

from which we obtain the least squares solutions

$$\hat{\mu} = \bar{y}_{..},$$
$$\hat{\tau}_i = \bar{y}_{i.} - \bar{y}_{..}, \quad i = 1, \ldots, v.$$

Still another solution, used, for example, by SAS computer software, is obtained by adding the equation $\hat{\tau}_v = 0$. Then the solutions to the normal equations are

$$\hat{\mu} = \bar{y}_{v.},$$
$$\hat{\tau}_i = \bar{y}_{i.} - \bar{y}_{v.}, \quad i = 1, \ldots, v.$$

In each of the three sets of solutions just obtained, it is always true that

$$\hat{\mu} + \hat{\tau}_i = \bar{y}_{i.}.$$

No matter which extra equation is added to the normal equations, $\bar{y}_{i.}$ will *always* be the least squares solution for $\mu + \tau_i$. Thus, although it is not possible to obtain unique least squares solutions for μ and τ_i separately, the least squares solution for the estimable true treatment mean $\mu + \tau_i$ *is* unique. We call $\bar{y}_{i.}$ the *least squares estimate* and $\bar{Y}_{i.}$ the *least squares estimator* of $\mu + \tau_i$. The notation $\mu + \tau_i$ is used somewhat ambiguously to mean both the least squares estimator and estimate. It should be clear from the context which of these is meant.

3.4.4 Properties of Least Squares Estimators

An important property of a least squares estimator is that

the least squares estimator of any estimable function of the parameters is the unique best linear unbiased estimator.

This statement, called the *Gauss–Markov Theorem*, is true for all linear models whose error variables are independent and have common variance σ^2. The theorem tells us that for the one-way analysis of variance model (3.3.1), the least squares estimator $\sum b_i \overline{Y}_{i.}$ of the estimable function $\sum b_i(\mu + \tau_i)$ is unique, is unbiased and has smallest variance. The theorem also tells us that τ_i cannot be estimable, since we have three different solutions for τ_i and none of the corresponding estimators has expected value equal to τ_i.

For the one-way analysis of variance model, Y_{it} has a normal distribution with mean $\mu + \tau_i$ and variance σ^2 (see Section 3.3), so $E[\overline{Y}_{i.}] = \mu + \tau_i$ and $\text{Var}(\overline{Y}_{i.}) = \sigma^2/r_i$. Therefore, the distribution of the least squares estimator $\overline{Y}_{i.}$ of $\mu + \tau_i$ is

$$\overline{Y}_{i.} \sim N(\mu + \tau_i, \sigma^2/r_i).$$

The $\overline{Y}_{i.}$'s are independent, since they are based on different Y_{it}'s. Consequently, the distribution of the least squares estimator $\sum c_i \overline{Y}_{i.}$ of the contrast $\sum c_i \tau_i$, with $\sum c_i = 0$, is

$$\sum c_i \overline{Y}_{i.} \sim N(\Sigma c_i \tau_i, \, \Sigma \frac{c_i^2}{r_i} \sigma^2).$$

Example 3.4.1 Heart–lung pump experiment

The following experiment was run by Richard Davis at The Ohio State University in 1987 to determine the effect of the number of revolutions per minute (rpm) of the rotary pump head of an Olson heart–lung pump on the fluid flow rate. The rpm was set directly on the tachometer of the pump console and PVC tubing of size 3/8" by 3/32" was used. The flow rate was measured in liters per minute. Five equally spaced levels of the treatment factor "rpm" were selected, namely, 50, 75, 100, 125, and 150 rpm, and these were coded as 1, 2, 3, 4, 5, respectively. The experimental design was a completely randomized design with $r_1 = r_3 = r_5 = 5, r_2 = 3$, and $r_4 = 2$. The data, in the order collected, are given in Table 3.2, and the summary information is

3.4 Estimation of Parameters

Table 3.2 Fluid flow obtained from the rotary pump head of an Olson heart–lung pump

Observation	rpm	Level	Liters/minute
1	150	5	3.540
2	50	1	1.158
3	50	1	1.128
4	75	2	1.686
5	150	5	3.480
6	150	5	3.510
7	100	3	2.328
8	100	3	2.340
9	100	3	2.298
10	125	4	2.982
11	100	3	2.328
12	50	1	1.140
13	125	4	2.868
14	150	5	3.504
15	100	3	2.340
16	75	2	1.740
17	50	1	1.122
18	50	1	1.128
19	150	5	3.612
20	75	2	1.740

$$y_{1.} = 5.676, \quad r_1 = 5, \quad \bar{y}_{1.} = 1.1352,$$
$$y_{2.} = 5.166, \quad r_2 = 3, \quad \bar{y}_{2.} = 1.7220,$$
$$y_{3.} = 11.634, \quad r_3 = 5, \quad \bar{y}_{3.} = 2.3268,$$
$$y_{4.} = 5.850, \quad r_4 = 2, \quad \bar{y}_{4.} = 2.9250,$$
$$y_{5.} = 17.646, \quad r_5 = 5, \quad \bar{y}_{5.} = 3.5292.$$

The least squares estimate of the mean fluid flow rate when the pump is operating at 150 rpm is

$$(\hat{\mu} + \hat{\tau}_5) = \bar{y}_{5.} = 3.5292$$

liters per minute. The other mean fluid flow rates are estimated in a similar way. The experimenter expected the flow rate to increase as the rpm of the pump head was increased. Figure 3.1 supports this expectation.

Since the variance of the least squares estimator $\bar{Y}_{i.}$ of $\mu + \tau_i$ is σ^2/r_i, the first, third, and fifth treatment means are more precisely measured than the second and fourth.

The least squares estimate of the difference in fluid flow rate between 50 rpm and 150 rpm is

$$(\hat{\tau}_5 - \hat{\tau}_1) = (\hat{\mu} + \hat{\tau}_5) - (\hat{\mu} + \hat{\tau}_1) = \bar{y}_{5.} - \bar{y}_{1.} = 2.394$$

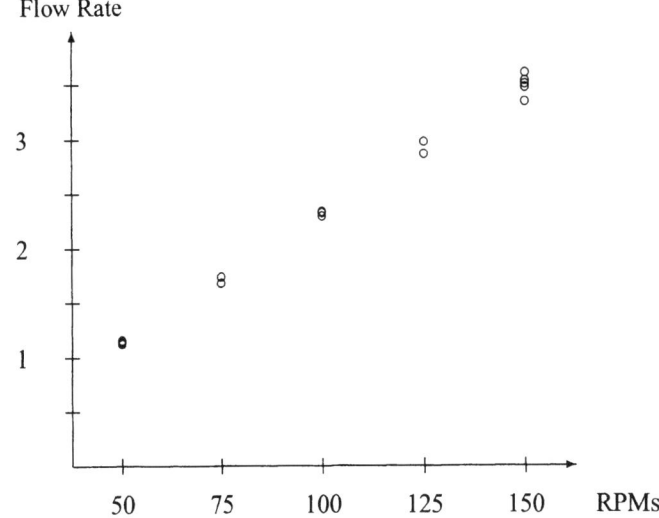

Figure 3.1
Plot of data for the heart–lung pump experiment

liters per minute. The associated variance is

$$\sum_{i=1}^{5} \frac{c_i^2}{r_i} \sigma^2 = \left(\frac{1}{5} + \frac{1}{5}\right)\sigma^2 = 0.4\,\sigma^2.$$

3.4.5 Estimation of σ^2

The least squares estimates $\hat{\mu} + \hat{\tau}_i = \bar{y}_{i.}$ of $\mu + \tau_i$ ($i = 1, \ldots, v$) minimize the sum of squared errors. Therefore, for the one-way analysis of variance model (3.3.1), the minimum possible value of the sum of squared errors (3.4.2), which we write as ssE, is equal to

$$ssE = \sum_i \sum_t \hat{e}_{it}^2 = \sum_i \sum_t (y_{it} - \hat{\mu} - \hat{\tau}_i)^2.$$

Here, $\hat{e}_{it} = (y_{it} - \hat{\mu} - \hat{\tau}_i)$ is the deviation of the tth observation on the ith treatment from the estimated ith treatment mean. This is called the (it)th *residual*. Substituting the least squares estimates $\hat{\mu} + \hat{\tau}_i = \bar{y}_{i.}$ into the formula for ssE, we have

$$ssE = \sum_i \sum_t (y_{it} - \bar{y}_{i.})^2. \tag{3.4.4}$$

The minimum sum of squared errors, ssE, is called the *sum of squares for error* or the *error sum of squares*, and is used below to find an unbiased estimate of the error variance σ^2.

Useful computational formulae for ssE, obtained by multiplying out the quantity in parentheses in (3.4.4), are

$$ssE = \sum_i \sum_t y_{it}^2 - \sum_i r_i \bar{y}_{i.}^2 \tag{3.4.5}$$

$$= \sum_i \sum_t y_{it}^2 - \sum_i (y_{i.}^2 / r_i). \tag{3.4.6}$$

3.4 Estimation of Parameters

Now, the random variable *SSE* corresponding to the minimum sum of squared errors *ssE* in (3.4.4) is

$$SSE = \sum_i \sum_t (Y_{it} - \overline{Y}_{i.})^2 = \sum_i (r_i - 1) S_i^2 , \qquad (3.4.7)$$

where $S_i^2 = \sum_{t=1}^{r_i} (Y_{it} - \overline{Y}_{i.})^2 / (r_i - 1))$ is the sample variance for the ith treatment. In Exercise 11, the reader is asked to verify that S_i^2 is an unbiased estimator of the error variance σ^2. Then, the expected value of *SSE* is

$$E(SSE) = \sum_i (r_i - 1) E(S_i^2) = (n - v)\sigma^2 ,$$

giving an unbiased estimator of σ^2 as

$$\hat{\sigma}^2 = SSE/(n - v) = MSE. \qquad (3.4.8)$$

The corresponding unbiased estimate of σ^2 is the observed value of *MSE*, namely $msE = ssE/(n - v)$. Both *MSE* and *msE* are called the *mean square for error* or *error mean square*. The estimate *msE* is sometimes called the "within groups (or within treatments) variation."

3.4.6 Confidence Bound for σ^2

If an experiment were to be repeated in the future, the estimated value of σ^2 obtained from the current experiment could be used at step (h) of the checklist to help calculate the number of observations that should be taken in the new experiment (see Sections 3.6.2 and 4.5). However, the error variance in the new experiment is unlikely to be exactly the same as that in the current experiment, and in order not to underestimate the number of observations needed, it is advisable to use a larger value of σ^2 in the sample size calculation. One possibility is to use the upper limit of a one-sided confidence interval for σ^2.

It can be shown that the distribution of SSE/σ^2 is chi-squared with $n - v$ degrees of freedom, denoted by χ^2_{n-v}. Consequently,

$$P\left(\frac{SSE}{\sigma^2} \geq \chi^2_{n-v, 1-\alpha}\right) = 1 - \alpha , \qquad (3.4.9)$$

where $\chi^2_{n-v, 1-\alpha}$ is the percentile of the chi-squared distribution with $n - v$ degrees of freedom and with probability of $1 - \alpha$ in the right-hand tail.

Manipulating the inequalities in (3.4.9), and replacing *SSE* by its observed value *ssE*, gives a one-sided $100(1 - \alpha)\%$ confidence bound for σ^2 as

$$\sigma^2 \leq \frac{ssE}{\chi^2_{n-v, 1-\alpha}} . \qquad (3.4.10)$$

This upper bound is called a $100(1 - \alpha)\%$ *upper confidence limit* for σ^2.

Table 3.3 Data for the battery experiment

Battery Type	Life per unit cost (minutes per dollar)				$\bar{y}_{i.}$
1	611	537	542	593	570.75
2	923	794	827	898	860.50
3	445	490	384	413	433.00
4	476	569	480	460	496.25

Example 3.4.2 Battery experiment, continued

The data of the battery experiment (Section 2.5.2, page 26) are summarized in Table 3.3. The sum of squares for error is obtained from (3.4.5); that is,

$$ssE = \sum_i \sum_t y_{it}^2 - \sum_i r_i \bar{y}_{i.}^2$$
$$= 6,028,288 - 4(570.75^2 + 860.50^2 + 433.00^2 + 496.25^2)$$
$$= 28,412.5.$$

An unbiased estimate of the error variance is then obtained as

$$msE = ssE/(n-v) = 28,412.5/(16-4) = 2367.71.$$

A 95% upper confidence limit for σ^2 is given by

$$\sigma^2 \leq \frac{ssE}{\chi^2_{12, 0.95}} = \frac{28,412.5}{5.23} = 5432.60,$$

and taking the square root of the confidence limit, a 95% upper confidence limit for σ is 73.71 minutes per dollar. If the experiment were to be repeated in the future, the calculation for the number of observations at step (h) of the checklist might take the largest likely value for σ to be around 70–75 minutes per dollar. □

3.5 One-Way Analysis of Variance

3.5.1 Testing Equality of Treatment Effects

In an experiment involving v treatments, an obvious question is whether or not the treatments differ at all in terms of their effects on the response variable. Thus one may wish to test the null hypothesis

$$H_0 : \{\tau_1 = \tau_2 = \cdots = \tau_v\}$$

that the treatment effects are all equal against the alternative hypothesis

$$H_A : \{\text{at least two of the } \tau_i\text{'s differ}\}.$$

At first glance, the null hypothesis appears to involve nonestimable parameters. However, we can easily rewrite it in terms of $v-1$ estimable contrasts, as follows:

$$H_0 : \{\tau_1 - \tau_2 = 0 \text{ and } \tau_1 - \tau_3 = 0 \text{ and } \cdots \text{ and } \tau_1 - \tau_v = 0\}.$$

3.5 One-Way Analysis of Variance

This is not the only way to rewrite H_0 in terms of estimable contrasts. For example, we could use the contrasts $\tau_i - \bar{\tau}_.$ (where $\bar{\tau}_. = \sum \tau_i/v$) and write the null hypothesis as follows:

$$H_0 : \{\tau_1 - \bar{\tau}_. = 0 \text{ and } \tau_2 - \bar{\tau}_. = 0 \text{ and } \cdots \text{ and } \tau_v - \bar{\tau}_. = 0\}.$$

Now $\bar{\tau}_.$ is the average of the τ_i's, so the $\tau_i - \bar{\tau}_.$'s add to zero. Consequently, if $\tau_i - \bar{\tau}_. = 0$ for $i = 1, \ldots, v - 1$, then $\tau_v - \bar{\tau}_.$ must also be zero. Thus, this form of the null hypothesis could be written in terms of just the first $v - 1$ estimable functions $\tau_1 - \bar{\tau}_., \ldots, \tau_{v-1} - \bar{\tau}_.$.

Any way that we rewrite H_0 in terms of estimable functions of the parameters, it will always depend on $v - 1$ distinct contrasts. The number $v - 1$ is called the *treatment degrees of freedom*.

The basic idea behind an analysis of variance test is that the sum of squares for error measures how well the model fits the data. Consequently, a way of testing H_0 is to compare the sum of squares for error under the original one-way analysis of variance model (3.3.1), known as the *full model*, with that obtained from the modified model, which assumes that the null hypothesis is true. This modified model is called the *reduced model*.

Under H_0, the τ_i's are equal, and we can write the common value of $\tau_1, \ldots, \tau_v$ as τ. If we incorporate this into the one-way analysis of variance model, we obtain the reduced model

$$Y_{it} = \mu + \tau + \epsilon_{it}^0,$$
$$\epsilon_{it}^0 \sim N(0, \sigma^2),$$
ϵ_{it}^0's are mutually independent,
$$t = 1, \ldots, r_i, \quad i = 1, \ldots, v,$$

where we write ϵ_{it}^0 for the (it)th error variable in the reduced model. To calculate the sum of squares for error, ssE_0, we need to determine the value of $\mu + \tau$ that minimizes the sum of squared errors

$$\sum_i \sum_t (y_{it} - \mu - \tau)^2.$$

Using calculus, the reader is asked to show in Exercise 7 that the unique least squares estimate of $\mu + \tau$ is the sample mean of all the observations; that is, $\hat{\mu} + \hat{\tau} = \bar{y}_{..}$. Therefore, the error sum of squares for the reduced model is

$$ssE_0 = \sum_i \sum_t (y_{it} - \bar{y}_{..})^2$$
$$= \sum_i \sum_t y_{it}^2 - n\bar{y}_{..}^2. \qquad (3.5.11)$$

If the null hypothesis $H_0 : \{\tau_1 = \tau_i = \ldots = \tau_v\}$ is false, and the treatment effects differ, the sum of squares for error ssE under the full model (3.3.1) is considerably smaller than the sum of squares for error ssE_0 for the reduced model. This is depicted in Figure 3.2. On the other hand, if the null hypothesis is true, then ssE_0 and ssE will be very similar. The analysis of variance test is based on the difference $ssE_0 - ssE$, relative to the size of ssE; that is, the test is based on $(ssE_0 - ssE)/ssE$. We would want to reject H_0 if this quantity is large.

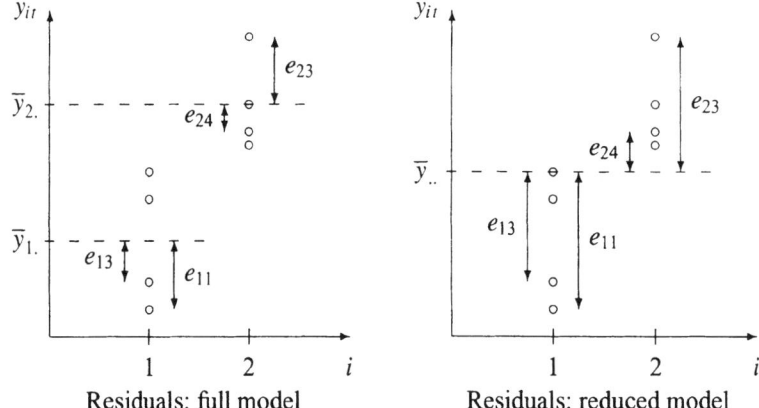

Figure 3.2
Residuals under the full and reduced models when H_0 is false

We call $ssT = ssE_0 - ssE$ the *sum of squares for treatments* or the *treatment sum of squares*,

since its value depends on the differences between the treatment effects. Using formulas (3.5.11) and (3.4.5) for ssE_0 and ssE, the treatment sum of squares is

$$ssT = ssE_0 - ssE \tag{3.5.12}$$

$$= \left(\sum_i \sum_t y_{it}^2 - n\bar{y}_{..}^2\right) - \left(\sum_i \sum_t y_{it}^2 - \sum_i r_i \bar{y}_{i.}^2\right)$$

$$= \sum_i r_i \bar{y}_{i.}^2 - n\bar{y}_{..}^2. \tag{3.5.13}$$

Since $\bar{y}_{i.} = y_{i.}/r_i$ and $\bar{y}_{..} = y_{..}/n$, we can also write ssT as

$$ssT = \frac{\sum_i y_{i.}^2}{r_i} - \frac{y_{..}^2}{n} \tag{3.5.14}$$

A third, equivalent, formulation is

$$ssT = \sum_i r_i (\bar{y}_{i.} - \bar{y}_{..})^2. \tag{3.5.15}$$

The reader is invited to multiply out the parentheses in (3.5.15) and verify that (3.5.13) is obtained. There is a shortcut method of expanding (3.5.15) to obtain (3.5.13). First write down each term in y and square it. Then associate with each squared term the signs in (3.5.15). Finally, precede each term with the summations and constant outside the parentheses in (3.5.15). This quick expansion will work for all terms like (3.5.15) in this book. Formula (3.5.15) is probably the easiest form of ssT to remember. The best form for computations is (3.5.14), since it is the least sensitive to rounding error, while (3.5.13) is the easiest to manipulate for theoretical work.

Since we will reject H_0 if ssT/ssE is large, we need to know what "large" means. This in turn means that we need to know the distribution of the corresponding random variable SST/SSE when H_0 is true, where

$$SST = \sum_i r_i (\bar{Y}_{i.} - \bar{Y}_{..})^2 \quad \text{and} \quad SSE = \sum_i \sum_t (Y_{it} - \bar{Y}_{i.})^2. \tag{3.5.16}$$

3.5   One-Way Analysis of Variance

Now, as mentioned in Section 3.4.6, it can be shown that SSE/σ^2 has a chi-squared distribution with $n - v$ degrees of freedom, denoted by χ^2_{n-v}. Similarly, it can be shown that when H_0 is true, SST/σ^2 has a χ^2_{v-1} distribution, and that SST and SSE are independent. The ratio of two independent chi-squared random variables, each divided by their degrees of freedom, has an F distribution. Therefore, if H_0 is true, we have

$$\frac{SST/\sigma^2(v-1)}{SSE/\sigma^2(n-v)} \sim F_{v-1,n-v}.$$

We now know the distribution of SST/SSE multiplied by the constant $(n-v)/(v-1)$, and we want to reject the null hypothesis $H_0 : \{\tau_1 = \cdots = \tau_v\}$ in favor of the alternative hypothesis H_A : {at least two of the treatment effects differ} if this ratio is large. Thus, if we write $msT = ssT/(v-1)$, $msE = ssE/(n-v)$, where ssT and ssE are the observed values of the treatment sum of squares and error sum of squares, respectively, our decision rule is to

$$\text{reject } H_0 \text{ if } \frac{msT}{msE} > F_{v-1,n-v,\alpha}, \qquad (3.5.17)$$

where $F_{v-1,n-v,\alpha}$ is the critical value from the F distribution with $v-1$ and $n-v$ degrees of freedom with α in the right-hand tail. The probability α is often called the *significance level* of the test and is the probability of rejecting H_0 when in fact it is true (a Type I error). Thus, α should be selected to be small if it is important not to make a Type I error ($\alpha = 0.01$ and 0.001 are typical choices); otherwise, α can be chosen to be a little larger ($\alpha = 0.10$ and 0.05 are typical choices). Critical values $F_{v-1,n-v,\alpha}$ for the F distribution are given in Table A.6. Due to lack of space, only a few typical values of α have been tabulated.

The calculations involved in the test of the hypothesis H_0 against H_A are usually written as an *analysis of variance table* as shown in Table 3.4. The last line shows the *total sum of squares* and *total degrees of freedom*. The total sum of squares, $sstot$, is $(n-1)$ times the sample variance of all of the data values. Thus,

$$sstot = \sum_i \sum_t (y_{it} - \overline{y}_{..})^2 = \sum_i \sum_t y_{it}^2 - n\overline{y}_{..}^2. \qquad (3.5.18)$$

From (3.5.11), we see that $sstot$ happens to be equal to ssE_0 for the one-way analysis of variance model, and from (3.5.12) we see that

$$sstot = ssT + ssE.$$

Table 3.4   One-way analysis of variance table

Source of Variation	Degrees of Freedom	Sum of Squares	Mean Square	Ratio	Expected Mean Square
Treatments	$v-1$	ssT	$\frac{ssT}{v-1}$	$\frac{msT}{msE}$	$\sigma^2 + Q(\tau_i)$
Error	$n-v$	ssE	$\frac{ssE}{n-v}$		σ^2
Total	$n-1$	$sstot$			
Computational Formulae					
$ssT = \sum_i r_i \overline{y}_{i.}^2 - n\overline{y}_{..}^2$			$ssE = \sum_i \sum_t y_{it}^2 - \sum_i r_i \overline{y}_{i.}^2$		
$sstot = \sum_i \sum_t y_{it}^2 - n\overline{y}_{..}^2$					
$Q(\tau_i) = \sum_i r_i(\tau_i - \sum_h r_h \tau_h/n)^2/(v-1)$					

Thus, the total sum of squares consists of a part ssT that is explained by differences between the treatment effects and a part ssE that is not explained by any of the parameters in the model.

Example 3.5.1 Battery experiment, continued

Consider the battery experiment introduced in Section 2.5.2, page 26. The sum of squares for error was calculated in Example 3.4.4, page 44, to be $ssE = 28{,}412.5$. From the life per unit cost data given in Table 3.3, page 44, we have

$$y_{1.} = 2283, \ y_{2.} = 3442, \ y_{3.} = 1732, \ y_{4.} = 1985, \ y_{..} = 9442,$$

Also, $\Sigma\Sigma y_{it}^2 = 6{,}028{,}288$ and $r_i = 4$. Hence, the sums of squares ssT (3.5.14) and $sstot$ (3.5.18) are

$$ssT = \sum (y_{i.}^2/r_i) - y_{..}^2/n$$
$$= (2283^2/4 + 3442^2/4 + 1732^2/4 + 1985^2/4) - 9442^2/16$$
$$= 427{,}915.25,$$
$$sstot = ssE_0 = \sum\sum y_{it}^2 - (y_{..})^2/n$$
$$= 6{,}028{,}288 - 9442^2/16 = 456{,}327.75,$$

and we can verify that $sstot = ssT + ssE$.

The decision rule for testing the null hypothesis $H_0 : \{\tau_1 = \tau_2 = \tau_3 = \tau_4\}$ that the four battery types have the same average life per unit cost against the alternative hypothesis that at least two of the battery types differ, at significance level α, is

reject H_0 if $msT/msE = 60.24 > F_{3,12,\alpha}$.

From Table A.6, it can be seen that $60.24 > F_{3,12,\alpha}$ for any of the tabulated values of α. For example, if α is chosen to be 0.01, then $F_{3,12,0.01} = 5.95$. Thus, for any tabulated choice of α, the null hypothesis is rejected, and it is concluded that at least two of the battery types differ in mean life per unit cost. In order to investigate which particular pairs of battery types differ, we would need to calculate confidence intervals. This will be done in Chapter 4. □

3.5.2 Use of p-Values

The *p-value* of a test is the smallest choice of α that would allow the null hypothesis to be rejected. For convenience, computer packages usually print the p-value as well as the ratio msT/msE. Having information about the p-value saves looking up $F_{v-1,n-v,\alpha}$ in Table A.6.

Table 3.5 One-way analysis of variance table for the battery experiment

Source of Variation	Degrees of Freedom	Sum of Squares	Mean Square	Ratio	p-value
Type	3	427,915.25	142,638.42	60.24	0.0001
Error	12	28,412.50	2,367.71		
Total	15	456,327.75			

All we need to do is to compare the p-value with our selected value of α. Therefore, the decision rule for testing $H_0 : \{\tau_1 = \cdots \tau_v\}$ against $H_A : \{$not all of τ_i's are equal$\}$ can be written as

reject H_0 if $p < \alpha$.

Example 3.5.2 Battery experiment, continued

In the battery experiment of Example 3.5.1, the null hypothesis $H_0 : \{\tau_1 = \tau_2 = \tau_3 = \tau_4\}$ that the four battery types have the same average life per unit cost was tested against the alternative hypothesis that they do not. The p-value for the test is shown in Table 3.5 as $p = 0.0001$. A value of 0.0001 in the SAS computer output indicates that the p-value is less than or equal to 0.0001. Smaller values are not printed explicitly. If α were chosen to be 0.01, then the null hypothesis would be rejected, since $p < \alpha$. □

3.6 Sample Sizes

Before an experiment can be run, it is necessary to determine the number of observations that should be taken on each treatment. This forms step (h) of the checklist in Section 2.2. In order to make this determination, the experimenter must first ascertain the approximate cost, in both time and money, of taking each observation and whether the cost differs for different levels of the treatment factor(s). There will probably be a fixed budget for the entire experiment. Therefore, remembering to set aside sufficient resources for the analysis of the experimental data, a rough calculation can be made of the maximum number, N, of observations that can be afforded. After having worked through steps (a)–(g) of the checklist, the experimenter will have identified the objectives of the experiment and the type of analysis required. It must now be ascertained whether or not the objectives of the experiment can be achieved within the budget. The calculations at step (h) may show that it is unnecessary to take as many as N observations, in which case valuable resources can be saved. Alternatively, and unfortunately the more likely, it may be found that more than N observations are needed in order to fulfill all the experimenter's requirements of the experiment. In this case, the experimenter needs to go back and review the decisions made so far in order to try to relax some of the requirements. Otherwise, an increase in budget needs to be obtained. There is little point in running the experiment with smaller sample sizes than those required without finding out what effect this will have on the analysis. The following quotation from J. N. R. Jeffers in his article "Acid rain and tree roots: an analysis" in *The Statistical Consultant in Action* (1987) is worth careful consideration:

> There is a quite strongly held view among experimenters that statisticians always ask for more replication than can be provided, and hence jeopardize the research by suggesting that it is not worth doing unless sufficient replication can be provided. There is, of course, some truth in this allegation, and equally, some truth in the view that, unless an experiment can be done with adequate replication, and with due regard

to the size of the difference which it is important to be able to detect, the research may indeed not be worth doing.

We will consider two methods of determining the number of observations on each treatment (the sample sizes). One method, which involves specifying the desired length of confidence intervals, will be presented in Section 4.5. The other method, which involves specifying the power required of the analysis of variance, is the topic of this section. Since the method uses the expected value of the mean square for treatments, we calculate this first.

3.6.1 Expected Mean Squares for Treatments

The formula for SST, the treatment sum of squares, was given in (3.5.16) on page 46. Its expected value is

$$E[SST] = E[\sum r_i(\overline{Y}_{i.} - \overline{Y}_{..})^2]$$
$$= E[\Sigma r_i \overline{Y}_{i.}^2 - n\overline{Y}_{..}^2]$$
$$= \sum r_i E[\overline{Y}_{i.}^2] - n E[\overline{Y}_{..}^2].$$

From the definition of the variance of a random variable, we know that $\text{Var}(X) = E[X^2] - (E[X])^2$, so we can write $E[SST]$ as

$$E[SST] = \Sigma r_i[\text{Var}(\overline{Y}_{i.}) + (E[\overline{Y}_{i.}])^2] - n[\text{Var}(\overline{Y}_{..}) + (E[\overline{Y}_{..}])^2].$$

For the one-way analysis of variance model (3.3.1), the response variables Y_{it} are independent, and each has a normal distribution with mean $\mu + \tau_i$ and variance σ^2. So,

$$E[SST] = \sum r_i \left(\sigma^2/r_i + (\mu + \tau_i)^2 \right)$$
$$\quad - n \left(\sigma^2/n + \left(\mu + \sum r_i \tau_i / n \right)^2 \right)$$
$$= v\sigma^2 + n\mu^2 + 2\mu \sum r_i \tau_i + \sum r_i \tau_i^2$$
$$\quad - \sigma^2 - n\mu^2 - 2\mu \sum r_i \tau_i - (\sum r_i \tau_i)^2/n$$
$$= (v-1)[\sigma^2 + Q(\tau_i)],$$

where

$$Q(\tau_i) = \Sigma_i r_i \left(\tau_i - \Sigma_h r_h \tau_h/n \right)^2 /(v-1), \qquad (3.6.19)$$

and the expected value of the mean square for treatments $MST = SST/(v-1)$ is

$$E[MST] = \sigma^2 + Q(\tau_i),$$

which is the quantity we listed in the analysis of variance table, Table 3.4. We note that when the treatment effects are all equal, $Q(\tau_i) = 0$, and $E[MST] = \sigma^2$.

3.6.2 Sample Sizes Using Power of a Test

Suppose that one of the major objectives of an experiment is to examine whether or not the treatments all have a similar effect on the response. The null hypothesis is actually somewhat unrealistic. The effects of the treatments are almost certainly not *exactly* equal, and even if they were, the nuisance variability in the experimental data would mask this fact. In any case, if the different levels produce only a very small difference in the response variable, the experimenter may not be interested in discovering this fact. For example, a difference of 5 minutes in life per dollar in two different batteries would probably not be noticed by most users. However, a larger difference such as 60 minutes may well be noticed. Thus the experimenter might require H_0 to be rejected with high probability if $\tau_i - \tau_s > 60$ minutes per dollar for some $i \neq s$ but may not be concerned about rejecting the null hypothesis if $\tau_i - \tau_s \leq 5$ minutes per dollar for all $i \neq s$. In most experiments, there is some value Δ such that if the difference in the effects of any two of the treatments exceeds Δ, the experimenter would like to reject the null hypothesis in favor of the alternative hypothesis with high probability.

The *power* of the test at Δ, denoted by $\pi(\Delta)$, is the probability of rejecting H_0 when the effects of at least two of the treatments differ by Δ. The power of the test $\pi(\Delta)$ is a function of Δ and also of the sample sizes, the number of treatments, the significance level α, and the error variance σ^2. Consequently, the sample sizes can be determined if $\pi(\Delta)$, v, α, and σ^2 are known. The values of Δ, $\pi(\Delta)$, v, and α are chosen by the experimenter, but the error variance has to be guessed using data from a pilot study or another similar experiment. In general, the largest likely value of σ^2 should be used. If the guess for σ^2 is too small, then the power of the test will be lower than the specified $\pi(\Delta)$. If the guess for σ^2 is too high, then the power will be higher than needed, and differences in the τ_i's smaller than Δ will cause H_0 to be rejected with high probability.

The rule for testing the null hypothesis $H_0 : \{\tau_1 = \cdots = \tau_v\}$ against H_A: {at least two of the τ_i's differ}, given in (3.5.17), on page 47, is

$$\text{reject } H_0 \text{ if } \frac{msT}{msE} > F_{v-1, n-v, \alpha} .$$

As stated in Section 3.5.1, the test statistic *MST/MSE* has an F distribution if the null hypothesis is correct. But if the null hypothesis is false, then *MST/MSE* has a related distribution called a noncentral F distribution. The noncentral F distribution is denoted by $F_{v-1, n-v, \delta^2}$, where δ^2 is called the *noncentrality parameter* and is defined to be

$$\delta^2 = (v-1)Q(\tau_i)/\sigma^2 , \qquad (3.6.20)$$

where $Q(\tau_i)$ was calculated in (3.6.19). When $Q(\tau_i) = 0$, then $\delta^2 = 0$, and the distribution of *MST/MSE* becomes the usual F-distribution. Otherwise, δ^2 is greater than zero, and the mean and spread of the distribution of *MST/MSE* are larger than those of the usual F-distribution. For equal sample sizes $r_1 = r_2 = \cdots = r_v = r$, we see that δ^2 reduces to

$$\delta^2 = r \sum_i (\tau_i - \overline{\tau}_.)^2/\sigma^2.$$

The calculation of the sample size r required to achieve a power $\pi(\Delta)$ at Δ for given v, α, and σ^2 rests on the fact that the hardest situation to detect is that in which the effects of two of the factor levels (say, the first and last) differ by Δ, and the others are all equal and midway between; that is,

$$\mu + \tau_2 = \mu + \tau_3 = \cdots = \mu + \tau_{v-1} = c,$$

$$\mu + \tau_1 = c + \Delta/2, \quad \text{and} \quad \mu + \tau_v = c - \Delta/2,$$

for some constant c. In this case,

$$\delta^2 = r \sum_i \frac{(\tau_i - \bar{\tau}.)^2}{\sigma^2} = \frac{r\Delta^2}{2\sigma^2}. \tag{3.6.21}$$

The power of the test depends on the sample size r through the distribution of MST/MSE, which depends on δ^2. Since the power of the test is the probability of rejecting H_0, we have

$$\pi(\Delta) = P\left(\frac{MST}{MSE} > F_{v-1, n-v, \alpha}\right).$$

The noncentral F distribution is tabulated in Table A.7, with power π given as a function of $\phi = \delta/\sqrt{v}$ for various values of $v_1 = v - 1$, $v_2 = n - v$, and α. Using (3.6.21),

$$\phi^2 = \frac{\delta^2}{v} = \frac{r\Delta^2}{2v\sigma^2},$$

so

$$r = \frac{2v\sigma^2\phi^2}{\Delta^2}. \tag{3.6.22}$$

Hence, given α, Δ, v, and σ^2, the value of r can be determined from Table A.7 to achieve a specified power $\pi(\Delta)$. The determination has to be done iteratively, since the denominator degrees of freedom, $v_2 = n - v = v(r - 1)$, depend on the unknown r. The procedure is as follows:

(a) Find the section of Table A.7 for the numerator degrees of freedom $v_1 = v - 1$ and the specified α (only $\alpha = 0.01$ and $\alpha = 0.05$ are shown).

(b) Calculate the denominator degrees of freedom using $v_2 = 1000$ in the first iteration and $v_2 = n - v = v(r - 1)$ in the following iterations, and locate the appropriate row of the table.

(c) For the required power $\pi(\Delta)$, use interpolation to determine the corresponding value of ϕ.

(d) Calculate $r = 2v\sigma^2\phi^2/\Delta^2$, rounding up to the nearest integer.

(e) Repeat steps (b)–(d) until the value of r is unchanged or alternates between two values. Select the larger of alternating values.

Example 3.6.1 Soap experiment, continued

The first part of the checklist for the soap experiment is given in Section 2.5.1, page 22, and is continued in Section 3.7, below. At step (h), the experimenter calculated the number of observations needed on each type of soap as follows.

The error variance was estimated to be about 0.007 square grams from the pilot experiment. In testing the hypothesis $H_0 : \{\tau_1 = \tau_2 = \tau_3\}$, the experimenter deemed it important to be able to detect a difference in weight loss of at least $\Delta = 0.25$ grams between any two soap types, with a probability 0.90 of correctly doing so, and a probability 0.05 of a Type I error. This difference was considered to be the smallest discrepancy in the weight loss of soaps that would be noticeable.

Using a one-way analysis of variance model, for $v = 3$ treatments, with $\Delta = 0.25$, $r = 2v\sigma^2\phi^2/\Delta^2 = 0.672\phi^2$, and $v_2 = v(r - 1) = 3(r - 1)$, r was calculated as follows. Using Table A.7 for $v_1 = v - 1 = 2$, $\alpha = 0.05$, and $\pi(\Delta) = 0.90$:

r	$v_2 = 3(r-1)$	ϕ	$r = 0.672\phi^2$	Action
	1000	2.33	3.65	Round up to $r = 4$
4	9	2.48	4.13	Round up to $r = 5$
5	12	2.33	3.65	Stop, and use $r = 4$ or 5.

The experimenter decided to take $r = 4$ observations on each soap type. □

3.7 A Real Experiment—Soap Experiment, Continued

The objective of the soap experiment described in Section 2.5.1, page 22, was to compare the extent to which three different types of soap dissolve in water. The three soaps selected for the experiment were a regular soap, a deodorant soap, and a moisturizing soap from a single manufacturer, and the weight-loss after 24 hours of soaking and 4 days drying is reproduced in Table 3.6. Steps (a)–(d) of the checklist were given in Section 2.5.1. The remaining steps and part of the analysis of the experimental data are described below. The first part of the description is based on the written report of the experimenter, Suyapa Silvia.

3.7.1 Checklist, Continued

(e) **Run a pilot experiment.**

A pilot experiment was run and used for two purposes. First, it helped to identify the difficulties listed at step (d) of the checklist. Secondly, it provided an estimate of σ^2 for step (h). The error variance was estimated to be about 0.007 grams^2. The value 0.007

Table 3.6 Data for the soap experiment

Soap	Weight-loss (grams)				$\bar{y}_{i.}$
1	−0.30	−0.10	−0.14	0.40	−0.0350
2	2.63	2.61	2.41	3.15	2.7000
3	1.86	2.03	2.26	1.82	1.9925

gm² was the value of *msE* in the pilot experiment. In fact, this is an underestimate, and it would have been better to have used the one-sided confidence bound (3.4.10) for σ^2.

(f) **Specify the model.**
Since care will be taken to control all extraneous sources of variation, it is assumed that the following model will be a reasonable approximation to the true model.

$$Y_{it} = \mu + \tau_i + \epsilon_{it},$$
$$\epsilon_{it} \sim N(0, \sigma^2),$$
ϵ_{it}'s are mutually independent
$$i = 1, 2, 3; \quad t = 1, \ldots r_i;$$

where τ_i is the (fixed) effect on the response of the ith soap, μ is a constant, Y_{it} is the weight loss of the tth cube of the ith soap, and ϵ_{it} is a random error.

Before analyzing the experimental data, the assumptions concerning the distribution of the error variables will be checked using graphical methods. (Assumption checking will be discussed in Chapter 5).

(g) **Outline the analysis.**
In order to address the question of differences in weights, a one-way analysis of variance will be computed at $\alpha = 0.05$ to test
$$H_0 : \{\tau_1 = \tau_2 = \tau_3\}$$
versus
$$H_A : \{ \text{ the effects of at least two pairs of soap types differ}\}.$$
To find out more about the differences among pairs of treatments, 95% confidence intervals for the pairwise differences of the τ_i will be calculated using Tukey's method (Tukey's method will be discussed in Section 4.4.4).

(h) **Calculate the number of observations that need to be taken.**
Four observations will be taken on each soap type. (See Example 3.6.2, page 53, for the calculation.)

(i) **Review the above decisions. Revise if necessary.**
It is not difficult to obtain 4 observations on each of 3 soaps, and therefore the checklist does not need revising. Small adjustments to the experimental procedure that were found necessary during the pilot experiment have already been incorporated into the checklist.

3.7.2 Data Collection and Analysis

The data collected by the experimenter are plotted in Figure 2.2, page 25, and reproduced in Table 3.6. The assumptions that the error variables are independent and have a normal distribution with constant variance were checked (using methods to be described in Chapter 5) and appear to be satisfied. The least squares estimates, $\hat{\mu} + \hat{\tau}_i = \bar{y}_{i.}$, of the average weight loss values (in grams) are

$$\bar{y}_{1.} = -0.0350, \quad \bar{y}_{2.} = 2.7000, \quad \bar{y}_{3.} = 1.9925.$$

3.7 A Real Experiment—Soap Experiment, Continued

Table 3.7 One-way analysis of variance table for the soap experiment

Source of Variation	Degrees of Freedom	Sum of Squares	Mean Square	Ratio	p-value
Soap	2	16.1220	8.0610	104.45	0.0001
Error	9	0.6945	0.0772		
Total	11	16.8166			

The hypothesis of no differences in weight loss due to the different soap types is tested below using an analysis of variance test.

Using the values $\bar{y}_{i.}$ given above, together with $\sum\sum y_{it}^2 = 45.7397$ and $r_1 = r_2 = r_3 = 4$, the sums of squares for Soap and Total are calculated using (3.5.14) and (3.5.18), as

$$ssT = \sum_i r_i \bar{y}_{i.}^2 - y_{..}^2/n$$
$$= \left[4(-0.0350)^2 + 4(2.7000)^2 + 4(1.9925)^2\right] - \left[(18.63)^2/12\right] = 16.1221,$$
$$sstot = ssE_0 = \sum\sum y_{it}^2 - (y_{..})^2/n$$
$$= 45.7397 - (18.63)^2/12 = 16.8166.$$

The sum of squares for error can be calculated by subtraction, giving $ssE = sstot - ssT = 0.6946$, or directly from (3.4.5), as

$$ssE = \sum_i \sum_t y_{it}^2 - \sum_i r_i \bar{y}_{i.}^2$$
$$= 45.7397 - \left[4(-0.0350)^2 + 4(2.7000)^2 + 4(1.9925)^2\right] = 0.6946.$$

The estimate of error variability is then

$$\hat{\sigma}^2 = msE = ssE/(n-v) = 0.6945/(12-3) = 0.0772.$$

The sums of squares and mean squares are shown in the analysis of variance table, Table 3.7. Notice that the estimate of σ^2 is ten times larger than the estimate of 0.007 grams^2 provided by the pilot experiment. This suggests that the pilot experiment was not sufficiently representative of the main experiment. As a consequence, the power of detecting a difference of $\Delta = 0.25$ grams between the weight losses of the soaps is, in fact, somewhat below the desired probability 0.90 (see Exercise 17).

The decision rule for testing $H_0 : \{\tau_1 = \tau_2 = \tau_3\}$ against the alternative hypothesis, that at least two of the soap types differ in weight loss, using a significance level of $\alpha = 0.05$, is to reject H_0 if $msT/msE = 104.45 > F_{2,9,0.05}$. From Table A.6, $F_{2,9,0.05} = 4.26$. Consequently, the null hypothesis is rejected, and it is concluded that at least two of the soap types do differ in their weight loss after 24 hours in water (and 4 days drying time). This null hypothesis would have been rejected for most practical choices of α. If α had been chosen to be as small as 0.005, $F_{2,9,\alpha}$ is still only 10.1. Alternatively, if the analysis is done by computer, the p-value would be printed in the computer output. Here the p-value is less than 0.0001, and H_0 would be rejected for any choice of α above this value.

The experimenter was interested in estimating the contrasts $\tau_i - \tau_u$ for all $i \neq u$, that is, she was interested in comparing the effects on weight loss of the different types of soaps. For the one-way analysis of variance model (3.3.1) and a completely randomized design, all contrasts are estimable, and the least squares estimate of $\tau_i - \tau_u$ is

$$\hat{\tau}_i - \hat{\tau}_u = (\hat{\mu} + \hat{\tau}_i) - (\hat{\mu} + \hat{\tau}_u) = \overline{y}_{i.} - \overline{y}_{u.}.$$

Hence, the least square estimates of the differences in the treatment effects are

$$\hat{\tau}_2 - \hat{\tau}_3 = 0.7075, \qquad \hat{\tau}_2 - \hat{\tau}_1 = 2.7350, \qquad \hat{\tau}_3 - \hat{\tau}_1 = 2.0275.$$

Confidence intervals for the differences will be evaluated in Example 4.4.5.

3.7.3 Discussion by the Experimenter

The results of this experiment were unexpected in that the soaps reacted with the water in very different ways, each according to its ingredients. An examination of the soap packages showed that for the deodorant soap and the moisturizing soap, water is listed as the third ingredient, whereas the regular soap claims to be 99.44% pure soap. Information on the chemical composition of soaps revealed that soaps are sodium and/or potassium salts of oleic, palmitic, and coconut oils and therefore in their pure form (without water) should float as the regular soap bars do. The other two soaps under discussion contain water and therefore are more dense and do not float.

One possible reason for the regular soap's actual increase in weight is that this "dry" soap absorbed and retained the water and dissolved to a lesser extent during the soaking period. The deodorant soap and the moisturizing soap, on the other hand, already contained water and did not absorb as much as the regular soap. They dissolved more easily during the soaking phase as a consequence. This is somewhat supported by the observation that the dissolved soap gel that formed extensively around the deodorant soap and the moisturizing soap did not form as much around the regular soap. Furthermore, the regular soap appeared to increase in size and remain larger, even at the end of the drying period.

3.7.4 Further Observations by the Experimenter

The soaps were weighed every day for one week after the experimental data had been collected in order to see what changes continued to occur. The regular soap eventually lost most of the water it retained, and the average loss of weight (due to dissolution) was less than that for the other two soaps.

If this study were repeated, with a drying period of at least one week, I believe that the results would indicate that regular soap loses less weight due to dissolution than either of the deodorant soap or the moisturizing soap

3.8 Using SAS Software

3.8.1 Randomization

A simple procedure for randomizing a completely randomized design was given in Section 3.2, on page 34. This procedure is easily implemented using the SAS software, as we now illustrate. Consider a completely randomized design for two treatments and $r = 3$ observations on each, giving a total of $n = 6$ observations. The following SAS statements create and print a data set named DESIGN, which includes the lists of values of the two variables TREATMNT and RANNO as required by steps 1 and 2 of the randomization procedure in Section 3.2. The statements INPUT and LINES are instructions to SAS that the values of TREATMNT are being input on the lines that follow rather than from an external data file. Inclusion of "@@" in the INPUT statement allows the levels of TREATMNT to be entered on one line as opposed to one per line. For each treatment label entered for the variable TREATMNT, a corresponding value of RANNO is generated using the SAS random number generating function RANUNI.

```
DATA DESIGN;
  INPUT TREATMNT @@;
  RANNO=RANUNI(0);
  LINES;
  1 1 1 2 2 2
PROC PRINT;
```

The statement PROC PRINT then prints the following output. The column labeled OBS (observation) is given by SAS as reference numbers.

	The SAS System	
OBS	TREATMNT	RANNO
1	1	0.37590
2	1	0.12212
3	1	0.74290
4	2	0.53347
5	2	0.95505
6	2	0.74718

The following statements, which follow PROC PRINT, sort the data set by the values of RANNO, as required by step 3 of the randomization procedure, and print the randomized design along with the ordered experimental unit labels 1–6 under the heading OBS.

```
PROC SORT;
  BY RANNO;
PROC PRINT;
```

The resulting output is as follows.

```
         The SAS System
OBS    TREATMNT    RANNO
 1        1        0.12212
 2        1        0.37590
 3        2        0.53347
 4        1        0.74290
 5        2        0.74718
 6        2        0.95505
```

Experimental units 1, 2, and 4 are assigned to treatment 1, and experimental units 3, 5, and 6 are assigned to treatment 2.

3.8.2 Analysis of Variance

In this section we illustrate how SAS software can be used to conduct a one-way analysis of variance test for equality of the treatment effects, assuming that model (3.3.1) is appropriate. We use the data in Table 2.7, page 24, from the soap experiment.

A sample SAS program to analyze the data is given in Table 3.8. Line numbers have been included for reference, but the line numbers are not part of the SAS program and if included would cause SAS software to generate error messages.

The option LINESIZE=72 in the OPTIONS statement in line 1 of the program causes all output generated by the program to be restricted to 72 characters per line. This is convenient

Table 3.8 Sample SAS program for the soap experiment

Line	SAS Program
1	OPTIONS LINESIZE=72;
2	DATA;
3	INPUT WTLOSS SOAP;
4	LINES;
5	-0.30 1
6	-0.10 1
7	-0.14 1
8	: :
9	1.82 3
10	;
11	PROC PRINT;
12	;
13	PROC PLOT;
14	PLOT WTLOSS*SOAP
15	/ VPOS=11 HPOS=40;
16	;
17	PROC GLM;
18	CLASS SOAP;
19	MODEL WTLOSS=SOAP;
20	MEANS SOAP;
21	LSMEANS SOAP;

3.8 Using SAS Software 59

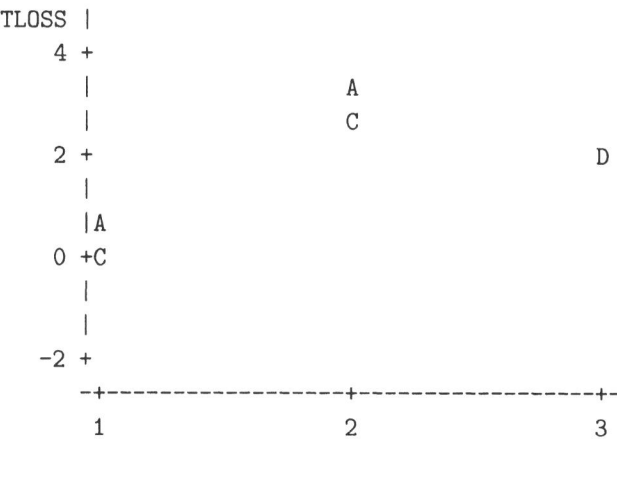

Figure 3.3
SAS data plot for the soap experiment

for viewing output on a terminal or printing output on 8.5 by 11 inch paper in the portrait orientation. We have included this option when running all of our programs, though it will not be shown henceforth in any programs.

Lines 2–10 of the program create a data set that includes as variables the response variable WTLOSS and the corresponding level of the treatment factor SOAP. Line 8 must be replaced by the additional data not shown here. The PRINT procedure (line 11) is used to print the data. The PLOT procedure (lines 13–15) generates the data plot shown in Figure 3.3, with WTLOSS on the vertical axis and SOAP on the horizontal axis, where A indicates a single data point, B indicates two data values very close together, C indicates three similar data values, and so on. Options VPOS and HPOS in the PLOT statement control the size of the plot (vertical and horizontal). If no size options were specified, then line 15 would be unnecessary, but line 14 would need to end with a semicolon. Lines 12 and 16 have no purpose except to separate the SAS procedures.

The General Linear Models procedure PROC GLM (lines 17–21) generates an analysis of variance table and calculates sample means. The CLASS statement identifies SOAP as a major source of variation whose levels are coded. The MODEL statement defines the response variable as WTLOSS, and the only source of variation included in the model is SOAP. The parameter μ and the error variables are automatically included in the model. The MODEL statement causes the analysis of variance table shown in Table 3.9 to be calculated. The F Value is the value of the ratio msT/msE for testing the null hypothesis that the three treatment effects are all equal. The value Pr > F is the p-value of the test to be compared with the chosen significance level. When the p-value is listed as 0.0001, it is actually less

Table 3.9 Sample SAS output from PROC GLM for the soap experiment.

```
                       The SAS System
                 General Linear Models Procedure

Dependent Variable: WTLOSS
                         Sum of        Mean
Source            DF    Squares      Square    F Value    Pr > F
Model              2   16.122050    8.061025    104.45    0.0001
Error              9    0.694575    0.077175
Corrected Total   11   16.816625
```

Table 3.10 Output from MEANS and LSMEANS for the soap experiment

```
                     The SAS System
               General Linear Models Procedure

Level of           ------------WTLOSS----------
SOAP       N          Mean              SD
1          4       -0.03500000       0.30259985
2          4        2.70000000       0.31601688
3          4        1.99250000       0.20022904

                Least Squares Means

              SOAP          WTLOSS
                            LSMEAN
               1         -0.03500000
               2          2.70000000
               3          1.99250000
```

than or equal to 0.0001. The null hypothesis is rejected for any chosen significance level larger than this.

The MEANS statement (line 20 of Table 3.8) causes the sample means $\bar{y}_{i.}$ to be printed, and the LSMEANS statement (line 21) requests printing of the least squares means, $\hat{\mu} + \hat{\tau}_i$. In the one-way analysis of variance model these are identical. The output from these two statements is shown in Table 3.10.

A calculation of the number of observations required for a future experiment cannot be done using PROC GLM. The value of *msE* from a pilot experiment can be calculated via an analysis of variance table, and then the sample size calculation needs to be done by hand as in Section 3.6.

Exercises

1. Suppose that you are planning to run an experiment with one treatment factor having four levels and no blocking factors. Suppose that the calculation of the required number of observations has given $r_1 = r_2 = r_3 = r_4 = 5$. Assign at random 20 experimental units to the $v = 4$ levels of the treatments, so that each treatment is assigned 5 units.

2. Suppose that you are planning to run an experiment with one treatment factor having three levels and no blocking factors. It has been determined that $r_1 = 3, r_2 = r_3 = 5$. Assign at random 13 experimental units to the $v = 3$ treatments, so that the first treatment is assigned 3 units and the other two treatments are each assigned 5 units.

3. Suppose that you are planning to run an experiment with three treatment factors, where the first factor has two levels and the other two factors have three levels each. Write out the coded form of the 18 treatment combinations. Assign 36 experimental units at random to the treatment combinations so that each treatment combination is assigned two units.

4. For the one-way analysis of variance model (3.3.1), page 36, the solution to the normal equations used by the SAS software is $\hat{\tau}_i = \bar{y}_{i.} - \bar{y}_{v.}$ ($i = 1, \ldots, v$) and $\hat{\mu} = \bar{y}_{v.}$.

    (a) Is τ_i estimable? Explain.

    (b) Calculate the expected value of the least squares estimator for $\tau_1 - \tau_2$ corresponding to the above solution. Is $\tau_1 - \tau_2$ estimable? Explain.

5. Consider a completely randomized design with observations on three treatments (coded 1, 2, 3). For the one-way analysis of variance model (3.3.1), page 36, determine which of the following are estimable. For those that are estimable, state the least squares estimator.

    (a) $\tau_1 + \tau_2 - 2\tau_3$.

    (b) $\mu + \tau_3$.

    (c) $\tau_1 - \tau_2 - \tau_3$.

    (d) $\mu + (\tau_1 + \tau_2 + \tau_3)/3$.

6. (requires calculus) Show that the normal equations for estimating $\mu, \tau_1, \ldots, \tau_v$ are those given in equation (3.4.3) on page 38.

7. (requires calculus) Show that the least squares estimator of $\mu + \tau$ is $\bar{Y}_{..}$ for the linear model $Y_{it} = \mu + \tau + \epsilon_{it}^0$ ($t = 1, \ldots, r_i; i = 1, 2, \ldots, v$), where the ϵ_{it}^0's are independent random variables with mean zero and variance σ^2. (This is the reduced model for the one-way analysis of variance test, Section 3.5.1, page 44.)

8. For the model in the previous exercise, find an unbiased estimator for σ^2. (Hint: first calculate $E[ssE_0]$ in (3.5.11), page 45.)

9. (requires calculus) Find the least squares estimates of $\mu_1, \mu_2, \ldots, \mu_v$ for the linear model $Y_{it} = \mu_i + \epsilon_{it}$ ($t = 1, \ldots, r_i; i = 1, 2, \ldots, v$), where the ϵ_{it}'s are independent

random variables with mean zero and variance σ^2. Compare these estimates with the least squares estimates of $\mu + \tau_i$ ($i = 1, 2, \ldots, v$) in model (3.3.1), page 36.

10. For the model in the previous exercise, find an unbiased estimator for σ^2. Compare the estimator with that in (3.4.8), page 43.

11. Verify, for the one-way analysis of variance model (3.3.1), page 36, that each treatment sample variance S_i^2 is an unbiased estimator of the error variance σ^2, so that

$$E(SSE) = \sum_i (r_i - 1)E(S_i^2) = (n - v)\sigma^2.$$

12. **Balloon experiment** (Meily Lin, 1985)

    Prior to 1985, the experimenter had observed that some colors of birthday balloons seem to be harder to inflate than others. She ran this experiment to determine whether balloons of different colors are similar in terms of the time taken for inflation to a diameter of 7 inches. Four colors were selected from a single manufacturer. An assistant blew up the balloons and the experimenter recorded the times (to the nearest 1/10 second) with a stop watch. The data, in the order collected, are given in Table 3.11, where the codes 1, 2, 3, 4 denote the colors pink, yellow, orange, blue, respectively.

    (a) Plot inflation time versus color and comment on the results.

    (b) Estimate the mean inflation time for each balloon color, and add these estimates to the plot from part (a).

    (c) Construct an analysis of variance table and test the hypothesis that color has no effect on inflation time.

    (d) Plot the data for each color in the order that it was collected. Are you concerned that the assumptions on the model are not satisfied? If so, why? If not, why not?

    (e) Is the analysis conducted in part (c) satisfactory?

13. **Heart–lung pump experiment, continued**

    The heart–lung pump experiment was described in Example 3.4.4, page 40, and the data were shown in Table 3.2, page 41.

Table 3.11 Times (in seconds) for the balloon experiment

Time Order	1	2	3	4	5	6	7	8
Coded color	1	3	1	4	3	2	2	2
Inflation Time	22.4	24.6	20.3	19.8	24.3	22.2	28.5	25.7
Time Order	9	10	11	12	13	14	15	16
Coded color	3	1	2	4	4	4	3	1
Inflation Time	20.2	19.6	28.8	24.0	17.1	19.3	24.2	15.8
Time Order	17	18	19	20	21	22	23	24
Coded color	2	1	4	3	1	4	4	2
Inflation Time	18.3	17.5	18.7	22.9	16.3	14.0	16.6	18.1
Time Order	25	26	27	28	29	30	31	32
Coded color	2	4	2	3	3	1	1	3
Inflation Time	18.9	16.0	20.1	22.5	16.0	19.3	15.9	20.3

Table 3.12 Times (in seconds) for the "walk" sign to appear in the pedestrian light experiment

\multicolumn{4}{c}{Number of pushes}
0	1	2	3
38.14	38.28	38.17	38.14
38.20	38.17	38.13	38.30
38.31	38.08	38.16	38.21
38.14	38.25	38.30	38.04
38.29	38.18	38.34	38.37
38.17	38.03	38.34	
38.20	37.95	38.17	
	38.26	38.18	
	38.30	38.09	
	38.21	38.06	

(a) Calculate an analysis of variance table and test the null hypothesis that the number of revolutions per minute has no effect on the fluid flow rate.

(b) Are you happy with your conclusion? Why or why not?

(c) Calculate a 90% upper confidence limit for the error variance σ^2.

14. **Pedestrian light experiment** (Larry Lesher, 1985)

Larry Lesher ran this experiment in order to determine whether pushing a certain pedestrian light button had an effect on how long he had to wait before the pedestrian light showed "walk." The treatment factor of interest was the number of pushes of the button, and 13 observations were taken for each of 0, 1, 2, and 3 pushes of the button. The waiting times for the "walk" sign for the first 32 observations in the order collected are shown in Table 3.12, giving $r_0 = 7, r_1 = r_2 = 10, r_3 = 5$ (where the levels of the treatment factor are coded as 0, 1, 2, 3 for simplicity). The observations were taken as close together as possible on a Saturday in February 1985.

(a) Plot the waiting times against the number of pushes of the button. What does the plot show?

(b) Construct an analysis of variance table and test the null hypothesis that the number of pushes of the pedestrian button has no effect on the waiting time for the "walk" sign.

(c) Estimate the mean waiting time for each number of pushes and the corresponding variance.

(d) Estimate the contrast $\tau_0 - (\tau_1 + \tau_2 + \tau_3)/3$, which compares the effect of no pushes of the button with the average effect of pushing the button once, twice, or three times.

(e) Calculate the variance associated with the least squares estimator of the contrast in part (d). How does the value of the variance compare with the variance σ^2 of the random error variables?

Table 3.13 Data for the Trout Experiment

Code	Hemoglobin (grams per 100 ml)									
1	6.7	7.8	5.5	8.4	7.0	7.8	8.6	7.4	5.8	7.0
2	9.9	8.4	10.4	9.3	10.7	11.9	7.1	6.4	8.6	10.6
3	10.4	8.1	10.6	8.7	10.7	9.1	8.8	8.1	7.8	8.0
4	9.3	9.3	7.2	7.8	9.3	10.2	8.7	8.6	9.3	7.2

Source: Gutsell, J. S. (1951). Copyright © 1951 International Biometric Society. Reprinted with permission.

15. **Trout experiment** (Gutsell, 1951, Biometrics)

    The data in Table 3.13 show the measurements of hemoglobin (grams per 100 ml) in the blood of brown trout. The trout were placed at random in four different troughs. The fish food added to the troughs contained, respectively, 0, 5, 10, and 15 grams of sulfamerazine per 100 pounds of fish (coded 1, 2, 3, 4). The measurements were made on ten randomly selected fish from each trough after 35 days.

    (a) Plot the data and comment on the results.

    (b) Write down a suitable model for this experiment.

    (c) Calculate the least squares estimate of the mean response for each treatment. Show these estimates on the plot obtained in (a). Can you draw any conclusions from these estimates?

    (d) Test the hypothesis that sulfamerazine has no effect on the hemoglobin content of trout blood.

    (e) Calculate a 95% upper confidence limit for σ^2.

16. **Trout experiment, continued**

    Suppose the trout experiment of Exercise 15 is to be repeated with the same $v = 4$ treatments, and suppose that the same hypothesis, that the treatments have no effect on hemoglobin content, is to be tested.

    (a) For calculating the number of observations needed on each treatment, what would you use as a guess for σ^2?

    (b) Calculate the sample sizes needed for an analysis of variance test with $\alpha = 0.05$ to have power 0.95 if (i) $\Delta = 1.5$. (ii) $\Delta = 1.0$. (iii) $\Delta = 2.0$.

17. **Soap experiment, continued**

    In Example 3.6.2, page 53, a sample size calculation was made for the number of observations needed to detect, with probability $\pi(0.25) = 0.90$, a difference in weight loss of at least $\Delta = 0.25$ grams in $v = 3$ difference types of soap, using an analysis of variance with a probability of $\alpha = 0.05$ of a Type I error. The calculation used an estimate of 0.007 grams^2 for σ^2 and showed that $r = 4$ observations were needed on each type of soap. The experiment was run with $r = 4$, and the least squares estimate for σ^2 was 0.0772. If the true value for σ^2 was, in fact, 0.08, what power did the test actually have for detecting a difference of $\Delta = 0.25$ grams in the weight loss of the three soaps?

Exercises 65

18. The diameter of a ball bearing is to be measured using three different calipers. How many observations should be taken on each caliper type if the null hypothesis H_0:{effects of the calipers are the same} is to be tested against the alternative hypothesis that the three calipers give different average measurements. It is required to detect a difference of 0.01 mm in the effects of the caliper types with probability 0.98 and a Type I error probability of $\alpha = 0.05$. It is thought that σ is about 0.03 mm.

19. An experiment is to be run to determine whether or not time differences in performing a simple manual task are caused by different types of lighting. Five levels of lighting are selected ranging from dim colored light to bright white light. The one-way analysis of variance model (3.3.1), page 36 is thought to be a suitable model, and $H_0 : \{\tau_1 = \tau_2 = \tau_3 = \tau_4 = \tau_5\}$ is to be tested against the alternative hypothesis H_A:{the τ_i's are not all equal} at significance level 0.05. How many observations should be taken at each light level given that the experimenter wishes to reject H_0 with probability 0.90 if the difference in the effects of any two light levels produces a 4.5-second time difference in the task? It is thought that σ is at most 3.0 seconds.

4. Inferences for Contrasts and Treatment Means

4.1 Introduction
4.2 Contrasts
4.3 Individual Contrasts and Treatment Means
4.4 Methods of Multiple Comparisons
4.5 Sample Sizes
4.6 Using SAS Software
Exercises

4.1 Introduction

The objective of an experiment is often much more specific than merely determining whether or not all of the treatments give rise to similar responses. For example, a chemical experiment might be run primarily to determine whether or not the yield of the chemical process increases as the amount of the catalyst is increased. A medical experiment might be concerned with the efficacy of each of several new drugs as compared with a standard drug. A nutrition experiment may be run to compare high fiber diets with low fiber diets. Such treatment comparisons are formalized in Section 4.2. The purpose of this chapter is to provide confidence intervals and hypothesis tests about treatment comparisons and treatment means. We start, in Section 4.3, by considering a single treatment comparison or mean, and then, in Section 4.4, we develop the techniques needed when more than one treatment comparison or mean is of interest. The number of observations required to achieve confidence intervals of given lengths is calculated in Section 4.5. SAS commands for confidence intervals and hypothesis tests are provided in Section 4.6.

4.2 Contrasts

In Chapter 3, we defined a contrast to be a linear combination of the parameters $\tau_1, \tau_2, \ldots, \tau_v$ of the form

$$\sum c_i \tau_i, \quad \text{with} \quad \sum c_i = 0.$$

For example, $\tau_u - \tau_s$ is the contrast that compares the effects (as measured by the response variable) of treatments u and s. If $\tau_u - \tau_s = 0$, then treatments u and s affect the response in exactly the same way, and we say that these treatments do not differ. Otherwise, the treatments do differ in the way they affect the response. We showed in Section 3.4 that for a completely randomized design and the one-way analysis of variance model (3.3.1), every contrast $\sum c_i \tau_i$ is estimable with least squares estimate

$$\sum c_i \hat{\tau}_i = \sum c_i (\hat{\mu} + \hat{\tau}_i) = \sum c_i \overline{y}_{i.} \quad (4.2.1)$$

and corresponding least squares estimator $\sum c_i \overline{Y}_{i.}$. The variance of the least squares estimator is

$$\text{Var}\left(\sum c_i \overline{Y}_{i.}\right) = \sum c_i^2 \text{Var}(\overline{Y}_{i.}) = \sum c_i^2 (\sigma^2 / r_i) = \sigma^2 \sum (c_i^2 / r_i). \quad (4.2.2)$$

The first equality uses the fact that the treatment sample means $\overline{Y}_{i.}$ involve different response variables, which in model (3.3.1) are independent. The error variance σ^2 is generally unknown and is estimated by the unbiased estimate msE, giving the estimated variance of the contrast estimator as

$$\widehat{\text{Var}}\left(\sum c_i \overline{Y}_{i.}\right) = msE \sum (c_i^2 / r_i).$$

The *estimated standard error* of the estimator is the square root of this quantity, namely,

$$\sqrt{\widehat{\text{Var}}\left(\sum c_i \overline{Y}_{i.}\right)} = \sqrt{msE \sum (c_i^2 / r_i)}. \quad (4.2.3)$$

Normalized contrasts When several contrasts are to be compared, it is sometimes helpful to be able to measure them all on the same scale. A contrast is said to be *normalized* if it is scaled so that its least squares estimator has variance σ^2. From (4.2.2), it can be seen that a contrast $\Sigma c_i \tau_i$ is normalized by dividing it by $\sqrt{\Sigma c_i^2 / r_i}$. If we write $h_i = c_i / \sqrt{\Sigma c_i^2 / r_i}$, then the least squares estimator $\Sigma h_i \overline{Y}_{i.}$ of the normalized contrast $\Sigma h_i \tau_i$ has the following distribution:

$$\sum h_i \overline{Y}_{i.} \sim N\left(\sum h_i \tau_i, \sigma^2\right), \quad \text{where } h_i = \frac{c_i}{\sqrt{\sum c_i^2 / r_i}}.$$

Normalized contrasts will be used for hypothesis testing (Section 4.3.3).

Contrast coefficients It is convenient to represent a contrast by listing only the coefficients of the parameters $\tau_1, \tau_2, \ldots, \tau_v$. Thus, $\sum c_i \tau_i = c_1 \tau_1 + c_2 \tau_2 + \cdots + c_v \tau_v$ would be represented by the list of *contrast coefficients*

$$[c_1, c_2, \ldots, c_v].$$

4.2.1 Pairwise Comparisons

As the name suggests, *pairwise comparisons* are simple differences $\tau_u - \tau_s$ of pairs of parameters τ_u and τ_s ($u \neq s$). These are of interest when the experimenter wishes to compare each treatment with every other treatment. The list of contrast coefficients for the pairwise difference $\tau_u - \tau_s$ is

$$[0, 0, 1, 0, \ldots, 0, -1, 0, \ldots, 0],$$

where the 1 and -1 are in positions u and s, respectively. The least squares estimate of $\tau_u - \tau_s$ is obtained from (4.2.1) by setting $c_u = 1$, $c_s = -1$, and all other c_i equal to zero, giving

$$\hat{\tau}_u - \hat{\tau}_s = \bar{y}_{u.} - \bar{y}_{s.},$$

and the corresponding least squares estimator is $\bar{Y}_{u.} - \bar{Y}_{s.}$. Its estimated standard error is obtained from (4.2.3) and is equal to

$$\sqrt{\widehat{\text{Var}}(\bar{Y}_{u.} - \bar{Y}_{s.})} = \sqrt{msE\left((1/r_u) + (1/r_s)\right)}.$$

Example 4.2.1 Battery experiment, continued

Details for the battery experiment were given in Section 2.5.2 (page 26). The experimenter was interested in comparing the life per unit cost of each battery type with that of each of the other battery types. The average lives per unit cost (in minutes/dollar) for the four batteries, calculated from the data in Table 2.8, page 28, are

$$\bar{y}_{1.} = 570.75, \quad \bar{y}_{2.} = 860.50, \quad \bar{y}_{3.} = 433.00, \quad \bar{y}_{4.} = 496.25.$$

The least squares estimates of the pairwise differences are, therefore,

$$\hat{\tau}_1 - \hat{\tau}_2 = -289.75, \quad \hat{\tau}_1 - \hat{\tau}_3 = 137.75, \quad \hat{\tau}_1 - \hat{\tau}_4 = 74.50,$$
$$\hat{\tau}_2 - \hat{\tau}_3 = 427.50, \quad \hat{\tau}_2 - \hat{\tau}_4 = 364.25, \quad \hat{\tau}_3 - \hat{\tau}_4 = -63.25.$$

The estimated pairwise differences suggest that battery type 2 (alkaline, store brand) is vastly superior to the other three battery types in terms of the mean life per unit cost. Battery type 1 (alkaline, name brand) appears better than types 3 and 4, and battery type 4 (heavy duty, store brand) better than type 3 (heavy duty, name brand). We do, however, need to investigate whether or not these perceived differences might be due only to random fluctuations in the data.

In Example 3.4.4 (page 44), the error variance was estimated to be $msE = 2367.71$. The sample sizes were $r_1 = r_2 = r_3 = r_4 = 4$, and consequently, the estimated standard error for each pairwise comparison is equal to

$$\sqrt{2367.71 \left(\tfrac{1}{4} + \tfrac{1}{4}\right)} = 34.41 \text{ min/\$}.$$

$ese = \sqrt{mse\left(\frac{1}{r_1} + \frac{1}{r_2} + \cdots + \frac{1}{r_i}\right)}$

It can be seen that all of the estimated pairwise differences involving battery type 2 are bigger than four times their estimated standard errors. This suggests that the perceived differences in battery type 2 and the other batteries are of sizeable magnitudes and are unlikely to be due to random error. We shall formalize these comparisons in terms of confidence intervals in Example 4.4.4 later in this chapter. □

4.2.2 Treatment Versus Control

If the experimenter is interested in comparing the effects of one special treatment with the effects of each of the other treatments, then the special treatment is called the *control*. For example, a pharmaceutical experiment might involve one or more experimental drugs together with a standard drug that has been on the market for some years. Frequently, the objective of such an experiment is to compare the effect of each experimental drug with that of the standard drug but not necessarily with the effects of any of the other experimental drugs. The standard drug is then the control. If we code the control as level 1, and the experimental drugs as levels $2, 3, \ldots, v$, respectively, then the contrasts of interest are $\tau_2-\tau_1, \tau_3-\tau_1, \ldots, \tau_v-\tau_1$. These contrasts are known as *treatment versus control* contrasts. They form a subset of the pairwise differences, so we can use the same formulae for the least squares estimate and the estimated standard error. The contrast coefficients for the contrast $\tau_i - \tau_1$ are $[-1, 0, \ldots, 0, 1, 0, \ldots, 0]$, where the 1 is in position i.

4.2.3 Difference of Averages

Sometimes the levels of the treatment factors divide naturally into two or more groups, and the experimenter is interested in the *difference of averages* contrast that compares the average effect of one group with the average effect of the other group(s). For example, consider an experiment that is concerned with the effect of different colors of exam paper (the treatments) on students' exam performance (the response). Suppose that treatments 1 and 2 represent the pale colors, white and yellow, whereas treatments 3, 4, and 5 represent the darker colors, blue, green and pink. The experimenter may wish to compare the effects of light and dark colors on exam performance. One way of measuring this is to estimate the contrast $\frac{1}{2}(\tau_1 + \tau_2) - \frac{1}{3}(\tau_3 + \tau_4 + \tau_5)$, which is the difference of the average effects of the light and dark colors. The corresponding contrast coefficients are

$$\left[\tfrac{1}{2}, \tfrac{1}{2}, -\tfrac{1}{3}, -\tfrac{1}{3}, -\tfrac{1}{3}\right].$$

From (4.2.1) and (4.2.3), the least squares estimate would be

$$\tfrac{1}{2}\bar{y}_{1.} + \tfrac{1}{2}\bar{y}_{2.} - \tfrac{1}{3}\bar{y}_{3.} - \tfrac{1}{3}\bar{y}_{4.} - \tfrac{1}{3}\bar{y}_{5.}$$

with estimated standard error

$$\sqrt{msE\left(\frac{1}{4r_1} + \frac{1}{4r_2} + \frac{1}{9r_3} + \frac{1}{9r_4} + \frac{1}{9r_5}\right)}.$$

Example 4.2.2 Battery experiment, continued

In the battery experiment of Section 2.5.2, page 26, battery types 1 and 2 were alkaline batteries, while types 3 and 4 were heavy duty. In order to compare the running time per unit cost of these two types of batteries, we examine the contrast $\frac{1}{2}(\tau_1 + \tau_2) - \frac{1}{2}(\tau_3 + \tau_4)$. The least squares estimate is

$$\tfrac{1}{2}(570.75 + 860.50) - \tfrac{1}{2}(433.00 + 496.25) = 251.00 \text{ min/\$},$$

suggesting that the alkaline batteries are more economical (on average by over two hours per dollar spent). The associated standard error is $\sqrt{msE(4/16)} = 24.32$ min/$, so the estimated difference in running time per unit cost is over ten times larger than the standard error, suggesting that the observed difference is not just due to random fluctuations in the data. □

4.2.4 Trends

Trend contrasts may be of interest when the levels of the treatment factor are quantitative and have a natural ordering. For example, suppose that the treatment factor is temperature and its selected levels are 50°C, 75°C, 100°C, coded as 1, 2, 3, respectively. The experimenter may wish to know whether the value of the response variable increases or decreases as the temperature increases and, if so, whether the rate of change remains constant. These questions can be answered by estimating linear and quadratic trends in the response.

The trend contrast coefficients for v *equally spaced* levels of a treatment factor and *equal sample sizes* are listed in Table A.2 for values of v between 3 and 7. For v treatments, trends up to $(v-1)$th order can be measured. Experimenters rarely use more than four levels for a quantitative treatment factor, since it is unusual for strong quartic and higher-order trends to occur in practice, especially within the narrow range of levels considered in a typical experiment.

Table A.2 does not tabulate contrast coefficients for unequally spaced levels or for unequal sample sizes. The general method of obtaining the coefficients of the trend contrasts involves fitting a regression model to the noncoded levels of the treatment factor. It can be shown that the linear trend contrast coefficients can easily be calculated as

$$c_i = r_i(nx_i - \Sigma r_i x_i), \qquad (4.2.4)$$

where r_i is the number of observations taken on the ith uncoded level x_i of the treatment factor, and $n = \Sigma r_i$ is the total number of observations. We are usually interested only in whether or not the linear trend is likely to be negligible, and to make this assessment, the contrast estimate is compared with its standard error. Consequently, we may multiply or divide the calculated coefficients by any integer without losing any information. When the r_i are all equal, the coefficients listed in Appendix A.2 are obtained, possibly multiplied or divided by an integer. Expressions for quadratic and higher-order trend coefficients are more complicated (see Draper and Smith, 1981, Sections 5.6 and 5.7).

Example 4.2.3 Heart–lung pump experiment, continued

The experimenter who ran the heart–lung pump experiment of Example 3.4.4, page 40, expected to see a linear trend in the data, since he expected the flow rate to increase as the number of revolutions per minute (rpm) of the pump head was increased. The plot of the data in Figure 3.1 (page 42) shows the observed flow rates at the five different levels of rpm. From the figure, it might be anticipated that the linear trend is large but higher-order trends are very small.

The five levels of rpm observed were 50, 75, 100, 125, 150, which are equally spaced. Had there been equal numbers of observations at each level, then we could have used the contrast coefficients $[-2, -1, 0, 1, 2]$ for the linear trend contrast and $[2, -1, -2, -1, 2]$ for the quadratic trend contrast as listed in Table A.2 for $v = 5$ levels of the treatment factor. However, here the sample sizes were $r_1 = r_3 = r_5 = 5, r_2 = 3$ and $r_4 = 2$. The coefficients for the linear trend are calculated via (4.2.4). Now $n = \Sigma r_i = 20$, and

$$\Sigma r_i x_i = 5(50) + 3(75) + 5(100) + 2(125) + 5(150) = 1975.$$

So, we have

x_i	$r_i(20 x_i - \Sigma r_i x_i)$
50	$5(1000 - 1975) = -4875$
75	$3(1500 - 1975) = -1425$
100	$5(2000 - 1975) = 125$
125	$2(2500 - 1975) = 1050$
150	$5(3000 - 1975) = 5125$

The coefficients are each divisible by 25, so rather than using the calculated coefficients $[-4875, -1425, 125, 1050, 5125]$, we can divide them by 25 and use the linear trend coefficients $[-195, -57, 5, 42, 205]$. The average flow rates (liters/minute) were calculated as

$$\bar{y}_{1.} = 1.1352, \quad \bar{y}_{2.} = 1.7220, \quad \bar{y}_{3.} = 2.3268, \quad \bar{y}_{4.} = 2.9250, \quad \bar{y}_{5.} = 3.5292.$$

The least squares estimate $\Sigma c_i \bar{y}_{i.}$ of the linear contrast is then

$$-195 \bar{y}_{1.} - 57 \bar{y}_{2.} + 5 \bar{y}_{3.} + 42 \bar{y}_{4.} + 205 \bar{y}_{5.} = 538.45$$

liters per minute. The linear trend certainly appears to be large. However, before drawing conclusions, we need to compare this trend estimate with its corresponding estimated standard error. The data give $\Sigma \Sigma y_{it}^2 = 121.8176$, and we calculate the error sum of squares (3.4.5), page 42, as $ssE = 0.0208$, giving an unbiased estimate of σ^2 as

$$msE = ssE/(n - v) = 0.0208/(20 - 5) = 0.001387.$$

The estimated standard error of the linear trend estimator is then

$$\sqrt{msE \left(\frac{(-195)^2}{5} + \frac{(-57)^2}{3} + \frac{(5)^2}{5} + \frac{(42)^2}{2} + \frac{(205)^2}{5} \right)} = 4.988.$$

Clearly, the estimate of the linear trend is extremely large compared with its standard error.

Had we normalized the contrast, the linear contrast coefficients would each have been divided by

$$\sqrt{\sum c_i^2/r_i} = \sqrt{\frac{(-195)^2}{5} + \frac{(-57)^2}{3} + \frac{(5)^2}{5} + \frac{(42)^2}{2} + \frac{(205)^2}{5}} = 134.09,$$

and the normalized linear contrast estimate would have been 4.0156. The estimated standard error of all normalized contrasts is $\sqrt{msE} = 0.03724$ for this experiment, so the normalized linear contrast estimate remains large compared with the standard error. □

4.3 Individual Contrasts and Treatment Means

4.3.1 Confidence Interval for a Single Contrast

In this section, we obtain a formula for a confidence interval for an individual contrast. If confidence intervals for more than one contrast are required, then the multiple comparison methods of Section 4.4 should be used instead. We give the formula first, and the derivation afterwards. A $100(1-\alpha)\%$ confidence interval for the contrast $\Sigma c_i \tau_i$ is

$$\sum c_i \bar{y}_{i.} - t_{n-v,\alpha/2} \sqrt{msE \sum c_i^2/r_i} \leq \sum c_i \tau_i \qquad (4.3.5)$$

$$\leq \sum c_i \bar{y}_{i.} + t_{n-v,\alpha/2} \sqrt{msE \sum c_i^2/r_i}.$$

We can write this more succinctly as

$$\sum c_i \tau_i \in \left(\sum c_i \bar{y}_{i.} \pm t_{n-v,\alpha/2} \sqrt{msE \sum c_i^2/r_i}\right), \qquad (4.3.6)$$

where the symbol $\pm$, which is read as "plus or minus," denotes that the upper limit of the interval is calculated using $+$ and the lower limit using $-$. The symbols "$\Sigma c_i \tau_i \in$" mean that the interval includes the true value of the contrast $\Sigma c_i \tau_i$ with $100(1-\alpha)\%$ confidence. For future reference, we note that the general form of the above confidence interval is

$$\sum c_i \tau_i \in \left(\sum c_i \hat{\tau}_i \pm t_{df,\alpha/2} \sqrt{\widehat{\text{Var}}(\Sigma c_i \hat{\tau}_i)}\right), \qquad (4.3.7)$$

where df is the number of degrees of freedom for error.

To derive the confidence interval (4.3.5), we will need to use some results about normally distributed random variables. As we saw in the previous section, for the completely randomized design and one-way analysis of variance model (3.3.1), the least squares estimator of the contrast $\sum c_i \tau_i$ is $\sum c_i \bar{Y}_{i.}$, which has variance $\text{Var}(\Sigma c_i \bar{Y}_{i.}) = \sigma^2 \sum c_i^2/r_i$. This estimator is a linear combination of normally distributed random variables and therefore also has a normal distribution. Subtracting the mean and dividing by the standard deviation gives us a random variable

$$\frac{D}{\sigma} = \frac{\sum c_i \bar{Y}_{i.} - \sum c_i \tau_i}{\sigma \sqrt{\sum c_i^2/r_i}}, \qquad (4.3.8)$$

which has a $N(0, 1)$ distribution. We estimate the error variance, σ^2, by msE, and from Section 3.4.6, page 43, we know that

$$MSE/\sigma^2 = SSE/(n-v)\sigma^2 \sim \chi^2_{n-v}/(n-v).$$

It can be shown that the random variables D and MSE are independent (see Graybill, 1976), and the ratio of a normally distributed random variable and a chi-squared random variable that are independent has a t-distribution with the same number of degrees of freedom as the chi-squared distribution. Hence, the ratio $D/\sqrt{MSE}$ has a t distribution with $n-v$ degrees of freedom. Using the expression (4.3.8), we can now write down the following probability statement about $D/\sqrt{MSE}$:

$$P\left(-t_{n-v,\alpha/2} \leq \frac{\sum c_i \overline{Y}_{i.} - \sum c_i \tau_i}{\sqrt{MSE \sum c_i^2/r_i}} \leq t_{n-v,\alpha/2}\right) = 1-\alpha,$$

where $t_{n-v,\alpha/2}$ is the percentile of the t_{n-v} distribution corresponding to a probability of $\alpha/2$ in the right-hand-tail, the value of which can be obtained from Table A.4. Manipulating the two inequalities, the probability statement becomes

$$P\left(\sum c_i \overline{Y}_{i.} - t_{n-v,\alpha/2}\sqrt{MSE \sum c_i^2/r_i} \leq \sum c_i \tau_i \right. \tag{4.3.9}$$

$$\left. \leq \sum c_i \overline{Y}_{i.} + t_{n-v,\alpha/2}\sqrt{MSE \sum c_i^2/r_i}\right) = 1-\alpha.$$

Then replacing the estimators by their observed values in this expression gives a $100(1-\alpha)\%$ confidence interval for $\sum c_i \tau_i$ as in (4.3.5).

Example 4.3.1 Heart–lung pump experiment, continued

Consider the heart–lung pump experiment of Examples 3.4.4 and 4.2.4, pages 40 and 72. The least squares estimate of the difference in fluid flow at 75 rpm and 50 rpm (levels 2 and 1 of the treatment factor, respectively) is

$$\Sigma c_i \overline{y}_{i.} = \overline{y}_{2.} - \overline{y}_{1.} = 0.5868$$

liters per minute. Since there were $r_2 = 5$ observations at 75 rpm and $r_1 = 3$ observations at 50 rpm, and $msE = 0.001387$, the estimated standard error of this contrast is

$$\sqrt{msE \, \Sigma c_i^2/r_i} = \sqrt{0.001387 \left(\tfrac{1}{3} + \tfrac{1}{5}\right)} = 0.0272 \text{ liters/minute}.$$

Using this information, together with $t_{15,0.025} = 2.131$, we obtain from (4.3.6) a 95% confidence interval (in units of liters per minute) for $\tau_2 - \tau_1$ as

$$(0.5868 \pm (2.131)(0.0272)) = (0.5288, 0.6448).$$

This tells us that with 95% confidence, the fluid flow at 75 rpm of the pump is between 0.53 and 0.64 liters per minute greater than at 50 rpm. □

Confidence bounds, or one-sided confidence intervals, can be derived in the same manner as two-sided confidence intervals. For the completely randomized design and one-way analysis of variance model (3.3.1), a $100(1 - \alpha)\%$ *upper confidence bound* for $\sum c_i \tau_i$ is

$$\sum c_i \tau_i < \sum c_i \bar{y}_{i.} + t_{df,\alpha} \sqrt{msE \sum c_i^2/r_i}, \qquad (4.3.10)$$

and a $100(1 - \alpha)\%$ *lower confidence bound* for $\sum c_i \tau_i$ is

$$\sum c_i \tau_i > \sum c_i \bar{y}_{i.} - t_{df,\alpha} \sqrt{msE \sum c_i^2/r_i}, \qquad (4.3.11)$$

where $t_{df,\alpha}$ is the percentile of the t distribution with df degrees of freedom and probability α in the right-hand tail.

4.3.2 Confidence Interval for a Single Treatment Mean

For the one-way analysis of variance model (3.3.1), the true mean response $\mu + \tau_s$ of the sth level of a treatment factor was shown in Section 3.4 to be estimable with least squares estimator $\bar{Y}_{s.}$. Although one is unlikely to be interested in only one of the treatment means, we can obtain a confidence interval as follows.

Since $\bar{Y}_{s.} \sim N(\mu + \tau_s, \sigma^2/r_s)$ for model (3.3.1), we can follow the same steps as those leading to (4.3.6) and obtain a $100(1 - \alpha)\%$ confidence interval for $\mu + \tau_s$ as

$$\mu + \tau_s \in (\bar{y}_{s.} \pm t_{df,\alpha/2} \sqrt{msE/r_s}). \qquad (4.3.12)$$

Example 4.3.2 Heart–lung pump experiment, continued

Suppose that the experimenter had required a 99% confidence interval for the true average fluid flow ($\mu + \tau_3$) for the heart–lung pump experiment of Example 3.4.4, page 40, when the revolutions per minute of the pump are set to 100 rpm. Using (4.3.12) and $r_3 = 5$, $\bar{y}_{3.} = 2.3268$, $msE = 0.001387$, $n - v = 20 - 5$, and $t_{15,0.005} = 2.947$, the 99% confidence interval for $\mu + \tau_3$ is

$$\mu + \tau_3 \in (2.3268 \pm (2.947)(0.01666)) = (2.2777, 2.3759).$$

So, with 99% confidence, the true average flow rate at 100 rpm of the pump is believed to be between 2.28 liters per minute and 2.38 liters per minute. □

4.3.3 Hypothesis Test for a Single Contrast or Treatment Mean

The outcome of a hypothesis test can be deduced from the corresponding confidence interval in the following way. The null hypothesis $H_0 : \Sigma c_i \tau_i = h$ will be rejected at significance level α in favor of the two-sided alternative hypothesis $H_A : \Sigma c_i \tau_i \neq h$ if the corresponding confidence interval for $\Sigma c_i \tau_i$ fails to contain h. For example, the 95% confidence interval for $\tau_2 - \tau_1$ in Example 4.3.1 does not contain zero, so the hypothesis $H_0 : \tau_2 - \tau_1 = 0$ (that the flow rates are the same at 50 rpm and 75 rpm) would be rejected at significance level $\alpha = 0.05$ in favor of the alternative hypothesis (that the flow rates are not equal).

We can make this more explicit, as follows. Suppose we wish to test the hypothesis $H_0 : \Sigma c_i \tau_i = 0$ against the alternative hypothesis $H_A: \Sigma c_i \tau_i \neq 0$. The interval (4.3.6) fails to contain 0 if the absolute value of $\Sigma c_i \bar{y}_{i.}$ is bigger than $t_{n-v,\alpha/2}\sqrt{msE\ \Sigma c_i^2/r_i}$. Therefore, the rule for testing the null hypothesis against the alternative hypothesis is

$$\text{reject } H_0 \text{ if } \left|\frac{\sum c_i \bar{y}_{i.}}{\sqrt{msE \sum c_i^2/r_i}}\right| > t_{n-v,\alpha/2}, \quad (4.3.13)$$

where $|\ |$ denotes absolute value. We call such rules *decision rules*. If H_0 is rejected, then H_A is automatically accepted. The test statistic can be squared, so that the decision rule becomes

$$\text{reject } H_0 \text{ if } \frac{(\sum c_i \bar{y}_{i.})^2}{msE \sum c_i^2/r_i} > t_{n-v,\alpha/2}^2 = F_{1,n-v,\alpha}, $$

and the F distribution can be used instead of the t distribution. Notice that the test statistic is the square of the normalized contrast estimate divided by msE. We call the quantity

$$ssc = \frac{(\sum c_i \bar{y}_{i.})^2}{\sum c_i^2/r_i} \quad (4.3.14)$$

the *sum of squares for the contrast*, or *contrast sum of squares* (even though it is the "sum" of only one squared term). The decision rule can be more simply expressed as

$$\text{reject } H_0 \text{ if } \frac{ssc}{msE} > F_{1,n-v,\alpha}. \quad (4.3.15)$$

For future reference, we can see that the general form of ssc/msE is

$$\frac{ssc}{msE} = \frac{(\sum c_i \hat{\tau}_i)^2}{\widehat{Var}(\sum c_i \hat{\tau}_i)}. \quad (4.3.16)$$

The above test is a two-tailed test, since the null hypothesis will be rejected for both large and small values of the contrast. One-tailed tests can be derived also, as follows.

The decision rule for the test of $H_0 : \Sigma c_i \tau_i = 0$ against the one-sided alternative hypothesis $H_A : \sum c_i \tau_i > 0$ is

$$\text{reject } H_0 \text{ if } \frac{\sum c_i \bar{y}_{i.}}{\sqrt{msE \sum c_i^2/r_i}} > t_{n-v,\alpha}. \quad (4.3.17)$$

Similarly, for the one-sided alternative hypothesis $H_A : \sum c_i \tau_i < 0$, the decision rule is

$$\text{reject } H_0 \text{ if } \frac{\sum c_i \bar{y}_{i.}}{\sqrt{msE \sum c_i^2/r_i}} < -t_{n-v,\alpha}. \quad (4.3.18)$$

If the hypothesis test concerns a single treatment mean, for example, $H_0 : \mu + \tau_s = 0$, then the decision rules (4.3.13)–(4.3.18) are modified by setting c_s equal to one and all the other c_i equal to zero.

Example 4.3.3 Filter experiment

Lorenz, Hsu, and Tuovinen (1982) describe an experiment that was carried out to determine the relative performance of seven membrane filters in supporting the growth of bacterial colonies. The seven filter types are regarded as the seven levels of the treatment factor and are coded 1, 2, ..., 7. Filter types 1, 4, and 7 were received presterilized. Several different types of data were collected, but the only data considered here are the colony counts of fecal coliforms from a sample of Olentangy River water (August 1980) that grew on each filter. Three filters of each type were observed and the average colony counts* were

$$\bar{y}_{1.} = 36.0, \ \bar{y}_{2.} = 18.0, \ \bar{y}_{3.} = 27.7, \ \bar{y}_{4.} = 28.0, \ \bar{y}_{5.} = 28.3, \ \bar{y}_{6.} = 37.7, \ \bar{y}_{7.} = 30.3.$$

The mean squared error was $msE = 21.6$. Suppose we wish to test the hypothesis that the presterilized filters do not differ from the nonpresterilized filters in terms of the average colony counts, against a two-sided alternative hypothesis that they do differ. The hypothesis of interest involves a difference of averages contrast, that is,

$$H_0 : \tfrac{1}{3}(\tau_1 + \tau_4 + \tau_7) - \tfrac{1}{4}(\tau_2 + \tau_3 + \tau_5 + \tau_6) = 0.$$

From (4.3.15), the decision rule is to reject H_0 if

$$\frac{ssc}{msE} = \frac{\left[\tfrac{1}{3}(\bar{y}_{1.} + \bar{y}_{4.} + \bar{y}_{7.}) - \tfrac{1}{4}(\bar{y}_{2.} + \bar{y}_{3.} + \bar{y}_{5.} + \bar{y}_{6.})\right]^2}{msE \left[\frac{3(\tfrac{1}{3})^2}{3} + \frac{4(-\tfrac{1}{4})^2}{3}\right]} > F_{1,14,\alpha}.$$

Selecting a probability of a Type I error equal to $\alpha = 0.05$, this becomes

$$\text{reject } H_0 \text{ if } \frac{(3.508)^2}{(21.6)(0.1944)} = 2.931 > F_{1,14,0.05}.$$

Since $F_{1,14,0.05} = 4.6$, there is not sufficient evidence to reject the null hypothesis, and we conclude that the presterilized filters do not differ significantly from the nonpresterilized filters when α is set at 0.05.

Notice that the null hypothesis would be rejected if the probability of a Type I error is set a little higher than $\alpha = 0.10$, since $F_{1,14,0.10} = 3.10$. Thus, if these experimenters are willing to accept a high risk of incorrectly rejecting the null hypothesis, they would be able to conclude that there is a difference between the presterilized and the nonpresterilized filters.

A 95% confidence interval for this difference can be obtained from (4.3.6) as follows:

$$\tfrac{1}{3}(\tau_1 + \tau_4 + \tau_7) - \tfrac{1}{4}(\tau_2 + \tau_3 + \tau_5 + \tau_6) \in \left(3.675 \pm t_{14,0.025}\sqrt{(21.6)(0.1944)}\right),$$

and since $t_{14,0.025} = 2.145$, the interval becomes

$$(3.508 \pm (2.145)(2.0492)) = (-0.888, 7.904),$$

where the measurements are average colony counts. The interval contains zero, which agrees with the hypothesis test at $\alpha = 0.05$. □

---

*Reprinted from Journal AWWA, Vol. 74, No. 8 (August 1982), by permission. Copyright © 1982, American Water Works Association.

4.4 Methods of Multiple Comparisons

4.4.1 Multiple Confidence Intervals

Often, the most useful analysis of experimental data involves the calculation of a number of different confidence intervals, one for each of several contrasts or treatment means. The confidence level for a single confidence interval is based on the probability, like (4.3.9), that the random interval will be "correct" (meaning that the random interval will contain the true value of the contrast or function).

It is shown below that when several confidence intervals are calculated, the probability that they are all simultaneously correct can be alarmingly small. Similarly, when several hypotheses are to be tested, the probability that at least one hypothesis is incorrectly rejected can be uncomfortably high. Much research has been done over the years to find ways around these problems. The resulting techniques are known as *methods of multiple comparison*, the intervals are called *simultaneous confidence intervals*, and the tests are called *simultaneous hypothesis tests*.

Suppose an experimenter wishes to calculate m confidence intervals, each having a $100(1 - \alpha^*)\%$ confidence level. Then each interval will be individually correct with probability $1 - \alpha^*$. Let S_j be the event that the jth confidence interval will be correct and $\overline{S}_j$ the event that it will be incorrect ($j = 1, \ldots, m$). Then, using the standard rules for probabilities of unions and intersections of events, it follows that

$$P(S_1 \cap S_2 \cap \cdots \cap S_m) = 1 - P(\overline{S}_1 \cup \overline{S}_2 \cup \cdots \cup \overline{S}_m).$$

This says that the probability that all of the intervals will be correct is equal to one minus the probability that at least one will be incorrect. If $m = 2$,

$$P(\overline{S}_1 \cup \overline{S}_2) = P(\overline{S}_1) + P(\overline{S}_2) - P(\overline{S}_1 \cap \overline{S}_2)$$
$$\leq P(\overline{S}_1) + P(\overline{S}_2).$$

A similar result, which can be proved by mathematical induction, holds for any number m of events, that is,

$$P(\overline{S}_1 \cup \overline{S}_2 \cup \cdots \cup \overline{S}_m) \leq \sum_j P(\overline{S}_j),$$

with equality if the events $\overline{S}_1, \overline{S}_2, \ldots, \overline{S}_m$ are mutually exclusive. Consequently,

$$P(S_1 \cap S_2 \cap \cdots \cap S_m) \geq 1 - \sum_j P(\overline{S}_j) = 1 - m\alpha^*; \quad (4.4.19)$$

that is, the probability that the m intervals will simultaneously be correct is at least $1 - m\alpha^*$. The probability $m\alpha^*$ is called the *overall significance level* or *experimentwise error rate*. A typical value for α^* for a single confidence interval is 0.05, so the probability that six confidence intervals each calculated at a 95% individual confidence level will simultaneously be correct is at least 0.7. Although "at least" means "bigger than or equal to," it is not known in practice how much bigger than 0.7 the probability might actually be. This is because the degree of overlap between the events $\overline{S}_1, \overline{S}_2, \ldots, \overline{S}_m$ is generally unknown. The probability "at least 0.7" translates into an *overall confidence level* of "at least 70%" when the responses are observed. Similarly, if an experimenter calculates ten confidence intervals each having individual confidence level 95%, then the simultaneous confidence level for the ten intervals is at least 50%, which is not very informative. As m becomes larger the problem becomes worse, and when $m \geq 20$, the overall confidence level is at least 0%, clearly a useless assertion!

Similar comments apply to the hypothesis testing situation. If m hypotheses are to be tested, each at significance level α^*, then the probability that at least one hypothesis is incorrectly rejected is at most $m\alpha^*$.

Various methods have been developed to ensure that the overall confidence level is not too small and the overall significance level is not too high. Some methods are completely general, that is, they can be used for any set of estimable functions, while others have been developed for very specialized purposes such as comparing each treatment with a control. Which method is best depends on which contrasts are of interest and the number of contrasts to be investigated. In this section, five methods are discussed that control the overall confidence level and overall significance level. The terms *preplanned contrasts* and *data snooping* occur in the summary of methods and the subsequent subsections. These have the following meanings. Before the experiment commences, the experimenter will have written out a checklist, highlighted the contrasts and/or treatment means that are of special interest, and designed the experiment in such a way as to ensure that these are estimable with as small variances as possible. These are the preplanned contrasts and means. After the data have been collected, the experimenter usually looks carefully at the data to see whether anything unexpected has occurred. One or more unplanned contrasts may turn out to be the most interesting, and the conclusions of the experiment may not be as anticipated. Allowing the data to suggest additional interesting contrasts is called data snooping.

The following summary is written in terms of confidence intervals, but it also applies to hypothesis tests. A shorter confidence interval corresponds to a more powerful hypothesis test. The block designs mentioned in the summary will be discussed in Chapters 10 and 11.

Summary of Multiple Comparison Methods

1. **Bonferroni method for preplanned comparisons**

   Applies to any m preplanned estimable contrasts or functions of the parameters. Gives shorter confidence intervals than the other methods listed if m is small. Can be used for any design. Cannot be used for data snooping.

2. **Scheffé method for all comparisons**

   Applies to any m estimable contrasts or functions of the parameters. Gives shorter intervals than Bonferroni's method if m is large. Allows data snooping. Can be used for any design.

3. **Tukey method for all pairwise comparisons**

   Best for all pairwise comparisons. Can be used for completely randomized designs, randomized block designs, and balanced incomplete block designs. Is believed to be applicable (conservative) for other designs as well. Can be extended to include all contrasts, but Scheffé's method is generally better for these.

4. **Dunnett method for treatment-versus-control comparisons**
Best for all treatment-versus-control contrasts. Can be used for completely randomized designs, randomized block designs, and balanced incomplete block designs.

5. **Hsu method for multiple comparisons with the best treatment**
Selects the best treatment and identifies those treatments that are significantly worse than the best. Can be used for completely randomized designs, randomized block designs, and balanced incomplete block designs.

Details of confidence intervals obtained by each of the above methods are given in Sections 4.4.2–4.4.7. The terminology "a set of simultaneous $100(1-\alpha)\%$ confidence intervals" will always refer to the fact that the *overall* confidence level for a set of contrasts or treatment means is (at least) $100(1-\alpha)\%$. Each of the five methods discussed gives confidence intervals of the form

$$\sum_i c_i \tau_i \in \left(\sum_i c_i \hat{\tau}_i \pm w \sqrt{\widehat{\text{Var}}(\Sigma c_i \hat{\tau}_i)} \right), \qquad (4.4.20)$$

where w, which we call the *critical coefficient*, depends on the method, on v, on the number of confidence intervals calculated, and on the number of error degrees of freedom. The term

$$msd = w \sqrt{\widehat{\text{Var}}(\Sigma c_i \hat{\tau}_i)},$$

which is added and subtracted from the least squares estimate in (4.4.20), is called the *minimum significant difference*, because if the estimate is larger than *msd*, the confidence interval excludes zero, and the contrast is significantly different from zero.

4.4.2 Bonferroni Method for Preplanned Comparisons

The inequality (4.4.19) shows that if m simultaneous confidence intervals are calculated for preplanned contrasts, and if each confidence interval has confidence level $100(1-\alpha^*)\%$, then the overall confidence level is greater than or equal to $100(1-m\alpha^*)\%$. Thus, an experimenter can ensure that the overall confidence level is at least $100(1-\alpha)\%$ by setting $\alpha^* = \alpha/m$. This is known as the Bonferroni method for simultaneous confidence intervals. Replacing α by α/m in the formula (4.3.6), page 73, for an individual confidence interval, we obtain a formula for a set of simultaneous $100(1-\alpha)\%$ confidence intervals for m preplanned contrasts $\Sigma c_i \tau_i$ in a completely randomized design with the one-way analysis of variance model (3.3.1), as

$$\sum_i c_i \tau_i \in \left(\sum_i c_i \bar{y}_{i.} \pm t_{n-v,\alpha/(2m)} \sqrt{msE \sum_i c_i^2/r_i} \right), \qquad (4.4.21)$$

where the critical coefficient, w_B, is

$$w_B = t_{n-v,\alpha/(2m)}.$$

Since $\alpha/(2m)$ is likely to be an atypical value, the percentiles $t_{n-v,\alpha/(2m)}$ may need to be obtained by use of a computer package, or by approximate interpolation between values in

Table A.4, or by using the following approximate formula due to Peiser (1943):

$$t_{df,\alpha/(2m)} \approx z_{\alpha/(2m)} + (z^3_{\alpha/(2m)} + z_{\alpha/(2m)})/(4(df)), \quad (4.4.22)$$

where df is the error degrees of freedom (equal to $n - v$ in the present context), and where $z_{\alpha/(2m)}$ is the percentile of the standard normal distribution corresponding to a probability of $\alpha/(2m)$ in the right hand tail. The standard normal distribution is tabulated in Table A.3 and covers the entire range of values for $\alpha/(2m)$. When m is very large, $\alpha/(2m)$ is very small, possibly resulting in extremely wide simultaneous confidence intervals. In this case, the Scheffé or Tukey methods described in the following subsections would be preferred.

If some of the m simultaneous intervals are for true mean responses $\mu + \tau_s$, then the required intervals are of the form (4.3.12), page 75, with α replaced by α/m, that is,

$$\mu + \tau_s \in \left(\bar{y}_{s.} \pm t_{n-v,\alpha/(2m)} \sqrt{msE/r_s} \right). \quad (4.4.23)$$

Similarly, replacing α by α/m in (4.3.15), a set of m null hypotheses, each of the form

$$H_0 : \sum_{i=1}^{v} c_i \tau_i = 0,$$

can be tested against their respective two-sided alternative hypotheses at overall significance level α using the set of decision rules each of the form

$$\text{reject } H_0 \text{ if } \frac{ssc}{msE} > F_{1,df,\alpha/m}. \quad (4.4.24)$$

Note that Bonferroni's method can be use only for *preplanned* contrasts and means. An experimenter who looks at the data and then proceeds to calculate simultaneous confidence intervals for the few contrasts that look interesting has effectively calculated a very large number of intervals. This is because the interesting contrasts are usually those that seem to be significantly different from zero, and a rough mental calculation of the estimates of a large number of contrasts has to be done to identify these interesting contrasts. Scheffé's method should be used for contrasts that were selected after the data were examined.

Example 4.4.1 Filter experiment, continued

The filter experiment was described in Example 4.3.3, page 76. Suppose that before the data had been collected, the experimenters had planned to calculate a set of simultaneous 90% confidence intervals for the following $m = 3$ contrasts. These contrasts have been selected based on the details of the original study described by Lorenz, Hsu, and Tuovinen (1982).

(i) $\frac{1}{3}(\tau_1 + \tau_4 + \tau_7) - \frac{1}{4}(\tau_2 + \tau_3 + \tau_5 + \tau_6)$. This contrast measures the difference in the average effect of the presterilized and the nonpresterilized filter types. This was used in Example 4.3.3 to illustrate a hypothesis test for a single contrast.

(ii) $\frac{1}{2}(\tau_1 + \tau_7) - \frac{1}{5}(\tau_2 + \tau_3 + \tau_4 + \tau_5 + \tau_6)$. This contrast measures the difference in the average effects of two filter types with gradated pore size and five filter types with uniform pore size.

(iii) $\frac{1}{6}(\tau_1 + \tau_2 + \tau_4 + \tau_5 + \tau_6 + \tau_7) - \tau_3$. This contrast is the difference in the average effect of the filter types that are recommended by their manufacturers for bacteriologic analysis of water and the single filter type that is recommended for sterility testing of pharmaceutical or cosmetic products.

From Example 4.3.3, we know that

$$\bar{y}_{1.} = 36.0, \quad \bar{y}_{2.} = 18.0, \quad \bar{y}_{3.} = 27.7, \quad \bar{y}_{4.} = 28.0, \quad \bar{y}_{5.} = 28.3,$$

$$\bar{y}_{6.} = 37.7, \quad \bar{y}_{7.} = 30.3, \quad r_i = 3, \quad msE = 21.6.$$

The formula for each of the three preplanned simultaneous 90% confidence intervals is given by (4.4.21) and involves the critical coefficient $w_B = t_{14,(0.1)/6} = t_{14,0.0167}$, which is not available in Table A.4. Either the value can be calculated from a computer program, or an approximate value can be obtained from formula (4.4.22) as

$$t_{14,0.0167} \approx 2.128 + (2.128^3 + 2.128)/(4 \times 14) = 2.338.$$

The minimum significant difference for each of the three simultaneous 90% confidence intervals is

$$msd = 2.338\sqrt{(21.6)\sum c_i^2/3} = 6.2735\sqrt{\sum c_i^2}.$$

Thus, for the first interval, we have

$$msd = 6.2735\sqrt{3\left(\frac{1}{9}\right) + 4\left(\frac{1}{16}\right)} = 4.791,$$

giving the interval as

$$\tfrac{1}{3}(\tau_1 + \tau_4 + \tau_7) - \tfrac{1}{4}(\tau_2 + \tau_3 + \tau_5 + \tau_6) \in (3.508 \pm 4.791) = (-1.283, 8.299).$$

Calculating the minimum significant differences separately for the other two confidence intervals leads to

$$\tfrac{1}{2}(\tau_1 + \tau_7) - \tfrac{1}{5}(\tau_2 + \tau_3 + \tau_4 + \tau_5 + \tau_6) \in (-0.039, 10.459);$$

$$\tfrac{1}{6}(\tau_1 + \tau_2 + \tau_4 + \tau_5 + \tau_6 + \tau_7) - \tau_3 \in (-4.759, 8.793).$$

Notice that all three intervals include zero, although the second is close to excluding it. Thus, at overall significance level $\alpha = 0.10$, we would fail to reject the hypothesis that there is no difference in average colony counts between the presterilized and nonpresterilized filters, nor between filter 3 and the others, nor between filters with gradated and uniform pore sizes. At a slightly higher significance level, we would reject the hypothesis that the filters with gradated pore size have the same average colony counts as those with uniform pore size. The same conclusion would be obtained if (4.4.24) were used to test simultaneously, at overall level $\alpha = 0.10$, the hypotheses that each of the three contrasts is zero. The confidence interval has the added benefit that we can say with overall 90% confidence that on average, the filters with gradated pore size give rise to colony counts up to 10.4 greater than the filters with uniform pore sizes. □

4.4.3 Scheffé Method of Multiple Comparisons

The main drawbacks of the Bonferroni method of multiple comparisons are that the m contrasts to be examined must be preplanned and the confidence intervals can become very wide if m is large. Scheffé's method, on the other hand, provides a set of simultaneous $100(1-\alpha)\%$ confidence intervals whose widths are determined only by the number of treatments and the number of observations in the experiment, no matter how many contrasts are of interest. The two methods are compared directly later in this section.

Scheffé's method is based on the fact that every possible contrast $\sum c_i \tau_i$ can be written as a linear combination of the set of $(v-1)$ treatment versus control contrasts, $\tau_2 - \tau_1, \tau_3 - \tau_1, \ldots, \tau_v - \tau_1$. (We leave it to the reader to check that this is true.) Once the experimental data have been collected, it is possible to find a $100(1-\alpha)\%$ *confidence region* for these $v-1$ treatment-versus-control contrasts. The confidence region not only determines confidence bounds for each treatment-versus-control contrast, it determines bounds for *every* possible contrast $\sum c_i \tau_i$ and, in fact, for *any number* of contrasts, while the overall confidence level remains fixed. The mathematical details are given by Scheffé (1959).

For v treatments in a completely randomized design and the one-way analysis of variance model (3.3.1), a set of simultaneous $100(1-\alpha)\%$ confidence intervals for all contrasts $\sum c_i \tau_i$ is given by

$$\sum_i c_i \tau_i \in \left(\sum_i c_i \bar{y}_{i.} \pm \sqrt{(v-1)F_{v-1, n-v, \alpha}} \sqrt{msE \sum_i c_i^2 / r_i} \right). \quad (4.4.25)$$

Notice that this is the same form as the general formula (4.4.20), page 80, where the critical coefficient w is

$$w_S = \sqrt{(v-1)F_{v-1, n-v, \alpha}} \, .$$

If confidence intervals for the treatment means $\mu + \tau_i$ are also of interest, the critical coefficient w_S needs to be replaced by

$$w_S^* = \sqrt{v F_{v, n-v, \alpha}} \, .$$

The reason for the increase in the numerator degrees of freedom is that any of the functions $\mu + \tau_i$ can be written as a linear combination of the $v-1$ treatment versus control contrasts and one additional function $\mu + \tau_1$. For the completely randomized design and model (3.3.1), a set of simultaneous $100(1-\alpha)\%$ confidence intervals for any number of true mean responses and contrasts is therefore given by

$$\sum_i c_i \tau_i \in \left(\sum_i c_i \bar{y}_{i.} \pm \sqrt{v F_{v, n-v, \alpha}} \sqrt{msE \sum_i c_i^2 / r_i} \right)$$

together with

$$\mu + \tau_s \in \left(\bar{y}_{i.} \pm \sqrt{v F_{v, n-v, \alpha}} \sqrt{msE / r_s} \right) . \quad (4.4.26)$$

Example 4.4.2 Filter experiment, continued

If we look at the observed average colony counts,

$$\bar{y}_{1.} = 36.0, \quad \bar{y}_{2.} = 18.0, \quad \bar{y}_{3.} = 27.7, \quad \bar{y}_{4.} = 28.0,$$
$$\bar{y}_{5.} = 28.3, \quad \bar{y}_{6.} = 37.7, \quad \bar{y}_{7.} = 30.3,$$

for the filter experiment of Examples 4.3.3 and 4.4.2 (pages 76 and 81), filter type 2 appears to give a much lower count than the other types. One may wish to recalculate each of the three intervals in Example 4.4.2 with filter type 2 excluded. It might also be of interest to compare the filter types 1 and 6, which showed the highest average colony counts, with the other filters. These are *not preplanned* contrasts. They have become interesting only after the data have been examined, and therefore we need to use Scheffé's method of multiple comparisons. In summary, we are interested in the following twelve contrasts:

$$\tfrac{1}{3}(\tau_1 + \tau_4 + \tau_7) - \tfrac{1}{3}(\tau_3 + \tau_5 + \tau_6), \quad \tfrac{1}{2}(\tau_1 + \tau_7) - \tfrac{1}{4}(\tau_3 + \tau_4 + \tau_5 + \tau_6),$$
$$\tfrac{1}{5}(\tau_1 + \tau_4 + \tau_5 + \tau_6 + \tau_7) - \tau_3,$$
$$\tau_1 - \tau_3, \quad \tau_1 - \tau_4, \quad \tau_1 - \tau_5, \quad \tau_1 - \tau_6, \quad \tau_1 - \tau_7,$$
$$\tau_6 - \tau_3, \quad \tau_6 - \tau_4, \quad \tau_6 - \tau_5, \quad \tau_6 - \tau_7.$$

The formula for a set of Scheffé 90% simultaneous confidence intervals is given by (4.4.25) with $\alpha = 0.10$. Since $v = 7$, $n = 21$, and $msE = 21.6$ for the filter experiment, the minimum significant difference for each interval becomes

$$msd = \sqrt{6F_{6,14,0.10}}\sqrt{21.6 \, \Sigma c_i^2/3} = 9.837\sqrt{\Sigma c_i^2}.$$

The twelve simultaneous 90% confidence intervals are then

$$\tfrac{1}{3}(\tau_1 + \tau_4 + \tau_7) - \tfrac{1}{3}(\tau_3 + \tau_5 + \tau_6)$$
$$\in \left((31.43 - 31.23) \pm 9.837\sqrt{3\left(\tfrac{1}{9}\right) + 3\left(\tfrac{1}{9}\right)}\right)$$
$$= (-7.83, 8.23),$$

$$\tfrac{1}{2}(\tau_1 + \tau_7) - \tfrac{1}{4}(\tau_3 + \tau_4 + \tau_5 + \tau_6) \in (-5.79, 11.24),$$
$$\tfrac{1}{5}(\tau_1 + \tau_4 + \tau_5 + \tau_6 + \tau_7) - \tau_3 \in (-6.42, 15.14),$$

$\tau_1 - \tau_3$	$\in$	$(-5.61, 22.21),$	$\tau_6 - \tau_3$	$\in$	$(-3.91, 23.91),$
$\tau_1 - \tau_4$	$\in$	$(-5.91, 21.91),$	$\tau_6 - \tau_4$	$\in$	$(-4.21, 23.61),$
$\tau_1 - \tau_5$	$\in$	$(-6.21, 21.61),$	$\tau_6 - \tau_5$	$\in$	$(-4.51, 23.31),$
$\tau_1 - \tau_6$	$\in$	$(-15.61, 12.21),$	$\tau_6 - \tau_7$	$\in$	$(-6.51, 21.31).$
$\tau_1 - \tau_7$	$\in$	$(-8.21, 19.61).$			

These intervals are all fairly wide and all include zero. Consequently, at overall error rate $\alpha = 0.1$, we are unable to infer that any of the contrasts are significantly different from zero. □

Relationship between analysis of variance and the Scheffé method The analysis of variance test and the Scheffé method of multiple comparisons are equivalent in the following sense. The analysis of variance test will reject the null hypothesis $H_0: \tau_1 = \tau_2 = \cdots = \tau_v$ at significance level α if there is at least one confidence interval among the infinite number of Scheffé simultaneous $100(1-\alpha)\%$ confidence intervals that excludes zero. However, the intervals that exclude zero may not be among those for the interesting contrasts being examined.

Other methods of multiple comparisons do not relate to the analysis of variance test in this way. It is possible when using one of the other multiple comparison methods that one or more intervals in a simultaneous $100(1-\alpha)\%$ set may exclude 0, while the analysis of variance test of H_0 is *not* rejected at significance level α. Hence, if specific contrasts of interest have been identified in advance of running the experiment and a method of multiple comparisons other than Scheffé's method is to be used, then it is sensible to analyze the data using only the multiple comparison procedure.

4.4.4 Tukey Method for All Pairwise Comparisons

In some experiments, confidence intervals may be required only for pairwise difference contrasts. Tukey, in 1953, proposed a method that is specially tailored to handle this situation and that gives shorter intervals for pairwise differences than do the Bonferroni and Scheffé methods.

For the completely randomized design and the one-way analysis of variance model (3.3.1), Tukey's simultaneous confidence intervals for all pairwise comparisons $\tau_i - \tau_s$, $i \neq s$, with overall confidence level at least $100(1-\alpha)\%$ is given by

$$\tau_i - \tau_s \in \left((\bar{y}_{i.} - \bar{y}_{s.}) \pm w_T \sqrt{msE \left(\frac{1}{r_i} + \frac{1}{r_s} \right)} \right), \qquad (4.4.27)$$

where the critical coefficient w_T is

$$w_T = q_{v, n-v, \alpha}/\sqrt{2},$$

and where $q_{v,n-v,\alpha}$ is tabulated in Appendix A.8. When the sample sizes are equal ($r_i = r$; $i = 1, \ldots, v$), the overall confidence level is *exactly* $100(1-\alpha)\%$. When the sample sizes are unequal, the confidence level is *at least* $100(1-\alpha)\%$.

The derivation of (4.4.27) is as follows. For equal sample sizes, the formula for Tukey's simultaneous confidence intervals is based on the distribution of the statistic

$$Q = \frac{\max\{T_i\} - \min\{T_i\}}{\sqrt{MSE/r}},$$

where $T_i = \bar{Y}_{i.} - (\mu + \tau_i)$ for the one-way analysis of variance model (3.3.1), and where $\max\{T_i\}$ is the maximum value of the random variables $T_1, T_2, \ldots, T_v$ and $\min\{T_i\}$ the minimum value. Since the $\bar{Y}_{i.}$'s are independent, the numerator of Q is the range of v independent $N(0, \sigma^2/r)$ random variables, and is standardized by the estimated standard deviation. The distribution of Q is called the *Studentized range distribution*. The percentile corresponding to a probability of α in the right-hand tail of this distribution is denoted by

$q_{v,n-v,\alpha}$, where v is the number of treatments being compared, and $n - v$ is the number of degrees of freedom for error. Therefore,

$$P\left(\frac{\max\{T_i\} - \min\{T_i\}}{\sqrt{MSE/r}} \leq q_{v,n-v,\alpha}\right) = 1 - \alpha.$$

Now, if $\max\{T_i\} - \min\{T_i\}$ is less than or equal to $q_{v,n-v,\alpha}\sqrt{MSE/r}$, then it must be true that $|T_i - T_s| \leq q_{v,n-v,\alpha}\sqrt{MSE/r}$ for *every* pair of random variables $T_i, T_s, i \neq s$. Using this fact and the above definition of T_i, we have

$$1 - \alpha = P\left(-q_{v,n-v,\alpha}\sqrt{MSE/r} \leq (\overline{Y}_{i.} - \overline{Y}_{s.}) - (\tau_i - \tau_s)\right.$$
$$\left. \leq q_{v,n-v,\alpha}\sqrt{MSE/r}, \text{ for all } i \neq s\right).$$

Replacing $\overline{Y}_{i.}$ by its observed value $\overline{y}_{i.}$, and MSE by the observed value msE, a set of simultaneous $100(1 - \alpha)\%$ confidence intervals for all pairwise differences $\tau_i - \tau_s, i \neq s$, is given by

$$\tau_i - \tau_s \in \left((\overline{y}_{i.} - \overline{y}_{s.}) \pm q_{v,n-v,\alpha}\sqrt{msE/r}\right),$$

which can be written in terms of the critical coefficient as

$$\tau_i - \tau_s \in \left((\overline{y}_{i.} - \overline{y}_{s.}) \pm w_T\sqrt{msE\left(\frac{1}{r} + \frac{1}{r}\right)}\right). \tag{4.4.28}$$

More recently, Hayter (1984) showed that the same form of interval can be used for unequal sample sizes as in (4.4.27), and that the overall confidence level is then at least $100(1 - \alpha)\%$.

Example 4.4.3   Battery experiment, continued

In the battery experiment of Example 4.2.1 (page 69), we considered the pairwise differences in the life lengths per unit cost of $v = 4$ different battery types, and we obtained the least squares estimates

$\hat{\tau}_1 - \hat{\tau}_2 = -289.75$, $\hat{\tau}_1 - \hat{\tau}_3 = 137.75$, $\hat{\tau}_1 - \hat{\tau}_4 = 74.50$,
$\hat{\tau}_2 - \hat{\tau}_3 = 427.50$, $\hat{\tau}_2 - \hat{\tau}_4 = 364.25$, $\hat{\tau}_3 - \hat{\tau}_4 = -63.25$.

The standard error was $\sqrt{msE(\frac{1}{4} + \frac{1}{4})} = 34.41$, and the number of error degrees of freedom was $n - v = (16 - 4) = 12$. From Table A.8, $q_{4,12,0.05} = 4.20$, so $w_T = 4.20/\sqrt{2}$, and the minimum significant difference is

$$msd = (4.20/\sqrt{2})(34.41) = 102.19.$$

Therefore, the simultaneous 95% confidence intervals for the pairwise comparisons of lifetimes per unit cost of the different battery types are

$\tau_1 - \tau_2 \in (-289.75 \pm 102.19) = (-391.94, -187.56)$,
$\tau_1 - \tau_3 \in (137.75 \pm 102.19) = (35.56, 239.94)$,
$\tau_1 - \tau_4 \in (-27.69, 176.69)$, $\tau_2 - \tau_3 \in (325.31, 529.69)$,

$\tau_2 - \tau_4 \in (262.06, 466.44), \qquad \tau_3 - \tau_4 \in (-165.44, 38.94).$

Four of these intervals exclude zero, and one can conclude (at an overall 95% confidence level) that battery type 2 (alkaline, store brand) has the highest lifetime per unit cost, and battery type 3 (heavy duty, name brand) has lower lifetime per unit cost than does battery type 1 (alkaline, name brand). The intervals show us that with overall 95% confidence, battery type 2 is between 188 and 391 minutes per dollar better than battery type 1 (the name brand alkaline battery) and even more economical than the heavy-duty brands. □

Example 4.4.4 Bonferroni, Scheffé and Tukey methods compared

Suppose that $v = 5, n = 35$, and $\alpha = 0.05$, and that only the 10 pairwise comparisons $\tau_i - \tau_s, i \neq s$, are of interest to the experimenter and these were specifically selected prior to the experiment (i.e., were preplanned). If we compare the critical coefficients for the three methods, we obtain

$$\text{Bonferroni:} \quad w_B = t_{30,.025/10} = 3.02,$$
$$\text{Scheffé:} \quad w_S = \sqrt{4\,F_{4,30,.05}} = 3.28,$$
$$\text{Tukey:} \quad w_T = \frac{1}{\sqrt{2}} q_{5,30,.05} = 2.91.$$

Since w_T is less than w_B, which is less than w_S for this example, the Tukey intervals will be shorter than the Bonferroni intervals, which will be shorter than the Scheffé intervals. □

4.4.5 Dunnett Method for Treatment-Versus-Control Comparisons

In 1955, Dunnett developed a method of multiple comparisons that is specially designed to provide a set of simultaneous confidence intervals for preplanned treatment-versus-control contrasts $\tau_i - \tau_1$ ($i = 2, \ldots, v$), where level 1 corresponds to the control treatment. The intervals are shorter than those given by the Scheffé, Tukey, and Bonferroni methods, but the method should not be used for any other type of contrasts.

The formulae for the simultaneous confidence intervals are based on the joint distribution of the estimators $\overline{Y}_{i.} - \overline{Y}_{1.}$ of $\tau_i - \tau_1$ ($i = 2, \ldots, v$). This distribution is a special case of the multivariate t distribution and depends on the correlation between $\overline{Y}_{i.} - \overline{Y}_{1.}$ and $\overline{Y}_{s.} - \overline{Y}_{1.}$. For the completely randomized design, with equal numbers of observations $r_2 = \cdots = r_v = r$ on the experimental treatments and $r_1 = c$ observations on the control treatment, the correlation is

$\rho = r/(c + r).$

In many experiments, the same number of observations will be taken on the control and experimental treatments, in which case $\rho = 0.5$. However, the shortest confidence intervals for comparing $v - 1$ experimental treatments with a control treatment are generally obtained when c/r is chosen to be close to $\sqrt{v - 1}$. Since we have tabulated the multivariate t-distribution only with correlation $\rho = 0.5$, we will discuss only the case $c = r$. Other tables can be found in the book of Hochberg and Tamhane (1987), and intervals can also be obtained via some computer packages (see Section 4.6.2 for the SAS package).

If the purpose of the experiment is to determine which of the experimental treatments give a significantly higher response than the control treatment, then one-sided confidence bounds should be used. For a completely randomized design with equal sample sizes and the one-way analysis of variance model (3.3.1), Dunnett's simultaneous one-sided $100(1-\alpha)\%$ confidence bounds for treatment-versus-control contrasts $\tau_i - \tau_1$ ($i = 2, 3, \ldots, v$) are

$$\tau_i - \tau_1 \geq (\overline{y}_{i.} - \overline{y}_{1.}) - w_{D1} \sqrt{msE\left(\frac{2}{r}\right)}, \qquad (4.4.29)$$

where the critical coefficient is

$$w_{D1} = t^{(0.5)}_{v-1, n-v, \alpha}$$

and where $t^{(0.5)}_{v-1, n-v, \alpha}$ is the percentile of the maximum of a multivariate t-distribution with common correlation 0.5 and $n - v$ degrees of freedom, corresponding to a Type I error probability of α in the right-hand tail. The critical coefficient is tabulated in Table A.9. If the right hand side of (4.4.29) is negative, we infer that the control gives a larger response than the ith experimental treatment.

If the purpose is to determine which of the experimental treatments give a significantly lower response than the control, then the inequality is reversed, and the confidence bound becomes

$$\tau_i - \tau_1 \leq (\overline{y}_{i.} - \overline{y}_{1.}) + w_{D1} \sqrt{msE\left(\frac{2}{r}\right)}. \qquad (4.4.30)$$

If the right-hand side is negative, we infer that the control gives a smaller response than the experimental treatment.

To determine which experimental treatments are better than the control *and* which ones are worse, two-sided intervals of the general form (4.4.20) are used as for the other multiple comparison methods. For the completely randomized design, one-way analysis of variance model (3.3.1), and equal sample sizes, the formula is

$$\tau_i - \tau_1 \in \left(\overline{y}_{i.} - \overline{y}_{1.} \pm w_{D2} \sqrt{msE\left(\frac{2}{r}\right)}\right), \qquad (4.4.31)$$

where the critical coefficient is

$$w_{D2} = |t|^{(0.5)}_{v-1, n-v, \alpha}$$

and is the upper critical value for the maximum of the absolute values of a multivariate t-distribution with correlation 0.5 and $n - v$ error degrees of freedom, corresponding to the chosen value of α in the right-hand tail. The critical coefficients for equal sample sizes are provided in Table A.10.

For future reference, the general formula for Dunnett's two-sided simultaneous $100(1-\alpha)\%$ confidence intervals for treatment versus control contrasts $\tau_i - \tau_1$ ($i = 2, 3, \ldots, v$) is

$$\tau_i - \tau_1 \in \left((\hat{\tau}_i - \hat{\tau}_1) \pm w_{D2} \sqrt{\widehat{\text{Var}}(\hat{\tau}_i - \hat{\tau}_1)} \right), \quad (4.4.32)$$

and, for one-sided confidence bounds, we replace w_{D2} by w_{D1} and replace "$\in$" by "$\leq$" or "$\geq$." The critical coefficients are

$$w_{D2} = |t|_{v-1,df,\alpha}^{(0.5)} \text{ and } w_{D1} = t_{v-1,df,\alpha}^{(0.5)}$$

for two-sided and one-sided intervals, respectively, where df is the number of error degrees of freedom.

Example 4.4.5 Soap experiment, continued

Suppose that as a preplanned objective of the soap experiment of Section 2.5.1, page 22, the experimenter had wanted simultaneous 99% confidence intervals comparing the weight losses of the deodorant and moisturizing soaps (levels 2 and 3) with that of the regular soap (level 1). Then it is appropriate to use Dunnett's method as given in (4.4.31). From Section 3.7.2, $r_1 = r_2 = r_3 = 4$, $msE = 0.0772$, $\hat{\tau}_2 - \hat{\tau}_1 = 2.7350$, and $\hat{\tau}_3 - \hat{\tau}_1 = 2.0275$. From Table A.10, $w_{D2} = |t|_{v-1,n-v,\alpha}^{(0.5)} = |t|_{2,9,0.01}^{(0.5)} = 3.63$, so the minimum significant difference is

$$msd = 3.63 \sqrt{msE(2/4)} = 0.713.$$

Hence, the simultaneous 99% confidence intervals are

$$\tau_2 - \tau_1 \in (2.7350 \pm 0.713) \approx (2.022, 3.448)$$

and

$$\tau_3 - \tau_1 \in (2.0275 \pm 0.713) \approx (1.314, 2.741).$$

One can conclude from these intervals (with overall 99% confidence) that the deodorant soap (soap 2) loses between 2 and 3.4 grams more weight on average than does the regular soap, and the moisturizing soap loses between 1.3 and 2.7 grams more weight on average than the regular soap. We leave it to the reader to verify that neither the Tukey nor the Bonferroni method would have been preferred for these contrasts (see Exercise 7). □

4.4.6 Hsu Method for Multiple Comparisons with the Best Treatment

"Multiple comparisons with the best treatment" is similar to multiple comparisons with a control, except that since it is unknown prior to the experiment which treatment is the best, a control treatment has not been designated. Hsu (1984) developed a method in which each treatment sample mean is compared with the best of the others, allowing some treatments to be eliminated as worse than best, and allowing one treatment to be identified as best if all others are eliminated. Hsu calls this method RSMCB, which stands for *Ranking, Selection, and Multiple Comparisons with the Best treatment.*

Suppose, first, that the best treatment is the treatment that gives the largest response on average. Let $\tau_i - \max(\tau_j)$ denote the effect of the ith treatment minus the effect of the best of the *other* $v - 1$ treatments. When the ith treatment is the best, $\max(\tau_j)$ $(j \neq i)$ will be the effect of the second-best treatment. So, $\tau_i - \max(\tau_j)$ will be positive if treatment i is the best, zero if the ith treatment is tied for being the best, or negative if the treatment i is worse than best.

Hsu's procedure provides a set of v simultaneous $100(1 - \alpha)\%$ confidence intervals for $\tau_i - \max(\tau_j)$, $j \neq i$ $(i = 1, 2, \ldots, v)$. For an equireplicate completely randomized design and model (3.3.1), Hsu's formula for the simultaneous $100(1 - \alpha)\%$ confidence intervals is

$$\tau_i - \max_{j \neq i}(\tau_j) \in \left((\bar{y}_{i.} - \max_{j \neq i} \bar{y}_{j.}) \pm w_H \sqrt{msE(2/r)} \right), \quad (4.4.33)$$

where a negative *upper* bound is set to zero and the ith treatment is declared *not* to be the best, while a positive *lower* bound is set to zero and the ith treatment is selected as the best. The critical coefficient

$$w_H = w_{D1} = t_{v-1, n-v, \alpha}^{(0.5)}$$

is the same as that for Dunnett's one-sided confidence bound and is tabulated in Table A.9. If more than one of the intervals has a positive upper bound, then the corresponding set of treatments contains the best treatment with $100(1 - \alpha)\%$ confidence.

Hsu's procedure can also be used in the case of unequal sample sizes. However, the formulas for the intervals are complicated, and their calculation is best left to a computer package (see Section 4.6.2).

If the best treatment factor level is the level that gives the smallest response rather than the largest, then Hsu's procedure has to be modified, and the simultaneous $100(1 - \alpha)\%$ confidence intervals for $\tau_i - \min(\tau_j)$ are given by

$$\tau_i - \min_{j \neq i}(\tau_j) \in \left((\bar{y}_{i.} - \min_{j \neq i}(\bar{y}_{j.})) \pm w_H \sqrt{msE(2/r)} \right), \quad (4.4.34)$$

where a positive lower bound is set to zero and treatment i is declared not to be the best, and a negative upper bound is set to zero and treatment i is selected as the best treatment.

Example 4.4.6 Filter experiment, continued

Example 4.3.3, page 76, described an experiment to determine the relative performance of seven membrane filters in supporting the growth of bacterial colonies. The summary information is

$$\bar{y}_{1.} = 36.0, \quad \bar{y}_{2.} = 18.0, \quad \bar{y}_{3.} = 27.7, \quad \bar{y}_{4.} = 28.0, \quad \bar{y}_{5.} = 28.3,$$

$$\bar{y}_{6.} = 37.7, \quad \bar{y}_{7.} = 30.3, \quad n = 21, \quad v = 7, \quad r = 3, \quad msE = 21.6.$$

A completely randomized design and model (3.3.1) was used. For a set of simultaneous 95% confidence intervals, the minimum significant difference using Hsu's method is obtained from (4.4.33) as

$$msd = w_H \sqrt{msE(2/r)} = t_{6, 14, 0.05}^{(0.5)} \sqrt{21.6(2/3)} = 2.54\sqrt{14.4} \approx 9.7,$$

Table 4.1 Multiple comparisons with the best for the filter data

i	$\overline{y}_{i.} - \max_{j \neq i} \overline{y}_{j.}$	Confidence Interval
1	36.0 − 37.7 = −1.7	(−11.4, 8.0)
2	18.0 − 37.7 = −19.7	(−29.4, 0.0)
3	27.7 − 37.7 = −10.0	(−19.7, 0.0)
4	28.0 − 37.7 = −9.7	(−19.4, 0.0)
5	28.3 − 37.7 = −9.4	(−19.1, 0.3)
6	37.7 − 36.0 = 1.7	(−8.0, 11.4)
7	30.3 − 37.7 = −7.4	(−17.1, 2.3)

where the critical coefficient w_H is the same as w_{D1} and is obtained by interpolation from Table A.9, and msd has been rounded to the same level of accuracy as the treatment sample means. Further computations and the resulting simultaneous 95% confidence intervals are shown in Table 4.1. Filters 2, 3, and 4 are declared not to be the best, since their corresponding intervals have negative upper bounds, which are then set to zero. No interval has a positive lower bound, so there is no selection of a single filter as being the best. The sixth filter yielded the highest sample mean, but any one of filters 1, 5, 6, or 7 could be the best (with 95% overall confidence). □

4.4.7 Combination of Methods

The Bonferroni method is based on the fact that if m individual confidence intervals are obtained, each with confidence level $100(1 - \alpha^*)\%$, then the overall confidence level is at least $100(1 - m\alpha^*)\%$. The same fact can be used to combine the overall confidence levels arising from more than one multiple comparison procedure.

In Example 4.4.2 (page 81), the Bonferroni method was used to calculate simultaneous 90% confidence intervals for $m = 3$ preplanned contrasts. In Example 4.4.3 (page 84), the analysis was continued by calculating simultaneous 90% Scheffé intervals for twelve other contrasts. The overall error rate for these two sets of intervals combined is therefore at most $0.1 + 0.1 = 0.2$, giving an overall, or "experimentwise," confidence level of at least $100(1 - 0.2)\% = 80\%$ for all fifteen intervals together.

Different possible strategies for multiple comparisons should be examined when outlining the analysis at step (g) of the checklist (Section 2.2, page 7). Suppose that in the above example the overall level for all intervals (both planned and otherwise) had been required to be at least 90%. We examine two possible strategies that could have been used. First, the confidence levels for the Bonferroni and Scheffé contrasts could have been adjusted, dividing $\alpha = 0.10$ into two pieces, α_1 for the preplanned contrasts and α_2 for the others, where $\alpha_1 + \alpha_2 = 0.10$. This strategy would have resulted in intervals that were somewhat wider than the above for all of the contrasts. Alternatively, Scheffé's method could have been used with $\alpha = 0.10$ for all of the contrasts including the three preplanned contrasts. This strategy would have resulted in wider intervals for the three preplanned contrasts but not for the others. Both strategies would result in an overall, or experimentwise, confidence level of 90% instead of 80%.

4.4.8 Methods Not Controlling Experimentwise Error Rate

We have introduced five methods of multiple comparisons, each of which allows the experimenter to control the overall confidence level, and the same methods can be used to control the experimentwise error rate when multiple hypotheses are to be tested. There exist other multiple comparison procedures that are more powerful (i.e., that more easily detect a nonzero contrast) but do not control the overall confidence level nor the experimentwise error rate. While some of these are used quite commonly, we do not advocate their use. Such procedures include Duncan's multiple range test, Fisher's protected LSD procedure, and the Newman–Keuls method. (For more details, see Hsu's 1996 text.)

4.5 Sample Sizes

Before an experiment can be run, it is necessary to determine the number of observations that should be taken on each level of each treatment factor (step (h) of the checklist in Section 2.2, page 7). In Section 3.6.2, a method was presented to calculate the sample sizes needed to achieve a specified power of the test of the hypothesis $H_0 : \tau_1 = \cdots = \tau_v$. In this section we show how to determine the sample sizes to achieve confidence intervals of specified lengths.

The lengths of confidence intervals decrease as sample sizes increase. Consequently, if the length of an interval is specified, it should be possible to calculate the required sample sizes, especially when these are equal. However, there is a problem. Since the experimental data have not yet been collected, the value of the mean squared error is not known. As in Section 3.6.2, if the value of the mean squared error can be reasonably well be guessed at, either from previous experience or from a pilot study, then a trial and error approach to the problem can be followed, as illustrated in the next example.

Example 4.5.1 Bean-soaking experiment

Suppose we were to plan an experiment to compare the effects of $v = 5$ different soaking times on the growth rate of mung bean seeds. The response variable will be the length of a shoot of a mung bean seed 48 hours after soaking. Suppose that a pilot experiment has indicated that the mean square for error is likely to be not more than 10 mm^2, and suppose that we would like a set of 95% simultaneous confidence intervals for pairwise differences of the soaking times, with each interval no wider than 6 mm (that is, the half width or minimum significant difference should be no greater than 3 mm).

The formula for each of the simultaneous confidence intervals for pairwise comparisons using Tukey's method of multiple comparisons is given by (4.4.27) page 85. For equal sample sizes, the interval half width, or minimum significant difference, is required to be at

4.5 Sample Sizes

most 3 mm; that is, we require

$$msd = w_T \sqrt{10\left(\frac{1}{r} + \frac{1}{r}\right)} \leq 3,$$

where $w_T = q_{5, 5r-5, .05}/\sqrt{2}$ or, equivalently,

$$q^2_{5, 5r-5, .05} \leq 0.9r.$$

Adopting a trial-and-error approach, we guess a value for r, say $r = 10$. Then, from Table A.8, we find $q^2_{5, 45, .05} \approx 4.03^2 = 16.24$, which does not satisfy the requirement that $q^2 \leq 0.9r = 9$. A larger value for r is needed, and we might try $r = 20$ next. The calculations are most conveniently laid out in table form, as follows.

r	$5r - 5$	$q^2_{5, 5r-5, 0.05}$	$0.9r$	Action
10	45	$4.03^2 = 16.24$	9.00	Increase r
20	95	$3.95^2 = 15.60$	18.00	Decrease r
15	70	$3.97^2 = 15.76$	13.50	Increase r
18	85	$3.96^2 = 15.68$	16.20	Decrease r
17	80	$3.96^2 = 15.68$	15.30	

If $r = 17$ observations are taken on each of the five soaking times, and if the mean square for error is approximately 10 mm² in the main experiment, then the 95% Tukey simultaneous confidence intervals for pairwise comparisons will be a little over the required 6 mm in length. If $r = 18$ observations are taken, the interval will be a little shorter than the 6 mm required. If the cost of the experiment is high, then $r = 17$ would be selected; otherwise, $r = 18$ might be preferred. □

Trial and error procedures such as that illustrated in Example 4.5 for Tukey's method of multiple comparisons can be used for any of the other multiple comparison methods to obtain the approximate sample sizes required to meet the objectives of the experiment. The same type of calculation can be done for unequal sample sizes, provided that the relative sizes are specified, for example $r_1 = 2r_2 = 2r_3 = 2r_4$.

Unless more information is desired on some treatments than on others, or unless costs or variances are unequal, it is generally advisable to select equal sample sizes whenever possible. Choosing equal sample sizes produces two benefits: Confidence intervals for pairwise comparisons are all the same length, which makes them easier to compare, and the multiple comparison and analysis of variance procedures are less sensitive to an incorrect assumption of normality of the error variables.

Quite often, the sample size calculation will reveal that the required number of observations is too large to meet the budget or the time restrictions of the experiment. There are several possible remedies:

(a) Refine the experimental procedure to reduce the likely size of msE,

(b) Omit one or more treatments,

(c) Allow longer confidence intervals,

(d) Allow a lower confidence level.

Table 4.2 Sample SAS program for the battery experiment

Line	SAS Program
1	DATA BATTERY;
2	INPUT TYPEBAT LIFEUC ORDER;
3	LINES;
4	1 611 1
5	2 923 2
6	1 537 3
7	: : :
8	3 413 16
9	;
10	PROC PRINT;
11	;
12	PROC PLOT;
13	PLOT LIFEUC*TYPEBAT
14	/ VPOS=19 HPOS=50;
15	;
16	PROC GLM;
17	CLASSES TYPEBAT;
18	MODEL LIFEUC = TYPEBAT;
19	ESTIMATE 'DUTY'
20	TYPEBAT 1 1 -1 -1 /DIVISOR = 2;
21	ESTIMATE 'BRAND'
22	TYPEBAT 1 -1 1 -1 /DIVISOR = 2;
23	ESTIMATE 'INTERACTN'
24	TYPEBAT 1 -1 -1 1 /DIVISOR = 2;
25	CONTRAST 'BRAND'
26	TYPEBAT 1 -1 1 -1;
27	MEANS TYPEBAT / TUKEY CLDIFF ALPHA=0.01;

4.6 Using SAS Software

In this section we illustrate how to use the SAS software to generate information for confidence intervals and hypothesis tests for individual contrasts and means and also for the multiple comparison procedures. We use the data from the battery experiment of Section 2.5.2 (page 26). The treatment factor is type of battery TYPEBAT, the response variable is the life per unit cost LIFEUC, and the one-way analysis of variance model (3.3.1) was used for the analysis.

A sample SAS program to analyze the data is given in Table 4.2. Line numbers have been included for the sake of reference but are not part of the SAS program. A data set is created by lines 1–9, line 10 causes the data to be printed, lines 12–14 generate a plot of the data, and lines 16–18 generate the analysis of variance table shown in the top of Table 4.3.

4.6.1 Inferences on Individual Contrasts

The SAS statements ESTIMATE and CONTRAST, part of the GLM procedure, are used for making inferences concerning specific contrasts. The ESTIMATE statements (lines 19–24

Table 4.3 Analysis of variance table and output from the ESTIMATE and CONTRASTS statements

```
                      The SAS System
              General Linear Models Procedure

Dependent Variable: LIFEUC
                          Sum of         Mean
Source             DF    Squares    Square    F Value   Pr > F
Model               3   427915.25  142638.42    60.24   0.0001
Error              12    28412.50    2367.71
Corrected Total    15   456327.75


Contrast           DF   Contrast SS  Mean Square  F Value   Pr > F
BRAND               1    124609.00    124609.00    52.63    0.0001


                                T for H0:    Pr > |T|   Std Error of
Parameter       Estimate      Parameter=0                 Estimate
DUTY           251.000000         10.32       0.0001    24.3295516
BRAND         -176.500000         -7.25       0.0001    24.3295516
INTERACTN     -113.250000         -4.65       0.0006    24.3295516
```

of Table 4.2) generate information for constructing confidence intervals or conducting hypothesis tests for individual contrasts.

Each of the three ESTIMATE statements includes a contrast name in single quotes, together with the name of the factor for which the effects of levels are to be compared, and the coefficients of the contrast to be estimated. If the contrast is to be divided by a constant, this is indicated by means of the DIVISOR option. The information generated by these statements is shown in the bottom half of Table 4.3. The columns show the contrast name, the contrast estimate $\Sigma c_i \bar{y}_{i.}$, the value of the test-statistic for testing the null hypothesis that the contrast is zero (see (4.3.13) page 76, the corresponding p-value for a two-tailed test, and the standard error $\sqrt{msE(\Sigma c_i^2/r_i)}$ of the estimate. For each of the contrasts shown in Table 4.3, the p-value is less than 0.0006, indicating that all three contrasts are significantly different from zero for any choice of *individual* significance level α^* greater than 0.0006. The overall and individual significance levels should be selected prior to analysis. The parameter estimates and standard errors can be used to construct confidence intervals by hand, using the critical coefficient for the selected multiple comparison methods (see also Section 4.6.2).

The CONTRAST statement in lines 25–26 of Table 4.2 generates the information shown in Table 4.3 that is needed in (4.3.15), page 76, for testing the single null hypothesis that the brand contrast is zero versus the alternative hypothesis that it is not zero. The "F Value" of 52.63 is the square of the "T for H0" value of −7.25 (up to rounding error) for the brand contrast generated by the ESTIMATE statement, the two tests (4.3.13) and (4.3.15) being equivalent.

Table 4.4 Tukey's method for the battery experiment

```
                        The SAS System
                  General Linear Models Procedure
         Tukey's Studentized Range (HSD) Test for variable: LIFEUC
         NOTE: This test controls the type I experimentwise error rate.
             Alpha= 0.01  Confidence= 0.99  df= 12  MSE= 2367.708
                    Critical Value of Studentized Range= 5.502
                      Minimum Significant Difference= 133.85
         Comparisons significant at the 0.01 level are indicated by '***'.

                         Simultaneous              Simultaneous
                            Lower      Difference     Upper
              TYPEBAT     Confidence    Between    Confidence
            Comparison      Limit        Means        Limit
            2    - 1        155.90      289.75       423.60    ***
            2    - 4        230.40      364.25       498.10    ***
            2    - 3        293.65      427.50       561.35    ***
            1    - 4        -59.35       74.50       208.35
            1    - 3          3.90      137.75       271.60    ***
            4    - 3        -70.60       63.25       197.10
```

4.6.2 Multiple Comparisons

The MEANS statement (line 27) in the GLM procedure in Table 4.2 can be used to generate the observed means $\bar{y}_{i.}$ for each level of a factor. Inclusion of the TUKEY option causes the SAS software to use the Tukey method to compare the effects of each pair of levels. The option CLDIFF asks that the results of Tukey's method be presented in the form of confidence intervals. The option ALPHA=0.01 indicates that 99% simultaneous confidence intervals are to be generated. The results generated by SAS software are given in Table 4.4. The three asterisks highlight those intervals that do not contain zero.

In order to show the output for Tukey's method when the sample sizes are unequal, we deleted lines 4 and 5 from the SAS program in Table 4.2, leaving battery types 1 and 2 with only 3 observations each and battery types 3 and 4 with the original 4 observations each. The output is shown in Table 4.5. We see that no value for the minimum significant difference is given. This is because the intervals are of different lengths depending upon the numbers of observations on the treatments involved in the intervals. Otherwise, the output is similar to the equal-observation case.

Other methods of multiple comparisons can also be requested as options in the MEANS statement of the GLM procedure. For example, the options BON and SCHEFFE request all pairwise comparisons using the methods of Bonferroni and Scheffé, respectively. The option DUNNETT('1') requests Dunnett's 2-sided method of comparing all treatments with a control, specifying level 1 as the control treatment. Any other level can be specified as the control by changing "('1')" to the specified level. The option DUNNETTU('1') requests upper bounds for the treatment-versus-control contrasts $\tau_i - \tau_1$ by Dunnett's method and is useful for showing which treatments have a larger effect than the control treatment (coded 1). Similarly, the

Table 4.5 Tukey's method for the battery experiment, unequal sample sizes

```
                        The SAS System
                 General Linear Models Procedure
         Tukey's Studentized Range (HSD) Test for variable: LIFEUC
      NOTE: This test controls the type I experimentwise error rate.
          Alpha= 0.01  Confidence= 0.99  df= 10  MSE= 2104.408
                Critical Value of Studentized Range= 5.769
     Comparisons significant at the 0.01 level are indicated by '***'.

                       Simultaneous              Simultaneous
                          Lower      Difference     Upper
            TYPEBAT     Confidence    Between     Confidence
          Comparison      Limit        Means        Limit
           2   - 1        129.55       282.33       435.12    ***
           2   - 4        200.50       343.42       486.33    ***
           2   - 3        263.75       406.67       549.58    ***
           1   - 4        -81.83        61.08       204.00
           1   - 3        -18.58       124.33       267.25
           4   - 3        -69.06        63.25       195.56
```

option DUNNETTL('1') provides lower bounds useful for showing which treatments have a smaller effect than the control treatment (coded 1).

Hsu's method of multiple comparisons with the best can be executed using a SAS macro that is available in the SAS/STAT® sample library starting with release 6.12. The method is also implemented in the SAS JMP® software beginning in version 2 under the "Fit Y by X" platform and in the Minitab software beginning in version 8 using the MCB subcommand of the ONEWAY command.

SAS does not give confidence bounds for σ^2, but the value of msE is, of course, available from the analysis of variance table.

Exercises

1. **Buoyancy experiment**
   Consider conducting an experiment to investigate the question, "Is the buoyancy of an object in water affected by different concentrations of salt in the water?"

   (a) Complete steps (a)–(d) of the checklist (page 7) in detail. Specify any preplanned contrasts or functions that should be estimated. State, with reasons, which, if any, methods of multiple comparisons will be used.

   (b) Run a small pilot experiment to obtain a preliminary estimate of σ^2.

   (c) Finish the checklist.

2. **Cotton-spinning experiment, continued**

For the cotton-spinning experiment of Section 2.3, page 14, identify any contrasts or functions that you think might be interesting to estimate. For any contrasts that you have selected, list the corresponding contrast coefficients.

3. **Pedestrian light experiment, continued**

The pedestrian light experiment was described in Exercise 14 of Chapter 3, and the data were given in Table 3.12, page 63.

(a) Test the hypothesis that pushing the button does not lessen the waiting time for the "walk" signal; that is, test the null hypothesis

$$H_0 : \tau_0 - (\tau_1 + \tau_2 + \tau_3)/3 = 0$$

against the one-sided alternative hypothesis

$$H_A : \tau_0 - (\tau_1 + \tau_2 + \tau_3)/3 < 0.$$

(b) Using "no pushes of the button" as the control treatment, give a set of simultaneous 95% confidence intervals for the treatment-versus-control contrasts. State your conclusions in a form that can be understood by all users of the pedestrian light.

4. **Reaction time experiment**

(L. Cai, T. Li, Nishant, and A. van der Kouwe, 1996)

The experiment was run to compare the effects of auditory and visual cues on speed of response of a human subject. A personal computer was used to present a "stimulus" to a subject, and the reaction time required for the subject to press a key was monitored. The subject was warned that the stimulus was forthcoming by means of an auditory or a visual cue. The experimenters were interested in the effects on the subjects' reaction time of the auditory and visual cues and also in different elapsed times between cue and stimulus. Thus, there were two different treatment factors: "cue stimulus" at two levels "auditory" or "visual," and "elapsed time between cue and stimulus" at three levels "five," "ten," or "fifteen" seconds. This gave a total of six treatment combinations, which can be coded as

1 = auditory, 5 seconds   4 = visual, 5 seconds
2 = auditory, 10 seconds   5 = visual, 10 seconds
3 = auditory, 15 seconds   6 = visual, 15 seconds

The results of a pilot experiment, involving only one subject, are shown in Table 4.6. The reaction times were measured by the computer and are shown in seconds. The order of observation is shown in parentheses.

(a) Identify a set of contrasts that you would find particularly interesting in this experiment. (Hint: A comparison between the auditory treatments and the visual treatments might be of interest). These are your preplanned contrasts.

(b) Plot the data. What does the plot suggest about the treatments?

(c) Test the hypothesis that the treatments do not have different effects on the reaction time against the alternative hypothesis that they do have different effects.

Table 4.6 Reaction times, in seconds, for the reaction time experiment—(order of collection in parentheses)

Treatments					
1	2	3	4	5	6
0.204 (9)	0.167 (3)	0.202 (13)	0.257 (7)	0.283 (6)	0.256 (1)
0.170 (10)	0.182 (5)	0.198 (16)	0.279 (14)	0.235 (8)	0.281 (2)
0.181 (18)	0.187 (12)	0.236 (17)	0.269 (15)	0.260 (11)	0.258 (4)

(d) Calculate a set of simultaneous 90% confidence intervals for your preplanned contrasts, using a method or methods of your choice. State your conclusions.

(e) Ignoring the previous parts of this exercise, use Hsu's method of multiple comparisons with the best to determine the best/worst treatment or treatments. Define "best" to be the treatment that produces the quickest response (that is, the smallest value of the response variable).

5. **Trout experiment, continued**

Exercise 15 of Chapter 3 (page 64) concerns a study of the effects of four levels of sulfamerazine (0, 5, 10, 15 grams per 100 lb of fish) on the hemoglobin content of trout blood. An analysis of variance test rejected the hypothesis that the four treatment effects are the same at significance level $\alpha = 0.01$.

(a) Compare the four treatments using Tukey's method of pairwise comparisons and a 99% overall confidence level.

(b) Compare the effect of no sulfamerazine on the hemoglobin content of trout blood with the average effect of the other three levels. The overall confidence level of all intervals in parts (a) and (b) should be at least 98%.

6. **Battery experiment, continued**

In Example 4.4.4 (page 87), Tukey's method is used to obtain a set of 95% simultaneous confidence intervals for the pairwise differences $\tau_i - \tau_s$. Verify that this method gives shorter confidence intervals than would either of the Bonferroni or Scheffé methods (for $v = 4$ and $r = 4$).

7. **Soap experiment, continued**

The soap experiment was described in Section 2.5.1, page 22, and an analysis was given in Section 3.7.2, page 54.

(a) Suppose that the experimenter had been interested only in the contrast $\tau_1 - \frac{1}{2}(\tau_2 + \tau_3)$, which compares the weight loss for the regular soap with the average weight loss for the other two soaps. Calculate a confidence interval for this single contrast.

(b) Test the hypothesis that the regular soap has the same average weight loss as the average of the other two soaps. Do this via your confidence interval in part (a) and also via (4.3.13) and (4.3.15).

(c) In Example 4.4.5 (page 89), Dunnett's method was used for simultaneous 99% confidence intervals for two preplanned treatment-versus-control contrasts. Would either or both of the Bonferroni and Tukey methods have given shorter intervals?

100   Exercises

(d) Which method would be the best if all pairwise differences are required? Calculate a set of simultaneous 99% confidence intervals for all of the pairwise differences. Why are the intervals longer than those in part (c)?

8. **Trout experiment, continued**

   (a) For the trout experiment in Exercise 15 of Chapter 3 (see page 64), test the hypotheses that the linear and quadratic trends in hemoglobin content of trout blood due to the amount of sulfamerazine added to the diet is negligible. State the overall significance level of your tests.

   (b) Regarding the absence of sulfamerazine in the diet as the control treatment, calculate simultaneous 99% confidence intervals for the three treatment-versus-control comparisons. Which method did you use and why?

   (c) What is the overall confidence level of the intervals in part (b) together with those in Exercise 5? Is there a better strategy than using three different procedures for the three sets of intervals? Explain.

9. **Battery experiment, continued**

   Suppose the battery experiment of Section 2.5.2 (page 26) is to be repeated. The experiment involved four treatments, and the error standard deviation is estimated from that experiment to be about 48.66 minutes per dollar (minutes/dollar).

   (a) Calculate a 90% upper confidence limit for the error variance σ^2.

   (b) How large should the sample sizes be in the new experiment if Tukey's method of pairwise comparisons is to be used and it is desired to obtain a set of 95% simultaneous confidence intervals of length at most 100 minutes per dollar?

   (c) How large should the sample sizes be in the new experiment if Scheffé's method is to be used to obtain a set of 95% simultaneous confidence intervals for various contrasts and if the confidence interval for the duty contrast is to be of length at most 100 minutes per dollar?

10. **Trout experiment, continued**

    Consider again the trout experiment in Exercise 15 of Chapter 3.

    (a) Suppose the experiment were to be repeated. Suggest the largest likely value for the error mean square *msE*.

    (b) How many observations should be taken on each treatment so that the length of each interval in a set of simultaneous 95% confidence intervals for pairwise comparisons should be at most 2 grams per 100 ml?

11. **Pedestrian light experiment, continued**

    (a) Suppose that you are planning to repeat the pedestrian light experiment (Exercise 14 of Chapter 3) at a pedestrian crossing of your choosing. Select $v = 4$ levels for the treatment factor "number of pushes," including the level "no pushes." Give reasons for your selection.

    (b) Using "no pushes" as the control treatment, write down the formula for a set of 95% simultaneous confidence intervals for treatment-versus-control contrasts.

(c) How many observations would you need to ensure that your treatment-versus-control confidence intervals are of length less than 0.1 seconds? What value are you going to use for *msE* and why?

(d) If you had selected $v = 6$ levels of the treatment factor instead of $v = 4$ levels, would you have required more observations per treatment or fewer or the same number?

5. Checking Model Assumptions

5.1 Introduction
5.2 Strategy for Checking Model Assumptions
5.3 Checking the Fit of the Model
5.4 Checking for Outliers
5.5 Checking Independence of the Error Terms
5.6 Checking the Equal Variance Assumption
5.7 Checking the Normality Assumption
5.8 Using SAS Software
Exercises

5.1 Introduction

Throughout the two previous chapters, we discussed experiments whose data could be described by the one-way analysis of variance model (3.3.1), that is,

$$Y_{it} = \mu + \tau_i + \epsilon_{it},$$
$$\epsilon_{it} \sim N(0, \sigma^2),$$
ϵ_{it}'s are mutually independent,
$t = 1, \ldots, r_i, \quad i = 1, \ldots, v.$

This model implies that the response variables Y_{it} are mutually independent and have a normal distribution with mean $\mu + \tau_i$ and variance σ^2, that is, $Y_{it} \sim N(\mu + \tau_i, \sigma^2)$. For a given experiment, the model is selected in step (f) of the checklist using any available knowledge about the experimental situation, including the anticipated major sources of variation, the measurements to be made, the type of experimental design selected, and the results of any pilot experiment. However, it is not until the data have been collected that the adequacy of the model can be checked. Even if a pilot experiment has been used to help select the model, it is still important to check that the chosen model is a reasonable description of the data arising from the main experiment.

Methods of checking the model assumptions form the subject of this chapter, together with some indications of how to proceed if the assumptions are not valid. We begin by presenting a general strategy, including the order in which model assumptions should be checked. For checking model assumptions, we rely heavily on residual plots. We do so because while examination of residual plots is more subjective than would be testing for model lack-of-fit, the plots are often more informative about the nature of the problem, the consequences, and the corrective action.

5.2 Strategy for Checking Model Assumptions

In this section we discuss strategy and introduce the notions of residuals and residual plots. A good strategy for checking the assumptions about the model is to use the following sequence of checks.

- *Check the form of the model*—are the mean responses for the treatments adequately described by $E(Y_{it}) = \mu + \tau_i, i = 1, \ldots, v$?

- *Check for outliers*—are there any unusual observations (outliers)?

- *Check for independence*—do the error variables ϵ_{it} appear to be independent?

- *Check for constant variance*—do the error variables ϵ_{it} have similar variances for each treatment?

- *Check for normality*—do the error variables ϵ_{it} appear to be a random sample from a normal distribution?

For all of the fixed-effects models considered in this book, these same assumptions should be checked, except that $E(Y_{it})$ differs from model to model. The assumptions of independence, equal variance, and normality are the error assumptions mentioned in Chapter 3.

5.2.1 Residuals

The assumptions on the model involve the error variables, $\epsilon_{it} = Y_{it} - E(Y_{it})$, and can be checked by examination of the *residuals*. The itth residual $\hat{e}_{it}$ is defined as the observed value of $Y_{it} - \hat{Y}_{it}$, where $\hat{Y}_{it}$ is the least squares estimator of $E[Y_{it}]$, that is,

$$\hat{e}_{it} = y_{it} - \hat{y}_{it}.$$

For the one-way analysis of variance model (3.3.1), $E[Y_{it}] = \mu + \tau_i$, so the (it)th residual is

$$\hat{e}_{it} = y_{it} - (\hat{\mu} + \hat{\tau}_i) = y_{it} - \bar{y}_{i.}.$$

5.2 Strategy for Checking Model Assumptions

Table 5.1 Residuals and standardized residuals for the trout experiment

Treatment	Residuals				
1	−0.50	0.60	−1.70	1.20	−0.20
	0.60	1.40	0.20	−1.40	−0.20
2	0.57	−0.93	1.07	−0.03	1.37
	2.57	−2.23	−2.93	−0.73	1.27
3	1.37	−0.93	1.57	−0.33	1.67
	0.07	−0.23	−0.93	−1.23	−1.03
4	0.61	0.61	−1.49	−0.89	0.61
	1.51	0.01	−0.09	0.61	−1.49
Treatment	Standardized Residuals				
1	−0.42	0.50	−1.41	1.00	−0.17
	0.50	1.16	0.17	−1.16	−0.17
2	0.47	−0.77	0.89	−0.02	1.14
	2.14	−1.85	−2.43	−0.61	1.06
3	1.14	−0.77	1.30	−0.27	1.39
	0.06	−0.19	−0.77	−1.02	−0.86
4	0.51	0.51	−1.24	−0.74	0.51
	1.25	0.01	−0.07	0.51	−1.24

We prefer to work with the *standardized residuals* rather than the residuals themselves, since standardization facilitates the identification of outliers. The standardization is achieved by dividing the residuals by their standard deviation, that is, by $\sqrt{ssE/(n-1)}$. The standardized residuals,

$$z_{it} = \frac{\hat{e}_{it}}{\sqrt{ssE/(n-1)}},$$

then have sample variance equal to 1.0.

If the assumptions on the model are correct, the standardized error variables ϵ_{it}/σ are independently distributed with a $N(0, 1)$ distribution, so the observed values $e_{it}/\sigma = (y_{it} - (\mu + \tau_i))/\sigma$ would constitute independent observations from a standard normal distribution. Although the standardized residuals are dependent and involve estimates of both e_{it} and σ, their behavior should be similar. Consequently, methods for evaluating the model assumptions using the standardized residuals look for deviations from patterns that would be expected of independent observations from a standard normal distribution.

5.2.2 Residual Plots

A *residual plot* is a plot of the standardized residuals z_{it} against the levels of another variable, the choice of which depends on the assumption being checked. In Example 5.2.2, we show a plot of the standardized residuals against the levels of the treatment factor for the trout experiment. Plots like this are useful for evaluating the assumption of constant error variance as well as the adequacy of the model.

Table 5.2 Data for the trout experiment

Code	Hemoglobin (grams per 100 ml)	$\bar{y}_{i.}$
1	6.7 7.8 5.5 8.4 7.0 7.8 8.6 7.4 5.8 7.0	7.20
2	9.9 8.4 10.4 9.3 10.7 11.9 7.1 6.4 8.6 10.6	9.33
3	10.4 8.1 10.6 8.7 10.7 9.1 8.8 8.1 7.8 8.0	9.03
4	9.3 9.3 7.2 7.8 9.3 10.2 8.7 8.6 9.3 7.2	8.69

Source: Gutsell, J. S. (1951). Copyright © 1951 International Biometric Society. Reprinted with permission.

Example 5.2.1 Constructing a residual plot: trout experiment

The trout experiment was described in Exercise 15 of Chapter 3. There was one treatment factor (grams of sulfamerazine per 100 lb of fish) with four levels coded 1, 2, 3, 4, each observed $r = 10$ times. The response variable was grams of hemoglobin per 100 ml of trout blood. The $n = 40$ data values are reproduced in Table 5.2 together with the treatment means.

Using the one-way analysis of variance model (3.3.1), it can be verified that $ssE = 56.471$. The residuals $\hat{e}_{it} = y_{it} - \bar{y}_{i.}$ and the standardized residuals $z_{it} = \hat{e}_{it}/\sqrt{ssE/(n-1)}$ are shown in Table 5.1. For example, the observation $y_{11} = 6.7$ yields the residual

$$\hat{e}_{11} = 6.7 - 7.2 = -0.5$$

and the standardized residual

$$z_{11} = -0.5/\sqrt{56.471/39} = -0.42$$

to two decimal places.

A plot of the standardized residuals against treatments is shown in Figure 5.1. The residuals sum to zero for each treatment since $\Sigma_t(y_{it} - \bar{y}_{i.}) = 0$ for each $i = 1, \ldots, v$. The standardized residuals seem fairly well scattered around zero, although the spread of the

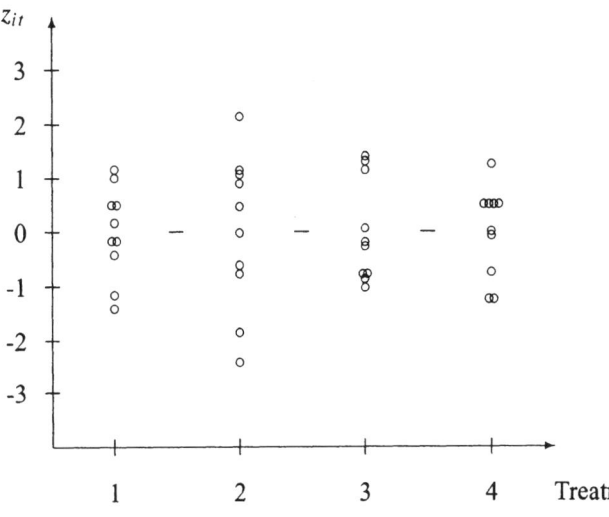

Figure 5.1 Plot of standardized residuals for the trout experiment

residuals for treatment 2 seems a little larger than the spread for the other three treatments. This could be interpreted as a sign of unequal variances of the error variables or that the data values having standardized residuals 2.14 and −2.43 are outliers, or it could be attributed to chance variation. Methods for checking for outliers and equality of variances will be discussed in Sections 5.4 and 5.6, respectively. □

5.3 Checking the Fit of the Model

The first assumption to be checked is the assumption that the model $E(Y_{it})$ for the mean response is correct. One purpose of running a pilot experiment is to choose a model that is a reasonable description of the data. If this is done, the model assumption checks for the main experiment should show no problems. If the model for mean response does not adequately fit the data, then there is said to be model *lack of fit*. If this occurs and if the model is changed accordingly, then any stated confidence levels and significance levels will only be approximate. This should be taken into account when decisions are to be made based on the results of the experiment.

In general, the fit of the model is checked by plotting the standardized residuals versus the levels of each independent variable (treatment factor, block factor, or covariate) included in the model. Lack of fit is indicated if the residuals exhibit a nonrandom pattern about zero in any such plot, being too often positive for some levels of the independent variable and too often negative for others.

For model (3.3.1), the only independent variable included in the model is the treatment factor. Since the residuals sum to zero for each level of the treatment factor, lack of fit would only be detected if there were a number of unusually large or small observations. However, lack of fit can also be detected by plotting the standardized residuals against the levels of factors that were omitted from the model. For example, for the trout experiment, if the standardized residuals were plotted against the age of the corresponding fish and if the plot were to show a pattern, then it would indicate that age should have been included in the model as a covariate. A similar idea is discussed in Section 5.5 with respect to checking for independence.

5.4 Checking for Outliers

An *outlier* is an observation that is much larger or much smaller than expected. This is indicated by a residual that has an unusually large positive or negative value. Outliers are fairly easy to detect from a plot of the standardized residuals versus the levels of the treatment factors. Any outlier should be investigated. Sometimes such investigation will reveal an error in recording the data, and this can be corrected. Otherwise, outliers may be due to the error variables not being normally distributed, or having different variances, or an incorrect specification of the model.

If all of the model assumptions hold, including normality, then approximately 68% of the standardized residuals should be between −1 and +1, approximately 95% between −2

Figure 5.2
Original residual plot for the battery experiment

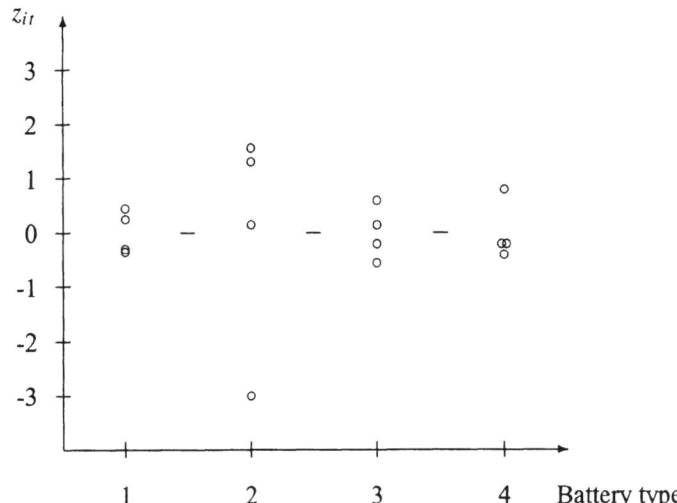

and +2, and approximately 99.7% between −3 and +3. If there are more outliers than expected under normality, then the true confidence levels are lower than stated and the true significance levels are higher.

Example 5.4.1 Checking for outliers: battery experiment

In the battery experiment of Example 2.5.2 (page 26), four observations on battery life per unit cost were collected for each of four battery types. Figure 5.2 shows the standardized residuals plotted versus battery type for the data as originally entered into the computer for analysis using model (3.3.1). This plot shows two related anomalies. There is one apparent outlier for battery type 2, the residual value being −2.98. Also, *all* of the standardized residuals for the other three battery types are less than one in magnitude. This is many more than the 68% expected.

An investigation of the outlier revealed a data entry error for the corresponding observation—a life length of 473 minutes was typed, but the recording sheet for the experiment showed the correct value to be 773 minutes. The unit cost for battery type 2 was $0.935 per battery, yielding the erroneous value of 506 minutes per dollar for the life per unit cost, rather than the correct value of 827. After correcting the error, the model was fitted again and the standardized residuals were replotted, as shown in Figure 5.3.

Observe how correcting the single data entry error corrects both problems observed in Figure 5.2. Not only is there no outlier, but the distribution of the 16 standardized residuals about zero is as one might anticipate for independent observations from a standard normal distribution—about a third of the standardized residuals exceed one in magnitude, and all are less than two in magnitude. The two anomalies are related, since correcting the data entry error makes ssE smaller and the standardized residuals correspondingly larger. □

For an outlier like that shown in Figure 5.2, the most probable cause of the problem is a measurement error, a recording error, or a transcribing error. When an outlier is detected, the experimenter should look at the original recording sheet to see whether the original data

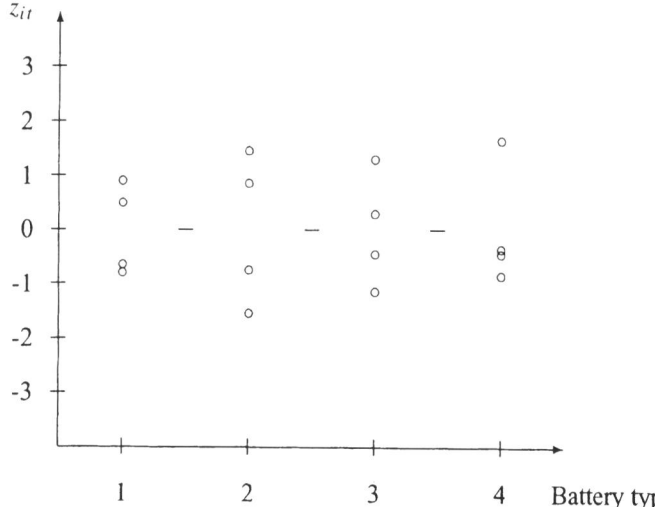

Figure 5.3
Residual plot after data correction for the battery experiment

value has been copied incorrectly at some stage. If the error can be found, then it can be corrected. When no obvious cause can be found for an outlier, the data value should not automatically be discarded, since it may be an indication of an occasional erratic behavior of a treatment. For example, had it not been due to a typographical error, the outlier for battery type 2 in the previous example might have been due to a larger variability in the responses for battery type 2.

The experimenter has to decide whether to include the unusual value in the analysis or whether to omit it. First, the data should be reanalyzed without the outlying value. If the conclusions of the experiment remain the same, then the outlier can safely be left in the analysis. If the conclusions change dramatically, then the outlier is said to be *influential*, and the experimenter must make a judgment as to whether the outlying observation is likely to be an experimental error or whether unusual observations do occur from time to time. If the experimenter decides on the former, then the analysis should be reported without the outlying observation. If the experimenter decides on the latter, then the model is not adequate to describe the experimental situation, and a more complicated model would be needed.

5.5 Checking Independence of the Error Terms

Since the checks for the constant variance and normality assumptions assume that the error terms are independent, a check for independence should be made next. The most likely cause of nonindependence in the error variables is the similarity of experimental units close together in time or space. The independence assumption is checked by plotting the standardized residuals against the order in which the corresponding observations were collected and against any spatial arrangement of the corresponding experimental units. If the independence assumption is satisfied, the residuals should be randomly scattered around zero with no discernible pattern. Such is the case for Figure 5.4 for the battery experiment. If

Figure 5.4
Residual plot for the battery experiment

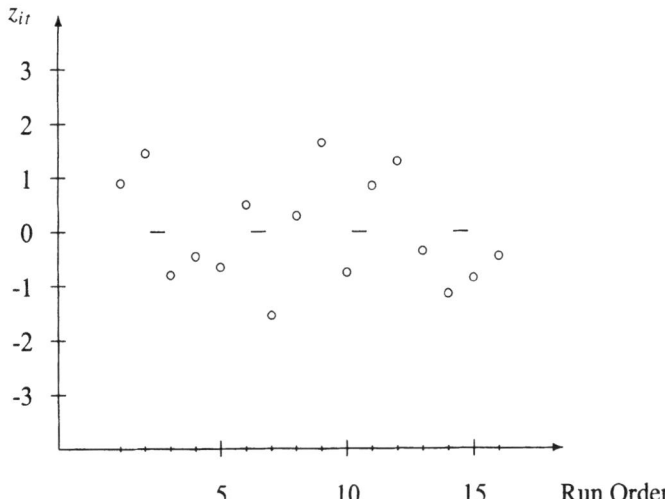

the plot were to exhibit a strong pattern, then this would indicate a serious violation of the independence assumption, as illustrated in the following example.

Example 5.5.1 Checking independence: balloon experiment

The experimenter who ran the balloon experiment in Exercise 12 of Chapter 3 was concerned about lack of independence of the observations. She had used a single subject to blow up all the balloons in the experiment, and the subject had become an expert balloon blower before the experiment was finished! Having fitted the one-way analysis of variance model (3.3.1) to the data (Table 3.11), she plotted the standardized residuals against the time order in which the balloons were inflated. The plot is shown in Figure 5.5. There appears to be a strong downward drift in the residuals as time progresses. The observations are clearly *dependent*.

Figure 5.5
Residual plot for the balloon experiment

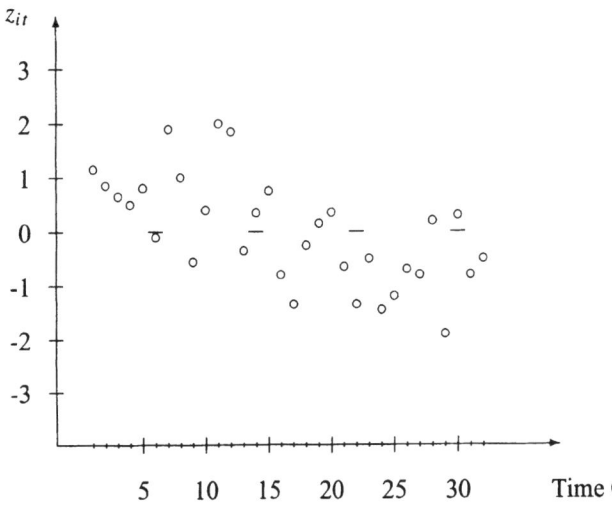

If an analysis is conducted under the assumptions of model (3.3.1) when, in fact, the error variables are dependent, the true significance levels of hypothesis tests can be much higher than stated and the true confidence levels and powers of tests can be much lower than stated. The problem of dependent errors can be difficult to correct. If there is a clear trend in the residual plot, such as the linear trend in Figure 5.5, it may be possible to add terms into the model to represent a time or space effect. For example, a more complex model that might be adequate for the balloon experiment is

$$Y_{it} = \mu + \tau_i + \gamma x_{it} + \epsilon_{it}$$
$$\epsilon_{it} \sim N(0, \sigma^2)$$

ϵ_{it}'s are mutually independent

$t = 1, 2, \ldots, r_i; \quad i = 1, \ldots, v,$

where the variable x_{it} denotes the time at which the observation was taken and γ is a linear time trend parameter that must be estimated. Such a model is called an *analysis of covariance* model and will be studied in Chapter 9. The assumptions for analysis of covariance models are checked using the same types of plots as discussed in this chapter. In addition, the standardized residuals should also be plotted against the values of x_{it}.

Had the experimenter in the balloon experiment anticipated a run order effect, she could have selected an analysis of covariance model prior to the experiment. Alternatively, she could have grouped the observations into blocks of, say, eight observations. Notice that each group of eight residuals in Figure 5.5 looks somewhat randomly scattered. As mentioned earlier in this chapter, when the model is changed after the data have been examined, then stated confidence levels and significance levels using that same data are inaccurate.

If a formal test of independence is desired, the most commonly used test is that of Durbin and Watson (1951) for time-series data (see Neter, Kutner, Nachtsheim, and Wasserman, 1996, pages 504–510).

5.6 Checking the Equal Variance Assumption

If the independence assumption appears to be satisfied, then the equal-variance assumption should be checked. Studies have shown that if the sample sizes $r_1, \ldots, r_v$ are chosen to be equal, then unless one variance is considerably larger than the others, the significance level of hypothesis tests and confidence levels of the associated confidence intervals remain close to the stated values. However, if the sample sizes are unequal, and if the treatment factor levels which are more highly variable in response happen to have been observed fewer times (i.e. if smaller r_i coincide with larger $\text{Var}(\epsilon_{it}) = \sigma_i^2$), then the statistical procedures are generally quite liberal, and the experimenter has a greater chance of making a Type I error in testing than anticipated, and also, the true confidence level of a confidence interval is lower than intended. On the other hand, if the large r_i coincide with large σ_i^2, then the procedures are conservative (significance levels are lower than stated and confidence levels

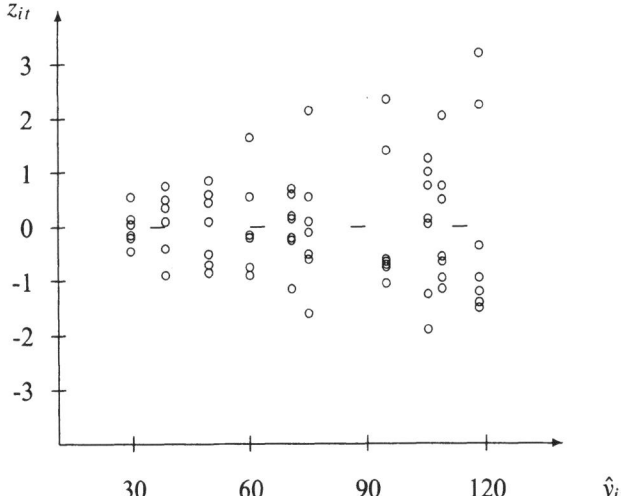

Figure 5.6
Megaphone-shaped residual plot

are higher). Thus, unless there is good knowledge of which treatment factor levels are the more variable, an argument can be made that *the sample sizes should be chosen to be equal.*

5.6.1 Detection of Unequal Variances

The most common pattern of nonconstant variance is that in which the error variance increases as the mean response increases. This situation is suggested when the plot of the standardized residuals versus the fitted values resembles a megaphone in shape, as in Figure 5.6. In such a case, one can generally find a transformation of the data, known as a variance-stabilizing transformation, which will correct the problem (see Section 5.6.2).

If the residual plot indicates unequal variances but not the pattern of Figure 5.6 (or its mirror image), then a variance-stabilizing transformation is generally not available. Approximate and somewhat less powerful methods of data analysis such as those discussed in Section 5.6.3 must then be applied.

An unbiased estimate of the error variance σ_i^2 for the ith treatment is the sample variance of the residuals for the ith treatment, namely

$$s_i^2 = \frac{1}{r_i - 1} \sum_{t=1}^{r_i} \hat{e}_{it}^2 = \frac{1}{r_i - 1} \sum_{t=1}^{r_i} (y_{it} - \hat{\mu} - \hat{\tau}_i)^2 \qquad (5.6.1)$$

$$= \frac{1}{r_i - 1} \sum_{t=1}^{r_i} (y_{it} - \bar{y}_{i.})^2 .$$

There do exist tests for the equality of variances, but they tend to have low power unless there are large numbers of observations on each treatment factor level. Also, the tests tend to be very sensitive to nonnormality. (The interested reader is referred to Neter, Kutner, Nachtsheim, and Wasserman, 1996, page 763).

A rule of thumb that we shall apply is that the usual analysis of variance F-test and the methods of multiple comparisons discussed in Chapter 4 are appropriate, provided that the

5.6 Checking the Equal Variance Assumption

ratio of the largest of the v treatment variance estimates to the smallest, $s^2_{\max}/s^2_{\min}$, does not exceed three. The rule of thumb is suggested by simulation studies in which the true variances σ_i^2 are specified. Since these are unknown in practice, we are basing our rule of thumb on the estimates s_i^2 of the variances. Be aware, however, that *it is possible for the ratio of extreme variance estimates $s^2_{\max}/s^2_{\min}$ to exceed three, even when the model assumptions are correct.*

Example 5.6.1 Comparing variances: trout experiment

Figure 5.1 (page 106) shows a plot of the standardized residuals against the levels of the treatment factor for the trout experiment. The plot suggests that the variance of the error variables for treatment 2 might be larger than the variances for the other treatments. Using the data in Table 3.13, we obtain

i	1	2	3	4
$\bar{y}_{i.}$	7.20	9.33	9.03	8.69
s_i^2	1.04	2.95	1.29	1.00

so $s^2_{\max}/s^2_{\min} = 2.95$, which satisfies our rule of thumb, but only just. Both the standard analysis using model (3.3.1) and an approximate analysis that does not require equal variances will be discussed in Example 5.6.3. □

5.6.2 Data Transformations to Equalize Variances

Finding a transformation of the data to equalize the variances of the error variables involves finding some function $h(y_{it})$ of the data so that the model

$$h(Y_{it}) = \mu^* + \tau_i^* + \epsilon_{it}^*$$

holds and $\epsilon_{it}^* \sim N(0, \sigma^2)$ and the ϵ_{it}^*'s are mutually independent for all $t = 1, \ldots, r_i$ and $i = 1, \ldots, v$. An appropriate transformation can generally be found if there is a clear relationship between the error variance $\sigma_i^2 = \text{Var}(\epsilon_{it})$ and the mean response $E[Y_{it}] = \mu + \tau_i$, for $i = 1, \ldots, v$. If the variance and the mean increase together, as suggested by the megaphone-shaped residual plot in Figure 5.6, or if one increases as the other decreases, then the relationship between σ_i^2 and $\mu + \tau_i$ is often of the form

$$\sigma_i^2 = k(\mu + \tau_i)^q, \tag{5.6.2}$$

where k and q are constants. In this case, the function $h(y_{it})$ should be chosen to be

$$h(y_{it}) = \begin{cases} (y_{it})^{1-(q/2)} & \text{if } q \neq 2, \\ \ln(y_{it}) & \text{if } q = 2 \text{ and all } y_{it}\text{'s are nonzero}, \\ \ln(y_{it} + 1) & \text{if } q = 2 \text{ and some } y_{it}\text{'s are zero.} \end{cases} \tag{5.6.3}$$

Here "ln" denotes the natural logarithm, which is the logarithm to the base e. Usually, the value of q is not known, but a reasonable approximation can be obtained empirically as follows. Substituting the least squares estimates for the parameters into equation (5.6.2) and taking logs of both sides gives

$$\ln(s_i^2) = \ln(k) + q(\ln(\bar{y}_{i.})).$$

Therefore, the slope of the line obtained by plotting $\ln(s_i^2)$ against $\ln(\bar{y}_{i.})$ gives an estimate for q. This will be illustrated in Example 5.6.2.

The value of q is sometimes suggested by theoretical considerations. For example, if the normal distribution assumed in the model is actually an approximation to the Poisson distribution, then the variance would be equal to the mean, and $q = 1$. The square-root transformation $h(y_{it}) = (y_{it})^{1/2}$ would then be appropriate. The binomial distribution provides another commonly occurring case for which an appropriate transformation can be obtained theoretically. If each Y_{it} has a binomial distribution with mean mp and variance $mp(1-p)$, then a variance-stabilizing transformation is

$$h(y_{it}) = \sin^{-1}\sqrt{y_{it}/m} = \arcsin\left(\sqrt{y_{it}/m}\right).$$

When a transformation is found that equalizes the variances, then it is necessary to check or recheck the other model assumptions, since a transformation that cures one problem could cause others. If there are no problems with the other model assumptions, then analysis can proceed using the techniques of the previous two chapters, but using the transformed data $h(y_{it})$.

Example 5.6.2 Choosing a transformation: battery life experiment

In Section 2.5.2, the response variable considered for the battery experiment was "battery life per unit cost," and a plot of the residuals versus the fitted values looks similar to Figure 5.3 and shows fairly constant error variances.

Suppose, however, that the response variable of interest had been "battery life" regardless of cost. The corresponding data are given in Table 5.3. The battery types are

1 = alkaline, name brand
2 = alkaline, store brand
3 = heavy duty, name brand
4 = heavy duty, store brand

Figure 5.7 shows a plot of the standardized residuals versus the fitted values. Variability seems to be increasing modestly with mean response, suggesting that a transformation can be found to stabilize the error variance. The ratio of extreme variance estimates is $s_{max}^2/s_{min}^2 = s_2^2/s_3^2 = 3151.70/557.43 \approx 5.65$. Hence, based on the rule of thumb, a variance stabilizing transformation should be used. Using the treatment sample means and

Table 5.3 Data for the battery lifetime experiment

Battery	Lifetime (minutes)				$\bar{y}_{i.}$	s_i^2
1	602	529	534	585	562.50	1333.71
2	863	743	773	840	804.75	3151.70
3	232	255	200	215	225.50	557.43
4	235	282	238	228	245.75	601.72

5.6 Checking the Equal Variance Assumption

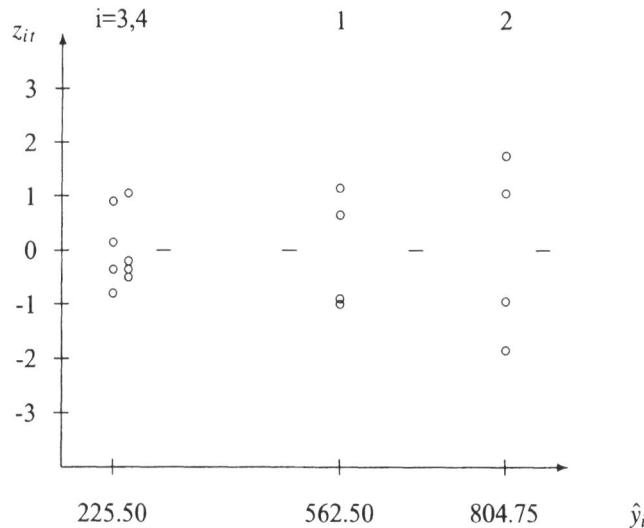

Figure 5.7 Residual plot for the battery lifetime data

variances from Table 5.3, we have

i	$\bar{y}_{i.}$	$\ln(\bar{y}_{i.})$	s_i^2	$\ln(s_i^2)$
1	562.50	6.3324	1333.71	7.1957
2	804.75	6.6905	3151.70	8.0557
3	225.50	5.4183	557.43	6.3233
4	245.75	5.5043	601.72	6.3998

Figure 5.8 shows a plot of $\ln(s_i^2)$ against $\ln(\bar{y}_{i.})$. This plot is nearly linear, so the slope will provide an estimate of q in (5.6.2). A line can be drawn by eye or by the regression methods of Chapter 8. Both methods give a slope approximately equal to $q = 1.25$. From equation (5.6.3) a variance-stabilizing transformation is

$$h(y_{it}) = (y_{it})^{0.375}.$$

Since $(y_{it})^{0.375}$ is close to $(y_{it})^{0.5}$, and since the square root of the data values is perhaps more meaningful than $(y_{it})^{0.375}$, we will try taking the square root transformation. The square roots of the data are shown in Table 5.4.

The transformation has stabilized the variances considerably, as evidenced by $s_{max}^2/s_{min}^2 = 0.982/0.587 \approx 1.67$. Checks of the other model assumptions for the transformed data also reveal no severe problems. The analysis can now proceed using the transformed data. The stated significance level and confidence levels will now be approximate, since the model has

Table 5.4 Transformed data $\sqrt{y_{it}}$ for the battery lifetime experiment

Brand	$x_{it} = h(y_{it}) = \sqrt{y_{it}}$				$\bar{x}_{i.}$	s_i^2
1	24.536	23.000	23.108	24.187	23.708	0.592
2	29.377	27.258	27.803	28.983	28.355	0.982
3	15.232	15.969	14.142	14.663	15.001	0.614
4	15.330	16.793	15.427	15.100	15.662	0.587

been changed based on the data. For the transformed data, $msE = 0.6936$. Using Tukey's method of multiple comparisons to compare the lives of the four battery types (regardless of cost) at an overall confidence level of 99%, the minimum significant difference obtained from equation (4.4.28) is

$$msd = q_{4,12,0.01}\sqrt{msE/4} = 5.50\sqrt{0.6936/4} = 2.29.$$

Comparing msd with the differences in the sample means $\bar{x}_{i.}$ of the transformed data in Table 5.4, we can conclude that at an overall 99% level of confidence, all pairwise differences are significantly different from zero except for the comparison of battery types 3 and 4. Furthermore, it is reasonable to conclude that type 2 (alkaline, store brand) is best, followed by type 1 (alkaline, name brand). However, any more detailed interpretation of the results is muddled by use of the transformation, since the comparisons use mean values of $\sqrt{\text{life}}$. A more natural transformation, which also provided approximately equal error variances, was used in Section 2.5.2. There, the response variable was taken to be "life per unit cost," and confidence intervals were able to be calculated in meaningful units. □

5.6.3 Analysis with Unequal Error Variances

An alternative to transforming the data to equalize the error variances is to use a method of data analysis that is designed for nonconstant variances. Such a method will be presented for constructing confidence intervals. The method is approximate and tends to be less powerful than the methods of Chapter 4 with transformed data. However, the original data units are maintained, and the analysis can be used whether or not a variance-stabilizing transformation is available.

Without the assumption of equal variances for all treatments, the one-way analysis of variance model (3.3.1) is

$$Y_{it} = \mu + \tau_i + \epsilon_{it},$$

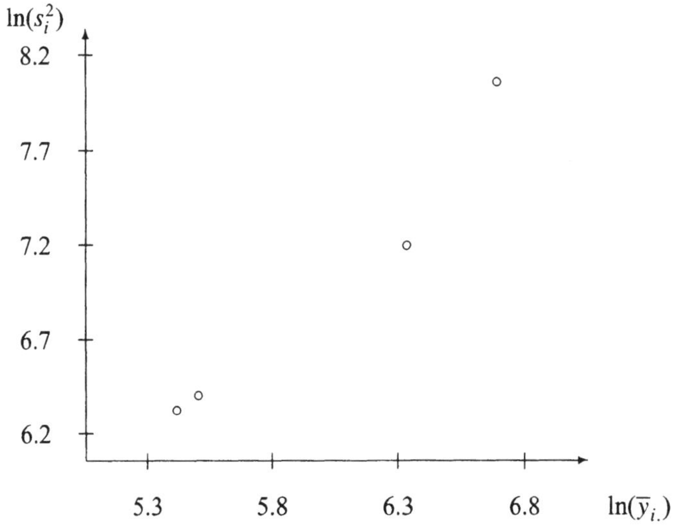

Figure 5.8
Plot of $\ln(s_i^2)$ versus $\ln(\bar{y}_{i.})$ for the battery lifetime experiment

5.6 Checking the Equal Variance Assumption

$$\epsilon_{it} \sim N(0, \sigma_i^2),$$

ϵ_{it}'s are mutually independent,

$t = 1, \ldots, r_i, \quad i = 1, \ldots, v.$

For this model, each contrast $\Sigma c_i \tau_i$ in the treatment parameters remains estimable, but the least squares estimator $\Sigma c_i \hat{\tau}_i = \Sigma c_i \overline{Y}_{i.}$ now has variance $\text{Var}(\Sigma c_i \overline{Y}_{i.}) = \Sigma c_i^2 \sigma_i^2 / r_i$. If we estimate σ_i^2 by s_i^2 as given in (5.6.1), then

$$\frac{\Sigma c_i \hat{\tau}_i - \Sigma c_i \tau_i}{\sqrt{\widehat{\text{Var}}(\Sigma c_i \hat{\tau}_i)}}$$

has approximately a t-distribution with df degrees of freedom, where

$$\widehat{\text{Var}}(\Sigma c_i \hat{\tau}_i) = \sum \frac{c_i^2}{r_i} s_i^2 \quad \text{and} \quad df = \frac{(\Sigma c_i^2 s_i^2 / r_i)^2}{\sum \frac{(c_i^2 s_i^2 / r_i)^2}{(r_i - 1)}}. \tag{5.6.4}$$

Then an approximate $100(1-\alpha)\%$ confidence interval for a single treatment contrast $\Sigma c_i \tau_i$ is

$$\sum c_i \tau_i \in \left(\sum c_i \hat{\tau}_i \pm w \sqrt{\widehat{\text{Var}}(\Sigma c_i \hat{\tau}_i)} \right), \tag{5.6.5}$$

where $w = t_{df, \alpha/2}$ and $\Sigma c_i \hat{\tau}_i = \Sigma c_i \overline{y}_{i.}$, all sums being from $i = 1$ to $i = v$. The formulae for $\widehat{\text{Var}}(\Sigma c_i \hat{\tau}_i)$ and df in (5.6.4), often called *Satterthwaite's approximation*, are due to Smith (1936), Welch (1938), and Satterthwaite (1946). The approximation is best known for use in inferences on a pairwise comparison $\tau_h - \tau_i$ of the effects of two treatments, in which case, for samples each of size r, (5.6.4) reduces to

$$\widehat{\text{Var}}(\hat{\tau}_h - \hat{\tau}_i) = \frac{s_h^2}{r} + \frac{s_i^2}{r} \quad \text{and} \quad df = \frac{(r-1)(s_h^2 + s_i^2)^2}{s_h^4 + s_i^4}. \tag{5.6.6}$$

Satterthwaite's approach can be extended to multiple comparison procedures by changing the critical coefficient w appropriately and computing $\Sigma c_i \hat{\tau}_i$ and df separately for each contrast. For Tukey's method, for example, the critical coefficient in (5.6.5) is $w_T = q_{v, df, \alpha}/\sqrt{2}$. Simulation studies by Dunnett (1980) have shown this variation on Tukey's method to be conservative (true α smaller than or equal to that stated).

Example 5.6.3 Satterthwaite's approximation: trout experiment

In Example 5.6.1, it was shown that the ratio of the maximum to the minimum error variance for the trout experiment satisfies the rule of thumb, but only just. The standardized residuals are plotted against the fitted values in Figure 5.9. The data for treatment 2 are the most variable and have the highest mean response, but there is no clear pattern of variability increasing as the mean response increases. In fact, it can be verified that a plot of $\ln(s_i^2)$ against $\ln(\overline{y}_{i.})$ is not very close to linear, suggesting that a transformation will not be successful in stabilizing the variances.

To obtain simultaneous approximate 95% confidence intervals for pairwise comparisons in the treatment effects by Tukey's method using Satterthwaite's approximation, we use

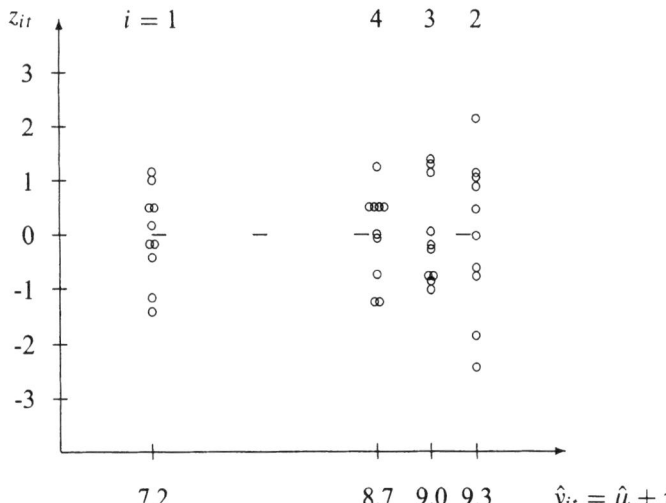

Figure 5.9
Residual plot for the trout experiment

equations (5.6.5) and (5.6.6) with $r = 10$. The minimum significant difference for pairwise comparison $\tau_h - \tau_i$ is

$$msd = \frac{1}{\sqrt{2}} q_{4,df,0.05} \sqrt{\frac{s_h^2}{r} + \frac{s_i^2}{r}},$$

the size of which depends upon which pair of treatments is being compared. From Example 5.6.1, we have

$$s_1^2 = 1.04, \quad s_2^2 = 2.95, \quad s_3^2 = 1.29, \quad s_4^2 = 1.00.$$

The values of $\sqrt{\widehat{\text{Var}}(\hat{\tau}_h - \hat{\tau}_i)} = \sqrt{s_h^2/r + s_i^2/r}$ are listed in Table 5.5. Comparing the values of *msd* with the values of $\bar{y}_{h.} - \bar{y}_{i.}$ in Table 5.5, we can conclude with simultaneous approximate 95% confidence that each of treatments 2, 3, and 4 yields statistically significantly higher mean response than does treatment 1.

Since $s_{max}^2 / s_{min}^2 = 2.95$, we could accept the rule of thumb and apply Tukey's method (4.4.28) for equal variances. The minimum significant difference for each pairwise

Table 5.5 Approximate values for Tukey's multiple comparisons for the trout experiment

(h, i)	$\sqrt{s_h^2/r + s_i^2/r}$	df	$q_{4,df,0.05}$	msd	$\bar{y}_{h.} - \bar{y}_{i.}$
(2, 3)	0.651	$15.6 \approx 16$	4.05	1.86	0.30
(2, 4)	0.629	$14.5 \approx 15$	4.08	1.82	0.64
(2, 1)	0.631	$14.6 \approx 15$	4.08	1.82	2.13
(3, 4)	0.478	$17.7 \approx 18$	4.00	1.35	0.34
(3, 1)	0.483	$17.8 \approx 18$	4.00	1.37	1.83
(4, 1)	0.452	$18.0 \approx 18$	4.00	1.28	1.49

comparison would then be

$$msd = q_{4,36,0.05}\sqrt{msE/10} = 3.82\sqrt{1.5685/10} \approx 1.51 \, .$$

Comparing this with the values of $\bar{y}_{h.} - \bar{y}_{i.}$ in Table 5.5, the same conclusion is obtained as in the analysis using Satterthwaite's approximation, namely, treatment 1 has significantly lower mean response than do treatments 2, 3, and 4. The three confidence intervals involving treatment 2, having length $2(msd)$, would be slightly wider using Satterthwaite's approximation, and the other three confidence intervals would be slightly narrower. Where there is so little difference in the two methods of analysis, the standard analysis would usually be preferred. □

5.7 Checking the Normality Assumption

The assumption that the error variables have a normal distribution is checked using a *normal probability plot*, which is a plot of the standardized residuals against their normal scores. *Normal scores* are percentiles of the standard normal distribution, and we will show how to obtain them after providing motivation for the normal probability plot.

If a given linear model is a reasonable description of a set of data without any outliers, and if the error assumptions are satisfied, then the standardized residuals would look similar to n independent observations from the standard normal distribution. In particular, the qth smallest standardized residual would be approximately equal to the $100[q/(n+1)]$th percentile of the standard normal distribution. Consequently, when the model assumptions hold, a plot of the qth smallest standardized residual against the $100[q/(n+1)]$th percentile of the standard normal distribution for each $q = 1, 2, \ldots, n$ would show points roughly on a straight line through the origin with slope equal to 1.0. However, if any of the model assumptions fail, and in particular if the normality assumption fails, then the normal probability plot shows a nonlinear pattern.

Blom, in 1958, recommended that the standardized residuals be plotted against the $100[(q - 0.375)/(n + 0.25)]$th percentiles of the standard normal distribution rather than the $100[q/(n+1)]$th percentiles, since this gives a slightly straighter line. These percentiles are called *Blom's normal scores*.

Blom's qth normal score is the value ξ_q for which

$$P(Z \leq \xi_q) = (q - 0.375)/(n + 0.25),$$

where Z is a standard normal random variable. Hence, Blom's qth normal score is

$$\xi_q = \Phi^{-1}[(q - 0.375)/(n + 0.25)], \quad (5.7.7)$$

where Φ is the cumulative distribution function (cdf) of the standard normal distribution. The normal scores possess a symmetry about zero, that is, the jth smallest and the jth largest scores are always equal in magnitude but opposite in sign.

The normal scores are easily obtained using most statistical packages. Use of the SAS software to obtain the normal scores and generate a normal probability plot will be illustrated

Table 5.6 Normal scores: battery experiment

z_{it}	ξ_q	$\sqrt{y_{it}}$	Battery
−1.47	−1.77	27.258	2
−1.15	−1.28	14.142	3
−0.95	−0.99	23.000	1
−0.80	−0.76	23.108	1
−0.76	−0.57	15.100	4
−0.74	−0.40	27.803	2
−0.45	−0.23	14.663	3
−0.45	−0.08	15.330	4
−0.32	0.08	15.427	4
0.31	0.23	15.232	3
0.64	0.40	24.187	1
0.84	0.57	28.983	2
1.11	0.76	24.536	1
1.30	0.99	15.969	3
1.37	1.28	29.377	2
1.52	1.77	16.793	4

in Section 5.8. Alternatively, the normal scores can be calculated as shown in Example 5.7 using Table A.3 for the standard normal distribution.

Example 5.7.1 Computing normal scores: battery life experiment

To illustrate the normal probability plot and the computation of normal scores, consider the battery life data (regardless of cost) that were transformed in Example 5.6.2 to equalize the variances. The transformed observations, standardized residuals, and normal scores are listed in Table 5.6, in order of increasing size of the residuals. In the battery life experiment there were $n = 16$ observations in total. The first normal score that corresponds to the smallest residual ($q = 1$) is

$$\xi_1 = \Phi^{-1}[(1 - 0.375)/(16 + 0.25)] = \Phi^{-1}(0.0385).$$

Thus, the area under the standard normal curve to the left of ξ_1 is 0.0385. Using a table for the standard normal distribution or a computer program, this value is

$$\Phi^{-1}(0.0385) = -1.77.$$

By symmetry, the largest normal score is 1.77. The other normal scores are calculated in a similar fashion, and the corresponding normal probability plot is shown in Figure 5.10. We discuss the interpretation of this plot below. □

For inferences concerning treatment means and contrasts, the assumption of normality needs only to be approximately satisfied. Interpretation of a normal probability plot, such as that in Figure 5.10, requires some basis of comparison. The plot is not completely linear. Such plots always exhibit some sampling variation even if the normality assumption is satisfied. Since it is difficult to judge a straight line for small samples, normal probability plots are useful only if there are at least 15 standardized residuals being plotted. A plot

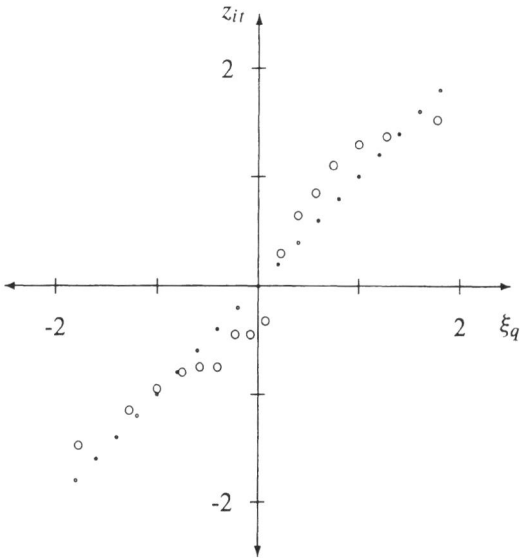

Figure 5.10
Normal probability plot for the square root battery life data

for 50 standardized residuals that are known to have a normal distribution is shown in plot (a) of Figure 5.11 and can be used as a benchmark of what might be expected when the assumption of normality is satisfied.

Small deviations from normality do not badly affect the stated significance levels, confidence levels, or power. If the sample sizes are equal, the main case for concern is that in which the distribution has heavier tails than the normal distribution, as in plot (b) of Figure 5.11. The apparent outliers are caused by the long tails of the nonnormal distribution, and a model based on normality would not be adequate to represent such a set of data. If this is the case, then use of nonparametric methods of analysis should be considered (as described, for example, by Hollander and Wolfe, 1973). Sometimes, a problem of nonnormality can be cured by taking a transformation of the data, such as $\ln(y_{it})$. However, it should be remembered that any transformation could cause a problem of unequal variances where none existed before. If the equal-variance assumption does not hold for a given set of data, then a separate normal probability plot should be generated for each treatment instead of one plot using all n residuals (provided that there are sufficient data values).

The plot for the transformed battery lifetime experiment shown in Figure 5.10 is less linear than the benchmark plot, but it does not exhibit the extreme behavior of plot (b) of Figure 5.11 for the heavy-tailed nonnormal distribution. Consequently, the normality assumption can be taken to be approximately satisfied, and the stated confidence and significance levels will be approximately correct.

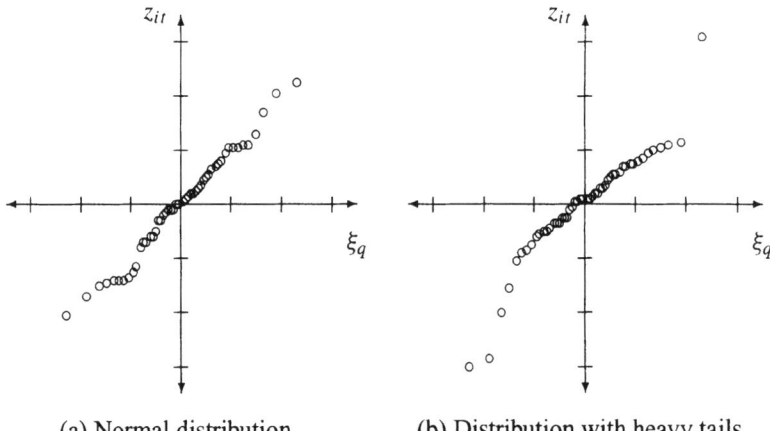

Figure 5.11
Normal probability plots for two distributions

(a) Normal distribution   (b) Distribution with heavy tails

5.8 Using SAS Software

5.8.1 Using SAS to Generate Residual Plots

We now illustrate use of the SAS software to generate the various plots used in this chapter. In the following sections, we will check the assumptions on the one-way analysis of variance model (3.3.1) for the data of the mung bean experiment described in Example 5.8.1 below.

Example 5.8.1 Mung bean experiment

An experiment was run in 1993 by K. H. Chen, Y. F. Kuo, R. Sengupta, J. Xu, and L. L. Yu to compare watering schedules and growing mediums for mung bean seeds. There were two treatment factors: "amount of water" with three levels (1, 2, and 3 teaspoons of water per day) and "growing medium" having two levels (tissue and paper towel, coded 1 and 2). We will recode the six treatment combinations as $1 = 11, 2 = 12, 3 = 21, 4 = 22, 5 = 31, 6 = 32$.

Table 5.7 Data for the mung bean experiment

Treat-ment	Shoot length in mm (Order of observation in parentheses)			
1	1.5 (14)	1.1 (15)	1.3 (18)	0.9 (30)
	8.5 (35)	10.6 (39)	3.5 (42)	7.4 (43)
2	0.0 (3)	0.6 (4)	9.5 (7)	11.3 (12)
	12.6 (17)	8.1 (27)	7.8 (29)	7.3 (37)
3	5.2 (16)	0.4 (23)	3.6 (31)	2.8 (36)
	12.3 (45)	14.1 (46)	0.3 (47)	1.8 (48)
4	13.2 (1)	14.8 (11)	10.7 (13)	13.8 (20)
	9.6 (24)	0.0 (34)	0.6 (40)	8.2 (44)
5	5.1 (5)	3.3 (21)	0.2 (26)	3.9 (28)
	7.0 (32)	9.5 (33)	11.1 (38)	6.2 (41)
6	11.6 (2)	2.3 (6)	6.7 (8)	2.5 (9)
	10.6 (10)	10.8 (19)	15.9 (22)	9.0 (25)

5.8 Using SAS Software 123

Table 5.8 SAS program to generate residual plots: mung bean experiment

```
DATA GROW;
  INPUT ORDER WATER MEDIUM LENGTH;
  TRTMT = 2*(WATER-1) + MEDIUM;
  LINES;
    1  2  1  13.2
    2  3  2  11.6
    3  1  2   0.0
    :  :  :   :
   48  2  1   1.8
  ;
PROC GLM;
  CLASS TRTMT;
  MODEL LENGTH = TRTMT;
  OUTPUT OUT=GROW2 PREDICTED=YPRED RESIDUAL=Z;
PROC STANDARD STD=1.0;
  VAR Z;
PROC RANK NORMAL=BLOM ;
  VAR Z;
  RANKS NSCORE;
PROC PRINT;
  ;
PROC PLOT;
  PLOT Z*TRTMT Z*ORDER Z*YPRED / VREF=0 VPOS=19 HPOS=50;
  PLOT Z*NSCORE / VREF=0 HREF=0 VPOS=19 HPOS=50;
```

Forty-eight beans of approximately equal weights were randomly selected for the experiment. These were all soaked in water in a single container for two hours. After this time, the beans were placed in separate containers and randomly assigned to a treatment (water/medium) combination in such a way that eight containers were assigned to each treatment combination. The 48 containers were placed on a table in a random order. The shoot lengths of the beans were measured (in mm) after one week. The data are shown in Table 5.7 together with the order in which they were collected. □

A SAS program that generates the residual plots for the mung bean experiment is shown in Table 5.8. The program uses the SAS procedures GLM, PRINT, and PLOT, all of which were introduced in Section 3.8.

The values of the factors ORDER (order of observation), WATER, MEDIUM, and the response variable LENGTH are entered into the data set GROW using the INPUT statement. The treatment combinations are then recoded, with the levels of TRTMT representing the recoded levels 1–6.

The OUTPUT statement in the GLM procedure calculates and saves the predicted values $\widehat{y}_{it}$ and the residuals $\hat{e}_{it}$ as the variables YPRED and Z, respectively, in a new data set named GROW2. The data set GROW2 also contains all of the variables in the original data set GROW. The residuals stored as the variable Z are then standardized using the procedure STANDARD. The residuals need to be standardized by dividing each residual by $\sqrt{ssE/(n-1)}$. This is

Table 5.9 Output from PROC PRINT

```
                        The SAS System
OBS   ORDER   WATER   MEDIUM   LENGTH   TRTMT   YPRED      Z         NSCORE
 1      1       2        2      13.2      4     8.8625    0.98205    0.92011
 2      2       3        2      11.6      6     8.6750    0.66224    0.57578
 3      3       1        2       0.0      2     7.1500   -1.61882   -1.60357
 :      :       :        :        :       :       :          :          :
48     48       2        1       1.8      3     5.0625   -0.73866   -0.77149
```

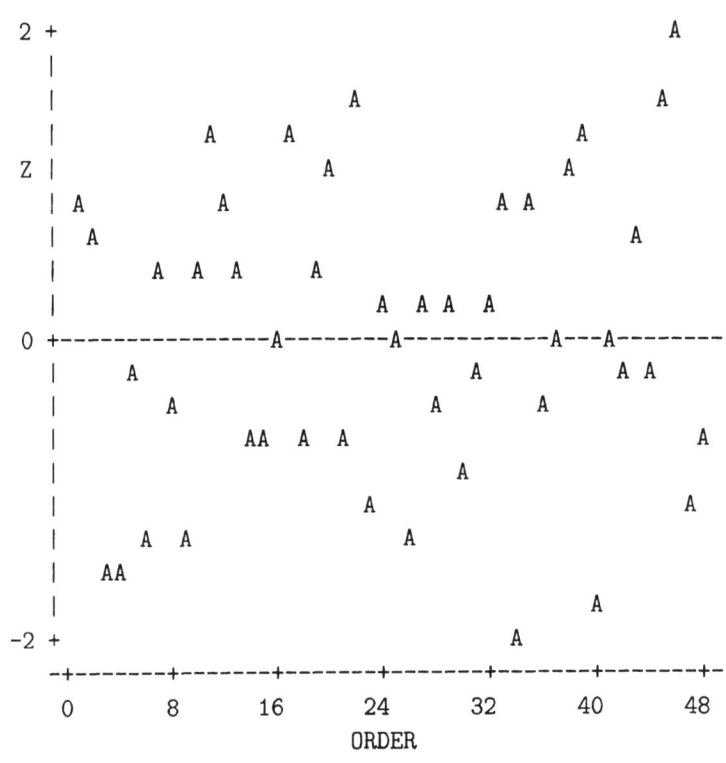

Figure 5.12
Plot of z_{it} versus order: mung bean experiment

done by requesting the procedure STANDARD to achieve a standard deviation of 1.0. The variable Z then represents the standardized residuals.

The procedure RANK is used to compute Blom's normal scores. The procedure orders the standardized residuals from smallest to largest and calculates their ranks. (The qth smallest residual has rank q.) The values of the variable NSCORE calculated by this procedure are the normal scores for the values of Z. The PRINT procedure prints all the values of the variables

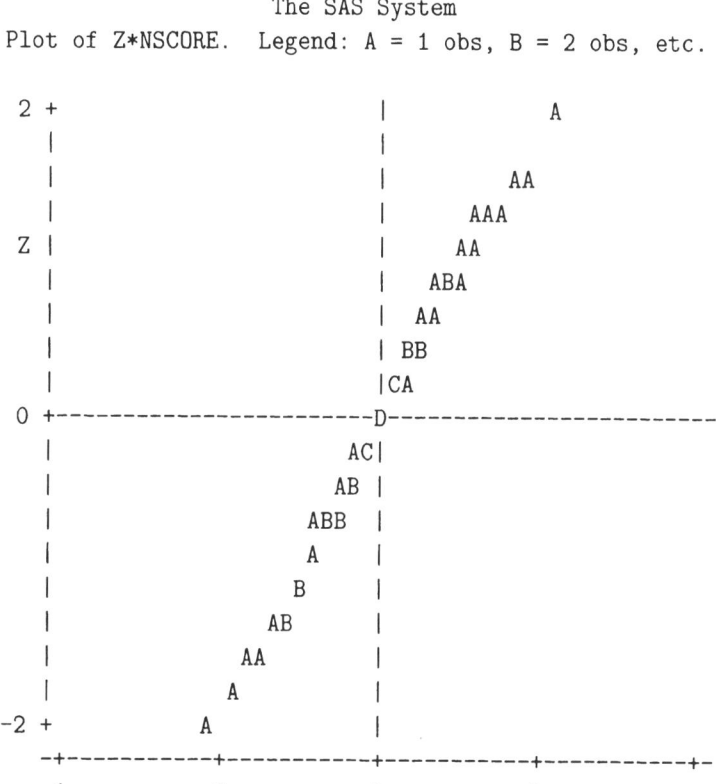

Figure 5.13
Plot of z_{it} versus normal score: mung bean experiment

created so far. Some representative output is shown in Table 5.9. The PRINT statement can be omitted if this information is not wanted.

Plots of the standardized residuals z_{it} against treatments, run order, predicted values, and normal scores are requested in the PLOT procedure. Vertical and horizontal reference lines at zero are included as appropriate via the VREF and HREF statements. The number of vertical and horizontal printing positions can be controlled through the VPOS and HPOS options. These have been set at 19 and 50, respectively, to give the plots shown in Figures 5.12–5.13. Plots with too few vertical printing positions will often produce points that are too close to be distinguished, indicated by B for two points, C for three points, and so on.

For the mung bean experiment, a plot of the standardized residuals against the order in which the observations are collected is shown in Figure 5.12, and a plot of standardized residuals against normal scores is shown in Figure 5.13. Neither of these plots indicates any serious problems with the assumptions on the model.

A plot of the standardized residuals against the predicted values (not shown) suggests that treatment variances are not too unequal, but that there could be outliers associated with one or two of the treatments. Lines 1–9 of the SAS program in Table 5.10 produced the first four columns of output of Table 5.11. From this, the rule of thumb can be checked that the

Table 5.10 SAS program to plot $\ln(s_i^2)$ against $\ln(\bar{y}_{i.})$: mung bean experiment

Line	SAS Program
1	DATA; SET GROW;
2	PROC SORT;
3	BY TRTMT;
4	PROC MEANS NOPRINT MEAN VAR;
5	VAR LENGTH;
6	BY TRTMT;
7	OUTPUT OUT=GROW3 MEAN=AVLNTH VAR=VARLNTH;
8	PROC PRINT;
9	VAR TRTMT AVLNTH VARLNTH;
10	DATA; SET GROW3;
11	LN_AV=LOG(AVLNTH); LN_VAR=LOG(VARLNTH);
12	PROC PRINT;
13	VAR TRTMT AVLNTH VARLNTH LN_AV LN_VAR;
14	PROC PLOT;
15	PLOT LN_VAR*LN_AV / VPOS=19 HPOS=50;

sample variances should not differ by more than a factor of 3. It can be verified that the ratio of the maximum and minimum variances is under 2.7 for this experiment.

When the equal-variance assumption does not appear to be valid, and when the experimenter chooses to use an analysis based on Satterthwaite's approximation, formulas need to be calculated by hand using sample variances such as those in Table 5.11. A normal probability plot such as that of Figure 5.13 would not be relevant, but the normality assumption needs to be checked for each treatment separately. This can be done by generating a separate normal probability plot for each treatment (provided that the sample sizes are sufficiently large). The plots can be obtained by including a BY TRTMT statement in the RANK and PLOT procedures. Sample program lines are:

```
PROC SORT; BY TRTMT;
PROC RANK NORMAL=BLOM;
  VAR Z; RANKS NSCORE;
  BY TRTMT;
PROC PLOT; BY TRTMT;
  PLOT Z*ORDER / VREF=0 VPOS=19 HPOS=50;
  PLOT Z*NSCORE / VREF=0 HREF=0 VPOS=19 HPOS=50;
```

5.8.2 Transforming the Data

If a variance-stabilizing transformation is needed, a plot of $\ln(s_i^2)$ against $\ln(\bar{y}_{i.})$ can be achieved via lines 1–7 and 10–15 in Table 5.10 (shown for the mung bean experiment). These statements can be added to the statements in Table 5.8 either before the GLM procedure or at the end of the program.

The SORT procedure and the BY statement sort the observations in the original data set GROW using the values of the variable TRTMT. This is required by the subsequent MEANS procedure with the NOPRINT option, which computes the mean and variance of the variable

Table 5.11 Treatment averages and variances for the mung bean experiment

```
                    The SAS System
OBS    TRTMT    AVLNTH    VARLNTH    LN_AV    LN_VAR
 1       1      4.3500    15.1714    1.47018   2.71941
 2       2      7.1500    21.1171    1.96711   3.05009
 3       3      5.0625    28.0570    1.62186   3.33424
 4       4      8.8625    32.8027    2.18183   3.49051
 5       5      5.7875    12.1555    1.75570   2.49778
 6       6      8.6750    21.6793    2.16045   3.07636
```

LENGTH separately for each treatment, without printing the results. The OUTPUT statement creates a data set named GROW3, with one observation for each treatment, and with the two variables AVLNTH and VARLNTH containing the sample mean lengths and sample variances for each treatment. Two new variables LN_AV and LN_VAR are created.

These are the natural logarithm, or log base e, of the average length and the variance for each treatment. The PRINT procedure requests the values of the variables TRTMT, AVLNTH, VARLNTH, LN_AV, LN_VAR to be printed. The output is in Table 5.11.

Finally, the PLOT procedure generates the plot of $\ln(s_i^2)$ against $\ln(\bar{y}_{i.})$, shown in Figure 5.14. The values do not fall along a straight line, so a variance-stabilizing transformation of the type given in equation (5.6.3) does not exist for this data set. However, since the ratio of the maximum to the minimum variance is less than 3.0, a transformation is not vital, according to our rule of thumb.

If an appropriate transformation is identified, then the transformed variable can be created from the untransformed variable in a DATA step of a SAS program, just as the variables LN_AV and LN_VAR were created in the data set GROW3 by transforming the variables AVLNTH and VARLNTH, respectively. Alternatively, the transformation can be achieved after the INPUT statement in the same way as the factor TRTMT was created. SAS statements useful for the variance-stabilizing transformations of equation (5.6.3) include:

Transformation	SAS Statement
$h = \ln(y)$	H = LOG(Y);
$h = \sin^{-1}(y)$	H = ARSIN(Y);
$h = y^p$	H = Y**P;

Exercises

1. **Pedestrian light experiment, continued**

   Check the assumptions on the one-way analysis of variance model (3.3.1) used for analyzing the data of the pedestrian light experiment in Exercise 14 of Chapter 3. The data are reproduced, together with their order of observation, in Table 5.12. Are the model assumptions approximately satisfied for these data?

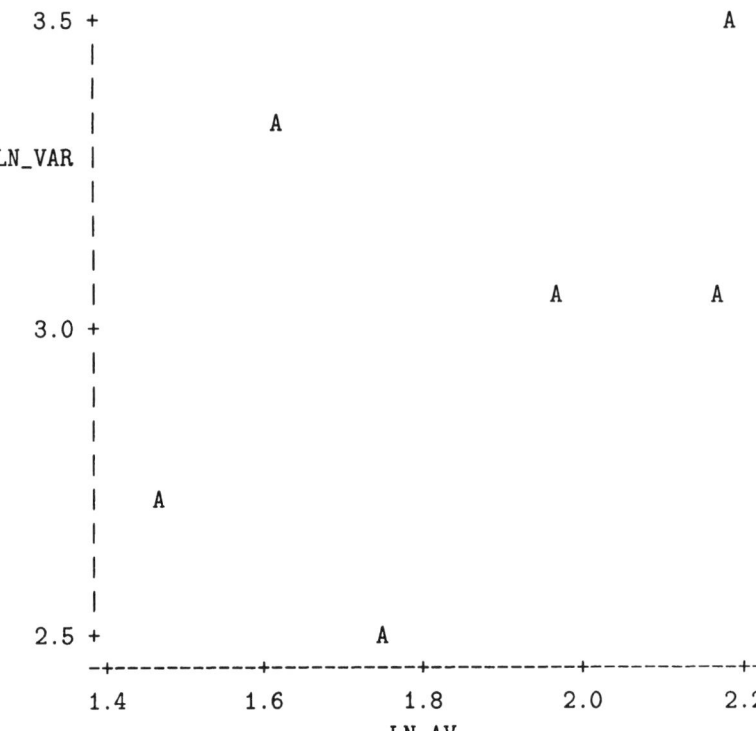

Figure 5.14
Plot of $\ln(s_i^2)$ against $\ln(\bar{y}_{i.})$; mung bean experiment

Table 5.12 Times (in seconds) for the "walk" sign to appear in the pedestrian light experiment

Time Order	1	2	3	4	5	6	7	8
Pushes	2	2	1	3	0	0	1	0
Waiting Time	38.17	38.13	38.28	38.14	38.14	38.20	38.17	38.31
Time Order	9	10	11	12	13	14	15	16
Pushes	1	2	0	1	2	2	1	2
Waiting Time	38.08	38.16	38.14	38.25	38.30	38.34	38.18	38.34
Time Order	17	18	19	20	21	22	23	24
Pushes	1	2	0	1	3	2	1	2
Waiting Time	38.03	38.17	38.29	37.95	38.30	38.18	38.26	38.09
Time Order	25	26	27	28	29	30	31	32
Pushes	2	3	3	0	1	0	3	1
Waiting Time	38.06	38.21	38.04	38.17	38.30	38.20	38.37	38.21

Table 5.13 Weight loss for the soap experiment

Soap	Weight Loss				$\bar{y}_{i.}$	s_i^2
1	−0.30	−0.10	−0.14	0.40	−0.0350	0.09157
2	2.63	2.61	2.41	3.15	2.7000	0.09986
3	1.72	2.07	2.17	2.01	1.9925	0.03736

Table 5.14 Melting times for margarine in seconds

Brand	Times	$\bar{y}_{i.}$	s_i
1	167, 171, 178, 175, 184, 176, 185, 172, 178, 178	176.4	5.56
2	231, 233, 236, 252, 233, 225, 241, 248, 239, 248	238.6	8.66
3	176, 168, 171, 172, 178, 176, 169, 164, 169, 171	171.4	4.27
4	201, 199, 196, 211, 209, 223, 209, 219, 212, 210	208.9	8.45

2. **Soap experiment, continued**

    Check the assumptions on the one-way analysis of variance model (3.3.1) for the soap experiment, which was introduced in Section 2.5.1. The data are reproduced in Table 5.13 (the order of collection of observations is not available).

3. **Margarine experiment** (Amy L. Phelps, 1987)

    The data in Table 5.14 are the melting times in seconds for three different brands of margarine (coded 1–3) and one brand of butter (coded 4). The butter was used for comparison purposes. The sizes and shapes of the initial margarine/butter pats were as similar as possible, and these were melted one by one in a clean frying pan over a constant heat.

    (a) Check the equal-variance assumption on model (3.3.1) for these data. If a transformation is required, choose the best transformation of the form (5.6.3), and recheck the assumptions.

    (b) Using the transformed data, compute a 95% confidence interval comparing the average melting times for the margarines with the average melting time for the butter.

    (c) Repeat part (b) using the untransformed data and Satterthwaite's approximation for unequal variances. Compare the results with those of part (b).

    (d) For this set of data, which analysis do you prefer? Why?

4. **Reaction time experiment, continued**

    The reaction time pilot experiment was described in Exercise 4 of Chapter 4. The experimenters were interested in the different effects on the reaction time of the aural and visual cues and also in the different effects of the elapsed time between the cue and the stimulus. There were six treatment combinations:

    1 = aural, 5 seconds   4 = visual, 5 seconds
    2 = aural, 10 seconds   5 = visual, 10 seconds
    3 = aural, 15 seconds   6 = visual, 15 seconds

Table 5.15 Reaction times (in seconds) for the reaction time experiment

Time Order	1	2	3	4	5	6
Coded treatment	6	6	2	6	2	5
Reaction Time	0.256	0.281	0.167	0.258	0.182	0.283
Time Order	7	8	9	10	11	12
Coded treatment	4	5	1	1	5	2
Reaction Time	0.257	0.235	0.20	0.170	0.260	0.187
Time Order	13	14	15	16	17	18
Coded treatment	3	4	4	3	3	1
Reaction Time	0.202	0.279	0.269	0.198	0.236	0.181

The data are reproduced, together with their order of observation, in Table 5.15.

The pilot experiment employed a single subject. Of concern to the experimenters was the possibility that the subject may show signs of fatigue. Consequently, fixed rest periods were enforced between every pair of observations.

(a) Check whether or not the assumptions on the one-way analysis of variance model (3.3.1) are approximately satisfied for these data. Pay particular attention to the experimenter's concerns about fatigue.

(b) Suggest a way to design the experiment using more than one subject. (Hint: consider using subjects as blocks in the experiment).

5. **Catalyst experiment**

   H. Smith, in the 1969 volume of *Journal of Quality Control*, described an experiment that investigated the effect of four reagents and three catalysts on the production rate in a catalyst plant. He coded the reagents as A, B, C, and D, and the catalysts as X, Y, and Z, giving twelve treatment combinations, coded as $AX, AY, \ldots, DZ$. Two observations were taken on each treatment combination, and these are shown in Table 5.16, together with the order in which the observations were collected.

Table 5.16 Production rates for the catalyst experiment

Time Order	1	2	3	4	5	6	7	8
Treatment	CY	AZ	DX	AY	CX	DZ	AX	CZ
Yield	9	5	12	7	13	7	4	13
Time Order	9	10	11	12	13	14	15	16
Treatment	BY	CZ	BZ	DX	BX	CX	DY	BZ
Yield	13	13	7	12	4	15	12	9
Time Order	17	18	19	20	21	22	23	24
Treatment	BX	DY	AY	DZ	BY	AX	CY	AZ
Yield	6	14	11	9	15	6	15	9

Source: Smith, H. Jr. (1969). Copyright © 1997 American Society for Quality. Reprinted with Permission.

Table 5.17 Data for the bicycle experiment

Code	Treatment	Crank Rates		
1	5 mph	15	19	22
2	10 mph	32	34	27
3	15 mph	44	47	44
4	20 mph	59	61	61
5	25 mph	75	73	75

Are the assumptions on the one-way analysis of variance model (3.3.1) approximately satisfied for these data? If not, can you suggest what needs to be done in order to be able to analyze the experiment?

6. **Bicycle experiment** (Debra Schomer, 1987)

   The bicycle experiment was run to compare the crank rates required to keep a bicycle at certain speeds, when the bicycle was in twelfth gear on flat ground. The speeds chosen were 5, 10, 15, 20, and 25 mph, (coded 1–5). The data are given in Table 5.17.

   The experimenter fitted the one-way analysis of variance model (3.3.1) and plotted the standardized residuals. She commented in her report:

   > Note the larger spread of the data at lower speeds. This is due to the fact that in such a high gear, to maintain such a low speed consistently for a long period of time is not only bad for the bike, it is rather difficult to do.

   Thus the experimenter was not surprised to find a difference in the variances of the error variables at different levels of the treatment factor.

   (a) Plot the standardized residuals against $\widehat{y}_{it}$, compare the sample variances, and evaluate equality of the error variances for the treatments.

   (b) Choose the best transformation of the data of the form (5.6.3), and test the hypotheses that the linear and quadratic trends in crank rates due to the different speeds are negligible, using an overall significance level of 0.01.

   (c) Repeat part (b), using the untransformed data and Satterthwaite's approximation for unequal variances,

   (d) Discuss the relative merits of the methods applied in parts (b) and (c).

7. **Dessert experiment**

   (P. Clingan, Y. Deng, M. Geil, J. Mesaros, and J. Whitmore, 1996)

   The experimenters were interested in whether the melting rate of a frozen orange dessert would be affected (and, in particular, slowed down) by the addition of salt and/or sugar. At this point, they were not interested in taste testing. Six treatments were selected, as follows:

   1 = 1/8 tsp salt, 1/4 cup sugar   4 = 1/4 tsp salt, 1/4 cup sugar
   2 = 1/8 tsp salt, 1/2 cup sugar   5 = 1/4 tsp salt, 1/2 cup sugar
   3 = 1/8 tsp salt, 3/4 cup sugar   6 = 1/4 tsp salt, 3/4 cup sugar

   For each observation of each treatment, the required amount of sugar and salt was added to the contents of a 12-ounce can of frozen orange juice together with 3 cups of

Table 5.18 Percentage melting of frozen orange cubes for the dessert experiment

Position	1	2	3	4	5	6
Treatment	2	5	5	1	4	3
% melt	12.06	9.66	7.96	9.04	10.17	7.86
Position	7	8	9	10	11	12
Treatment	4	1	3	1	2	4
% melt	8.14	9.52	4.28	8.32	10.74	5.98
Position	13	14	15	16	17	18
Treatment	2	6	6	3	6	5
% melt	9.84	7.58	6.65	9.26	8.46	12.83

water. The orange juice mixes were frozen in ice cube trays and allocated to random positions in a freezer. After 48 hours, the cubes were removed from the freezer, placed on half-inch mesh wire grid and allowed to melt into a container in the laboratory (which was held at 24.4°C) for 30 mins. The percentage melting (by weight) of the cubes are recorded in Table 5.18. The coded position on the table during melting is also recorded.

(a) Plot the data. Does it appear that the treatments have different effects on the melting of the frozen orange dessert?

(b) Check whether the assumptions on the one-way analysis of variance model (3.3.1) are satisfied for these data. Pay particular attention to the equal-variance assumption.

(c) Use Satterthwaite's method to compare the pairs of treatments.

(d) What conclusions can you draw about the effects of the treatments on the melting of the frozen orange dessert? If your concern was to produce frozen dessert with a long melting time, which treatment would you recommend? What other factors should be taken into account before production of such a dessert?

8. **Wildflower experiment** (Barbra Foderaro, 1986)

An experiment was run to determine whether or not the germination rate of the endangered species of Ohio plant *Froelichia floridana* is affected by storage temperature or storage method. The two levels of the factor "temperature" were "spring temperature, 14°C–24°C" and "summer temperature, 18°C–27°C." The two levels of the factor "storage" were "stratified" and "unstratified." Thus, there were four treatment combinations in total. Seeds were divided randomly into sets of 20 and the sets assigned at random to the treatments. Each stratified set of seeds was placed in a mesh bag, spread out to avoid overlapping, buried in two inches of moist sand, and placed in a refrigeration unit for two weeks at 50°F. The unstratified sets of seeds were kept in a paper envelope at room temperature. After the stratification period, each set of seeds was placed on a dish with 5 ml of distilled deionized water, and the dishes were put into one of two growth chambers for two weeks according to their assigned level of temperature. At the end of this period, each dish was scored for the number of germinated seeds. The resulting data are given in Table 5.19.

Table 5.19 Data for the wildflower experiment

Treatment Combination	Number Germinating					$\bar{y}_{i.}$	s_i
1: Spring/Stratified	12	13	2	7	19	8.4	6.995
	0	0	3	17	11		
2: Spring/Unstratified	6	2	0	2	4	2.5	3.308
	1	0	10	0	0		
3: Summer/Stratified	6	4	5	7	6	5.0	1.633
	5	7	5	2	3		
4: Summer/Unstratified	0	6	2	5	1	3.6	2.271
	5	2	3	6	6		

(a) For the original data, evaluate the constant-variance assumption on the one-way analysis of variance model (3.3.1) both graphically and by comparing sample variances.

(b) It was noted by the experimenter that since the data were the numbers of germinated seeds out of a total of 20 seeds, the observations Y_{it} should have a binomial distribution. Does the corresponding transformation help to stabilize the variances?

(c) Plot $\ln(s_i^2)$ against $\ln(\bar{y}_{i.})$ and discuss whether or not a power transformation of the form given in equation (5.6.3) might equalize the variances.

(d) Use Scheffé's method of multiple comparisons, in conjunction with Satterthwaite's approximation, to construct 95% confidence intervals for all pairwise comparisons and for the two contrasts

$$\tfrac{1}{2}[1, 1, -1, -1] \quad \text{and} \quad \tfrac{1}{2}[1, -1, 1, -1],$$

which compare the effects of temperature and storage methods, respectively.

9. **Spaghetti sauce experiment**
(K. Brewster, E. Cesmeli, J, Kosa, M. Smith, and M. Soliman, 1996)
The spaghetti sauce experiment was run to compare the thicknesses of three particular brands of spaghetti sauce, both when stirred and unstirred. The six treatments were:

1 = store brand, unstirred   2 = store brand, stirred
3 = national brand, unstirred   4 = national brand, stirred
5 = gourmet brand, unstirred   6 = gourmet brand, stirred

Part of the data collected is shown in Table 5.20. There are three observations per treatment, and the response variable is the weight (in grams) of sauce that flowed through a colander in a given period of time. A thicker sauce would give rise to smaller weights.

(a) Check the assumptions on the one-way analysis of variance model (3.3.1).

(b) Use Satterthwaite's method to obtain simultaneous confidence intervals for the six preplanned contrasts

$$\tau_1 - \tau_2, \quad \tau_3 - \tau_4, \quad \tau_5 - \tau_6, \quad \tau_1 - \tau_5, \quad \tau_1 - \tau_3, \quad \tau_3 - \tau_5,$$

Select an overall confidence level of at least 94%.

Table 5.20 Weights (in grams) for the spaghetti sauce experiment

Time order	1	2	3	4	5	6	7	8	9
Treatment	3	2	4	3	4	5	1	6	6
Weight	14	69	26	15	20	12	55	14	16
Time order	10	11	12	13	14	15	16	17	18
Treatment	5	1	2	4	6	3	5	2	1
Weight	16	66	64	23	17	22	18	64	53

6
Experiments with Two Crossed Treatment Factors

 6.1 Introduction
 6.2 Models and Factorial Effects
 6.3 Contrasts
 6.4 Analysis of the Two-Way Complete Model
 6.5 Analysis of the Two-Way Main-Effects Model
 6.6 Calculating Sample Sizes
 6.7 Small Experiments
 6.8 Using SAS Software
 Exercises

6.1 Introduction

In this chapter, we discuss the use of completely randomized designs for experiments that involve two treatment factors. We label the treatment factors as A and B, where factor A has a levels coded $1, 2, \ldots, a$, and factor B has b levels coded $1, 2, \ldots, b$. Every level of A is observed with every level of B, so the factors are *crossed*. In total, there are $v = ab$ treatments (treatment combinations), and these are coded as $11, 12, \ldots, 1b, 21, 22, \ldots, 2b, \ldots, ab$.

In the previous three chapters, we recoded the treatment combinations as $1, 2, \ldots, v$ and used the one-way analysis of variance for comparing their effects. In this chapter, we investigate the contributions that each of the factors make individually to the response, and it is more convenient to retain the a 2-digit code ij for a treatment combination in which factor A is at level i and factor B is at level j. In Section 6.2.1, we define the "interaction" of two treatment factors. Allowing for the possibility of interaction leads one to select a "two-way complete model" to model the data (Section 6.4). However, if it is known in advance that the factors do not interact, a "two-way main-effects model" would be selected (Section 6.5). Estimation of contrasts, confidence intervals, and analysis of variance techniques are described for these basic models. The calculation of sample sizes is also discussed (Section 6.6). The corresponding SAS commands are described in Section 6.8.

If each of the two factors has a large number of levels, the total number of treatment combinations could be quite large. When observations are costly, it may be necessary to limit the number of observations to one per treatment combination. Analysis for this situation is discussed in Section 6.7.

6.2 Models and Factorial Effects

6.2.1 The Meaning of Interaction

In order to understand the meaning of the interaction between two treatment factors, it is helpful to look at possible data sets from a hypothetical experiment. In 1994, the Statistics Department at The Ohio State University introduced a self-paced version of a data analysis course that is usually taught by lecture. Although no experiment was run in this particular instance, suppose that a hypothetical department wishes to know to what extent student performance in an introductory course is affected by the teaching method used (lecture-based or self-paced) and also by the particular instructor teaching the course. Suppose that the instructors to be used in the experiment are the three professors who will be teaching the course for the foreseeable future. The self-paced system uses the same text, same instructors, additional notes, and extended instructor office hours, but no lectures.

There are two treatment factors of interest, namely "instructor," which has three levels, coded 1, 2, and 3, and "teaching method," which has two levels, coded 1 and 2. Both of the treatment factors are fixed effects, since their levels have been specifically chosen (see Section 2.2, page 7, step (f)). The students who enroll in the introductory course are the experimental units and are allocated at random to one of the six treatment combinations in such a way that approximately equal numbers of students are enrolled in each section. Student performance is to be measured by means of a computer-graded multiple-choice examination, and an average exam score $\bar{y}_{ij.}$ for each treatment combination will be obtained.

There are eight different types of situations that could occur, and these are depicted in Figure 6.1, where the plotted character indicates the teaching method used. The plots are called *interaction plots* and give an indication of how the different instructor–teaching method combinations affect the average exam score.

In plots (a)–(d) of Figure 6.1, the dotted lines joining the average exam scores for the two teaching methods are parallel (and sometimes coincide). In plot (b), all the instructors have obtained higher exam scores with teaching method 1 than with method 2, but the instructors themselves look very similar in terms of the average exam scores obtained. Thus there is an effect on the average exam score of teaching method (M) but no effect of instructor (I). Below the plot this is highlighted by the notation "I = no, M = yes." The notation "IM = no" refers to the fact that the lines are parallel, indicating that there is no interaction (see below). In plot (c), no difference can be seen in the average scores obtained from the two teaching methods for any instructor, although the instructors themselves appear to have achieved different average scores. Thus, the instructors have an effect on the average exam score, but

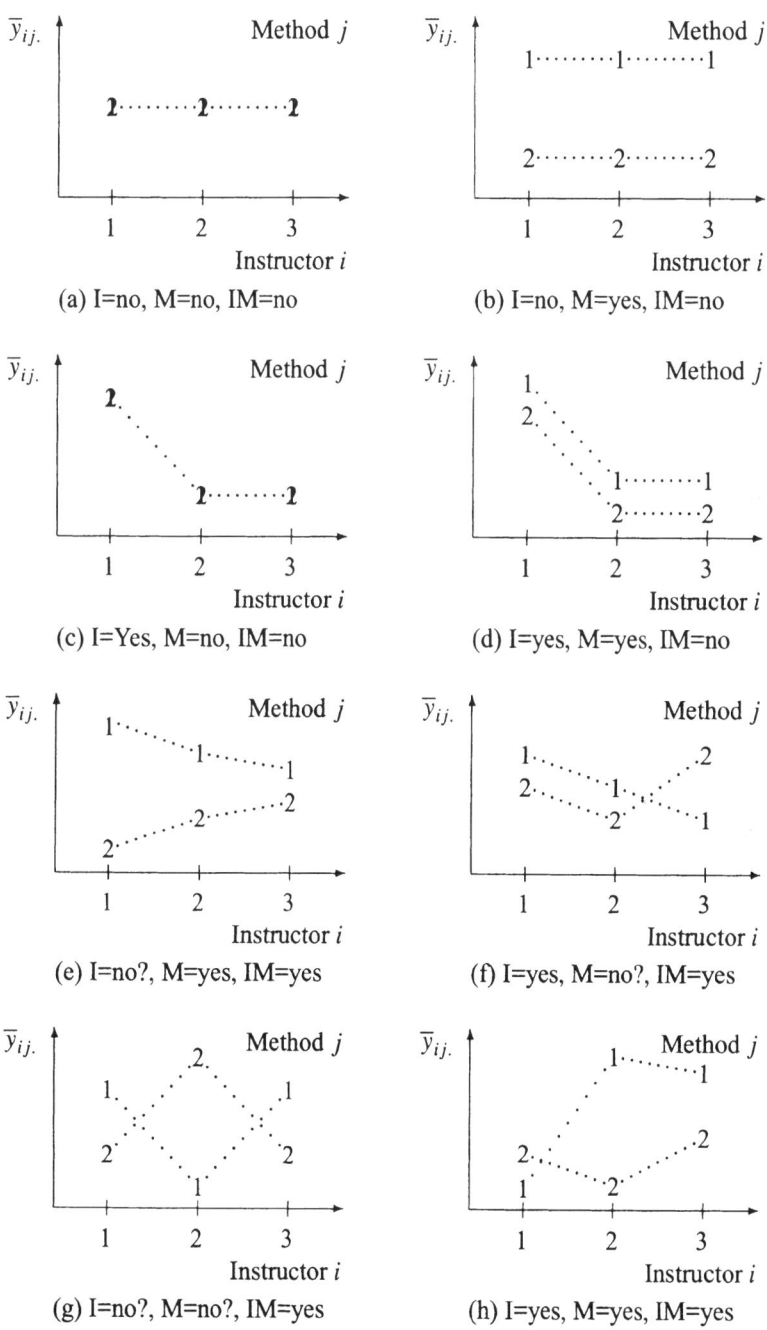

Figure 6.1 Possible configurations of effects present for two factors, Instructor (I) and teaching Method (M). Legend is level j of Method.

the teaching methods do not (I = yes, M = no). Plot (d) shows the type of plot that might be obtained if there is both an instructor effect and a teaching-method effect. The plot shows that all three instructors have obtained higher average exam scores using method 1 than using method 2. But also, instructor 1 has obtained higher average scores than the other two instructors. The individual teaching method effects and instructor effects are known as *main effects*.

In plots (e)–(h) of Figure 6.1, the dotted lines are not parallel. This means that more is needed to explain the differences in exam scores than just teaching method and instructor effects. For example, in plot (e), all instructors have obtained higher exam scores using method 1 than using method 2, but the difference is very small for instructor 3 and very large for instructor 1. In plot (h), instructor 1 has obtained higher exam scores with method 2, while the other two instructors have done better with method 1. In all of plots (e)–(h) the instructors have performed differently with the different methods. This is called an effect of *interaction* between instructor and teaching method.

In plot (g), the instructors clearly differ. Two do better with method 1 and one with method 2. However, if we ignore teaching methods, the instructors appear to have achieved very similar average exam scores overall. So, averaged over the methods, there is little difference between them. In such a case, a standard computer analysis will declare that there is no difference between instructors, which is somewhat misleading. We use the notation "IM = yes" to denote an interaction between Instructor and Method, and "I = no?" to highlight the fact that a conclusion of no difference between instructors should be interpreted with caution in the presence of interaction. In general, *if there is an interaction between two treatment factors, then it may not be sensible to examine either of the main effects separately*. Instead, it will often be preferable to compare the effects of the treatment combinations themselves.

While interaction plots are extremely helpful in interpreting the analysis of an experiment, they give no indication of the size of the experimental error. Sometimes a perceived interaction in the plot will not be distinguishable from error variability in the analysis of variance. On the other hand, if the error variability is very small, then an interaction effect may be statistically significant in the analysis, even if it appears negligible in the plot.

6.2.2 Models for Two Treatment Factors

If we use the two-digit codes ij for the treatment combinations in the one-way analysis of variance model (3.3.1), we obtain the model

$$Y_{ijt} = \mu + \tau_{ij} + \epsilon_{ijt}, \qquad (6.2.1)$$
$$\epsilon_{ijt} \sim N(0, \sigma^2),$$
$$\epsilon_{ijt}\text{'s independent},$$
$$t = 1, \ldots, r_{ij}; \quad i = 1, \ldots, a; \quad j = 1, \ldots, b,$$

where i and j are the levels of A and B, respectively. This model is known as the *cell-means model*. The "cell" refers to the cell of a table whose rows represent the levels of A and whose columns represent the levels of B.

6.2 Models and Factorial Effects

Since the interaction plot arising from a two-factor experiment could be similar to any of plots (a)–(h) of Figure 6.1, it is often useful to model the effect on the response of treatment combination ij to be the sum of the individual effects of the two factors, together with their interaction; that is,

$$\tau_{ij} = \alpha_i + \beta_j + (\alpha\beta)_{ij}.$$

Here, α_i is the effect (positive or negative) on the response due to the fact that the ith level of factor A is observed, and β_j is the effect (positive or negative) on the response due to the fact that the jth level of factor B is observed, and $(\alpha\beta)_{ij}$ is the extra effect (positive or negative) on the response of observing levels i and j of factors A and B together. The corresponding model, which we call the *two-way complete model*, or the *two-way analysis of variance model*, is as follows:

$$Y_{ijt} = \mu + \alpha_i + \beta_j + (\alpha\beta)_{ij} + \epsilon_{ijt}, \qquad (6.2.2)$$
$$\epsilon_{ijt} \sim N(0, \sigma^2),$$
ϵ_{ijt}'s are mutually independent,
$$t = 1, \ldots, r_{ij}; \quad i = 1, \ldots, a; \quad j = 1, \ldots, b.$$

The phrase "two-way" refers to the fact that there are two primary sources of variation, namely, the two treatment factors. Model (6.2.2) is equivalent to model (6.2.1), since all we have done is to express the effect of the treatment combination in terms of its constituent parts.

Occasionally, an experimenter has sufficient knowledge about the two treatment factors being studied to state with reasonable certainty that the factors do not interact and that an interaction plot similar to one of plots (a)–(d) of Figure 6.1 will occur. This knowledge may be gleaned from previous similar experiments or from scientific facts about the treatment factors. If this is so, then the interaction term can be dropped from model (6.2.2), which then becomes

$$Y_{ijt} = \mu + \alpha_i + \beta_j + \epsilon_{ijt}, \qquad (6.2.3)$$
$$\epsilon_{ijt} \sim N(0, \sigma^2),$$
ϵ_{ijt}'s are mutually independent,
$$t = 1, \ldots, r_{ij}; \quad i = 1, \ldots, a; \quad j = 1, \ldots, b.$$

Model (6.2.3) is a "submodel" of the two-way complete model and is called a *two-way main-effects model*, or *two-way additive model*, since the effect on the response of treatment combination ij is modeled as the sum of the individual effects of the two factors. If an additive model is used when the factors really do interact, then inferences on main effects can be very misleading. Consequently, if the experimenter does not have reasonable knowledge about the interaction, then the two-way complete model (6.2.2) or the equivalent cell-means model (6.2.1) should be used.

6.2.3 Checking the Assumptions on the Model

The assumptions implicit in both the two-way complete model (6.2.2) and the two-way main-effects model (6.2.3) are that the error random variables have equal variances, are mutually independent, and are normally distributed. The strategy and methods for checking the error assumptions are the same as those in Chapter 5. The standardized residuals are calculated as

$$z_{ijt} = (y_{ijt} - \hat{y}_{ijt})/\sqrt{ssE/(n-1)}$$

with

$$\hat{y}_{ijt} = \hat{\tau}_{ij} = \hat{\alpha}_i + \hat{\beta}_j + \widehat{(\alpha\beta)}_{ij}$$

or

$$\hat{y}_{ijt} = \hat{\tau}_{ij} = \hat{\alpha}_i + \hat{\beta}_j,$$

depending upon which model is selected, where the "hat" denotes a least squares estimate. The residuals are plotted against

(i) the order of observation to check independence,

(ii) the levels of each factor and $\hat{y}_{ijt}$ to check for outliers and for equality of variances,

(iii) the normal scores to check the normality assumption.

When the main-effects model is selected, interaction plots of the data, such as those in Figure 6.1, can be used to check the assumption of no interaction. An alternative way to check for interaction is to plot the standardized residuals against the levels of one of the factors with the plotted labels being the levels of the second factor. An example of such a plot is shown in Figure 6.2. (For details of the original experiment, see Exercise 17.9, page 630.) If the main-effects model had represented the data well, then the residuals would have been randomly scattered around zero. However, a pattern can be seen that is reminiscent of the interaction plot (f) of Figure 6.1 suggesting that a two-way complete model would have been a much better description of the data. If the model is changed based on the data, subsequent stated confidence levels and significance levels will be inaccurate, and analyses must be interpreted with caution.

If there is some doubt about the equality of the variances, the rule of thumb $s^2_{max}/s^2_{min} < 3$ can be employed, where s^2_{max} is the maximum of the variances of the data values within the cells, and s^2_{min} is the minimum (see Section 5.6.1). In a two-way layout, however, there may not be sufficient observations per cell to allow this calculation to be made. Nevertheless, we can at least check that the error variances are the same for each level of any given factor by employing the rule of thumb for the variances of the nonstandardized residuals calculated at each level of the factor.

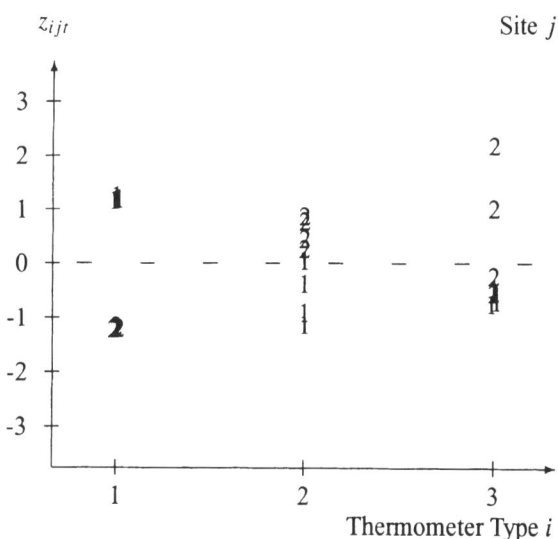

Figure 6.2
Residual plot for the temperature experiment

6.3 Contrasts

6.3.1 Contrasts for Main Effects and Interactions

Since the cell-means model (6.2.1) is equivalent to the one-way analysis of variance model, we know that all contrasts in the treatment effects τ_{ij} are estimable (cf. Section 3.4.1, page 37). Contrasts of interest for a cell-means model are typically of three main types: treatment contrasts, interaction contrasts, and main-effect contrasts.

Treatment contrasts $\sum_i \sum_j d_{ij} \tau_{ij}$ are no different from the types of contrasts described in Chapter 4. For example, $\tau_{ij} - \tau_{sh}$ is a pairwise difference between treatment combinations ij and sh. All the confidence interval methods of Chapter 4 are directly applicable.

Interaction contrasts are the contrasts that we use in order to measure whether or not the lines on the interaction plots (cf. Figure 6.1) are parallel. An example of an interaction contrast is

$$(\tau_{sh} - \tau_{(s+1)h}) - (\tau_{sq} - \tau_{(s+1)q}). \tag{6.3.4}$$

We can verify that this is, indeed, an interaction contrast by using the equivalent two-way complete model notation with $\tau_{ij} = \alpha_i + \beta_j + (\alpha\beta)_{ij}$. Substituting this into (6.3.4) gives the contrast

$$\left((\alpha\beta)_{sh} - (\alpha\beta)_{(s+1)h}\right) - \left((\alpha\beta)_{sq} - (\alpha\beta)_{(s+1)q}\right), \tag{6.3.5}$$

which is a function of interaction parameters only. Interaction contrasts are always of the form

$$\sum_i \sum_j d_{ij} \tau_{ij} = \sum_i \sum_j d_{ij} (\alpha\beta)_{ij}, \tag{6.3.6}$$

where

$$\sum_i d_{ij} = 0 \text{ for each } j \quad \text{and} \quad \sum_j d_{ij} = 0 \text{ for each } i.$$

Some, but not all, interaction contrasts have coefficients $d_{ij} = c_i k_j$. For example, if we take $c_s = k_h = 1$ and $c_{s+1} = k_q = -1$ and all other c_i and k_j zero, then, setting $d_{ij} = c_i k_j$ in (6.3.6), we obtain the coefficients in contrast (6.3.5).

If the interaction effect is very small, then the lines on an interaction plot are almost parallel (as in plots (a)–(d) of Figure 6.1). We can then compare the average effects of the different levels of A (averaging over the levels of B). Thus, contrasts of the form $\Sigma c_i \bar{\tau}_{i.}$, with $\Sigma c_i = 0$, would be of interest. However, if there is an interaction (as in plot (g) of Figure 6.1), such an average may make little sense. This becomes obvious when we use the two-way complete model formulation, since a main effect contrast in A is

$$\sum_i c_i \bar{\tau}_{i.} = \sum_i c_i (\alpha_i + \overline{(\alpha\beta)}_{i.}) \tag{6.3.7}$$

where $\overline{(\alpha\beta)}_{i.} = \frac{1}{b} \sum_j (\alpha\beta)_{ij}$, and we can see clearly that we have averaged over any interaction effect that might be present. We will often write

$$\alpha_i^* = \alpha_i + \overline{(\alpha\beta)}_{i.} \quad \text{and} \quad \beta_j^* = \beta_j + \overline{(\alpha\beta)}_{.j}$$

for convenience. A contrast in the main effect of A for the two-way complete model is then written as $\Sigma c_i \alpha_i^* (\Sigma c_i = 0)$, and a contrast in the main effect of B is

$$\sum_j k_j \bar{\tau}_{.j} = \sum_j k_j (\beta_j + \overline{(\alpha\beta)}_{.j}) = \sum_j k_j \beta_j^*, \tag{6.3.8}$$

where $\Sigma k_j = 0$ and $\overline{(\alpha\beta)}_{.j} = \frac{1}{a} \sum_i (\alpha\beta)_{ij}$.

Sometimes, it is of interest to compare the effects of the levels of one factor separately at each level of the other factor. For example, in the hypothetical experiment in Section 6.2.1, a natural objective might be to choose a best teaching method for each instructor separately. If comparison of the effects of levels of factor B for each level of factor A is required, then contrasts of the form

$$\sum_j c_j \tau_{ij}, \quad \text{with} \quad \sum_j c_j = 0 \quad \text{for each } i = 1, 2, \ldots, a,$$

are of interest. We call such contrasts *simple contrasts* in the levels of B. As a special case, we have the *simple pairwise differences* of factor B:

$$\tau_{ih} - \tau_{ij}, \quad \text{for each } i = 1, \ldots, a.$$

These are a subset of the pairwise comparison contrasts. Simple contrasts and simple pairwise differences of factor A are defined in an analogous way.

When it is known in advance of the experiment that factors A and B do not interact, the two-way main-effects model (6.2.3) would normally be used. In this model, there is no interaction term, so $\tau_{ij} = \alpha_i + \beta_j$. The main-effects contrasts for A and B are respectively of the form

$$\sum c_i \bar{\tau}_{i.} = \sum c_i \alpha_i \quad \text{and} \quad \sum k_j \bar{\tau}_{.j} = \sum k_j \beta_j,$$

6.3.2 Writing Contrasts as Coefficient Lists

with $\sum c_i = 0$ and $\sum k_j = 0$.

Instead of writing out a contrast explicitly, it is sometimes sufficient, and more convenient, to list the contrast coefficients only. For the two-way complete model, we have a choice. We can refer to contrasts as either a list of coefficients of the parameters α_i^*, β_j^*, and $(\alpha\beta)_{ij}$ or as a list of coefficients of the τ_{ij}'s. This is illustrated in the following example.

Example 6.3.1 Battery experiment, continued

The four treatment combinations in the battery experiment of Section 2.5.2, page 26, involved two treatment factors, "duty" and "brand," each having two levels (1 for alkaline and 2 for heavy duty; 1 for name brand and 2 for store brand), giving treatment combinations 11, 12, 21, and 22. (These were coded in previous examples as 1, 2, 3, and 4, respectively.) There were $r = 4$ observations on each treatment combination.

The interaction plot in Figure 6.3 shows a possible interaction between the two factors, since the dotted lines on the plot are not close to parallel. However, we should remember that we cannot be certain whether the nonparallel lines are due to an interaction or to inherent variability in the data, and we will need to investigate the cause in more detail later.

The interaction is measured by the contrast

$$\tau_{11} - \tau_{12} - \tau_{21} + \tau_{22} = (\alpha\beta)_{11} - (\alpha\beta)_{12} - (\alpha\beta)_{21} + (\alpha\beta)_{22},$$

which can be written in terms of the coefficient list $[\,1, -1, -1, 1\,]$.

The contrast that compares the average lifetimes of heavy duty and alkaline batteries (averaged across brands) is

$$\overline{\tau}_{2.} - \overline{\tau}_{1.} = \tfrac{1}{2}(\tau_{21} + \tau_{22}) - \tfrac{1}{2}(\tau_{11} + \tau_{12}) = \alpha_2^* - \alpha_1^*.$$

This has coefficient list $[-1, 1\,]$ in terms of the effects α_1^*, α_2^* of the levels of duty, but coefficient list $\tfrac{1}{2}[-1, -1, 1, 1\,]$ in terms of the effects $\tau_{11}, \tau_{12}, \tau_{21}, \tau_{22}$ of the treatment combinations. Similarly, the contrast that compares the average life of store brand with that of name brand (averaged over duty) has coefficient list $[-1, 1\,]$ in terms of the effects β_j^* of brand, but coefficient list $\tfrac{1}{2}[-1, 1, -1, 1\,]$ in terms of the τ_{ij}'s.

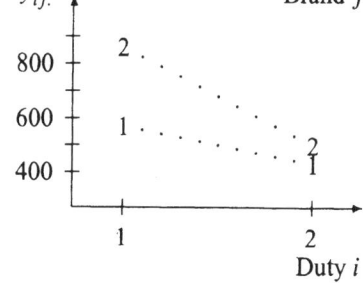

Figure 6.3 Plot of average life per unit cost against "Duty" level i by "Brand" level j for the battery experiment

Since the main-effect contrasts each have divisor 2, the interaction contrast is often divided by 2 also. This has the effect that the least squares estimators of all three contrasts have the same variances (see Example 6.4.1), and their magnitudes are more directly comparable. An alternative way to achieve equal variances is to normalize the contrasts (see Section 4.2), in which case all three contrasts would all be divided by $\sqrt{\Sigma c_i^2 / r}$. □

Contrast coefficients are often listed as columns in a table. For example, the contrast coefficients of the τ_{ij}'s for the main effect and interaction contrasts of Example 6.3.2 are written as below, with ±1's in the body of the table, and the constants listed as divisors in the last row.

ij	A	B	AB
11	−1	−1	1
12	−1	1	−1
21	1	−1	−1
22	1	1	1
Divisor	2	2	2

The benefit of this representation is that we can see easily that each AB interaction coefficient can be obtained by multiplying the corresponding A and B main-effect coefficients. Most of the interaction contrasts that we shall use have this product form. We will mention the exceptions when they arise.

Example 6.3.2 Trend contrasts

Suppose that the two factors, A and B, have $a = 3$ and $b = 6$ equally spaced quantitative levels, respectively, and that the sample sizes are equal. From Table A.2, we see that A_L, the linear trend contrast for A, has contrast coefficient list $[-1, 0, 1]$ in terms of the α_i^*'s, and A_Q, the quadratic trend contrast for A, has contrast coefficient list $[1, -2, 1]$; that is

$$A_L = -\alpha_1^* + \alpha_3^*,$$
$$A_Q = \alpha_1^* - 2\alpha_2^* + \alpha_3^*.$$

Similarly, in terms of the β_j^*'s, the coefficient lists for the linear and quadratic trends in the effects of the six levels of B are also obtained from Table A.2 as $[-5, -3, -1, 1, 3, 5]$ and $[5, -1, -4, -4, -1, 5]$, respectively; that is,

$$B_L = -5\beta_1^* - 3\beta_2^* - \beta_3^* + \beta_4^* + 3\beta_5^* + 5\beta_6^*,$$
$$B_Q = 5\beta_1^* - \beta_2^* - 4\beta_3^* - 4\beta_4^* - \beta_5^* + 5\beta_6^*.$$

Now,

$$\alpha_i^* = \overline{\tau}_{i.}, \text{ giving } \Sigma_i c_i \alpha_i^* = \Sigma_i c_i \left(\tfrac{1}{6} \Sigma_j \tau_{ij}\right) = \tfrac{1}{6} \Sigma_i \Sigma_j c_i \tau_{ij},$$

and

$$\beta_j^* = \overline{\tau}_{.j}, \text{ giving } \Sigma_j k_j \beta_j^* = \Sigma_j k_j \left(\tfrac{1}{3} \Sigma_i \tau_{ij}\right) = \tfrac{1}{3} \Sigma_i \Sigma_j k_j \tau_{ij},$$

Table 6.1 Trend contrasts when A and B have 3 and 6 equally spaced levels, respectively

ij	A_L	A_Q	B_L	B_Q	B_C	B_{qr}	B_{qn}	$A_L B_L$	$A_L B_{qn}$
11	−1	1	−5	5	−5	1	−1	5	1
12	−1	1	−3	−1	7	−3	5	3	−5
13	−1	1	−1	−4	4	2	−10	1	10
14	−1	1	1	−4	−4	2	10	−1	−10
15	−1	1	3	−1	−7	−3	−5	−3	5
16	−1	1	5	5	5	1	1	−5	−1
21	0	−2	−5	5	−5	1	−1	0	0
22	0	−2	−3	−1	7	−3	5	0	0
23	0	−2	−1	−4	4	2	−10	0	0
24	0	−2	1	−4	−4	2	10	0	0
25	0	−2	3	−1	−7	−3	−5	0	0
26	0	−2	5	5	5	1	1	0	0
31	1	1	−5	5	−5	1	−1	−5	−1
32	1	1	−3	−1	7	−3	5	−3	5
33	1	1	−1	−4	4	2	−10	−1	−10
34	1	1	1	−4	−4	2	10	1	10
35	1	1	3	−1	−7	−3	−5	3	−5
36	1	1	5	5	5	1	1	5	1
Divisor	6	6	3	3	3	3	3	1	1

and we can write all of the above trends in terms of contrasts in τ_{ij}, as shown in the columns of Table 6.1. Contrast coefficients are also listed for cubic, quartic, and quintic trends for B. If we wish to compare the A and B trends on the same scale, we can normalize the contrasts (see Section 4.2).

In order to model a three-dimensional surface, we need to know not only how the response is affected by the levels of each factor averaged over the levels of the other factor, but also how the response changes as the levels of A and B change together. The linear A × linear B trend ($A_L B_L$) measures whether or not the linear trend in A changes in a linear fashion as the levels of B are increased, and vice versa. This is an interaction contrast whose coefficients are of the form $d_{ij} = c_i k_j$, where c_i are the contrast coefficients for A, and k_j are the contrast coefficients for B. The $A_L B_L$ contrast coefficients are shown in Table 6.1, and it can be verified that they are obtained by multiplying together corresponding main-effect linear trend coefficients in the same row. Coefficients for the linear A × quintic B ($A_L B_{qn}$) contrast is also shown for use later in this chapter. □

6.4 Analysis of the Two-Way Complete Model

In the analysis of an experiment with two treatment factors that possibly interact, we may proceed with the analysis in two equivalent ways. We may use the cell-means model (6.2.1) together with all the analysis techniques of Chapters 3 and 4, or we may use the two-way complete model (6.2.2) and isolate the contributions to the response made by each of the two factors and their interaction separately.

A sensible strategy is to start with the two-way complete model and test a hypothesis of no interaction. If the hypothesis is not rejected, we may then continue with the analysis by examining the main effects under the same two-way complete model. We would not change to the two-way main-effects model, since this is not an equivalent model. However, if the hypothesis of no interaction is rejected, then we would normally prefer to change to the equivalent cell-means model and examine differences in the effects of the treatment combinations. We would also use the cell-means model when the objective of the experiment is to find the best treatment combination.

6.4.1 Least Squares Estimators for the Two-Way Complete Model

As in Section 3.4.3, page 38, the least squares estimator of $\mu + \tau_{ij}$ is $\overline{Y}_{ij.}$, so the least squares estimators of the parameters in the cell-means model (6.2.1) and the equivalent two-way complete model (6.2.2) are

$$\hat{\mu} + \hat{\tau}_{ij} = \hat{\mu} + \hat{\alpha}_i + \hat{\beta}_j + \widehat{(\alpha\beta)}_{ij} = \overline{Y}_{ij.} ,$$

and the corresponding variance is σ^2/r_{ij}. Any interaction contrast of the form $\Sigma\Sigma d_{ij}\tau_{ij}$ (with $\Sigma_i d_{ij} = 0$ and $\Sigma_j d_{ij} = 0$) has least squares estimator and associated variance equal to

$$\sum_i \sum_j d_{ij} \overline{Y}_{ij.} \text{ and } \sigma^2 \sum_i \sum_j \left(\frac{d_{ij}^2}{r_{ij}}\right).$$

In particular, the least squares estimator of the interaction contrast

$$(\tau_{sh} - \tau_{uh}) - (\tau_{sq} - \tau_{uq})$$

is

$$\overline{Y}_{sh.} - \overline{Y}_{uh.} - \overline{Y}_{sq.} + \overline{Y}_{uq.} \tag{6.4.9}$$

with variance

$$\sigma^2 \left(\frac{1}{r_{sh}} + \frac{1}{r_{uh}} + \frac{1}{r_{sq}} + \frac{1}{r_{uq}}\right). \tag{6.4.10}$$

The least squares estimators of main-effect contrasts $\Sigma c_i \alpha_i^*$ and $\Sigma k_j \beta_j^*$ are

$$\sum_i c_i \hat{\alpha}_i^* = \sum_i c_i \left(\frac{1}{b} \sum_j \overline{Y}_{ij.}\right) \text{ and } \sum_j k_j \hat{\beta}_j^* = \sum_j k_j \left(\frac{1}{a} \sum_i \overline{Y}_{ij.}\right) \tag{6.4.11}$$

with variances

$$\text{Var}(\Sigma c_i \hat{\alpha}_i^*) = \sigma^2 \left(\sum_i \sum_j \frac{c_i^2}{b^2 r_{ij}}\right) \text{ and } \text{Var}(\Sigma k_j \hat{\beta}_j^*) = \sigma^2 \left(\sum_i \sum_j \frac{k_j^2}{a^2 r_{ij}}\right), \tag{6.4.12}$$

respectively. If the sample sizes are equal, the least squares estimators of $\sum c_i \alpha_i^*$ and $\sum k_j \beta_j^*$ reduce to

$$\sum_i c_i \hat{\alpha}_i^* = \sum_i c_i \overline{Y}_{i..} \quad \text{and} \quad \sum_j k_j \hat{\beta}_j^* = \sum_j k_j \overline{Y}_{.j.}, \qquad (6.4.13)$$

where $\overline{Y}_{i..} = \sum_j \sum_t Y_{ijt}/br$ and $\overline{Y}_{.j.} = \sum_i \sum_t Y_{ijt}/ar$. Thus, for equal sample sizes,

$$\hat{\alpha}_i^* - \hat{\alpha}_s^* = \overline{Y}_{i..} - \overline{Y}_{s..} \quad \text{and} \quad \hat{\beta}_j^* - \hat{\beta}_q^* = \overline{Y}_{.j.} - \overline{Y}_{.q.} \qquad (6.4.14)$$

with associated variances $2\sigma^2/(br)$ and $2\sigma^2/(ar)$, respectively.

Example 6.4.1 Battery experiment

The four treatment combinations in the battery experiment of Section 2.5.2, page 26, involved two treatment factors, "duty" and "brand," each having two levels (1 for alkaline and 2 for heavy duty; 1 for name brand and 2 for store brand), giving treatment combinations 11, 12, 21, and 22. There were $r = 4$ observations on each treatment combination. The observed average lifetimes per unit cost for the treatment combinations were

$$\overline{y}_{11.} = 570.75, \quad \overline{y}_{12.} = 860.50, \quad \overline{y}_{21.} = 433.00, \quad \overline{y}_{22.} = 496.25.$$

The interaction contrast

$$\tfrac{1}{2}(\tau_{11} - \tau_{12} - \tau_{21} + \tau_{22}) = \tfrac{1}{2}((\alpha\beta)_{11} - (\alpha\beta)_{12} - (\alpha\beta)_{21} + (\alpha\beta)_{22})$$

has least squares estimate

$$\tfrac{1}{2}(\overline{y}_{11.} - \overline{y}_{12.} - \overline{y}_{21.} + \overline{y}_{22.}) = -113.25,$$

with associated variance

$$\sigma^2 \left(\sum\sum d_{ij}^2/r\right) = \sigma^2 \left((\tfrac{1}{2})^2 + (-\tfrac{1}{2})^2 + (-\tfrac{1}{2})^2 + (\tfrac{1}{2})^2\right)/4 = \sigma^2/4.$$

The duty contrast,

$$\alpha_1^* - \alpha_2^* = (\alpha_1 + \overline{(\alpha\beta)}_{1.}) - (\alpha_2 + \overline{(\alpha\beta)}_{2.}) = \tfrac{1}{2}(\tau_{11} + \tau_{12} - \tau_{21} - \tau_{22}),$$

has least squares estimate $\overline{y}_{1..} - \overline{y}_{2..} = 251.00$ and associated variance $\sigma^2/4$. The brand contrast,

$$\beta_1^* - \beta_2^* = (\beta_1 + \overline{(\alpha\beta)}_{.1}) - (\beta_2 + \overline{(\alpha\beta)}_{.2}) = \tfrac{1}{2}(\tau_{11} - \tau_{12} + \tau_{21} - \tau_{22}),$$

has least squares estimate $\overline{y}_{.1.} - \overline{y}_{.2.} = -176.50$ and associated variance $\sigma^2/4$. □

6.4.2 Estimation of σ^2 for the Two-Way Complete Model

Since the two-way complete model (6.2.2) is equivalent to the cell-means model (6.2.1), an unbiased estimate of σ^2 is the same as that for the one-way analysis of variance model, apart from an extra subscript j. Thus, the error sum of squares ssE can be obtained from (3.4.4)–(3.4.6), page 42, that is,

$$ssE = \sum_i \sum_j \sum_t (y_{ijt} - \overline{y}_{ij.})^2 \qquad (6.4.15)$$

$$= \sum_i \sum_j \sum_t y_{ijt}^2 - \sum_i \sum_j r_{ij} \bar{y}_{ij.}^2. \qquad (6.4.16)$$

$$= \sum_i \sum_j \sum_t y_{ijt}^2 - \sum_i \sum_j y_{ij.}^2/r_{ij}. \qquad (6.4.17)$$

An unbiased estimate for σ^2 is obtained as $msE = ssE/(n-v)$, with $v = ab$. An upper $100(1-\alpha)\%$ confidence bound for σ^2 is given by (3.4.10), page 43, that is,

$$\sigma^2 \leq \frac{ssE}{\chi^2_{n-ab, 1-\alpha}}. \qquad (6.4.18)$$

Example 6.4.2 Reaction time experiment, continued

The reaction time pilot experiment, run in 1996 by Liming Cai, Tong Li, Nishant, and Andre van der Kouwe, was described in Exercise 4 of Chapter 4. The experiment was run to compare the speed of response of a human subject to audio and visual stimuli. A personal computer was used to present a "stimulus" to a subject, and the time that the subject took to press a key in response was monitored. The subject was warned that the stimulus was forthcoming by means of an auditory or a visual cue. The two treatment factors were "Cue Stimulus" at two levels, "auditory" and "visual" (Factor A, coded 1, 2), and "Cue Time" at three levels, 5, 10, and 15 seconds between cue and stimulus (Factor B, coded 1, 2, 3), giving a total of $v = 6$ treatment combinations (coded 11, 12, 13, 21, 22, 23). Three observations were taken on each treatment combination for a single subject. The reaction times are shown in Table 6.2. It can be verified that $\sum\sum\sum y_{ijt}^2 = 0.96519$. Using (6.4.17) and the sums in Table 6.2, the sum of squares for error is

$$ssE = \sum_i \sum_j \sum_t y_{ijt}^2 - 3\sum_i \sum_j \bar{y}_{ij.}^2$$
$$= 0.96519 - 3(0.32057) = 0.00347,$$

and an unbiased estimate of σ^2 is $msE = ssE/(18-6) = 0.000289$ seconds^2. An upper 95% confidence bound for σ^2 is

$$\frac{ssE}{\chi^2_{12,.95}} = \frac{0.00347}{5.226} = 0.000664 \text{ seconds}^2,$$

Table 6.2 Data (in seconds) for the reaction time experiment

Cue Stimulus	Cue Time	Treatment Combination	Reaction Time y_{ijt}			Sums $y_{ij.}$
1	1	11	0.204	0.170	0.181	0.555
1	2	12	0.167	0.182	0.187	0.536
1	3	13	0.202	0.198	0.236	0.636
2	1	21	0.257	0.279	0.269	0.805
2	2	22	0.283	0.235	0.260	0.778
2	3	23	0.256	0.281	0.258	0.795

and taking square roots, an upper 95% confidence bound for σ is 0.0257 seconds. □

6.4.3 Multiple Comparisons for the Complete Model

In outlining the analysis at step (g) of the checklist of Chapter 2, the experimenter should specify which treatment contrasts are of interest, together with overall error rates for hypothesis tests and overall confidence levels for confidence intervals. If the two-way complete model has been selected, comparison of treatment combinations, comparison of main effects of A, and comparison of main effects of B may all be of interest. A possibility in outlining the analysis is to select error rates of α_1, α_2, and α_3 for the three sets of inferences. Then, by the Bonferroni method, the *experimentwise* simultaneous error rate is at most $\alpha_1 + \alpha_2 + \alpha_3$, and the experimentwise confidence level is at least $100(1 - \alpha_1 - \alpha_2 - \alpha_3)\%$. If interaction contrasts are also of interest, then the overall α-level can be divided into four parts instead of three.

Comparing treatment combinations When comparison of treatment combinations is of most interest, the cell-means model (6.2.1) is used. The formulae for the Bonferroni, Scheffé, Tukey, Dunnett, and Hsu methods can all be used in the same way as was done in Chapter 4, but with ssE given by (6.4.17) and with $v = ab$.

The best treatment combination can be found using either Tukey's or Hsu's method of multiple comparisons. The best treatment combination may not coincide with the apparent best levels of A and B separately. For example, in Figure 6.1(h), page 137, the apparent best treatment combination occurs with instructor 2 and method 1, whereas the best instructor, on average, appears to be number 3.

Comparing main effects Main-effect contrasts compare the effects of the levels of one factor *averaging* over the levels of the other factor and may not be of interest if the two factors interact. If main-effect contrasts are to be examined, then the Bonferroni, Scheffé, Tukey, Dunnett, and Hsu methods can be used for each factor separately. The general formula is equivalent to (4.4.20), page 80. For factor A and *equal sample sizes* the formula is

$$\sum_i c_i \bar{\tau}_{i.} = \sum_i c_i \alpha_i^* \in \left(\sum_i c_i \bar{y}_{i..} \pm w \sqrt{msE \sum_i c_i^2 / br} \right), \qquad (6.4.19)$$

where the critical coefficient w for each of the five methods is, respectively,

$$w_B = t_{n-ab, \alpha/2m} \; ; \; w_S = \sqrt{(a-1)F_{a-1,n-ab,\alpha}} \; ; \; w_T = q_{a,n-ab,\alpha}/\sqrt{2} \; ;$$

$$w_{D1} = w_H = t^{(0.5)}_{a-1,n-ab,\alpha} \; ; \; w_{D2} = |t|^{(0.5)}_{a-1,n-ab,\alpha} \, .$$

The general formula for a confidence interval for a contrast in factor B is

$$\sum_j k_j \bar{\tau}_{.j} = \sum_j k_j \beta_j^* \in \left(\sum_j k_j \bar{y}_{.j.} \pm w \sqrt{msE \sum_j k_j^2 / (ar)} \right) \qquad (6.4.20)$$

with critical coefficients as above but interchanging a and b. The error variance estimate is $msE = ssE/(n - ab)$, where ssE is obtained from (6.4.17).

For *unequal sample sizes*, the Bonferroni and Scheffé methods can be used, but the least squares estimates and variances must be replaced by (6.4.11) and (6.4.12), respectively. It has not yet been proved that the other three methods retain an overall confidence level of at least $100(1-\alpha)\%$ for unequal sample sizes, although this is widely believed to be the case for Tukey's method.

Example 6.4.3 Reaction time experiment, continued

Suppose the preplanned analysis for the reaction time experiment of Example 6.4.2 (page 148) had been to use the two-way complete model and to test the null hypothesis of no interaction. If the hypothesis were to be rejected, then the plan was to use Tukey's method at level 99% for the pairwise comparisons of the treatment combinations. Otherwise, Tukey's method would be used at level 99% for the pairwise comparison of the levels of B (cue time), and a single 99% confidence interval would be obtained for comparing the two levels of A (cue stimulus). Then the experimentwise confidence level for the three sets of intervals would have been at least 97%.

After looking at the data plotted in Figure 6.4, the experimenters might decide that comparison of the levels of cue stimulus (averaged over cue time) is actually the only comparison of interest. However, the experimentwise confidence level remains at least 97%, because two other sets of intervals were planned ahead of time and only became uninteresting after the data were examined.

The sample mean weights for the two cue stimuli (averaged over cue times) are

$$\bar{y}_{1..} = 0.1919, \quad \bar{y}_{2..} = 0.2642.$$

The mean square for error was calculated in Example 6.4.2 to be $msE = 0.000289$ sec^2. The formula for a 99% confidence interval for the comparison of $a = 2$ treatments and $br = 9$ observations on each treatment is obtained from (6.4.19) with $w = t_{18-6, 0.01} = 2.681$, giving

$$\alpha_2^* - \alpha_1^* \in \left(\bar{y}_{2..} - \bar{y}_{1..} \pm w_t \sqrt{msE\,(1/br + 1/br)}\right)$$
$$= 0.0723 \pm (2.681)\sqrt{0.000289(2/9)} = (0.0508, 0.0937).$$

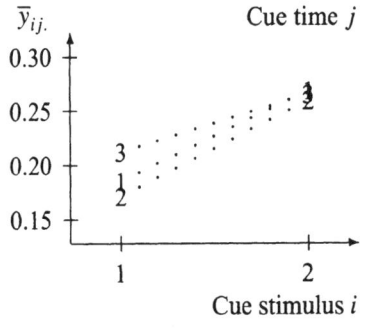
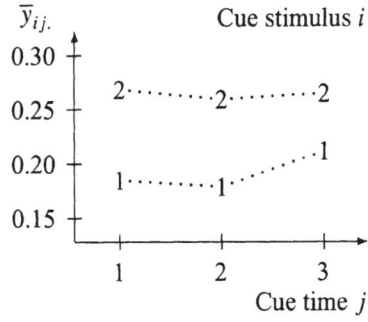

Figure 6.4 Average times for the reaction time experiment

Thus, at an experimentwise confidence level of at least 97%, we can conclude that the average reaction time with an auditory cue is between 0.0508 and 0.0937 seconds faster than with a visual cue. □

Multiple comparisons when variances are unequal When the variances of the error variables are unequal, and no transformation can be found to remedy the problem, Satterthwaite's approximation, introduced in Section 5.6.3 (page 116), can be used. This is illustrated in Example 6.4.3.

Example 6.4.4 Bleach experiment

The bleach experiment was run by Annie Autret in 1986 to study the effect of different bleach concentrations (factor A) and the effect of the type of stain (factor B) on the speed of stain removal from a piece of cloth. The bleach concentration was to be observed at levels 3, 5, and 7 teaspoonfuls of bleach per cup of water (coded 1, 2, 3), and three types of stain (blue ink, jam, tomato sauce; coded 1, 2, 3) were of interest, giving $v = 9$ treatment combinations in total. The experimenter calculated that she needed $r = 5$ observations per treatment combination in order to be able to detect, with probability 0.9, a difference of 5 minutes in the time of stain removal between the treatment combinations.

The data are shown in Table 6.3 together with the sample mean and standard deviation for each treatment combination. The maximum sample standard deviation is about 8.9 times the size of the minimum sample standard deviation, so the ratio of the maximum to the minimum variance is about 80, and a transformation of the data should be contemplated. The reader can verify, using the technique described in Section 5.6.2, that a plot of $\ln(s_{ij}^2)$ against $\ln(\bar{y}_{ij.})$ is not linear, so no transformation of the form $h(y_{ijt}) = y_{ijt}^{1-(q/2)}$ will adequately equalize the error variances.

An alternative is to apply Satterthwaite's approximation. The plan of the analysis was to use Tukey's method with an error rate of 0.01 for each of the main-effect comparisons and for the pairwise differences of the treatment combinations, giving an experimentwise confidence level of at least 97%. The methodology will be illustrated for main effects of B, following Section 5.6.3, page 116.

Table 6.3 Data for the bleach experiment, with treatment factors "concentration" (A) and "stain type" (B)

ij	Time for stain removal (in seconds)					$\bar{y}_{ij.}$	s_{ij}
11	3600	3920	3340	3173	2452	3297.0	550.27
12	495	236	515	573	555	474.8	137.04
13	733	525	793	1026	510	717.4	212.85
21	2029	2271	2156	2493	2805	2350.8	305.94
22	428	432	335	288	376	371.8	61.60
23	880	759	1138	780	1625	1036.4	361.91
31	3660	4105	4545	3569	3342	3844.2	479.85
32	410	225	437	350	140	312.4	126.32
33	539	1354	347	584	781	721.0	386.02

A pairwise comparison of levels u and h of factor B is of the form

$$\beta_u^* - \beta_h^* = \overline{\tau}_{.u} - \overline{\tau}_{.h} = \tfrac{1}{3}(\tau_{1u} + \tau_{2u} + \tau_{3u} - \tau_{1h} - \tau_{2h} - \tau_{3h}),$$

which has least squares estimate

$$\widehat{\beta}_u^* - \widehat{\beta}_h^* = \overline{y}_{.u.} - \overline{y}_{.h.} = \tfrac{1}{3}(\overline{y}_{1u.} + \overline{y}_{2u.} + \overline{y}_{3u.} - \overline{y}_{1h.} - \overline{y}_{2h.} - \overline{y}_{3h.}).$$

If s_{ij}^2 denotes the sample variance of the data for treatment combination ij, the estimated variance of this estimator, as in (5.6.4), page 117, is

$$\widehat{\text{Var}}\left(\widehat{\beta}_u^* - \widehat{\beta}_h^*\right) = \sum_i \sum_j d_{ij} \frac{s_{ij}^2}{r_{ij}} = \frac{1}{9 \times 5}(s_{1u}^2 + s_{2u}^2 + s_{3u}^2 + s_{1h}^2 + s_{2h}^2 + s_{3h}^2),$$

and since $r = 5$, the approximate number of degrees of freedom for error is

$$df = \frac{(s_{1u}^2 + s_{2u}^2 + s_{3u}^2 + s_{1h}^2 + s_{2h}^2 + s_{3h}^2)^2}{(s_{1u}^4/4) + (s_{2u}^4/4) + (s_{3u}^4/4) + (s_{1h}^4/4) + (s_{2h}^4/4) + (s_{3h}^4/4)}$$

after canceling the factor $r^2 = 25$ in the numerator and denominator.

For Tukey's method of pairwise comparisons for factor B with $b = 3$ levels, the minimum significant difference is

$$msd = w_T \sqrt{\widehat{\text{Var}}\left(\widehat{\beta}_u^* - \widehat{\beta}_h^*\right)},$$

with $w_T = q_{3,df,.01}/\sqrt{2}$. For measurements in seconds, we have the following values:

(u, h)	df	$q_{3,df,0.01}$	$\widehat{\text{Var}}\left(\widehat{\beta}_u^* - \widehat{\beta}_h^*\right)$	msd	$\overline{y}_{.u.} - \overline{y}_{.h.}$
(1, 2)	11.5	5.09	14,780.6	437.57	2,777.67
(1, 3)	18.6	4.68	21,153.5	481.31	2,339.07
(3, 2)	12.6	4.99	8,084.7	317.26	438.60

The set of 99% simultaneous Tukey confidence intervals for pairwise differences is then

$$\beta_1^* - \beta_2^* \in (2777.67 \pm 437.57) = (2340.10, 3215.24),$$

$$\beta_1^* - \beta_3^* \in (1857.76, 2820.38), \quad \beta_1^* - \beta_3^* \in (121.34, 755.86).$$

Since none of the intervals contains zero, we can state that all pairs of levels of B (stain types) have different effects on the speed of stain removal, averaged over the three concentrations of bleach. With experimentwise confidence level at least 97%, the mean time to remove blue ink (level 1) is between 1857 and 2820 seconds longer than that for tomato sauce (level 3), and the mean time to remove tomato sauce is between 121 and 755 seconds longer than that for jam (level 2). □

6.4.4 Analysis of Variance for the Complete Model

There are three standard hypotheses that are usually examined when the two-way complete model is used. The first hypothesis is that the interaction between treatment factors A and

6.4 Analysis of the Two-Way Complete Model

B is negligible; that is,

$$H_0^{AB} : \{(\alpha\beta)_{ij} - (\alpha\beta)_{iq} - (\alpha\beta)_{sj} + (\alpha\beta)_{sq} = 0 \text{ for all } i \neq s, j \neq q\},$$

which occurs when the interaction plots show parallel lines. Notice that if all of the contrasts $(\alpha\beta)_{ij} - (\alpha\beta)_{iq} - (\alpha\beta)_{sj} + (\alpha\beta)_{sq}$ are zero, then their averages over s and q are also zero. This leads to an equivalent way to write H_0^{AB} as

$$H_0^{AB} : \{(\alpha\beta)_{ij} - \overline{(\alpha\beta)}_{i.} - \overline{(\alpha\beta)}_{.j} + \overline{(\alpha\beta)}_{..} = 0 \text{ for all } ij\}.$$

In this form, it appears that H_0^{AB} is based on ab estimable contrasts, but in fact, some of them are redundant, since the ab contrasts add to zero over the subscript $i = 1, 2, \ldots, a$ and also over the subscript $j = 1, 2, \ldots, b$. Consequently, H_0^{AB} is actually based on $(a - 1)(b - 1)$ estimable contrasts, and the test is based on $(a - 1)(b - 1)$ degrees of freedom.

The other two standard hypotheses are the main-effect hypotheses

$$H_0^A : \{\alpha_1^* = \alpha_2^* = \ldots = \alpha_a^*\} \text{ and } H_0^B : \{\beta_1^* = \beta_2^* = \ldots = \beta_b^*\},$$

where $\alpha_i^* = \alpha_i + \overline{(\alpha\beta)}_{i.}$ and $\beta_j^* = \beta_j + \overline{(\alpha\beta)}_{.j}$. However, these main-effect hypotheses may not be of interest if there is a sizable interaction. Each of the main-effect hypotheses can be rephrased in terms of estimable contrasts in the parameters, and so can be tested. As in Chapter 3, the tests will be based on $(a - 1)$ and $(b - 1)$ degrees of freedom, respectively.

When the sample sizes are unequal, there are no neat algebraic formulae for the decision rules of the hypothesis tests. Therefore, we will obtain the tests for equal sample sizes and postpone discussion of the unequal sample size case to Section 6.8, where analysis will be done by computer.

Testing interactions—equal sample sizes Since tests for main effects may not be relevant if the two factors interact, the hypothesis of negligible interaction should be tested first. As in Section 3.5.1, page 44, in order to test

$$H_0^{AB} : \{(\alpha\beta)_{ij} - \overline{(\alpha\beta)}_{i.} - \overline{(\alpha\beta)}_{.j} + \overline{(\alpha\beta)}_{..} = 0 \text{ for all } ij\}$$

against the alternative hypothesis H_A^{AB}:{ the interaction is not negligible}, we compare the sum of squares for error ssE under the two-way complete model (6.2.2) with the sum of squares for error ssE_0^{AB} under the reduced model obtained when H_0^{AB} is true. The difference

$$ssAB = ssE_0^{AB} - ssE$$

is called the *sum of squares for the interaction AB*, and the test rejects H_0^{AB} in favor of H_A^{AB} if $ssAB$ is large relative to ssE.

We can rewrite the two-way complete model as

$$\begin{aligned} y_{ijt} &= \mu + \alpha_i + \beta_j + (\alpha\beta)_{ij} + \epsilon_{ijt} \\ &= \mu^* + \alpha_i^* + \beta_j^* + [(\alpha\beta)_{ij} - \overline{(\alpha\beta)}_{i.} - \overline{(\alpha\beta)}_{.j} + \overline{(\alpha\beta)}_{..}] + \epsilon_{ijt}, \end{aligned}$$

where μ^* is the constant $\mu - \overline{(\alpha\beta)}_{..}$. So, when H_0^{AB} is true, the reduced model is

$$y_{ijt} = \mu^* + \alpha_i^* + \beta_j^* + \epsilon_{ijt},$$

which has the same form as the two-way main-effects model.

We will show in Section 6.5.1 that the least squares estimate of $\mu + \alpha_i + \beta_j$ for the two-way main-effects model is $\bar{y}_{i..} + \bar{y}_{.j.} - \bar{y}_{...}$, for equal sample sizes. Similarly, the least squares estimate of $\mu^* + \alpha_i^* + \beta_j^*$ in the above reduced model is also $\bar{y}_{i..} + \bar{y}_{.j.} - \bar{y}_{...}$. Hence, the sum of squares for error for the reduced model is

$$ssE_0^{AB} = \sum_i \sum_j \sum_t \left(y_{ijt} - \hat{\mu}^* - \hat{\alpha}_i^* - \hat{\beta}_j^*\right)^2$$

$$= \sum_i \sum_j \sum_t (y_{ijt} - \bar{y}_{i..} - \bar{y}_{.j.} + \bar{y}_{...})^2 .$$

Adding and subtracting a term $\bar{y}_{ij.}$ to this expression, we have

$$ssE_0^{AB} = \sum_i \sum_j \sum_t \left((y_{ijt} - \bar{y}_{ij.}) + (\bar{y}_{ij.} - \bar{y}_{i..} - \bar{y}_{.j.} + \bar{y}_{...})\right)^2$$

$$= \sum_i \sum_j \sum_t (y_{ijt} - \bar{y}_{ij.})^2 + \sum_i \sum_j \sum_t (\bar{y}_{ij.} - \bar{y}_{i..} - \bar{y}_{.j.} + \bar{y}_{...})^2 .$$

But the first term is just ssE given in (6.4.17). So, for equal sample sizes,

$$ssAB = ssE_0^{AB} - ssE$$

$$= r \sum_i \sum_j (\bar{y}_{ij.} - \bar{y}_{i..} - \bar{y}_{.j.} + \bar{y}_{...})^2 \qquad (6.4.21)$$

$$= \sum_i \sum_j y_{ij.}^2/r - \sum_i y_{i..}^2/(br) - \sum_j y_{.j.}^2/(ar) + y_{...}^2/(abr) .$$

It can be shown that when H_0^{AB} is true, the corresponding random variable $SS(AB)/\sigma^2$ has a chi-squared distribution with $(a-1)(b-1)$ degrees of freedom. Also, $SSE/\sigma^2 \sim \chi^2_{n-ab}$ and SSE can be shown to be independent of $SS(AB)$. So, when H_0^{AB} is true,

$$\frac{SS(AB)/(a-1)(b-1)\sigma^2}{SSE/(n-ab)\sigma^2} = \frac{MS(AB)}{MSE} \sim F_{(a-1)(b-1), n-ab} .$$

We reject H_0^{AB} for large values of the ratio $msAB/msE$. Thus, the rule for testing the hypothesis H_0^{AB} against the alternative hypothesis that the interaction is not negligible is

$$\text{reject } H_0^{AB} \text{ if } \frac{msAB}{msE} > F_{(a-1)(b-1), n-ab, \alpha} , \qquad (6.4.22)$$

where $msAB = ssAB/(a-1)(b-1)$, $msE = ssE/(n-ab)$, $ssAB$ is given in (6.4.21), and ssE is

$$ssE = \sum_i \sum_j \sum_t y_{ijt}^2 - \sum_i \sum_j y_{ij.}^2/r .$$

If H_0^{AB} is rejected, it is often preferable to use the equivalent cell-means model and look at contrasts in the treatment combinations. If H_0^{AB} is not rejected, then tests and contrasts for main effects are usually of interest, and the two-way complete model is retained. (We do not change to the inequivalent main-effects model.)

6.4 Analysis of the Two-Way Complete Model

Testing main effects of A—equal sample sizes In testing the hypothesis that factor A has no effect on the response, one can either test the hypothesis that the levels of A (averaged over the levels of B) have the same average effect on the response, that is,

$$H_0^A : \{\alpha_1^* = \alpha_2^* = \cdots = \alpha_a^*\},$$

or one can test the hypothesis that the response depends only on the level of B, that is

$$H_0^{A+AB} : \{H_0^A \text{ and } H_0^{AB} \text{ are both true}\}.$$

The traditional test, which is produced automatically by many computer packages, is a test of the former, and the sum of squares for error ssE under the two-way complete model is compared with the sum of squares for error ssE_0^A under the reduced model

$$Y_{ijt} = \mu^{**} + \beta_j^* + \left((\alpha\beta)_{ij} - \overline{(\alpha\beta)}_{i.} - \overline{(\alpha\beta)}_{.j} + \overline{(\alpha\beta)}_{..}\right) + \epsilon_{ijt}.$$

It is, perhaps, more intuitively appealing to test H_0^{A+AB} rather than H_0^A, since the corresponding reduced model is

$$Y_{ijt} = \mu^* + \beta_j^* + \epsilon_{ijt},$$

suggesting that A has no effect on the response whatsoever.

In this book, we take the view that the main effect of A would not be tested unless the hypothesis of no interaction were first accepted. If it is true that there is no interaction, then the two hypotheses and corresponding reduced models are the same, and the results of the two tests should be similar. Consequently, we will derive the test of the standard hypothesis H_0^A.

It can be shown that if the sample sizes are equal, the least squares estimate of $E[Y_{ijt}]$ for the reduced model under H_0^A is

$$\overline{y}_{ij.} - \overline{y}_{i..} + \overline{y}_{...},$$

and so the sum of squares for error for the reduced model is

$$ssE_0^A = \sum_i \sum_j \sum_t (y_{ijt} - \overline{y}_{ij.} + \overline{y}_{i..} - \overline{y}_{...})^2.$$

Taking the terms in pairs and expanding the terms in parentheses, we obtain

$$ssE_0^A = \sum_{i=1}^a \sum_{j=1}^b \sum_{t=1}^r (y_{ijt} - \overline{y}_{ij.})^2 - br \sum_{i=1}^a (\overline{y}_{i..} - \overline{y}_{...})^2.$$

Since the first term is the formula (6.4.17) for ssE, the *sum of squares for treatment factor A* is

$$ssA = ssE_0^A - ssE = br \sum_{i=1}^a (\overline{y}_{i..} - \overline{y}_{...})^2 = \sum_{i=1}^a y_{i..}^2/(br) - y_{...}^2/(abr). \quad (6.4.23)$$

Notice that this formula for ssA is similar to the formula (3.5.12), page 46, for ssT used to test the hypothesis $H_0:\{\tau_1 = \tau_2 = \cdots = \tau_a\}$ in the one-way analysis of variance.

We write *SSA* for the random variable corresponding to *ssA*. It can be shown that if H_0^A is true, SSA/σ^2 has a chi-squared distribution with $a-1$ degrees of freedom, and that *SSA* and *SSE* are independent. So, writing $MSA = SSA/(a-1)$, we have that MSA/MSE has an F-distribution when H_0^A is true, and the rule for testing $H_0^A : \{\alpha_1^* = \cdots = \alpha_a^*\}$ against $H_0^A : \{$ not all of the α_i^*'s are equal$\}$ is

$$\text{reject } H_0^A \text{ if } \frac{msA}{msE} > F_{a-1, n-ab, \alpha}, \tag{6.4.24}$$

where $msA = ssA/(a-1)$ and $msE = ssE/(n-ab)$.

Testing main effects of B—equal sample sizes Analogous to the test for main effects of A, we can show that the rule for testing $H_0^B : \{\beta_1^* = \beta_2^* = \cdots = \beta_b^*\}$ against $H_A^B : \{$ not all of the β_j^*'s are equal$\}$ is

$$\text{reject } H_0^B \text{ if } \frac{msB}{msE} > F_{b-1, n-ab, \alpha}, \tag{6.4.25}$$

where $msB = ssB/(b-1)$, $msE = ssE/(n-ab)$, and

$$ssB = ar \sum_j (\bar{y}_{.j.} - \bar{y}_{...})^2 = \sum_j y_{.j.}^2/(ar) - y_{...}^2/(abr). \tag{6.4.26}$$

Analysis of variance table The tests of the three hypotheses are summarized in a two-way analysis of variance table, shown in Table 6.4. The computational formulae are given for equal sample sizes. The last line of the table is $sstot = \sum_i \sum_j \sum_t (y_{ijt} - \bar{y}_{...})^2$, which is the total sum of squares similar to (3.5.18). It can be verified that

$$ssA + ssB + ssAB + ssE = sstot.$$

When the sample sizes are not equal, the formulae for *ssA*, *ssB*, and *ssAB* are more complicated, the corresponding random variables *SSA*, *SSB*, and *SS(AB)* are not independent,

Table 6.4 Two-way ANOVA, crossed fixed effects with interaction

Source of Variation	Degrees of Freedom	Sum of Squares	Mean Square	Ratio
Factor A	$a-1$	ssA	$\frac{ssA}{a-1}$	$\frac{msA}{msE}$
Factor B	$b-1$	ssB	$\frac{ssB}{b-1}$	$\frac{msB}{msE}$
AB	$(a-1)(b-1)$	$ssAB$	$\frac{ssAB}{(a-1)(b-1)}$	$\frac{msAB}{msE}$
Error	$n-ab$	ssE	$\frac{ssE}{n-ab}$	
Total	$n-1$	$sstot$		

Computational Formulae for Equal Sample Sizes	
$ssE = \sum_i \sum_j \sum_t y_{ijt}^2$	$ssA = \sum_i y_{i..}^2/(br) - y_{...}^2/n$
$\quad - \sum_i \sum_j y_{ij.}^2/r$	$ssB = \sum_j y_{.j.}^2/(ar) - y_{...}^2/n$
$sstot = \sum_i \sum_j \sum_t y_{ijt}^2 - y_{...}^2/n$	$ssAB = \sum_i \sum_j y_{ij.}^2/r - \sum_i y_{i..}^2/(br)$
$n = abr$	$\quad - \sum_j y_{.j.}^2/(ar) + y_{...}^2/n$

and

$$ssA + ssB + ssAB + ssE \neq sstot.$$

The analysis of experiments with unequal sample sizes will be discussed in Section 6.8, page 175, using the SAS computer package.

Example 6.4.5 Reaction time experiment, continued

The reaction time experiment was described in Example 6.4.2, page 148. There were $a = 2$ levels of cue stimulus and $b = 3$ levels of cue time, and $r = 3$ observations per treatment combination. Using the data in Table 6.2, we have

$$sstot = \sum_i \sum_j \sum_t y_{ijt}^2 - y_{...}^2/(abr)$$
$$= 0.96519 - 0.93617 = 0.02902,$$
$$ssA = \sum_i y_{i..}^2/(br) - y_{...}^2/(abr)$$
$$= (1.727^2 + 2.378^2)/9 - 0.93617 = 0.02354,$$
$$ssB = \sum_j y_{.j.}^2/(ar) - y_{...}^2/(abr)$$
$$= (1.360^2 + 1.314^2 + 1.431^2)/6 - 0.93617 = 0.00116$$
$$ssAB = \sum_i \sum_j y_{ij.}^2/r - \sum_i y_{i..}^2/(br) - \sum_j y_{.j.}^2/(ar) + y_{...}^2/(abr)$$
$$= 0.96172 - 0.95971 - 0.93733 + 0.93617 = 0.00085,$$

and in Example 6.4.2, ssE was calculated to be 0.00347. It can be seen that $sstot = ssA + ssB + ssAB + ssE$. The analysis of variance table is shown in Table 6.5. The mean squares are the sums of squares divided by their degrees of freedom.

There are three hypotheses to be tested. If the Type I error probability α is selected to be 0.01 for each test, then the probability of incorrectly rejecting at least one hypothesis when it is true is at most 0.03. The interaction plots in Figure 6.4, page 150, suggest that there is no interaction between cue stimulus (A) and cue time (B). To test this hypothesis, we obtain from the analysis of variance table

$$msAB/msE = 0.00043/0.00029 = 1.5,$$

which is less than $F_{2,12,.01} = 6.93$. Therefore, at individual significance level, $\alpha = 0.01$, there is not sufficient evidence to reject the null hypothesis H_0^{AB} that the interaction is negligible. This agrees with the interaction plot.

Now consider the main effects. Looking at Figure 6.4, if we average over cue stimulus, there does not appear to be much difference in the effect of cue time. If we average over cue time, then auditory cue stimulus (level 1) appears to produce a shorter reaction time than a visual cue stimulus (level 2). From the analysis of variance table, $msA/msE = 0.02354/0.00029 = 81.4$. This is larger than $F_{1,12,.01} = 9.33$, so we reject $H_0^A:\{\alpha_1^* = \alpha_2^*\}$, and we would conclude that there is a difference in cue stimulus averaged over the cue times.

Table 6.5 Two-way ANOVA for the reaction time experiment

Source of Variation	Degrees of Freedom	Sum of Squares	Mean Square	Ratio	p-value
Cue stimulus	1	0.02354	0.02354	81.38	0.0001
Cue time	2	0.00116	0.00058	2.00	0.1778
Interaction	2	0.00085	0.00043	1.46	0.2701
Error	12	0.00347	0.00029		
Total	17	0.02902			

On the other hand, $msB/msE = 0.00058/0.00029 = 2.0$, which is less than $F_{2,12,.01} = 6.93$. Consequently, we do not reject $H_0^B : \{\beta_1^* = \beta_2^* = \beta_3^*\}$ and conclude that there is no evidence for a difference in the effects of the cue times averaged over the two cue stimuli.

If the analysis were done by a computer program, the p-values in Table 6.5 would be printed. We would reject any hypothesis whose corresponding p-value is less than the selected individual α^* level. In this example, we selected $\alpha^* = 0.01$, and we would fail to reject H_0^{AB} and H_0^B, but we would reject H_0^A, as in the hand calculations.

This was a pilot experiment, and since the experimenters already believed that cue stimulus and cue time really do not interact, they selected the two-way main-effects model in planning the main experiment. □

6.5 Analysis of the Two-Way Main-Effects Model

6.5.1 Least Squares Estimators for the Main-Effects Model

The two-way main-effects model (6.2.3) is

$$Y_{ijt} = \mu + \alpha_i + \beta_j + \epsilon_{ijt},$$
$$\epsilon_{ijt} \sim N(0, \sigma^2),$$
ϵ_{ijt}'s are mutually independent,
$t = 1, \ldots, r_{ij}; \quad i = 1, \ldots, a; \quad j = 1, \ldots, b.$

This model is a submodel of the two-way complete model (6.2.2) in the sense that it can only describe situations similar to those depicted in plots (a)–(d) of Figure 6.1 and cannot describe plots (e)–(h). When the sample sizes are unequal, the least squares estimators of the parameters in the main-effects model are not easy to obtain, and calculations are best left to a computer (see Section 6.8). In the optional subsection below, we show that when the sample sizes are all equal to r, the least squares estimator of $E[Y_{ijt}] = (\mu + \alpha_i + \beta_j)$ is

$$\hat{\mu} + \hat{\alpha}_i + \hat{\beta}_j = \overline{Y}_{i..} + \overline{Y}_{.j.} - \overline{Y}_{...} . \qquad (6.5.27)$$

The least squares estimator for the estimable main-effect contrast $\sum_i c_i \alpha_i$ with $\sum_i c_i = 0$ is then

$$\sum_i c_i \hat{\alpha}_i = \sum_i c_i (\hat{\mu} + \hat{\alpha}_i + \hat{\beta}_j) = \sum_i c_i \left(\overline{Y}_{i..} + \overline{Y}_{.j.} - \overline{Y}_{...} \right)$$

$$= \sum_i c_i \overline{Y}_{i..},$$

which has variance

$$\text{Var}\left(\sum_i c_i \hat{\alpha}_i\right) = \text{Var}\left(\sum_i c_i \overline{Y}_{i..}\right) = \frac{\sigma^2}{br} \sum_i c_i^2. \tag{6.5.28}$$

For example, $\alpha_p - \alpha_s$, the pairwise comparison of levels p and s of A, has least squares estimator and associated variance

$$\hat{\alpha}_p - \hat{\alpha}_s = \overline{Y}_{p..} - \overline{Y}_{s..} \quad \text{with} \quad \text{Var}(\overline{Y}_{p..} - \overline{Y}_{s..}) = \frac{2\sigma^2}{br}.$$

These are exactly the same formulas as for the two-way complete model and similar to those for the one-way model. Likewise for B, a main-effect contrast $\sum k_j \beta_j$ with $\sum_j k_j = 0$ has least squares estimator and associated variance

$$\sum_j k_j \hat{\beta}_j = \sum_j k_j \overline{Y}_{.j.} \quad \text{and} \quad \text{Var}\left(\sum_j k_j \overline{Y}_{.j.}\right) = \frac{\sigma^2}{ar} \sum_j k_j^2, \tag{6.5.29}$$

and the least squares estimator and associated variance for the pairwise difference $\beta_h - \beta_q$ is

$$\hat{\beta}_h - \hat{\beta}_q = \overline{Y}_{.h.} - \overline{Y}_{.q.} \quad \text{with} \quad \text{Var}(\overline{Y}_{.h.} - \overline{Y}_{.q.}) = \frac{2\sigma^2}{ar}.$$

Example 6.5.1 Nail varnish experiment

An experiment on the efficacy of nail varnish solvent in removing nail varnish from cloth was run by Pascale Quester in 1986. Two different brands of solvent (factor A) and three different brands of nail varnish (factor B) were investigated. One drop of nail varnish was applied to a piece of cloth (dropped from the applicator 20 cm above the cloth). The cloth was immersed in a bowl of solvent and the time measured (in minutes) until the varnish completely dissolved. There were six treatment combinations 11, 12, 13, 21, 22, 23, where the first digit represents the brand of solvent and the second digit represents the brand of nail varnish used in the experiment. The design was a completely randomized design with $r = 5$ observations on each of the six treatment combinations. The data are listed in Table 6.6 in the order in which they were collected.

The experimenter had run a pilot experiment to estimate the error variance σ^2 and to check that the experimental procedure was satisfactory. The pilot experiment indicated that the interaction between nail varnish and solvent was negligible. The similarity of the chemical composition of the varnishes and solvents, and the verification from the pilot experiment, suggest that the main-effects model (6.2.3) will be a satisfactory model for the main experiment. The data from the main experiment give the interaction plots in Figure 6.5. Although the lines are not quite parallel, the selected main-effects model would not be a severely incorrect representation of the data.

Table 6.6 Data (minutes) for the nail varnish experiment

Solvent	2	1	1	2	2	2	1	2
Varnish	3	3	3	3	2	2	2	2
Time	32.50	30.20	27.25	24.25	34.42	26.00	22.50	31.08
Solvent	1	2	1	1	2	1	2	2
Varnish	2	1	1	1	1	3	3	2
Time	25.17	29.17	27.58	28.75	31.75	29.75	30.75	29.17
Solvent	1	1	2	1	2	2	1	2
Varnish	2	1	2	2	1	3	3	1
Time	27.75	25.83	24.75	21.50	32.08	29.50	24.50	28.50
Solvent	2	1	1	2	1	1		
Varnish	3	3	1	1	1	2		
Time	28.75	22.75	29.25	31.25	22.08	25.00		

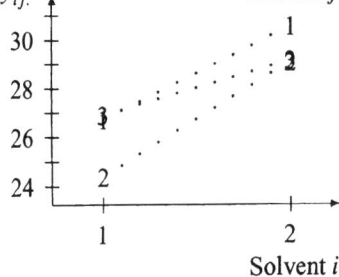
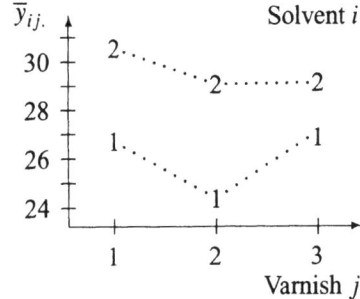

Figure 6.5 Average dissolving times for the nail varnish experiment

Using the data in Table 6.6, the average dissolving time (in minutes) for the two brands of solvent are

$$\bar{y}_{1..} = 25.9907 \quad \text{and} \quad \bar{y}_{2..} = 29.5947.$$

So the least squares estimate of the difference in the dissolving times for the two solvents is

$$\hat{\alpha}_1 - \hat{\alpha}_2 = \bar{y}_{1..} - \bar{y}_{2..} = -3.6040,$$

and the variance of the estimator is $2\sigma^2/(rb) = 2\sigma^2/15$. A difference of 3.6 minutes seems quite substantial, but this needs to be compared with the experimental error via a confidence interval to see whether such a difference could have occurred by chance (see Examples 6.5.2 and 6.5.3).

The average dissolving times for the three brands of nail varnish are

$$\bar{y}_{.1.} = 28.624, \quad \bar{y}_{.2.} = 26.734, \quad \text{and} \quad \bar{y}_{.3.} = 28.020,$$

and the least squares estimates of the pairwise comparisons are

$$\hat{\beta}_1 - \hat{\beta}_2 = 1.890, \quad \hat{\beta}_1 - \hat{\beta}_3 = 0.604, \quad \text{and} \quad \hat{\beta}_2 - \hat{\beta}_3 = -1.286,$$

each with associated variance $2\sigma^2/10$. Since levels 1 and 2 of the nail varnish represented French brands, while level 3 represented an American brand, the difference of averages

6.5 Analysis of the Two-Way Main-Effects Model

contrast

$$\tfrac{1}{2}(\beta_1 + \beta_2) - \beta_3$$

would also be of interest. The least squares estimate of this contrast is

$$\tfrac{1}{2}(\hat{\beta}_1 + \hat{\beta}_2) - \hat{\beta}_3 = \tfrac{1}{2}(\overline{y}_{.1.} + \overline{y}_{.2.}) - \overline{y}_{.3.} = -0.341,$$

with associated variance $6\sigma^2/40$. □

Deriving least squares estimators for equal sample sizes (optional) We now sketch the derivation (using calculus) of the least squares estimators for the parameters of the two-way main-effects model (6.2.3), when the sample sizes are all equal to r. A reader without knowledge of calculus may jump to Section 6.5.2, page 162.

As in Section 3.4.3, the least squares estimates of the parameters in a model are those estimates that give the minimum value of the sum of squares of the estimated errors. For the two-way main-effects model (6.2.3), the sum of squared errors is

$$\sum_{i=1}^{a}\sum_{j=1}^{b}\sum_{t=1}^{r} \hat{e}_{ijt}^2 = \sum_{i=1}^{a}\sum_{j=1}^{b}\sum_{t=1}^{r}\left(y_{ijt} - (\mu + \alpha_i + \beta_j)\right)^2.$$

The least squares estimates are obtained by differentiating the sum of squared errors with respect to each of the parameters μ, α_i ($i = 1, \ldots, a$), and β_j ($j = 1, \ldots, b$) in turn and setting the derivatives equal to zero. The resulting set of normal equations is as follows.

$$y_{...} - abr\hat{\mu} - br\sum_i \hat{\alpha}_i - ar\sum_j \hat{\beta}_j = 0, \tag{6.5.30}$$

$$y_{i..} - br\hat{\mu} - br\hat{\alpha}_i - r\sum_j \hat{\beta}_j = 0, \quad i = 1, \ldots, a, \tag{6.5.31}$$

$$y_{.j.} - ar\hat{\mu} - r\sum_i \hat{\alpha}_i - ar\hat{\beta}_j = 0, \quad j = 1, \ldots, b. \tag{6.5.32}$$

There are $1+a+b$ normal equations in $1+a+b$ unknowns. However, the equations are not all distinct (linearly independent), since the sum of the a equations (6.5.31) is equal to the sum of the b equations (6.5.32), which is equal to the equation (6.5.30). Consequently, there are at most, and, in fact, exactly, $1+a+b-2$ distinct equations, and two extra equations are needed in order to obtain a solution. Many computer packages, including the SAS software, use the extra equations $\hat{\alpha}_a = 0$ and $\hat{\beta}_b = 0$. However, when working by hand, it is easier to use the equations $\sum_i \hat{\alpha}_i = 0$ and $\sum_j \hat{\beta}_j = 0$, in which case (6.5.30)–(6.5.32) give the following least squares solutions:

$$\hat{\mu} = \overline{y}_{...}, \tag{6.5.33}$$

$$\hat{\alpha}_i = \overline{y}_{i..} - \overline{y}_{...}, \quad i = 1, \ldots, a,$$

$$\hat{\beta}_j = \overline{y}_{.j.} - \overline{y}_{...}, \quad j = 1, \ldots, b.$$

Then the least squares estimate of $\mu + \alpha_i + \beta_j$ is

$$\hat{\mu} + \hat{\alpha}_i + \hat{\beta}_j = \overline{y}_{i..} + \overline{y}_{.j.} - \overline{y}_{...}, \quad i = 1, \ldots, a, \quad j = 1, \ldots, b.$$

Deriving least squares estimators for unequal sample sizes (optional) If the sample sizes are not equal, then the normal equations for the two-way main-effects model become

$$y_{...} - n\hat{\mu} - \sum_{p=1}^{a} r_{p.}\hat{\alpha}_p - \sum_{q=1}^{b} r_{.q}\hat{\beta}_q = 0, \tag{6.5.34}$$

$$y_{i..} - r_{i.}\hat{\mu} - r_{i.}\hat{\alpha}_i - \sum_{q=1}^{b} r_{iq}\hat{\beta}_q = 0, \quad i = 1, \ldots, a, \tag{6.5.35}$$

$$y_{.j.} - r_{.j}\hat{\mu} - \sum_{p=1}^{a} r_{pj}\hat{\alpha}_p - r_{.j}\hat{\beta}_j = 0, \quad j = 1, \ldots, b, \tag{6.5.36}$$

where $n = \sum_i \sum_j r_{ij}$, $r_{p.} = \sum_j r_{pj}$, and $r_{.q} = \sum_i r_{iq}$. As in the equal sample size case, the normal equations represent $a+b-1$ distinct equations in $1+a+b$ unknowns, and two extra equations are needed to obtain a particular solution. Looking at (6.5.34), a sensible choice might be $\sum_p r_{p.}\hat{\alpha}_p = 0$ and $\sum_q r_{.q}\hat{\beta}_q = 0$. Then $\hat{\mu} = \overline{y}_{...}$ as in the equal sample size case. However, obtaining solutions for the $\hat{\alpha}_i$'s and $\hat{\beta}_j$'s is not so easy. One can solve for $\hat{\beta}_j$ in (6.5.36) and substitute this into (6.5.35), which gives the following equations in the $\hat{\alpha}_i$'s:

$$\hat{\alpha}_i - \sum_{q=1}^{b} \frac{r_{iq}}{r_{.q}r_{i.}} \sum_{p=1}^{a} r_{pq}\hat{\alpha}_p = \overline{y}_{i..} - \sum_{q=1}^{b} \frac{r_{iq}}{r_{.q}r_{i.}} y_{.q}, \quad \text{for } i = 1, \ldots, a. \tag{6.5.37}$$

Equations in the $\hat{\beta}_j$'s can be obtained similarly. Algebraic expressions for the individual parameter estimates are generally complicated, and we will leave the unequal sample size case to a computer analysis (Section 6.8).

6.5.2 Estimation of σ^2 in the Main-Effects Model

The minimum value of the sum of squares of the estimated errors for the two-way main-effects model is

$$ssE = \sum_{i=1}^{a} \sum_{j=1}^{b} \sum_{t=1}^{r} (y_{ijt} - \hat{\mu} - \hat{\alpha}_i - \hat{\beta}_j)^2 \tag{6.5.38}$$

$$= \sum_{i=1}^{a} \sum_{j=1}^{b} \sum_{t=1}^{r} (y_{ijt} - \overline{y}_{i..} - \overline{y}_{.j.} + \overline{y}_{...})^2.$$

Expanding the terms in parentheses in (6.5.38) yields the following formulae useful for direct hand calculation of *ssE*:

$$ssE = \sum_i \sum_j \sum_t y_{ijt}^2 - br \sum_i \overline{y}_{i..}^2 - ar \sum_j \overline{y}_{.j.}^2 + abr\overline{y}_{...}^2 \tag{6.5.39}$$

$$= \sum_i \sum_j \sum_t y_{ijt}^2 - \frac{1}{br} \sum_i y_{i..}^2 - \frac{1}{ar} \sum_j y_{.j.}^2 + \frac{1}{abr} y_{...}^2.$$

Now, ssE is the observed value of

$$SSE = \sum_i \sum_j \sum_t (Y_{ijt} - \overline{Y}_{i..} - \overline{Y}_{.j.} + \overline{Y}_{...})^2.$$

In Exercise 19, the reader will be asked to prove, for the equal sample size case, that

$$E[SSE] = (n - a - b + 1)\sigma^2,$$

where $n = abr$, so an unbiased estimator of σ^2 is

$$MSE = SSE/(n - a - b + 1).$$

It can be shown that SSE/σ^2 has a chi-squared distribution with $(n - a - b + 1)$ degrees of freedom. An upper $100(1 - \alpha)$% confidence bound for σ^2 is therefore given by

$$\sigma^2 \leq \frac{ssE}{\chi^2_{n-a-b+1,1-\alpha}}.$$

Example 6.5.2 Nail varnish experiment

The data for the nail varnish experiment are given in Table 6.6 of Example 6.5.1, page 159, and $a = 2, b = 3, r = 5, n = 30$. It can be verified that

$$\sum_i \sum_j \sum_t y_{ijt}^2 = 23,505.7976, \quad \overline{y}_{...} = 27.7927,$$

and

$$\overline{y}_{1..} = 25.9907, \quad \overline{y}_{2..} = 29.5947,$$

$$\overline{y}_{.1.} = 28.624, \quad \overline{y}_{.2.} = 26.734, \quad \overline{y}_{.3.} = 28.020.$$

Thus, from (6.5.39),

$$ssE = 23,505.7976 - 23,270.3857 - 23,191.6053 + 23,172.9696$$
$$= 216.7762,$$

and an unbiased estimate of σ^2 is

$$msE = 216.7762/(30 - 2 - 3 + 1) = 8.3375 \text{ minutes}^2.$$

A 95% upper confidence bound for σ^2 is

$$\frac{ssE}{\chi^2_{26,.95}} = \frac{216.7762}{15.3791} = 14.096 \text{ minutes}^2,$$

and taking square roots, a 95% upper confidence limit for σ is 3.7544 minutes. □

6.5.3 Multiple Comparisons for the Main-Effects Model

When the sample sizes are equal, the Bonferroni, Scheffé, Tukey, Dunnett, and Hsu methods described in Section 4.4 can all be used for obtaining simultaneous confidence intervals for sets of contrasts comparing the levels of A or of B. A set of $100(1 - \alpha)$% simultaneous confidence intervals for contrasts comparing the levels of factor A is of the form (4.4.20), which for the two-way model becomes

$$\sum c_i \alpha_i \in \left(\sum c_i \bar{y}_{i..} \pm w \sqrt{msE \sum c_i^2 / br} \right) , \qquad (6.5.40)$$

where the critical coefficients for the various methods are, respectively,

$$w_B = t_{n-a-b+1, \alpha/2m} \; ; \; w_S = \sqrt{(a-1) F_{a-1, n-a-b+1, \alpha}} \; ;$$

$$w_T = q_{a, n-a-b+1, \alpha} / \sqrt{2} \; ;$$

$$w_H = w_{D1} = t_{a-1, n-a-b+1, \alpha}^{(0.5)} \; ; \; w_{D2} = |t|_{a-1, n-a-b+1, \alpha}^{(0.5)} .$$

Similarly, a set of $100(1 - \alpha)$% confidence intervals for contrasts comparing the levels of factor B is of the form

$$\sum k_j \beta_j \in \left(\sum k_j \bar{y}_{.j.} \pm w \sqrt{msE \sum k_j^2 / ar} \right) , \qquad (6.5.41)$$

and the critical coefficients are as above after interchanging a and b.

We can also obtain confidence intervals for the treatment means $\mu + \alpha_i + \beta_j$ using the least squares estimators $\bar{Y}_{i..} + \bar{Y}_{.j.} - \bar{Y}_{...}$, each of which has a normal distribution and variance $\sigma^2 (a + b - 1)/(abr)$. We obtain a set of $100(1 - \alpha)$% simultaneous confidence intervals for the ab treatment means as

$$\mu + \alpha_i + \beta_j \in \left\{ (\bar{y}_{i..} + \bar{y}_{.j.} - \bar{y}_{...}) \pm w \sqrt{msE \left(\frac{a+b-1}{abr} \right)} \right\} , \qquad (6.5.42)$$

with critical coefficient

$$w_{BM} = t_{\alpha/(2ab), (n-a-b+1)} \quad \text{or} \quad w_{SM} = \sqrt{(a+b-1) F_{a+b-1, n-a-b+1, \alpha}}$$

for the Bonferroni and Scheffé methods, respectively.

When confidence intervals are calculated for treatment means and for contrasts in the main effects of factors A and B, an experimentwise confidence level should be calculated. For example, if intervals for contrasts for factor A have overall confidence level $100(1 - \alpha_1)$%, and intervals for B have overall confidence level $100(1 - \alpha_2)$%, and intervals for means have overall confidence level $100(1 - \alpha_3)$%, the experimentwise confidence level for all the intervals combined is at least $100(1 - (\alpha_1 + \alpha_2 + \alpha_3))$%. Alternatively, w_{SM} could be used in (6.5.40) and (6.5.42), and the overall level for all three sets of intervals together would be $100(1 - \alpha)$%.

Example 6.5.3 Nail varnish experiment

The least squares estimates for the differences in the effects of the two nail varnish solvents and for the pairwise differences in the effects of the three nail varnishes were calculated in Example 6.5.1, page 159. From Table 6.8, $msE = 8.3375$ with error degrees of freedom $n - a - b + 1 = 26$. There is only $m = 1$ contrast for factor A, and a simple 99% confidence interval of the form (6.5.40) can be used to give

$$\alpha_2 - \alpha_1 \in \left(\overline{y}_{2..} - \overline{y}_{1..} \pm t_{n-a-b+1,\alpha/2}\sqrt{msE(2/br)}\right)$$
$$= \left(3.6040 \pm t_{26,0.005}\sqrt{(8.3375/15)}\right).$$

From Table A.4, $t_{26,0.005} = 2.779$, so a 99% confidence interval for $\alpha_2 - \alpha_1$ is

$$1.5321 \leq \alpha_2 - \alpha_1 \leq 5.6759.$$

The confidence interval indicates that solvent 2 takes between 1.5 and 5.7 minutes longer, on average, in dissolving the three nail varnishes than does solvent 1.

To compare the nail varnishes in terms of their speed of dissolving, confidence intervals are required for the three pairwise comparisons $\beta_1 - \beta_2$, $\beta_1 - \beta_3$, and $\beta_2 - \beta_3$. If an overall confidence level of 99% is required, Tukey's method gives confidence intervals of the form

$$\beta_j - \beta_p \in \left(\overline{y}_{.j.} - \overline{y}_{.p.} \pm (q_{b,df,0.01}/\sqrt{2})\sqrt{msE(2/(ar))}\right).$$

From Table A.8, $q_{3,26,0.01} = 4.54$. Using the least squares estimates computed in Example 6.5.1, page 159, and $msE = 8.3375$ with $n - a - b + 1 = 26$ as above, the minimum significant difference is $msd = (4.54/\sqrt{2})\sqrt{8.3375(2/10)} = 4.145$. A set of 99% confidence intervals for the pairwise comparisons for factor B is

$$\beta_1 - \beta_2 \in (1.890 \pm 4.145) = (-2.255, 6.035),$$

$$\beta_1 - \beta_3 \in (-3.541, 4.749), \quad \beta_2 - \beta_3 \in (-5.431, 2.859).$$

Each of these intervals includes zero, indicating insufficient evidence to conclude a difference in the speed at which the nail varnishes dissolve. The overall confidence level for the four intervals for factors A and B together is at least 98%. Bonferroni's method could have been used instead for all four intervals. To have obtained an overall level of at least 98%, we could have set $\alpha^* = \alpha/m = 0.02/4 = 0.005$ for each of the four intervals. The critical coefficient in (6.5.40) would then have been $w_B = t_{0.0025,26} = 3.067$. So the Bonferroni method would have given a longer interval for $\alpha_1 - \alpha_2$ but shorter intervals for $\beta_j - \beta_p$. □

6.5.4 Unequal Variances

When the variances of the error variables are unequal and no equalizing transformation can be found, Satterthwaite's approximation can be used. Since the approximation uses the sample variances of the observations for each treatment combination individually, and since the least squares estimates of the main-effect contrasts are the same whether or not interaction terms are included in the model, the procedure is exactly the same as that illustrated for the bleach experiment in Example 6.4.3, page 151.

6.5.5 Analysis of Variance for Equal Sample Sizes

Testing main effects of B—equal sample sizes The hypothesis that the levels of B all have the same effect on the response is $H_0^B : \{\beta_1 = \beta_2 = \cdots = \beta_b\}$, which can be written in terms of estimable contrasts as $H_0^B:\{\beta_j - \bar{\beta}_. = 0, \text{ for all } j = 1, \ldots, b\}$. To obtain a test of H_0^B against the alternative hypothesis H_A^B : { at least two of the β_j's differ}, the sum of squares for error for the two-way main-effects model is compared with the sum of squares for error for the reduced model

$$Y_{ijt} = \mu + \alpha_i + \epsilon_{ijt} \,. \tag{6.5.43}$$

This is identical to the one-way analysis of variance model (3.3.1) with μ replaced by $\mu* = \mu + \bar{\beta}_.$ and with br observations on the ith level of treatment factor A. Thus ssE_0^B is the same as the sum of squares for error in a one-way analysis of variance, and can be obtained from equation (3.4.4), page 42, by replacing the subscript j by the pair of subscripts jt, yielding

$$ssE_0^B = \sum_i \sum_j \sum_t (y_{ijt} - \bar{y}_{i..})^2 \,. \tag{6.5.44}$$

The sum of squares for testing H_0^B is $ssE_0^B - ssE$, where ssE was derived in (6.5.38), page 162. So,

$$ssB = \sum_{i=1}^{a}\sum_{j=1}^{b}\sum_{t=1}^{r}(y_{ijt} - \bar{y}_{i..})^2 - \sum_{i=1}^{a}\sum_{j=1}^{b}\sum_{t=1}^{r}\left((y_{ijt} - \bar{y}_{i..}) - (\bar{y}_{.j.} - \bar{y}_{...})\right)^2$$
$$= ar\sum_j (\bar{y}_{.j.} - \bar{y}_{...})^2$$
$$= \sum_j y_{.j.}^2/(ar) - y_{...}^2/(abr) \,. \tag{6.5.45}$$

Notice that the formula for ssB is identical to the formula (6.4.26) for testing the equivalent main-effect hypothesis in the two-way complete model. It can be shown that when H_0^B is true, the corresponding random variable SSB/σ^2 has a chi-squared distribution with $(b-1)$ degrees of freedom, and SSB and SSE are independent. Therefore, when H_0^B is true,

$$\frac{SSB/(b-1)\sigma^2}{SSE/(n-a-b+1)\sigma^2} = \frac{MSB}{MSE} \sim F_{b-1, n-a-b+1} \,,$$

and the decision rule for testing H_0^B against H_A^B is

$$\text{reject } H_0^B \text{ if } \frac{msB}{msE} > F_{b-1,n-a-b+1,\alpha} \,. \tag{6.5.46}$$

Testing main effects of A—equal sample sizes A similar rule is obtained for testing $H_0^A : \{\alpha_1 = \alpha_2 = \cdots = \alpha_a\}$ against the alternative hypothesis H_A^A : {at least two of the α_i's differ}. The decision rule is

$$\text{reject } H_0^A \text{ if } \frac{msA}{msE} > F_{a-1,n-a-b+1,\alpha} \,, \tag{6.5.47}$$

where $msA = ssA/(a-1)$, and

$$ssA = rb\sum_i (\bar{y}_{i..} - \bar{y}_{...})^2 = \sum_{i=1}^{a} y_{i..}^2/(rb) - y_{...}^2/(abr) \,. \tag{6.5.48}$$

6.5 Analysis of the Two-Way Main-Effects Model

Table 6.7 Two-Way ANOVA, negligible interaction, equal sample sizes

Source of Variation	Degrees of Freedom	Sum of Squares	Mean Square	Ratio
Factor A	$a-1$	ssA	$\frac{ssA}{a-1}$	$\frac{msA}{msE}$
Factor B	$b-1$	ssB	$\frac{ssB}{b-1}$	$\frac{msB}{msE}$
Error	$n-a-b+1$	ssE	$\frac{ssE}{n-a-b+1}$	
Total	$n-1$	sstot		

Computational Formulae for Equal Sample Sizes	
$ssA = \sum_i y_{i..}^2/(br) - y_{...}^2/n$	$ssB = \sum_j y_{.j.}^2/(ar) - y_{...}^2/n$
$sstot = \sum_i \sum_j \sum_t y_{ijt}^2 - y_{...}^2/n$	$ssE = sstot - ssA - ssB$
$n = abr$	

similar to the formula (6.4.23) for testing the equivalent hypothesis in the two-way complete model.

Analysis of variance table The information for testing H_0^A and H_0^B is summarized in the analysis of variance table shown in Table 6.7. When sample sizes are equal, $ssE = sstot - ssA - ssB$. When the sample sizes are not equal, the formulae for the sums of squares are complicated, and the analysis should be done by computer (Section 6.8).

Example 6.5.4 Nail varnish experiment

The analysis of variance table for the nail varnish experiment of Example 6.5.1, page 159, is given in Table 6.8. The experimenter selected the Type I error probability as 0.05 for testing each of H_0^A and H_0^B, giving an overall error rate of at most 0.1. The ratio $msA/msE = 11.68$ is larger than $F_{1,26,0.05} \approx 4.0$, and therefore, the null hypothesis can be rejected. It can be concluded at individual significance level 0.05 that there is a difference in dissolving times for the two solvents.

The ratio $msB/msE = 1.12$ is smaller than $F_{2,26,0.05} \approx 3.15$. Therefore, the null hypothesis H_0^B cannot be rejected at individual significance level 0.05, and it is not possible to conclude that there is a difference in dissolving time among the three varnishes. □

Table 6.8 Analysis of variance for the nail varnish experiment

Source of Variation	Degrees of Freedom	Sum of Squares	Mean Square	Ratio	p-value
Solvent	1	97.4161	97.4161	11.68	0.0021
Varnish	2	18.6357	9.3178	1.12	0.3423
Error	26	216.7761	8.3375		
Total	29	332.8279			

6.5.6 Model Building

In some experiments, the primary objective is to find a model that gives an adequate representation of the experimental data. Such experiments are called experiments for *model building*. If there are two crossed, fixed treatment factors, it is legitimate to use the two-way complete model (6.2.2) as a preliminary model. Then, if H_0^{AB} fails to be rejected, the two-way main effects model (6.2.3) can be accepted as a reasonable model to represent the same type of experimental data in *future* experiments.

Note that it is *not legitimate* to adopt the two-way main effects model and to use the corresponding analysis of variance table, Table 6.7, to test further hypotheses or calculate confidence intervals using the *same* set of data. If this is done, the model is changed based on the data, and the quoted significance levels and confidence levels associated with further inferences will not be correct. Model building should be regarded as a completely different exercise from confidence interval calculation. *They should be done using different experimental data.*

6.6 Calculating Sample Sizes

In Chapter 3, we showed two methods of calculating sample sizes. The method of Section 3.6 aims to achieve a specified power of a hypothesis test, and the method of Section 4.5 aims to achieve a specified length of a confidence interval. Both of these techniques rely on knowledge of the largest likely value of σ^2 or msE and can also be used for the two-way complete model.

Alternatively, sample sizes can be calculated to ensure that confidence intervals for main-effect contrasts are no longer than a stated size, using the formulae (6.4.19) and (6.4.20) or, for the two-way main-effects model, the formulae (6.5.40) and (6.5.41).

Similarly, the method of Section 3.6 for choosing the sample size to achieve the required power of a hypothesis test can be used for each factor separately, with the modification that the sample size calculation is based on

$$r = 2a\sigma^2\phi^2/(b\Delta_A^2) \tag{6.6.49}$$

for factor A and

$$r = 2b\sigma^2\phi^2/(a\Delta_B^2)$$

for factor B, where Δ_A is the smallest difference in the α_i's (or α_i^*'s) and Δ_B is the smallest difference in the β_j's (or β_j^*'s) that are of interest. The calculation procedure is identical to that in Section 3.6, except that the error degrees of freedom are $v_2 = n - v$ for the complete model and $v_2 = n - a - b + 1$ for the main-effects model (with $n = abr$), and the numerator degrees of freedom are $v_1 = a - 1$ for factor A and $v_1 = b - 1$ for factor B.

If several different calculations are done and the calculated values of r differ, then the largest value should be selected.

6.7 Small Experiments

6.7.1 One Observation per Cell

When observations are extremely time-consuming or expensive to collect, an experiment may be designed to have $r = 1$ observation on each treatment combination. Such experiments are called *experiments with one observation per cell* or *single replicate experiments*. Since the ability to choose the sample sizes is lost, it should be recognized that confidence intervals may be wide and hypothesis tests not very powerful.

If it is known in advance that the interaction between the two treatment factors is negligible, then the experiment can be analyzed using the two-way main-effects model (6.2.3). If this information is not available, then the two-way complete model (6.2.2) needs to be used. However, there is a problem. Under the two-way complete model, the number of degrees of freedom for error is $ab(r - 1)$. If $r = 1$, then this number is zero, and σ^2 cannot be estimated.

Thus, a single replicate experiment with a possible interaction between the two factors can be analyzed only if one of the following is true:

(i) σ^2 is known in advance.

(ii) The interaction is expected to be of a certain form that can be modeled with fewer than $(a - 1)(b - 1)$ degrees of freedom.

(iii) The number of treatment combinations is large, and only a few contrasts are likely to be nonnegligible (*effect sparsity*).

If σ^2 is known in advance, formulae for confidence intervals would be based on the normal distribution, and hypothesis tests would be based on the chi-squared distribution. However, this situation is unlikely to occur, and we will not pursue it. The third case tends to occur when the experiment involves a large number of treatment factors and will be discussed in detail in Chapter 7. Here, we look at the second situation and consider two methods of analysis, the first based on orthogonal contrasts, and the second known as Tukey's test for additivity.

6.7.2 Analysis Based on Orthogonal Contrasts

Two estimable contrasts are called *orthogonal contrasts* if and only if their least squares estimators are uncorrelated. For the moment, we recode the treatment combinations to obtain a single-digit code, as we did in Chapter 3. Two contrasts $\Sigma c_i \tau_i$ and $\Sigma k_s \tau_s$ are orthogonal if and only if

$$0 = \text{Cov}\left(\sum_{i=1}^{v} c_i \overline{Y}_{i.}, \sum_{s=1}^{v} k_s \overline{Y}_{s.}\right) = \sum_{i=1}^{v} \sum_{s=1}^{v} c_i k_s \text{Cov}(\overline{Y}_{i.}, \overline{Y}_{s.})$$
$$= \sum_{i} c_i k_i \text{Cov}(\overline{Y}_{i.}, \overline{Y}_{i.}) + \sum_{i} \sum_{s \neq i} c_i k_s \text{Cov}(\overline{Y}_{i.}, \overline{Y}_{s.})$$

$$= \sum_i c_i k_i \text{Var}(\overline{Y}_{i.}) + 0$$

$$= \sigma^2 \sum_i c_i k_i / r_i .$$

In the above calculation $\text{Cov}(\overline{Y}_{i.}, \overline{Y}_{s.})$ is zero when $s \neq i$, because all the Y_{it}'s are independent of each other in the cell-means model. Thus, two contrasts $\Sigma c_i \tau_i$ and $\Sigma k_i \tau_i$ are orthogonal if and only if

$$\sum_i c_i k_i / r_i = 0 . \quad (6.7.50)$$

If the sample sizes are equal, then this reduces to

$$\sum_i c_i k_i = 0 .$$

Changing back to two subscripts, we have that two contrasts $\Sigma\Sigma d_{ij}\tau_{ij}$ and $\Sigma\Sigma h_{ij}\tau_{ij}$ are orthogonal if and only if

$$\sum_{i=1}^{a}\sum_{j=1}^{b} d_{ij} h_{ij} / r_{ij} = 0 , \quad (6.7.51)$$

or, for equal sample sizes, the contrasts are orthogonal if and only if

$$\sum_{i=1}^{a}\sum_{j=1}^{b} d_{ij} h_{ij} = 0 . \quad (6.7.52)$$

For equal sample sizes, the trend contrasts provide an illustration of orthogonal contrasts. For example, it can be verified that any pair of trend contrasts in Table 6.1, page 145, satisfy (6.7.52). For the models considered in this book, the contrast estimators are normally distributed, so orthogonality of contrasts implies that their least squares estimators are independent.

For v treatments, or treatment combinations, a set of $v - 1$ orthogonal contrasts is called a *complete set of orthogonal contrasts*. It is not possible to find more than $v - 1$ contrasts that are mutually orthogonal. We write the sum of squares for the qth orthogonal contrast in a complete set as ssc_q, where

$$ssc_q = (\Sigma\Sigma c_{ij}\overline{y}_{ij.})^2 / (\Sigma\Sigma c_{ij}^2 / r_{ij})$$

is the square of the normalized contrast estimator (see page 76). The sum of squares for treatments, ssT, can be partitioned into the sums of squares for the $v - 1$ orthogonal contrasts in a complete set; that is,

$$ssT = ssc_1 + ssc_2 + \ldots + ssc_{v-1} . \quad (6.7.53)$$

Example 6.7.1 Battery experiment, continued

Main effect and interaction contrasts for the battery experiment were examined in Example 6.3.2, page 143 and, following that example, were written as columns in a table. Since

6.7 Small Experiments

Table 6.9 Three orthogonal contrasts for the battery experiment

Contrast	Coefficients	$\sum c_i \bar{y}_{i.}$	$\sum c_i^2/r_i$	ssc
Duty	$\frac{1}{2}[\,1,\ \ 1,-1,-1\,]$	251.00	$\frac{1}{4}$	252,004.00
Brand	$\frac{1}{2}[\,1,-1,\ \ 1,-1\,]$	−176.50	$\frac{1}{4}$	124,609.00
Interaction	$\frac{1}{2}[\,1,-1,-1,\ \ 1\,]$	−113.25	$\frac{1}{4}$	51,302.25

the sample sizes are all equal, we need only check that (6.7.52) holds by multiplying corresponding coefficients for any two contrasts and adding their products. The duty, brand, and interaction contrasts form a complete set of $v - 1 = 3$ orthogonal contrasts.

The sums of squares for the three contrasts are shown in Table 6.9. It can be verified that they add to the treatment sum of squares $ssT = 427{,}915.25$ that was calculated in Example 3.5.1, page 48. □

We can use the same idea to split the interaction sum of squares $ssAB$ into independent pieces. For the two-way complete model (6.2.2) with $r = 1$ observation per cell, the sum of squares for testing the null hypothesis that a particular interaction contrast, say $\sum_i \sum_j d_{ij}(\alpha\beta)_{ij}$ (with $\sum_i d_{ij} = 0$ and $\sum_j d_{ij} = 0$), is negligible, against the alternative hypothesis that the contrast is not negligible, is

$$ssc = \frac{(\sum_i \sum_j d_{ij} y_{ij})^2}{\sum_i \sum_j d_{ij}^2}. \tag{6.7.54}$$

The interaction has $(a-1)(b-1)$ degrees of freedom. Consequently, there are $(a-1)(b-1)$ orthogonal interaction contrasts in a complete set, and their corresponding sums of squares add to $ssAB$, that is,

$$ssAB = \sum_{h=1}^{(a-1)(b-1)} ssc_h,$$

where ssc_h is the sum of squares for the hth such contrast.

Suppose it is known *in advance* that e specific orthogonal interaction contrasts are likely to be negligible. Then the sums of squares for these e negligible contrasts can be pooled together to obtain an estimate of error variance, based on e degrees of freedom,

$$ssE = \sum_{h=1}^{e} ssc_h \quad \text{and} \quad msE = ssE/e.$$

The sums of squares for the remaining interaction contrasts can be used to test the contrasts individually or added together to obtain an interaction sum of squares

$$ssAB_m = \sum_{h=e+1}^{(a-1)(b-1)} ssc_h.$$

Then the decision rule for testing the hypothesis H_0^{AB}:{ the interaction AB is negligible} against the alternative hypothesis that the interaction is not negligible is

$$\text{reject } H_0^{AB} \text{ if } \frac{ssAB_m/m}{ssE/e} > F_{m,e,\alpha},$$

Table 6.10 Two-way ANOVA, one observation per cell, e negligible interaction contrasts, and $m = (a - 1)(b - 1) - e$ interaction degrees of freedom

Source of Variation	Degrees of Freedom	Sum of Squares	Mean Square
Factor A	$a - 1$	ssA	msA
Factor B	$b - 1$	ssB	msB
Interaction	m	$ssAB_m$	$msAB$
Error	e	ssE	msE
Total	$ab - 1$	$sstot$	

where $m = (a - 1)(b - 1) - e$. Likewise, the main effect test statistics have denominator ssE/e and error degrees of freedom $df = e$. The tests are summarized in Table 6.10, which shows a modified form of the analysis of variance table for the two-way complete model. A worked example is given in Section 6.7.4.

To save calculating the sums of squares for all of the contrasts, the error sum of squares is usually obtained by subtraction, that is,

$$ssE = sstot - ssA - ssB - ssAB_m.$$

The above technique is most often used when the factors are quantitative, since higher-order interaction trends are often likely to be negligible. The information about the interaction effects must be known prior to running the experiment. If this information is not available, then one of the techniques discussed in Section 7.5 must be used instead.

6.7.3 Tukey's Test for Additivity

Tukey's test for additivity uses only one degree of freedom to measure the interaction. It tests the null hypothesis $H_0^\gamma : \{(\alpha\beta)_{ij} = \gamma\alpha_i\beta_j$ for all $i, j\}$ against the alternative hypothesis that the interaction is not of this form. The test is appropriate only if the size of the interaction effect is expected to increase proportionally to each of the main effects, and it is not designed to measure any other form of interaction. The test requires that the normality assumption be well satisfied. The decision rule is

$$\text{reject } H_0^\gamma \text{ if } \frac{ssAB^*}{ssE/e} > F_{1,e,\alpha}, \tag{6.7.55}$$

where

$$ssAB^* = \frac{ab\left[\sum_i \sum_j y_{ij}\bar{y}_{i.}\bar{y}_{.j} - (ssA + ssB + ab\bar{y}_{..}^2)\bar{y}_{..}\right]^2}{(ssA)(ssB)}$$

and

$$ssE = sstot - ssA - ssB - ssAB^*.$$

The analysis of variance table is as in Table 6.10 with $m = 1$ and with $e = (a-1)(b-1)-1$.

Table 6.11 Data for the air velocity experiment, with factors Rib Height (A) and Reynolds Number (B)

		\multicolumn{6}{c}{Reynolds Number, j}	
	i	1	2	3	4	5	6	$\bar{y}_{i.}$
Rib	1	−24	−23	1	8	29	23	2.333
Height	2	33	28	45	57	74	80	52.833
	3	37	79	79	95	101	111	83.667
	$\bar{y}_{.j}$	15.333	28.00	41.667	53.333	68.000	71.333	$46.278 = \bar{y}_{..}$

Source: Willke, D. (1962). Copyright © 1962 Blackwell Publishers. Reprinted with permission.

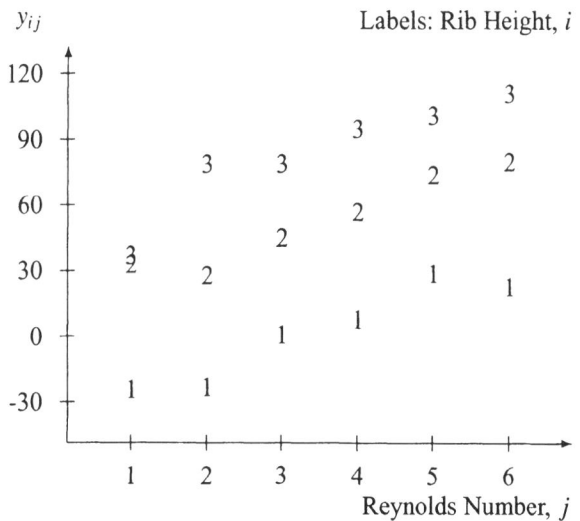

Figure 6.6 Data for the air velocity experiment

6.7.4 A Real Experiment—Air Velocity Experiment

The data given in Table 6.11, and plotted in Figure 6.6, form part of an experiment described by D. Wilkie in the 1962 issue of *Applied Statistics* (volume 11, pages 184–195). The experiment was designed to examine the position of maximum velocity of air blown down the space between a roughened rod and a smooth pipe surrounding it. The treatment factors were the height of ribs on the roughened rod (factor A) at equally spaced heights 0.010, 0.015, and 0.020 inches (coded 1, 2, 3) and Reynolds number (factor B) at six levels (coded 1–6) equally spaced logarithmically over the range 4.8 to 5.3. The responses were measured as $y = (d - 1.4) \times 10^3$, where d is the distance in inches from the center of the rod.

Figure 6.6 shows very little interaction between the factors. However, prior to the experiment, the investigators had thought that the factors would interact to some extent. They wanted to use the set of orthogonal polynomial trend contrasts for the AB interaction and were reasonably sure that the contrasts $A_Q B_{qr}$, $A_L B_{qn}$, $A_Q B_{qn}$ would be negligible. Thus the sum of squares for these three contrasts could be used to estimate σ^2 with 3 degrees of freedom. We are using "L, Q, C, qr, qn" as shorthand notation for linear, quadratic, cubic, quartic, and quintic contrasts, respectively. The coefficients for these three orthogonal polynomial trend contrasts can be obtained by multiplying the corresponding main-effect coefficients shown in Table 6.1, page 145. The coefficients for $A_L B_{qn}$ are shown in the table as an example. Also shown are the contrast coefficients for the linear$A \times$linear$B$ contrast, $A_L B_L$. These are

$$[5, 3, 1, -1, -3, -5, 0, 0, 0, 0, 0, 0, -5, -3, -1, 1, 3, 5].$$

The estimate of $A_L B_L$ is then

$$\Sigma\Sigma d_{ij} y_{ij} = 5(-24) + 3(-23) + \cdots + 3(101) + 5(111) = 54.$$

Now,

$$\Sigma\Sigma d_{ij}^2 = (5^2 + 3^2 + \cdots + 3^2 + 5^2) = 140,$$

so the corresponding sum of squares is

$$ss(A_L B_L) = \frac{54^2}{140} = 20.829.$$

The sums of squares for the other contrasts are computed similarly, and the error sum of squares is calculated as the sum of the sums of squares of the three negligible contrasts. The analysis of variance table is given in Table 6.12.

The hypotheses that the individual contrasts are zero can be tested using Scheffé's procedure or Bonferroni's procedure. If Bonferroni's procedure is used, each of the 14 hypotheses should be tested at a very small α-level. Taking $\alpha = 0.005$, so that the overall level is at most 0.07, we have $F_{1,3,0.005} = 55.6$, and only the linear A and linear B contrasts appear to

Table 6.12 Analysis of variance for the air velocity experiment

Source of Variation	Degrees of Freedom	Sum of Squares	Mean Square	Ratio	p-value
Rib height (A)	2	20232.111			
A_L	1	19845.333	19845.333	338.77	0.0003
A_Q	1	386.778	386.778	6.60	0.0825
Reynolds number (B)	5	7386.944			
B_L	1	7262.976	7262.976	123.98	0.0016
B_Q	1	65.016	65.016	1.11	0.3695
B_C	1	36.296	36.296	0.62	0.4887
B_{qr}	1	13.762	13.762	0.23	0.6611
B_{qn}	1	8.894	8.894	0.15	0.7228
Interaction (AB)	7	616.817			
$A_L B_L$	1	20.829	20.829	0.36	0.5930
$A_L B_Q$	1	47.149	47.149	0.80	0.4358
$A_L B_C$	1	265.225	265.225	4.53	0.1233
$A_L B_{qr}$	1	33.018	33.018	0.56	0.5073
$A_Q B_L$	1	15.238	15.238	0.26	0.6452
$A_Q B_Q$	1	170.335	170.335	2.91	0.1867
$A_Q B_C$	1	65.023	65.023	1.11	0.3694
Error	3	175.739	58.580		
Total	17	28411.611			

6.8 Using SAS Software

Table 6.13 contains a sample SAS program for analysis of the two-way complete model (6.2.2). For illustration, we use the data of the reaction time experiment shown in Table 4.6, page 99, but with the last four observations missing, so that $r_{11} = r_{21} = 2, r_{12} = r_{22} = r_{23} = 3, r_{13} = 1$. In the data input lines, the levels of each of the two treatment factors A and B are shown together with the response and the order in which the observations were collected. A two-digit code for each treatment combination TC can very easily be generated by the statement TC = 10*A + B following the INPUT statement. This way of coding the treatment combinations works well for all applications except for drawing plots with TC on one axis. Such a plot would not show codes 11, 12, and 21 as equally spaced. An alternative way of coding the treatment combinations for generating plots will be given in Section 6.8.2.

The GLM procedure is used to generate the analysis of variance table and to estimate and to test contrasts. As in the one-way analysis of variance, the treatment factors must be declared as class variables using a CLASSES statement. The two-way complete model is represented as

    MODEL Y = A B A*B;

with the main effects listed in either order, but before the interaction. The two-way main-effects model (6.2.3) would be represented as

    MODEL Y = A B;

The program also shows the cell-means model (6.2.1) in a second GLM procedure, using

    MODEL Y = TC;

The output from the first GLM procedure is shown in Table 6.14. The analysis of variance table is organized differently from that in Table 6.4, page 156. The five "model" degrees of freedom are the treatment degrees of freedom corresponding to the six treatment combinations. Information concerning main effects and interactions is provided underneath the table under the heading "Type I" and "Type III" sums of squares.

The *Type III sums of squares* are the values ssA, ssB, and $ssAB$ and are used for hypothesis testing whether or not the sample sizes are equal. They are calculated by comparing the sum of squares for error in the full and reduced models as in Section 6.4.4. The sums of squares listed in the output are always in the same order as the effects in the MODEL statement, but the hypothesis of no interaction should be tested first.

The *Type I sum of squares* for an effect is the additional variation in the data that is explained by adding that effect to a model containing the previously listed sources of variation. For example, in the program output, the Type I sum of squares for A is the reduction in the error sum of squares that is achieved by adding the effect of factor A to a model containing only an intercept term. The reduction in the error sum of squares is equivalent to the extra variation in the data that is explained by adding A to the model. Here, the "full model" contains A and the intercept, while the "reduced model" contains only the intercept. The

Table 6.13 SAS program to illustrate aspects of analysis of a two-way complete model (reaction time experiment)

```
DATA  DATA1;
  INPUT ORDER A B Y;
  *** code the treatment combinations 11, 12, ..., 23;
  TC = 10*A + (B/5);
  LINES;
    1 2 15 0.256
    2 2 15 0.281
    3 1 10 0.167
    : :  :   :
   13 1 15 0.202
   14 2  5 0.279
  ;
PROC PRINT;
PROC GLM;
  CLASSES A B;
  MODEL Y = A B A*B;
     * Use MEANS only if sample sizes are equal;
  MEANS A  B / TUKEY CLDIFF ALPHA=0.01;
     * If sample sizes are unequal, use LSMEANS;
  LSMEANS A B/PDIFF=ALL CL ADJUST=TUKEY ALPHA=0.01;
  CONTRAST '11-13-21+23'  A*B 1  0 -1 -1  0  1;
  CONTRAST 'B1-B2' B 1 -1  0;
  ESTIMATE 'B1-B2' B 1 -1  0;
  ESTIMATE 'B1-B3' B 1  0 -1;
  ESTIMATE 'B2-B3' B 0  1 -1;
PROC GLM;
  CLASSES TC;
  MODEL Y = TC;
  MEANS TC / TUKEY CLDIFF ALPHA=0.01;
  LSMEANS TC/PDIFF=CONTROL CL ADJUST=DUNNETT ALPHA=0.01;
  LSMEANS TC/PDIFF=CONTROLL CL ADJUST=DUNNETT ALPHA=0.01;
  LSMEANS TC/PDIFF=CONTROLU CL ADJUST=DUNNETT ALPHA=0.01;
  CONTRAST '11-13-21+23'  TC  1  0 -1  -1  0  1;
  CONTRAST 'B1-B2'        TC  1 -1  0   1 -1  0;
```

Type I sum of squares for B is the additional variation in the data that is explained by adding the effect of factor B to a model that already contains the intercept and the effect of A (so that the "full model" contains A, B and the intercept, while the "reduced model" contains only the A and the intercept). The Type I sums of squares (also known as *sequential sums of squares*) depend upon the order in which the effects are listed in the MODEL statement. Type I sums of squares are used for model building, not for hypothesis testing under an assumed model. Consequently, we will use only the Type III sums of squares.

The Type I and Type III sums of squares are identical when the sample sizes are equal, since the factorial effects are then estimated independently of one another. But when the

Table 6.14 Some output for the SAS program for a two-way complete model with unequal sample sizes (reaction time experiment)

```
                         The SAS System
                  General Linear Models Procedure

Dependent Variable: Y
                          Sum of        Mean
Source            DF     Squares       Square     F Value   Pr > F
Model              5    0.02153160    0.00430632   13.38    0.0010
Error              8    0.00257533    0.00032192
Corrected Total   13    0.02410693

Source            DF    Type I SS    Mean Square   F Value   Pr > F
A                  1    0.02101572    0.02101572   65.28    0.0001
B                  2    0.00033302    0.00016651    0.52    0.6148
A*B                2    0.00018286    0.00009143    0.28    0.7600

Source            DF    Type III SS  Mean Square   F Value   Pr > F
A                  1    0.01682504    0.01682504   52.27    0.0001
B                  2    0.00045773    0.00022887    0.71    0.5198
A*B                2    0.00018286    0.00009143    0.28    0.7600

Contrast          DF    Contrast SS  Mean Square   F Value   Pr > F
11-13-21+23        1    0.00013886    0.00013886    0.43    0.5298
B1-B2              1    0.00017340    0.00017340    0.54    0.4839

                                 T for H0:     Pr > |T|    Std Error of
Parameter       Estimate        Parameter=0                 Estimate
B1-B2           0.00850000          0.73         0.4839     0.01158153
B1-B3          -0.00600000         -0.44         0.6731     0.01370346
B2-B3          -0.01450000         -1.14         0.2861     0.01268694
```

sample sizes are unequal, as in the illustrated data set, the Type I and Type III sums of squares differ. In the absence of a sophisticated computer package, each Type I and Type III sum of squares can be calculated as the difference of the error sums of squares obtained from two analysis of variance tables, one for the full model and one for the reduced model.

6.8.1 Contrasts and Multiple Comparisons

In the first GLM procedure in Table 6.13, the two-way complete model is used, and the coefficient lists are entered for each factor separately, rather than for the treatment combinations. The first CONTRAST statement is used to test the hypothesis that the interaction contrast $(\alpha\beta)_{11} - (\alpha\beta)_{13} - (\alpha\beta)_{21} + (\alpha\beta)_{23}$ is negligible, and the second CONTRAST statement is used to test the hypothesis that $\beta_1^* - \beta_2^*$ is negligible. These same contrasts are entered as coefficient lists for the treatment combinations in the second GLM procedure. In either case, the contrast sum of squares is as shown under Contrast SS in Table 6.14, and the *p*-value for the test is as shown under Pr > F.

Options such as TUKEY, SCHEFFE, BON, and DUNNETT in the MEANS statement of the first GLM procedure can be used for the generation of all pairwise comparisons, provided that the sample sizes are equal. The statement

```
MEANS A B / TUKEY CLDIFF ALPHA=0.01;
```

causes generation of simultaneous 99% confidence intervals for the main effects of A averaged over the levels of B, and separate simultaneous 99% confidence intervals for the main effects of B averaged over the levels of A, using the method of Tukey for each. The multiple comparisons obtained via a MEANS statement use differences in sample means rather than least squares estimates to calculate the pairwise difference estimates. These coincide only if the sample sizes are equal.

For unequal sample sizes, the LSMEANS option is used instead. The statement requesting simultaneous confidence intervals for pairwise comparisons is a little different from the MEANS option, as follows:

```
LSMEANS B /PDIFF=ALL CL ADJUST=TUKEY ALPHA=0.01;
```

This statement requests Tukey's simultaneous confidence intervals for pairwise comparisons in the levels of B via the ADJUST=TUKEY option. The option PDIFF=ALL requests intervals for *all* of the pairwise comparisons, and CL asks for the comparisons to be displayed as confidence intervals. The output for the reaction time experiment with unequal sample sizes, shown in Table 6.15, includes not only the confidence intervals for pairwise comparisons, but also *p*-values for simultaneous hypothesis tests using the Tukey method. Also given are confidence intervals for the B means $\mu + \overline{\alpha}_. + \beta_j + \overline{(\alpha\beta)}_{.j}$. If CL is omitted, then only the simultaneous tests and intervals for means are printed. The request TUKEY can be replaced by BON or SCHEFFE as appropriate. Notice that for the pairwise comparisons, SAS recodes the levels 5, 10, 15 of B as 1, 2, 3.

For treatment-versus-control comparisons, the option PDIFF=ALL is replaced by the option PDIFF=CONTROL for two-sided confidence intervals, and by the option PDIFF=CONTROLL or PDIFF=CONTROLU for upper or lower confidence bounds on the treatment−control differences, as follows:

```
LSMEANS TC/PDIFF=CONTROL   CL ADJUST=DUNNETT ALPHA=0.01;
LSMEANS TC/PDIFF=CONTROLL  CL ADJUST=DUNNETT ALPHA=0.01;
LSMEANS TC/PDIFF=CONTROLU  CL ADJUST=DUNNETT ALPHA=0.01;
```

The treatments are again renumbered by SAS in numerical order. In our program, in Table 6.13, we have requested the treatment-versus-control contrasts be done for the treatment combinations 11, 12, 13, 21, 22, 23. SAS recodes these as 1–6, and treatment 1 (our treatment combination 11) is taken as the control. The output for the simultaneous confidence intervals is shown in Table 6.16. The first set of intervals is for the two-sided treatment-versus-control comparisons. The second set arises from the CONTROLL option, which gives upper bounds for the "treatment−control" comparisons. The third set arises from the CONTROLU option, which gives lower bounds for the "treatment−control" comparisons. We have shown only

6.8 Using SAS Software

Table 6.15 LSMEANS output for a two-way complete model with unequal sample sizes (reaction time experiment)

```
                        The SAS System
                 General Linear Models Procedure
                      Least Squares Means

      B           Y      Pr > |T| H0: LSMEAN(i)=LSMEAN(j)
               LSMEAN   i/j      1        2        3

      5      0.22750000   1     .       0.7512   0.9010
     10      0.21900000   2    0.7512    .       0.5166
     15      0.23350000   3    0.9010   0.5166    .

                 99%                           99%
                Lower                         Upper
              Confidence                    Confidence
      B         Limit      Y LSMEAN           Limit
      5       0.197399    0.227500          0.257601
     10       0.194422    0.219000          0.243578
     15       0.198742    0.233500          0.268258

        Adjustment for multiple comparisons: Tukey-Kramer
              Least Squares Means for effect B
          99% Confidence Limits for LSMEAN(i)-LSMEAN(j)

              Simultaneous                  Simultaneous
                 Lower       Difference        Upper
              Confidence      Between       Confidence
      i   j      Limit         Means           Limit
      1   2    -0.037650      0.008500       0.054650
      1   3    -0.060605     -0.006000       0.048605
      2   3    -0.065054     -0.014500       0.036054
```

the simultaneous confidence intervals, but simultaneous tests are also given by SAS in each case.

We remind the reader that for unequal sample sizes, it has not yet been proved that the overall confidence levels achieved by the Tukey and Dunnett methods are at least as great as those stated, except in some special cases such as the one-way layout.

An alternative method of obtaining simultaneous confidence intervals for pairwise comparisons can be obtained from the output of the ESTIMATE statement for each contrast. The corresponding confidence intervals are of the form

Estimate $\pm w$ (Std Error of Estimate),

where w is the critical coefficient given in (6.4.19) for the complete model and in (6.5.40) for the main-effects model.

Table 6.16 LSMEANS output for a two-way complete model with unequal sample sizes (reaction time experiment)

```
                       The SAS System
                 General Linear Models Procedure
                     Least Squares Means
             Adjustment for multiple comparisons: Dunnett
                 Least Squares Means for effect TC
             99% Confidence Limits for LSMEAN(i)-LSMEAN(j)

                    Simultaneous              Simultaneous
                      Lower       Difference     Upper
                    Confidence     Between     Confidence
          i    j      Limit         Means        Limit
          2    1    -0.078427     -0.008333     0.061760
          3    1    -0.079040      0.015000     0.109040
          4    1     0.004217      0.081000     0.157783
          5    1     0.002240      0.072333     0.142427
          6    1     0.007907      0.078000     0.148093

             Adjustment for multiple comparisons: Dunnett
                 Least Squares Means for effect TC
             99% Confidence Limits for LSMEAN(i)-LSMEAN(j)

                    Simultaneous              Simultaneous
                      Lower       Difference     Upper
                    Confidence     Between     Confidence
          i    j      Limit         Means        Limit
          2    1        .         -0.008333     0.053393
          3    1        .          0.015000     0.097814
          4    1        .          0.081000     0.148617
          5    1        .          0.072333     0.134059
          6    1        .          0.078000     0.139726

             Adjustment for multiple comparisons: Dunnett
                 Least Squares Means for effect TC
             99% Confidence Limits for LSMEAN(i)-LSMEAN(j)

                    Simultaneous              Simultaneous
                      Lower       Difference     Upper
                    Confidence     Between     Confidence
          i    j      Limit         Means        Limit
          2    1    -0.070059     -0.008333        .
          3    1    -0.067814      0.015000        .
          4    1     0.013383      0.081000        .
          5    1     0.010607      0.072333        .
          6    1     0.016274      0.078000        .
```

6.8 Using SAS Software 181

Table 6.17 Input statements to obtain treatment combination codes as names, not numbers

```
DATA DATA1;
  INPUT ORDER A$ B$ Y;
  TC=trim(A)||trim(B);   *name the trt comb as  11, 12, ... , 23;
  LINES;
    1  2  3  0.256
    :  :  :   :
   14  2  1  0.279
;
PROC GLM;
  CLASSES TC;
  MODEL Y = TC;
  OUTPUT OUT=RESIDS PREDICTED=PDY RESIDUAL=Z;
;
PROC PLOT;
  PLOT Y*TC Z*TC / VPOS=19 HPOS=50;
```

6.8.2 Plots

Residual plots for checking the error assumptions on the model are generated in the same way as shown in Chapter 5. If the two-way main-effects model (6.2.3) is used, the assumption of additivity should also be checked. For this purpose it is useful to plot the standardized residuals against the level of one factor, using the levels of the other factor for plotting labels (see, for example, Figure 6.2, page 141, for the temperature experiment). Once the standardized residuals z have been obtained, a plot of these against the levels of factor A using the labels of factor B can be generated using the following SAS program lines:

```
PROC PLOT;
  PLOT Z*A=B / VPOS=19 HPOS=50;
```

An interaction plot can be obtained by adding the following statements to the end of the program in Table 6.13:

```
PROC SORT DATA=DATA1;
  BY A B;
PROC MEANS DATA=DATA1 NOPRINT MEAN VAR;
  VAR Y;
  BY A B;
  OUTPUT OUT=DATA2 MEAN=AV_Y VAR=VAR_Y;
PROC PRINT;
  VAR A B AV_Y VAR_Y;
PROC PLOT;
  PLOT AV_Y*A=B / VPOS=19 HPOS=50;
```

The PROC PRINT statement following PROC MEANS also gives the information about the variances that would be needed to check the rule of thumb that $s^2_{max}/s^2_{min} \leq 3$.

In order to check for equal error variances, the observations or the residuals may be plotted against the treatment combinations. If the treatment combination codes are formed as TC = 10*A + B as in Table 6.13, they will not be equally spaced along the axis. This is because the codes 11, 12, 21, etc. are regarded by SAS as numerical values, and as 2-digit numbers, 12 is closer to 11 than to 21. One way around this problem is as shown in Table 6.17. Notice that A and B have each been followed by a $ sign in the INPUT statement. This tells SAS to regard the codes as names rather than numbers. The statement TC = trim(A)||trim(B) forms two-digit name codes for the corresponding treatment combinations, and a plot of the residuals against TC will show the codes evenly spaced along the axis.

When there are not sufficient observations to be able to check equality of error variances for all the cells, the standardized residuals should be plotted against the levels of each factor. The rule of thumb may be checked for the levels of each factor by comparing the maximum and minimum variances of the (nonstandardized) residuals. This is done for factor A by augmenting the statements in Table 6.17 with the following lines.

```
PROC SORT DATA=RESIDS;
  BY A;
PROC MEANS DATA=RESIDS NOPRINT VAR;
  VAR Z;
  BY A;
  OUTPUT OUT=DATA2 VAR=VAR_Z;
PROC PRINT;
  VAR A VAR_Z;
```

6.8.3 One Observation per Cell

In order to split the interaction sum of squares into parts corresponding to negligible and nonnegligible orthogonal contrasts, we can enter the data in the usual manner and obtain the sums of squares for all of the contrasts via CONTRAST statements in the procedure PROC GLM. The analysis of variance table can then be constructed with the error sum of squares being the sum of the contrast sums of squares for the negligible contrasts. It is possible, however, to achieve this in a more direct way, as follows.

First, enter the contrast coefficients as part of the input data as shown in Table 6.18 for the air velocity experiment. In the air velocity experiment, factor A had $a = 3$ levels and factor B had $b = 6$ levels. The main-effect trend contrast coefficients are entered via the INPUT statement line by line directly from Table 6.1, page 145, and the interaction trend contrast coefficients are obtained by multiplication following the INPUT statement. In the PROC GLM statement, the CLASSES designation is omitted. If it were included, then Aln, for example, would be interpreted as one factor with three coded levels $-1, 0, 1$, and Aqd as a second factor with two coded levels $1, -2$, and so on. The model is fitted using those contrasts that have not been declared to be negligible. The error sum of squares will be based on the three omitted contrasts AlnBqn, AqdBqr, and AqdBqn, and the resulting analysis of variance table will be equivalent to that in Table 6.12, page 174.

It is not necessary to input the levels of A and B separately as we have done in columns 2 and 3 of the data, but these would be needed if plots of the data were required.

Table 6.18 Fitting a model in terms of contrasts (air velocity experiment)

```
DATA AIR;
  INPUT Y A B Aln Aqd Bln Bqd Bcb Bqr Bqn;
    AlnBln  =  Aln*Bln;
    AlnBqd  =  Aln*Bqd;
    AlnBcb  =  Aln*Bcb;
    AlnBqr  =  Aln*Bqr;
    AqdBln  =  Aqd*Bln;
    AqdBqd  =  Aqd*Bqd;
    AqdBcb  =  Aqd*Bcb;
  LINES;
   -24      1   1     -1    1    -5    5   -5    1   -1
   -23      1   2     -1    1    -3   -1    7   -3    5
     1      1   3     -1    1    -1   -4    4    2  -10
     8      1   4     -1    1     1   -4   -4    2   10
    29      1   5     -1    1     3   -1   -7   -3   -5
    23      1   6     -1    1     5    5    5    1    1
    33      2   1      0   -2    -5    5   -5    1   -1
    28      2   2      0   -2    -3   -1    7   -3    5
    45      2   3      0   -2    -1   -4    4    2  -10
    57      2   4      0   -2     1   -4   -4    2   10
    74      2   5      0   -2     3   -1   -7   -3   -5
    80      2   6      0   -2     5    5    5    1    1
    37      3   1      1    1    -5    5   -5    1   -1
    79      3   2      1    1    -3   -1    7   -3    5
    79      3   3      1    1    -1   -4    4    2  -10
    95      3   4      1    1     1   -4   -4    2   10
   101      3   5      1    1     3   -1   -7   -3   -5
   111      3   6      1    1     5    5    5    1    1
;
PROC PRINT;
PROC GLM;    *  omit the class statement;
  MODEL Y = Aln Aqd Bln Bqd Bcb Bqr Bqn AlnBln AlnBqd
                  AlnBcb AlnBqr AqdBln AqdBqd AqdBcb;
```

Exercises

1. Under what circumstances should the two-way main effects model (6.2.3) be used rather than the two-way complete model (6.2.2)? Discuss the interpretation of main effects in each model.

2. Verify that $(\tau_{ij} - \bar{\tau}_{i.} - \bar{\tau}_{.j} + \bar{\tau}_{..})$ is an interaction contrast for the two-way complete model. Write down the list of contrast coefficients in terms of the τ_{ij}'s when factor A has $a = 3$ levels and factor B has $b = 4$ levels.

3. Consider the functions $\{\alpha_1^* - \alpha_2^*\}$ and $\{(\alpha\beta)_{11} - (\alpha\beta)_{21} - (\alpha\beta)_{12} + (\alpha\beta)_{22}\}$ under the two-way complete model (6.2.2).

(a) Verify that the functions are estimable contrasts.

(b) Discuss the meaning of each of these contrasts for plots (d) and (g) of Figure 6.1 (page 137).

(c) If $a = b = 3$, give the list of contrast coefficients for each contrast, first for the parameters involved in the contrast, and then in terms of the parameters τ_{ij} of the equivalent cell-means model.

4. Show that when the parentheses is expanded in formula (6.4.15) for ssE on page 147, the computational formula (6.4.17) is obtained.

5. **Weight Lifting Experiment** (Gary Mirka, 1986)

The experimenter was interested in the effect on pulse rate (heart rate) of lifting different weights with legs either straight or bent (factor A, coded 1, 2). The selected weights were 50 lb, 75 lb, 100 lb (factor B, coded 1, 2, 3). He expected to see a higher pulse rate when heavier weights were lifted. He also expected that lifting with legs bent would result in a higher pulse rate than lifting with legs straight.

(a) Write out a detailed checklist for running an experiment similar to this. In the calculation of the number of observations needed, either run your own pilot experiment or use the information that for a single subject in the above study, the error sum of squares was $ssE = 130.909$ bpfs^2 based on $df=60$ error degrees of freedom (where bpfs is beats per 15 seconds).

(b) The data collected for a single subject by the above experimenter are shown in Table 6.19 in the order collected. The experimenter wanted to use a two-way complete model. Check the assumptions on this model, paying particular attention to the facts that (i) these are count data and may not be approximately normally

Table 6.19 Data (beats per 15 seconds) for the weightlifting experiment

A	2	1	1	1	2	2	1	2	1	2	1
B	2	1	3	1	2	3	3	2	2	2	1
Rate	31	27	37	28	32	32	35	30	32	31	27
A	2	1	1	1	2	1	1	2	2	2	2
B	2	3	3	2	1	1	3	1	3	3	3
Rate	34	33	34	31	26	25	35	24	33	31	36
A	1	1	1	1	1	1	1	1	2	1	2
B	3	1	1	2	1	2	2	3	3	2	1
Rate	36	27	30	33	29	32	34	37	32	34	27
A	2	1	1	2	2	1	1	2	2	1	2
B	2	1	3	1	2	1	2	1	2	1	3
Rate	31	27	38	27	30	29	34	25	34	28	34
A	2	1	2	1	1	2	1	1	2	2	2
B	2	1	3	2	2	1	3	2	1	1	1
Rate	31	30	34	35	34	24	35	31	27	26	25
A	2	2	2	1	2	2	2	2	2	1	1
B	2	3	1	2	1	2	3	3	3	3	3
Rate	32	35	24	33	23	30	34	32	33	37	38

distributed, and (ii) the measurements were made in groups of ten at a time in order to reduce the fatigue of the subject.

(c) Taking account of your answer to part (a), analyze the experiment, especially noting any trends in the response.

6. **Battery experiment, continued**

   Consider the battery experiment introduced in Section 2.5.2, for which $a = b = 2$ and $r = 4$. Suppose it is of interest to calculate confidence intervals for the four simple effects $\tau_{11} - \tau_{12}$, $\tau_{21} - \tau_{22}$, $\tau_{11} - \tau_{21}$, $\tau_{12} - \tau_{22}$, with an overall confidence level of 95%.

   (a) Determine whether the Tukey or Bonferroni method of multiple comparisons would provide shorter confidence intervals.

   (b) Apply the better of the methods from part (a) and comment on the results. (The data give $\bar{y}_{11.} = 570.75$, $\bar{y}_{12.} = 860.50$, $\bar{y}_{21.} = 433.00$, and $\bar{y}_{22.} = 496.25$ minutes per unit cost and $msE = 2,367.71$.)

   (c) Discuss the practical meaning of the contrasts estimated in (b) and explain what you have learned from the confidence intervals.

7. **Weld strength experiment**

   The data shown in Table 6.20 are a subset of the data given by Anderson and McLean (1974) and show the strength of a weld in a steel bar. Two factors of interest were gage bar setting (the distance the weld die travels during the automatic weld cycle) and time of welding (total time of the automatic weld cycle). Assume that the levels of both factors were selected to be equally spaced.

   (a) Using the cell-means model (6.2.1) for these data, test the hypothesis that there is no difference in the effects of the treatment combinations on weld strength against the alternative hypothesis that at least two treatment combinations have different effects.

   (b) Suppose the experimenters had planned to calculate confidence intervals for all pairwise comparisons between the treatment combinations, and also to look at the confidence interval for the difference between gage bar setting 3 and the average of the other two. Show what these contrasts look like in terms of the parameters τ_{ij} of the cell-means model, and suggest a strategy for calculating all intervals at overall level "at least 98%."

Table 6.20 Strength of weld

		\multicolumn{5}{c}{Time of welding (j)}				
	i	1	2	3	4	5
Gage	1	10, 12	13, 17	21, 30	18, 16	17, 21
bar	2	15, 19	14, 12	30, 38	15, 11	14, 12
setting	3	10, 8	12, 9	10, 5	14, 15	19, 11

Source: Reprinted from Anderson, V. L. and McLean, R. A. (1974), pp. 62–63, by courtesy of Marcel Dekker, Inc.

(c) Give formulae for the intervals in part (b). As an example, calculate the actual interval for $\tau_{13} - \tau_{15}$ (the difference in the true mean strengths at the 3rd and 5th times of welding for the first gage bar setting). Explain what this interval tells you.

(d) Calculate an upper 90% confidence limit for σ^2.

(e) If the experimenters were to repeat this experiment and needed the pairwise comparison intervals in (b) to be of width at most 8, how many observations should they take on each treatment combination? How many observations is this in total?

8. **Weld strength experiment, continued**

For the experiment described in Exercise 7, use the two-way complete model instead of the equivalent cell means model.

(a) Test the hypothesis of no interaction between gage bar setting and time of weld and state your conclusion.

(b) Draw an interaction plot for the two factors Gage bar setting and Time of welding. Does your interaction plot support the conclusion of your hypothesis test? Explain.

(c) In view of your answer to part (b), is it sensible to investigate the differences between the effects of gage bar setting? Why or why not? Indicate on your plot what would be compared.

(d) Regardless of your answer to (c), suppose the experimenters had decided to look at the linear trend in the effect of gage bar settings. Test the hypothesis that the linear trend in gage setting is negligible (against the alternative hypothesis that it is not negligible).

9. **Sample size calculation**

An experiment is to be run to examine three levels of factor A and four levels of factor B, using the two-way complete model (6.2.2). Determine the required sample size if the error variance σ^2 is expected to be less than 15 and simultaneous 99% confidence intervals for pairwise comparisons between treatment combinations should have length at most 10 to be useful.

10. **Bleach experiment, continued**

Use the data of the bleach experiment of Example 6.4.3, on page 151.

(a) Evaluate the effectiveness of a variance-equalizing transformation.

(b) Apply Satterthwaite's approximation to obtain 99% confidence intervals for the pairwise comparisons of the main effects of factor A using Tukey's method of multiple comparisons.

11. **Bleach experiment, continued**

The experimenter calculated that she needed $r = 5$ observations per treatment combination in order to be able to detect a difference in the treatment combinations of 5 minutes (300 seconds) with probability 0.9. Verify that her calculations were correct. She obtained a mean squared error of 43220.8 seconds^2 in her pilot experiment.

12. **Memory experiment** (James Bost, 1987)

    The memory experiment was planned in order to examine the effects of external distractions on short-term memory and also to examine whether some types of words were easier to memorize than others. Consequently, the experiment involved two treatment factors, "word type" and "type of distraction." The experimenter selected three levels for each factor. The levels of "word type" were

    Level 1 (fruit): words representing fruits and vegetables commonly consumed;

    Level 2 (nouns): words selected at random from Webster's pocket dictionary, representing tangible (i.e., visualizable) items;

    Level 3 (mixed): words of any description selected at random from Webster's pocket dictionary.

    A list of 30 words was prepared for each level of the treatment factor, and the list was not altered throughout the experiment.

    The levels of "type of distraction" were

    Level 1: No distraction other than usual background noise;

    Level 2: Constant distraction, supplied by a regular banging of a metal spoon on a metal pan;

    Level 3: Changing distraction, which included vocal, music, banging and motor noise, and varying lighting.

    The response variable was the number of words remembered (by a randomly selected subject) for a given treatment combination. The response variable is likely to have approximately a binomial distribution, with variance $30p(1-p)$ where p is the probability that a subject remembers a given word and 30 is the number of words on the list. It is unlikely that p is constant for all treatment combinations or for all subjects. However, since $np(1-p)$ is less than $30(0.5)(0.5) = 7.5$, a reasonable guess for the variance σ^2 is that it is less than 7.5.

    The experimenter wanted to reject each of the main-effect hypotheses H_0^A:{the memorization rate for the three word lists is the same} and H_0^B:{the three types of distraction have the same effect on memorization} with probability 0.9 if there was a difference of four words in memorization rates between any two word lists or any two distractions (that is $\Delta_A = \Delta_B = 4$), using a significance level of $\alpha = 0.05$. Calculate the number of subjects that are needed if each subject is to be assigned to just one treatment combination and measured just once.

13. **Memory experiment, continued**

    (a) Write out a checklist for the memory experiment of Exercise 12. Discuss how you would obtain the subjects and how applicable the experiment would be to the general population.

    (b) Consider the possibility of using each subject more than once (i.e., consider the use of a blocking factor). Discuss whether or not an assumption of independent observations is likely to be valid.

Table 6.21 Data and standardized residuals for the memory experiment

Word list	Distraction								
	None (1)			Constant (2)			Changing (3)		
Fruit (1)	20	14	24	15	22	17	17	13	12
	0.27	−2.16	1.89	−1.21	1.62	−0.40	1.21	−0.40	−0.81
Nouns (2)	19	14	19	12	11	14	12	15	8
	0.67	−1.35	0.67	−0.13	−0.54	0.67	0.13	1.35	−1.48
Mixed (3)	11	12	15	8	8	9	12	7	10
	−0.67	−0.27	0.94	−0.13	−0.13	0.27	0.94	−1.08	0.13

14. **Memory experiment, continued**

    The data for the memory experiment of Exercise 12 are shown in Table 6.21, with three observations per treatment combination.

    (a) The experimenter intended to use the two-way complete model. Check the assumptions on the model for the given data, especially the equal-variance assumption.

    (b) Analyze the experiment. A transformation of the data or use of the Satterthwaite approximation may be necessary.

15. **Ink experiment**

    (M. Chambers, Y.-W. Chen, E. Kurali, R. Vengurlekar, 1996)

    Teaching associates who give classes in computer labs at the Ohio State University are required to write on white boards with "dry markers" rather than on chalk boards with chalk. The ink from these dry markers can stain rather badly, and four graduate students planned an experiment to determine which type of cloth (factor A, 1= cotton/polyester, 2=rayon, 3=polyester) was most difficult to clean, and whether a detergent plus stain remover was better than a detergent without stain remover (factor B, levels 1, 2) for washing out such a stain.

    Pieces of cloth were to be stained with 0.1 ml of dry marker ink and allowed to air dry for 24 hours. The cloth pieces were then to be washed in a random order in the detergent to which they were allocated. The stain remaining on a piece of cloth after washing and drying was to be compared with a 19 point scale and scored accordingly, where 1=black and 19=white.

    (a) Make a list of the difficulties that might be encountered in running and analyzing an experiment of this type. Give suggestions on how these difficulties might be overcome or their effects reduced.

    (b) Why should each piece of cloth be washed separately? (Hint: think about the error variability.)

Table 6.22 Data for the ink experiment in the order of collection

Cloth Type	3	1	3	1	2	1	2	2	2	3	3	1
Stain Remover	2	2	2	2	1	1	1	2	2	1	1	1
Stain Score	1	6	1	5	11	9	9	8	6	3	4	8

Table 6.23 Data for the survival experiment (units of 10 hours)

Poison	Treatment			
	1	2	3	4
I	0.31	0.82	0.43	0.45
	0.45	1.10	0.45	0.71
	0.46	0.88	0.63	0.66
	0.43	0.72	0.76	0.62
II	0.36	0.92	0.44	0.56
	0.29	0.61	0.35	1.02
	0.40	0.49	0.31	0.71
	0.23	1.24	0.40	0.38
III	0.22	0.30	0.23	0.30
	0.21	0.37	0.25	0.36
	0.18	0.38	0.24	0.31
	0.23	0.29	0.22	0.33

Source: Box, G. E. P. and Cox, D. R. (1964). Copyright 1964 Blackwell Publishers. Reprinted with permission.

(c) The results of a small pilot study run by the four graduate students are shown in Table 6.22. Plot the data against the levels of the two treatment factors. Can you learn anything from this plot? Which model would you select for the main experiment? Why?

(d) Calculate the number of observations that you would need to take on each treatment combination in order to try to ensure that the lengths of confidence intervals for pairwise differences in the effects of the levels of each of the factors were no more than 2 points (on the 19-point scale).

16. **Survival experiment** (G.E.P. Box and D.R. Cox)
The data in Table 6.23 show survival times of animals to whom a poison and a treatment have been administered. The data were presented by G. E. P. Box and D. R. Cox in an article in the *Journal of the Royal Statistical Society* in 1964. There were three poisons (factor A at $a = 3$ levels), four treatments (factor B at $b = 4$ levels), and $r = 4$ animals (experimental units) assigned at random to each treatment combination.

(a) Check the assumptions on a two-way complete model for these data. If the assumptions are satisfied, then analyze the data and discuss your conclusions.

(b) Take a reciprocal transformation (y^{-1}) of the data. The transformed data values then represent "rates of dying." Check the assumptions on the model again. If the assumptions are satisfied, then analyze the data and discuss your conclusions.

(c) Draw an interaction plot for both the original and the transformed data. Discuss the interaction between the two factors in each of the measurement scales.

17. Use the two-way main-effects model (6.2.3) with $a = b = 3$.

(a) Which of the following are estimable?

(i) $\mu + \alpha_1 + \beta_2$.

(ii) $\mu + \alpha_1 + \frac{1}{2}(\beta_1 + \beta_2)$.

(iii) $\beta_1 - \frac{1}{3}(\beta_2 + \beta_3)$.

(b) Show that $\overline{Y}_{i..} + \overline{Y}_{.j.} - \overline{Y}_{...}$ is an unbiased estimator of $\mu + \alpha_i + \beta_j$ with variance, $\sigma^2(a+b-1)/(abr)$.

(c) Show that $\sum_i c_i \overline{Y}_{i..}$ is an unbiased estimator of the contrast $\sum_i c_i \alpha_i$.

18. **Nail varnish experiment, continued**

    The nail varnish experiment, introduced in Example 6.5.1, page 159, concerned the dissolving time of three brands of nail varnish in two brands of solvent. There were $r = 5$ observations on each treatment combination, and the two-way main-effects model (6.2.3) was used. The least squares estimate of the contrast comparing brand 3 with the average of the other two brands was $\frac{1}{2}(\widehat{\beta}_1 + \widehat{\beta}_2) - \widehat{\beta}_3 = -0.341$, with associated variance $3\sigma^2/20$. From Example 6.5.2, $msE = 8.3375$.

    (a) Verify that the assumptions on the two-way main effects model hold for this experiment.

    (b) Verify that $\frac{1}{2}(\overline{Y}_{.1.} + \overline{Y}_{.2.}) - \overline{Y}_{.3.}$ is an unbiased estimator of the difference of averages contrast $\frac{1}{2}(\beta_1 + \beta_2) - \beta_3$.

    (c) Obtain a 95% confidence interval for $\frac{1}{2}(\beta_1 + \beta_2) - \beta_3$, as though this were preplanned and the only contrast of interest.

19. For the two-way main-effects model (6.2.3) with equal sample sizes,

    (a) verify the computational formulae for ssE given in (6.5.39),

    (b) and, if SSE is the corresponding random variable, show that $E[SSE]$ is $(n - a - b + 1)\sigma^2$. [Hint: $E[X^2] = \text{Var}(X) + E[X]^2$.]

20. An experiment is to be run to compare the two levels of factor A and to examine the pairwise differences between the four levels of factor B, with a simultaneous confidence level of 90%. The experimenter is confident that the two factors will not interact. Find the required sample size if the error variance will be at most 25 and the confidence intervals should have length at most 10 to be useful.

21. **Water boiling experiment** (Kate Ellis, 1986)

    The experiment was run in order to examine the amount of time taken to boil a given amount of water on the four different burners of her stove, and with 0, 2, 4, or 6 teaspoons of salt added to the water. Thus the experiment had two treatment factors with four levels each. The experimenter ran the experiment as a completely randomized design by taking $r = 3$ observations on each of the 16 treatment combinations in a random order. The data are shown in Table 6.24. The experimenter believed that there would be no interaction between the two factors.

    (a) Check the assumptions on the two-way main-effects model.

    (b) Calculate a 99% set of Tukey confidence intervals for pairwise differences between the levels of salt, and calculate separately a 99% set of intervals for pairwise differences between the levels of burner.

Table 6.24 Data for the water boiling experiment, in minutes.

	Salt (teaspoons)			
Burner	0	2	4	6
Right back	7(7)	4(13)	7(24)	5(15)
	8(21)	7(25)	7(34)	7(33)
	7(30)	7(26)	7(41)	7(37)
Right front	4(6)	4(36)	4(1)	4(28)
	4(20)	5(44)	4(14)	4(31)
	4(27)	4(45)	5(18)	4(38)
Left back	6(9)	6(46)	7(8)	5(35)
	7(16)	6(47)	6(12)	6(39)
	6(22)	5(48)	7(43)	6(40)
Left front	9(29)	8(5)	8(3)	8(2)
	9(32)	8(10)	9(19)	8(4)
	9(42)	8(11)	10(23)	7(17)

(c) Test a hypothesis that there is no linear trend in the time to boil water due to the level of salt. Do a similar test for a quadratic trend.

(d) The experimenter believed that observation number 13 was an outlier, since it has a large standardized residual and it was an observation taken late on a Friday evening. Repeat the analysis in (b) and (c) removing this observation. Which analysis do you prefer? Why?

22. For $v = 5$ and $r = 4$, show that the first three "orthogonal polynomial contrasts" listed in Table A.2 are mutually orthogonal. (In fact all four are.) Find a pair of orthogonal contrasts that are not orthogonal polynomial contrasts. Can you find a third contrast that is orthogonal to each of these? How about a fourth? (This gets progressively harder!)

23. **Air velocity experiment, continued**
   (a) For the air velocity experiment introduced in Section 6.7.4 (page 173), calculate the sum of squares for each of the three interaction contrasts assumed to be negligible, and verify that these add to the value $ssE = 175.739$, as in Table 6.12.
   (b) Check the assumptions on the model by plotting the standardized residuals against the predicted responses, the treatment factor levels, and the normal scores. State your conclusions.

7 Several Crossed Treatment Factors

7.1 Introduction
7.2 Models and Factorial Effects
7.3 Analysis—Equal Sample Sizes
7.4 A Real Experiment—Popcorn–Microwave Experiment
7.5 One Observation per Cell
7.6 Design for the Control of Noise Variability
7.7 Using SAS Software
Exercises

7.1 Introduction

Experiments that involve more than two treatment factors are designed and analyzed using many of the same principles that were discussed in Chapter 6 for two-factor experiments. We continue to label the factors with uppercase Latin letters and their numbers of levels with the corresponding lowercase letters. An experiment that involves four factors, A, B, C, and D, having a, b, c, and d levels, respectively, for example, is known as an "$a \times b \times c \times d$ factorial experiment" (read "a by b by c by d") and has a total of $v = abcd$ treatment combinations.

There are several different models that may be appropriate for analyzing a factorial experiment with several treatment factors, depending on which interactions are believed to be negligible. These models, together with definitions of interaction between three or more factors, and estimation of contrasts, form the topic of Section 7.2. General rules are given in Section 7.3 for writing down confidence intervals and hypothesis tests when there are equal numbers of observations on all treatment combinations. In Section 7.5, methods are investigated for analyzing small experiments where there is only one observation per treatment combination.

The difference between a "design factor" and a "noise factor" is highlighted in Section 7.6, and the concept of selecting design factor levels that are least affected by the settings of noise factors is introduced. Finally, SAS commands for analyzing experiments with several treatment factors are given in Section 7.7 and can be used for unequal sample sizes. Problems caused by empty cells are also investigated.

7.2 Models and Factorial Effects

7.2.1 Models

One of a number of different models may be appropriate for describing the data from an experiment with several treatment factors. The selection of a suitable model prior to the experiment depends upon available knowledge about which factors do and do not interact. We take as an example an experiment with three factors. Our first option is to use the *cell-means model*, which is similar to the one-way analysis of variance model (3.3.1), page 36. For example, the cell-means model for three treatment factors is

$$Y_{ijkt} = \mu + \tau_{ijk} + \epsilon_{ijkt}, \qquad (7.2.1)$$

$$\epsilon_{ijkt} \sim N(0, \sigma^2),$$

ϵ_{ijkt}'s mutually independent,

$t = 1, \ldots, r_{ijk}; \ i = 1, \ldots, a; \ j = 1, \ldots, b; \ k = 1, \ldots, c.$

If there are more than three factors, the cell-means model has correspondingly more subscripts. As in Chapter 6, use of this model allows all of the formulae presented in Chapters 3 and 4 for one treatment factor to be used to compare the effects of the treatment combinations.

Alternatively, we can model the effect on the response of treatment combination ijk to be

$$\tau_{ijk} = \alpha_i + \beta_j + \gamma_k + (\alpha\beta)_{ij} + (\alpha\gamma)_{ik} + (\beta\gamma)_{jk} + (\alpha\beta\gamma)_{ijk},$$

where $\alpha_i, \beta_j, \gamma_k$ are the effects (positive or negative) on the response of factors A, B, C at levels i, j, k, respectively, $(\alpha\beta)_{ij}, (\alpha\gamma)_{ik}$, and $(\beta\gamma)_{jk}$ are the additional effects of the pairs of factors together at the specified levels, and $(\alpha\beta\gamma)_{ijk}$ is the additional effect of all three factors together at levels i, j, k. The three sets of factorial effects are called the main-effect parameters, the two-factor interaction parameters, and the three-factor interaction parameter, respectively. The interpretation of a three-factor interaction is discussed in the next section. If we replace τ_{ijk} in model (7.2.1) by the main-effect and interaction parameters, we obtain the equivalent *three-way complete model*; that is,

$$Y_{ijkt} = \mu + \alpha_i + \beta_j + \gamma_k + (\alpha\beta)_{ij} + (\alpha\gamma)_{ik} + (\beta\gamma)_{jk} + (\alpha\beta\gamma)_{ijk} + \epsilon_{ijkt},$$

$$\epsilon_{ijkt} \sim N(0, \sigma^2), \qquad (7.2.2)$$

ϵ_{ijkt}'s mutually independent,

$t = 1, \ldots, r_{ijk}; \ i = 1, \ldots, a; \ j = 1, \ldots, b; \ k = 1, \ldots, c.$

This form of the model extends in an obvious way to more than three factors by including a main-effect parameter for every factor and an interaction effect parameter for every combination of two factors, three factors, etc.

If prior to the experiment certain interaction effects are known to be negligible, the corresponding parameters can be removed from the complete model to give a submodel. For example, if the factors A and B are known not to interact in a three-factor experiment, then the AB and ABC interaction effects are negligible, so the terms $(\alpha\beta)_{ij}$ and $(\alpha\beta\gamma)_{ijk}$ are excluded from model (7.2.2). In the extreme case, if no factors are expected to interact, then a *main-effects model* (which includes no interaction terms) can be used.

When a model includes an interaction between a specific set of m factors, then all interaction terms involving subsets of those m factors should be included in the model. For example, a model that includes the effect of the three-factor interaction ABC would also include the effects of the AB, AC, and BC interactions as well as the main effects A, B, and C.

Use of a submodel, when appropriate, is advantageous, because simpler models generally yield tighter confidence intervals and more powerful tests of hypotheses. However, if interaction terms are removed from the model when the factors do, in fact, interact, then the resulting analysis and conclusions may be totally incorrect.

7.2.2 The Meaning of Interaction

The same type of interaction plot as that used in Section 6.2.1, page 136, can be used to evaluate interactions between pairs of factors in an experiment involving three or more factors. The graphical evaluation of three-factor interactions can be done by comparing separate interaction plots at the different levels of a third factor. Such plots will be illustrated for experiments that involve only three factors, but the methods are similar for experiments with four or more factors, except that the sample means being plotted would be averages over the levels of all the other factors.

The following sample means are for a hypothetical $3 \times 2 \times 2$ experiment involving the factors A, B, and C at 3, 2, and 2 levels, respectively.

ijk :	111	112	121	122	211	212	221	222	311	312	321	322
$\overline{y}_{ijk.}$:	3.0	4.0	1.5	2.5	2.5	3.5	3.0	4.0	3.0	4.0	1.5	2.5

An AB-interaction plot for these hypothetical data is shown in Figure 7.1. As in the previous chapter, we must remember that interaction plots give no indication of the size of the experimental error and must be interpreted with a little caution. The lines of the plot in Figure 7.1 are not parallel, indicating that the factors possibly interact. For factor A, level 2 appears to be the best on average, but not consistently the best at each level of B. Likewise, level 1 of factor B appears to be better on average, but not consistently better at each level of A. The perceived AB interaction is averaged over the levels of C and may have no practical meaning if there is an ABC interaction. Consequently, the three-factor interaction should be investigated first.

A three-factor interaction would be indicated if the interaction effect between any pair of factors were to change as the level of the third factor changes. In Figure 7.2, a separate

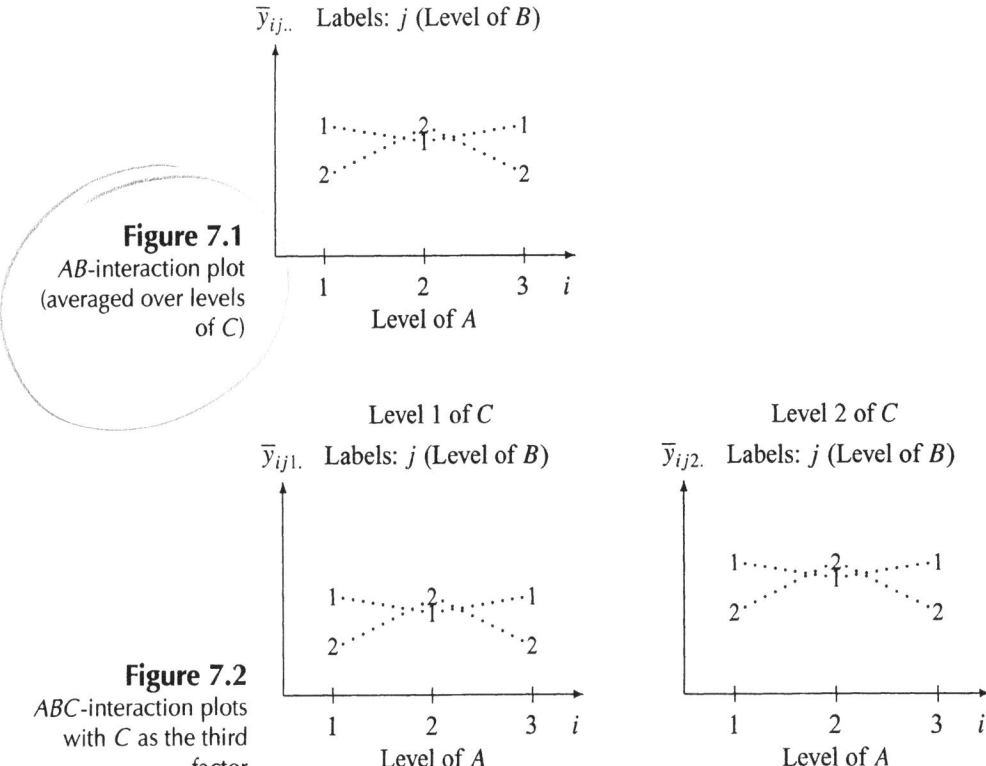

Figure 7.1
AB-interaction plot (averaged over levels of C)

Figure 7.2
ABC-interaction plots with C as the third factor

AB-interaction plot is shown for each level of factor C. Each of the two plots suggests the presence of an AB-interaction effect, but the patterns in the two plots are the same. In other words, the factors A and B apparently interact in the same way at each level of factor C. This indicates a negligible ABC-interaction effect. The shift in the interaction plot as the level of C changes from one plot to the other indicates a possible main effect of factor C. The AB interaction plot in Figure 7.1 is the average of the two plots in Figure 7.2, showing the AB interaction averaged over the two levels of C.

Other three-factor interaction plots can be obtained by interchanging the roles of the factors. For example, Figure 7.3 contains plots of $\bar{y}_{ijk.}$ against the levels i of A for each level j of factor B, using the levels k of C as labels and the same hypothetical data. Lines are parallel in each plot, indicating no AC-interaction at either level of B. Although the patterns differ from plot to plot, if there is no AC-interaction at either of the levels of B, there is no change in the AC-interaction from one level of B to the other. So, again the ABC-interaction effect appears to be negligible. An AC interaction plot would show the average of the two plots in Figure 7.3, and although the plot would again look different, the lines would still be parallel.

To see what the plots might look like when there is an ABC-interaction present, we look at the following second set of hypothetical data.

ijk :	111	112	121	122	211	212	221	222	311	312	321	322
$\bar{y}_{ijk.}$:	3.0	2.0	1.5	4.0	2.5	3.5	3.0	4.0	3.0	5.0	3.5	6.0

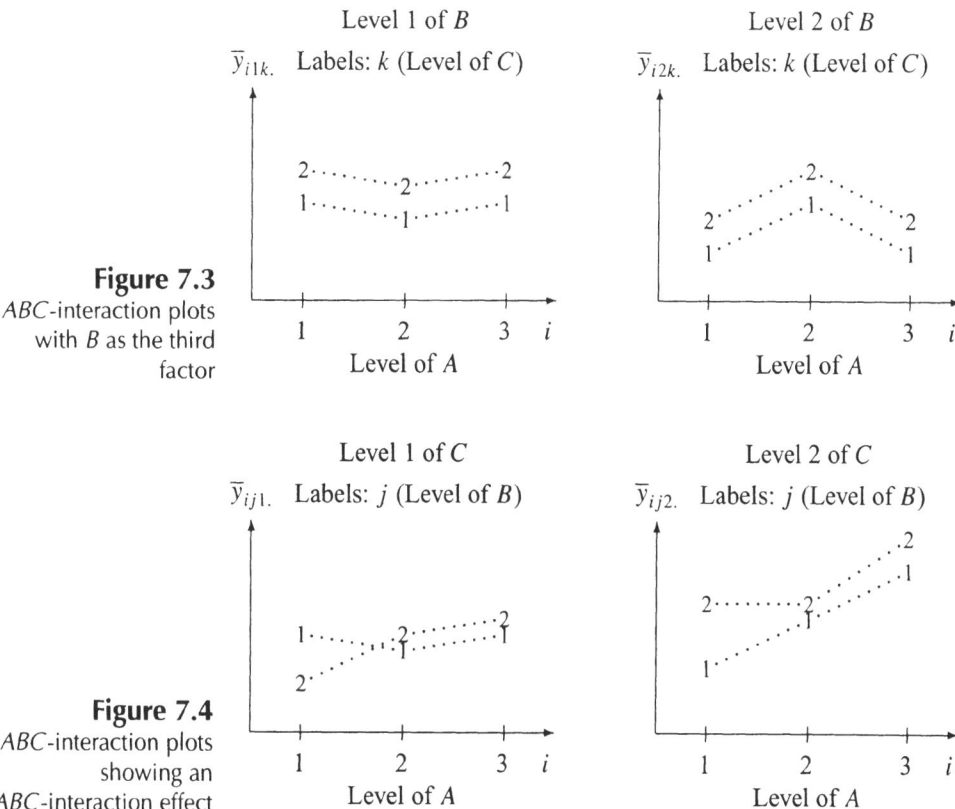

Figure 7.3
ABC-interaction plots with B as the third factor

Figure 7.4
ABC-interaction plots showing an ABC-interaction effect

Figure 7.4 shows plots of $\bar{y}_{ijk.}$ against the level i of factor A for each level k of factor C, using the level j of factor B as the plotting label. In each plot, corresponding lines are not all parallel, *and* the pattern changes from one plot to the next. In other words, the interaction effect between factors A and B apparently changes with the level of C, so there appears to be an ABC-interaction effect.

Four-factor interactions can be evaluated graphically by comparing the pairs of plots representing a three-factor interaction for the different levels of a fourth factor. Clearly, higher-order interactions are harder to envisage than those of lower order, and we would usually rely solely on the analysis of variance table for evaluating the higher-order interactions. In general, one should examine the higher-order interactions first, and work downwards. In many experiments, high-order interactions do tend to be small, and lower-order interactions can then be interpreted more easily.

7.2.3 Separability of Factorial Effects

In an experiment involving three factors, A, B, and C, for which it is known in advance that factor C will not interact with factor A or B, the AC, BC, and ABC interaction effects can be excluded from the model. Interpretation of the results of the experiment is simplified, because a specific change in the level of factor C has the same effect on the mean response

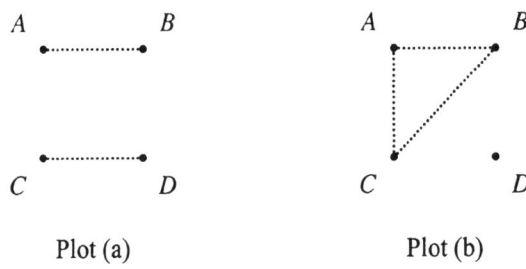

Figure 7.5
Separability plots

Plot (a)   Plot (b)

for every combination ij of levels of A and B. Likewise, a specific change in the combination of levels of A and B has the same effect on the mean response regardless of the level of C. If the objective is to find the best treatment combination ijk, then the task is reduced to two smaller problems involving fewer comparisons, namely choice of the best combination ij of levels of A and B and, separately, choice of the best level k of C.

When there is such separability of effects, the experimenter should generally avoid the temptation to run separate experiments, one to determine the best combination ij of levels of factors A and B and another to determine the best level k of C. A single factorial experiment involving n observations provides the same amount of information on the A, B, C, and AB effects as would two separate experiments—a factorial experiment for factors A and B and another experiment for factor C—*each* involving n observations!

One way to determine an appropriate model for an experiment is as follows. Suppose that the experiment involves p factors. Draw p points, labeling one point for each factor (see, for example, Figure 7.5 for four factors A–D). Connect each pair of factors that might conceivably interact with a line to give a *line graph*. For every pair of factors that are joined by a line in the line graph, a two-factor interaction should be included in the model. If three factors are joined by a triangle, then it may be appropriate to include the corresponding three-factor interaction in the model as well as the three two-factor interactions. Similarly, if four factors are joined by all six possible lines, it may be appropriate to include the corresponding four-factor interaction as well as the three-factor and two-factor interactions.

The line graphs in Figure 7.5 fall into two pieces. Line graph (a) represents the situation where A and B are thought to interact, as are C and D. The model would include the AB and CD interaction effects, in addition to all main effects. Line graph (b) represents an experiment in which it is believed that A and B interact and also A and C and also B and C. An appropriate model would include all main effects and the AC, AB, and BC interactions. The three-factor ABC interaction effect might also be included in the model depending upon the type of interaction plots expected by the experimenter. Thus, a possible model would be

$$Y_{ijklt} = \mu + \alpha_i + \beta_j + \gamma_k + \delta_l + (\alpha\beta)_{ij}$$
$$+ (\alpha\gamma)_{ik} + (\beta\gamma)_{jk} + (\alpha\beta\gamma)_{ijk} + \epsilon_{ijklt}.$$

7.2.4 Estimation of Factorial Contrasts

For an $a \times b \times c$ factorial experiment and the three-way complete model, all treatment contrasts are of the form

$$\sum_i \sum_j \sum_k h_{ijk} \tau_{ijk} \quad \text{with} \quad \sum_i \sum_j \sum_k h_{ijk} = 0,$$

and are estimable when there is at least one observation per treatment combination.

A contrast in the main effect of A is any treatment contrast for which the coefficients h_{ijk} depend only on the level i of A. For example, if we set h_{ijk} equal to $p_i/(bc)$, with $\Sigma p_i = 0$, then the contrast $\Sigma\Sigma\Sigma h_{ijk}\tau_{ijk}$ becomes

$$\sum_i p_i \bar{\tau}_{i..} = \sum_i p_i [\alpha_i + \overline{(\alpha\beta)}_{i.} + \overline{(\alpha\gamma)}_{i.} + \overline{(\alpha\beta\gamma)}_{i..}] = \sum_i p_i \alpha_i^*.$$

We notice that a main-effect contrast for factor A can be interpreted only as an average over all of the interaction effects involving A in the model and, consequently, may not be of interest if any of these interactions are nonnegligible. The B and C main-effect contrasts are defined in similar ways.

An *AB interaction contrast* is any treatment contrast for which the coefficients h_{ijk} depend only on the combination ij of levels of A and B, say $h_{ijk} = d_{ij}/c$, and for which $\sum_i d_{ij} = 0$ for all j and $\sum_j d_{ij} = 0$ for all i. An AB interaction contrast can be expressed as

$$\sum_i \sum_j d_{ij} \bar{\tau}_{ij.} = \sum_i \sum_j d_{ij}[(\alpha\beta)_{ij} + \overline{(\alpha\beta\gamma)}_{ij.}] = \sum_i \sum_j d_{ij}(\alpha\beta)_{ij}^*.$$

Thus, the AB interaction contrast can be interpreted only as an average over the ABC interaction and may not be of interest if the ABC interaction is nonnegligible. The AC and BC interaction contrasts are defined in similar ways.

An ABC *interaction contrast* is any contrast of the form

$$\sum_i \sum_j \sum_k h_{ijk} \tau_{ijk} = \sum_i \sum_j \sum_k h_{ijk}(\alpha\beta\gamma)_{ijk}$$

for which $\sum_i h_{ijk} = 0$ for all jk, $\sum_j h_{ijk} = 0$ for all ik, and $\sum_k h_{ijk} = 0$ for all ij. When we investigated the interaction plot for ABC using Figures 7.2–7.4, we compared the AB interaction at two different levels of factor C. In other words, we looked at contrasts of the type

$$(\tau_{112} - \tau_{122} - \tau_{212} + \tau_{222}) - (\tau_{111} - \tau_{121} - \tau_{211} + \tau_{221}).$$

If the levels 1 and 2 of A and B interact in the same way at each level of C, then this ABC interaction contrast is zero. If all interaction contrasts of this type (for all levels of A, B, and C) are zero, then the ABC interaction is negligible.

When a sub-model, rather than a complete model, is used, parameters for the negligible interactions are removed from the above expressions. If the experiment involves more than three factors, then the above definitions can be generalized by including the additional subscripts on $h_{ijk}\tau_{ijk}$ and averaging over the appropriate higher-order interactions.

As in Chapter 6, all contrasts can be represented by coefficient lists in terms of the main-effect and interaction parameters or in terms of the treatment combination parameters. This is illustrated in the next example.

Example 7.2.1 Coefficient lists for contrasts

Suppose that we have an experiment that involves four factors, A, B, C, and D, each to be examined at two levels (so that $a = b = c = d = 2$ and $v = 16$). Suppose that a model that includes AB, BC, BD, CD, and BCD interactions is expected to provide a good description of the data; that is,

$$Y_{ijklt} = \mu + \alpha_i + \beta_j + \gamma_k + \delta_l + (\alpha\beta)_{ij}$$
$$+ (\beta\gamma)_{jk} + (\beta\delta)_{jl} + (\gamma\delta)_{kl} + (\beta\gamma\delta)_{jkl} + \epsilon_{ijklt},$$
$$\epsilon_{ijklt} \sim N(0, \sigma^2),$$

ϵ_{ijklt}'s are mutually independent,

$$t = 1, \ldots, r_{ijkl}, \quad i = 1, 2, \quad j = 1, 2, \quad k = 1, 2, \quad l = 1, 2.$$

The contrast that compares the two levels of C is

$$\overline{\tau}_{..2.} - \overline{\tau}_{..1.} = \gamma_2^* - \gamma_1^*,$$

where $\gamma_k^* = \gamma_k + \overline{(\beta\gamma)}_{.k} + \overline{(\gamma\delta)}_{k.} + \overline{(\beta\gamma\delta)}_{.k.}$. This contrast can be represented as a coefficient list $[-1, \ 1\ ]$ in terms of the parameters γ_1^* and γ_2^* or as

$$\tfrac{1}{8}[-1, -1, \ 1, \ 1, -1, -1, \ 1, \ 1, -1, -1, \ 1, \ 1, -1, -1, \ 1, \ 1\ ]$$

in terms of the τ_{ijkl}. These are listed under the heading C in Table 7.1. Coefficient lists for the other main-effect contrasts in terms of the τ_{ijkl} are also shown in Table 7.1. The treatment combinations in the table are listed in ascending order when regarded as 4-digit numbers. The main-effect contrast coefficients are -1 when the corresponding factor is at level 1, and the coefficients are $+1$ when the corresponding factor is at level 2, although these can be interchanged if contrasts such as $\gamma_1^* - \gamma_2^*$ are required rather than $\gamma_2^* - \gamma_1^*$. The divisor shown in the table is the number of observations taken on each level of the factor.

The two-factor interaction contrast for CD is

$$(\gamma\delta)_{11}^* - (\gamma\delta)_{12}^* - (\gamma\delta)_{21}^* + (\gamma\delta)_{22}^*,$$

where $(\gamma\delta)_{kl}^* = (\gamma\delta)_{kl} + \overline{(\beta\gamma\delta)}_{.kl}$. This has coefficient list $[\ 1, -1, -1, \ 1\ ]$ in terms of the interaction parameters $(\gamma\delta)_{kl}^*$ but has coefficient list

$$\tfrac{1}{4}[\ 1, -1, -1, \ 1, \ 1, -1, -1, \ 1, \ 1, -1, -1, \ 1, \ 1, -1, -1, \ 1\ ]$$

in terms of the treatment combination parameters τ_{ijkl}. The coefficients are $+1$ when C and D are at the same level and -1 when they are at different levels. Notice that these coefficients can easily be obtained by multiplying together the C and D coefficients in the same rows of Table 7.1. The coefficient lists for some of the other interaction contrasts are also shown in the table, and it can be verified that their coefficients are also products of the corresponding main-effect coefficients. The divisors are the numbers of observations on each pair of levels of C and D. To obtain the same precision (estimator variance) as a main

Table 7.1 Contrast coefficient lists in terms of treatment combination parameters

Treatment combination	A	B	C	D	AB	BC	BD	CD	BCD
1111	−1	−1	−1	−1	1	1	1	1	−1
1112	−1	−1	−1	1	1	1	−1	−1	1
1121	−1	−1	1	−1	1	−1	1	−1	1
1122	−1	−1	1	1	1	−1	−1	1	−1
1211	−1	1	−1	−1	−1	−1	−1	1	1
1212	−1	1	−1	1	−1	−1	1	−1	−1
1221	−1	1	1	−1	−1	1	−1	−1	−1
1222	−1	1	1	1	−1	1	1	1	1
2111	1	−1	−1	−1	−1	1	1	1	−1
2112	1	−1	−1	1	−1	1	−1	−1	1
2121	1	−1	1	−1	−1	−1	1	−1	1
2122	1	−1	1	1	−1	−1	−1	1	−1
2211	1	1	−1	−1	1	−1	−1	1	1
2212	1	1	−1	1	1	−1	1	−1	−1
2221	1	1	1	−1	1	1	−1	−1	−1
2222	1	1	1	1	1	1	1	1	1
Divisor	8	8	8	8	4	4	4	4	2

effect contrast, the divisor would need to be changed to 8 (or all contrasts would need to be normalized).

Contrast coefficients are also shown for the BCD interaction. These are the products of the main-effect coefficients for B, C, and D. This contrast compares the CD interaction at the two levels of B (or, equivalently, the BC interaction at the two levels of D, or the BD interaction at the two levels of C). □

7.3 Analysis—Equal Sample Sizes

For an experiment involving p factors, we can select a cell-means model or the equivalent p-way complete model, or any of the possible submodels. When the sample sizes are equal, the formulae for the degrees of freedom, least squares estimates, and sums of squares for testing hypotheses follow well-defined patterns. We saw in Chapter 6 that for an experiment with two factors, we obtain similar formulae for the least squares estimates of the contrasts $\sum c_i \alpha_i$ in the two-way main-effects model and $\sum c_i \alpha_i^*$ in the two-way complete model. Similarly, the sum of squares for A was of the same form in both cases. This is also true for experiments with more than two factors.

We now give a series of rules that can be applied to any complete model with equal sample sizes. The rules are illustrated for the ABD interaction in an experiment involving four treatment factors A, B, C, and D, with corresponding symbols α, β, γ, and δ and subscripts i, j, k, and l to represent their effects on the response in a four-way complete model with r observations per treatment combination. The corresponding rules for submodels are obtained by dropping the relevant interaction terms from the complete model. When the sample sizes are not equal, the formulae are more complicated, and we will analyze such experiments only via a computer package (see Section 7.7).

Rules for estimation and hypothesis testing—equal sample sizes

1. Write down the name of the main effect or interaction of interest and the corresponding numbers of levels and subscripts.
   Example: ABD; numbers of levels a, b, and d; subscripts i, j, and l.

2. The number of degrees of freedom ν for a factorial effect is the product of the "number of levels minus one" for each of the factors included in the effect.
   Example: For ABD, $\nu = (a-1)(b-1)(d-1)$.

3. Multiply out the number of degrees of freedom and replace each letter with the corresponding subscript.
   Example: For ABD, $df = abd - ab - ad - bd + a + b + d - 1$, which gives $ijl - ij - il - jl + i + j + l - 1$.

4. The sum of squares for testing the hypothesis that a main effect or an interaction is negligible is obtained as follows. Use each group of subscripts in rule 3 as the subscripts of a term $\bar{y}$, averaging over all subscripts not present and keeping the same signs. Put the resulting estimate in brackets, square it, and sum over all possible subscripts. To expand the parentheses, square each term in the parentheses, keep the same signs, and sum over all possible subscripts.
   Example:
   $$ss(ABD) = rc \sum_i \sum_j \sum_l (\bar{y}_{ij.l.} - \bar{y}_{ij...} - \bar{y}_{i.l.} - \bar{y}_{.j.l.}$$
   $$+ \bar{y}_{i....} + \bar{y}_{.j...} + \bar{y}_{...l.} - \bar{y}_{.....})^2$$
   $$= rc \sum_i \sum_j \sum_l \bar{y}^2_{ij.l.} - rcd \sum_i \sum_j \bar{y}^2_{ij...} - rbc \sum_i \sum_l \bar{y}^2_{i.l.}$$
   $$- rac \sum_j \sum_l \bar{y}^2_{.j.l.} + rbcd \sum_i \bar{y}^2_{i....} + racd \sum_j \bar{y}^2_{.j...}$$
   $$+ rabc \sum_l \bar{y}^2_{...l.} - rabcd \bar{y}^2_{.....}\,.$$

5. The total sum of squares $sstot$ is the sum of all the squared deviations of the data values from their overall mean. The total degrees of freedom is $n-1$, where n is the total number of observations.
   Example:
   $$sstot = \sum_i \sum_j \sum_k \sum_l \sum_t (y_{ijklt} - \bar{y}_{.....})^2$$
   $$= \sum_i \sum_j \sum_k \sum_l \sum_t y^2_{ijklt} - n\bar{y}^2_{.....}\,,$$
   $n - 1 = abcdr - 1$.

6. The error sum of squares *ssE* is *sstot* minus the sums of squares for all other effects in the model. The error degrees of freedom df is $n - 1$ minus the degrees of freedom for all of the effects in the model. For a complete model, $df = n - v$, where v is the total number of treatment combinations.
Example:
$$ssE = sstot - ssA - ssB - ssC - ssD$$
$$- ss(AB) - ss(AC) - \cdots - ss(BCD) - ss(ABCD),$$
$$df = (n-1) - (a-1) - (b-1) - \cdots - (a-1)(b-1)(c-1)(d-1).$$

7. The mean square for an effect is the corresponding sum of squares divided by the degrees of freedom.
Example: $\quad ms(ABD) = ss(ABD)/((a-1)(b-1)(d-1))$,
$$msE = ssE/df$$

8. The decision rule for testing the null hypothesis that an effect is zero against the alternative hypothesis that the effect is nonzero is
$$\text{reject } H_0 \text{ if } \frac{ss/v}{ssE/df} = \frac{ms}{msE} > F_{v,df,\alpha},$$
where ss is the sum of squares calculated in rule 4, v is the degrees of freedom in rule 2, and $ms = ss/v$.

Example: To test H_0^{ABD} : {the interaction *ABD* is negligible}
against H_A^{ABD} : {the interaction *ABD* is not negligible},
the decision rule is
$$\text{reject } H_0^{ABD} \quad \text{if} \quad \tfrac{ms(ABD)}{msE} > F_{(a-1)(b-1)(d-1),df,\alpha}.$$

9. An estimable contrast for an interaction or main effect is a linear combination of the corresponding parameters (averaged over all higher-order interactions in the model), where the coefficients add to zero over each of the subscripts in turn.
Example: All estimable contrasts for the *ABD* interaction are of the form
$$\sum_i \sum_j \sum_l h_{ijl}(\alpha\beta\delta)^*_{ijl}$$
where
$$\sum_i h_{ijl} = 0 \text{ for all } j,l; \quad \sum_j h_{ijl} = 0 \text{ for all } i,l; \quad \sum_l h_{ijl} = 0 \text{ for all } i,j,$$
and where $(\alpha\beta\delta)^*$ is the parameter representing the *ABD* interaction averaged over all the higher-order interactions in the model.

10. If the sample sizes are equal, the least squares estimate of an estimable contrast in rule 9 is obtained by replacing each parameter with $\bar{y}$ having the same subscripts and averaging over all subscripts not present.

Example: The least squares estimate of the ABD contrast in rule 9 is

$$\sum_i \sum_j \sum_l h_{ijl}(\widehat{\alpha\beta\delta})^*_{ijl} = \sum_i \sum_j \sum_l h_{ijl}\,\bar{y}_{ij.l.}\,.$$

11. The variance of an estimable contrast for a factorial effect is obtained by adding the squared contrast coefficients, dividing by the product of r and the numbers of levels of all factors not present in the effect, and multiplying by σ^2.

    Example: $\text{Var}\left(\sum_i \sum_j \sum_l h_{ijl}(\widehat{\alpha\beta\delta})^*_{ijl}\right) = \left(\sum_i \sum_j \sum_l h_{ijl}^2/(cr)\right)\sigma^2$.

12. The "sum of squares" for testing the null hypothesis H_0^c that a contrast is zero is the square of the normalized contrast estimate.

    Example: The sum of squares for testing the null hypothesis that the contrast $\sum\sum\sum h_{ijl}(\alpha\beta\delta)^*_{ijl}$ is zero against the alternative hypothesis that the contrast is nonzero is the square of the least squares estimate of the normalized contrast $\sum\sum\sum h_{ijl}(\alpha\beta\delta)^*_{ijl}/\sqrt{\sum\sum\sum h_{ijl}^2/(cr)}$; that is,

    $$ssc = \frac{\left(\sum_i \sum_j \sum_l h_{ijl}\bar{y}_{ij.l.}\right)^2}{\sum_i \sum_j \sum_l h_{ijl}^2/(cr)}.$$

13. The decision rule for testing the null hypothesis H_0^c that an estimable contrast is zero, against the alternative hypothesis that the contrast is nonzero, is

    $$\text{reject } H_0^c \text{ if } \frac{ssc}{msE} > F_{1,df,\alpha/m},$$

    where ssc is the square of the normalized contrast estimate, as in rule 12; msE is the error mean square; df is the number of error degrees of freedom; α is the overall Type I error probability; and m is the number of preplanned hypotheses being tested.

14. Simultaneous confidence intervals for contrasts in the treatment combinations can be obtained from the general formula (4.4.20), page 80, with the appropriate critical coefficients for the Bonferroni, Scheffé, Tukey, Dunnett, and Hsu methods.

    Example: For the four-way complete model, the general formula for simultaneous $100(1-\alpha)\%$ confidence intervals for a set of contrasts of the form $\sum\sum\sum\sum c_{ijkl}\tau_{ijkl}$ is

    $$\sum\sum\sum\sum c_{ijkl}\tau_{ijkl} \in \left(\sum\sum\sum\sum c_{ijkl}\bar{y}_{ijkl.} \pm w\sqrt{msE\left(\sum\sum\sum\sum c_{ijkl}^2/r\right)}\right),$$

    where the critical coefficient, w, is

    $$w_B = t_{df,\alpha/2m}\,;\quad w_S = \sqrt{(v-1)F_{v-1,df,\alpha}}\,;\quad w_T = q_{v,df,\alpha}/\sqrt{2}\,;$$

    $$w_{D1} = w_H = t_{v-1,df,\alpha}^{(0.5)}\,;\quad w_{D2} = |t|_{v-1,df,\alpha}^{(0.5)}\,;$$

    for the five methods, respectively, and v is the number of treatment combinations, and df is the number of error degrees of freedom.

15. Simultaneous confidence intervals for the true mean effects of the treatment combinations in the complete model can be obtained from the general formula (4.3.12), page 75, with the appropriate critical coefficients for the Bonferroni or Scheffé methods.
Example: For the four-way complete model, the general formula for simultaneous $100(1 - \alpha)\%$ confidence intervals for true mean effects of the treatment combinations $\mu + \tau_{ijkl}$ is

$$\mu + \tau_{ijkl} \in \left(\bar{y}_{ijkl.} \pm w \sqrt{msE/r}\right),$$

where the critical coefficient, w, is

$$w_B = t_{df,\alpha/(2v)} \quad \text{or} \quad w_s = \sqrt{v\, F_{v,df,\alpha}}$$

for the Bonferroni and Scheffé methods, respectively.

16. Simultaneous $100(1 - \alpha)\%$ confidence intervals for contrasts in the levels of a single factor can be obtained by modifying the formulae in rule 14. Replace v by the number of levels of the factor of interest, and r by the number of observations on each level of the factor of interest.
Example: For the four-way complete model, the general formula for simultaneous confidence intervals for contrasts $\sum_i c_i \bar{\tau}_{i...} = \sum_i c_i \alpha_i^*$ in A is

$$\sum_i c_i \alpha_i^* \in \left(\sum_i c_i \bar{y}_{i....} \pm w \sqrt{msE \left(\sum_i c_i^2/(bcdr)\right)}\right), \quad (7.3.3)$$

where the critical coefficients for the five methods are, respectively,

$$w_B = t_{df,\alpha/(2m)} \;;\; w_S = \sqrt{(a-1)F_{a-1,df,\alpha}} \;;\; w_T = q_{a,df,\alpha}/\sqrt{2} \;;$$

$$w_{D1} = w_H = t_{a-1,df,\alpha}^{(0.5)} \;;\; w_{D2} = |t|_{a-1,df,\alpha}^{(0.5)} \;;$$

where df is the number of error degrees of freedom.

7.4 A Real Experiment—Popcorn–Microwave Experiment

The experiment described in this section was run by Jianjian Gong, Chongqing Yan, and Lihua Yang in 1992 to compare brands of microwave popcorn. The details in the following checklist have been extracted from the experimenters' report.

The design checklist

(a) **Define the objectives of the experiment.**
The objective of the experiment was to find out which brand gives rise to the best popcorn in terms of the proportion of popped kernels. The experiment was restricted to popcorn produced in a microwave oven.

(b) **Identify all sources of variation.**

(i) Treatment factors and their levels.

The first treatment factor was "brand." Three levels were selected, including two national brands (levels 1 and 2) and one local brand (level 3). These brands were the brands most commonly used by the experimenters and their colleagues. All three brands are packaged for household consumers in boxes of 3.5 ounce packages, and a random selection of packages was used in this experiment.

Power of the microwave oven was identified as a possible major source of variation and was included as a second treatment factor. Three available microwave ovens had power ratings of 500, 600, and 625 watts. The experimenters used only one oven for each power level. This means that their conclusions could be drawn only about the three ovens in the study and not about power levels in general.

Popping time was taken as a third treatment factor. The usual instructions provided with microwave popcorn are to microwave it until rapid popping slows to 2 to 3 seconds between pops. Five preliminary trials using brand 3, a 600 watt microwave oven, and times equally spaced from 3 to 5 minutes suggested that the best time was between 4 and 5 minutes. Hence, time levels of 4, 4.5, and 5 minutes were selected for the experiment and coded 1–3, respectively.

(ii) Experimental units

The experiment was to be run sequentially over time. The treatment combinations were to be examined in a completely random order. Consequently, the experimental units were the time slots that were to be assigned at random to the treatment combinations.

(iii) Blocking factors, noise factors, and covariates.

Instead of randomly ordering the observations on all of the treatment combinations, it might have been more convenient to have taken the observations oven by oven. In this case, the experiment would have been a "split-plot design" (see Section 2.4.4) with ovens representing the blocks. In this experiment, no blocking factors or covariates were identified by the experimenters. The effects of noise factors, such as room temperature, were thought to be negligible and were ignored.

(c) **Choose a rule by which to assign the experimental units to the treatments.**

A completely randomized design was indicated. The time-slots were randomly assigned to the brand–power–time combinations. Popcorn packages were selected at random from a large batch purchased by the experimenters to represent each brand. Changes in quality, if any, of the packaged popcorn over time could not be detected by this experiment.

(d) **Specify measurements to be made, the experimental procedure, and the anticipated difficulties.**

A main difficulty for the experimenters was to choose the response variable. They considered weight, volume, number, and percentage of successfully popped kernels as possible response variables. In each case, they anticipated difficulty in consistently being able to classify kernels as popped or not. To help control such variation or inconsistency in the measurement process, a single experimenter made all measurements. For measuring weight, the experimenters needed a more accurate scale than was available, since popcorn is very light. They decided against measuring volume, since brands with smaller kernels would appear to give less volume, as the popcorn would pack more easily into a measuring cylinder. The percentage, rather than number, of successfully popped kernels for each package was selected as the response variable.

(e) **Run a pilot experiment.**

The experimenters ran a very small pilot experiment to check their procedure and to obtain a rough estimate of the error variance. they collected observations on only 9 treatment combinations. Using a three-way main-effects model, they found that the overall average popping rate was about 70% and the error standard deviation was a little less than 10.7%. The highest popping rate occurred when the popping time was at its middle level (4.5 minutes), suggesting that the range of popping times under consideration was reasonable. Results for 600 and 625 watt microwave ovens were similar, with lower response rates for the 500 watt microwave oven. However, since all possible interactions had been ignored for this preliminary analysis, the experimenters were cautious about drawing any conclusions from the pilot experiment.

(f) **Specify the model.**

For their main experiment, the experimenters selected the three-way complete model, which includes all main effects and interactions between the three treatment factors. They assumed that the packages selected to represent each brand would be very similar, and package variability for each brand could be ignored.

(a) — **revisited. Define the objectives of the experiment.**

Having identified the treatment factors, response variables, etc., the experimenters were able to go back to step (a) and reformalize the objectives of the experiment. They decided that the three questions of most interest were:

- Which combination of brand, power, and time will produce the highest popping rate? (Thus, pairwise comparisons of all treatment combinations were required.)

- Which brand of popcorn performs best overall? (Pairwise comparison of the levels of brand, averaging over the levels of power and time, was required.)

- How do time and power affect response? (Pairwise comparison of time–power combinations, averaging over brands, was required. Also, main-effect comparisons of power and time were required.)

(g) **Outline the analysis.**

Tukey's method of simultaneous confidence intervals for pairwise comparisons was to be used separately at level 99% for each of the above five sets of contrasts, giving an experimentwise confidence level of at least 95%.

(h) **Calculate the number of observations that need to be taken.**

The data from the pilot study suggested that 10.7% would be a reasonable guess for the error standard deviation. This was calculated using a main-effects model rather than the three-way complete model, but we would expect a model with more terms to reduce the estimated error variance, not to enlarge it. Consequently, the value $msE = 10.7^2$ was used in the sample-size calculations. The experimenters decided that their confidence intervals for any pairwise main-effect comparison should be no wider than 15% (that is, the half-width, or minimum significant difference, should be less than 7.5%). Using rule 16, page 205, for Tukey's pairwise comparisons, a set of 99% simultaneous confidence intervals for the pairwise differences between the brands (factor A) is

$$\bar{y}_{i\ldots} - \bar{y}_{s\ldots} \pm w_T \sqrt{msE\,(2/(bcr))}\,,$$

where the critical coefficient is $w_T = q_{3,27r-27,0.01}/\sqrt{2}$. The error degrees of freedom are calculated for a complete model as $df = n - v = 27r - 27$. Consequently, using 10.7^2 as the rough estimate of msE, we need to solve

$$msd = \left(q_{3,27(r-1),.01}/\sqrt{2}\right)\sqrt{(10.7)^2(2/(9r))} \leq 7.5\,.$$

Trial and error shows that $r = 4$ is adequate. Thus a total of $n = rv = 108$ observations would be needed.

(i) **Review the above decisions. Revise, if necessary.**

The experimenters realized that it would not be possible to collect 108 observations in the time they had available. Since the effects of power levels of 600 and 625 watts were comparable in the pilot study, they decided to drop consideration of the 600 watt microwave and to include only power levels of 500 watts (level 1) and 625 watts (level 2) in the main experiment. Also, they decided to take only $r = 2$ observations (instead of the calculated $r = 4$) on each of the $v = 18$ remaining treatment combinations. The effect of this change is to widen the proposed confidence intervals. A set of 99% simultaneous confidence intervals for pairwise comparisons in the brands using Tukey's method and $msE = 10.7^2$ would have half-width

$$msd = (q_{3,18,.01}/\sqrt{2})\sqrt{(10.7)^2(2/(6 \times 2))} = 14.5\,,$$

about twice as wide as in the original plan. It was important, therefore, to take extra care in running the experiment to try to reduce the error variability.

The experiment was run, and the resulting data are shown in Table 7.2. Unfortunately, the error variance does not seem to be much smaller than in the pilot experiment, since the mean squared error was reduced only to $(9.36)^2$. A plot of the standardized residuals against fitted values did not show any pattern of unequal variances or outliers. Likewise, a plot of the standardized residuals against the normal scores was nearly linear, giving no reason to question the model assumption of normality. Unfortunately, the experimenters did not keep information concerning the order of observations, so the independence assumption cannot be checked.

Data analysis Table 7.3 contains the analysis of variance for investigating the three-way complete model. If an overall significance level of $\alpha \leq 0.07$ is selected, allowing each hypothesis to be tested at level $\alpha^* = 0.01$, the only null hypothesis that would be rejected would be H_0^T :{popping time has no effect on the proportion of popped kernels}. However, at a slightly higher significance level, the brand–time interaction also appears to have an effect on the proportion of popped kernels.

7.4 A Real Experiment—Popcorn–Microwave Experiment

Table 7.2 Percentage y_{ijkl} of kernels popped—popcorn–microwave experiment

Brand (i)	Power (j)	Time (k) 1	2	3
1	1	73.8, 65.5	70.3, 91.0	72.7, 81.9
1	2	70.8, 75.3	78.7, 88.7	74.1, 72.1
2	1	73.7, 65.8	93.4, 76.3	45.3, 47.6
2	2	79.3, 86.5	92.2, 84.7	66.3, 45.7
3	1	62.5, 65.0	50.1, 81.5	51.4, 67.7
3	2	82.1, 74.5	71.5, 80.0	64.0, 77.0
		$\bar{y}_{..1.} = 72.9000$	$\bar{y}_{..2.} = 79.8667$	$\bar{y}_{..3.} = 63.8167$

Table 7.3 Three-way ANOVA for the popcorn–microwave experiment

Source of Variation	Degrees of Freedom	Sum of Squares	Mean Square	Ratio	p-value
B	2	331.1006	165.5503	1.89	0.1801
P	1	455.1111	455.1111	5.19	0.0351
T	2	1554.5756	777.2878	8.87	0.0021
B*P	2	196.0406	98.0203	1.12	0.3485
B*T	4	1433.8578	358.4644	4.09	0.0157
P*T	2	47.7089	23.8544	0.27	0.7648
B*P*T	4	47.3344	11.8336	0.13	0.9673
Treatments	17	4065.7289	239.1605	2.73	0.0206
Error	18	1577.8700	87.6594		
Total	35	5643.5989			

If the equivalent cell-means model is used, the null hypothesis of no difference between the treatment combinations would be rejected at significance level $\alpha = 0.07$. This is shown in the row of Table 7.3 labeled "Treatments." Since the design is equireplicate, the main effects and interactions are estimated independently, and their sums of squares add to the treatment sum of squares. The corresponding numbers of degrees of freedom likewise add up.

Figure 7.6 shows an interaction plot for the factors "Brand" and "Time." The plot suggests that use of time level 2, namely 4.5 minutes, generally gives a higher popping rate for all three brands. Using level 2 of time, brands 1 and 2 appear to be better than brand 3. The two national brands thus appear to be better than the local brand. Brand 1 appears to be less sensitive than brand 2 to the popping time. (We say that brand 1 appears to be *more robust* to changes in popping time.) Unless this perceived difference is due to error variability, which does not show on the plot, brand 1 is the brand to be recommended.

Having examined the analysis of variance table and Figure 7.6, the most interesting issue seems to be that the differences in the brands might not be the same at the different popping times. This is not one of the comparisons that had been preplanned at step (g) of the checklist. It is usually advisable to include in the plan of the analysis the use of Scheffé's multiple comparisons for all contrasts that look interesting after examining the data. If we had done this at overall 99% confidence level, then the experimentwise error rate would have been

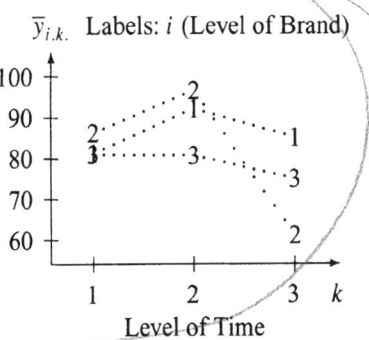

Figure 7.6
Interaction plot for factors "brand" and "time" averaged over "power" for the popcorn–microwave experiment

at least 94%. Interaction contrasts and their least squares estimates are defined in rules 9 and 10, page 203. The interaction contrast of most interest is, perhaps,

$$\bar{\tau}_{1.2} - \bar{\tau}_{1.3} - \bar{\tau}_{2.2} + \bar{\tau}_{2.3},$$

which compares the differences in brands 1 and 2 at popping times 2 and 3. This has least squares estimate

$$\bar{y}_{1.2.} - \bar{y}_{1.3.} - \bar{y}_{2.2.} + \bar{y}_{2.3.} = 82.175 - 75.20 - 86.65 + 51.225 = -28.45.$$

The importance of preplanning will now become apparent. Using Scheffé's method (rule 14) at overall level 99%, a confidence interval for this contrast is given by

$$\sum_i \sum_k c_{ik} \bar{y}_{i.k.} \pm w_S \sqrt{msE \left(\sum_i \sum_k c_{ik}^2/(br) \right)}$$

$$= -28.45 \pm \sqrt{17 F_{17,18,.01}} \sqrt{87.6594 \, (4/4)}$$

$$= -28.45 \pm 69.69 \quad = (-96.41, \, 39.52).$$

Our popping rates are percentages, so our minimum significant difference is 69%. This is far too large to give any useful information. The resulting interval gives the value of the interaction contrast as being between −96.4% and 39.5%! Had this contrast been preplanned for at individual confidence level 99%, we would have used the critical value $w_B = t_{18,0.005} = 2.878$ instead of $w_S = 7.444$, and we would have obtained a minimum significant difference of about 30%, leading to the interval $(-55.40, -1.50)$. Although still wide, this interval would have given more information, and in particular, it would have indicated that the interaction contrast was significantly different from zero.

The other important effect that showed up in the analysis of variance table was the effect of the different popping times. Comparisons of popping times did feature as one of the preplanned sets of multiple comparisons, and consequently, we use Tukey's method (rule 16) for pairwise differences $\gamma_k - \gamma_u$ at overall level 99%. The minimum significant difference is

$$msd = w_T \sqrt{msE \, \Sigma c_k^2/(abr)} = (q_{3,18,.01}/\sqrt{2}) \sqrt{(87.6594)(2/12)} = 12.703.$$

The average percentages of popped kernels for the three popping times are shown in Table 7.2 as

$$\bar{y}_{..1.} = 72.9000, \quad \bar{y}_{..2.} = 79.8667, \quad \bar{y}_{..3.} = 63.8167,$$

so the three confidence intervals are

$$\begin{aligned}
\gamma_1 - \gamma_2 &\in (-6.9667 \pm 12.7030) = (-5.736, 19.670), \\
\gamma_1 - \gamma_3 &\in (9.0833 \pm 12.7030) = (-3.620, 21.786), \\
\gamma_2 - \gamma_3 &\in (16.0500 \pm 12.7030) = (3.347, 28.753).
\end{aligned}$$

We see that at an experimentwise confidence level of at least 94%, use of popping time 2 (4.5 min) produces on average between 3.35% and 28.75% more popcorn than use of popping time 3 (5 min).

The other questions asked by the experimenters appear to be of less interest, and we will omit these. The experimentwise confidence level is still at least 94%, even though we have chosen not to calculate all of the preplanned intervals.

7.5 One Observation per Cell

If the complete model is used for a factorial experiment with one observation per cell, then there are no degrees of freedom available to estimate the error variance. This problem was discussed in Section 6.7, where one possible method of analysis was described. The method relies on being able to identify a number of negligible contrasts, which are then excluded from the model. The corresponding sums of squares and degrees of freedom are used to estimate the error variance. With this approach, confidence intervals can be constructed and hypothesis tests conducted. An example with four treatment factors that are believed not to interact with each other is presented in the next section.

Two alternative approaches for the identification of nonnegligible contrasts are provided in the subsequent sections. In Section 7.5.2 we show an approach based on the evaluation of a normal probability plot of a set of contrast estimates, and in Section 7.5.3 we discuss a more formalized approach. These two approaches work well under effect sparsity, that is, when most of the treatment contrasts under examination are negligible.

7.5.1 Analysis Assuming That Certain Interaction Effects Are Negligible

For a single replicate factorial experiment, if the experimenter knows ahead of time that certain interactions are negligible, then by excluding those interactions from the model, the corresponding degrees of freedom can be used to estimate the error variance. It must be recognized, however, that if interactions are incorrectly assumed to be negligible, then msE will be inflated, in which case the results of the experiment may be misleading.

Example 7.5.1 Drill advance experiment

Daniel (1976) described a single replicate $2 \times 2 \times 2 \times 2$ experiment to study the effects of four treatment factors on the rate of advance of a small stone drill. The treatment factors were "load on the drill" (A), "flow rate through the drill" (B), "speed of rotation" (C), and "type of mud used in drilling" (D). Each factor was observed at two levels, coded 1 and 2. The author examined several different transformations of the response and concluded that the log transform was one of the more satisfactory ones. In the rest of our discussion, y_{ijkl} represents the log (to the base 10) of the units of drill advance, as was illustrated in the original paper.

In many experiments with a number of treatment factors, experimenters are willing to believe that some or all of the interactions are very small. Had that been the case here, the experimenter would have used the four-way main-effects model. (Analysis of this experiment without assuming negligible interactions is discussed in Example 7.5.2, page 214.)

Degrees of freedom and sums of squares are given by rules 2 and 4 in Section 7.3. For example, the main effect of B has $b - 1$ degrees of freedom and

$$ssB = acd \sum_i (\bar{y}_{.j..} - \bar{y}_{....})^2 = acd \sum_i \bar{y}_{.j..}^2 - acbd\bar{y}_{....}^2 .$$

The sums of squares for the other effects are calculated similarly and are listed in the analysis of variance table, Table 7.5. The error sum of squares shown in Table 7.5 is the total of all the eleven (negligible) interaction sums of squares and can be obtained by subtraction, as in rule 6, page 203:

$$ssE = sstot - ssA - ssB - ssC - ssD = 0.01998 .$$

Similarly, the number of error degrees of freedom is the total of the $15 - 4 = 11$ interaction degrees of freedom. An estimate of σ^2 is therefore $msE = ssE/11 = 0.0018$. Since $F_{1,11,.01} = 9.65$, the null hypotheses of no main effects of B, C, and D would all have been rejected at overall significance level $\alpha \leq 0.04$. Alternatively, from a computer analysis we would see that the p-values for B, C, and D are each less than or equal to an individual significance level of $\alpha^* = 0.01$.

Table 7.4 Data for the drill advance experiment

ABCD	advance	$y = \log(\text{advance})$	ABCD	advance	$y = \log(\text{advance})$
1111	1.68	.2253	2111	1.98	.2967
1112	2.07	.3160	2112	2.44	.3874
1121	4.98	.6972	2121	5.70	.7559
1122	7.77	.8904	2122	9.43	.9745
1211	3.28	.5159	2211	3.44	.5366
1212	4.09	.6117	2212	4.53	.6561
1221	9.97	.9987	2221	9.07	.9576
1222	11.75	1.0700	2222	16.30	1.2122

Source: *Applications of Statistics to Industrial Experimentation*, by C. Daniel, Copyright © 1976, John Wiley & Sons, New York. Reprinted by permission of John Wiley & Sons, Inc.

Table 7.5 Analysis of variance for the drill advance experiment

Source of Variation	Degrees of Freedom	Sum of Squares	Mean Square	Ratio	p-value
A	1	0.01275	0.01275	7.02	0.0226
B	1	0.25387	0.25387	139.74	0.0001
C	1	1.00550	1.00550	553.46	0.0001
D	1	0.08045	0.08045	44.28	0.0001
Error	11	0.01998	0.00182		
Total	15	1.37254			

Confidence intervals for the $m = 4$ main-effect contrasts using Bonferroni's method at an overall level of at least 95% can be calculated from rule 16. From rules 10 and 11 on page 203, the least squares estimate for the contrast that compares the effects of the high and low levels of B is $\bar{y}_{.2..} - \bar{y}_{.1..}$, with variance $\sigma^2(2/8)$, giving the confidence interval

$$\left(\bar{y}_{.2..} - \bar{y}_{.1..} \pm w_B \sqrt{msE\,(2/8)}\right),$$

where the critical coefficient is

$$w_B = t_{11,.025/4} = t_{11,.00625} \approx z + (z^3 + z)/((4)(11)) \approx 2.911,$$

from (4.4.22), page 81, and

$$msd = w_B \sqrt{msE\,(2/8)} = 2.911\sqrt{(0.00182)/4} = 0.062.$$

Now, $\bar{y}_{.2..} = 0.820$, $\bar{y}_{.1..} = 0.568$, so the confidence interval for the B contrast is

$$(0.252 \pm 0.062) \approx (0.190, 0.314),$$

where the units are units of log drill advance. Confidence intervals for the other three main effects comparing high with low levels can be calculated similarly as

$$\begin{aligned} A: &\quad 0.056 \pm 0.062 &=&\quad (-0.006, 0.118),\\ C: &\quad 0.501 \pm 0.062 &=&\quad (0.439, 0.563),\\ D: &\quad 0.142 \pm 0.062 &=&\quad (0.080, 0.204).\end{aligned}$$

We see that the high levels of B, C, D give a somewhat higher response in terms of log drill advance (with overall confidence level at least 95%), whereas the interval for A includes zero. □

7.5.2 Analysis Using Normal Probability Plot of Effect Estimates

For a single replicate factorial experiment, with v treatment combinations, one can find a set of $v - 1$ orthogonal contrasts. When these are normalized, the contrast estimators all have variance σ^2. If the assumptions of normality, equal variances, and independence of the response variables are approximately satisfied, the estimates of negligible contrasts are like independent observations from a normal distribution with mean zero and variance σ^2. If we plot the normalized contrast estimates against their normal scores (in the same way that we checked for normality of the error variables in Chapter 5), the estimates of negligible effects tend to fall nearly on a straight line. Any contrast for which the corresponding estimate appears to be far from the line is considered to be nonnegligible. Provided that there is effect sparsity (all but a few contrast estimates are expected to be negligible), it is not difficult to pick out the nonnegligible contrasts.

Example 7.5.2 Drill advance experiment, continued

The data for the drill advance experiment were given in Table 7.4 in Example 7.5.1. The experiment involved treatment factors "load" (A), "flow" (B), "speed" (C), and "mud" (D) and response "log(drill advance)" (Y). If we have no information about which factors are likely to be negligible, we would use the four-way complete model or the equivalent cell-means model:

$$Y_{ijkl} = \mu + \tau_{ijkl} + \epsilon_{ijkl},$$
$$\epsilon_{ijkl} \sim N(0, \sigma^2),$$
ϵ_{ijkl}'s mutually independent,
$i = 1, 2; \ j = 1, 2; \ k = 1, 2; \ l = 1, 2.$

The contrast coefficients for the four main effects and some of the interactions in such a cell-means model were listed in Table 7.1, page 201. The contrast coefficients for the interactions can be obtained by multiplying together the corresponding main-effect coefficients. Each contrast is normalized by dividing the coefficients by $\sqrt{\Sigma c_{ijkl}^2 / r} = \sqrt{16}$ rather than by the divisors of Table 7.1. For example, the normalized BCD interaction contrast has coefficient list

$$\tfrac{1}{4}[-1, \ 1, \ 1, -1, \ 1, -1, -1, \ 1, -1, \ 1, \ 1, -1, \ 1, -1, -1, \ 1\,].$$

The least squares estimate of the normalized BCD interaction contrast is then

$$\tfrac{1}{4}[-(0.2253) + (0.3160) + \cdots - (0.9576) + (1.2122)] \ = \ -0.0300\,.$$

The 15 normalized factorial contrast estimates are given in Table 7.6, and the normal probability plot of these estimates is shown in Figure 7.7, with the main effect estimates labeled. Observe that all the estimates fall roughly on a straight line, except for the estimates for the main-effects of factors D, B, and C. Hence, these three main effects appear to be nonnegligible. □

Table 7.6 Normalized contrast estimates for the drill advance experiment

Effect:	A	B	C	D		
Estimate:	0.1129	0.5039	1.0027	0.2836		
Effect:	AB	AC	AD	BC	BD	CD
Estimate:	−0.0298	0.0090	0.0581	−0.0436	−0.0130	0.0852
Effect:	ABC	ABD	ACD	BCD	$ABCD$	
Estimate:	0.0090	0.0454	0.0462	−0.0300	0.0335	

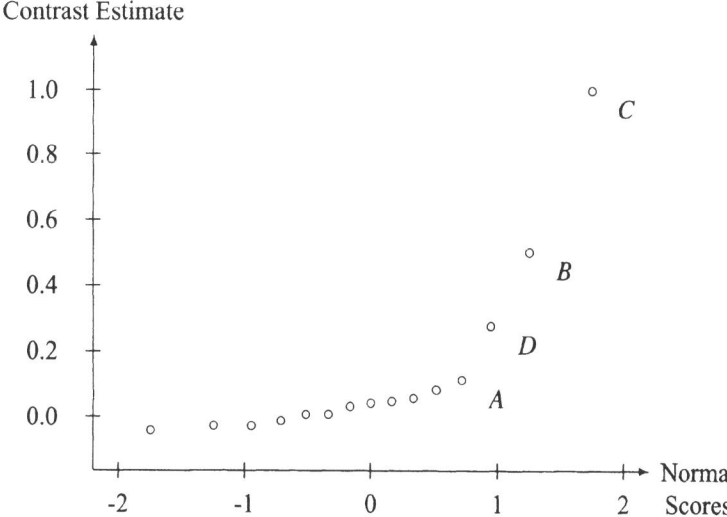

Figure 7.7
Normal probability plot of contrast estimates for the drill advance experiment

In the construction of the normal probability plot, the contrasts must be scaled to have the same variance, and normalization is one way to achieve this. When *all factors* have two levels, and when the contrasts are written in terms of the treatment combination parameters as in Table 7.1, their least squares estimators will all have the same variance, as long as the *same divisor* is used for every contrast. A popular selection for divisor is $v/2$, which is the natural divisor for main-effect contrasts comparing the average treatment combination effects at the two levels of a factor. Thus, rather than using divisor $\sqrt{16}$ in Example 7.5.2, we could have used divisor $v/2 = 8$. If the divisor $v/2$ is used, the estimators all have variance $4\sigma^2/v$. If no divisor is used, the estimators all have variance $v\sigma^2$. As long as the variances are all equal, the normal probability plot can be used to identify the important contrasts. In *all other* sizes of experiment, the contrast coefficients are not all ± 1, and we recommend that all contrasts be normalized so that their estimators all have variance σ^2.

7.5.3 Analysis Using Confidence Intervals

In this section, an alternative to the normal probability plot is presented for the analysis of a single replicate factorial experiment. As with the normal probability plot, we require a set of m orthogonal contrasts and effect sparsity, and we make no assumptions as to *which* effects are negligible. The procedure provides confidence intervals for the m contrasts with a simultaneous confidence level of at least $100(1-\alpha)\%$. For the moment, we recode the treatment combinations as $1, 2, \ldots, v$, their effects as $\tau_1, \tau_2, \ldots \tau_v$, and we generically denote each of the m contrasts by $\sum c_i \tau_i$.

First, let d equal the integer part of $(m+1)/2$, which is $m/2$ if m is even and is $(m+1)/2$ if m is odd. The method requires that there be at least d negligible effects (effect sparsity). In general, this will be true if at least one of the factors has no effect on the response (and so does not interact with any of the other factors) or if most of the higher-order interactions are negligible.

216   Chapter 7   Several Crossed Treatment Factors

We take each of the m contrasts in turn. For the kth contrast $\sum c_i \tau_i$, we calculate its least squares estimate $\sum c_i y_i$ and its sum of squares ssc_k, using rules 10 and 12, page 204. We then calculate the *quasi mean squared error* msQ_k for the kth contrast by taking the average of the d smallest of $ssc_1, \ldots, ssc_{k-1}, ssc_{k+1}, \ldots, ssc_m$ (that is, the smallest d contrast sums of squares ignoring the kth).

The *Voss–Wang method* gives simultaneous $100(1-\alpha)\%$ confidence intervals for the m contrasts, the confidence interval for the kth contrast being

$$\sum c_i \tau_i \;\in\; \left(\sum c_i y_i \pm w_V \sqrt{msQ_k \sum_i c_i^2} \right). \tag{7.5.4}$$

The critical coefficients $w_V = v_{m,d,\alpha}$ are provided in Appendix A.11. The critical values $v_{m,d,\alpha}$ were obtained by Voss and Wang (1997) as the square root of the percentile corresponding to α in the right-hand tail of the distribution of

$$V^2 = \max \{ SSC_k / MSQ_k \},$$

where the maximum is over $k = 1, 2, \ldots, m$.

Example 7.5.3   Drill advance experiment, continued

Consider again the single replicate drill advance experiment of Examples 7.5.1 and 7.5.2 with four factors having two levels each. We can find $m = 15$ orthogonal factorial contrasts, nine of which are shown in Table 7.1, page 201. The Voss–Wang method of simultaneous confidence intervals, described above, is reasonably effective as long as there are at least $d = 8$ negligible contrasts in this set.

For an overall 95% confidence level, the critical coefficient is obtained from Appendix A.11 as $w_V = v_{15,8,0.05} = 9.04$. Selecting divisors $v/2 = 8$ for each contrast, we obtain the least squares estimates in Table 7.7.

The sums of squares for the 15 contrasts are also listed in Table 7.7 in descending order. For the contrasts corresponding to each of the seven largest sums of squares, the quasi mean squared error is composed of the eight smallest contrast sums of squares; that is,

$$msQ_k = (0.0000808 + \cdots + 0.0020571)/8 = 0.0009004,$$

and the minimum significant difference for each of these seven contrasts is

$$msd_k = v_{15,8,0.05} \sqrt{msQ_k \, (16/(8 \times 8))} = (9.04)\sqrt{0.0009004 \times 0.25} \approx 0.1356.$$

The quasi mean squared errors for the contrasts corresponding to the eight smallest sums of squares are modestly larger, leading to slightly larger minimum significant differences and correspondingly wider intervals. All contrast estimates and minimum significant differences are summarized in Table 7.7.

Table 7.7 Confidence interval information for the drill advance experiment

Effect	ssc_k	msQ_k	Estimate	msd_k
C	1.0054957	0.0009004	0.5014	0.1356
B	0.2538674	0.0009004	0.2519	0.1356
D	0.0804469	0.0009004	0.1418	0.1356
A	0.0127483	0.0009004	0.0565	0.1356
CD	0.0072666	0.0009004	0.0426	0.1356
AD	0.0033767	0.0009004	0.0291	0.1356
ACD	0.0021374	0.0009004	0.0231	0.1356
ABD	0.0020571	0.0009105	0.0227	0.1364
BC	0.0019016	0.0009299	−0.0218	0.1378
ABCD	0.0011250	0.0010270	0.0168	0.1449
BCD	0.0008986	0.0010553	−0.0150	0.1468
AB	0.0008909	0.0010563	−0.0149	0.1469
BD	0.0001684	0.0011466	−0.0065	0.1531
ABC	0.0000812	0.0011575	0.0045	0.1538
AC	0.0000808	0.0011575	0.0045	0.1537

The four largest contrast estimates in absolute value are 0.5014 for C, 0.2519 for B, 0.1418 for D, and 0.0565 for A, giving the intervals

For C: 0.5014 ± 0.1356 = $(0.3658, 0.6370)$,

For B: 0.2519 ± 0.1356 = $(0.1163, 0.3875)$,

For D: 0.1418 ± 0.1356 = $(0.0062, 0.2774)$,

For A: 0.0565 ± 0.1356 = $(-0.0791, 0.1921)$.

Thus, in the 95% simultaneous set, the intervals for the main-effect contrasts of C, B, and D exclude zero and are declared to be the important effects. The intervals for A and for all of the interaction contrasts include zero, so we conclude that these contrasts are not significantly different from zero. Notice that our conclusion agrees with that drawn from the normal probability plot. The benefit of the Voss–Wang method is that we no longer need to guess which contrast estimates lie on the straight line, and also that we have explicit confidence intervals for the magnitudes of the nonnegligible contrasts. □

7.6 Design for the Control of Noise Variability

Design for the control of noise variability is sometimes known as *robust design* or *parameter design* and refers to the procedure of developing or designing a product in such a way that it performs consistently as intended under the variety of conditions of its use throughout its life. The ideas apply equally well to the design of manufacturing and other organizational processes. Factors included in experimentation are categorized as either design factors or noise factors. *Design factors* are factors that are easy and inexpensive to control in the design of the product—these are also known as *control factors* or *design parameters*. Factors that may affect the performance of a product but that are difficult or impossible to control when the product is in use are called *noise factors*. Noise factors can be internal or external sources of noise. For example, the climate in which a product is used is an external source of noise. Internal sources of noise include variation in the materials used to make a product as well as the wear of the components and materials over the life of the product.

Each combination of levels of the design factors is a potential product design and is said to be *robust* if the product functions consistently well despite uncontrolled variation in the levels of the noise factors. A general philosophy of robust design has been espoused by Dr. Genichi Taguchi. Dr. Taguchi is a Japanese quality consultant who has advocated the use of quality improvement techniques, including the design of experiments, to the Japanese engineering and industrial communities since the 1950s. One of his fundamental contributions is the principle that reduction of variation is generally the most difficult task from an engineering perspective and so should be the focus of attention during product design. In contrast, adjustment of the mean level of some response variable to a target level is a relatively easy engineering task. In 1980, Dr. Taguchi traveled to the U.S.A. for the first time. He gave a series of talks and visited a number of companies. Acceptance of his ideas began slowly, but is gaining momentum, especially in the major electronic and automobile industries in the United States and Europe.

In the previous sections and chapters, we have focused on how the mean response changes as the factor levels change. In robust design, we pay considerable attention to how the variability of the response changes as the factor levels change. Two approaches to designing for the control of noise variability will be discussed in the following two subsections. The fundamental ideas are those proposed by Dr. Taguchi, but the methods of analysis discussed are those preferred by many statisticians.

7.6.1 Analysis of Design-by-Noise Interactions

In this subsection, the application of traditional statistical methods of analysis of a factorial experiment is illustrated for a robust product design experiment. Treatment combinations are combinations of levels of both design and noise factors. This is sometimes referred to as a *mixed array*. The interactions between design factors and noise factors are exploited to obtain the robust settings of the design factors. We illustrate the ideas via the following experiment.

Example 7.6.1 Torque optimization experiment

Rich Bigham (1987) reported on several experiments conducted at the Elsie Division of ITT Automotive to maximize the operating efficiency of car seat tracks produced by the division. One of the purposes of one of those experiments was to stabilize frame torque of a car seat track, with a target value of 14 ± 4 inch-pounds.

The experiment involved two design factors: "anvil type" (factor A) with levels "coined," "flat," and "crowned," coded 1–3, respectively, and "rivet diameter" (factor B) with levels 7.0, 7.5, and 8.0 mm, also coded 1–3, respectively. These are design factors because the best combination of their levels is desired for future production. Measurements were taken

7.6 Design for the Control of Noise Variability

Table 7.8 Torque optimization experiment data: Torque (inch-pounds)

AB	Machine $k=1$				Machine $k=2$			
	y_{ij11}	y_{ij12}	$\bar{y}_{ijk.}$	s^2_{ijk}	y_{ij21}	y_{ij21}	$\bar{y}_{ijk.}$	s^2_{ijk}
11	16	21	18.5	12.5	24	18	21.0	18.0
12	38	40	39.0	2.0	36	38	37.0	2.0
13	48	60	54.0	72.0	42	40	41.0	2.0
21	8	10	9.0	2.0	16	12	14.0	8.0
22	22	28	25.0	18.0	16	18	17.0	2.0
23	28	34	31.0	18.0	16	16	16.0	0.0
31	8	14	11.0	18.0	8	6	7.0	2.0
32	18	24	21.0	18.0	8	14	11.0	18.0
33	20	14	17.0	18.0	16	14	15.0	2.0

Source: Bigham, R. (1987). © Copyright, American Supplier Institute, Inc., Livonia, Michigan (U.S.A.). Reproduced by permission under License No. 980701.

on two different machines. "Machine" (factor M) is regarded as a noise factor because it is desirable to use both machines at the same settings of the design factors in the production process. In such a case, settings of the design factors are needed that give a nonvarying response across machines.

Two observations were collected for each of the 18 treatment combinations, as shown in Table 7.8. We use the three-way complete model to analyze the data; that is,

$$Y_{ijkt} = \mu + \tau_{ijk} + \epsilon_{ijkt}$$
$$= \mu + \alpha_i + \beta_j + \gamma_k + (\alpha\beta)_{ij} + (\alpha\gamma)_{ik} + (\beta\gamma)_{jk} + (\alpha\beta\gamma)_{ijk} + \epsilon_{ijkt}.$$

Plots of the standardized residuals against predicted values and normal scores do not reveal any concerns about the model, except that the two most extreme standardized residuals are ± 2.33, corresponding to treatment combination 131, and the two residuals for treatment combination 232 are zero. It is impossible, with such a small amount of data, to determine whether or not the error variances are unequal for different treatment combinations. Since this is an experiment for robust product design, confidence intervals and hypothesis tests (which rely heavily on the model assumptions) are not of great interest. As long as the model is a reasonable description of the data, the model assumptions do not need to be satisfied exactly.

The analysis of variance table is shown in Table 7.9. Of particular interest are the interactions between design factors and the noise factor. Although the model assumptions may not be valid, we can still gain an impression of the important effects from the list of p-values. For example, the AM interaction appears to be negligible. However, the BM interaction effect does appear to be important, and the ABM interaction should probably be considered also.

By appropriate choice of the levels of A and B, it may be possible to exploit these interactions to dampen the effects of the two machines (noise factor M) on response variability. To investigate the possibilities, an ABM interaction plot with the treatment means plotted against the levels of M is given in Figure 7.8, with AB combinations as labels. The objective is to choose a combination of levels of the design factors A and B for which two conditions

Table 7.9 Analysis of variance for the torque optimization experiment

Source of Variation	Degrees of Freedom	Sum of Squares	Mean Square	Ratio	p-value
A	2	3012.7222	1506.3611	116.62	0.0001
B	2	1572.0555	786.0277	60.85	0.0001
AB	4	470.4444	117.6111	9.11	0.0003
M	1	240.2500	240.2500	18.60	0.0004
AM	2	5.1666	2.5833	0.20	0.8205
BM	2	197.1666	98.5833	7.63	0.0040
ABM	4	170.6666	42.6666	3.30	0.0339
Error	18	232.5000	12.9166		
Total	35	5900.9722			

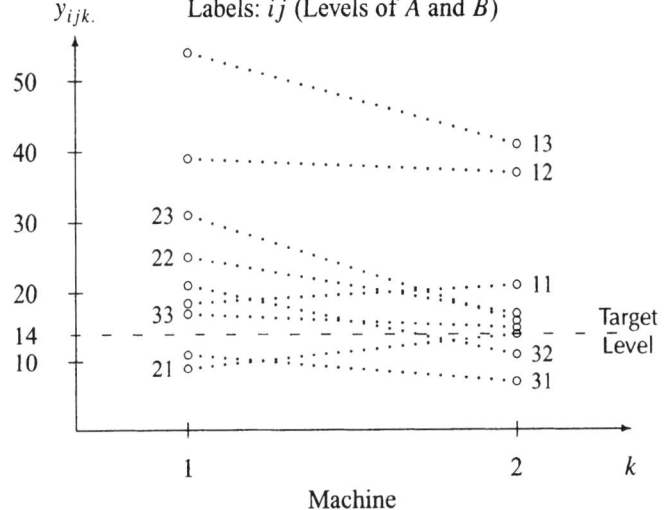

Figure 7.8 Interaction plot for the torque optimization experiment

are satisfied: First, the line connecting the two corresponding points should be relatively flat, so that there is little variation from machine to machine. Secondly, the mean response should be within 4 inch-pounds of the target value of 14 inch-pounds. Observe from Figure 7.8 that the treatment combination 33 satisfies both conditions rather well. If we had ignored variability, we might have been tempted to say that since $\bar{y}_{32.} = 16.0$, the average response for treatment combination 32 is approximately on target. However, the difference in mean response from one machine to the other results in much variation in the resulting torques, making treatment combination 32 an unsuitable choice.

Notice that the response is very consistent from machine to machine for treatment combination 12, while the mean response is way too high. In some experiments, this would not necessarily be bad. If there were another design factor, called an *adjustment factor*, that affected the mean response but not the variability, then that factor could be used to adjust the mean response on target while maintaining small response variability.

The writer of the original article selected treatment combination 33, since not only was it one of the most suitable treatment combinations in the study we have described, but it also appeared to be the best in a study that examined the effects on a different response variable. A followup study was done, examining the levels of B in more detail. The experimenters finally settled on a level of B that was halfway between levels 2 and 3, that is, 7.75 mm. □

7.6.2 Analyzing the Effects of Design Factors on Variability

The approach of the previous subsection, which relied on the subjective examination of an interaction plot to identify a robust design factor combination, can become complicated if an experiment involves many factors. Instead, we can calculate the sample variance of all the observations (for all the noise factor combinations) for each design factor combination. We then do two analyses, one examining the effects of the design factors on the observed sample variances and the other examining the effects on the sample means.

We design the experiment so that the same noise factor combinations are observed for each design factor combination. The list of design factor combinations in the experiment is called the *design array*. The list of noise factor combinations is called the *noise array*. Since the same noise array is used for each design factor combination, such an experimental design is called a *product array*.

There are two ways in which the design can be randomized. First, the experimental units can be assigned completely at random to the design–noise combinations, giving the usual completely randomized design. Alternatively, the randomization can be done as for a split-plot design (Section 2.4.4, page 21), where randomization is done for the design array first (the whole-plots) and then for the noise array (split-plots) separately for each design factor combination.

Suppose there are v design factor combinations and u noise factor combinations in the design. Denote a particular design factor combination by w and a noise factor combination by x, and the corresponding observation by y_{wx}. For each design factor combination w, there are u observations corresponding to the u noise factor combinations. Denote the sample mean and sample variance of these u observations by

$$\bar{y}_{w.} = \sum_x y_{wx}/u \quad \text{and} \quad s_w^2 = \sum_x (y_{wx} - \bar{y}_{w.})^2/(u-1).$$

Under the cell-means model,

$$Y_{wx} = \mu + \tau_{wx} + \epsilon_{wx},$$ (7.6.5)

$$\epsilon_{wx} \sim N(0, \sigma^2),$$

ϵ_{wx}'s are mutually independent,

$$w = 1, \ldots, v; \ x = 1, \ldots, u;$$

the sample means $\bar{Y}_{w.}$ are independently and normally distributed with common variance σ^2/u. Therefore, $\bar{Y}_{w.}$ can be used as the dependent variable in a cell-means model

$$\bar{Y}_{w.} = \mu + \alpha_w + \epsilon_w,$$

$$\epsilon_w \sim N(0, \sigma^2/u),$$

ϵ_w's are mutually independent,

$$w = 1, \ldots, v;$$

and we can analyze the effects of the design factors averaged over the noise factors via the usual analysis of variance, but with $\bar{y}_{w.}$, $w = 1, \ldots, v$, as the v observations.

The more difficult problem concerns how best to analyze the sample variances. The noise factor levels are systematically, rather than randomly, varied in the experiment. Nevertheless, s_w^2 can be used as a measure of the variability of the design factor combination w calculated over the levels of the noise factors.

The usual analysis of variance using s_w^2 as the response variable would not be appropriate, because $\mathrm{Var}(S_w^2)$ is not constant. It can be shown that

$$E[S_w^2] = \sigma^2 + Q(\tau_{wx}) \quad \text{and} \quad \mathrm{Var}[S_w^2] = \frac{2\sigma^4}{u-1} + \frac{4\sigma^2 Q(\tau_{wx})}{u-1},$$

where

$$Q(\tau_{wx}) = \sum_x (\tau_{wx} - \bar{\tau}_{w.})^2/(u-1) \text{ and } \bar{\tau}_{w.} = \sum_x \tau_{wx}/u.$$

It follows that

$$E[S_w^2](2\sigma^2/(u-1)) \leq \mathrm{Var}[S_w^2] \leq E[S_w^2]^2(2/(u-1)).$$

This is true even if σ^2 depends on w. So, $\mathrm{Var}[S_w^2]$ increases with $E[S_w^2]$. If $\mathrm{Var}[S_w^2]$ achieves it's upper bound, then a log transformation is appropriate to stabilize $\mathrm{Var}[S_w^2]$ (see Section 5.6.2, page 113). If it achieves its lower bound, then a square-root transformation is appropriate. Hence, provided that the distribution of $\ln(S_w^2)$ is approximately normal, we can use it as a response variable in an analysis of variance to analyze the effect of the design factors on the variability of the response as the levels of the noise factors change.

Since there is only one value of $\bar{y}_{w.}$ and one value of $\ln(s_w^2)$ for each design factor combination w, the effects of the design factors on $\bar{y}_{w.}$ and $\ln(s_w^2)$ must be analyzed like a single replicate design.

Example 7.6.2 Torque optimization experiment, continued

For illustration of the above methodology, we use the torque optimization experiment introduced in Example 7.6.1. The design array consisted of $v = 9$ combinations $w = ij$ of levels of two design factors, A and B, each having three levels. The noise array consisted of the $u = 4$ combinations $x = kt$, for two levels ($k = 1, 2$) of the noise factor, M, and for two observations ($t = 1, 2$) on each treatment combination. The corresponding values of $\bar{y}_{ij.}$, s_{ij}^2, and $\ln(s_{ij}^2)$, computed from the data in Table 7.8, are shown in Table 7.10.

Consider first the effects of the design factors on response variability. Figure 7.9(a) is an AB interaction plot for these data, with $\ln(s_{ij}^2)$ plotted against the level of factor A, and the level of factor B as the label. There appears to be a strong AB-interaction effect for the response variable $\ln(s_{ij}^2)$. This is the equivalent conclusion to that of observing an ABM interaction in Example 7.6.1, page 218.

The interaction plot for the effects of A and B on mean response is shown in Figure 7.9(b), and suggests that there is an interaction between the two factors when averaged over the

Table 7.10 Torque optimization experiment sample means and variances

AB	$\bar{y}_{ij.}$	s^2_{ij}	$\ln(s^2_{ij})$
11	19.75	12.2500	2.5055
12	38.00	2.6667	0.9808
13	47.50	81.0000	4.3945
21	11.50	11.6667	2.4567
22	21.00	28.0000	3.3322
23	23.50	81.0000	4.3945
31	9.00	12.0000	2.4849
32	16.00	45.3333	3.8140
33	16.00	8.0000	2.0794

noise factors. This is the equivalent conclusion to observing an AB interaction averaged over M in Example 7.6.1.

Since both design factors apparently affect both the mean response and the response variability, the experimenter needs to choose a design factor combination that is satisfactory with respect to both. Since the target response is 14 ± 4 inch-pounds, we see from Table 7.10 that acceptable levels of mean response are obtained only for treatment combinations 21, 32, and 33, with treatment combination 33 having yielded a much lower sample variance. This conclusion is in accordance with the conclusion from Example 7.6.1. Some additional observations should be taken to confirm that the design factor combination 33 is indeed a good choice, since the analysis leading to it was subjective—no inferential statistical methods were used. □

7.7 Using SAS Software

The analysis of experiments with three or more factors and at least one observation per cell uses the same types of SAS commands as illustrated for two factors in Section 6.8. In Section 7.7.1, we illustrate the additional commands needed to obtain a normal probability plot of the normalized contrast estimates in the drill advance experiment. In Section 7.7.3,

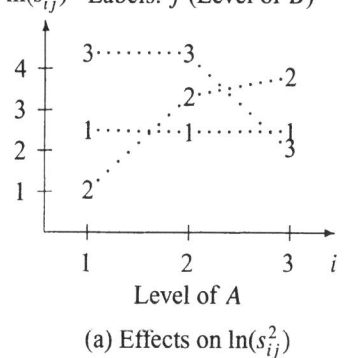
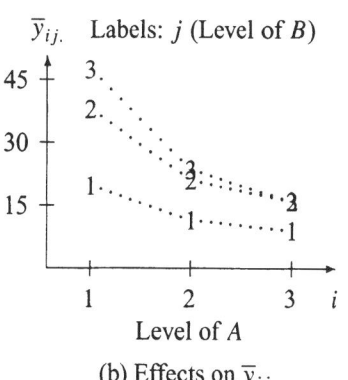

Figure 7.9 Interaction plots for the effects of Anvil Type (factor A) and Rivet Diameter (factor B) on $\ln(s^2_{ij})$ and $\bar{y}_{ij..}$ for the torque optimization experiment

(a) Effects on $\ln(s^2_{ij})$

(b) Effects on $\bar{y}_{ij..}$

we show how to calculate and plot the sample means and variances for robust design, and in Section 7.7.4, we show the complications that can arise when one or more cells are empty.

7.7.1 Normal Probability Plots of Contrast Estimates

In Table 7.11, we show a SAS program for producing a normal probability plot similar to that of Figure 7.7, page 215, for the contrast estimates of the drill advance experiment. The levels of A, B, C, and D together with the responses ADV are entered via the INPUT statement as usual. A log transformation is then taken so that the response Y used in the analysis is the log of the units of drill advance. Note that Y = LOG10(ADV) calculates log to the base 10, whereas Y = LOG(ADV) would calculate log to the base e, which is the more usual transformation. In the subsequent DATA statement, which creates data set DRILL2, the contrast coefficients are calculated for the main effects (for example, A = 2*A - 3), so that level 1 of a factor becomes contrast coefficient -1 (e.g., 2*1 - 3 = -1), and level 2 of a factor becomes coefficient $+1$. These coefficients could have been entered directly via the INPUT statement, as shown in Table 6.18, page 183. The interaction coefficients are obtained by multiplying together the main-effect coefficients (e.g., AB = A*B). The contrast coefficients are printed as columns similar to those in Table 7.1, page 201.

The contrast coefficients need to be divided by the selected divisor, multiplied by the corresponding responses, and then summed. In the data set DRILL3, we have calculated $2c_i y_i$ (for example, A = 2*A*Y). The subsequent PROC MEANS procedure calculates the averages of these; that is, $\Sigma_i (2c_i y_i)/v = \Sigma(c_i y_i)/(v/2)$. Thus each contrast effectively has divisor $v/2$. If a different divisor is required, then the 2 in the expressions 2*A*Y etc. would be adjusted accordingly.

In order to be able to plot the estimates, we need to gather them into the different values of a single variable. This is achieved by PROC TRANSPOSE, which turns the rows of the data set into columns. After deleting the first two values via the statement IF _N_>2 (these values are merely information for SAS), the resulting least squares estimates are listed as values of the variable EST1. The normal scores corresponding to the values of EST1 are calculated as in Chapter 5 using PROC RANK NORMAL=BLOM, and then printed. Finally, the last statement in Table 7.11 draws a plot of the least squares estimates versus the normal scores. Assuming effect sparsity, the nonnegligible contrasts are those whose estimates do not lie on the straight line.

7.7.2 Voss–Wang Confidence Interval Method

For analysis of a single-replicate experiment by the Voss–Wang method of simultaneous confidence intervals (Section 7.5.3), we first fit a full model. In the SAS output, all of the contrast sums of squares are calculated, and these can be rank ordered by hand. Notice the shortcut for fitting the full model in the first call of PROC GLM in Table 7.12. The four factors are listed in the MODEL statement, separated by vertical lines. SAS automatically adds all possible interactions to the model.

The quasi mean squared error msQ_k is most easily calculated as the error mean square obtained from the submodel that omits the terms corresponding to the d smallest contrast

Table 7.11 SAS program for a normal probability plot for the drill advance 2^4 experiment

```
DATA DRILL;
  INPUT A B C D  ADV;
  Y=LOG10(ADV);   *  log to base 10;
  LINES;
  1 1 1 1   1.68
  : : : :    :
  2 2 2 2   16.30
;
*  Calculate contrast coefficients for m contrasts and print them;
DATA DRILL2;    SET DRILL;
    A=2*A-3;    B=2*B-3;    C=2*C-3;    D=2*D-3;
    AB=A*B;    AC=A*C;    AD=A*D;    BC=B*C;    BD=B*D;    CD=C*D;
    ABC=AB*C;  ABD=AB*D;  ACD=AC*D;  BCD=BC*D;  ABCD=ABC*D;
PROC PRINT;
  VAR Y A B C D AB AC AD BC BD CD ABC ABD ACD BCD ABCD;
;
*  calculate  2*sum(c_iy_i)/v  which gives the contrast least
*       squares estimates with divisor v/2.  Then print these;
*       change the 2 to square root of v for normalized contrasts;
DATA DRILL3;    SET DRILL2;
    A=2*A*Y;       B=2*B*Y;       C=2*C*Y;      D=2*D*Y;
    AB=2*AB*Y;     AC=2*AC*Y;     AD=2*AD*Y;
    BC=2*BC*Y;     BD=2*BD*Y;     CD=2*CD*Y;
    ABC=2*ABC*Y;   ABD=2*ABD*Y;   ACD=2*ACD*Y;  BCD=2*BCD*Y;
  ABCD=2*ABCD*Y;
PROC MEANS NOPRINT;
  VAR A B C D AB AC AD BC BD CD ABC ABD ACD BCD ABCD;
  OUTPUT OUT=ESTIMATS
    MEAN=A B C D AB AC AD BC BD CD ABC ABD ACD BCD ABCD;
PROC PRINT;
*  At this point the data set has m least squares estimates
*  Turn the data set so that these form 2+m observations on a
*  single variable.  The first 2 are merely headings.  Remove them;
PROC TRANSPOSE PREFIX=EST OUT=ESTS;
DATA ESTS;    SET ESTS;    IF _N_>2;
;
*  Calculate the normal scores corresponding to the contrast estimates;
PROC RANK NORMAL=BLOM OUT=PLT;
  VAR EST1;
  RANKS NSCORE;
PROC PRINT;
*  Plot the contrast estimates against the normal scores;
PROC PLOT;
  PLOT EST1*NSCORE / VPOS=19 HPOS=50;
```

Table 7.12 SAS program for the Voss–Wang method for the drill advance experiment

```
* Data set DRILL contains the original data;
DATA DRILL4;  SET DRILL;
* Fit complete model including all main effects and interactions;
PROC GLM;
  CLASSES A B C D;
  MODEL Y = A | B | C | D;
;
* Data set DRILL2 contains the contrast coefficients and observations;
DATA DRILL5;  SET DRILL2;
* Calculate quasi mean squares for the Voss-Wang method as follows;
* Omit the d=8 smallest contrasts.  The resulting mean squared error
* is the msQ for confidence intervals for the m-d=7 largest contrasts;
PROC GLM;
  CLASSES A B C D;
  MODEL Y = C B D A CD AD ACD;
;
* Omit the d=8 smallest contrasts apart from ABD. The resulting mean
* squared error is the msQ for confidence intervals for ABD;
PROC GLM;
  CLASSES A B C D;
  MODEL Y = C B D A CD AD ABD;
;
* Omit the d=8 smallest contrasts apart from BC. The resulting mean
* squared error is the msQ for confidence intervals for BC;
PROC GLM;
  CLASSES A B C D;
  MODEL Y = C B D A CD AD BC;
;
* etc for each contrast in turn;
```

sums of squares (not counting ssc_k). A second run of the program is required in order to fit the submodels for each contrast in turn. Some of these submodels are shown in the SAS program in Table 7.12. The models must be fitted using the contrast coefficients. This is because a model of the form MODEL Y = C B D A C*D A*D A*C*D used in PROC GLM would result in 2 degrees of freedom being assigned to ACD, since the "subinteraction" AC of ACD is not in the model, and its information is assigned to ACD. We can calculate the contrast coefficients as described in the previous subsection and shown in the data set DRILL2 of Table 7.11, or we can enter them directly as in Table 6.18, page 183.

The critical values $v_{m,d,\alpha}$ for the Voss–Wang method are not obtainable through SAS, so the intervals must be completed by hand.

7.7.3 Identification of Robust Factor Settings

A SAS program for analyzing the torque optimization experiment of Examples 7.6.1 and 7.6.2 both as a mixed array and as a product array is shown in Table 7.13. A data set TORQUE2 containing the mean and variances of the responses at each combination of the A, B, and M levels is created via the PROC MEANS statement and printed as in Table 7.8, page 219. An interaction plot for ABM, such as that in Figure 7.8, page 220, is provided via the PROC PLOT statement. Notice that we have coded the treatment combinations as $1, 2, \ldots, 9$, rather than $11, 12, \ldots, 33$, since the labels on the SAS plot are single digits. The first call of PROC GLM analyzes the experiment as a mixed array with the noise factor M included in the model as a third treatment factor. The output gives the information in Table 7.9, page 220.

Some SAS commands for analysis of the experiment as a product array utilizing the mean and variance of the response is shown next in Table 7.13. The mean and variance of the responses at each combination of the A and B levels are created via the PROC MEANS statement, stored in data set TORQUE3 and printed as in Table 7.10, page 223. The second call of PROC GLM analyzes the mean response AV_Y, giving an analysis of variance table with no degrees of freedom for interaction, and PROC PLOT gives an interaction plot similar to that of Figure 7.9(a), page 223. The third calls of PROC GLM and PROC PLOT give the equivalent output for the log variance LVY of the response. The interpretation of all of these analyses was given in the examples of Section 7.6.

7.7.4 Experiments with Empty Cells

We now illustrate the use of SAS software for the analysis of an experiment with empty cells. No new SAS procedures or commands are introduced, but the empty cells can cause complications. For illustration, we use the following experiment.

Example 7.7.1 Rail weld experiment

S. M. Wu (1964) illustrated the usefulness of two-level factorial designs using the data listed in the SAS program of Table 7.14. Under investigation were the effects of three factors—ambient temperature (T), wind velocity (V), and rail steel bar size (S)—on the ultimate tensile strength of welds. The factor levels were $0°$ and $70°$ Fahrenheit for temperature, 0 and 20 miles per hour for wind velocity, and 4/11 and 11/11 inches for bar size, each coded as levels 1 and 2, respectively. Only six of the possible eight treatment combinations were observed, but $r = 2$ observations were taken on each of these six.

Some SAS commands for analyzing the rail weld experiment are presented in Table 7.14. Notice that rather than listing the observations for each treatment combination on separate lines, we have listed them as Y1 and Y2 on the same line. We have then combined the observations into the response variable Y. The new variable REP, which will be ignored in the model, is merely a device to keep the observations distinct. This method of input is often useful if the data have been stored in a table, with the observations for the same treatment combinations listed side by side, as in Table 7.2, page 209.

The three-way complete model is requested in the first call of PROC GLM in Table 7.14. The output is shown in Table 7.15. With two cells empty, there are data on only six treatment combinations, so there are only five degrees of freedom available for comparing treatments. This is not enough to measure the three main effects, the three two-factor interactions,

Table 7.13 SAS program for analysis of a mixed array and a product array

```
DATA TORQUE;
  INPUT A B M Y;
  TC = (3*(A-1)) + B;
  LINES;
  : : : (input data lines here)
;
*  calculate the average data values for each ABM combination;
PROC SORT;    BY A B M;
PROC MEANS  NOPRINT MEAN VAR;
  VAR Y;    BY A B M ;
  OUTPUT OUT=TORQUE2 MEAN=AV_Y VAR=VAR_Y;
DATA TORQUE2;    SET TORQUE2;
  TC = (3*(A-1)) + B;
PROC PRINT;
  VAR A B TC M AV_Y VAR_Y;
PROC PLOT;
  PLOT AV_Y*M=TC AV_Y*M=B/VPOS=19 HPOS=50;
;
*  fit model for mixed array analysis with three factors;
DATA TORQUE;    SET TORQUE;
PROC GLM;
  CLASSES A B M;
  MODEL Y = A B A*B M A*M B*M A*B*M;
;
*  calculate  average and log var of data for each AB combination;
PROC SORT;    BY A B ;
PROC MEANS NOPRINT MEAN VAR;
  VAR Y;    BY A B ;
  OUTPUT OUT=TORQUE3 MEAN=AV_Y VAR=VAR_Y;
DATA TORQUE3; SET TORQUE3;
  TC = (3*(A-1)) + B;
  LVY = LOG(VAR_Y);
PROC PRINT;
  VAR A B TC AV_Y VAR_Y LVY;
;
*  analysis as a product array with response AV_Y;
PROC GLM;
  CLASSES A B ;
  MODEL AV_Y = A B A*B  ;
PROC PLOT;
  PLOT AV_Y*A=B/VPOS=19 HPOS=50;
*  analysis as a product array with response log VAR_Y;
PROC GLM;
  CLASSES A B ;
  MODEL LVY = A B A*B;
PROC PLOT;
  PLOT LVY*A=B/VPOS=19 HPOS=50;
```

7.7 Using SAS Software

Table 7.14 SAS program for the rail weld experiment with two empty cells

```
DATA;
  INPUT T V S Y1 Y2;
  REP=1; Y=Y1; OUTPUT; * create SAS observation for y=y1;
  REP=2; Y=Y2; OUTPUT; * create SAS observation for y=y2;
  LINES;
1 1 1 84.0 91.0
1 1 2 77.7 80.5
2 1 1 95.5 84.0
2 1 2 99.7 95.4
2 2 1 76.0 98.0
2 2 2 93.7 81.7
PROC PRINT;
  VAR T V S REP Y;
* try to fit a 3-way complete model;
PROC GLM;
  CLASS T V S;
  MODEL Y = T | V | S;
  ESTIMATE 'TEMPERATURE' T  -1 1;
* fit a sub-model using 5 degrees of freedom;
PROC GLM;
  CLASS T V S;
  MODEL Y = T V S T*S V*S;
  ESTIMATE 'TEMPERATURE' T -1 1;
  ESTIMATE 'VELOCITY'    V -1 1;
  ESTIMATE 'SIZE'        S -1 1;
  ESTIMATE 'TEMPERATURE*SIZE' T*S 1 -1 -1 1/DIVISOR=2;
  ESTIMATE 'VELOCITY*SIZE'    V*S 1 -1 -1 1/DIVISOR=2;
```

Source: Data is from Wu, S. M. (1964). Copyright © 1964 American Welding Society. Reprinted with permission. (Reprinted University of Wisconsin Engineering Experiment Station, Reprint 684.)

and the three-factor interaction. This is indicated in the output, since two effects have zero degrees of freedom. The ESTIMATE statement for the contrast under the first call of PROC GLM generates no output. Instead, it generates a note in the SAS log indicating that the contrast is not estimable.

The only model that can be used is one that uses at most five degrees of freedom. Of course, this should be anticipated ahead of time during step (g) of the checklist (Chapter 2). Figure 7.10 illustrates with a solid ball at the corresponding corners of the cube the treatment combinations for which data are collected. One might guess that the TV interaction effect is not estimable, since data are only collected at three of the four combinations of levels of these two factors.

One possibility is to exclude from the complete model those interactions for which the Type I degrees of freedom are zero, namely the TV and TVS interaction effects. The contrast coefficient lists for the seven factorial effects are shown in Table 7.16. It is clear

Table 7.15 Output from the first call of PROC GLM for the rail weld experiment

```
                          The SAS System
                     General Linear Models Procedure

Dependent Variable: Y
                          Sum of           Mean
Source         DF        Squares         Square     F Value    Pr > F
Model           5       349.510000      69.902000      1.00    0.4877
Error           6       417.790000      69.631667
C.Total        11       767.300000


Source         DF       Type I SS    Mean Square    F Value    Pr > F
T               1       138.240000     138.240000      1.99    0.2085
V               1        79.380000      79.380000      1.14    0.3267
T*V             0         0.000000         .             .        .
S               1         0.003333       0.003333      0.00    0.9947
T*S             1       106.681667     106.681667      1.53    0.2620
V*S             1        25.205000      25.205000      0.36    0.5694
T*V*S           0         0.000000         .             .        .
```

Table 7.16 Contrast coefficients for the observed treatment combinations (T.C.) in the rail weld experiment

T.C.	T	V	TV	S	TS	VS	TVS
111	−1	−1	1	−1	1	1	−1
112	−1	−1	1	1	−1	−1	1
211	1	−1	−1	−1	−1	1	1
212	1	−1	−1	1	1	−1	−1
221	1	1	1	−1	−1	−1	−1
222	1	1	1	1	1	1	1

that the T and V contrasts are not orthogonal to the TV interaction contrast, and that the S, TS, and VS contrasts are not orthogonal to the TVS interaction contrast. Consequently, the incorrect omission of TV and TVS from the model will bias the estimates of all the other contrasts. If we do decide to exclude both the TV and TVS interaction effects, then the model is of the form

$$Y_{ijkl} = \mu + \alpha_i + \beta_j + \gamma_k + (\alpha\gamma)_{ik} + (\beta\gamma)_{ji} + \epsilon_{ijkl}.$$

We illustrate analysis of this model using the second call of PROC GLM in Table 7.14. Some of the output is shown in Table 7.17. The contrasts for T and V are not orthogonal to each other, but they can be estimated (although with a small positive correlation). Similar comments apply to the S, TS, and VS contrasts. None of the factorial effects appears particularly strong in Table 7.17.

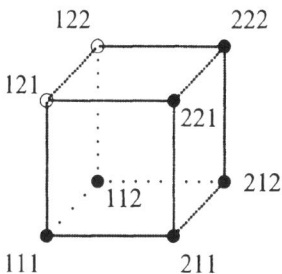

Figure 7.10 Treatment combinations included in the design of the rail weld experiment

Table 7.17 Output from the second call of PROC GLM

```
                      The SAS System

Source       DF    Type III SS    Mean Square    F Value    Pr > F
T             1     214.245000     214.245000       3.08    0.1300
V             1      79.380000      79.380000       1.14    0.3267
S             1      29.645000      29.645000       0.43    0.5383
T*S           1     131.220000     131.220000       1.88    0.2189
V*S           1      25.205000      25.205000       0.36    0.5694

                                    T for H0:   Pr > |T|   Std Error of
Parameter          Estimate       Parameter=0                Estimate
Temperature       10.3500000          1.75        0.1300     5.90049433
Velocity          -6.3000000         -1.07        0.3267     5.90049433
Size              -3.8500000         -0.65        0.5383     5.90049433
Temperature*Size   8.1000000          1.37        0.2189     5.90049433
Velocity*Size     -3.5500000         -0.60        0.5694     5.90049433
```

The ESTIMATE statements under the second call of PROC GLM generate the information shown in Table 7.17 for testing or constructing confidence intervals for the usual main effects and two-factor interaction effects under the given model. □

Exercises

1. For the following hypothetical data sets of Section 7.2.2 reproduced below, draw interaction plots to evaluate the BC and ABC interaction effects, with levels of B on the horizontal axis and levels of C for labels. In each case, comment on the apparent presence or absence of BC and ABC interaction effects.

   (a) ijk : 111 112 121 122 211 212 221 222 311 312 321 322
        $\bar{y}_{ijk.}$: 3.0 4.0 1.5 2.5 2.5 3.5 3.0 4.0 3.0 4.0 1.5 2.5

   (b) ijk : 111 112 121 122 211 212 221 222 311 312 321 322
        $\bar{y}_{ijk.}$: 3.0 2.0 1.5 4.0 2.5 3.5 3.0 4.0 3.0 5.0 3.5 6.0

2. In planning a five-factor experiment, it is determined that the factors A, B, and C might interact and the factors D and E might interact but that no other interaction effects should be present. Draw a line graph for this experiment and give an appropriate model.

3. Consider an experiment with four treatment factors, A, B, C, and D, at a, b, c, and d levels, respectively, with r observations per treatment combination. Assume that the four-way complete model is a valid representation of the data. Use the rules of Section 7.3 to answer the following.
   (a) Find the number of degrees of freedom associated with the AC interaction effect.
   (b) Obtain an expression for the sum of squares for AC.
   (c) Give a rule for testing the hypothesis that the AC interaction is negligible against the alternative hypothesis that it is not negligible. How should the results of the test be interpreted, given the other terms in the model?
   (d) Write down a contrast for measuring the AC interaction. Give an expression for its least squares estimate and associated variance.
   (e) Give a rule for testing the hypothesis that your contrast in part (d) is negligible.

4. **Popcorn–microwave experiment, continued**
   In the popcorn–microwave experiment of Section 7.4 (page 205), the experimenters studied the effects of popcorn brand, microwave oven power, and cooking time on the percentage of popped kernels in packages of microwave popcorn. Suppose that, rather than using a completely randomized design, the experimenters first collected all the observations for one microwave oven, followed by all observations for the other microwave oven. Would you expect the assumptions on the three-way complete model to be satisfied? Why or why not?

5. **Weathering experiment**
   An experiment is described in the paper "Accelerated weathering of marine fabrics" (Moore, M. A. and Epps, H. H., *Journal of Testing and Evaluation* 20, 1992, 139–143). The purpose of the experiment was to compare the effects of different types of weathering on the breaking strength of marine fabrics used for sails. The factors of interest were

   F: Fabric at 3 levels (1 = polyester, 2 = acrylic, 3 = nylon).

   E: Exposure conditions (1 = continuous light at 62.7°C, 2 = alternating 30 minutes light and 15 minutes condensation).

   A: Exposure levels (1 = 1200 AFU, 2 = 2400 AFU, 3 = 3600 AFU).

   D: Direction of cut of the fabric (1 = warp direction, 2 = filling direction).

   In total there were $v = 3 \times 2 \times 3 \times 2 = 36$ treatment combinations, and $r = 2$ observations were taken on each. The response variable was "percent change in breaking strength of fabric after exposure to weathering conditions." The average response for each of the 36 treatment combinations is shown in Table 7.18.
   (a) How would you decide whether or not the error variables have approximately the same variance for each fabric?

Table 7.18 Percent change in breaking strength of fabrics after exposure

Exposure (E)	AFU (A)	Direction (D)	Fabric (F)		
			1	2	3
1	1	1	−43.0	−1.7	−74.7
		2	−46.1	+11.7	−86.7
	2	1	−45.3	−4.2	−87.9
		2	−51.3	+10.0	−97.9
	3	1	−53.3	−5.1	−98.2
		2	−54.5	+7.5	−100.0
2	1	1	−48.1	−6.8	−85.0
		2	−43.6	−3.3	−91.7
	2	1	−52.3	−4.2	−100.0
		2	−53.8	−3.3	−100.0
	3	1	−56.5	−5.9	−100.0
		2	−56.4	−6.7	−100.0

Source: Moore, M. A. and Epps H. H., (1992). Copyright © ASTM. Reprinted with permission.

(b) Using the cell-means model, test the hypothesis $H_0 : [\tau_1 = \cdots = \tau_{36}]$ against the alternative hypothesis H_A : [at least two τ_i's differ]. What can you conclude?

(c) Write down a contrast in the treatment combinations that compares the polyester fabric with the nylon fabric. Is your contrast estimable?

(d) If your contrast in (c) is estimable, give a formula for the least squares estimator and its variance. Otherwise, go to part (e).

(e) Assuming that you are likely to be interested in a very large number of contrasts and you want your overall confidence level to be 95%, calculate a confidence interval for any pairwise comparison of your choosing. What does the interval tell you?

(f) Calculate a 90% confidence bound for σ^2.

(g) If you were to repeat this experiment and you wanted your confidence interval in (d) to be of length at most 20%, how many observations would you take on each treatment combination?

6. **Weathering experiment, continued**

Suppose you were to analyze the weathering experiment described in Exercise 5 using a four-way complete model.

(a) What conclusions can you draw from the analysis of variance table? (Be explicit about your overall error rate.)

(b) Give an explicit formula for testing that the FA-interaction is negligible.

(c) Would confidence intervals for differences in fabrics be of interest? If not, why not? If so, how would they be interpreted? Give a formula for such confidence intervals assuming that these intervals are preplanned and are the only intervals envisaged, and the overall level is to be at least 99%.

Table 7.19 Data (ml) for the evaporation experiment in the order observed

A	2	1	2	1	2	1	1	2
B	1	2	1	2	1	1	2	2
C	1	2	2	1	2	1	1	1
y_{ijkt}	17.5	9.0	19.5	7.0	19.0	8.0	8.0	17.0
A	1	1	1	1	2	2	2	2
B	1	1	2	1	2	2	2	1
C	2	1	2	2	2	1	2	1
y_{ijkt}	7.5	7.5	7.5	8.0	16.0	16.0	16.0	18.0

(d) In the original paper, the authors write "Fabric direction (D) had essentially no effect on percent change in breaking strength for any of the fabrics." Do you agree with this statement? Explain.

7. **Evaporation experiment**
(L. Jen, S.-M. Hsieh, P.-C. Kao, and M. Prenger, 1990)
The experimenters were interested in the evaporation rate of water under different conditions. Either 1 or 3 teaspoons of salt (levels 1 or 2 of factor B) were added to 100 ml of water and thoroughly stirred. The water was placed in a cup or on a plate (levels 1 or 2 of factor A), which was then placed on a windowsill or on the floor of a particular closet (levels 1 or 2 of factor C). After 48 hours, the amount of water remaining in the container was measured, and the response was the amount of water evaporated (100 ml less the amount remaining). Two observations were taken on each of the treatment combinations n a random order, as indicated in Table 7.19.

(a) Choose a model and outline the analysis that you would wish to perform for such an experiment (step (g) of the checklist; see Chapter 2).

(b) Using your choice of model, carry out your analysis as outlined in part (a). Check the assumptions on your model, allowing for the fact that there are only two observations for each treatment combination.

8. **Paper towel strength experiment**
(Burt Beiter, Doug Fairchild, Leo Russo, and Jim Wirtley, 1990)
The experimenters compared the relative strengths of two similarly priced brands of paper towel under varying levels of moisture saturation and liquid type. The treatment factors were "amount of liquid" (factor A, with levels 5 and 10 drops coded 1 and 2), "brand of towel" (factor B, with levels coded 1 and 2), and "type of liquid" (factor C, with levels "beer" and "water" coded 1 and 2). A $2 \times 2 \times 2$ factorial experiment with $r = 3$ was run in a completely randomized design. The resulting data, including run order, are given in Table 7.20.

(a) The experimenters assumed only factors A and B would interact. Specify the corresponding model.

(b) List all treatment contrasts that are likely to be of primary interest to the experimenters.

Table 7.20 Data for paper towel strength experiment: A ="amount of liquid," B ="brand of towel," and C ="liquid type"

ABC	Strength	(Order)	Strength	(Order)	Strength	(Order)
111	3279.0	(3)	4330.7	(15)	3843.7	(16)
112	3260.8	(11)	3134.2	(20)	3206.7	(22)
121	2889.6	(5)	3019.5	(6)	2451.5	(21)
122	2323.0	(1)	2603.6	(2)	2893.8	(14)
211	2964.5	(4)	4067.8	(10)	3327.0	(18)
212	3114.2	(12)	3009.3	(13)	3242.0	(19)
221	2883.4	(9)	2581.4	(23)	2385.9	(24)
222	2142.3	(7)	2364.9	(8)	2189.9	(17)

(c) Using the data in Table 7.20, draw an interaction plot for each interaction, and use the plots to assess the separability of effects assumed by the experimenters.

(d) Use residual plots to evaluate the adequacy of the model specified in part (a).

(e) Provide an analysis of variance table for this experiment, test the various effects, and draw conclusions.

(f) Construct confidence intervals for each of the treatment contrasts that you listed in part (b), using an appropriate method of multiple comparisons. Discuss the results.

9. **Rocket experiment**

S. R. Wood and D. E. Hartvigsen describe an experiment in the 1964 issue of *Industrial Quality Control* on the testing of an auxiliary rocket engine. According to the authors, the rocket engine must be capable of satisfactory operation after exposure to environmental conditions encountered during storage, transportation, and the in-flight environment. Four environmental factors were deemed important. These were vibration (Factor A; absent, present, coded 0, 1), temperature cycling (Factor B; absent, present, coded 0, 1), altitude cycling (Factor C; absent, present, coded 0, 1) and firing temperature/altitude (Factor D, 4 levels, coded 0, 1, 2, 3). The response variable was "thrust duration," and the observations are shown in Table 7.21, where C_k and D_l denote the kth level of C and the lth level of D, respectively.

The experimenters were willing to assume that the 3-factor and 4-factor interactions were negligible.

Table 7.21 Thrust duration (in seconds) for the rocket experiment

		C_0				C_1			
A	B	D_0	D_1	D_2	D_3	D_0	D_1	D_2	D_3
0	0	21.60	11.54	19.09	13.11	21.60	11.50	21.08	11.72
0	1	21.09	11.14	21.31	11.26	22.17	11.32	20.44	12.82
1	0	21.60	11.75	19.50	13.72	21.86	9.82	21.66	13.03
1	1	19.57	11.69	20.11	12.09	21.86	11.18	20.24	12.29
Total		83.86	46.12	80.01	50.18	87.49	43.82	83.42	49.86

Source: Wood, S. R. and Hartvigsen, D. E. (1964). Copyright 1964 American Society for Quality. Reprinted with permission.

(a) State a reasonable model for this experiment, including any assumptions on the error term.

(b) How would you check the assumptions on your model?

(c) Calculate an analysis of variance table and test any relevant hypotheses, stating your choice of the overall level of significance and your conclusions.

(d) Levels 0 and 1 of factor D represent temperatures $-75°F$ and $170°F$, respectively at sea level. Level 2 of D represents $-75°F$ at 35,000 feet. Suppose the experimenters had been interested in two preplanned contrasts. The first compares the effects of levels 0 and 1 of D, and the second compares the effects the levels 0 and 2 of D. Using an overall level of at least 98%, give a set of simultaneous confidence intervals for these two contrasts.

(e) Test the hypotheses that each contrast identified in part (d) is negligible. Be explicit about which method you are using and your choice of the overall level of significance.

(f) If the contrasts in part (d) had not been preplanned, would your answer to (d) have been different? If so, give the new calculations.

(g) Although it may not be of great interest in this particular experiment, draw an interaction plot for the CD interaction and explain what it shows.

(h) If the experimenters had included the 3-factor and 4-factor interactions in the model, how could they have decided upon the important main effects and interactions?

10. **Spectrometer experiment**

A study to determine the causes of instability of measurements made by a Baird spectrometer during production at North Star Steel Iowa was reported by J. Inman, J. Ledolter, R. V. Lenth, and L. Niemi in the *Journal of Quality Technology* in 1992. A brainstorming session with members of the Quality Assurance and Technology Department of the company produced a list of five factors that could be controlled and could be the cause of the observed measurement variability. The factors and their selected experimental levels were:

A: Temperature of the lab. (67°, 72°, 77°).

B: Cleanliness of entrance window seal (clean, one week's use).

C: Placement of sample (sample edge tangential to edge of disk, sample completely covering disk, sample partially covering disk).

D: Wear of boron nitride disk (new, one month old).

E: Sharpness of counterelectrode tip (newly sharpened, one week's wear).

Spectrometer measurements were made on several different elements. The manganese measurements are shown in Table 7.22, where A_i and B_j denote the ith level of A and the jth level of B, respectively. The experimenters were willing to assume that the 4-factor and 5-factor interactions were negligible.

(a) Test any relevant hypotheses, at a 0.05 overall level of significance, and state your conclusions.

Exercises 237

Table 7.22 Manganese data for the spectrometer experiment

C	D	E	A_1		A_2		A_3	
			B_1	B_2	B_1	B_2	B_1	B_2
1	1	1	0.9331	0.9214	0.8664	0.8729	0.8711	0.8627
1	1	2	0.9253	0.9399	0.8508	0.8711	0.8618	0.8785
1	2	1	0.8472	0.8417	0.7948	0.8305	0.7810	0.8009
1	2	2	0.8554	0.8517	0.7810	0.7784	0.7887	0.7853
2	1	1	0.9253	0.9340	0.8879	0.8729	0.8618	0.8692
2	1	2	0.9301	0.9272	0.8545	0.8536	0.8720	0.8674
2	2	1	0.8435	0.8674	0.7879	0.8009	0.7904	0.7793
2	2	2	0.8463	0.8526	0.7784	0.7863	0.7939	0.7844
3	1	1	0.9146	0.9272	0.8769	0.8683	0.8591	0.8683
3	1	2	0.9399	0.9488	0.8739	0.8729	0.8729	0.8481
3	2	1	0.8499	0.8417	0.7893	0.8009	0.7893	0.7904
3	2	2	0.8472	0.8300	0.7913	0.7904	0.7956	0.7827
Total			10.6578	10.6836	9.9331	9.9991	9.9376	9.9172

Source: Inman, J., Ledolter, J., Lenth, R. V. and Niemi, L. (1992). Copyright © 1997 American Society for Quality. Reprinted with Permission.

(b) Draw an interaction plot for the AE interaction. Does the plot show what you expected it to show? Why or why not? (Mention AE, A, and E.)

(c) The spectrometer manual recommends that the placement of the sample be at level 2. Using level 2 as a control level, give confidence intervals comparing the other placements with the control placement. You may assume that these comparisons were preplanned. State which method you are using and give reasons for your choice. Use an overall confidence level of at least 98% for these two intervals.

(d) Test the hypotheses of no linear and quadratic trends in the manganese measurements due to temperature. Use a significance level of at least 0.01 for each test.

11. **Galling experiment**

    A. Ertas, H. J. Carper, and W. R. Blackstone (1992, *Experimental Mechanics*) described an experiment to study the effects of speed (factor A at 1.5 and 5.0 rpm), surface roughness (factor B at 1.9 and 3.8 microns), and axial load (factor C at 413.4 and 689 MPa) on the "galling" of a metal collar. Galling is the name given to the failure phenomenon of severe adhesive wear, and the amount of galling in the experimental observations was scored with 0 for no galling and 10 for severe galling. The data are shown in Table 7.23.

    (a) Would you expect the assumptions on the three-way complete model to hold for these data? Why or why not?

    (b) Calculate the least squares estimates for a set of seven orthogonal contrasts, measuring the main effects and interactions.

Table 7.23 Scores for the galling experiment

A	1	1	1	1	2	2	2	2
B	1	1	2	2	1	1	2	2
C	1	2	1	2	1	2	1	2
y_{ijk}	2	5	0	2	6	10	4	8

Source: Ertas, A., Carper, H. J., and Blackstone, W. R. (1992). Published by the Society for Experimental Mechanics. Reprinted with permission.

(c) Draw a normal probability plot of the seven contrast estimates. Although $m = 7$ contrasts is too few to be able to draw good conclusions about the plot, which contrasts should be investigated in more detail later?

(d) Use the Voss–Wang procedure to examine the seven contrasts used in part (c). What conclusions can you draw about this experiment?

12. **Washing power experiment**

   E. G. Schilling (1973, *Journal of Quality Technology*) illustrates the estimation of orthogonal trend contrasts using a set of data originally collected by Feuell and Wagg in 1949. The Feuell and Wagg experiment investigated the washing power of a solution as measured by the reflectance of pieces of cotton cloth after washing. Pieces of cloth were soiled with colloidal graphite and liquid paraffin and then washed for 20 minutes at 60°C followed by two rinses at 40°C and 30°C, respectively. The three factors in the washing solution of interest were

   "sodium carbonate" (Factor A, levels 0%, 0.05%, and 0.1%);
   "detergent" (Factor B, levels 0.05%, 0.1%, and 0.2%);
   "sodium carboxymethyl cellulose" (Factor C, levels 0%, 0.025%, 0.05%).

   We code the levels of each factor as 1, 2, and 3. One observation was taken per treatment combination, and the responses are shown in Table 7.24.

Table 7.24 Data for the washing power experiment

A	1	1	1	1	1	1	1	1	1
B	1	1	1	2	2	2	3	3	3
C	1	2	3	1	2	3	1	2	3
y_{ijk}	10.6	14.9	18.2	19.8	24.3	23.2	27.0	31.5	34.0
A	2	2	2	2	2	2	2	2	2
B	1	1	1	2	2	2	3	3	3
C	1	2	3	1	2	3	1	2	3
y_{ijk}	19.7	25.5	25.9	32.9	36.4	38.9	36.1	39.0	40.6
A	3	3	3	3	3	3	3	3	3
B	1	1	1	2	2	2	3	3	3
C	1	2	3	1	2	3	1	2	3
y_{ijk}	22.3	29.4	29.7	32.0	41.0	41.6	32.1	41.5	38.7

Source: Schilling, E. G. (1973). Copyright © 1997 American Society for Quality. Reprinted with Permission.

(a) Make a table similar to that of Table 7.1, page 201, with the first column containing the 27 treatment combinations for the washing power experiment in ascending order. List the contrast coefficients for the main effect trend contrasts: Linear A, Quadratic A, Linear C, and Quadratic C. Also list the contrast coefficients for the interaction trend contrasts Linear $A\times$ Linear C, Linear $A\times$ Quadratic C, Quadratic $A\times$ Linear C, Quadratic $A\times$ Quadratic C.

(b) What divisors are needed to normalize each of the contrasts? Calculate, by hand, the least squares estimates for the normalized contrasts Linear A and Quadratic A.

(c) The levels of B are not equally spaced. Select two orthogonal contrasts that compare the levels of B and add these to your table in part (a).

(d) Use a computer program (similar to that of Table 7.11) to calculate the least squares estimates of a complete set of 26 orthogonal normalized contrasts that measure the main effects of A, B, and C and their interactions. Prepare a normal probability plot of the 26 contrast estimates. Explain what you can conclude from the plot.

(e) Use the method of Voss and Wang (Sections 7.5.3 and 7.7.2) to examine a complete set of 26 orthogonal normalized contrasts that measure the main effects of A, B, and C and their interactions. Compare your conclusions with those obtained from part (d).

13. **Paper towel experiment, continued**

Consider the paper towel strength experiment of Exercise 8. Suppose that *only the first ten observations* had been collected. These are labeled (1)–(10) in Table 7.20, page 235

(a) Is it possible to perform an analysis of variance of these data, using a model that includes main effects and the AB interaction as required by the experimenters? If so, analyze the experiment.

(b) Use a computer program to fit a three-way complete model. Can all of the main effects and interactions be measured? If not, investigate which models could have been used in the analysis of such a design with two empty cells and unequal numbers of observations in the other cells.

14. **Popcorn–robust experiment**

(M. Busam, M. Cooper, H. Livatyali, T. Miller and V. Vazquez, 1996)

The experimenters were interested in examining three brands of popcorn (factor A) and two types of oil (factor B) in terms of the percentage of edible kernels obtained after popping 200 kernels of corn for a certain length of time. The popping time (factor T) was regarded as a noise factor. The experimenters wanted to discover whether some combinations of popcorn brand and oil not only give a higher percentage of popped kernels than others, but also whether some combinations are more robust than others to variations in popping times. The selected levels of the noise factor T were 1.5 min, 1.75 min, and 2 min (coded 1, 2, and 3). The three brands of popcorn consisted of one store brand (level 1) and two different name brands (levels 2 and 3). The two types of

Table 7.25 Percentage of popped kernels for the robust popcorn experiment

	Time $k=1$		Time $k=2$		Time $k=3$	
AB	y_{ij11}	y_{ij12}	y_{ij21}	y_{ij21}	y_{ij31}	y_{ij31}
11	45.5	57.0	73.5	83.5	62.5	80.0
12	59.0	52.0	76.0	84.5	83.0	86.0
21	76.0	53.0	58.5	64.5	57.5	51.0
22	86.0	74.0	77.0	69.5	61.0	62.5
31	83.5	67.5	85.5	84.0	81.5	78.5
32	51.0	69.0	66.5	76.0	85.5	78.0

oil were a store brand corn oil (level 1) and a specialized name brand popping oil (level 2).

Two observations were taken on each brand–oil combination for each of the popping times. The observations were collected in a random order and are summarized in Table 7.25.

(a) Analyze the experiment as a mixed array, using a three-way complete model. Draw an ABT interaction plot, similar to that of Figure 7.8, page 220. If the goal of the experiment is to find brand–oil combinations that give a high percentage of edible kernels and that are not too sensitive to the popping time, what recommendations would you make?

(b) Does the store brand of popcorn differ substantially in terms of percentage of edible kernels from the average of the name brands? Do the different types of oil differ? State your overall confidence levels or significance levels.

(c) Analyze the experiment as a product array, and calculate the sample average and the log sample variance percentage of popped kernels for each brand–oil combination. Draw AB interaction plots similar to those of Figure 7.9, page 223. If the goal of the experiment is still to find brand–oil combinations that give a high percentage of edible kernels and that are not too sensitive to the popping time, what recommendations would you make? How do your recommendations compare with those that you made in part (a)?

15. **Steel bar experiment**

W. D. Baten (1956, *Industrial Quality Control*) described an experiment that investigated the cause of variability of the length of steel bars in a manufacturing process. Each bar was processed with one of two different heat treatments (factor A, levels 1, 2) and was cut on one of four different screw machines (factor B, levels 1, 2, 3, 4) at one of three different times of day (factor C, levels 8 am, 11 am, 3 pm, coded 1, 2, 3). There were considerable differences in the lengths of the bars after cutting, and a purpose for this experiment was to try to determine whether there were assignable causes for this variation.

(a) Discuss possible ways to design and analyze this experiment, but assume that it needs to be run in a working factory. In your discussion, consider using

(i) a completely randomized design,

Table 7.26 Data for the steel bar experiment

ABC	y_{1jk1}	y_{1jk2}	y_{1jk3}	y_{1jk4}	ABC	y_{2jk1}	y_{2jk2}	y_{2jk3}	y_{2jk4}
111	6	9	1	3	211	4	6	0	1
112	6	3	1	−1	212	3	1	1	−2
113	5	4	9	6	213	6	0	3	7
121	7	9	5	5	221	6	5	3	4
122	8	7	4	8	222	6	4	1	3
123	10	11	6	4	223	8	7	10	0
131	1	2	0	4	231	−1	0	0	1
132	3	2	1	0	232	2	0	−1	1
133	−1	2	6	1	233	0	−2	4	−4
141	6	6	7	3	241	4	5	5	4
142	7	9	11	6	242	9	4	6	3
143	10	5	4	8	243	4	3	7	0

Source: Baten, W. D. (1956). Copyright 1956 American Society for Quality. Reprinted with permission.

(ii) a randomized block design,

(iii) a design with times of day (factor C) regarded as a block factor,

(iv) a design with times of day (factor C) regarded as a noise factor.

(b) The randomization employed by the experimenter is not specified in the published article, and we proceed as though it were run as a completely randomized design with the three factors A, B, and C described above. List some of the sources of variation that must have been deemed as minor and ignored.

(c) The data that were collected by the experimenter are shown in Table 7.26. There are $r = 4$ observations on each of the $v = 24$ treatment combinations. The data values are "y_{ijkt} = (length −4.38)× 100 inches." Check the assumptions on the three-way complete model for these data. (You may wish to remove an outlier). If the assumptions are satisfied, calculate an analysis of variance table. What are your conclusions?

(d) Draw a machine × time interaction plot. Which machine is most robust to the time of day at which the bars are cut? What possible causes could there be for this difference in machines?

(e) The desired length for each bar was 4.385 ± 0.005 inches, which means that the desired value for the response y_{ijkt} is 5 units. Calculate confidence intervals for the true mean lengths of the bars cut on the four machines. Which machines appear to give bars closest to specification?

(f) Calculate the log sample variance of the twelve observations at each heat/machine combination. Is one of the heat treatments more robust than the other to the time of day at which the bars are cut? How do the machines compare in terms of the variability of the steel bar lengths? Does your conclusion agree with the one you made in part (d)?

(g) Given all the information that you have gained about this steel cutting process, what recommendations would you make to the management of this manufacturing company?

16. **Paper towel strength experiment, continued**

    Consider the paper towel strength experiment of Exercise 8, page 234. The 24 observations are shown in Table 7.20.

    (a) Explain why we could have regarded this as an experiment for the control of noise variability, with brand of paper towel as the design factor and the amount and type of liquid as noise factors.

    (b) Use the complete model and construct an analysis of variance table. For each significant interaction effect involving brand (if any), draw the corresponding interaction plot. Discuss the overall performance and relative robustness of the two brands of paper towel.

    (c) For each level j of brand of paper towel, there are twelve combinations ikt ($i = 1, 2$, $k = 1, 2$, $t = 1, 2, 3$) of levels of the noise factors "amount," "type of liquid," and observation number. For each j, calculate the sample variance s_j^2 and log sample variance $\ln(s_j^2)$ of these twelve observations. Also calculate the sample mean strength $\overline{y}_{.j..}$.

    (d) The best brand is that which has the greatest average strength and the smallest variance in the strength (calculated over the noise factors). Which brand would you recommend? Does your recommendation agree with your discussion in part (b)?

8 Polynomial Regression

8.1 Introduction
8.2 Models
8.3 Least Squares Estimation (Optional)
8.4 Test for Lack of Fit
8.5 Analysis of the Simple Linear Regression Model
8.6 Analysis of Polynomial Regression Models
8.7 Orthogonal Polynomials and Trend Contrasts (Optional)
8.8 A Real Experiment—Bean-Soaking Experiment
8.9 Using SAS Software

8.1 Introduction

In each of the previous chapters we were concerned with experiments that were run as completely randomized designs for the purpose of investigating the effects of one or more treatment factors on a response variable. Analysis of variance and methods of multiple comparisons were used to analyze the data. These methods are applicable whether factor levels are qualitative or quantitative.

In this chapter, we consider an alternative approach for quantitative factors, when the set of possible levels of each factor is real-valued rather than discrete. We restrict attention to a single factor and denote its levels by x. The mean response $E[Y_{xt}]$ is modeled as a polynomial function of the level x of the factor, and the points $(x, E[Y_{xt}])$ are called the *response curve*. For example, if $E[Y_{xt}] = \beta_0 + \beta_1 x$ for unknown parameters β_0 and β_1, then the mean response is a linear function of x and the response curve is a line, called the *regression line*. Using data collected at various levels x, we can obtain estimates $\hat{\beta}_0$ and $\hat{\beta}_1$ of the intercept and slope of the line. Then $\hat{y}_x = \hat{\beta}_0 + \hat{\beta}_1 x$ provides an estimate of $E[Y_{xt}]$

as a function of x, and it can be used to estimate the mean response or to predict the values of new observations for any factor level x, including values for which no data have been collected. We call $\hat{y}_x$ the *fitted model* or the *estimated mean response* at the level x.

In Section 8.2, we look at polynomial regression and the fit of polynomial response curves to data. Estimation of the parameters in the model, using the method of least squares, is discussed in the optional Section 8.3. In Section 8.4, we investigate how well a regression model fits a given set of data via a "lack-of-fit" test. In Section 8.5, we look at the analysis of a simple linear regression model and test hypotheses about the values of the model parameters. Confidence intervals are also discussed. The general analysis of a higher-order polynomial regression model using a computer package is discussed in Section 8.6. Investigation of linear and quadratic trends in the data via orthogonal polynomials is the topic of optional Section 8.7. An experiment is examined in detail in Section 8.8, and analysis using the SAS computer package is done in Section 8.9.

Polynomial regression methods can be extended to experiments involving two or more quantitative factors. The mean response $E[Y_{xt}]$ is then a function of several variables and defines a *response surface* in three or more dimensions. Specialized designs are usually required for fitting response surfaces, and consequently, we postpone their discussion to Chapter 16.

8.2 Models

The standard model for polynomial regression is

$$Y_{xt} = \beta_0 + \beta_1 x + \beta_2 x^2 + \cdots + \beta_p x^p + \epsilon_{xt}, \quad (8.2.1)$$

$$\epsilon_{xt} \sim N(0, \sigma^2),$$

ϵ_{xt}'s are mutually independent

$$t = 1, \ldots, r_x; \quad x = x_1, \ldots, x_v.$$

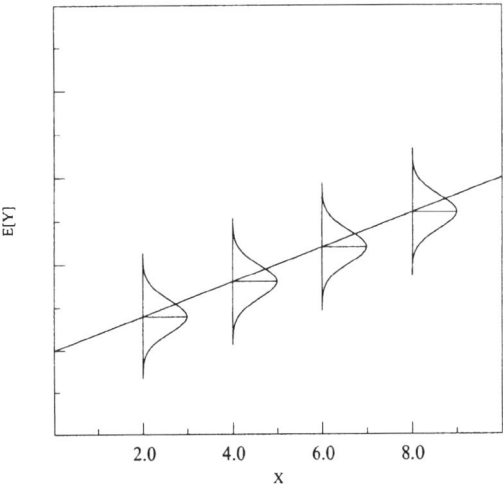

Figure 8.1
Simple linear regression model

8.2 Models

The treatment factor is observed at v different levels $x_1, \ldots, x_v$. There are r_x observations taken when the treatment factor is at level x, and Y_{xt} is the response for the tth of these. The responses Y_{xt} are modeled as independent random variables with mean

$$E[Y_{xt}] = \beta_0 + \beta_1 x + \beta_2 x^2 + \cdots + \beta_p x^p,$$

which is a pth-degree polynomial function of the level x of the treatment factor. Since $\epsilon_{xt} \sim N(0, \sigma^2)$, it follows that

$$Y_{xt} \sim N(\beta_0 + \beta_1 x + \beta_2 x^2 + \cdots + \beta_p x^p, \sigma^2).$$

Typically, in a given experiment, the exact functional form of the true response curve is unknown. In polynomial regression, the true response curve is assumed to be well approximated by a polynomial function. If the true response curve is relatively smooth, then a low-order polynomial function will often provide a good model, at least for a limited range of levels of the treatment factor.

If $p = 1$ in the polynomial regression function, we have the case known as *simple linear regression*, for which the mean response is

$$E[Y_{xt}] = \beta_0 + \beta_1 x,$$

which is a linear function of x. This model assumes that an increase of one unit in the level of x produces a mean increase of β_1 in the response, and is illustrated in Figure 8.1. At each value of x, there is a normal distribution of possible values of the response, the mean of which is the corresponding point, $E[Y_{xt}] = \beta_0 + \beta_1 x$, on the regression line and the variance of which is σ^2.

Consider now the data plotted in Figure 8.2, for which polynomial regression might be appropriate. Envisage a normal distribution of possible values of Y_{xt} for each level x, and a smooth response curve connecting the distribution of their means, $E[Y_{xt}]$. It would appear that a quadratic response curve may provide a good fit to these data. This case, for which

$$E[Y_{xt}] = \beta_0 + \beta_1 x + \beta_2 x^2,$$

is called *quadratic regression*. If this model is adequate, the fitted quadratic model can be used to estimate the value of x for which the mean response is maximized, even though it may not occur at one of the x values for which data have been collected.

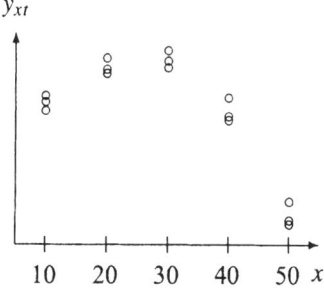

Figure 8.2 Three hypothetical observations y_{xt} at each of five treatment factor levels

Although regression models can be used to estimate the mean response at values of x that have not been observed, estimation outside the range of observed x values must be done with caution. There is no guarantee that the model provides a good fit outside the observed range.

If observations are collected for v distinct levels x of the treatment factor, then any polynomial regression model of degree $p \le v - 1$ (that is, with v or fewer parameters) can be fitted to the data. However, it is generally preferable to use the simplest model that provides an adequate fit. So for polynomial regression, lower-order models are preferred. Higher-order models are susceptible to *overfit*, a circumstance in which the model fits the data too well at the expense of having the fitted response curve vary or fluctuate excessively between data points. Over-fit is illustrated in Figure 8.3, which contains plots for a simple linear regression model and a sixth-degree polynomial regression model, each fitted to the same set of data. The sixth-degree polynomial model provides the better fit in the sense of providing a smaller value for the sum of squared errors. However, since we may be looking at natural fluctuation of data around a true linear model, it is arguable that the simple linear regression model is actually a better model—better for predicting responses at new values of x, for example. Information concerning the nature of the treatment factor and the response variable may shed light on which model is more likely to be appropriate.

Least squares estimates Once data are available, we can use the method of least squares to find estimates $\hat{\beta}_j$ of the parameters β_j of the chosen regression model. The fitted model is then

$$\hat{y}_x = \hat{\beta}_0 + \hat{\beta}_1 x + \hat{\beta}_2 x^2 + \cdots + \hat{\beta}_p x^p,$$

and the error sum of squares is

$$ssE = \sum_x \sum_t (y_{xt} - \hat{y}_x)^2.$$

The number of error degrees of freedom is the number of observations minus the number of parameters in the model; that is, $n - (p + 1)$. The mean squared error,

$$msE = \sum_x \sum_t (y_{xt} - \hat{y}_x)^2 / (n - p - 1),$$

provides an unbiased estimate of σ^2.

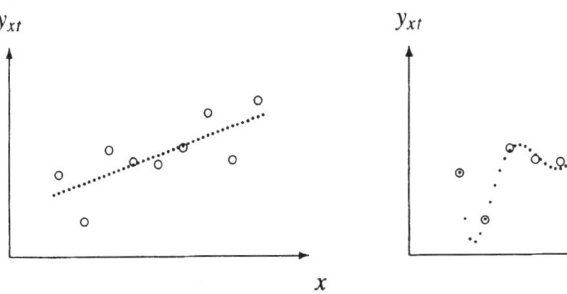

Figure 8.3 Data and fitted linear and sixth-degree polynomial regression models

(a) Simple linear regression model   (b) Degree 6 polynomial model

8.2 Models

In the following optional section, we obtain the least squares estimates of the parameters β_0 and β_1 in a simple linear regression model. However, in general we leave the determination of least squares estimates to a computer, since the formulae are not easily expressed without the use of matrices, and the hand computations are generally tedious. An exception to this occurs with the use of orthogonal polynomial models, discussed in Section 8.7.

Checking model assumptions Having made an initial selection for the degree of polynomial model required in a given scenario, the model assumptions should be checked. The first assumption to check is that the proposed polynomial model for $E[Y_{xt}]$ is indeed adequate. This can done either by examination of a plot of the residuals versus x or by formally testing for model lack of fit. The standard test for lack of fit is discussed in Section 8.4.

If no pattern is apparent in a plot of the residuals versus x, this indicates that the model is adequate. Lack of fit is indicated if there is a clear function-like pattern. For example, suppose a quadratic model is fitted but a cubic model is needed. Any linear or quadratic pattern in the data would then be explained by the model and would not be evident in the residual plot, but the residual plot would show the pattern of a cubic polynomial function unexplained by the fitted model (see Figure 8.4).

Residual plots can also be used to assess the assumptions on the random error terms in the model in the same way as discussed in Chapter 5. The residuals are plotted versus run order to evaluate independence of the error variables, plotted versus fitted values $\hat{y}_x$ to check the constant variance assumption and to check for outliers, and plotted versus the normal scores to check the normality assumption.

If the error assumptions are not valid, the fitted line still provides a model for mean response. However, the results of confidence intervals and hypothesis tests can be misleading. Departures from normality are generally serious problems only when the true error distribution has long tails or when prediction of a single observation is required. Nonconstant variance can sometimes be corrected via transformations, as in Chapter 5, but this may also change the order of the model that needs to be fitted.

If no model assumptions are invalidated, then analysis of variance can be used to determine whether or not a simpler model would suffice than the one postulated by the experimenter (see Section 8.6).

Figure 8.4
Plots for a quadratic polynomial regression model fitted to data from a cubic model

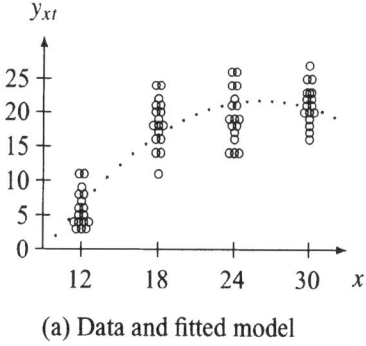
(a) Data and fitted model

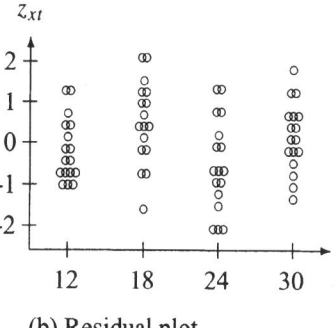
(b) Residual plot

8.3 Least Squares Estimation (Optional)

In this section, we derive the normal equations for a general polynomial regression model. These equations can be solved to obtain the set of least squares estimates $\hat{\beta}_j$ of the parameters β_j. We illustrate this for the case of simple linear regression.

8.3.1 Normal Equations

For the pth-order polynomial regression model (8.2.1), the normal equations are obtained by differentiating the sum of squared errors

$$\sum_x \sum_t e_{xt}^2 = \sum_x \sum_t (y_{xt} - \beta_0 - \beta_1 x - \cdots - \beta_p x^p)^2$$

with respect to each parameter and setting each derivative equal to zero. For example, if we differentiate with respect to β_j, set the derivative equal to zero, and replace each β_i with $\hat{\beta}_i$, we obtain the jth normal equation as

$$\sum_x \sum_t x^j y_{xt} = \sum_x \sum_t x^j \left(\hat{\beta}_0 + x\hat{\beta}_1 + \cdots + x^p \hat{\beta}_p \right). \tag{8.3.2}$$

We have one normal equation of this form for each value of j, $j = 0, 1, \ldots, p$. Thus, in total, we have $p+1$ equations in $p+1$ unknowns $\hat{\beta}_j$. Provided that the number of levels of the treatment factor exceeds the number of parameters in the model (that is, $v \geq p+1$), there is a unique solution to the normal equations giving a unique set of least squares estimates, with the result that all parameters are estimable.

8.3.2 Least Squares Estimates for Simple Linear Regression

For the simple linear regression model, we have $p = 1$, and there are two normal equations obtained from (8.3.2) with $j = 0, 1$. These are

$$\sum_x \sum_t y_{xt} = n\hat{\beta}_0 + \sum_x \sum_t x\hat{\beta}_1,$$

$$\sum_x \sum_t x y_{xt} = \sum_x \sum_t x\hat{\beta}_0 + \sum_x \sum_t x^2 \hat{\beta}_1,$$

where $n = \Sigma_x r_x$ denotes the total number of observations in the experiment. Dividing the first equation by n, we obtain

$$\hat{\beta}_0 = \overline{y}_{..} - \hat{\beta}_1 \overline{x}_{..}, \tag{8.3.3}$$

where $\overline{x}_{..} = \sum_x r_x x / n$. Substituting this into the second equation gives

$$\hat{\beta}_1 = \frac{\sum_x \sum_t x y_{xt} - n\overline{x}_{..}\overline{y}_{..}}{ss_{xx}}, \tag{8.3.4}$$

where $ss_{xx} = \sum_x r_x (x - \overline{x}_{..})^2$.

8.4 Test for Lack of Fit

We illustrate the lack-of-fit test via the quadratic regression model

$$E[Y_{xt}] = \beta_0 + \beta_1 x + \beta_2 x^2 .$$

If data have been collected for only three levels $x = x_1, x_2, x_3$ of the treatment factor, then the fitted model $\hat{y}_x = \hat{\beta}_0 + \hat{\beta}_1 x + \hat{\beta}_2 x^2$ will pass through the sample means $\bar{y}_{x.}$ computed at each value of x. This means that the predicted response $\hat{y}_x$ at the observed values of x is $\hat{y}_x = \bar{y}_{x.}$ (for $x = x_1, x_2, x_3$). This is the same fit as would be obtained using the one-way analysis of variance model, so we know that it is the best possible fit of a model to the data in the sense that no other model can give a smaller sum of squares for error, ssE.

If observations have been collected at more than three values of x, however, then the model is unlikely to fit the data perfectly, and in general, $\hat{y}_x \neq \bar{y}_{x.}$. If the values $\hat{y}_x$ and $\bar{y}_{x.}$ are too far apart relative to the amount of variability inherent in the data, then the model does not fit the data well, and there is said to be model *lack of fit*. In other words, in our example, the quadratic function is not sufficient to model the mean response $E[Y_{xt}]$.

If there is replication at one or more of the x-values, and if data are collected at more than three x-values, then it is possible to conduct a test for lack-of-fit of the quadratic model. The null hypothesis is that the quadratic model is adequate for modeling mean response; that is,

$$H_0^Q : E[Y_{xt}] = \beta_0 + \beta_1 x + \beta_2 x^2 .$$

The alternative hypothesis is that a more general model (the one-way analysis of variance model) is needed; that is,

$$H_A^Q : E[Y_{xt}] = \mu + \tau_x ,$$

where τ_x is the effect on the response of the treatment factor at level x. We fit the quadratic regression model and obtain ssE and $msE = ssE/(n-3)$. Now, MSE is an unbiased estimator of the error variance if the quadratic model is correct, but otherwise it has expected value larger than σ^2.

At each level x where more than one observation has been taken, we can calculate the sample variance s_x^2 of the responses. Each sample variance s_x^2 is an unbiased estimator of the error variance, σ^2, and these can be pooled to obtain the *pooled sample variance*,

$$s_p^2 = \left[\sum_x (r_x - 1)s_x^2 \right] / (n - v) . \tag{8.4.5}$$

Provided that the assumption of equal error variances is valid, the pooled sample variance is an unbiased estimator of σ^2 even if the model does not fit the data well. This pooled sample variance is called the *mean square for pure error* and denoted by $msPE$. An alternative way to compute $msPE$ is as the mean square for error obtained by fitting the one-way analysis of variance model.

The test of lack of fit, which is the test of H_0^Q versus H_A^Q, is based on a comparison of the two fitted models (the quadratic model and the one-way analysis of variance model), using the difference in the corresponding error sums of squares. We write ssE for the error sum of squares obtained from the quadratic regression model and $ssPE$ for the error sum of squares from the one-way analysis of variance model. Then the *sum of squares for lack of fit* is

$$ssLOF = ssE - ssPE.$$

The sum of squares for pure error has $n - v$ degrees of freedom associated with it, whereas the sum of squares for error has $n - (p+1) = n - 3$ (since there are $p + 1 = 3$ parameters in the quadratic regression model). The number of degrees of freedom for lack of fit is therefore $(n - 3) - (n - v) = v - 3$. The corresponding *mean square for lack of fit*,

$$msLOF = ssLOF/(v - 3),$$

measures model lack of fit because it is an unbiased estimator of σ^2 if the null hypothesis is true but has expected value larger than σ^2 otherwise.

Under the polynomial regression model (8.2.1) for $p = 2$, the decision rule for testing H_0^Q versus H_A^Q at significance level α is

reject H_0^Q if $msLOF/msPE > F_{v-3, n-v, \alpha}$.

In general, a polynomial regression model of degree p can be tested for lack of fit as long as $v > p + 1$ and there is replication for at least one of the x-levels. A test for lack of fit of the pth-degree polynomial regression model is a test of the null hypothesis

$$H_0^p : \{ E[Y_{xt}] = \beta_0 + \beta_1 x + \cdots + \beta_p x^p; \ x = x_1, \ldots, x_v \}$$

versus the alternative hypothesis

$$H_A^p : \{ E[Y_{xt}] = \mu + \tau_x; \ x = x_1, \ldots, x_v \}.$$

The decision rule at significance level α is

reject H_0^p if $msLOF/msPE > F_{v-p-1, n-v, \alpha}$,

where

$$msLOF = ssLOF/(v - p - 1) \quad \text{and} \quad ssLOF = ssE - ssPE.$$

Here, ssE is the error sum of squares obtained by fitting the polynomial regression model of degree p, and $ssPE$ is the error sum of squares obtained by fitting the one-way analysis of variance model.

Table 8.1 Hypothetical data for one continuous treatment factor

x	y_{xt}			$\bar{y}_{x.}$	s_x^2
10	69.42	66.07	71.70	69.0633	8.0196
20	79..91	81.45	85.52	82.2933	8.4014
30	88.33	82.01	84.43	84.9233	10.1681
40	62.59	70.98	64.12	65.8967	19.9654
50	25.86	32.73	24.39	27.6600	19.8189

Table 8.2 Test for lack of fit of quadratic regression model for hypothetical data

Source of Variation	Degrees of Freedom	Sum of Squares	Mean Square	Ratio	p-value
Lack of Fit	2	30.0542	15.0271	1.13	0.3604
Pure Error	10	132.7471	13.2747		
Error	12	162.8013			

Example 8.4.1 Lack-of-fit test for quadratic regression

In this example we conduct a test for lack of fit of a quadratic polynomial regression model, using the hypothetical data that were plotted in Figure 8.2 (page 245). Table 8.1 lists the $r = 3$ observations for each of $v = 5$ levels x of the treatment factor, together with the sample mean and sample variance. The pooled sample variance (8.4.5) is

$$s_p^2 = msPE = \sum_x 2s_x^2/(15-5) = 13.2747,$$

and the sum of squares for pure error is therefore

$$ssPE = (15-5)msPE = 132.7471.$$

Alternatively, this can be obtained as the sum of squares for error from fitting the one-way analysis of variance model.

The error sum of squares ssE is obtained by fitting the quadratic polynomial regression model using a computer program (see Section 8.9 for achieving this via SAS). We obtain $ssE = 162.8013$. Thus

$$ssLOF = ssE - ssPE = 162.8013 - 132.7471 = 30.0542$$

with

$$v - p - 1 = 5 - 2 - 1 = 2$$

degrees of freedom. The test for lack of fit is summarized in Table 8.2. Since the p-value is large, there is no significant lack of fit. The quadratic model seems to be adequate for these data. □

8.5 Analysis of the Simple Linear Regression Model

Suppose a linear regression model has been postulated for a given scenario, and a check of the model assumptions finds no significant violations including lack of fit. Then it is appropriate to proceed with analysis of the data.

It was shown in the optional Section 8.3 that the least squares estimates of the intercept and slope parameters in the simple linear regression model are

$$\hat{\beta}_0 = \bar{y}_{..} - \hat{\beta}_1 \bar{x}_{..} \quad \text{and} \quad \hat{\beta}_1 = \frac{\sum_x \sum_t xy_{xt} - n\bar{x}_{..}\bar{y}_{..}}{ss_{xx}}, \quad (8.5.6)$$

where $\bar{x}_{..} = \sum_x r_x x / n$ and $ss_{xx} = \sum_x r_x (x - \bar{x}_{..})^2$. The corresponding estimators (random variables), which we also denote by $\hat{\beta}_0$ and $\hat{\beta}_1$, are normally distributed, since they are linear combinations of the normally distributed random variables Y_{xt}. In Exercise 1, the reader is asked to show that the variances of $\hat{\beta}_0$ and $\hat{\beta}_1$ are equal to

$$\text{Var}(\hat{\beta}_0) = \sigma^2 \left(\frac{1}{n} + \frac{\bar{x}_{..}^2}{ss_{xx}} \right) \quad \text{and} \quad \text{Var}(\hat{\beta}_1) = \sigma^2 \left(\frac{1}{ss_{xx}} \right). \tag{8.5.7}$$

If we estimate σ^2 by

$$msE = \frac{\sum_x \sum_t (y_{xt} - (\hat{\beta}_0 + \hat{\beta}_1 x))^2}{n - 2}, \tag{8.5.8}$$

it follows that

$$\frac{\hat{\beta}_0 - \beta_0}{\sqrt{msE \left(\frac{1}{n} + \frac{\bar{x}_{..}^2}{ss_{xx}} \right)}} \sim t_{n-2} \quad \text{and} \quad \frac{\hat{\beta}_1 - \beta_1}{\sqrt{msE \left(\frac{1}{ss_{xx}} \right)}} \sim t_{n-2}.$$

Thus, the decision rule at significance level α for testing whether or not the intercept is equal to a specific value a ($H_0^{\text{int}} : \{\beta_0 = a\}$ versus $H_A^{\text{int}} : \{\beta_0 \neq a\}$) is

$$\text{reject } H_0^{\text{int}} \text{ if } \frac{\hat{\beta}_0 - a}{\sqrt{msE \left(\frac{1}{n} + \frac{\bar{x}_{..}^2}{ss_{xx}} \right)}} > t_{n-2,\alpha/2} \text{ or } < t_{n-2,1-\alpha/2}. \tag{8.5.9}$$

The decision rule at significance level α for testing whether or not the slope of the regression model is equal to a specific value b ($H_0^{\text{slp}} : \{\beta_1 = b\}$ versus $H_A^{\text{slp}} : \{\beta_1 \neq b\}$) is

$$\text{reject } H_0^{\text{slp}} \text{ if } \frac{\hat{\beta}_1 - b}{\sqrt{msE \left(\frac{1}{ss_{xx}} \right)}} > t_{n-2,\alpha/2} \text{ or } < t_{n-2,1-\alpha/2}. \tag{8.5.10}$$

Corresponding one-tailed tests can be constructed by choosing the appropriate tail of the t distribution and replacing $\alpha/2$ by α.

Confidence intervals at individual confidence levels of $100(1 - \alpha)\%$ for β_0 and β_1 are, respectively,

$$\hat{\beta}_0 \pm t_{n-2,\alpha/2} \sqrt{msE \left(\frac{1}{n} + \frac{\bar{x}_{..}^2}{ss_{xx}} \right)} \tag{8.5.11}$$

and

$$\hat{\beta}_1 \pm t_{n-2,\alpha/2} \sqrt{msE \left(\frac{1}{ss_{xx}} \right)}. \tag{8.5.12}$$

We can use the regression line to estimate the expected mean response $E[Y_{xt}]$ at any particular value of x, say x_a; that is,

$$\widehat{E}[Y_{x_a t}] = \hat{y}_{x_a} = \hat{\beta}_0 + \hat{\beta}_1 x_a.$$

8.5 Analysis of the Simple Linear Regression Model

The variance associated with this estimator is

$$\text{Var}(\hat{Y}_{x_a}) = \sigma^2 \left(\frac{1}{n} + \frac{(x_a - \bar{x}_{..})^2}{ss_{xx}} \right).$$

Since $\hat{Y}_{x_a}$ is a linear combination of the normally distributed random variables $\hat{\beta}_0$ and $\hat{\beta}_1$, it, too, has a normal distribution. Thus, if we estimate σ^2 by msE given in (8.5.8), we obtain a $100(1-\alpha)\%$ confidence interval for the expected mean response at x_a as

$$\hat{\beta}_0 + \hat{\beta}_1 x_a \pm t_{n-2,\alpha/2} \sqrt{msE \left(\frac{1}{n} + \frac{(x_a - \bar{x}_{..})^2}{ss_{xx}} \right)}. \qquad (8.5.13)$$

A confidence "band" for the entire regression line can be obtained by calculating confidence intervals for the mean response at all values of x. Since this is an extremely large number of intervals, we need to use Scheffé's method of multiple comparisons. So, a $100(1-\alpha)\%$ confidence band for the regression line is given by

$$\hat{\beta}_0 + \hat{\beta}_1 x_a \pm \sqrt{2 F_{2,n-2,\alpha}} \sqrt{msE \left(\frac{1}{n} + \frac{(x_a - \bar{x}_{..})^2}{ss_{xx}} \right)}. \qquad (8.5.14)$$

The critical coefficient here is $w = \sqrt{2\, F_{2,n-2,\alpha}}$ rather than the value $w = \sqrt{(v-1)\, F_{v-1,n-v,\alpha}}$ that we had in the one-way analysis of variance model, since there are only two parameters of interest in our model (instead of linear combinations of $v-1$ pairwise comparisons) and the number of error degrees of freedom is $n-2$ rather than $n-v$.

Finally, we note that it is also possible to use the regression line to predict a future observation at a particular value x_a of x. The predicted value $\hat{y}_{x_a}$ is the same as the estimated mean response at x_a obtained from the regression line; that is,

$$\hat{y}_{x_a} = \hat{\beta}_0 + \hat{\beta}_1 x_a.$$

The variance associated with this prediction is larger by an amount σ^2 than that associated with the estimated mean response, since the model acknowledges that the data values are distributed around their mean according to a normal distribution with variance σ^2. Consequently, we may adapt (8.5.13) to obtain a $100(1-\alpha)\%$ *prediction interval* for a future observation at x_a, as follows:

$$\hat{\beta}_0 + \hat{\beta}_1 x_a \pm t_{n-2,1-\alpha/2} \sqrt{msE \left(1 + \frac{1}{n} + \frac{(x_a - \bar{x}_{..})^2}{ss_{xx}} \right)}. \qquad (8.5.15)$$

Alternatively, the prediction interval follows, because

$$\frac{\hat{Y}_{x_a} - Y_{x_a}}{\sqrt{msE \left(1 + \frac{1}{n} + \frac{(x_a - \bar{x}_{..})^2}{ss_{xx}} \right)}} \sim t(n-2)$$

under our model.

254   Chapter 8   Polynomial Regression

Table 8.3 Fluid flow in liters/minute for the heart–lung pump experiment

rpm	Liters per minute				
50	1.158	1.128	1.140	1.122	1.128
75	1.740	1.686	1.740		
100	2.340	2.328	2.328	2.340	2.298
125	2.868	2.982			
150	3.540	3.480	3.510	3.504	3.612

Example 8.5.1   Heart–lung pump experiment, continued

In Example 4.2.4, page 72, a strong linear trend was discovered in the fluid flow rate as the number of revolutions per minute increases in a rotary pump head of an Olson heart–lung pump. Consequently, a simple linear regression model may provide a good model for the data. The data are reproduced in Table 8.3. It can be verified that

$$\bar{x}_{..} = \sum_x r_x x / n = [5(50) + 3(75) + 5(100) + 2(125) + 5(150)]/20 = 98.75,$$

and

$$\bar{y}_{..} = 2.2986 \quad \text{and} \quad \sum_x \sum_t x y_{xt} = 5212.8.$$

So,

$$ss_{xx} = [5(-48.75)^2 + 3(-23.75)^2 + 5(1.25)^2 + 2(26.25)^2 + 5(51.25)^2]$$
$$= 28,093.75,$$

giving

$$\hat{\beta}_1 = [5212.8 - 20(98.75)(2.2986)]/[28,093.75]$$
$$= 673.065/28,093.75 = 0.02396.$$

The mean square for error (8.5.8) for the regression model is best calculated by a computer package. It is equal to $msE = 0.001177$, so the estimated variance of $\hat{\beta}_1$ is

$$\text{Var}(\hat{\beta}_1) = msE/ss_{xx} = (0.001177)/28,093.75 = 0.000000042.$$

A 95% confidence interval for β_1 is then given by (8.5.12), as

$$0.02396 \pm t_{18,.025}\sqrt{0.000000042},$$
$$0.02396 \pm (2.101)(0.00020466),$$
$$(0.02353, 0.02439).$$

To test the null hypothesis $H_0^{slp} : \{\beta_1 = 0\}$, against the one-sided alternative hypothesis $H_A^{slp} : \{\beta_1 > 0\}$ that the slope is greater than zero at significance level $\alpha = 0.01$, we use a one-sided version of the decision rule (8.5.10) and calculate

$$\frac{\hat{\beta}_1 - 0}{\sqrt{msE\left(\frac{1}{ss_{xx}}\right)}} = \frac{0.02396}{0.00020466} = 117.07,$$

and since this is considerably greater than $t_{18,0.01} = 2.552$, we reject H_0^{slp}. We therefore conclude that the slope of the regression line is greater than zero, and the fluid flow increases as the revolutions per minute increase. □

8.6 Analysis of Polynomial Regression Models

8.6.1 Analysis of Variance

Suppose a polynomial regression model has been postulated for a given experiment, and the model assumptions appear to be satisfied, including no significant lack of fit. Then it is appropriate to proceed with analysis of the data. A common objective of the analysis of variance is to determine whether or not a lower-order model might suffice. One reasonable approach to the analysis, which we demonstrate for the quadratic model ($p = 2$), is as follows.

First, test the null hypothesis $H_0^L : \beta_2 = 0$ that the highest-order term $\beta_2 x^2$ is not needed in the model so that the simple linear regression model is adequate. If this hypothesis is rejected, then the full quadratic model is needed. Otherwise, testing continues and attempts to assess whether an even simpler model is suitable. Thus, the next step is to test the hypothesis $H_0 : \beta_1 = \beta_2 = 0$. If this is rejected, the simple linear regression model is needed and adequate. If it is not rejected, then x is apparently not useful in modeling the mean response.

Each test is constructed in the usual way, by comparing the error sum of squares of the full (quadratic) model with the error sum of squares of the reduced model corresponding to the null hypothesis being true. For example, to test the null hypothesis $H_0^L : \beta_2 = 0$ that the simple linear regression model is adequate versus the alternative hypothesis H_A^L that the linear model is not adequate, the decision rule at significance level α is

reject H_0^L if $ms(\beta_2)/msE > F_{1,n-v,\alpha}$,

where the mean square $ms(\beta_2) = ss(\beta_2)/1$ is based on one degree of freedom, and

$$ss(\beta_2) = ssE_1 - ssE_2,$$

where ssE_1 and ssE_2 are the error sums of squares obtained by fitting the models of degree one and two, respectively.

Similarly, the decision rule at significance level α for testing $H_0 : \beta_1 = \beta_2 = 0$ versus the alternative hypothesis that H_0 is false is

reject H_0 if $ms(\beta_1, \beta_2)/msE > F_{2,n-v,\alpha}$,

Table 8.4 Analysis of variance table for polynomial regression model of degree p. Here ssE_b denotes the error sum of squares obtained by fitting the polynomial regression model of degree b.

Source of Variation	Degrees of Freedom	Sum of Squares	Mean Square	Ratio
β_p	1	$ssE_{p-1} - ssE$	$ms(\beta_p)$	$ms(\beta_p)/msE$
β_{p-1}, β_p	2	$ssE_{p-2} - ssE$	$ms(\beta_{p-1}, \beta_p)$	$ms(\beta_{p-1}, \beta_p)/msE$
$\vdots$	$\vdots$	$\vdots$	$\vdots$	$\vdots$
$\beta_2, \ldots, \beta_p$	$p-1$	$ssE_1 - ssE$	$ms(\beta_2, \ldots, \beta_p)$	$ms(\beta_2, \ldots, \beta_p)/msE$
Model	p	$ssE_0 - ssE$	$ms(\beta_1, \ldots, \beta_p)$	$ms(\beta_1, \ldots, \beta_p)/msE$
Error	$n - p - 1$	ssE	msE	
Total	$n - 1$	$sstot$		

where the mean square $ms(\beta_1, \beta_2) = ss(\beta_1, \beta_2)/2$ is based on 2 degrees of freedom, and

$$ss(\beta_1, \beta_2) = (ssE_0 - ssE_2)/2 \,,$$

and ssE_0 and ssE_2 are the error sums of squares obtained by fitting the models of degree zero and two, respectively.

The tests are generally summarized in an analysis of variance table, as indicated in Table 8.4 for the polynomial regression model of degree p. In the table, under sources of variability, "Model" is listed rather than "$\beta_1, \ldots, \beta_p$" for the test of $H_0 : \beta_1 = \cdots = \beta_p = 0$, since this is generally included as standard output in a computer package. Also, to save space, we have written the error sum of squares as ssE for the full model, rather than indicating the order of the model with a subscript p. Analysis of variance for quadratic regression ($p = 2$) is illustrated in the following example.

Example 8.6.1 Analysis of variance for quadratic regression

Consider the hypothetical data in Table 8.1, page 250, with three observations for each of the levels $x = 10, 20, 30, 40, 50$. For five levels, the quartic model is the highest-order polynomial model that can be fitted to the data. However, a quadratic model was postulated for these data, and a test for lack of fit of the quadratic model, conducted in Example 8.4, suggested that this model is adequate.

The analysis of variance for the quadratic model is given in Table 8.5. The null hypothesis $H_0^L : \{\beta_2 = 0\}$ is rejected, since the p-value is less than 0.0001. So, the linear model is not adequate, and the quadratic model is needed. This is no surprise, based on the plot of the data shown in Figure 8.5.

Now, suppose the objective of the experiment was to determine how to maximize mean response. From the data plot, it appears that the maximum response occurs within the range of the levels x that were observed. The fitted quadratic regression model can be obtained from a computer program, as illustrated in Section 8.9 for the program SAS. The fitted

8.6 Analysis of Polynomial Regression Models

Table 8.5 Analysis of variance for the quadratic model

Source of Variation	Degrees of Freedom	Sum of Squares	Mean Square	Ratio	p-value
β_2	1	3326.2860	3326.2860	245.18	0.0001
Model	2	6278.6764	3139.3382	231.40	0.0001
Error	12	162.8013	13.5668		
Total	14	6441.4777			

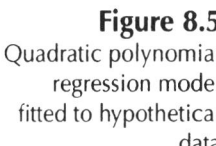

Figure 8.5
Quadratic polynomial regression model fitted to hypothetical data

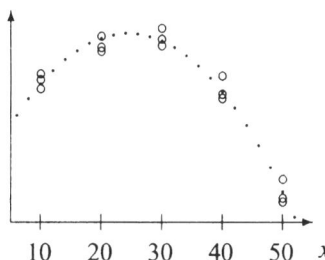

model is

$$\hat{y}_x = 33.43333 + 4.34754x - 0.08899x^2,$$

and is plotted in Figure 8.5 along with the raw data. The fitted curve achieves its maximum value when x is around 24.4, which should provide a good estimate of the level x that maximizes mean response. Further experimentation involving levels around this value could now be done. □

The adequacy of a regression model is sometimes assessed in terms of the proportion of variability in the response variable that is explained by the model. This proportion, which is the ratio of the model sum of squares to the sum of squares total, is called the *coefficient of multiple determination*, or the R^2-value. In the notation of Table 8.4,

$$R^2 = (ssE_0 - ssE)/sstot = ss(\beta_1, \ldots, \beta_p)/sstot.$$

For simple linear regression,

$$R^2 = ss(\beta_1)/sstot$$

is called the *coefficient of determination*, and in this case $R^2 = r^2$, where

$$r = ss_{xy}/\sqrt{ss_{xx} ss_{yy}}$$

is the usual *sample correlation*, or *Pearson product-moment correlation*.

8.6.2 Confidence Intervals

When the model is fitted via a computer program, the least squares estimates of $\hat{\beta}_j$ and their corresponding standard errors (estimated standard deviations) usually form part of the standard computer output. If the model assumptions are satisfied, then

$$\frac{\hat{\beta}_j - \beta_j}{\sqrt{\widehat{\text{Var}}(\hat{\beta}_j)}} \sim t_{n-p-1} .$$

Individual confidence intervals can be obtained for the model parameters, as we illustrated in Section 8.5 for the simple linear regression model. The general form is

$$\hat{\beta}_j \ \pm \ t_{n-2,\alpha/2} \sqrt{\widehat{\text{Var}}(\hat{\beta}_j)} .$$

Most programs will also allow calculation of the estimated mean response at any value of $x = x_a$ together with its standard error, and also calculation of the predicted response at $x = x_a$ plus its standard error. Confidence and prediction intervals for these can again be calculated using the t_{n-p-1} distribution. The confidence interval formula for mean response at $x = x_a$ is

$$\hat{\beta}_0 + \hat{\beta}_1 x_a + \cdots + \hat{\beta}_p x_a^p \ \pm \ t_{n-p-1,\alpha/2} \sqrt{\widehat{\text{Var}}(\hat{Y}_{x_a})}$$

and the prediction interval formula for a new observation at $x = x_a$ is

$$\hat{\beta}_0 + \hat{\beta}_1 x_a + \cdots + \hat{\beta}_p x_a^p \ \pm \ t_{n-p-1,\alpha/2} \sqrt{\hat{\sigma}^2 + \widehat{\text{Var}}(\hat{Y}_{x_a})} .$$

The overall confidence level for all the intervals combined should be computed via the Bonferroni method as usual. A confidence band for the regression line is obtained by calculating confidence intervals for the estimated mean response at all values of x, using the critical coefficient for Scheffé's method; that is,

$$\hat{\beta}_0 + \hat{\beta}_1 x + \cdots + \hat{\beta}_p x^p \ \pm \ \sqrt{(p+1) F_{p+1,n-p-1,\alpha}} \sqrt{\widehat{\text{Var}}(\hat{Y}_x)} .$$

8.7 Orthogonal Polynomials and Trend Contrasts (Optional)

The normal equations for polynomial regression were presented in equation (8.3.2). It was noted that solving the equations can be tedious. However, the factor levels can be transformed in such a way that the least squares estimates have a simple algebraic form and are easily computed. Furthermore, the parameter estimators become uncorrelated and are multiples of the corresponding trend contrast estimators. This transformation is illustrated in this section for simple linear regression and for quadratic regression, when the factor levels x are equally spaced with equal numbers r of observations per level.

8.7.1 Simple Linear Regression

Consider the simple linear regression model, for which

$$Y_{xt} = \beta_0 + \beta_1 x + \epsilon_{xt} ; \quad x = x_1, \ldots, x_v; \quad t = 1, \ldots, r . \tag{8.7.16}$$

8.7 Orthogonal Polynomials and Trend Contrasts (Optional)

When there are r observations on each of the v quantitative levels x of the treatment factor, the average value of x is $\bar{x}_{..} = r\sum_x x/n = \sum_x x/v$. The transformation $z_x = x - \bar{x}_{..}$ centers the levels x at zero, so that $\sum_x z_x = 0$. This makes the estimates of the slope and intercept parameters uncorrelated (or orthogonal). We can replace x in model (8.7.16) by z_x, so that the "centered" form of the model is

$$Y_{xt} = \beta_0^* + \beta_1^* z_x + \epsilon_{xt}; \quad x = x_1, \ldots, x_v; \quad t = 1, \ldots, r. \tag{8.7.17}$$

A transformation of the independent variable changes the interpretation of some of the parameters. For example, in the simple linear regression model (8.7.16), β_0 denotes mean response when $x = 0$, whereas in the transformed model (8.7.17), β_0^* denotes mean response when $z_x = 0$, which occurs when $x = \bar{x}_{..}$.

The normal equations corresponding to $j = 0$ and $j = 1$ for the centered model are obtained from (8.3.2) with z_x in place of x. Thus, we have

$$\sum_x \sum_t y_{xt} = \sum_x \sum_t \left(\hat{\beta}_0^* + z_x \hat{\beta}_1^*\right), \quad = vr\hat{\beta}_0^*$$

$$\sum_x \sum_t z_x y_{xt} = \sum_x \sum_t z_x \left(\hat{\beta}_0^* + z_x \hat{\beta}_1^*\right) = \sum_x r z_x^2 \hat{\beta}_1^*.$$

Solving these equations gives the least squares estimates as

$$\hat{\beta}_0^* = \bar{y}_{..} \quad \text{and} \quad \hat{\beta}_1^* = \frac{1}{r\sum_x z_x^2} \sum_x \sum_t z_x y_{xt}.$$

Now,

$$\text{Cov}\left(Y_{..}, \sum_x \sum_t z_x Y_{xt}\right) = \sum_x \sum_t z_x \text{Cov}(Y_{xt}, Y_{xt}) = r\sigma^2 \sum_x z_x = 0,$$

so the estimators $\hat{\beta}_0^*$ and $\hat{\beta}_1^*$ are uncorrelated.

We now consider a special case to illustrate the relationship of the slope estimator with the linear trend contrast that we used in Section 4.2.4. Suppose equal numbers of observations are collected at the three equally spaced levels

$$x_1 = 5, \quad x_2 = 7, \quad \text{and} \quad x_3 = 9.$$

Then $\bar{x}_{..} = 7$, so

$$z_5 = -2, \quad z_7 = 0, \quad \text{and} \quad z_9 = 2.$$

These values are twice the corresponding linear trend contrast coefficients $(-1, 0, 1)$ listed in Appendix A.2. Now, $r = 2$, so $r \sum_x z_x^2 = 8r$, and

$$\hat{\beta}_1^* = \frac{1}{r\sum_x z_x^2} \sum \sum z_x y_{z_x t} = \frac{1}{8r}(2y_{9.} - 2y_{5.})$$

$$= \frac{1}{4}(\bar{y}_{9.} - \bar{y}_{5.}),$$

which is a quarter of the value of the linear trend contrast estimate. It follows that β_1^* and the linear trend contrast have the same normalized estimate and hence also the same sum of squares. Thus, testing $H_0 : \beta_1^* = 0$ under model (8.7.17) is analogous to testing the hypothesis $H_0 : \tau_3 - \tau_1 = 0$ of no linear trend effect under the one-way analysis of variance model

$$Y_{it} = \mu + \tau_i + \epsilon_{it}; \quad i = 1, 2, 3; \quad t = 1, 2,$$

where τ_i is the effect on the response of the ith coded level of the treatment factor. The one difference is that in the first case, the model is the linear regression model ($p = 1$), while in the second case, the model is the one-way analysis of variance model, which is equivalent to a model of order $p = v - 1 = 2$. Thus the two models will not yield the same mean squared error, so the F-statistics will not be identical.

8.7.2 Quadratic Regression

Consider the quadratic regression model, for which

$$Y_{xt} = \beta_0 + \beta_1 x + \beta_2 x^2 + \epsilon_{xt}. \tag{8.7.18}$$

Assume that the treatment levels $x = x_1, \ldots, x_v$ are equally spaced, with r observations per level. To achieve orthogonality of estimates, it is necessary to transform both the linear and the quadratic independent variables.

Let $z_x = x - \bar{x}_{..}$ as in the case of simple linear regression, so that again $\sum_x z_x = 0$. Similarly, define

$$z_x^{(2)} = z_x^2 - \sum_x z_x^2 / v.$$

Then $\sum_x z_x^{(2)} = 0$. Also, writing z_i for the ith value of z_x in rank order, we note that since the levels x are equally spaced,

$$z_i = -z_{v+1-i} \quad \text{and} \quad z_i^{(2)} = z_{v+1-i}^{(2)},$$

so $\sum_x z_x z_x^{(2)} = 0$. These conditions give uncorrelated parameter estimators. To see this, consider the transformed model

$$Y_{xt} = \beta_0^* + \beta_1^* z_x + \beta_2^* z_x^{(2)} + \epsilon_{xt}. \tag{8.7.19}$$

The normal equations (8.3.2) become

$$\sum_x \sum_t y_{xt} = \sum_x \sum_t \left(\hat{\beta}_0^* + z_x \hat{\beta}_1^* + z_x^{(2)} \hat{\beta}_2^*\right) = vr\hat{\beta}_0^*,$$

$$\sum_x \sum_t z_x y_{xt} = \sum_x \sum_t z_x \left(\hat{\beta}_0^* + z_x \hat{\beta}_1^* + z_x^{(2)} \hat{\beta}_2^*\right) = r\sum_x z_x^2 \hat{\beta}_1^*,$$

$$\sum_x \sum_t z_x^{(2)} y_{xt} = \sum_x \sum_t z_x^{(2)} \left(\hat{\beta}_0^* + z_x \hat{\beta}_1^* + z_x^{(2)} \hat{\beta}_2^*\right) = r\sum_x (z_x^{(2)})^2 \hat{\beta}_2^*.$$

The least squares estimates, obtained by solving the normal equations, are

$$\hat{\beta}_0^* = \bar{y}_{..}, \quad \hat{\beta}_1^* = \frac{\sum_x \sum_t z_x y_{xt}}{r \sum_x z_x^2}, \quad \hat{\beta}_2^* = \frac{\sum_x \sum_t z_x^{(2)} y_{xt}}{r \sum_x (z_x^{(2)})^2}.$$

The estimators $\hat{\beta}_0^*$ and $\hat{\beta}_1^*$ are unchanged from the simple linear regression model (8.7.17), so they remain uncorrelated. Similarly, $\hat{\beta}_0^*$ and $\hat{\beta}_2^*$ are uncorrelated, because

$$\text{Cov}\left(Y_{..}, \sum_x \sum_t z_x^{(2)} Y_{xt}\right) = r\sigma^2 \sum_x z_x^{(2)} = 0.$$

Observe that $\text{Cov}(\hat{\beta}_1^*, \hat{\beta}_2^*)$ is also zero, since it is proportional to

$$\text{Cov}\left(\sum_x \sum_t z_x Y_{xt}, \sum_x \sum_t z_x^{(2)} Y_{xt}\right) = r\sigma^2 \sum_x z_x z_x^{(2)} = 0.$$

The transformed variables z_x and $z_x^{(2)}$ are called *orthogonal polynomials*, because they are polynomial functions of the levels x and give rise to uncorrelated parameter estimators $\hat{\beta}_0^*$, $\hat{\beta}_1^*$, and $\hat{\beta}_2^*$. It was illustrated in the previous subsection on simple linear regression that the values z_x are multiples of the coefficients of the linear trend contrast. Likewise, the values $z_x^{(2)}$ are multiples of the coefficients of the quadratic trend contrast. For example, suppose we have $r = 17$ observations on the equally spaced levels

$$x_1 = 12, \quad x_2 = 18, \quad x_3 = 24, \quad x_4 = 30.$$

Then $z_x = x - \bar{x}_{..}$, so

$$z_{12} = -9, \quad z_{18} = -3, \quad z_{24} = 3, \quad z_{30} = 9.$$

These are 3 times the linear trend contrast coefficients listed in Appendix A.2. Also, $\sum_x z_x^2/v = 45$, so

$$z_{12}^{(2)} = 36, \quad z_{18}^{(2)} = -36, \quad z_{24}^{(2)} = -36, \quad z_{30}^{(2)} = 36,$$

which are 36 times the quadratic trend contrasts.

As in the simple linear regression case, one can likewise show that the least squares estimates $\hat{\beta}_1^*$ and $\hat{\beta}_2^*$ are constant multiples of the corresponding linear and quadratic trend contrast estimates $\hat{\tau}_3 - \hat{\tau}_1$ and $\hat{\tau}_1 - 2\hat{\tau}_2 + \hat{\tau}_3$ that would be used in the one-way analysis of variance model. Consequently, the sums of squares for testing no quadratic trend and no linear trend are the same, although again, the error mean square will differ.

8.7.3 Comments

We have illustrated via two examples the equivalence between the orthogonal trend contrasts in analysis of variance and orthogonal polynomials in regression analysis for the case of equispaced, equireplicated treatment levels. While both are convenient tools for data analysis, identification of orthogonal trend contrasts and orthogonal polynomials can be rather complicated for higher-order trends, unequally spaced levels, or unequal numbers of observations per level. Fortunately, analogous testing information can also be generated by fitting appropriate full and reduced models, as was discussed in Section 8.6.1. This is easily accomplished using computer regression software. Use of the SAS software for such tests will be illustrated in Section 8.9.

8.8 A Real Experiment—Bean-Soaking Experiment

The bean-soaking experiment was run by Gordon Keeler in 1984 to study how long mung bean seeds ought to be soaked prior to planting in order to promote early growth of the bean sprouts. The experiment was run using a completely randomized design, and the experimenter used a one-way analysis of variance model and methods of multiple comparisons to analyze the data. In Section 8.8.2, we present the one-way analysis of variance, and then in Section 8.8.3, we reanalyze the data using polynomial regression methods.

8.8.1 Checklist

The following checklist has been drawn from the experimenter's report.

(a) **Define the objectives of the experiment.**
The objective of the experiment is to determine whether the length of the soaking period affects the rate of growth of mung bean seed sprouts. The directions for planting merely advise soaking overnight, and no further details are given.
As indicated in Figure 8.6, I expect to see no sprouting whatsoever for short soaking times, as the water does not have sufficient time to penetrate the bean coat and initiate sprouting. Then, as the soaking time is increased, I would expect to see a transition period of sprouting with higher rates of growth as water begins to penetrate the bean coat. Eventually, the maximum growth rate would be reached due to complete saturation of the bean. A possible decrease in growth rates could ensue from even longer soaking times due to bacterial infection and "drowning" the bean.

(b) **Identify all sources of variation.**
(i) Treatment factors and their levels.
There is just one treatment factor in this experiment, namely soaking time. A pilot experiment was run to obtain an indication of suitable times to be examined in the main experiment. The pilot experiment examined soaking times from 0.5 hour to 16 hours. Many beans that had been soaked for less than 6 hours failed to germinate, and at 16 hours the saturation point had not yet been reached. Consequently, the five equally spaced soaking times of 6, 12, 18, 24 and 30 hours will be selected as treatment factor levels for the experiment.
(ii) Experimental units.

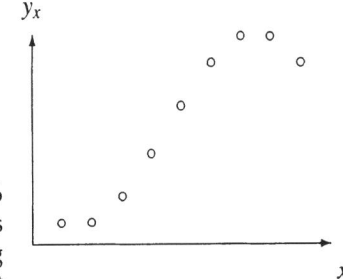

Figure 8.6
Anticipated results from the bean-soaking experiment

8.8 A Real Experiment—Bean-Soaking Experiment

The experimental units are the mung bean seeds selected at random from a large sack of approximately 10,000 beans.

(iii) Blocking factors, noise factors, and covariates.

Sources of variation that could affect growth rates include: individual bean differences; protozoan, bacterial, fungal, and viral parasitism; light; temperature; humidity; water quality.

Differences between beans will hopefully balance out in the random assignment to soaking times. Light, temperature, humidity, and water quality will be kept constant for all beans in the experiment. Thus, no blocking factors or covariates will be needed in the model.

Bacterial infection could differ from one treatment factor level to another due to soaking the beans in different baths. However, if the beans assigned to different treatment factor levels are soaked in the same bath, this introduces the possibility of a chemical signal from beans ready to germinate to the still dormant beans that sprouting conditions are prime. Consequently, separate baths will be used.

(c) **Choose a rule by which to assign experimental units to treatments.**

A completely randomized design will be used with an equal number of beans assigned to each soaking time.

(d) **Specify the measurements to be made, the experimental procedure, and the anticipated difficulties.**

The soaking periods will be started at 6-hour intervals, so that the beans are removed from the water at the same time. They will then be allowed to grow in the same environmental conditions for 48 hours, when the lengths of the bean sprouts will be measured (in millimeters).

The main difficulty in running the experiment is in controlling all the factors that affect growth. The beans themselves will be randomly selected and randomly assigned to soaking times. Different soaking dishes for the different soaking times will be filled at the same time from the same source.

On removal from the soaking dishes, the beans will be put in a growth chamber with no light but high humidity. During the pilot experiment, the beans were rinsed after 24 hours to keep them from dehydrating. However, the procedure cannot be well controlled from treatment to treatment, and will not be done in the main experiment.

A further difficulty is that of accurately measuring the shoot length.

(e) **Run a pilot experiment.**

A pilot study was run and the rest of the checklist was completed. As indicated in step (b), the results were used to determine the soaking times to be included in the experiment.

(f) **Specify the model.**

The one-way analysis of variance model (3.3.1) will be used, and the assumptions will be checked after the data are collected.

(g) **Outline the analysis.**
Confidence intervals for the pairwise differences in the effects of soaking time on the 48-hour shoot lengths will be calculated. Also, in view of the expected results, linear, quadratic and cubic trends in the shoot length will be examined. Tukey's method will be used for the pairwise comparisons with $\alpha_1 = 0.01$, and Bonferroni's method will be used for the three trend contrasts with overall level $\alpha_2 \leq 0.01$. The experimentwise error rate will then be at most 0.02.

(h) **Calculate the number of observations that need to be taken.**
Using the results of the pilot experiment, a calculation showed that 17 observations should be taken on each treatment (see Example 4.5, page 92).

(i) **Review the above decisions. Revise, if necessary.**
Since 17 observations could easily be taken for the soaking time, there was no need to revise the previous steps of the checklist.

The experiment was run, and the resulting data are shown in Table 8.6. The data for soaking time 6 hours have been omitted from the table, since none of these beans germinated.

The data are plotted in Figure 8.7 and show that the trend expected by the experimenter is approximately correct. For the soaking times included in the study, sprout length appears to increase with soaking time, with soaking times of 18, 24, and 30 hours yielding similar results, but a soaking of time of only 12 hours yielding consistently shorter sprouts.

8.8.2 One-Way Analysis of Variance and Multiple Comparisons

The experimenter used Tukey's method with a 99% simultaneous confidence level to compare the effects of soaking the beans for 12, 18, 24, or 30 hours. The formula for Tukey's

Table 8.6 Length of shoots of beans after 48 hours for the bean-soaking experiment

Soaking Time (hours)	r	Length (mm)						Average length	Sample variance
12	17	5	11	8	11	4	4	5.9412	7.0596
		8	3	6	4	7	3		
		5	4	6	9	3			
18	17	11	16	18	24	18	18	18.4118	12.6309
		21	14	21	19	17	24		
		14	20	16	20	22			
24	17	17	16	26	18	14	24	19.5294	15.6420
		18	14	24	26	21	21		
		22	19	14	19	19			
30	17	20	18	22	20	21	17	21.2941	8.5966
		16	23	25	19	21	20		
		27	25	22	23	23			

8.8 A Real Experiment—Bean-Soaking Experiment

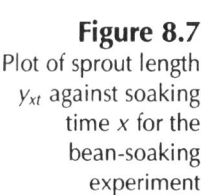

Figure 8.7
Plot of sprout length y_{xt} against soaking time x for the bean-soaking experiment

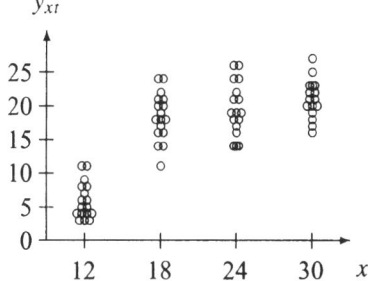

method for the one-way analysis of variance model was given in (4.4.28) as

$$\tau_i - \tau_s \in \left(\bar{y}_{i.} - \bar{y}_{s.} \pm w_T \sqrt{\left(\frac{2}{r}\right) msE} \right),$$

where $w_T = q_{v,n-v,\alpha}/\sqrt{2}$.

The treatment sample means are shown in Table 8.6. There are $r = 17$ observations on each of the $v = 4$ levels of the treatment factor. The formula for the sum of squares for error in the one-way analysis of variance model was given in (3.4.5), page 42. Using the data in Table 8.6 we have

$$msE = ssE/(n-v) = 10.9816.$$

From Table A.8, $q_{4,64,0.01} = 4.60$. Thus, in terms of the coded factor levels, the 99% simultaneous confidence intervals for pairwise comparisons are

$$\tau_4 - \tau_3 \in (-1.93, \ 5.46), \qquad \tau_3 - \tau_2 \in (-2.58, \ 4.81),$$
$$\tau_4 - \tau_2 \in (-0.81, \ 6.58), \qquad \tau_3 - \tau_1 \in (9.89, \ 17.29),$$
$$\tau_4 - \tau_1 \in (11.66, \ 19.05), \qquad \tau_2 - \tau_1 \in (8.77, \ 16.17).$$

From these, we can deduce that soaking times of 18, 24, and 30 hours yield significantly longer sprouts on average after 48 hours than does a soaking time of only 12 hours. The three highest soaking times are not significantly different in their effects on the sprout lengths, although the plot (Figure 8.7) suggests that the optimum soaking time might approach or even exceed 30 hours.

The one-way analysis of variance for the data is given in Table 8.7 and includes the information for testing for linear, quadratic, and cubic trends. The coefficients for the trend contrasts, when there are $v = 4$ equally spaced levels and equal sample sizes, are listed in Table A.2. The linear contrast is $[-3, -1, 1, 3]$, and the hypothesis of no linear trend is $H_0^L : \{-3\tau_1 - \tau_2 + \tau_3 + 3\tau_4 = 0\}$. Obtaining the treatment sample means from Table 8.6, the estimate of the linear trend is

$$\sum_i c_i \bar{y}_i = -3\bar{y}_{1.} - \bar{y}_{2.} + \bar{y}_{3.} + 3\bar{y}_{4.} = 47.1765,$$

with associated variance

$$\Sigma_i (c_i^2/r)\sigma^2 = (1/17)(9 + 1 + 1 + 9)\sigma^2 = (20/17)\sigma^2.$$

The sum of squares is calculated from (4.3.14), page 76; that is,

$$ssc = \left(\sum_i c_i \bar{y}_{i.}\right)^2 / \left(\sum_i c_i^2 / 17\right).$$

So, the sum of squares for the linear trend is

$$ssc = (47.1765)^2/(20/17) = 1891.78.$$

The quadratic and cubic trends correspond to the contrasts [1, −1, −1, 1] and [−1, 3, −3, 1], respectively, and their corresponding sums of squares are calculated in a similar way and are listed in Table 8.7. If we test the hypotheses that each of these three trends is zero with an overall significance level of $\alpha = 0.01$ using the Bonferroni method, then, using (4.4.24) on page 81 for each trend, the null hypothesis that the trend is zero is rejected if $ssc/msE > F_{1,64,0.01/3}$. This critical value is not tabulated, but since $F_{1,64,0.0033} = t^2_{1,64,0.00166}$, it can be approximated using (4.4.22) as follows:

$$t_{1,64,0.00166} \approx 2.935 + (2.935^3 + 2.935)/(4 \times 64) = 3.0454,$$

so the critical value is $F_{1,64,0.0033} \approx 9.2747$. (Alternatively, the critical value could be obtained from a computer packge using the "inverse cumulative distribution function" of the F-distribution.)

To test the null hypothesis H_0^L that the linear trend is zero against the alternative hypothesis $H_A^L : -3\tau_1 - \tau_2 + \tau_3 + 3\tau_4 \neq 0$ that the linear trend is nonzero, the decision rule is to

reject H_0 if $ssc/msE = 172.27 > F_{1,64,.0033} \approx 9.2747$.

Thus, using a simultaneous significance level $\alpha = 0.01$ for the three trends, the linear trend is determined to be nonzero.

The corresponding test ratios for the quadratic and cubic trends are given in Table 8.7. There is sufficient evidence to conclude that the linear, quadratic, and cubic trends are all significantly different from zero. The probability that one or more of these hypotheses would be incorrectly rejected by this procedure is at most $\alpha = 0.01$.

Table 8.7 One-way ANOVA for the bean-soaking experiment

Source of Variation	Degrees of Freedom	Sum of Squares	Mean Square	Ratio	p-value
Soaking Time	3	2501.29	833.76	75.92	0.0001
Linear Trend	1	1891.78	1891.78	172.27	0.0001
Quadratic Trend	1	487.12	487.12	44.36	0.0001
Cubic Trend	1	122.40	122.40	11.15	0.0014
Error	64	702.82	10.98		
Total	67	3204.12			

8.8.3 Regression Analysis

In the previous subsection, the bean-soaking experiment was analyzed using the one-way analysis of variance and multiple comparison methods. In this subsection, we reanalyze the experiment using regression analysis. Since there are four levels of the treatment factor "soaking time," the highest-order polynomial regression model that can be (uniquely) fitted to the data is the cubic regression model, namely,

$$Y_{xt} = \beta_0 + \beta_1 x + \beta_2 x^2 + \beta_3 x^3 + \epsilon_{xt},$$
$$\epsilon_{xt} \sim N(0, \sigma^2),$$

ϵ_{xt}'s are mutually independent,

$$x = 12, 18, 24, 30; \quad t = 1, \ldots, 17.$$

Using the data given in Table 8.6, the fitted model can be obtained from a computer program (see Section 8.9) as

$$\hat{y}_x = -101.058824 + 15.475490x - 0.657680x^2 + 0.009259x^3.$$

Table 8.8 contains the analysis of variance for the bean experiment data based on the cubic regression model. The cubic model provides the same fit as does the one-way analysis of variance model, since $p + 1 = v = 4$. Thus, $\hat{y}_x = \bar{y}_{x.}$ for $x = 12, 18, 24, 30$, and the number of degrees of freedom, the sum of squares, and the mean square for the major sources of variation—the treatment factor ("Model"), error, and total—are the same in the regression analysis of variance as in the one-way analysis of variance. It is not possible to test for model lack of fit, since the postulated model is of order $p = 3 = v - 1$. We can, however, test to see whether a lower-order model would suffice.

We first test the null hypothesis $H_0^Q : \beta_3 = 0$, or equivalently, that the quadratic regression model $E[Y_{xt}] = \beta_0 + \beta_1 x + \beta_2 x^2$ would provide an adequate fit to the data. The result of the test is summarized in Table 8.8. The test ratio is 11.15 with a p-value of 0.0014. So, we reject H_0^Q and conclude that the cubic model is needed. Since the cubic regression model provides the same fit as the analysis of variance model, this test is identical to the test that the cubic trend contrast is zero in the one-way analysis of variance, shown in Table 8.7.

If $H_0^Q : \beta_3 = 0$ had not been rejected, then the next step would have been to have tested the null hypothesis $H_0^L : \beta_2 = \beta_3 = 0$, or equivalently, that the simple linear regression model is adequate. If neither $H_0^Q : \beta_3 = 0$ nor $H_0^L : \beta_2 = \beta_3 = 0$ had been rejected, the next step would have been to have tested $H_0 : \beta_1 = \beta_2 = \beta_3 = 0$.

Table 8.8 Cubic regression ANOVA for the bean-soaking experiment

Source of Variation	Degrees of Freedom	Sum of Squares	Mean Square	Ratio	p-value
β_3	1	122.40	122.40	11.15	0.0014
β_2, β_3	2	609.52	304.76	27.76	0.0001
Model	3	2501.29	833.76	75.92	0.0001
Error	64	702.82	10.98		
Total	67	3204.12			

Figure 8.8
Plot of data and fitted cubic polynomial regression model for the bean-soaking experiment

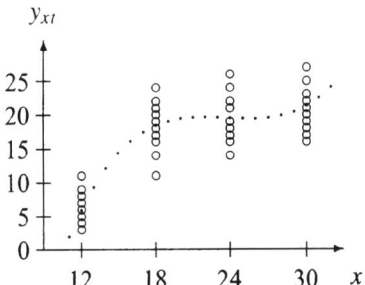

Based on the previous analysis, the cubic model is needed to provide an adequate fit to the data. Figure 8.8 illustrates the cubic model fitted to the data. We may now see the dangers of using a model to predict the value of the response beyond the range of observed x values. The cubic model predicts that mean sprout length will increase rapidly as soaking time is increased beyond 30 hours! Clearly, this model is extremely unlikely to be reliable for extrapolation beyond 30 hours.

Recall that Tukey's method of multiple comparisons did not yield any significant differences in mean response between the soaking times of 18, 24, and 30 hours. Yet the plot of the data in Figure 8.8 suggests that a trend over these levels might well exist. There is a lot of variability inherent in the data that prevents significant differences between the soaking times from being detected. Nevertheless, a followup experiment examining soaking times from 18 to, say, 48 hours might provide the information needed to determine the best range of soaking times.

8.9 Using SAS Software

Polynomial regression models can be fitted using the SAS regression procedure PROC REG. The procedure provides least squares estimates of the regression parameters. Predicted (fitted) values and residuals can be saved to an output data set, as can 95% confidence limits for mean response, 95% prediction limits for new observations for given treatment levels x, and corresponding standard errors.

A sample SAS program to analyze the data from the bean-soaking experiment of Section 8.8 is shown in Table 8.9. In the first DATA statement, the variables x^2 and x^3 are created for the cubic regression model. PROC REG is used to fit the cubic regression model, and the output is shown in Table 8.10.

An analysis of variance table is automatically generated and includes information needed for testing the hypothesis that the treatment factor "soaking time" has no predictive value for mean growth length, namely, $H_0 : \{\beta_1 = \beta_2 = \beta_3 = 0\}$. The information for this test is listed with source of variation "Model." We see that the p-value is less than 0.0001, so H_0 would be rejected.

Below the analysis of variance table, parameter estimates for the fitted model are given. Using these, we have the fitted cubic regression model

$$\hat{y}_x = -101.058824 + 15.475490x - 0.657680x^2 + 0.009259x^3 .$$

Table 8.9 SAS program for analysis of the bean-soaking experiment

```
OPTIONS LS=72 NOOVP;
DATA BEAN;
  INPUT X LENGTH;
  X2=X**2; X3=X**3;
  LINES;
  12   5
  12  11
  12   8
  12  11
   :   :
  30  23
  30  23
;
PROC PRINT;
;
* create extra x-values for plotting the fitted curve;
DATA TOPLOT;
  DO X=8 TO 34;  X2=X**2; X3=X**3;
    LENGTH=.; * "." denotes a missing value;
    OUTPUT;
    END; * X loop;
;
* concatenate data sets BEAN and TOPLOT;
DATA; SET BEAN TOPLOT;
;
* do the analysis;
PROC REG;  MODEL LENGTH = X X2 X3 / SS1;
  QUAD:    TEST X3=0; * test adequacy of quadratic model;
  LINEAR:  TEST X2=0, X3=0; * test adequacy of linear model;
  OUTPUT PREDICTED=LHAT RESIDUAL=E
     L95M=L95M U95M=U95M STDP=STDM   L95=L95I U95=U95I STDI=STDI;
;
* plot the data and fitted model, overlayed on one plot;
PROC PLOT; PLOT LENGTH*X LHAT*X='*' / OVERLAY VPOS=20 HPOS=58;
;
* 95% confidence intervals and standard errors for mean response;
PROC PRINT; VAR X L95M LHAT U95M STDM;
* 95% prediction intervals and standard errors for new observations;
PROC PRINT; VAR X L95I LHAT U95I STDI;
;
* generate residual plots;
PROC RANK NORMAL=BLOM; VAR E; RANKS NSCORE;
PROC PLOT;   PLOT E*X E*LHAT / VREF=0 VPOS=19 HPOS=50;
             PLOT E*NSCORE / VREF=0 HREF=0 VPOS=19 HPOS=50;
```

270   Chapter 8   Polynomial Regression

Table 8.10 Output generated by PROC REG

```
                         The SAS System
Model: MODEL1
Dependent Variable: LENGTH

                      Analysis of Variance

                    Sum of         Mean
  Source    DF     Squares       Square     F Value    Prob>F
  Model      3   2501.29412    833.76471    75.924     0.0001
  Error     64    702.82353     10.98162
  C Total   67   3204.11765

      Root MSE     3.31385    R-square     0.7806
      Dep Mean    16.29412    Adj R-sq     0.7704
      C.V.        20.33772

                       Parameter Estimates

                  Parameter    Standard    T for H0:
  Variable   DF    Estimate       Error   Parameter=0   Prob > |T|
  INTERCEP    1  -101.058824  21.87851306     -4.619      0.0001
  X           1    15.475490   3.49667261      4.426      0.0001
  X2          1    -0.657680   0.17508291     -3.756      0.0004
  X3          1     0.009259   0.00277344      3.339      0.0014

  Variable   DF    Type I SS
  INTERCEP    1       18054
  X           1   1891.776471
  X2          1    487.117647
  X3          1    122.400000

Dependent Variable: LENGTH
Test: QUAD    Numerator:    122.4000  DF:   1   F value:   11.1459
              Denominator:   10.98162  DF:  64   Prob>F:     0.0014

Dependent Variable: LENGTH
Test: LINEAR  Numerator:    304.7588  DF:   2   F value:   27.7517
              Denominator:   10.98162  DF:  64   Prob>F:     0.0001
```

The standard error of each estimate is also provided, together with the information for conducting a t-test of each individual hypothesis $H_0 : \{\beta_i = 0\}, i = 1, 2, 3$.

Inclusion of the option SS1 in the MODEL statement of PROC REG causes printing of the Type I (sequential) sums of squares in the output. Each Type I sum of squares is the variation explained by entering the corresponding variable into the model, given that the previously listed variables are already in the model. For example, the Type I sum of squares for X is

$ssE_0 - ssE_1$, where ssE_0 is the error sum of squares for the model with $E[Y_{xt}] = \beta_0$, and ssE_1 is the error sum of squares for the simple linear regression model $E[Y_{xt}] = \beta_0 + \beta_1 x$; that is,

$$ss(\beta_1|\beta_0) = ssE_0 - ssE_1 = 1891.776471.$$

Likewise, the Type I sum of squares for X2 is the difference in error sums of squares for the linear and quadratic regression models; that is,

$$ss(\beta_2|\beta_0, \beta_1) = ssE_1 - ssE_2 = 487.117647,$$

and for X3, the Type I sum of squares is the difference in error sums of squares for the quadratic and cubic regression models; that is,

$$ss(\beta_3|\beta_0, \beta_1, \beta_2) = ssE_2 - ssE = 122.400000,$$

where we have written ssE for the error sum of squares for the full cubic model (rather than ssE_3). Thus, the ratio used to test the null hypothesis $H_0^Q : \{\beta_3 = 0\}$ versus $H_A^Q : \{\beta_3 \neq 0\}$ is

$$ss(\beta_3)/msE = ss(\beta_3|\beta_0, \beta_1, \beta_2)/msE = 122.4/10.98162 = 11.1459.$$

The output of the TEST statement labeled QUAD provides the same information, as well as the p-value 0.0014. The null hypothesis H_0^Q is thus rejected, so the quadratic model is not adequate—the cubic model is needed. Hence, there is no reason to test further reduced models, but the information for such tests will be discussed for illustrative purposes.

To test $H_0^L : \beta_2 = \beta_3 = 0$, the full model is the cubic model and the reduced model is the linear model, so the numerator sum of squares of the test statistic is

$$\begin{aligned} ss(\beta_2, \beta_3) = ssE_1 - ssE &= ss(\beta_2|\beta_0, \beta_1) + ss(\beta_3|\beta_0, \beta_1, \beta_2) \\ &= 487.117647 + 122.400000 = 609.517647, \end{aligned}$$

and the decision rule for testing H_0^L against the alternative hypothesis H_A^L that the cubic model is needed is

reject H_0^L if $ms(\beta_2, \beta_3)/msE > F_{2,64,\alpha}$,

where

$$ms(\beta_2, \beta_3) = ss(\beta_2, \beta_3)/2.$$

The information for this test of adequacy of the linear model is also generated by the TEST statement labeled LINEAR.

The OUTPUT statement in PROC REG saves into an output data set the upper and lower 95% confidence limits for mean response and the corresponding standard error under the variable names L95M, U95M and STDM. This is done for each x-value in the input data set for which all regressors are available. Similarly, the upper and lower 95% prediction limits for a new individual observation and the corresponding standard error are saved under the variable names L95I, U95I and STDI. These could be printed or plotted, though we do not do so here.

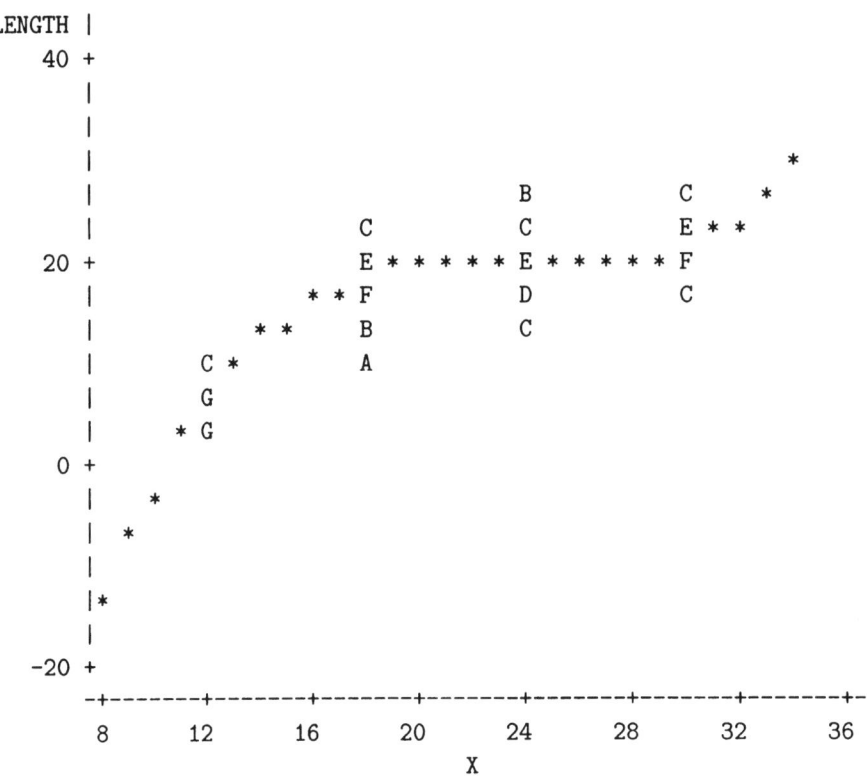

Figure 8.9
Plot of data and fitted response curve

The output of PROC PLOT is shown in Figure 8.9. Overlaid on the same axes are plots of the raw data and the fitted cubic polynomial regression curve. A trick was used to generate data to plot the fitted curve. Actual x values range from 12 to 30. In the DATA TOPLOT step in Table 8.9, additional observations were created in the data set corresponding to the integer x values ranging from 8 to 34 but with missing values for the dependent variable length. While observations with missing length values cannot be used to fit the model, the regression procedure does compute the corresponding predicted values LHAT. The OUTPUT statement includes these fitted values in the newly created output data set, so they can be plotted to show the fitted model. The option NOOVP was included in line 1 of the program so there is "no overprinting" of characters when the plots are overlaid.

In this example, it is not possible to test for lack of fit of the cubic model, since data were collected at only four x-levels. If we had been fitting a quadratic model, then a lack-of-fit test would have been possible. An easy way to generate the relevant output using the SAS

Exercises 273

software is as follows. In line 4 of the program, add a classification variable A, using the statement "A=X;". Then insert a PROC GLM procedure before PROC REG as follows.

PROC GLM;
CLASS A;
MODEL LENGTH = X X2 A;

Then the Type I sum of squares for A is the appropriate numerator $ssLOF$ for the test ratio.

Statements for generation of residual plots for checking the error assumptions are included in the sample SAS program in Table 8.9, but the output is not shown here.

Exercises

1. For the simple linear regression model

$$E[Y_{xt}] = \beta_0 + \beta_1 x,$$

the least squares estimators $\hat{\beta}_0$ and $\hat{\beta}_1$ for the parameters β_0 and β_1 are given in (8.5.6), page 251. Show that their variances are

$$\text{Var}(\hat{\beta}_0) = \sigma^2 \left(\frac{1}{n} + \frac{\bar{x}_{..}^2}{ss_{xx}}\right) \quad \text{and} \quad \text{Var}(\hat{\beta}_1) = \sigma^2 \left(\frac{1}{ss_{xx}}\right),$$

where $ss_{xx} = \sum_x r_x (x - \bar{x}_{..})^2$, as given in (8.5.7).

2. **Bicycle experiment, continued**

The bicycle experiment was run to compare the crank rates required to keep a bicycle at certain speeds, when the bicycle (a Cannondale SR400) was in twelfth gear on flat ground. The speeds chosen were $x = 5, 10, 15, 20,$ and 25 mph. The data are given in Table 8.11. (See also Exercise 6 of Chapter 5.)

(a) Fit the simple linear regression model to the data, and use residual plots to check the assumptions of the simple linear regression model.

(b) If a transformation of the data is needed, choose a transformation, refit the simple linear regression model, and check for lack of fit.

(c) Using your results from parts (a) and (b), select a model for the data. Use this model to obtain an estimate for the mean crank rate needed to maintain a speed of 18 mph in twelfth gear on level ground.

Table 8.11 Data for the bicycle experiment

Treatment	Crank Rates		
x	y_{xt}		
5 mph	15	19	22
10 mph	32	34	27
15 mph	44	47	44
20 mph	59	61	61
25 mph	75	73	75

(d) Calculate a 95% confidence interval for the mean crank rate needed to maintain a speed of 18 mph in twelfth gear on level ground.

(e) Find the 95% confidence band for the regression line. Draw a scatter plot of the data and superimpose the regression line and the confidence band on the plot.

(f) Would you be happy to use your model to estimate the mean crank rate needed to maintain a speed of 35 mph in twelfth gear on level ground. Why or why not?

3. **Systolic blood pressure experiment**

    A pilot experiment was run by John Spitak in 1987 to investigate the effect of jogging on systolic blood pressure. Only one subject was used in the pilot experiment, and a main experiment involving a random sample of subjects from a population of interest would need to be run in order to draw more general conclusions. The subject jogged in place for a specified number of seconds and then his systolic blood pressure was measured. The subject rested for at least 5 minutes, and then the next observation was taken.

    The data and their order of observation are given in Table 8.12.

    (a) Fit a simple linear regression model to the data and test for model lack of fit.

    (b) Use residual plots to check the assumptions of the simple linear regression model.

    (c) Give a 95% confidence interval for the slope of the regression line.

    (d) Using the confidence interval in part (c), test at significance level $\alpha = 0.05$ whether the linear term is needed in the model.

    (e) Repeat the test in part (d) but using the formula for the orthogonal polynomial linear trend coefficients for unequally spaced levels and unequal sample sizes given in Section 4.2.4. Do these two tests give the same information?

    (f) Estimate the mean systolic blood pressure of the subject after jogging in place for 35 seconds and calculate a 99% confidence interval.

    (g) The current experiment was only a pilot experiment. Write out a checklist for the main experiment.

4. **Trout experiment, continued**

    The data in Table 8.13 show the measurements of hemoglobin (grams per 100 ml) in the blood of brown trout. (The same data were used in Exercise 15 of Chapter 3.) The trout were placed at random in four different troughs. The fish food added to the troughs contained, respectively, $x = 0, 5, 10,$ and 15 grams of sulfamerazine per 100 pounds

Table 8.12 Systolic blood pressure measurements—(order of collection in parentheses)

Jogging time in seconds (order of collection)					
10	20	25	30	40	50
120 (1)	125 (2)	127 (10)	128 (3)	137 (5)	143 (6)
118 (9)	126 (4)		131 (7)		
	123 (8)				

Exercises 275

Table 8.13 Data for the trout experiment

x	Hemoglobin (grams per 100 ml)									
0	6.7	7.8	5.5	8.4	7.0	7.8	8.6	7.4	5.8	7.0
5	9.9	8.4	10.4	9.3	10.7	11.9	7.1	6.4	8.6	10.6
10	10.4	8.1	10.6	8.7	10.7	9.1	8.8	8.1	7.8	8.0
15	9.3	9.3	7.2	7.8	9.3	10.2	8.7	8.6	9.3	7.2

Source: Gutsell, J. S. (1951). Copyright © 1951 International Biometric Society. Reprinted with permission.

of fish. The measurements were made on ten randomly selected fish from each trough after 35 days.

(a) Fit a quadratic regression model to the data.

(b) Test the quadratic model for lack of fit.

(c) Use residual plots to check the assumptions of the quadratic model.

(d) Test whether the quadratic term is needed in the model.

(e) Use the fitted quadratic model to estimate the number of grams of sulfamerazine per 100 pounds of fish to maximize the mean amount of hemoglobin in the blood of the brown trout.

5. **Bean-soaking experiment, continued**
Use residual plots to check the assumptions of the cubic regression model for the data of the bean-soaking experiment. (The data are in Table 8.6, page 264).

6. **Bean-soaking experiment, continued**
Suppose the experimenter in the bean-soaking experiment of Section 8.8 had presumed that the quadratic regression model would be adequate for soaking times ranging from 12 to 30 hours.

(a) Figure 8.8, page 268, shows the fitted response curve and the standardized residuals each plotted against soaking time. Based on these plots, discuss model adequacy.

(b) Test the quadratic model for lack of fit.

7. **Orthogonal polynomials**
Consider an experiment in which an equal number of observations are collected for each of the treatment factor levels $x = 10, 20, 30, 40, 50$.

(a) Compute the corresponding values z_x for the linear orthogonal polynomial, and determine the rescaling factor by which the z_x differ from the coefficients of the linear trend contrast.

(b) Compute the values $z_x^{(2)}$ for the quadratic orthogonal polynomial, and determine the rescaling factor by which the $z_x^{(2)}$ differ from the coefficients of the quadratic trend contrast.

(c) Use the data of Table 8.1 and the orthogonal polynomial coefficients to test that the quadratic and linear trends are zero.

(d) Using the data of Table 8.1 and a statistical computing package, fit a quadratic model to the original values. Test the hypotheses

$$H_0^L : \{\beta_2 = 0\} \quad \text{and} \quad H_0 : \{\beta_1 = \beta_2 = 0\}$$

against their respective two-sided alternative hypotheses. Compare the results of these tests with those in (c).

8. **Orthogonal polynomials**

Consider use of the quadratic orthogonal polynomial regression model (8.7.19), page 260, for the data at levels 18, 24, and 30 of the bean-soaking experiment—the data are in Table 8.6, page 264.

(a) Compute the least squares estimates of the parameters.

(b) Why is it not possible to test for lack of fit of the quadratic model?

(c) Give an analysis of variance table and test the hypothesis that a linear model would provide an adequate representation of the data.

9. **Heart–lung pump experiment, continued**

In Example 8.5, page 254, we fitted a linear regression model to the data of the heart–lung pump experiment. We rejected the null hypothesis that the slope of the line is zero.

(a) Show that the numerator sum of squares for testing $H_0 : \{\beta_1 = 0\}$ against the alternative hypothesis $H_A : \{\beta_1 \neq 0\}$ is the same as the sum of squares ssc that would be obtained for testing that the linear trend is zero in the analysis of variance model (the relevant calculations were done in Example 4.2.4, page 72).

(b) Obtain a 95% confidence band for the regression line.

(c) Calculate a 99% prediction interval for the fluid flow rate at 100 revolutions per minute.

(d) Estimate the intercept β_0. This is not zero, which suggests that the fluid flow rate is not zero at 0 rpm. Since this should not be the case, explain what is happening.

9. Analysis of Covariance

9.1 Introduction
9.2 Models
9.3 Least Squares Estimates
9.4 Analysis of Covariance
9.5 Treatment Contrasts and Confidence Intervals
9.6 Using SAS Software
Exercises

9.1 Introduction

In Chapters 3–7, we used completely randomized designs and analysis of variance to compare the effects of one or more treatment factors on a response variable. If nuisance factors are expected to be a major source of variation, they should be taken into account in the design and analysis of the experiment. If the values of the nuisance factors can be measured in advance of the experiment or controlled during the experiment, then they can be taken into account at the design stage using blocking factors, as discussed in Chapter 10. Analysis of covariance, which is the topic of this chapter, is a means of adjusting the analysis for nuisance factors that cannot be controlled and that sometimes cannot be measured until the experiment is conducted. The method is applicable if the nuisance factors are related to the response variable but are themselves unaffected by the treatment factors.

For example, suppose an investigator wants to compare the effects of several diets on the weights of month-old piglets. The response (weight at the end of the experimental period) is likely to be related to the weight at the beginning of the experimental period, and these weights will typically be somewhat variable. To control or adjust for this prior weight variability, one possibility is to use a block design, dividing the piglets into groups (or blocks) of comparable weight, then comparing the effects of diets within blocks. A second possibility is to use a completely randomized design with response being the weight gain over the experimental period. This loses information, however, since heavier piglets may experience higher weight gain than lighter piglets, or vice versa. It is preferable to include the prior weight in the model as a variable, called a *covariate*, that helps to explain the final weight.

The model for a completely randomized design includes the effects of the treatment factors of interest, together with the effects of any nuisance factors (covariates). *Analysis of covariance* is the comparison of treatment effects, adjusting for one or more covariates. Standard analysis of covariance models and assumptions are discussed in Section 9.2. Least squares estimates are derived in Section 9.3. Sections 9.4 and 9.5 cover analysis of covariance tests and confidence interval methods for the comparison of treatment effects. Analysis using the SAS software is illustrated in Section 9.6.

9.2 Models

Consider an experiment conducted as a completely randomized design to compare the effects of the levels of v treatments on a response variable Y. Suppose that the response is also affected by a nuisance factor (covariate) whose value x can be measured during or prior to the experiment. Furthermore, suppose that there is a linear relationship between $E[Y]$ and x, with the same slope for each treatment. Then, if we plot $E[Y]$ versus x for each treatment separately, we would see parallel lines, as illustrated for two treatments in Figure 9.1(a). A comparison of the effects of the two treatments can be done by comparison of mean response at any value of x. The model that allows this type of analysis is the analysis of covariance model:

$$Y_{it} = \mu + \tau_i + \beta x_{it} + \epsilon_{it},$$ (9.2.1)
$$\epsilon_{it} \sim N(0, \sigma^2),$$

ϵ_{it}'s are mutually independent,

$t = 1, 2, \ldots, r_i; \quad i = 1, \ldots, v$.

In this model, the effect of the ith treatment is modeled as τ_i, as usual. If there is more than one treatment factor, then τ_i represents the effect of the ith treatment combination and could be replaced by main-effect and interaction parameters. The value of the covariate on the tth time that treatment i is observed is written as x_{it}, and the linear relationship between the response and the covariate is modeled as βx_{it} as in a regression model. It is important for the analysis that follows that the value x_{it} of the covariate not be affected by the treatment—otherwise, comparison of treatment effects at a common x-value would not be meaningful.

A common alternative form of the analysis of covariance model is

$$Y_{it} = \mu^* + \tau_i + \beta(x_{it} - \overline{x}_{..}) + \epsilon_{it},$$ (9.2.2)

in which the covariate values have been "centered." The two models are equivalent for comparison of treatment effects. The slope parameter β has the same interpretation in both

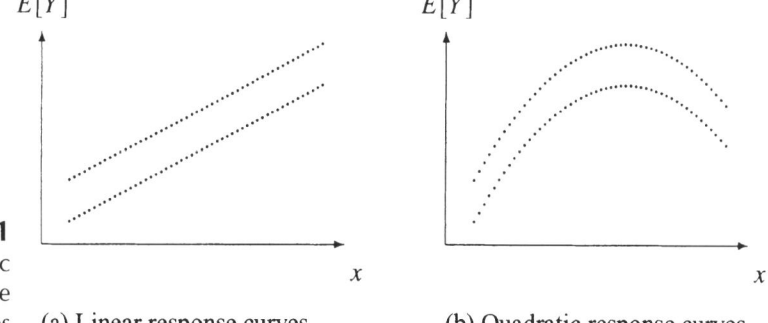

Figure 9.1
Linear and quadratic parallel response curves

(a) Linear response curves (b) Quadratic response curves

models. In model (9.2.2), $\mu + \tau_i$ denotes the mean response when $x_{it} = \bar{x}_{..}$, whereas in model (9.2.1), $\mu^* + \tau_i$ denotes the mean response when $x_{it} = 0$, with the parameter relationship $\mu^* = \mu - \beta \bar{x}_{..}$. Model (9.2.2) is often used to reduce computational problems and is a little easier to work with in obtaining least squares estimates.

9.2.1 Checking Model Assumptions and Equality of Slopes

In addition to the usual assumptions on the error variables, the analysis of covariance model (9.2.2) assumes a linear relationship between the covariate and the mean response, with the same slope for each treatment, as illustrated in Figure 9.1(a). It is appropriate to start by checking for model lack of fit.

Lack of fit can be investigated by plotting the residuals versus the covariate for each treatment on the same scale. If the plot looks nonlinear for any treatment, then a linear relationship between the response and covariate may not be adequate. If each plot does look linear, one can assess whether the slopes are comparable. A formal test of equality of slopes can be conducted by comparing the fit of the analysis of covariance model (9.2.2) with the fit of the corresponding model that does not require equal slopes, for which

$$Y_{it} = \mu + \tau_i + \beta_i(x_{it} - \bar{x}_{..}) + \epsilon_{it}. \qquad (9.2.3)$$

If there is no significant lack of fit of the model, then plots of the residuals versus run order, predicted values, and normal scores can be used as in Chapter 5 to assess the assumptions of independence, equal variances, and normality of the random error terms.

9.2.2 Model Extensions

The analysis of covariance model (9.2.1) can be generalized in various ways that we will mention here but not consider further.

If the effect of the covariate is not linear, then βx can be replaced with a higher-order polynomial function $\beta_1 x + \beta_2 x^2 + \cdots + \beta_p x^p$ to adequately model the common shape of the response curves for each treatment, analogous to the polynomial response curve models of Chapter 8. For example, parallel quadratic response curves for two treatments are shown in Figure 9.1(b).

If there is more than one covariate, the single covariate term can be replaced by an appropriate polynomial function of all the covariates. For example, for two covariates x_1 and x_2, the second-order function

$$\beta_1 x_1 + \beta_2 x_2 + \beta_{12} x_1 x_2 + \beta_{11} x_1^2 + \beta_{22} x_2^2$$

might be used, analogous to the polynomial response surface models of Chapter 16. Centered forms of these functions can also be obtained (see Section 8.7).

9.3 Least Squares Estimates

We now obtain the least squares estimates for the parameters in the analysis of covariance model, and then illustrate the need to use adjusted means to compare treatment effects.

9.3.1 Normal Equations (Optional)

To obtain the least squares estimates of the parameters in model (9.2.2), we need to minimize the sum of squared errors,

$$\sum_{i=1}^{v} \sum_{t=1}^{r_i} e_{it}^2 = \sum_{i=1}^{v} \sum_{t=1}^{r_i} (y_{it} - \mu - \tau_i - \beta(x_{it} - \overline{x}_{..}))^2 \ .$$

Differentiating this with respect to each parameter in turn and setting the corresponding derivative equal to zero gives the normal equations as

$$y_{..} = n\hat{\mu} + \sum_{i=1}^{v} r_i \hat{\tau}_i \ , \qquad (9.3.4)$$

$$y_{i.} = r_i(\hat{\mu} + \hat{\tau}_i) + \hat{\beta} \sum_{t=1}^{r_i} (x_{it} - \overline{x}_{..}), \quad i = 1, \ldots, v \ , \qquad (9.3.5)$$

$$\sum_{i=1}^{v} \sum_{t=1}^{r_i} y_{it}(x_{it} - \overline{x}_{..}) = \sum_{i=1}^{v} \sum_{t=1}^{r_i} (\hat{\mu} + \hat{\tau}_i)(x_{it} - \overline{x}_{..}) \qquad (9.3.6)$$

$$+ \sum_{i=1}^{v} \sum_{t=1}^{r_i} \hat{\beta}(x_{it} - \overline{x}_{..})^2 \ .$$

There are $v + 2$ normal equations and $v + 2$ unknown parameters. However, equation (9.3.4) is the sum of the v equations given in (9.3.5), since $\Sigma r_i = n$ and $\Sigma \Sigma (x_{it} - \overline{x}_{..}) = 0$. Thus, the normal equations are not linearly independent, and the equations do not have a unique solution. However, the $v + 1$ equations in (9.3.5) and (9.3.6) are linearly independent and can be solved to obtain the unique least squares estimates for β, $\mu + \tau_1, \ldots, \mu + \tau_v$ given in the next subsection.

9.3.2 Least Squares Estimates and Adjusted Treatment Means

Under model (9.2.2) the expected value

$$E[\overline{Y}_{i.}] = \mu + \tau_i + \beta(\overline{x}_{i.} - \overline{x}_{..})$$

is an estimate of the mean response of the ith treatment when the value of the covariate x_{it} is $\overline{x}_{i.}$. So, unless the covariate means $\overline{x}_{i.}$ all happen to be equal, the difference of response means $\overline{y}_{i.} - \overline{y}_{s.}$ does not estimate $\tau_i - \tau_s$ and cannot be used to compare treatment effects. The least squares estimates of the parameters in the model are obtained by solving the normal equations in optional Section 9.3.1 and are

$$\hat{\mu} + \hat{\tau}_i = \overline{y}_{i.} - \hat{\beta}(\overline{x}_{i.} - \overline{x}_{..}), \quad i = 1, \ldots, v, \tag{9.3.7}$$

$$\hat{\beta} = sp^*_{xy}/ss^*_{xx}, \tag{9.3.8}$$

where

$$sp^*_{xy} = \sum_{i=1}^{v}\sum_{t=1}^{r_i}(x_{it} - \overline{x}_{i.})(y_{it} - \overline{y}_{i.}) \quad \text{and} \quad ss^*_{xx} = \sum_{i=1}^{v}\sum_{t=1}^{r_i}(x_{it} - \overline{x}_{i.})^2.$$

In this notation, ss can be read as "sum of squares" and sp as "sum of products." In Exercise 2, the reader is asked to verify that $E[\hat{\beta}] = \beta$. Consequently,

$$E[\hat{\mu} + \hat{\tau}_i] = \mu + \tau_i.$$

The least squares estimators $\hat{\mu} + \hat{\tau}_i$ therefore estimate the mean response for the ith treatment at the value of the covariate equal to $\overline{x}_{..}$. We call the estimates $\hat{\mu} + \hat{\tau}_i$ the *adjusted means*, since they adjust the response mean $\overline{y}_{i.}$ by the amount $\hat{\beta}(\overline{x}_{i.} - \overline{x}_{..})$, which is equivalent to measuring the responses at the same point on the covariate scale. The need for this adjustment is illustrated in the following example.

Example 9.3.1 Adjusted versus unadjusted means

Table 9.1 contains hypothetical data arising from two treatments at various values of the covariate. Using equations (9.3.7) and (9.3.8), one can show that the corresponding fitted model is

$$\hat{y}_{it} = \hat{\mu} + \hat{\tau}_i + 0.5372(x_{ij} - 60),$$

where

$$\hat{\mu} + \hat{\tau}_1 = 62.5416 \quad \text{and} \quad \hat{\mu} + \hat{\tau}_2 = 48.2516.$$

The data and fitted model are plotted in Figure 9.2. Observe that treatment one has the larger effect, and correspondingly the higher fitted line. However if the treatment effects were estimated as $\overline{y}_{1.}$ and $\overline{y}_{2.}$, it would appear that treatment two has the larger effect, since it has the larger unadjusted mean:

$$\overline{y}_{2.} = 61.68 > \overline{y}_{1.} = 49.11.$$

This bias in the treatment effect estimates is due to the relative values of $\overline{x}_{1.}$ and $\overline{x}_{2.}$.

Table 9.1 Hypothetical analysis of covariance data

i	x_{it}		y_{it}		$\bar{x}_{i.}$	$\bar{y}_{i.}$
1	20	44.29	39.51	42.87	35	49.11
	30	44.48	48.39	49.14		
	40	50.24	51.63	46.55		
	50	57.75	59.23	55.23		
2	70	48.67	56.79	52.03	85	61.68
	80	57.68	67.25	52.88		
	90	62.04	66.12	64.39		
	100	63.54	72.49	76.33		

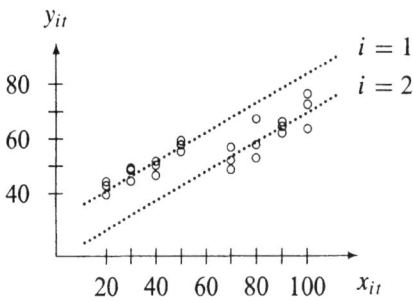

Figure 9.2 Illustration of bias if unadjusted means $\bar{y}_{i.}$ are compared

These data provide an exaggerated illustration of the need for adjustment of treatment sample means in analysis of covariance. □

9.4 Analysis of Covariance

For a completely randomized design and analysis of covariance model (9.2.2), a *one-way analysis of covariance* is used to test the null hypothesis $H_0 : \{\tau_1 = \tau_2 = \cdots = \tau_v\}$ against the alternative hypothesis H_A that at least two of the τ_i's differ. The test is based on the comparison of error sums of squares under the full and reduced models. If the null hypothesis is true with common treatment effect $\tau_i = \tau$, then the reduced model is

$$Y_{it} = \mu + \tau + \beta(x_{it} - \bar{x}_{..}) + \epsilon_{it}.$$

This is similar to a simple linear regression model, with constant $\beta_0 = \mu + \tau$, slope $\beta_1 = \beta$, and with regressor x_{it} centered. Thus, if we replace x by $x_{it} - \bar{x}_{..}$ in the formula (8.5.6), page 251, and the average $\bar{x}_{..}$ in (8.5.6) by the averaged centered value 0, the least squares estimates under the reduced model are

$$\hat{\mu} + \hat{\tau} = \bar{y}_{..} \quad \text{and} \quad \hat{\beta} = sp_{xy}/ss_{xx},$$

where

$$sp_{xy} = \sum_{i=1}^{v}\sum_{t=1}^{r_i}(x_{it} - \bar{x}_{..})y_{it} = \sum_{i=1}^{v}\sum_{t=1}^{r_i}(x_{it} - \bar{x}_{..})(y_{it} - \bar{y}_{..})$$

9.4 Analysis of Covariance

and

$$ss_{xx} = \sum_{i=1}^{v}\sum_{t=1}^{r_i}(x_{it}-\bar{x}_{..})^2.$$

So,

$$ssE_0 = \sum_i\sum_t \left(y_{it} - \hat{\mu} - \hat{\tau} - \hat{\beta}(x_{it}-\bar{x}_{..})\right)^2 \qquad (9.4.9)$$

$$= \sum_i\sum_t \left(y_{it} - \bar{y}_{..} - sp_{xy}(x_{it}-\bar{x}_{..})/ss_{xx}\right)^2$$

$$= ss_{yy} - (sp_{xy})^2/ss_{xx},$$

where $ss_{yy} = \sum_i\sum_t(y_{it}-\bar{y}_{..})^2$. The number of degrees of freedom for error is equal to the number of observations minus a degree of freedom for the constant $\mu + \tau$ and one for the slope β; that is, $n-2$.

Under the full analysis of covariance model (9.2.2), using the least squares estimates given in equations (9.3.7) and (9.3.8), the error sum of squares is

$$ssE = \sum_i\sum_t \left(y_{it} - \hat{\mu} - \hat{\tau}_i - \hat{\beta}(x_{it}-\bar{x}_{..})\right)^2$$

$$= \sum_i\sum_t \left(y_{it} - \bar{y}_{i.} + \hat{\beta}(\bar{x}_{i.}-\bar{x}_{..}) - \hat{\beta}(x_{it}-\bar{x}_{..})\right)^2$$

$$= \sum_i\sum_t \left((y_{it}-\bar{y}_{i.}) - \hat{\beta}(x_{it}-\bar{x}_{i.})\right)^2$$

$$= ss^*_{yy} - \hat{\beta}(sp^*_{xy}) \ = \ ss^*_{yy} - (sp^*_{xy})^2/ss^*_{xx}, \qquad (9.4.10)$$

where

$$ss^*_{xx} = \sum_i\sum_t(x_{it}-\bar{x}_{i.})^2,$$

$$ss^*_{yy} = \sum_i\sum_t(y_{it}-\bar{y}_{i.})^2,$$

$$sp^*_{xy} = \sum_i\sum_t(x_{it}-\bar{x}_{i.})(y_{it}-\bar{y}_{i.}).$$

The values ss^*_{xx} and ss^*_{yy} can be obtained from a computer program as the values of ssE fitting the one-way analysis of variance models with x_{it} and y_{it} as the response variables, respectively. The number of error degrees of freedom is $n-v-1$ (one less than the error degrees of freedom under the analysis of variance model due to the additional parameter β).

The sum of squares for treatments $ss(T|\beta)$ is the difference in the error sums of squares under the reduced and full models,

$$ss(T|\beta) = ssE_0 - ssE \qquad (9.4.11)$$

$$= \left(ss_{yy} - (sp_{xy})^2/ss_{xx}\right) - \left(ss^*_{yy} - (sp^*_{xy})^2/ss^*_{xx}\right).$$

The difference in the error degrees of freedom for the reduced and full models is

$$(n-2) - (n-v-1) = v-1.$$

Table 9.2 Analysis of covariance for one linear covariate

Source of Variation	Degrees of Freedom	Sum of Squares	Mean Square	Ratio
$T\|\beta$	$v-1$	$ss(T\|\beta)$	$\frac{ss(T\|\beta)}{v-1}$	$\frac{ms(T\|\beta)}{msE}$
$\beta\|T$	1	$ss(\beta\|T)$	$ss(\beta\|T)$	$\frac{ms(\beta\|T)}{msE}$
Error	$n-v-1$	ssE	msE	
Total	$n-1$	ss_{yy}		
Formulae				

$ss(T|\beta) = \left(ss_{yy} - (sp_{xy})^2/ss_{xx}\right) - \left(ss^*_{yy} - (sp^*_{xy})^2/ss^*_{xx}\right)$

$ss(\beta|T) = (sp^*_{xy})^2/ss^*_{xx} \qquad\qquad ssE = ss^*_{yy} - (sp^*_{xy})^2/ss^*_{xx}$

$ss_{xx} = \sum_i \sum_t (x_{it} - \overline{x}_{..})^2 \qquad\qquad ss_{yy} = \sum_i \sum_t (y_{it} - \overline{y}_{..})^2$

$sp_{xy} = \sum_i \sum_t (x_{it} - \overline{x}_{..})(y_{it} - \overline{y}_{..}) \qquad ss^*_{xx} = \sum_i \sum_t (x_{it} - \overline{x}_{i.})^2$

$sp^*_{xy} = \sum_i \sum_t (x_{it} - \overline{x}_{i.})(y_{it} - \overline{y}_{i.}) \qquad ss^*_{yy} = \sum_i \sum_t (y_{it} - \overline{y}_{i.})^2$

We denote the corresponding mean square by

$ms(T|\beta) = ss(T|\beta)/(v-1)$.

If the null hypothesis is true, then

$MS(T|\beta)/MSE \sim F_{v-1,n-v-1}$,

so we can obtain a decision rule for testing $H_0 : \{\tau_1 = \tau_2 = \cdots = \tau_v\}$ against $H_A : \{\tau_i \text{ not all equal}\}$ as

reject H_0 if $ms(T|\beta)/msE > F_{v-1,n-v-1,\alpha}$

at chosen significance level α. The information for testing equality of the treatment effects is typically summarized in an analysis of covariance table such as that shown in Table 9.2.

The table also includes information for testing the null hypothesis $H_0 : \{\beta = 0\}$ against the alternative hypothesis $H_A : \{\beta \neq 0\}$. The reduced model for this test is the one-way analysis of variance model (3.3.1), for which $Y_{it} = \mu + \tau_i + \epsilon_{it}$. From Chapter 3, the corresponding error sum of squares is

$$ssE_0 = \sum_i \sum_t (y_{it} - \overline{y}_{i.})^2 = ss^*_{yy},$$

and the number of error degrees of freedom is $n - v$. The error sum of squares for the full model is given in (9.4.10). Denoting the difference in error sums of squares by $ss(\beta|T)$, we have

$ss(\beta|T) = ssE_0 - ssE = (sp^*_{xy})^2/ss^*_{xx} = \hat{\beta}^2 ss^*_{xx}$.

The difference in the error degrees of freedom is $(n-v) - (n-v-1) = 1$, so the corresponding mean square, $ms(\beta|T)$, has the same value $ss(\beta|T)$. Under the assumptions of the analysis of covariance model (9.2.2), if $H_0 : \{\beta = 0\}$ is true, then

$MS(\beta|T)/MSE \sim F_{1,n-v-1}$.

9.4 Analysis of Covariance

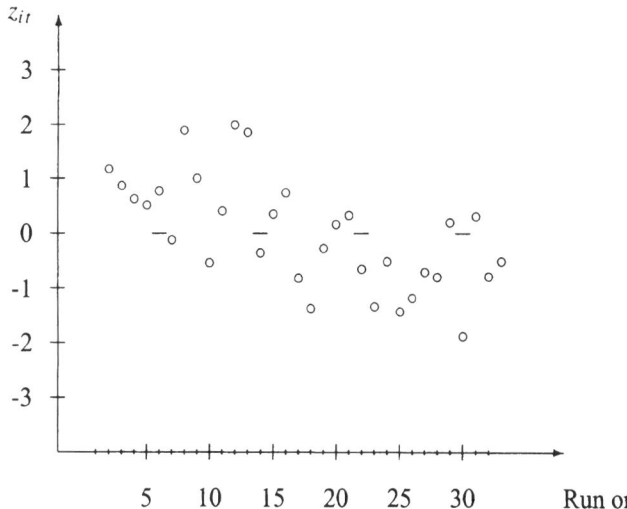

Figure 9.3
Residual plot for the balloon experiment

Thus, the decision rule for testing $H_0 : \{\beta = 0\}$ against $H_A : \{\beta \neq 0\}$, at significance level α, is

reject H_0 if $ms(\beta|T)/msE > F_{1,n-v-1,\alpha}$.

Example 9.4.1 Balloon experiment, continued

Consider the balloon experiment of Meily Lin, in which she compared the effects of four colors on balloon inflation time. In Example 5.5, page 110, the standardized residuals were plotted against the run order of the observations. The plot, reproduced in Figure 9.3, shows a clear linear decreasing trend in the residuals. This trend indicates a definite lack of independence in the error terms under the one-way analysis of variance model, but the trend can be eliminated by including the run order as a covariate in the model.

The analysis of covariance table for this experiment is shown in Table 9.3. Residual plots for checking the model assumptions will be discussed in Section 9.6 and reveal no anomalies.

The decision rule for testing equality of the treatment effects is to

reject $H_0 : \{\tau_1 = \cdots = \tau_v\}$ if $ms(T|\beta)/msE = 6.32 > F_{3,27,\alpha}$.

Since $F_{3,27,.01} = 4.60$, the null hypothesis is rejected at significance level $\alpha = 0.01$, and we can conclude that there is a difference in inflation times for the different colors of balloon.

Table 9.3 Analysis of covariance for the balloon experiment

Source of Variation	Degrees of Freedom	Sum of Squares	Mean Square	Ratio
$T\|\beta$	3	127.679	42.560	6.32
$\beta\|T$	1	120.835	120.835	17.95
Error	27	181.742	6.731	
Total	31	430.239		

Of secondary interest is the test of $H_0 : \{\beta = 0\}$ against $H_A : \{\beta \neq 0\}$. The decision rule is to reject the null hypothesis if $ms(\beta|T)/msE = 17.95 > F_{1,27,\alpha}$. Again, the null hypothesis is rejected at significance level $\alpha = 0.01$, since $F_{1,27,.01} = 7.68$. We may conclude that the apparent linear trend in the inflation times due to order is a real trend and not due to random error. □

9.5 Treatment Contrasts and Confidence Intervals

9.5.1 Individual Confidence Intervals

Since $\mu + \tau_i$ is estimable under model (9.2.2), any treatment contrast $\sum_i c_i \tau_i$ ($\sum_i c_i = 0$) is also estimable. From (9.3.7), $\sum_i c_i \tau_i$ has least squares estimator

$$\sum_i c_i \hat{\tau}_i = \sum_i c_i (\hat{\mu} + \hat{\tau}_i) = \sum_i c_i \left(\overline{Y}_{i.} - \hat{\beta}(\bar{x}_{i.} - \bar{x}_{..}) \right) = \sum_i c_i \left(\overline{Y}_{i.} - \hat{\beta} \bar{x}_{i.} \right). \quad (9.5.12)$$

(The term $\Sigma c_i \hat{\beta} \bar{x}_{..}$ is zero, since $\Sigma c_i = 0$.) Now, $\text{Var}(\overline{Y}_{i.}) = \sigma^2/r_i$, and it can be shown that $\text{Var}(\hat{\beta}) = \sigma^2/ss^*_{xx}$ and $\text{Cov}(\overline{Y}_{i.}, \hat{\beta}) = 0$. Using these results and (9.5.12), the variance of $\sum_i c_i \hat{\tau}_i$ is

$$\text{Var}\left(\sum_i c_i \hat{\tau}_i \right) = \text{Var}\left(\sum_i c_i \overline{Y}_{i.} \right) + \text{Var}\left(\hat{\beta} \sum_i c_i \bar{x}_{i.} \right)$$

$$= \sigma^2 \left(\sum_i \frac{c_i^2}{r_i} \right) + \left(\frac{\sigma^2}{ss^*_{xx}} \right) \left(\sum_i c_i \bar{x}_{i.} \right)^2 .$$

So, the estimated variance is

$$\widehat{\text{Var}}(\Sigma c_i \hat{\tau}_i) = msE \left(\sum_i \frac{c_i^2}{r_i} + \frac{(\sum_i c_i \bar{x}_{i.})^2}{ss^*_{xx}} \right). \quad (9.5.13)$$

From (9.5.12), the least squares estimator $\sum_i c_i \hat{\tau}_i$ is a function of $\overline{Y}_{i.}$ and $\hat{\beta}$. Since Y_{ij} has a normal distribution, both $\overline{Y}_{i.}$ and $\hat{\beta}$ are normally distributed. Consequently, $\sum_i c_i \hat{\tau}_i$ also has a normal distribution. Also, MSE/σ^2 has a chi-squared distribution with $n - v - 1$ degrees of freedom. Then, for any treatment contrast $\sum_i c_i \tau_i$, it follows that

$$\frac{\sum c_i \hat{\tau}_i - \sum c_i \tau_i}{\sqrt{\widehat{\text{Var}}(\sum c_i \hat{\tau}_i)}} \sim t_{n-v-1} .$$

So, a $100(1 - \alpha)\%$ confidence interval for $\sum_i c_i \tau_i$ is

$$\sum_i c_i \tau_i \in \left(\sum_i c_i \hat{\tau}_i \pm t_{n-v-1,\alpha/2} \sqrt{\widehat{\text{Var}}\left(\sum_i c_i \hat{\tau}_i \right)} \right). \quad (9.5.14)$$

9.5.2 Multiple Comparisons

The multiple comparison methods of Bonferroni and Scheffé are applicable in the analysis of covariance setting. However, since the adjusted treatment means $\hat{\mu} + \hat{\tau}_i = \overline{Y}_{i.} - \hat{\beta}(\overline{x}_{i.} - \overline{x}_{..})$ are not independent unless the $\overline{x}_{i.}$ are all equal, the methods of Tukey, Dunnett, and Hsu are not known to apply. It is believed that Tukey's method does still control the experimentwise confidence level in this case, but there is no known proof of this conjecture.

Confidence intervals are obtained as

$$\sum_i c_i \tau_i \in \left(\sum_i c_i \hat{\tau}_i \pm w \sqrt{\widehat{\text{Var}}\left(\sum_i c_i \hat{\tau}_i\right)} \right), \qquad (9.5.15)$$

where w is the appropriate critical coefficient. For the Bonferroni method and m predetermined treatment contrasts, $w = t_{n-v-1, \alpha/2m}$. For the Scheffé method for all treatment contrasts, $w = \sqrt{(v-1)F_{v-1, n-v-1, \alpha}}$. Formulae for the estimate $\sum c_i \hat{\tau}_i$ and the corresponding estimated variance are given in equations (9.5.12) and (9.5.13).

Example 9.5.1 Balloon experiment, continued

We now illustrate the Scheffé method of multiple comparisons to obtain simultaneous 95% confidence intervals for all pairwise treatment comparisons for the balloon experiment of Example 9.4. (The data are in Table 3.11, page 62.) For pairwise comparisons, the confidence intervals are obtained from (9.5.14),

$$\tau_i - \tau_s \in \left(\hat{\tau}_i - \hat{\tau}_s \pm \sqrt{3F_{3,27,.05}} \sqrt{\widehat{\text{Var}}(\hat{\tau}_i - \hat{\tau}_s)} \right),$$

where $\hat{\tau}_i - \hat{\tau}_s = (\overline{y}_{i.} - \overline{y}_{s.}) - \hat{\beta}(\overline{x}_{i.} - \overline{x}_{s.})$. The treatment and covariate means are

$\overline{y}_{1.} = 18.337, \quad \overline{y}_{2.} = 22.575, \quad \overline{y}_{3.} = 21.875, \quad \overline{y}_{4.} = 18.187,$
$\overline{x}_{1.} = 16.250, \quad \overline{x}_{2.} = 15.625, \quad \overline{x}_{3.} = 17.500, \quad \overline{x}_{4.} = 16.625,$

and from (9.3.8), we obtain

$$\hat{\beta} = sp_{xy}^* / ss_{xx}^* = -572.59/2713.3 = -0.21103.$$

Now, $msE = 6.731$ from Table 9.3, so

$$\widehat{\text{Var}}(\hat{\tau}_i - \hat{\tau}_s) = msE \left(\frac{1}{8} + \frac{1}{8} + \frac{(\overline{x}_{i.} - \overline{x}_{s.})^2}{ss_{xx}^*} \right)$$

$$= (6.731) \left(0.25 + \frac{(\overline{x}_{i.} - \overline{x}_{s.})^2}{2713.3} \right)$$

$$= 1.68275 + (0.00248)(\overline{x}_{i.} - \overline{x}_{s.})^2.$$

Using the critical coefficient $w = \sqrt{3F_{3,27,.05}} = \sqrt{3 \times 2.96}$, one can obtain the confidence interval information given in Table 9.4. The estimated difference exceeds the minimum significant difference

$$msd = w \sqrt{\widehat{\text{Var}}(\hat{\tau}_i - \hat{\tau}_s)} \quad \text{with} \quad w = \sqrt{3F_{3,27,.05}}$$

Table 9.4 Scheffé pairwise comparisons for the balloon experiment; overall confidence level is 95%

i	s	$\hat{\tau}_i - \hat{\tau}_s$	$\sqrt{\widehat{\mathrm{Var}}(\hat{\tau}_i - \hat{\tau}_s)}$	msd
1	2	−4.106	1.298	3.868
1	3	−3.801	1.299	3.871
1	4	0.071	1.297	3.865
2	3	0.304	1.301	3.877
2	4	4.176	1.298	3.868
3	4	3.872	1.298	3.868

for the first two and last two comparisons. One can conclude from the corresponding confidence intervals that the mean time to inflate balloons is longer for color 2 (yellow) than for colors 1 and 4 (pink and blue), and the mean inflation time for color 3 (orange) is longer than for color 4 (blue). At a slightly lower confidence level, we would also detect a difference in mean inflation times for colors 3 and 1 (orange and pink). The corresponding four intervals with overall confidence level 95% are

$$\tau_2 - \tau_1 \in (0.238, 7.974), \quad \tau_3 - \tau_1 \in (-0.070, 7.672),$$

$$\tau_2 - \tau_4 \in (0.308, 8.044), \quad \tau_3 - \tau_4 \in (\ 0.004, 7.740).$$

Whenever the data are used to determine or modify the model, the confidence levels and error rates associated with any subsequent analyses of the same data will not be exactly as stated. Such is the case for the analyses presented in Example 9.5.2 for the balloon experiment, since the covariate "run order" was included in the model as a result of a trend observed in the residuals from the original analysis of variance model. Thus, although Scheffé's method was used, we cannot be certain that the overall level of the confidence intervals in Example 9.5.2 is exactly 95%.

9.6 Using SAS Software

Table 9.5 contains a sample SAS program for performing a one-way analysis of covariance involving a single covariate with a linear effect. The program uses the data from the balloon experiment discussed in Examples 9.4 and 9.5.2. The data are given in Table 3.11, page 62. The experimenter was interested in comparing the effects of four colors (pink, yellow, orange, and blue) on the inflation time of balloons, and she collected eight observations per color. The balloons were inflated one after another by the same person. Residual analysis for the one-way analysis of variance model showed a linear trend in the residuals plotted against run order (Figure 9.3, page 285). Hence, run order is included in the model here as a linear covariate. To obtain the "centered" form of the model, as in model (9.2.2), a centered variable has been created immediately after the INPUT statement, using the SAS statement

```
X = RUNORDER - 16.5;
```

where 16.5 is the average value of RUNORDER.

9.6 Using SAS Software

Table 9.5 SAS program for analysis of covariance—Balloon experiment

```
DATA;
  INPUT RUNORDER COLOR INFTIME;
  X = RUNORDER - 16.5;
  LINES;
      1 1 22.0
      2 3 24.6
      3 1 20.3
      4 4 19.8
      : :  :
     30 1 19.3
     31 1 15.9
     32 3 20.3
;
PROC GLM;
  CLASS COLOR;
  MODEL INFTIME = COLOR X;
  ESTIMATE '1-2'   COLOR    1 -1  0  0;
  ESTIMATE '1-3'   COLOR    1  0 -1  0;
  ESTIMATE '1-4'   COLOR    1  0  0 -1;
  ESTIMATE '2-3'   COLOR    0  1 -1  0;
  ESTIMATE '2-4'   COLOR    0  1  0 -1;
  ESTIMATE '3-4'   COLOR    0  0  1 -1;
  ESTIMATE 'BETA'  X    1;
  OUTPUT OUT=B P=PRED R=Z;
PROC STANDARD STD=1;
  VAR Z;
PROC RANK NORMAL=BLOM OUT=C;
  VAR Z;
  RANKS NSCORE;
PROC PLOT;
  PLOT Z*RUNORDER Z*PRED Z*COLOR / VREF=0 VPOS=11 HPOS=50;
  PLOT Z*NSCORE / VREF=0 HREF=0 VPOS=11 HPOS=50;
```

In Table 9.5, PROC GLM is used to generate the analysis of covariance. The output is shown in Table 9.6. The treatment factor COLOR has been included in the CLASS statement to generate a parameter τ_i for each level of the treatment factor "color," while the covariate X has been excluded from the class statement so that it is included in the model as a regressor, or covariate, as in model (9.2.2). To obtain the "uncentered" form of the model, as in model (9.2.1), the variable RUNORDER would replace X throughout the program. The output in Table 9.6 would not change, since only the definition of the constant in the model has been altered.

The information for testing the null hypotheses $H_0^T : \{\tau_1 = \cdots = \tau_4\}$ against $H_A^T : \{H_0^T \text{ not true}\}$ and $H_0 : \{\beta = 0\}$ against $H_A : \{\beta \neq 0\}$ is in Table 9.6 under the heading Type III SS. Specifically, $ss(T|\beta) = 127.678829$ and $ss(\beta|T) = 120.835325$. The corresponding ratio statistics and p-values are listed under F Value and Pr > F, re-

Table 9.6 Output from SAS PROC GLM

```
                        The SAS System
                 General Linear Models Procedure

                    Class Level Information
                  Class    Levels    Values
                  COLOR       4      1 2 3 4

             Number of observations in data set = 32
```

Dependent Variable: INFTIME

Source	DF	Sum of Squares	Mean Square	F Value	Pr > F
Model	4	248.49657	62.12414	9.23	0.0001
Error	27	181.74218	6.73119		
Corrected Total	31	430.23875			

R-Square	C.V.	Root MSE	INFTIME Mean
0.577578	12.81607	2.5945	20.244

Source	DF	Type I SS	Mean Square	F Value	Pr > F
COLOR	3	127.66125	42.55375	6.32	0.0022
X	1	120.83532	120.83532	17.95	0.0002

Source	DF	Type III SS	Mean Square	F Value	Pr > F
COLOR	3	127.67883	42.55961	6.32	0.0022
X	1	120.83532	120.83532	17.95	0.0002

Parameter	Estimate	T for H0: Parameter=0	Pr > |T|	Std Error of Estimate
1-2	-4.10560387	-3.16	0.0038	1.29760048
1-3	-3.80129227	-2.93	0.0069	1.29872024
1-4	0.07086232	0.05	0.9568	1.29736147
2-3	0.30431160	0.23	0.8168	1.30058436
2-4	4.17646618	3.22	0.0034	1.29818288
3-4	3.87215459	2.98	0.0060	1.29795891
BETA	-0.21103382	-4.24	0.0002	0.04980823

spectively. Since the p-values are very small, both null hypotheses would be rejected for any reasonable overall significance level. Thus, there are significant differences in the effects of the four colors on inflation time after adjusting for linear effects of run order. Also, there is a significant linear trend in mean inflation as a function of run order after adjusting for the treatment effects. The least squares estimate for β is negative ($\hat{\beta} = -0.211$), so the trend is decreasing, as we saw in Figure 9.3.

The Type I and Type III sums of squares for color are similar but not quite equal, indicating that the treatment effects and the covariate effect are not independent. This is because the

9.6 Using SAS Software

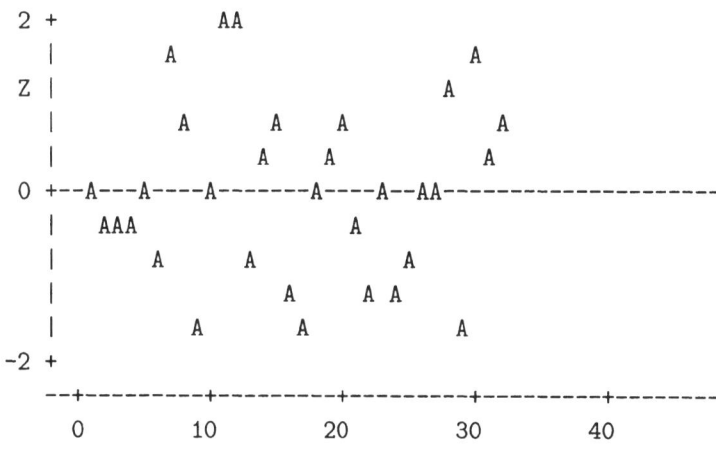

Figure 9.4
SAS plot of z_{it} against run order

comparison of treatment effects is a comparison of the adjusted means, which do depend on β, since the covariate means $\bar{x}_{i.}$ are not all equal for these data.

ESTIMATE statements under PROC GLM are used to generate the least squares estimate and estimated standard error for each pairwise comparison of treatment effects and for the coefficient β of the covariate. The standard errors of each $\hat{\tau}_i - \hat{\tau}_j$ are not quite equal but are all approximately 1.30. To compare all treatment effects pairwise using Scheffé's method and a simultaneous 95% confidence level, the calculations proceed as shown in Example 9.5.2.

The OUTPUT statement under PROC GLM and the procedures PROC STANDARD, PROC RANK, and PROC PLOT are used as they were in Chapter 5 to generate four residual plots. The resulting plots (not shown) show no problems with the model assumptions. Of special note, the plot of the residuals against run order in Figure 9.4 no longer shows any trend, so the linear run order covariate has apparently adequately modeled any run order dependence in the observations.

A test for equality of slopes as discussed in Section 9.2.1 can be generated using the SAS statements

    PROC GLM; CLASS COLOR;
      MODEL INFTIME = COLOR X COLOR*X;

The interaction term COLOR*X will be significantly different from zero if the linear run order trends are not the same for each color.

Exercises

1. Consider the hypothetical data of Example 9.3.2, in which two treatments are to be compared.

   (a) Fit the analysis of covariance model (9.2.1) or (9.2.2) to the data of Table 9.1, page 282.

   (b) Plot the residuals against the covariate, the predicted values, and normal scores. Use the plots to evaluate the model assumptions.

   (c) Test for inequality of slopes, using a level of significance $\alpha = 0.05$.

   (d) Test for equality of the treatment effects, using a significance level of $\alpha = 0.05$. Discuss the results.

   (e) Construct a 95% confidence interval for the difference in treatment effects. Discuss the results.

2. (optional) Assume that the analysis of covariance model (9.2.2) holds, so that $Y_{it} = \mu + \tau_i + \beta(x_{it} - \bar{x}_{..}) + \epsilon_{it}$.

   (a) Compute $E[Y_{it}]$.

   (b) Verify that $sp^*_{xY} = \sum_i \sum_t (x_{it} - \bar{x}_{i.}) Y_{it}$, given that $sp^*_{xY} = \sum_i \sum_t (x_{it} - \bar{x}_{i.})(Y_{it} - \bar{Y}_{i.})$.

   (c) Show that $E[\hat{\beta}] = \beta$, where $\hat{\beta} = sp^*_{xY}/ss^*_{xx}$ and $ss^*_{xx} = \sum_i \sum_t (x_{it} - \bar{x}_{i.})^2$.

   (d) Verify that $\text{Var}(\hat{\beta}) = \sigma^2/ss^*_{xx}$ and $\text{Cov}(\bar{Y}_{i.}, \hat{\beta}) = 0$.

   (e) Verify that $E[\hat{\mu} + \hat{\tau}_i] = \mu + \tau_i$, where $\hat{\mu} + \hat{\tau}_i = \bar{Y}_{i.} - \hat{\beta}(\bar{x}_{i.} - \bar{x}_{..})$.

   (f) Using the results of (c) and (e), argue that $\mu + \tau_i$ and β and all linear combinations of these are estimable.

3. **Zinc plating experiment**

   The following experiment was used by C. R. Hicks (1965), *Industrial Quality Control*, to illustrate the possible bias caused by ignoring an important covariate. The experimental units consisted of 12 steel brackets. Four steel brackets were sent to each of three vendors to be zinc plated. The response variable was the thickness of the zinc plating, in hundred-thousandths of an inch. The thickness of each bracket before plating was measured as a covariate. The data are reproduced in Table 9.7.

   (a) Plot y_{it} versus x_{it}, using the vendor index i as the plotting symbol. Discuss the relationship between plating thickness and bracket thickness before plating. Based on the plot, discuss appropriateness of the analysis of covariance model. Based on the plot, discuss whether there appears to be a vendor effect.

   (b) Fit the analysis of covariance model (9.2.1) or (9.2.2) to the data.

   (c) Plot the residuals against the covariate, predicted values, and normal scores. Use the plots to evaluate model assumptions.

   (d) Test for equality of slopes, using a level of significance $\alpha = 0.05$.

   (e) Test for equality of the vendor effects, using a significance level $\alpha = 0.05$.

Table 9.7 Bracket thickness x_{it} and plating thickness y_{it} in 10^{-5} inches for three vendors (Hicks, 1965)

	Vendor					
	1		2		3	
t	x_{1t}	y_{1t}	x_{2t}	y_{2t}	x_{3t}	y_{3t}
1	110	40	60	25	62	27
2	75	38	75	32	90	24
3	93	30	38	13	45	20
4	97	47	140	35	59	13

Source: Hicks, C. R. (1965). Copyright © 1965 American Society for Quality. Reprinted with permission.

(f) Fit the analysis of variance model to the data, ignoring the covariate.

(g) Using analysis of variance, ignoring the covariate, test for equality of the vendor effects using a significance level $\alpha = 0.05$.

(h) Compare and discuss the results of parts (e) and (g). For which model is msE smaller? Which model gives the greater evidence that vendor effects are not equal? What explanation can you offer for this?

4. **Paper towel absorbancy experiment**
   (S. Bortnick, M. Hoffman, K.K. Lewis, and C. Williams, 1996)
   Four students conducted a pilot experiment to compare the effects of two treatment factors, brand and printing, on the absorbancy of paper towels. Three brands of paper towels were compared (factor A at 3 levels). For each brand, both white and printed towels were evaluated (factor B, 1=white, 2=printed). For each observation, water was dripped from above a towel, which was horizontally suspended between two pairs of books on a flat surface, until the water began leaking through to the surface below. The time to collect each observation was measured in seconds. Absorbancy was measured as the number of water drops absorbed per square inch of towel. The rate at which the water droplets fell to the towel was measured (in drops per second) as a covariate. The data are reproduced in Table 9.8.

   (a) Plot absorbancy versus rate, using the treatment level as the plotting symbol. Based on the plot, discuss appropriateness of the analysis of covariance model, and discuss whether there appear to be treatment effects.

   (b) Fit the one-way analysis of covariance model to the data.

   (c) Plot the residuals against the covariate, run order, predicted values, and normal scores. Use the plots to evaluate model assumptions.

   (d) Test for equality of slopes, using a level of significance $\alpha = 0.05$.

   (e) Test for equality of treatment effects, using a significance level $\alpha = 0.05$.

   (f) Conduct a two-way analysis of covariance. Test the main effects and interactions for significance.

Table 9.8 Data for the paper tower absorbancy experiment

Run	Treatment	AB	Drops	Time	Area	Rate	Absorbancy
1	2	12	89	50	121.00	1.780	0.7355
2	4	22	28	15	99.00	1.867	0.2828
3	2	12	47	22	121.00	2.136	0.3884
4	1	11	82	42	121.00	1.952	0.6777
5	5	31	54	30	123.75	1.800	0.4364
6	1	11	74	37	121.00	2.000	0.6116
7	4	22	29	14	99.00	2.071	0.2929
8	6	32	80	41	123.75	1.951	0.6465
9	3	21	25	11	99.00	2.272	0.2525
10	3	21	27	12	99.00	2.250	0.2727
11	6	32	83	40	123.75	2.075	0.6707
12	5	31	41	19	123.75	2.158	0.3313

5. **Catalyst experiment, continued**

   The catalyst experiment was described in Exercise 5 of Chapter 5. The data were given in Table 5.16, page 130. There were twelve treatment combinations consisting of four levels of reagent, which we may recode as $A = 1$, $B = 2$, $C = 3$, $D = 4$, and three levels of catalyst, which we may recode as $X = 1$, $Y = 2$, $Z = 3$, giving the treatment combinations 11, 12, 13, 21, ..., 43.

   The order of observation of the treatment combinations is also given in Table 5.16.

   (a) Fit a two-way complete model to the data and plot the residuals against the time order. If you are happy about the independence of the error variables, then check the other assumptions on the model and analyze the data. Otherwise, go to part (b).

   (b) Recode the treatment combinations as 1, 2, ..., 12. Fit an analysis of covariance model (9.2.1) or (9.2.2) to the data, where the covariate x_{it} denotes the time in the run order at which the tth observation on the ith treatment combination was made. Check all of the assumptions on your model, and if they appear to be satisfied, analyze the data.

   (c) Plot the adjusted means of the twelve treatment combinations in such a way that you can investigate the interaction between the reagents and catalysts. Test the hypothesis that the interaction is negligible.

   (d) Check the model for lack of fit; that is, investigate the treatment × time interaction. State your conclusions.

10. Complete Block Designs

10.1 Introduction
10.2 Blocks, Noise Factors or Covariates?
10.3 Design Issues
10.4 Analysis of Randomized Complete Block Designs
10.5 A Real Experiment—Cotton-Spinning Experiment
10.6 Analysis of General Complete Block Designs
10.7 Checking Model Assumptions
10.8 Factorial Experiments
10.9 Using SAS Software
Exercises

10.1 Introduction

In step (b)(iii) of the checklist in Chapter 2, we raised the possibility that an experiment may involve one or more nuisance factors that although not of interest to the experimenter could have a major effect on the response. We classified these nuisance factors into three types: blocking factors, noise factors, and covariates. Different types of nuisance factors lead to different types of analyses, and the choice between these is revisited in Section 10.2.

The cotton-spinning experiment of Section 2.3, page 14, illustrates some of the considerations that might lead an experimenter to include a blocking factor in the model and to adopt a block design. The most commonly used block designs are the randomized complete block designs and the general complete block designs. These are defined in Section 10.3.2, and their randomization is illustrated in Section 10.3.3. Models, multiple comparisons, and analysis of variance for randomized complete block designs are given in Section 10.4 and those for general complete block designs in Section 10.6. Model assumption checks are outlined briefly in Section 10.7. An analysis of the cotton-spinning experiment is described in

Section 10.5, and in Section 10.8, we illustrate the analysis of a complete block design with factorial treatment combinations. Analysis of complete block designs by the SAS computer package is discussed in Section 10.9.

10.2 Blocks, Noise Factors or Covariates?

It is not always obvious whether to classify a nuisance factor as a blocking factor, a covariate, or a noise factor. The decision will often be governed by the goal of the experiment.

Nuisance factors are classified as *noise factors* if the objective of the experiment is to find settings of the treatment factors whose response is least affected by varying the levels of the nuisance factors. Settings of noise factors can usually be controlled during an experiment but are uncontrollable outside the laboratory. We gave an illustration of a noise factor in Section 7.6, page 217, and we will give some more examples in Chapter 15.

Covariates are nuisance factors that cannot be controlled but can be measured prior to, or during, the experiment. Sometimes covariates are of interest in their own right, but when they are included in the model as nuisance variables, their effects are used to adjust the responses so that treatments can be compared as though all experimental units were identical (see Chapter 9).

A block design is appropriate when the goal of the experiment is to compare the effects of different treatments averaged over a range of different conditions. The experimental units are grouped into sets in such a way that two experimental units in the same set are similar and can be measured under similar experimental conditions, but two experimental units in different sets are likely to give rise to quite different measurements even when assigned to the same treatment. The sets of similar experimental units are called *blocks*, and the conditions that vary from block to block form the levels of the *blocking factor*. The analysis of a block design involves the comparison of treatments applied to experimental units within the same block. Thus, the intent of blocking is to prevent large differences in the experimental units from masking differences between treatment factor levels, while at the same time allowing the treatments to be examined under different experimental conditions.

The levels of a blocking factor may be the values of a covariate that has been measured prior to the experiment and whose value is used to group the experimental units. More often, however, the levels of a blocking factor are groupings of characteristics that cannot be conveniently measured. For example, grouping the time slots in the same day into the same block, as was done for the cotton-spinning experiment in Section 2.3, ensures that environmental conditions within a block are fairly similar without the necessity of measuring them.

Since the levels of the blocking factor do not necessarily need to be measured, the block design is very popular. Agricultural experimenters may know that plots close together in a field are alike, while those far apart are not alike. Industrial experimenters may know that two items produced by one machine have similar characteristics, while those produced by two different machines are somewhat different. Medical experimenters may know that measurements taken on the same subject will be alike, while those taken on different subjects will not be alike. Consequently, blocks may be formed without actually knowing the precise levels of the blocking factor. Some more examples are given in the next section and throughout the chapter.

10.3 Design Issues

10.3.1 Block Sizes

Although it is perfectly possible for the numbers of experimental units in each block to be unequal, this is outside the scope of this book, and we will examine only block designs in which the block sizes are the same. We will use b to represent the number of blocks and k to represent the common block sizes.

Sometimes the block sizes are naturally defined, and sometimes they need to be specifically selected by the experimenter. In a bread-baking experiment, for example, the experimental units are the baking tins in different positions in the oven. If the temperature cannot be carefully controlled, there is likely to be a temperature gradient from the top shelf to the bottom shelf of the oven, although the temperature at all positions within a shelf may be more or less constant. If the measured response is affected by temperature, then experimental units on the same shelf are alike, but those on different shelves are different. There is a natural grouping of experimental units into blocks defined by the shelf of the oven. Thus, the shelves are the blocks of experimental units and represent the levels of the blocking factor "temperature." The number b of blocks is the number of shelves in the oven. The block size k is the number of baking tins that can be accommodated on each shelf.

It is not always the case that the block size is naturally defined by the experimental equipment. It will often need to be determined by the judgment of the experimenter. For example, the data in Figure 10.1 were gathered in a pilot experiment by Bob Belloto in the Department of Pharmacy at Ohio State University. The data show the readings given by a breathalyzer for a given concentration of alcohol. Notice how the readings decrease over time. Likely causes for this decrease include changes in atmospheric conditions, evaporation of alcohol, and deterioration of the breathalyzer filters. In other experiments, such trends in the data can be caused by equipment heating over time, by variability of batches of raw material, by experimenter fatigue, etc.

The block sizes for the breathalyzer experiment were chosen to be five, that is, the first five observations would be in one block, the next five in the next block, and so on. The reason for the choice was twofold. First, it can be seen from Figure 10.1 that the observations in the pilot experiment seem to be fairly stable in groups of five. Secondly, the experiment was to be run by two different technicians, who alternated shifts, and five observations could be taken per shift. Thus the blocking factor was factorial in nature, and its levels represented combinations of time and technicians.

It is not uncommon in industry for an experiment to be automatically divided into blocks according to time of day as a precaution against changing experimental conditions. A pilot experiment such as that in the breathalyzer experiment is the ideal way of determining the necessity for blocking. If blocks were to be created when they are not needed, hypothesis

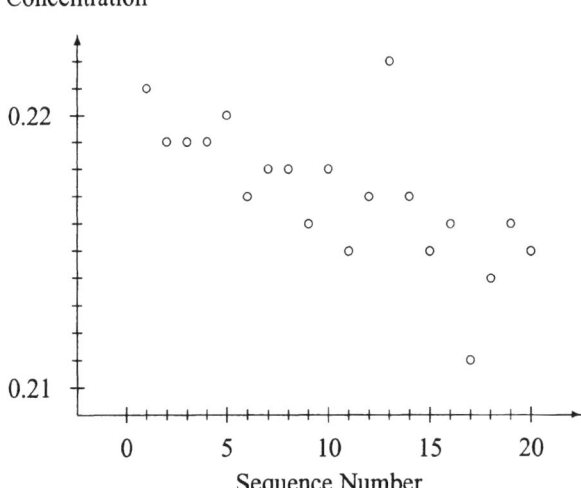

Figure 10.1
Pilot data for the breathalyzer experiment

tests would be less powerful and confidence intervals would be wider than those obtained via a completely randomized design.

10.3.2 Complete Block Design Definitions

Having decided on the block size and having grouped the experimental units into blocks of similar units, the next step is to assign the units to the levels of the treatment factors. The worst possible assignment of experimental units to treatments is to assign all the units within a block to one treatment, all units within another block to a different treatment, and so on. This assignment is bad because it does not allow the analysis to distinguish between differences in blocks and differences in treatments. The effects of the treatment factors and the effects of the blocking factor are said to be *confounded*.

The best possible assignment is one that allocates to every treatment the same number of experimental units per block. This can be achieved only when the block size k is a multiple of v, the number of treatments. Such designs are called *complete block designs*, and in the special case of $k = v$, they have historically been called *randomized complete block designs* or, simply, *randomized block designs*. The historical name is unfortunate, since all block designs need to be randomized. Nevertheless, we will retain the name randomized complete block design for block size $k = v$ and use the name *general complete block design* for block size a larger multiple of v.

If the block size is not a multiple of v, then the block design is known as an *incomplete block design*. This term is sometimes reserved for the smaller designs where $k < v$, but we will find it convenient to classify all designs as either complete or incomplete. Incomplete block designs are more complicated to design and analyze than complete block designs, and we postpone their discussion to Chapter 11.

We will continue to use r_i to mean the number of times that treatment i is observed in the experiment, and we introduce a new symbol n_{hi} to denote the number of times that treatment

Table 10.1 Randomization for block 1 of the bread-baking experiment

Unsorted Treatments	Unsorted Random Numbers	Sorted Treatments	Sorted Random Numbers	Experimental Unit
1	74	2	11	FL
2	11	3	39	FR
3	39	4	68	BL
4	68	1	74	BR

i is observed in block h. For complete block designs, all the n_{hi} are equal to the constant $s = k/v$ and all the r_i are equal to the constant $r = bs$.

10.3.3 The Randomized Complete Block Design

A *randomized complete block design* is a design with v treatments (which may be factorial treatment combinations) and with $n = bv$ experimental units grouped into b blocks of $k = v$ units in such a way that units within a block are alike and units in different blocks are substantially different. The $k = v$ experimental units within each block are randomly assigned to the v treatments so that each treatment is assigned one unit per block. Thus, each treatment appears once in every block ($s = 1$) and $r = b$ times in the design.

Example 10.3.1 Bread-baking experiment

An experimenter wishes to compare the shelf life of loaves made from $v = 4$ different bread doughs, coded 1, 2, 3, 4. An oven with three shelves will be used, and each shelf is large enough to take four baking tins. A temperature difference is anticipated between the different shelves but not in different positions within a shelf. The oven will be used twice, giving a total of six blocks defined by shelf/run of the oven, and the block size is $k = 4$ defined by the four positions on each shelf: FL, FR, BL, BR (front left, front right, back left, back right).

Since the block size is the same as the number of treatments, a randomized complete block design can be used. The experimental units (positions) in each block (shelf/run) are assigned at random to the four levels of the treatment factor (doughs) using the procedure described in Section 3.2, page 34, for each block separately. For example, suppose we obtain the four 2-digit random numbers 74, 11, 39, 68 from a computer random number generator, or from Table A.1, and associate them in this order with the four treatments to be observed once each in block 1. If we now sort the random numbers into ascending order, the treatment codes are randomly sorted into the order 2, 3, 4, 1 (see Table 10.1). If we allocate the experimental units in the order FL, FR, BL, BR to the randomly sorted treatments, we obtain the randomized block shown in the first row of Table 10.2. The other randomized blocks in Table 10.2 are obtained in a similar fashion.

Notice that the randomization that we have obtained in Table 10.2 has allowed bread dough 1 to be observed four times in the back right position, and that dough 2 is never observed in this position. If a temperature difference in positions is present, then this could cause problems in estimating treatment differences. If a temperature difference in positions is

Table 10.2 Example of a randomized complete block design

Block	Run	Shelf	FL	FR	BL	BR
1	1	1	2	3	4	1
2		2	1	2	3	4
3		3	4	3	2	1
4	2	1	2	4	3	1
5		2	2	4	1	3
6		3	3	2	4	1

likely, the randomized complete block design is not the correct design to use. A row–column design (Chapter 12) should be used instead. □

10.3.4 The General Complete Block Design

A *general complete block design* is a design with v treatments (which may be factorial treatment combinations) and with $n = bvs$ experimental units grouped into b blocks of $k = vs$ units in such a way that units within a block are alike and units in different blocks are substantially different. The $k = vs$ experimental units within each block are randomly assigned to the v treatments so that each treatment is assigned s units per block. Thus, each treatment appears s times in every block and $r = bs$ times in the design. In the special case of $s = 1$, the complete block design is called a randomized complete block design (Section 10.3.3).

Example 10.3.2 Light bulb experiment

An experiment was run by P. Bist, G. Deshpande, T.-W. Kung, R. Laifa, and C.-H. Wang in 1995 in order to compare the light intensities of different brands of light bulbs. Three brands were selected, and coded as 1, 2, and 3. Brand 3 was cheaper than brands 1 and 2. A further factor that was examined in the experiment was the effect of the percentage capacity of the bulb, which was controlled by the amount of current being passed through the bulb. Two levels of capacity were selected, 50% and 100%. Thus, there were $v = 6$ treatment combinations in total, which we will code as follows:

(50%, Brand 1) = 1,   (50%, Brand 2) = 2,   (50%, Brand 3) = 3,
(100%, Brand 1) = 4, (100%, Brand 2) = 5, (100%, Brand 3) = 6.

The experimenters wanted to compare the six treatment combinations for both 60 watt bulbs and 100 watt bulbs. Comparison of illumination between two different wattages was not of particular interest, since 100 watt bulbs should be brighter than 60 watt bulbs. The experiment needed to be run at two separate times, so it was convenient to examine the 60 watt bulbs on one day and the 100 watt bulbs on another. Therefore, two blocks were used, with the blocks being defined by the combinations of day and wattage. Four observations were taken on each of the $v = 6$ treatment combinations in each block.

The experiment was run in a dark room. Each light bulb was wired to a dimmer switch, and a digital multimeter was used to measure the amount of current flowing through the bulb. A photo resister was positioned one foot from the bulb, and the illumination was also

10.4 Analysis of Randomized Complete Block Designs

measured with a multimeter. The response variable was the observed resistance of the photo resister, where high illumination corresponds to low resistance. The data (resistances) are shown in Table 10.9, page 311, where the analysis is discussed. □

10.3.5 How Many Observations?

If the block size is pre-determined, we can calculate the number of blocks that are required to achieve a confidence interval of given length, or a hypothesis test of desired power, in much the same way as we calculated sample sizes in Chapter 6. If the number of blocks is limited, but the block sizes can be very large, the same techniques can be used to calculate the required block size for a general complete block design. A calculation of the required number of blocks using confidence intervals will be illustrated for a randomized complete block design in Section 10.5.2, and a calculation of the required block size using the power of a test will be done in Section 10.6.3 for a general complete block design.

10.4 Analysis of Randomized Complete Block Designs

10.4.1 Model and Analysis of Variance

The standard model for a randomized complete block design is

$$Y_{hi} = \mu + \theta_h + \tau_i + \epsilon_{hi}, \qquad (10.4.1)$$
$$\epsilon_{hi} \sim N(0, \sigma^2),$$

ϵ_{hi}'s are mutually independent,

$h = 1, \ldots, b; \quad i = 1, \ldots, v,$

where μ is a constant, θ_h is the effect of the hth block, τ_i is the effect of the ith treatment, Y_{hi} is the random variable representing the measurement on treatment i observed in block h, and ϵ_{hi} is the associated random error. We will call this standard model the *block–treatment model*.

Notice that the block–treatment model does not include a term for the interaction between blocks and treatments. If interaction effects were to be included in the model, there would be no degrees of freedom for error with which to estimate the error variance (cf. Section 6.7). In many blocked experiments, absence of block×treatment interaction is a reasonable assumption. However, if interaction is suspected in a given experiment, then the block size must be increased to allow its estimation.

The block–treatment model (10.4.1) looks similar to the two-way main-effects model (6.2.3) for two treatment factors in a completely randomized design with one observation per cell. Not surprisingly, then, the analysis of variance table in Table 10.3 for the randomized complete block design looks similar to the two-way analysis of variance table in Table 6.7, page 167, for two treatment factors and one observation per treatment combination. There is, however, an important difference. In a completely randomized design, the treatment combinations, and so the levels of *both* factors, are randomly assigned experimental units. On the other hand, in a block design, although observations are taken on all combinations of treatments and blocks, only the levels of the treatment factor are randomly assigned experimental units. The levels of the block factor represent intentional groupings of the experimental units. This leads to some controversy as to whether or not a test of equality of block effects is valid. In the present situation, where the blocks represent nuisance sources of variation, we do not need to know whether or not the block effects are truly equal. It is very unlikely that we can use the identical blocks again. So, rather than testing for equality of block effects, we will merely compare the block mean square $ms\theta$ with the error mean square msE to determine whether or not blocking was beneficial in the experiment at hand.

If $ms\theta$ is considerably larger than msE, this suggests that the creation of blocks was worthwhile in the sense of having reduced the size of the error mean square. If $ms\theta$ is less than msE, then the creation of blocks has lowered the power of hypothesis tests and increased the lengths of confidence intervals for treatment contrasts. The comparison of $ms\theta$ and msE is not a significance test. There is no statistical conclusion about the equality of block effects. The comparison is merely an assessment of the usefulness of having created blocks in this particular experiment and does provide some information for the planning of future, similar experiments. Of course, if $ms\theta$ is less than msE, it is *not* valid to pretend that the experiment was designed as a completely randomized design and to remove the block effects from the model—the randomization is not correct for a completely randomized design.

For testing hypotheses about treatment effects, we can use the analogy with the two-way main-effects model. The decision rule for testing the null hypothesis $H_0 : \{\tau_1 = \tau_2 = \cdots = \tau_v\}$, that the treatments have the same effect on the response, against the alternative hypothesis $H_A : \{$at least two of the τ_i differ$\}$ is

$$\text{reject } H_0 \text{ if } msT/msE > F_{v-1, bv-b-v+1, \alpha} \tag{10.4.2}$$

for some chosen significance level α, where msT and msE are defined in Table 10.3.

Example 10.4.1 Resting metabolic rate experiment

In the 1993 issue of *Annals of Nutrition and Metabolism*, R. C. Bullough and C. L. Melby describe an experiment that was run to compare the effects of inpatient and outpatient pro-

Table 10.3 Analysis of variance: randomized complete block design

Source of Variation	Degrees of Freedom	Sum of Squares	Mean Square	Ratio
Block	$b-1$	$ss\theta$	$ms\theta = \frac{ss\theta}{b-1}$	—
Treatment	$v-1$	ssT	$msT = \frac{ssT}{v-1}$	$\frac{msT}{msE}$
Error	$bv-b-v+1$	ssE	$msE = \frac{ssE}{bv-b-v+1}$	
Total	$bv-1$	$sstot$		
Formulae				
$ss\theta = \sum_h y_{h.}^2/v - y_{..}^2/(bv)$			$sstot = \sum_h \sum_i y_{hi}^2 - y_{..}^2/(bv)$	
$ssT = \sum_i y_{.i}^2/b - y_{..}^2/(bv)$			$ssE = sstot - ss\theta - ssT$	

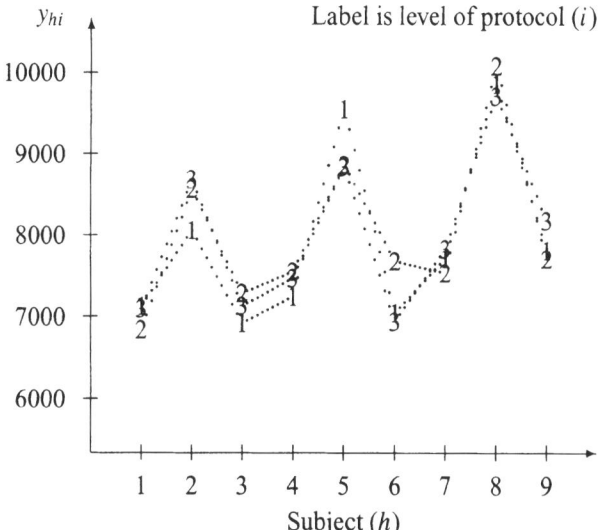

Figure 10.2
Resting metabolic rates by protocol (i) and subject (h)

tocols on the in-laboratory measurement of resting metabolic rate (RMR) in humans. A previous study had indicated measurements of resting metabolic rate on elderly individuals to be 8% higher using an outpatient protocol than with an inpatient protocol. If the measurements depend on the protocol, then comparison of the results of studies conducted by different laboratories using different protocols would be difficult. The experimenters hoped to show that the effects of protocol were negligible.

The experimental treatments consisted of three protocols: (1) an inpatient protocol in which meals were controlled—the patient was fed the evening meal and spent the night in the laboratory, then RMR was measured in the morning; (2) an outpatient protocol in which meals were controlled—the patient was fed the same evening meal at the laboratory but spent the night at home, then RMR was measured in the morning; and (3) an outpatient protocol in which meals were not strictly controlled—the patient was instructed to fast for 12 hours prior to measurement of RMR in the morning. The three protocols formed the $v = 3$ treatments in the experiment.

Since subjects tend to differ substantially from each other, error variability can be reduced by using the subjects as blocks and measuring the effects of all treatments for every subject. In this experiment, there were nine subjects (healthy, adult males of similar age), and they formed the $b = 9$ levels of a blocking factor "subject." Every subject was measured under all three treatments, so the blocks were of size $k = 3 = v$. RMR readings were taken over a one-hour period shortly after the subject arrived in the laboratory. The data collected during the second 30 minutes of testing are given in Table 10.4 and are plotted in Figure 10.2. The figure clearly suggests large subject differences, but no consistent treatment differences.

The analysis of variance is shown in Table 10.5. The value of $ms\theta$ is 37 times larger than msE, indicating that blocking by subject has greatly reduced the error variance estimate. So a block design was a good choice for this experiment.

The null hypothesis of no difference in the protocols cannot be rejected at any reasonable selection of α, since $msT/msE = 0.23$. The ratio tells us that the average variability of the

Table 10.4 Data for the resting metabolic rate experiment

	Protocol		
Subject	1	2	3
1	7131	6846	7095
2	8062	8573	8685
3	6921	7287	7132
4	7249	7554	7471
5	9551	8866	8840
6	7046	7681	6939
7	7715	7535	7831
8	9862	10087	9711
9	7812	7708	8179

Source: Bullough, R. C. and Melby, C. L. (1993). Copyright © 1993 Karger, Basel. Reprinted with permission.

measurements from one protocol to another was four times smaller than the measurement error variability. This is unusual, since measurements from one protocol to another must include measurement error. The p-value is 0.7950, indicating that there is only a 20% chance that we would see a value this small or smaller when there is no difference whatsoever in the effects of the protocols. Thus, we should ask how well the model fits the data—perhaps treatment–block interaction is missing from the model and has been included incorrectly in the error variability. Even if this were the case, however, there is still no indication that protocol 3 provides higher RMR readings than protocol 1—in fact, protocol 3 gave the lowest readings for four of the nine subjects.

It is not possible to check the model assumptions of equal error variances for each cell because of the small amount of data. But we can check the equal-variance assumptions for the different levels of the treatment factor. We find that the variances of the unstandardized residuals are very similar for the three protocols. The normality assumption seems to be reasonable. The only possible outlier is the observation for protocol 1, subject 5, but its removal does not change the above conclusions.

In their article, the experimenters discuss possible reasons for the fact that their conclusions differ from those of previous studies. Reasons included the different age of the subjects (27–29 years rather than 64–67 years) and the fact that they provided transport to the laboratory for the outpatients, whereas previous studies had not. □

Table 10.5 Analysis of variance for the resting metabolic rate experiment

Source of Variation	Degrees of Freedom	Sum of Squares	Mean Square	Ratio	p-value
Subject	8	23,117,462.30	2,889,682.79	–	–
Protocol	2	35,948.74	17,974.37	0.23	0.7950
Error	16	1,235,483.26	77217.70		
Total	26	24,388,894.30			

10.4.2 Multiple Comparisons

The block–treatment model (10.4.1) for the randomized complete block design is similar to the two-way main-effects model (6.2.3) for an experiment with two treatment factors and one observation per cell. Consequently, the least squares estimator for each $\mu + \theta_h + \tau_i$ ($h = 1, \ldots, b;\ i = 1, \ldots, v$) is similar to the estimator for each $\mu + \alpha_i + \beta_j$ ($i = 1, \ldots, a;\ j = 1, \ldots, b$) in (6.5.27), page 158, without the third subscript; that is,

$$\hat{\mu} + \hat{\theta}_h + \hat{\tau}_i = \overline{Y}_{h.} + \overline{Y}_{.i} - \overline{Y}_{..} \qquad (10.4.3)$$

It follows that any contrast $\Sigma c_i \tau_i$ (with $\Sigma c_i = 0$) in the treatment effects is estimable in the randomized complete block design and has least squares estimator

$$\Sigma c_i \hat{\tau}_i = \Sigma c_i \overline{Y}_{.i}\ .$$

The corresponding least squares estimate is $\Sigma c_i \overline{y}_{.i}$, and the corresponding variance is $\sigma^2(\Sigma c_i^2 / b)$. As for the two-way main-effects model, all of the multiple comparison procedures of Chapter 4 are valid for treatment contrasts in the randomized complete block design. The formulae, adapted from (6.5.40), page 164, are

$$\sum c_i \tau_i \in \left(\sum c_i \overline{y}_{.i} \pm w \sqrt{msE \sum c_i^2 / b} \right), \qquad (10.4.4)$$

where the critical coefficients for the Bonferroni, Scheffé, Tukey, Hsu, and Dunnett methods are, respectively,

$$w_B = t_{bv-b-v+1, \alpha/2m}\ ;\quad w_S = \sqrt{(v-1) F_{v-1, bv-b-v+1, \alpha}}\ ;$$

$$w_T = q_{v, bv-b-v+1, \alpha} / \sqrt{2}\ ;$$

$$w_H = w_{D1} = t^{(0.5)}_{v-1, bv-b-v+1, \alpha}\ ;\quad w_{D2} = |t|^{(0.5)}_{v-1, bv-b-v+1, \alpha}\ .$$

Example 10.4.2 Resting metabolic rate experiment, continued

In the resting metabolic rate experiment, described in Example 10.4.1, page 302, all three pairwise comparisons in the $v = 3$ protocol effects were of interest prior to the experiment, together with the contrast that compares the inpatient protocol with the two outpatient protocols. This latter contrast has coefficient list $[1, -\frac{1}{2}, -\frac{1}{2}]$. Suppose that the experimenters had wished to calculate simultaneous 95% confidence intervals for these four contrasts. From (10.4.4), the method of Scheffé uses critical coefficient

$$w_S = \sqrt{2 F_{2, 16, .05}} = \sqrt{2(3.63)} = 2.694\ ,$$

whereas the critical coefficient for the $m = 4$ confidence intervals using the Bonferroni method is

$$w_B = t_{16, .05/(2m)} = t_{16, .00625} \approx 2.5 + (2.5^3 + 2.5)/(4(16)) \approx 2.783\ ,$$

for $z_{.00625} = 2.5$ (see equation (4.4.22), page 81). Hence, the Scheffé method gives tighter intervals.

306   Chapter 10   Complete Block Designs

For each pairwise comparison $\tau_i - \tau_p$, we have $\sum c_i^2 = 2$, so using the Scheffé method of multiple comparisons and $msE = 77217.7$ from Table 10.5, the interval becomes

$$\tau_i - \tau_p \in \left(\bar{y}_{.i} - \bar{y}_{.p} \pm 2.694\sqrt{(77217.7)(2)/9}\right) = \left(\bar{y}_{.i} - \bar{y}_{.p} \pm 352.89\right).$$

The treatment sample means are obtained from the data in Table 10.4 as

$$\bar{y}_{.1} = 7927.7, \quad \bar{y}_{.2} = 8015.2, \quad \bar{y}_{.3} = 7987.0,$$

the biggest difference being $\bar{y}_{.2} - \bar{y}_{.1} = 87.5$. Since all three intervals contain zero, we can assert with 95% confidence that no two protocols differ.

Similarly, the Scheffé confidence interval for $\tau_1 - \frac{1}{2}(\tau_2 + \tau_3)$ is

$$\tau_1 - \frac{1}{2}(\tau_2 + \tau_3) \in \left(\bar{y}_{.1} - \frac{1}{2}(\bar{y}_{.2} + \bar{y}_{.3})\right) \pm 2.694\sqrt{(77217.7)(1.5)/9}$$
$$= (-73.44 \pm 305.62),$$

and again the interval contains zero. These results are expected in light of the failure in Example 10.4.1 to reject equality of treatment effects in the analysis of variance. □

10.5 A Real Experiment—Cotton-Spinning Experiment

10.5.1 Design Details

The checklist for the cotton-spinning experiment was given in Section 2.3, page 14. After considering several different possible designs, the experimenters settled on a randomized complete block design. Each experimental unit was the production of one full set of bobbins on a single machine with a single operator. A block consisted of a group of experimental units with the same machine, the same operator, and observed in the same week. Thus, the different levels of the blocking factor represented differences due to combinations of machines, operators, environmental conditions, and raw material. The block size was chosen to be six, as this was equal to the number of treatment combinations and also to the number of observations that could be taken on one machine in one week.

The treatment combinations were combinations of levels of two treatment factors, "flyer" and "degree of twist." Flyer had two levels, "ordinary" and "special." Twist had four levels, 1.63, 1.69, 1.78, and 1.90. For practical reasons, the combinations of flyer and twist equal to (ordinary, 1.63) and (special, 1.90) were not observed. We will recode the six treatment combinations that were observed as follows:

(ordinary, 1.69) = 1, (ordinary, 1.78) = 2, (ordinary, 1.90) = 3,
(special, 1.63) = 4,   (special, 1.69) = 5,   (special, 1.78) = 6.

The goal of the experiment was to investigate the effects of the flyers and degrees of twist on the breakage rate of cotton.

10.5.2 Sample-Size Calculation

Since the experimenters were interested in all pairwise comparisons of the effects of the treatment combinations, as well as some other special treatment contrasts, we will apply the Scheffé method of multiple comparisons at overall confidence level 95%. The experimenters initially wanted a confidence interval to indicate a difference in a pair of treatment combinations if the true difference in their effects was at least 2 breaks per 100 pounds of material. We will calculate the number of blocks that are needed to obtain a minimum significant difference of at most 2 for the Scheffé simultaneous confidence intervals for pairwise comparisons. Using (10.4.4) with $v = 6$, $\alpha = 0.05$, and $\Sigma c_i^2 = 2$, we need to find b such that

$$\sqrt{5 F_{5,5b-5,0.05}} \sqrt{msE\,(2/b)} \leq 2.$$

The error variability σ^2 was expected to be about 7 breaks^2, so we need to find the smallest value of b satisfying

$$F_{5,5b-5,0.05} \leq \frac{4 \times b}{5 \times 7 \times 2} = \frac{2b}{35}.$$

Trial and error shows that $b = 40$ will suffice.

Each block took a week to complete, and it was not clear how many machines would be available at any one time, so the experimenters decided that they would analyze the data after the first 13 blocks had been observed. With $b = 13$, $v = 6$, and a value of σ^2 expected to be about 7 breaks^2, the Scheffé 95% confidence intervals for pairwise comparisons have minimum significant difference equal to

$$msd = \sqrt{5 F_{5,5(13-1),0.05}} \sqrt{7 \times (2/13)} \;=\; 3.57.$$

A difference in treatment combinations i and p will be indicated if their observed average difference is more than 3.57 breaks per 100 pounds (with a probability of 0.95 of no false indications) rather than 2 breaks per 100 pounds.

10.5.3 Analysis of the Cotton-Spinning Experiment

The data for the first $b = 13$ blocks observed in the experiment were shown in Table 2.3 (page 17) and some of the data were plotted in Figure 2.1. There is an indication of block differences over time. The low number of breaks tend to be in block 1, and the high number of breaks in blocks 11, 12, and 13. This suggests that blocking was worthwhile. This is also corroborated by the fact that $ms\theta$ is nearly three times as large as msE.

The error assumptions for the block–treatment model (10.4.1) are satisfied apart from two outlying observations for treatment 1 (from blocks 5 and 10). The two outliers cause the variances of the unstandardized residuals to be unequal. Also, the normality assumption appears to be not quite satisfied. Since the experiment was run a long time ago, we are not able to investigate possible causes of the outliers. The best we can do is to run the analysis both with and without them. Here, we will continue the analysis including the outliers, and in Exercise 3, we ask the reader to verify that the model assumptions are approximately satisfied when the outliers are removed and that similar conclusions can be drawn.

Table 10.6 Analysis of variance for the cotton-spinning experiment

Source of Variation	Degrees of Freedom	Sum of Squares	Mean Square	Ratio	p-value
Block	12	177.155	14.763	–	
Treatment	5	231.034	46.207	9.05	0.0001
Error	60	306.446	5.107		
Total	77	714.635			

The analysis of variance table is shown in Table 10.6. Luckily, the error variance is smaller than that expected, and consequently, the confidence intervals will not be as wide as feared. The null hypothesis of equality of the treatment effects is rejected at significance level $\alpha = 0.01$, since the p-value is less than 0.01 and, equivalently,

$$msT/msE = 9.05 > F_{5,60,.01} = 3.34\,.$$

The treatment sample means are

i :	1	2	3	4	5	6
$\bar{y}_{.i}$:	10.8000	9.2769	7.1846	6.7538	7.0846	5.6538

With $b = 13$ blocks and $msE = 5.107$, the minimum significant difference for a set of Scheffé's simultaneous 95% confidence intervals is

$$msd = \sqrt{5F_{5,60,0.05}}\sqrt{msE\,\Sigma c_i^2/13} = \sqrt{5(2.37)}\sqrt{5.107\,\Sigma c_i^2/13}$$
$$= 2.158\sqrt{\Sigma c_i^2}\,.$$

For pairwise comparisons we have $\Sigma_i c_i^2 = 2$, so $msd = 3.052$. Comparing this value with differences in treatment sample means, we see that treatment 1 (ordinary flyer, 1.63 twist) yields significantly more breaks on average than all other treatment combinations except 2 (ordinary flyer, 1.69 twist), and 2 is significantly worse on average than 6 (special flyer, 1.90 twist). This might lead one to suspect that the special flyer might be better than the ordinary flyer.

In Figure 10.3, the treatment sample means $\bar{y}_{.i}$ are plotted against the uncoded twist levels, with the type of flyer as the label. This plot reveals informative patterns in the treatment means. In particular, it appears as if the mean number of breaks per 100 pounds decreases almost linearly as the amount of twist increases, with the special flyer (coded 2) yielding consistently smaller means for each amount of twist. Notice that the levels of twist are not equally spaced, so we cannot use the contrast coefficients in Appendix A.2 to measure trends in the breakage rate due to increasing twist. The linear trend is investigated in Section 10.9 using the SAS software.

The contrast $\frac{1}{2}(\tau_1 + \tau_2) - \frac{1}{2}(\tau_5 + \tau_6)$ compares the two flyers, averaging over the common levels (1.69 and 1.78) of twist. The corresponding confidence interval (still using Scheffé's method at an overall 95% confidence level) is

$$\left(\frac{1}{2}(\bar{y}_{.1}+\bar{y}_{.2}) - \frac{1}{2}(\bar{y}_{.5}+\bar{y}_{.6})\right) \pm 2.158\sqrt{\Sigma c_i^2} \approx (3.670 \pm 2.158)$$
$$= (1.512, 5.828)\,.$$

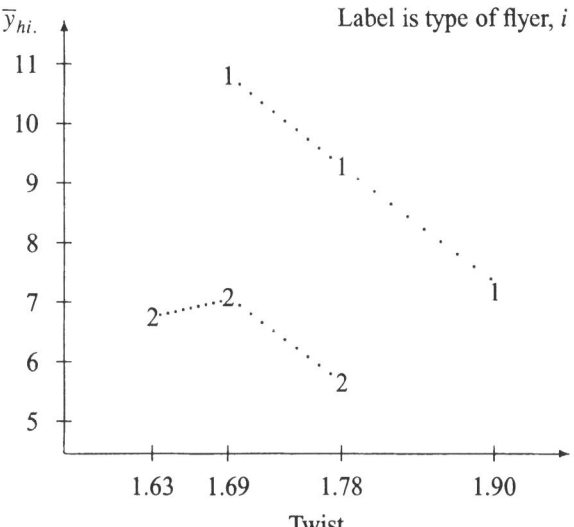

Figure 10.3
Mean number of breaks per 100 pounds for the cotton-spinning experiment

This confidence interval suggests that averaged over the middle two levels of twist, the ordinary flyer 1 is worse than the special flyer, producing on average between about 1.5 and 5.8 more breaks per 100 pounds.

10.6 Analysis of General Complete Block Designs

10.6.1 Model and Analysis of Variance

In this section we discuss general complete block designs with $s > 1$ observations on each treatment in each block. Having every level of the treatment factor observed more than once per block gives sufficient degrees of freedom to be able to measure a block×treatment interaction if one is anticipated. Therefore, there are two standard models for the general complete block design, the *block–treatment model* (without interaction)

$$Y_{hit} = \mu + \theta_h + \tau_i + \epsilon_{hit} \qquad (10.6.5)$$

and the *block–treatment interaction model*, which includes the effect of block–treatment interaction:

$$Y_{hit} = \mu + \theta_h + \tau_i + (\theta\tau)_{hi} + \epsilon_{hit}. \qquad (10.6.6)$$

In each case, the model includes the error assumptions

$$\epsilon_{hit} \sim N(0, \sigma^2),$$

ϵ_{hit}'s are mutually independent,

$t = 1, \ldots, s \, ; \, h = 1, \ldots, b \, ; \, i = 1, \ldots, v.$

The assumptions on these two models should be checked for any given experiment (see Section 10.7).

Table 10.7 Analysis of variance for the general complete block design with negligible block×treatment interaction and block size $k = vs$

Source of Variation	Degrees of Freedom	Sum of Squares	Mean Square	Ratio
Block	$b-1$	$ss\theta$	–	–
Treatment	$v-1$	ssT	$msT = \frac{ssT}{v-1}$	$\frac{msT}{msE}$
Error	$bvs - b - v + 1$	ssE	$msE = \frac{ssE}{bvs-b-v+1}$	
Total	$bvs - 1$	$sstot$		

Formulae

$ss\theta = \sum_h y_{h..}^2/vs - y_{...}^2/(bvs)$ $\quad sstot = \sum_h \sum_i \sum_t y_{hit}^2 - y_{...}^2/(bvs)$

$ssT = \sum_i y_{.i.}^2/bs - y_{...}^2/(bvs)$ $\quad ssE = sstot - ss\theta - ssT$

Table 10.8 Analysis of variance for the general complete block design with block×treatment interaction and block size $k = vs$

Source of Variation	Degrees of Freedom	Sum of Squares	Mean Square	Ratio
Block	$b-1$	$ss\theta$	–	–
Treatment	$v-1$	ssT	$msT = \frac{ssT}{(v-1)}$	$\frac{msT}{msE}$
Interaction	$(b-1)(v-1)$	$ss\theta T$	$ms\theta T = \frac{ss\theta T}{(b-1)(v-1)}$	$\frac{ms\theta T}{msE}$
Error	$bv(s-1)$	ssE	$msE = \frac{ssE}{bv(s-1)}$	
Total	$bvs - 1$	$sstot$		

Formulae

$ss\theta = \sum_h y_{h..}^2/(vs) - y_{...}^2/(bvs)$ $\quad ss\theta T = \sum_h \sum_i y_{hi.}^2/s - \sum_i y_{.i.}^2/(bs)$

$ssT = \sum_i y_{.i.}^2/(bs) - y_{...}^2/(bvs)$ $\qquad\qquad - \sum_h y_{h..}^2/(vs) + y_{...}^2/(bvs)$

$ssE = sstot - ss\theta - ssT - ss\theta T$ $\quad sstot = \sum_h \sum_i \sum_t y_{hit}^2 - y_{...}^2/(bvs)$

The block–treatment model (10.6.5) for a general complete block design is similar to the two-way main-effects model (6.2.3), and the block–treatment interaction model (10.6.6) is like the two-way complete model (6.2.2) for two treatment factors in a completely randomized design, each with s observations per cell. Analogously, the analysis of variance tables (Tables 10.7 and 10.8) for the block–treatment models, with and without interaction, look similar to those for the two-way main-effects and two-way complete models (see Tables 6.7 and 6.4, pages 167 and 156).

The decision rule for testing the null hypothesis $H_0^T : \{\tau_1 = \tau_2 = \cdots = \tau_v\}$ that the treatment effects are equal against the alternative hypothesis H_A^T that at least two of the treatment effects differ is given by the decision rule

reject H_0 if $msT/msE > F_{v-1, df, \alpha}$,   (10.6.7)

where α is the chosen significance level, and where msT, msE, and the error degrees of freedom, df, are obtained from Table 10.7 or 10.8 as appropriate.

Table 10.9 Resistances for the light bulb experiment. Low resistance implies high illumination. (Order of observations is shown in parentheses.)

Block	Treatments					
	1	2	3	4	5	6
I	314 (12)	285 (3)	350 (6)	523 (2)	460 (1)	482 (7)
(60 watt)	300 (13)	296 (9)	339 (8)	497 (4)	470 (5)	498 (11)
	310 (15)	301 (10)	360 (14)	520 (18)	488 (17)	505 (19)
	290 (22)	292 (24)	333 (16)	510 (20)	468 (21)	490 (23)
II	214 (28)	196 (27)	235 (42)	303 (26)	341 (32)	342 (25)
(100 watt)	205 (31)	201 (29)	247 (44)	319 (30)	350 (38)	347 (33)
	197 (35)	197 (39)	233 (46)	305 (34)	323 (41)	352 (37)
	204 (47)	215 (40)	244 (48)	316 (36)	343 (45)	323 (43)

If the block×treatment interaction term has been included in the model, a test of the hypothesis $H_0^{\theta T}: \{(\theta\tau)_{hi} - \overline{(\theta\tau)}_{h.} - \overline{(\theta\tau)}_{.i} + \overline{(\theta\tau)}_{..} = 0$ for all $h, i\}$ against the alternative hypothesis $H_A^{\theta T}$ that at least one interaction contrast is nonzero is given by

reject H_0 if $ms\theta T/msE > F_{(b-1)(v-1),bv(s-1),\alpha}$ (10.6.8)

for some chosen significance level α, where $ms\theta T$ and msE are obtained from Table 10.8. As usual, if the interaction is significantly different from zero, a test of equality of the treatment effects may not be of interest. An evaluation of the usefulness of blocking in the experiment at hand can be made by comparing $ms\theta$ with msE as in Section 10.4.1.

Example 10.6.1 Light bulb experiment, continued

The light bulb experiment, which was run in order to compare the light intensities of different brands of light bulbs, was described in Example 10.3.4, page 311. Three brands were selected and each was examined at 50% and 100% capacity. The $v = 6$ treatments were coded as

(Brand 1, 50%) = 1,   (Brand 2, 50%) = 2,   (Brand 3, 50%) = 3,
(Brand 1, 100%) = 4,   (Brand 2, 100%) = 5,   (Brand 3, 100%) = 6.

The experiment was run as a general complete block design with $b = 2$ blocks defined by combinations of days and wattages (60 watts and 100 watts). There were $s = 4$ observations per treatment per block. The data are given in Table 10.9 and are plotted in Figure 10.4. The figure suggests that there might be a small interaction between block (wattage) and treatment combination.

The analysis of variance table is shown in Table 10.10, and we see that $ms\theta T = 3500.69$ and $msE = 100.69$. Using (10.6.8), $H_0^{\theta T}$, the hypothesis of negligible interaction, is rejected if $ms\theta T/msE = 34.77$ is larger than $F_{5,36,\alpha}$ for some chosen significance level α. For $\alpha = 0.01$, $F_{5,36,.01} = 3.59$, so we reject $H_0^{\theta T}$ and conclude that the block×treatment interaction that appears in Figure 10.4 is significantly greater than zero. Notice that the p-value is at most 0.0001, so any choice of α greater than this would also lead to rejection of $H_0^{\theta T}$.

The purpose of the experiment was to determine the best brand (in terms of illumination). Consequently, even if there had not been a block×treatment interaction, a global test of equality of treatment effects is unlikely to be of interest. The experimenters were more

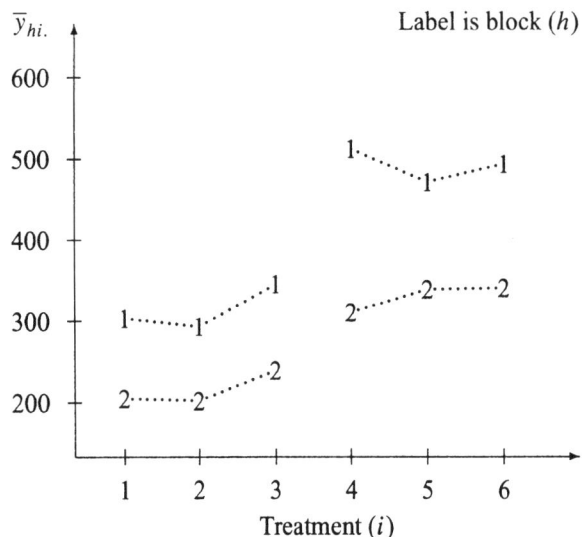

Figure 10.4
Plot of treatment averages for the light bulb experiment

Table 10.10 Analysis of variance for the light bulb experiment

Source of Variation	Degrees of Freedom	Sum of Squares	Mean Square	Ratio	p-value
Block (watts)	1	203971.688	203971.688	–	–
Treatment	5	267645.438	53529.088	531.64	0.0001
Block × Treatment	5	17503.438	3500.688	34.77	0.0001
Error	36	3624.750	100.688		
Total	47	492745.313			

interested in comparing brands averaged over capacities and comparing brands for each capacity separately. Since there is a block×treatment interaction, the experimenters decided to make these comparisons within each block separately. The response variable is resistance, and low resistance coincides with high illumination. Consequently, Figure 10.4 suggests that treatments 1 and 2 are better than treatment 3 at 50% capacity. These treatments correspond to brands 1 and 2, which are the more expensive brands. The position is less clear for 100% capacity. These contrasts will be examined in detail in Example 10.6.2.

We note that there is, as expected, a large difference in blocks. Not only is there a block×treatment interaction, but also the mean square for blocks (averaged over treatments) is 2000 times the size of the mean square for error. □

10.6.2 Multiple Comparisons for the General Complete Block Design

No interaction term in the model The Bonferroni, Scheffé, Tukey, Hsu, and Dunnett methods described in Section 4.4 can all be used for obtaining simultaneous confidence intervals for sets of treatment contrasts in a general complete block design. Since the block–treatment model (10.6.5), without interaction, is similar to the two-way main-effects

10.6 Analysis of General Complete Block Designs

model (6.2.3) with s observations per cell, formulae for multiple comparisons are similar to those given in (6.5.40), page 164, with a replaced by v and r replaced by s. Thus, a set of $100(1-\alpha)\%$ simultaneous confidence intervals for treatment contrast $\Sigma c_i \tau_i$ is of the form

$$\sum c_i \tau_i \in \left(\sum c_i \bar{y}_{.i.} \pm w \sqrt{msE \sum c_i^2 / bs} \right), \quad (10.6.9)$$

where the critical coefficients for the five methods are, respectively,

$$w_B = t_{n-b-v+1, \alpha/2m} \; ; \; w_S = \sqrt{(v-1) F_{v-1, n-b-v+1, \alpha}} \; ;$$

$$w_T = q_{v, n-b-v+1, \alpha}/\sqrt{2}$$

$$w_H = w_{D1} = t^{(0.5)}_{v-1, n-b-v+1, \alpha} \; ; \; w_{D2} = |t|^{(0.5)}_{v-1, n-b-v+1, \alpha},$$

where $n = bvs$.

Interaction term included in the model The block–treatment interaction model (10.6.6) for the general complete block design is similar to the two-way complete model (6.2.2), page 139, for two treatment factors with s observations per cell. Consequently, formulae for confidence intervals for treatment comparisons, averaging over the block×treatment interaction, are similar to those given in (6.4.19), page 149, with a replaced by v and r replaced by s.

The Bonferroni, Scheffé, Tukey, Hsu, and Dunnett methods can be used for comparing the treatments averaged over the block×treatment interaction. The general formula for a set of $100(1-\alpha)\%$ simultaneous confidence intervals for treatment contrasts is given by (10.6.9) above, where the error degrees of freedom in each of the critical coefficients is $n - bv$ instead of $n - b - v + 1$.

Treatment comparisons may not be of interest if treatments do interact with blocks. Instead, within-block comparisons are likely to be preferred. These are similar to the simple contrasts of Section 6.3.1 and are most easily calculated via a cell-means representation of the model. If we write

$$\eta_{hi} = \theta_h + \tau_i + (\theta\tau)_{hi},$$

then the comparison of treatments i and p in block h is the contrast

$$\eta_{hi} - \eta_{hp} = (\theta_h + \tau_i + (\theta\tau)_{hi}) - (\theta_h + \tau_p + (\theta\tau)_{hp}).$$

The following example illustrates the estimation of contrasts of this form.

Example 10.6.2 Light bulb experiment, continued

The data and analysis of variance for the light bulb experiment were given in Example 10.6.1, page 311. Due to the block×treatment interaction, the experimenters decided that they wanted to examine the following treatment comparisons, each of which involved within-block comparisons:

(i) pairwise differences between brands averaged over capacities for each block (wattage) separately, and

(ii) differences between brands for each capacity and each block separately.

In terms of the parameters in the block–treatment interaction model (10.6.6), the contrast that compares brand 1 with brand 2 averaged over capacities in block 1 is a comparison of treatments 1 and 4 with 2 and 5; that is,

$$\tfrac{1}{2}[(\tau_1 + (\theta\tau)_{11}) + (\tau_4 + (\theta\tau)_{14})] - \tfrac{1}{2}[(\tau_2 + (\theta\tau)_{12}) + (\tau_5 + (\theta\tau)_{15})].$$

If we rewrite the model as a cell-means model, we have

$$Y_{hit} = \eta_{hi} + \epsilon_{hit},$$

where $\eta_{hi} = \theta_h + \tau_i + (\theta\tau)_{hi}$ represents the effect on the response of treatment i in block h. The above contrast can then be expressed as

$$\tfrac{1}{2}(\eta_{11} + \eta_{14}) - \tfrac{1}{2}(\eta_{12} + \eta_{15}).$$

The least squares estimate for a parameter η_{hi} of a cell-means model is the cell average $\overline{y}_{hi.}$, so the least squares estimate of the contrast that compares brand 1 with brand 2 in block 1 is

$$\frac{1}{2}(\overline{y}_{11.} + \overline{y}_{14.}) - \frac{1}{2}(\overline{y}_{12.} + \overline{y}_{15.}) = \frac{1}{2}(303.5 + 512.5) - \frac{1}{2}(293.5 + 471.5) = 25.5.$$

The corresponding estimated variance is $(4/16)msE = 25.172$. The least squares estimates for the other contrasts, comparing brands i and j in either block, can be calculated in a similar way.

The second type of contrast of interest is a comparison of brands for each capacity separately in each block. For example,

$$\eta_{25} - \eta_{26}$$

compares brands 2 and 3 at 50% capacity in block 2. The least squares estimate is

$$\overline{y}_{25.} - \overline{y}_{26.} = 339.25 - 341.00 = -1.75,$$

with corresponding estimated variance of $(2/4)\, msE = 50.344$. Suppose we use Scheffé's multiple comparison procedure to calculate a set of 95% simultaneous confidence intervals for the 6 contrasts of the first type and for the 12 contrasts of the second type. The minimum significant difference for the first six contrasts is

$$msd = \sqrt{11\, F_{11,36,.05}}\, \sqrt{(4/16)\, msE} = 23.92,$$

and the minimum significant difference for the other twelve contrasts is

$$msd = \sqrt{11\, F_{11,36,.05}}\, \sqrt{(2/4)\, msE} = 33.86.$$

We find that averaged over capacities, in block 1 (60 watt bulbs, day 1), brands 1 and 2 were significantly different from each other, as were brands 2 and 3. Brand 2 had the lower resistance (higher illumination). In block 2 (100 watt bulbs, day 2), however, only brands 1 and 3 were significantly different, with brand 1 being better. This suggests that on average the cheaper brand (brand 3) was not as good as the more expensive brands.

Now, if we assume that most bulbs are used at 100% capacity and not on a dimmer switch, we find that brands 1 and 2 were each better than brand 3 in both blocks, but not significantly different from each other. At 50% capacity, the only difference that we find is a difference between brands 1 and 2 in block 1, with brand 2 being superior. Putting together all this information might lead one to select brand 2 for most illumination purposes, unless cost was an important factor. □

10.6.3 Sample-Size Calculations

A complete block design has $n = bvs$ experimental units divided into b blocks of size $k = vs$. The block size k and the number of blocks b must be chosen to accommodate the experimental conditions, the budget constraints, and the requirements on the lengths of confidence intervals or powers of hypothesis tests in the usual way. An example of the calculation of b to achieve confidence intervals of given length was given for the randomized complete block design in Section 10.5.2. In Example 10.6.3, we calculate, for a general complete block design, the block size required to achieve a given power of a hypothesis test.

Computing the number of observations s per treatment per block needed to achieve a prescribed power of a test of no treatment differences is analogous to the sample-size calculation for testing main effects of a factor in a two-way layout; that is, from equation (6.6.49), page 168, s must satisfy

$$s \geq \frac{2v\sigma^2\phi^2}{b\Delta^2}, \tag{10.6.10}$$

where Δ is the minimum difference between the treatment effects that is to be detected. The tables for power $\pi(\Delta)$ as a function of ϕ are in Appendix Table A.7. In (10.6.10), we can solve for either s or b, depending upon whether the block size or the number of blocks can be large.

Example 10.6.3 Colorfastness experiment

The colorfastness experiment was planned by D-Y Duan, H. Rhee, and C. Song in 1990 to investigate the linear and quadratic effects of the number of washes on the color change of a denim fabric. The experiment was to be carried out according to the guidelines of the American Association of Textile Chemists and Colorists Test 61-1980.

The experimenters anticipated that there would be systematic differences in the way they made their color determinations, and consequently, they grouped the denim fabric swatches into blocks according to which experimenter was to make the determination. Thus the levels of the blocking factor denoted the experimenter, and there were $b = 3$ blocks. They decided to use a general complete block design and allowed the block size to be $k = vs = 5s$, where s could be chosen. Rightly or wrongly, they did not believe that experimenter fatigue would have a large effect on the results, and they were happy for the block sizes to be large.

They planned to use a block–treatment interaction model (10.6.6), and they wanted to test the null hypothesis of no treatment differences whether or not there was block×treatment interaction. The test was to be carried out at significance level 0.01, and they wanted to reject the null hypothesis with probability 0.95 if there was a true difference of $\Delta = 0.5$ or more in the effect of the number of washes on color rating. They expected σ to be no larger than about 0.4.

We need to find the minimum value of s that satisfies equation (10.6.10); that is,

$$s \geq \frac{2v\sigma^2\phi^2}{b\Delta^2} = \frac{(2)(5)(0.4)^2\phi^2}{(3)(0.5)^2} = 2.13\phi^2.$$

The denominator (error) degrees of freedom for the block–treatment interaction model is $v_2 = bv(s-1) = 15(s-1)$. First we locate that portion of Appendix Table A.7 corresponding to numerator degrees of freedom $v_1 = v - 1 = 4$ and $\alpha = 0.01$. Then to achieve power $\pi = 0.95$, trial and error starting with $s = 100$ gives

s	$15(s-1)$	ϕ	$s = 2.13\phi^2$	Action
100	1485	2.33	11.56	Round up to $s = 12$
12	165	2.37	11.96	Round up to $s = 12$

So about $s = 12$ observations per treatment per block should be taken. □

10.7 Checking Model Assumptions

The assumptions on the block–treatment models (10.4.1) and (10.6.5) and on the block–treatment interaction model (10.6.6) for complete block designs need to be checked as usual. The assumptions on the error variables are that they have equal variances, are independent, and have a normal distribution. The form of the model must also be checked.

A visual check of an assumption of no block × treatment interaction can be made by plotting y_{hit} against the treatment factor levels i for each block h in turn. If the lines plotted for each block are parallel (as in plots (a)–(d) of Figure 6.1, page 137), then block × treatment interaction is likely to be absent, and error variability is small. If the lines are not parallel, then either block × treatment interaction is present or error variability is large.

For the block–treatment model (10.4.1) for the randomized complete block design, the (hi)th residual is

$$\hat{e}_{hi} = y_{hi} - \hat{y}_{hi} = y_{hi} - \bar{y}_{h.} - \bar{y}_{.i} + \bar{y}_{..}.$$

For the block–treatment model (10.6.5) for the general complete block design, the (hit)th residual is similar; that is,

$$\hat{e}_{hit} = y_{hit} - \hat{y}_{hit} = y_{hit} - \bar{y}_{h..} - \bar{y}_{.i.} + \bar{y}_{...}.$$

Table 10.11 Checking error assumptions for a complete block design

To check for:	Plot residuals against:
Independence	Order of observations (in space or time)
Equal variance and outliers	Predicted values $\hat{y}_{hit}$, levels of treatment factor, levels of block factor
Normality	Normal scores (also plot separately for each treatment if r is large and for each block if k is large)

For the block–treatment interaction model (10.6.6), the (hit)th residual is

$$\hat{e}_{hit} = y_{hit} - \hat{y}_{hit} = y_{hit} - \bar{y}_{hi.} \, .$$

The error assumptions are checked by residual plots, as summarized in Table 10.11 and described in Chapter 5.

10.8 Factorial Experiments

When the treatments are factorial in nature, the treatment parameter τ_i in the complete block design models (10.4.1), (10.6.5), and (10.6.6) can be replaced by main-effect and interaction parameters. Suppose, for example, we have an experiment with two treatment factors that is designed as a randomized complete block design—a situation similar to that of the cotton-spinning experiment of Section 10.5. In order not to confuse the number b of blocks with the number of levels of a treatment factor, we will label the two treatment factors as C and D with c and d levels respectively. If instead of recoding the treatment combinations as $1, 2, \ldots, v$ we retain the two digit codes, then the block–treatment model is

$$Y_{hijt} = \mu + \theta_h + \tau_{ij} + \epsilon_{hijt}$$

with the usual assumptions on the error variables. We can then express τ_{ij}, the effect of treatment combination ij, in terms of γ_i (the effect of C at level i), δ_j (the effect of D at level j), and, unless negligible, $(\gamma\delta)_{ij}$ (the effect of their interaction when C is at level i and D at level j); that is,

$$Y_{hijt} = \mu + \theta_h + \gamma_i + \delta_j + (\gamma\delta)_{ij} + \epsilon_{hijt}, \qquad (10.8.11)$$

$$\epsilon_{hijt} \sim N(0, \sigma^2),$$

ϵ_{hijt}'s are mutually independent,

$t = 1, \ldots, s \, ; \, h = 1, \ldots, b \, ; \, i = 1, \ldots, c \, ; \, j = 1, \ldots, d.$

In a general complete block design with $s > 1$ observations per treatment combination per block, we may include in the model some or all of the block×treatment interactions. For example, with two treatment factors, the block–treatment interaction model can be expressed as

$$\begin{aligned} Y_{hijt} &= \mu + \theta_h + \gamma_i + \delta_j + (\gamma\delta)_{ij} + (\theta\gamma)_{hi} \\ &\quad + (\theta\delta)_{hj} + (\theta\gamma\delta)_{hij} + \epsilon_{hijt}, \end{aligned} \qquad (10.8.12)$$

$$\epsilon_{hijt} \sim N(0, \sigma^2),$$

ϵ_{hijt}'s are mutually independent,

$t = 1, \ldots, s \, ; \, h = 1, \ldots, b \, ; \, i = 1, \ldots, c \, ; \, j = 1, \ldots, d.$

If there are more than two factors, the additional main effects and interactions can be added to the model in the obvious way.

Example 10.8.1 Banana experiment

The objectives section of the report of an experiment run in 1995 by K. Collins, D. Marriott, P. Kobrin, G. Kennedy, and S. Kini reads as follows:

> Recently a banana hanging device has been introduced in stores with the purpose of providing a place where bananas can be stored in order to slow the ripening process, thereby allowing a longer time over which the consumer has to ingest them. Commercially, bananas are picked from trees while they are fully developed but quite green and are artificially ripened under controlled conditions (hanging up) prior to transport (usually in boxes) to grocery stores. Once they are purchased and brought into the consumer's home, they are typically placed on a counter top and left there until they are either eaten or turn black, after which they can be thrown away or made into banana bread. Considering that the devices currently being marketed to hang bananas cost some money and take up counter space, it is of interest to us to determine whether or not they retard the ripening process.
>
> While there exist many ways to measure the degree of banana ripening, perhaps the simplest method is via visual inspection. The banana undergoes a predictable transition from the unripened green color to yellow then to yellow speckled with black and finally to fully black. The percentage of black color can be quantified through computer analysis of photographs of the skins of the bananas. Other methods to detect differences in the degree of banana ripening would require specialized instrumentation and techniques that are not available to us.
>
> The major objective of our experiment, then, is to determine whether or not any differences in the percent of black skin exist between bananas that are treated conventionally, i.e., placed on a counter, and bananas that are hung up. As a minor objective, we would like to determine whether or not any difference exists in the percentage of black skin between bananas allowed to ripen in a normal day/night cycle versus those ripening in the dark such as might occur if placed in a pantry.

The unripened bananas were bought as a single batch from a single store. They were assigned at random to four treatment combinations, consisting of combinations of levels of two factors at two levels each. Factor C was Lighting conditions (1= day/night cycle, 2 = dark closet). Factor D was Storage method (1 = hanging, 2 = counter-top). There were 12 bananas assigned to each treatment combination. After five days, the bananas were peeled and photographed. The images from the photographic slides were traced by hand, and the percentage of blackened skin was calculated using an image analyzer on a computer. Three of the experimenters prepared the images for the image analyzer and, since they were unskilled, they decided to regard themselves as blocks in order to remove experimenter differences from the comparisons of the treatment combinations. They selected a general complete block design and assigned the treated bananas in such a way that $s = 4$ observations on each treatment combination were obtained by each experimenter. The treatment combinations were observed in a random order, and the resulting data are shown in Table 10.12.

Since the experimenters did not anticipate a block×treatment interaction, they selected block–treatment model (10.8.11) to represent the data. The decision rule for testing the

10.8 Factorial Experiments

Table 10.12 Percentage blackened banana skin

Experimenter (Block)	Light C	Storage D	Percentage of blackened skin (y_{hijt})			
I	1	1	30	30	17	43
	1	2	43	35	36	64
	2	1	37	38	23	53
	2	2	22	35	30	38
II	1	1	49	60	41	61
	1	2	57	46	31	34
	2	1	20	63	64	34
	2	2	40	47	62	42
III	1	1	21	45	38	39
	1	2	42	13	21	26
	2	1	41	74	24	51
	2	2	38	22	31	55

Table 10.13 Analysis of variance for the banana experiment

Source of Variation	Degrees of Freedom	Sum of Squares	Mean Square	Ratio	p-value
Block (Experimenter)	2	1255.79	627.89	–	
C (Light)	1	80.08	80.08	0.42	0.5218
D (Storage)	1	154.08	154.08	0.80	0.3754
CD	1	24.08	24.08	0.13	0.7250
Error	42	8061.88	191.95		
Total	47	9575.92			

hypothesis H_0^{CD} of no interaction between the treatment factors Light and Storage, using a Type I error probability of $\alpha = 0.01$, is

$$\text{reject } H_0^{CD} \text{ if } \frac{ms(CD)}{msE} > F_{(c-1)(d-1), df, 0.01},$$

where $ms(CD) = ss(CD)/(c-1)(d-1)$ and df is the number of error degrees of freedom calculated below. Since there are equal numbers of observations per cell, we use rule 4 of Chapter 7, page 202; so

$$ss(CD) = bs \sum_i \sum_j \overline{y}_{.ij.}^2 - bds \sum_i \overline{y}_{.i..}^2 - bcs \sum_j \overline{y}_{..j.}^2 + bcds\, \overline{y}_{....}^2 = 24.0833.$$

Similarly,

$$ssC = bds \sum_j \overline{y}_{..j.}^2 - bcds\, \overline{y}_{....}^2 = 80.08,$$

$$ssD = bcs \sum_i \overline{y}_{.i..}^2 - bcds\, \overline{y}_{....}^2 = 154.08,$$

$$ss\theta = cds \sum_h \overline{y}_{h...}^2 - bcds\, \overline{y}_{....}^2 = 1255.79,$$

$$sstot = 9575.92.$$

So,

$$ssE = sstot - ss\theta - ssC - ssD - ss(CD) = 8061.88,$$

and

$$\begin{aligned}df &= (bcds - 1) - (b - 1) - (c - 1) - (d - 1) - (c - 1)(d - 1)\\&= 47 - 2 - 1 - 1 - 1 = 42.\end{aligned}$$

These values are shown in the analysis of variance table, Table 10.13. We can see that the mean square for blocks is much larger than the error mean square, so it was worthwhile designing this experiment as a block design. We also see that the mean square for the Light×Storage interaction is a lot smaller than the error mean square. As mentioned in the context of the resting metabolic rate experiment (Example 10.4.1, page 302), this is unusual when the model fits well, since the Light and Storage measurements include the error measurement. It suggests that the error mean square may have been inflated by some other source of variability, such as block×treatment interaction, that has been omitted from the model.

An interaction plot of the two factors Light and Storage (averaged over blocks and the Light×Storage interaction) is shown in Figure 10.5. There is no indication that hanging bananas (Storage level 1) might retard the ripening process. In fact, Storage level 1 seems to have given a higher percentage of blackened skin on average than Storage level 2. However, this apparent difference may be due to chance, as the treatment effects are not significantly different from each other. The experimenters commented that it was difficult to select the correct threshold levels for the image analysis and also that the bananas themselves seemed extremely variable. The experimenters felt that rather than draw firm conclusions at this stage, it might be worthwhile working to improve the experimental procedure to reduce variability and then to repeat the experiment. □

10.9 Using SAS Software

The analysis of variance table for a complete block design can be obtained from any computer package that has an analysis of variance routine or a regression routine. It is good practice

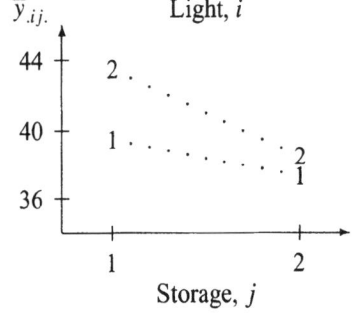
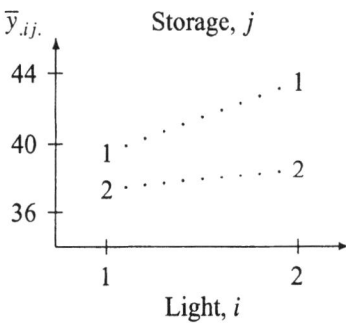

Figure 10.5 Interaction plot for the banana experiment

10.9 Using SAS Software

Table 10.14 A SAS program for analysis of the cotton-spinning experiment

```
DATA COTTON;
  INPUT BLOCK TRTMT FLYER TWIST BREAK;
  LINES;
    1   12 1 1.69  6.0
    2   12 1 1.69  9.7
    :   :  : :      :
   13   23 2 1.78  6.4
;
PROC PRINT;
;
* block--treatment model for a complete block design;
PROC GLM;
  CLASS BLOCK TRTMT;
  MODEL BREAK = BLOCK TRTMT;
;
* Factorial main effects model plus blocks;
DATA;    SET COTTON;
PROC GLM;
  CLASS BLOCK FLYER TWIST;
  MODEL BREAK = BLOCK FLYER TWIST ;
;
* Model with twist as a linear regressor variable;
DATA;    SET COTTON;
PROC GLM;
  CLASS BLOCK FLYER;
  MODEL BREAK = BLOCK FLYER TWIST / SOLUTION;
  ESTIMATE 'FLYER 1-2'  FLYER 1 -1;
```

to enter the block term in the model before the terms for the treatment factors. Although the order does not matter for complete block designs, it does matter for the incomplete block designs in the next chapter.

Computer programs do not distinguish between block and treatment factors, so a test for the hypothesis of no block effects will generally be listed. We suggest that the latter be ignored, and that blocking be considered to have been effective if $ms\theta$ exceeds msE (see Section 10.4.1).

The input statements of a computer program for analyzing a complete block design are similar to those described in Section 6.8 for a two-way model. The statements are illustrated in Table 10.14. The first call of PROC GLM shows a standard block–treatment model for the cotton-spinning experiment, (Section 2.3, page 14). If the block×treatment interaction term had been included in the model, this would have been entered in the usual way as BLOCK*TRTMT. Part of the corresponding output is shown in Table 10.15. The TYPE I and TYPE III sums of squares are equal, since there is one observation per block–treatment combination.

Table 10.15 SAS output for the block–treatment model; cotton-spinning experiment

```
                        The SAS System
                  General Linear Models Procedure

Dependent Variable: BREAK
                            Sum of        Mean
Source              DF     Squares       Square    F Value    Pr > F
Model               17    408.18923     24.01113     4.70     0.0001
Error               60    306.44615      5.10744
Corrected Total     77    714.63538


Source              DF   Type III SS   Mean Square   F Value   Pr > F
BLOCK               12    177.15538     14.76295      2.89     0.0033
TRTMT                5    231.03385     46.20677      9.05     0.0001
```

The second call of PROC GLM in Table 10.14 replaces the treatment parameter TRTMT with main-effect parameters for FLYER and TWIST. The TYPE III sums of squares are used for hypothesis testing, and ESTIMATE statements must now be used for confidence intervals, since not every combination of FLYER and TWIST was observed.

The plot of mean response against twist, with flyer type for labels, in Figure 10.3, page 309, suggested the possibility that the number of breaks per 100 pounds could be well modeled by a flyer effect and a linear twist effect. This can be evaluated by comparing the fit of the block–treatment model,

$$Y_{hi} = \mu + \theta_h + \tau_i + \epsilon_{hi}$$

($i = 12, 13, 14, 21, 22, 23$; $h = 1, \ldots, 13$), with the fit of the reduced model,

$$Y_{hjx} = \mu + \theta_h + \alpha_j + \gamma x + \epsilon_{hjx},$$

where α_j is the effect of flyer j ($j = 1, 2$) and x is the *uncoded* amount of twist. This can be done easily in the SAS software using two calls of the GLM procedure, one for each model. The first call of PROC GLM in Table 10.14 fitted the full block–treatment model, and the third call of PROC GLM fits the reduced model that includes the flyer effect and a linear regression in the levels of twist. Notice the similarity of the third call with the factorial main-effects model in the second call. The difference is that when TWIST is to be regarded as a linear regressor, it is omitted from the CLASS statement. The reduced model fits parallel linear regression lines, with intercepts adjusted for Block and Flyer effects. Again the TYPE I and TYPE III sums of squares are unequal, indicating that FLYER and TWIST cannot be estimated independently.

In the third call of PROC GLM, the SOLUTION option requests printing of the solution to the normal equations. The NOTE at the bottom of the SAS output regarding biased estimates means that the individual flyer effect parameters are not estimable, and the numbers given are nonunique solutions to the normal equations. The contrast representing the difference in the effects of the two flyers *is* estimable, and we can obtain its unique least squares estimate by taking the difference in the two values given for the individual flyers. This gives

Table 10.16 SAS output for the factorial main-effects model; cotton-spinning experiment

```
                        The SAS System
                 General Linear Models Procedure

Source           DF   Type I SS   Mean Square   F Value   Pr > F
BLOCK            12   177.15538     14.76295      2.94    0.0028
FLYER             1   130.78205    130.78205     26.03    0.0001
TWIST             3   100.22410     33.40803      6.65    0.0006

Source           DF Type III SS   Mean Square   F Value   Pr > F
BLOCK            12   177.15538     14.76295      2.94    0.0028
FLYER             1   175.02231    175.02231     34.84    0.0001
TWIST             3   100.22410     33.40803      6.65    0.0006
```

3.8587, which matches the value obtained from the ESTIMATE statement. The difference in the effects of the two flyers is declared to be significantly different from zero, since the corresponding p-value is at most 0.0001. The slope coefficient of TWIST is estimated to be -14.1003, which, being negative, suggests that the breakages decrease as the twist increases. This slope is declared to be significantly different from zero, since the test of $H_0: \gamma = 0$ versus $H_A: \gamma \neq 0$ has p-value at most 0.0001.

The difference in the error sum of squares for the full and reduced models divided by the difference in the error degrees of freedom is the mean square for lack of fit, $msLF$ (cf. Chapter 8). It provides the numerator of the test statistic for testing the null hypothesis H_0^R : {the reduced model is adequate} against the alternative hypothesis that the reduced model is not adequate. The decision rule is

reject H_0^R if $msLF/msE > F_{3,60,\alpha}$,

where

$$msLF = [ssE(\text{reduced}) - ssE(\text{full})] / [df(\text{reduced}) - df(\text{full})].$$

For the cotton-spinning experiment,

$$msLF = (319.854 - 306.446)/(63 - 60) = 4.469.$$

Since $msLF/msE = 4.469/5.10744 = 0.88 < 1$, we cannot reject H_0^R for any reasonable significance level α. Hence, the reduced model provides an adequate fit to the data, making interpretation of the parameters in the reduced model meaningful.

We note that this has been an exercise in model-building, and we can use the model to predict the breakage rates with either flyer over a range of values of twist. However, the model may not fit well for flyer 2 below a twist of 1.69 (see Figure 10.3).

Table 10.17 SAS program output for the reduced model for the cotton-spinning experiment

```
                        The SAS System
                  General Linear Models Procedure

Dependent Variable: Y
                        Sum of        Mean
Source          DF      Squares       Square      F Value    Pr > F
Model           14      394.78103     28.19864      5.55     0.0001
Error           63      319.85436      5.07705
Corrected Total 77      714.63538

Source          DF     Type I SS    Mean Square   F Value    Pr > F
BLOCK           12     177.15538      14.76295      2.91     0.0029
FLYER            1     130.78205     130.78205     25.76     0.0001
TWIST            1      86.84359      86.84359     17.11     0.0001

Source          DF    Type III SS   Mean Square   F Value    Pr > F
BLOCK           12     177.15538      14.76295      2.91     0.0029
FLYER            1     213.24566     213.24566     42.00     0.0001
TWIST            1      86.84359      86.84359     17.11     0.0001

                                  T for H0:     Pr > |T|    Std Error of
Parameter         Estimate       Parameter=0                  Estimate
FLYER 1-2        3.85876832          6.48         0.0001     0.59540773

                                  T for H0:     Pr > |T|    Std Error of
Parameter         Estimate       Parameter=0                  Estimate
FLYER   1        3.85876832 B        6.48         0.0001     0.59540773
        2        0.00000000 B         .             .            .
TWIST          -14.10027473          -4.14        0.0001     3.40929472
```

(Information has been deleted for intercept and blocks)

NOTE: The X'X matrix has been found to be singular and a generalized inverse was used to solve the normal equations. Estimates followed by the letter 'B' are biased, and are not unique estimators of the parameters.

Exercises

1. **Randomization**

   Conduct the randomization for a randomized complete block design with 3 treatments observed once in each of 5 blocks.

2. **Randomization**

   Conduct the randomization for a general complete block design for 3 treatments each observed twice in each of 4 blocks.

Table 10.18 Respiratory exchange ratio data

Subject	Protocol		
	1	2	3
1	0.79	0.80	0.83
2	0.84	0.84	0.81
3	0.84	0.93	0.88
4	0.83	0.85	0.79
5	0.84	0.78	0.88
6	0.83	0.75	0.86
7	0.77	0.76	0.71
8	0.83	0.85	0.78
9	0.81	0.77	0.72

Source: Bullough, R. C. and Melby, C. L. (1993). Copyright © 1993 Karger, Basel. Reprinted with permission.

3. **Cotton-spinning experiment**

   In the cotton-spinning experiment of Section 10.5, page 306, the two observations on treatment 1 (ordinary flier, 1.63 twist) arising from blocks 5 and 10 appear to be outliers.

   (a) Repeat the analysis of Sections 10.5 and 10.9 without these two observations.

   (b) Are the assumptions on the block–treatment model (10.4.1) approximately satisfied?

   (c) Draw conclusions about the fliers and degrees of twist from your analysis. Do any of your conclusions contradict those drawn when the outliers were included?

   (d) Which analysis do you prefer and why?

4. **Respiratory exchange ratio experiment**

   In the resting metabolic rate experiment introduced in Example 10.4.1, the experimenters also measured respiratory exchange ratio, which is another measure of energy expenditure. The data for the second 30 minutes of testing are given in Table 10.18.

   (a) Evaluate the assumptions of the block–treatment model (10.4.1) for these data.

   (b) Construct an analysis of variance table and test for equality of the effects of the protocols on respiratory exchange ratio.

   (c) Evaluate the usefulness of blocking.

   (d) Use the Scheffé method of multiple comparisons to construct simultaneous 99% confidence intervals for all pairwise comparisons of the protocols as well as the inpatient versus outpatient protocols corresponding to the contrast coefficient list $[\,1, -\frac{1}{2}, -\frac{1}{2}\,]$.

5. **Gasoline pilot experiment** (J. Stout and T. Houle, 1995)

   The experimenters were interested in determining whether the use of a higher-octane gasoline would increase the miles per gallon (mpg) achieved by an automobile. Due to the complexity of running such an experiment, they decided to run a pilot experiment

Table 10.19 Data (mpg) for the gasoline pilot experiment (order of observation within block is shown in parentheses)

Car/Driver (Block)	Octane		
	87	89	93
1	33.48 (2)	34.20 (3)	35.30 (1)
2	33.23 (2)	33.79 (3)	36.10 (1)
3	32.95 (3)	31.25 (1)	32.70 (2)

in order to determine the difficulties, decide on a model, and calculate the number of observations that they should take in the main experiment.

(a) Write out a checklist for such an experiment. Be careful to think about all sources of variation and how to control for them. In particular, be concerned about how to measure miles per gallon and whether it is possible to avoid mixing octane levels in the tank of a car.

(b) The experimenters selected a randomized complete block design for the pilot experiment, where each block was defined by a particular car with a particular driver. Three levels of octane (87, 89, 93) were selected from the same brand of gas. The driving was done mostly while commuting to and from work on the highway. The pilot experiment data are shown in Table 10.19. Using the block–treatment model (10.4.1), page 301, for a randomized complete block design, analyze the data.

(c) Investigate the error assumptions on the block–treatment model (10.4.1) for these data. Would you recommend that this model be used for the main experiment?

(d) Assuming that you could run a similar experiment, how many observations would you take on each octane level for each car/driver in order that simultaneous confidence intervals for pairwise differences in the mpg achieved by the different octane levels would be of width at most 2 mpg?

6. **Candle experiment**
An experiment to determine whether different colored candles (red, white, blue, yellow) burn at different speeds was conducted by Hsing-Chuan Tsai, Mei-Chiao Yang, Derek Wheeler, and Tom Schultz in 1989. Each experimenter collected four observations on each color in a random order, and "experimenter" was used as a blocking factor. Thus, the design was a general complete block design with $v = 4$, $k = 16$, $b = 4$, and $s = 4$. The resulting burning times (in seconds) are shown in Table 10.20. A pilot experiment indicated that treatments and blocks do interact. The candles used in the experiment were cake candles made by a single manufacturer.

(a) Explain what block × treatment interaction means in the context of this experiment. Can you think of any causes that might have led to the presence of interaction in the pilot experiment?

(b) Plot the data (Table 10.20) from the main experiment and interpret your plot(s).

Table 10.20 Data for the candle experiment (in seconds)

Block	Red		White		Blue		Yellow	
Tom	989	1032	1044	979	1011	951	974	998
	1077	1019	987	1031	928	1022	1033	1041
Derek	899	912	847	880	899	800	886	859
	911	943	879	830	820	812	901	907
Tsai	898	840	840	952	909	790	950	992
	955	1005	961	915	871	905	920	890
Yang	993	957	987	960	864	925	949	973
	1005	982	920	1001	824	790	978	938

(c) Complete an analysis of variance table for the data using the block–treatment interaction model (10.6.6) for a general complete block design.

(d) Test the null hypotheses of negligible block×treatment interaction and, if appropriate, test the null hypothesis of equality of treatment effects.

(e) Use an appropriate multiple comparisons procedure to evaluate which color of candle is best. Interpret the results.

(f) Discuss whether blocking was important in this experiment.

7. **Saltwater experiment**

An experiment to study the effect of salt on the boiling point of water was conducted by Alan Safer, Brian Jones, Carlos Blanco, and Yu-Hui Tao in 1989. The treatment factor was the amount of salt added to 500 ml of water. There were five levels—0, 8, 16, 24, and 32 grams of salt. The design used was a general complete block design. Three of the experimenters collected $s = 3$ observations each on each level of the treatment factor. "Experimenter" was used as a blocking factor. Thus, $v = 5$, $k = 15$, and $b = 3$. The resulting boiling point temperatures (in degrees C) are given in Table 10.21. The experimenters assumed that treatments and blocks do not interact.

(a) Plot the data and interpret the results.

Table 10.21 Data for the saltwater experiment

Block	Grams of Salt				
	0	8	16	24	32
1	97.9	97.0	97.8	98.6	98.8
	97.3	96.3	97.6	98.0	98.8
	97.1	97.3	97.9	98.4	98.9
2	97.5	97.1	97.5	98.7	98.8
	97.2	97.1	97.2	97.9	98.3
	97.7	97.4	97.6	98.4	98.8
3	97.4	97.8	98.0	98.6	98.9
	97.4	96.9	97.7	98.0	98.8
	97.5	97.6	97.6	98.1	98.9

(b) Complete an analysis of variance table for the data and test for equality of treatment effects.

(c) Evaluate whether blocking was worthwhile and whether the assumption of no treatment–block interaction looks reasonable.

(d) Compute sums of squares for orthogonal trend contrasts, and test the trends for significance, using a simultaneous error rate of 5%. Explain your results in terms of the effect of salt on the boiling point of water.

(e) Calculate the number of observations needed per treatment if a test of equality of treatment effects using $\alpha = 0.05$ is to have power 0.95 in detecting a difference of $1°C$ when $\sigma = 0.5°C$.

8. **Hypothetical chemical experiment**

   An experiment to examine the yield of a certain chemical was conducted in $b = 4$ different laboratories. The treatment factors of interest were

   A: acid strength (80% and 90%, coded 1, 2)

   B: time allowed for reaction (15 min and 30 min, coded 1, 2)

   C: temperature (50°C and 75°C, coded 1, 2)

   The experiment was run as a randomized complete block design with the laboratories as the levels of the blocking factor. The resulting data (in grams) are shown in Table 10.22. The goal of the experiment was to find the treatment combination(s) that give(s) the highest average yield.

   (a) Draw a graph of the data and comment on it.

   (b) State a possible model for this experiment. List any assumptions you have made and discuss the circumstances in which you would expect the assumptions to be valid.

   (c) Check the assumptions on your model.

   (d) Analyze the data and state your conclusions.

9. **Reaction time experiment, continued**

   The reaction time pilot experiment was described in Example 4, page 98, and analyzed in Examples 6.4.3 and 6.4.4, pages 150 and 157. The experiment was run to compare the speed of response of a human subject to audio and visual stimuli. A personal computer was used to present a "stimulus" to a subject and the time that the subject took to press a key in response was monitored. The subject was warned that the stimulus was forthcoming by means of an auditory or a visual cue. The two treatment factors were

Table 10.22 Data for the hypothetical chemical experiment

Lab (Block)	Treatment Combinations							
	111	112	121	122	211	212	221	222
1	7.3	9.5	13.8	15.4	16.0	18.7	11.3	14.5
2	8.8	11.3	15.3	17.7	17.9	20.8	12.0	15.4
3	11.7	14.1	17.2	22.3	22.6	24.8	16.9	18.5
4	6.2	8.3	11.2	15.4	16.8	17.4	8.2	12.5

"Cue Stimulus" at two levels "auditory" and "visual" (Factor A, coded 1, 2), and "Cue Time" at three levels 5, 10, and 15 seconds between cue and stimulus (Factor B, coded 1, 2, 3), giving a total of $v = 6$ treatment combinations (coded 11, 12, 13, 21, 22, 23). The pilot experiment used only one subject, for whom $msE = 0.00029$ seconds^2 based on 12 degrees of freedom. An upper 95% confidence bound for the error variance was calculated in Example 6.4.2, page 148, as $\sigma^2 \leq 0.000664$ seconds^2. To be able to draw conclusions about these six treatment combinations, it is important for the main experiment to use a random sample of subjects from the population. Consider using a randomized complete block design with b subjects representing blocks for the main experiment. Let the block sizes be $k = 6$, so that each treatment combination can be observed once for each subject.

How many subjects are needed if the simultaneous 95% confidence intervals for the pairwise comparisons of the treatment combinations need to be less than 0.1 seconds to be useful (that is, we require $msd < 0.05$ seconds)?

10. **Length perception experiment**

An experiment was run in 1996 by B. Millen, R. Shankar, K. Christoffersen, and P. Nevathia to explore subjects' ability to reproduce accurately a straight line of given length. A 5 cm line (1.9685 inches) was drawn horizontally in a 1-point width on an 11×8.5 in sheet of plain white paper. The sheet was affixed horizontally at eye level to a white projection screen located four feet in front of a table at which the subject was asked to sit. The subject was asked to reproduce the line on a sheet of white paper on which a border had been drawn. Subjects were selected from a population of university students, both male and female, between 20 and 30 years of age. The subjects were all right-handed and had technical backgrounds.

There were six different borders representing the combinations of three shapes—square, circle, equilateral triangle (levels of factor C, coded 1, 2, 3) and two areas—16 in^2 and

Table 10.23 Data for the length perception experiment

Subject	Treatment Combinations (shape, area)					
	11	12	21	22	31	32
1	0.20	−0.25	0.85	−0.50	0.40	0.05
2	1.70	0.30	1.80	0.40	1.40	1.80
3	−0.60	−0.90	−0.90	−0.50	−0.70	−0.50
4	0.60	0.10	0.70	0.20	0.70	0.60
5	0.50	0.40	0.30	0.70	0.50	0.60
6	0.20	−0.60	0.00	−1.40	−0.60	−1.20
7	1.30	−0.10	−0.40	0.50	−0.15	0.30
8	−0.85	−1.30	−0.40	−1.55	−0.85	−1.30
9	0.80	0.05	0.55	1.25	1.30	0.20
10	0.10	−0.10	−0.30	0.95	0.30	−0.95
11	−0.20	−0.40	−0.50	−0.30	−0.40	−0.40
12	0.05	−0.20	0.55	0.60	0.10	0.10
13	0.80	−0.60	0.20	−0.60	−0.60	−0.30
14	−0.25	−0.70	0.00	−0.70	−0.10	−0.95

9 in^2 (levels of factor D, coded 1, 2). The purpose of the experiment was not to see how close to the 5 cm that subjects could draw, but rather to compare the effects of the shape and area of the border on the length of the lines drawn. The subjects were all able to draw reasonably straight lines by hand, and the one of the experimenters measured, to the nearest half millimeter, the distance between the two endpoints of each line. Data from 14 of the subjects are shown as deviations from the target 5 cm in Table 10.23.

(a) Fit a block–treatment model to the data using subjects as blocks and with six treatments representing the shape–area combinations. Check the error assumptions on your model.

(b) Draw at least one graph and examine the data.

(c) Write down contrasts in the six treatment combinations representing the following comparisons:

   (i) differences in the effects of area for each shape separately,

   (ii) average difference in the effects of area,

   (iii) average difference in the effects of shape.

(d) Give a set of 99% simultaneous confidence intervals for the contrasts in (c)(i). State your conclusions.

(e) Under what conditions would the contrasts in (c)(ii) and (c)(iii) be of interest? Do these conditions hold for this experiment?

11. **Load-carrying experiment**

In 1993, an experiment was run by Mark Flannery, Chi-Chang Lee, Eric Nelson, and Pat Sparto to investigate the load-carrying capability of the human arm. Subjects were selected from a population of healthy males. The maximum torque generated at the elbow joint was measured (in newtons) using a dynamometer for each subject in a 5 minute exertion for nine different arm positions (in a random order). The nine arm positions were represented by $v = 9$ treatment combinations in a 3×3 factorial experiment. The first factor was "flex" with levels $0°, 45°, 90°$ of elbow flexion, coded 1, 2, 3. The second factor was "rotation" with levels $0°, 45°, 90°$ of shoulder rotation, coded 1, 2, 3.

Table 10.24 Data and order of collection for the load-carrying experiment

Order	1	2	3	4	5	6	7	8	9
Treat. Comb.	20	10	12	00	11	02	01	21	22
Subj 1	250	230	170	160	240	160	150	200	180
Treat. Comb.	00	11	20	21	12	10	01	22	02
Subj 2	230	260	260	220	250	270	230	190	210
Treat. Comb.	10	02	20	21	00	11	01	22	12
Subj 3	230	180	210	190	150	190	140	160	180
Treat. Comb.	20	00	22	11	02	21	10	01	12
Subj 4	360	200	380	290	240	310	280	350	210

Exercises

The experiment was run as a randomized complete block design with four blocks, each block being defined by a different subject. The subjects were selected from the populations of male students in the 20–30 year range in a statistics class.

(a) Identify contrasts that you think might be of interest in this experiment.

(b) Write down your model and identify your proposed analysis, including your overall significance level and confidence level.

(c) The experimenters decided that they required Scheffé's 95% confidence intervals for normalized contrasts in the main effects of each factor to be no wider than 10 newtons. How many subjects are needed to satisfy this requirement if the error variance is similar to the value $msE = 670$ newtons^2 that was obtained in the pilot experiment?

(d) The data are shown in the order collected in Table 10.24. Plot the data and identify which of your preplanned contrasts ought to be examined. Are there any other contrasts that you would like to examine?

(e) Are the assumptions on block–treatment model (10.4.1) approximately satisfied for this data? Pay particular attention to outliers. If the assumptions are satisfied, analyze the experiment. If they are not satisfied, what information can you gather from the data?

(f) Do your conclusions apply to the whole human population?

12. **Biscuit experiment**

An experiment to study how to make fluffy biscuits was conducted by Nathan Buurma, Kermit Davis, Mark Gross, Mary Krejsa, and Khaled Zitoun in 1994. The two treatment factors of interest were "height of uncooked biscuit" (0.25, 0.50, or 0.75 inches, coded 1, 2, and 3) and "kneading time" (7, 14, or 21 times, coded 1, 2, and 3). The design used was a general complete block design. The $b = 4$ blocks consisted of the four runs of the oven, and the experimental units consisted of $k = 18$ positions on a baking pan. The $v = 9$ treatment combinations were each observed $s = 2$ times per block and $r = bs = 8$ times in total. The resulting observations are "percentage of original height" and are shown in Table 10.25.

Table 10.25 Data for the biscuit experiment (percentage of original height)

Block	\multicolumn{9}{c}{Treatment Combination}								
	11	12	13	21	22	23	31	32	33
1	350.0	375.0	362.5	237.5	237.5	256.3	191.7	216.7	208.3
	300.0	362.5	312.5	231.3	231.3	243.8	200.0	212.5	225.0
2	362.8	350.0	367.5	250.0	262.5	250.0	245.8	212.5	241.7
	412.5	350.0	387.5	268.8	231.3	237.5	225.0	250.0	225.0
3	350.0	387.5	425.0	300.0	275.0	231.3	204.4	187.5	187.5
	337.5	362.5	400.0	262.5	206.3	262.5	204.2	204.2	208.3
4	375.0	362.5	400.0	318.8	250.0	243.8	200.0	216.7	212.5
	350.0	337.5	350.0	256.3	243.8	250.0	150.0	183.3	187.5

(a) State a suitable model for this experiment and check that the assumptions on your model hold for these data.

(b) Use an appropriate multiple comparisons procedure to evaluate which treatment combination is best.

(c) Evaluate whether blocking was worthwhile in this experiment.

13. **Algorithm experiment**

J. Pinheiro, R. Yao, and H. Ying, of the University of Wisconsin, ran an experiment in 1992 to compare the speeds of three computer algorithms for selecting a simple random sample of size m without replacement from a population of size N. If the population members are labeled from 1 to N, the naive algorithm of selecting a random number and checking to see whether or not it has been previously selected is rather slow. Consequently, the experimenters looked at two other possible algorithms.

Algorithm 1 selects sample members in turn, each one being selected from the part of the population that has not yet been included in the sample. This algorithm uses a lot of computer memory. Algorithm 2, which requires less memory, selects from the entire population, discarding repeats, and uses a list of indices to show whether an element has already been selected. These two algorithms (Factor A, levels 1 and 2) were compared for three equally spaced population sizes N (Factor C, levels 2000, 6000, 10000, coded 1, 2, 3), and for three equally spaced sampling fractions m/N (Factor B, levels 0.10, 0.15, 0.20, coded 1, 2, 3). The algorithms were written in Turbo Pascal 5.5. The execution times are variable, since each algorithm uses a random number generator.

The algorithms were each run on a PC according to a randomized complete block design, with blocks representing different days. The execution times (in 1/100 sec) from two of the blocks are shown in Table 10.26. The treatment combinations are listed in the order Algorithm, Sample fraction, Population size.

(a) Make a sketch of the interaction plot for Sampling fraction and Population size separately for each Algorithm. In your opinion, given the interaction effects, are the main effects of Algorithm and Sampling fraction worth investigating? Explain.

(b) Regardless of your answer in (a), test the hypothesis that the linear trend in Sampling fraction (averaged over Algorithm and Population size) is negligible against the alternative hypothesis that it is not negligible. Are you happy with your conclusion? Why or why not?

Table 10.26 Execution times (1/100 sec.) of simple random sample algorithms

Trt. Comb.	111	112	113	121	122	123	131	132	133
day 1	11	22	38	11	27	39	16	28	44
day 2	5	22	33	17	21	44	11	33	44
Trt. Comb.	211	212	213	221	222	223	231	232	233
day 1	17	49	77	22	66	116	33	88	148
day 2	17	44	82	22	72	116	28	88	148

(c) Calculate a confidence interval to compare the execution times of the two algorithms for the largest population size and largest sampling fraction. (Assume that you will want to calculate a large number of confidence intervals for various contrasts in the treatment combinations and that you want the overall level to be 95%.)

(d) If you were to repeat this experiment, would you run it as a block design or a completely randomized design? Justify your answer.

14. **Colorfastness experiment**

An experiment was run in 1990 by D-Y Duan, H. Rhee, and C. Song on the colorfastness of a blue cotton denim fabric. Swatches of the fabric were to be laundered several times, and a determination of fading of the material due to washing was to be made by the three experimenters. The experimenters anticipated that there would be systematic differences in the way they made their color determinations, and consequently, they grouped the fabric swatches into blocks according to which experimenter was to make the determination. Thus the levels of the blocking factor denoted the experimenter.

The levels of the treatment factor were the number of times of laundering, and these were selected to be 1, 2, 3, 4, and 5. Each experimenter was to make color determinations on a total of $k = vs = 60$ swatches of material. The swatches were the experimental units and were assigned at random to the five levels of the treatment factor in such a way that each treatment was assigned $s = 12$ swatches per experimenter. The randomization was done block by block as illustrated in Section 10.3.3 for the randomized complete block design.

The experiment was carried out according to the guidelines of the American Association of Textile Chemists and Colorists Test 61-1980. The measurements that are given in Table 10.27 were made using the Gray Scale for Color Change. This scale is measured using the integers 1–5, where 5 represents no change from the original color. Using their own continuous version of the Gray Scale, each of the $b = 3$ experimenters made color determinations on $s = 12$ swatches of fabric for each of the $v = 5$ treatments (numbers of washes). For each experimenter in turn, the 60 swatches were presented by the other two experimenters in a random order for assessment. The experimenter who was making the color assessment was not told which treatment she was observing—a "blind study." This was done to remove experimenter bias from the measurements.

(a) Plot the treatment averages for each block. Comment on a possible interaction between experimenter and number of washes, and also on any surprising features of the data.

(b) Fit a block–treatment interaction model (10.6.6) to these data for $b = 3$ blocks (experimenter) and $v = 5$ treatments (number of washes). Check the assumptions of normality, equal variance, and independence of the error variables.

(c) Using only the data from experimenters 1 and 2, repeat part (b). Under what circumstances could you justify ignoring the results of experimenter 3?

(d) Investigate the linear and quadratic trends in the effect on color of the number of washes. If necessary, use Satterthwaite's approximation for unequal variances.

Table 10.27 Data for the colorfastness experiment

Block (Experimenter)	Number of Washes	y_{hit} (Measurement on the Gray Scale)
1	1	3.8, 4.0, 4.0, 3.9, 3.8, 3.7, 3.9, 4.0, 4.0, 4.0, 3.9, 4.0
	2	3.0, 3.7, 3.8, 3.0, 3.7, 4.0, 2.9, 3.5, 3.2, 3.5, 4.0, 3.5
	3	3.7, 3.3, 3.5, 3.6, 3.1, 3.0, 3.2, 3.7, 3.8, 3.7, 3.6, 3.6
	4	3.0, 3.6, 3.9, 3.8, 3.8, 3.1, 3.6, 3.4, 4.0, 3.2, 3.0, 3.8
	5	3.6, 3.1, 3.8, 3.4, 3.9, 3.4, 3.5, 4.0, 3.4, 3.9, 3.0, 3.3
2	1	4.5, 3.8, 3.5, 3.5, 3.6, 3.8, 4.6, 3.9, 4.0, 3.9, 3.8, 4.2
	2	3.7, 3.6, 3.8, 3.5, 3.8, 4.0, 3.6, 3.6, 3.4, 3.7, 3.4, 3.3
	3	3.0, 3.7, 2.8, 3.0, 3.6, 3.4, 3.8, 3.6, 3.4, 3.7, 3.9, 3.8
	4	4.2, 3.8, 3.1, 2.8, 3.2, 3.0, 3.7, 3.0, 3.7, 3.5, 3.2, 3.9
	5	3.2, 3.5, 3.1, 3.3, 2.8, 3.5, 3.5, 3.2, 3.6, 3.7, 3.2, 3.2
3	1	4.0, 4.2, 3.8, 3.8, 4.2, 4.2, 3.8, 4.2, 4.2, 3.8, 4.2, 3.9
	2	3.2, 2.8, 2.8, 4.0, 3.0, 3.2, 3.8, 3.5, 4.0, 3.2, 3.5, 3.4
	3	3.8, 4.0, 3.8, 3.4, 4.2, 3.4, 4.0, 3.8, 4.2, 3.9, 3.9, 3.1
	4	4.2, 3.8, 3.5, 3.4, 4.2, 2.9, 3.5, 3.2, 3.5, 4.0, 3.2, 3.9
	5	3.5, 3.8, 2.8, 4.2, 4.0, 3.8, 3.9, 2.9, 3.9, 3.2, 3.5, 3.5

15. **Insole cushion experiment**

The insole experiment was run in the Gait Laboratory at The Ohio State University in 1995 by V. Agresti, S. Decker, T. Karakostas, E. Patterson, and S. Schwartz. The objective of the experiment was to compare the effect on the force with which the foot hits the ground of a regular shoe insole cushion and a heel cushion (factor C, coded 1, 2, respectively) available both as brand name and a store name (factor D, coded 1, 2, respectively).

Only one subject (and one pair of shoes) was used. A pilot experiment indicated that fatigue would not be a factor. The natural walking pace of the subject was measured before the experiment. This same pace was maintained throughout the experiment by use of a metronome.

The experiment was divided into two days (blocks). On one day the dominant leg (kicking leg) was examined, and on the second day the nondominant leg was examined. Each of the $v = 4$ treatment combinations were measured $s = 5$ times per block in a randomized order. For each treatment combination, the subject was instructed to walk naturally along the walkway of the laboratory without looking down. As the foot hit the "force plate," an analog signal was sent to a computer, which then converted the signal to a digital form. The response variable, shown in Table 10.28, is the maximum deceleration of the vertical component of the ground reaction force measured in newtons.

(a) Fit a model that includes a block×treatment interaction. Prepare an analysis of variance table. What can you conclude?

(b) Draw interaction plots for the CD interaction, C×block interaction, D×block interaction, and Treatment Combination×block interaction. Which contrasts would be of interest to examine?

Table 10.28 Data for the insole experiment

		Block I (Right Leg)				
C	D		Response in Newtons (order)			
1	1	899.99 (3)	910.81 (5)	927.79 (10)	888.77 (11)	911.93 (16)
1	2	924.92 (2)	900.10 (6)	923.55 (12)	891.56 (17)	885.73 (20)
2	1	888.09 (4)	954.11 (7)	937.41 (9)	911.85 (14)	908.41 (18)
2	2	884.01 (1)	918.36 (8)	880.23 (13)	891.16 (15)	917.16 (19)
		Block II (Left Leg)				
C	D		Response in Newtons (order)			
1	1	852.94 (22)	866.28 (27)	886.65 (28)	851.14 (33)	869.80 (34)
1	2	882.95 (21)	865.58 (24)	868.15 (25)	893.82 (37)	875.98 (38)
2	1	920.93 (26)	880.26 (31)	897.10 (35)	893.78 (39)	885.80 (40)
2	2	872.50 (23)	892.76 (29)	895.93 (30)	899.44 (32)	912.00 36)

(c) Calculate confidence intervals for any means or contrasts that you identified in part (b). Remember that these were selected after having looked at the data.

(d) Check the assumptions on the model. There are two possible outliers. Reexamine the data without either or both of these observations. Do any of your conclusions change? Which analysis would you report?

16. **Exam paper experiment**

At The Ohio State University, there are many large undergraduate courses with multiple sections. The midterm and final examinations for such courses are frequently given in one large room. Several versions of the exams are made up, and these are printed on different colored paper. In 1986, an experiment was run by Marbu Brown to see whether student average exam scores differed appreciably according to which version and color paper they were assigned.

The experiment was run with the cooperation of a professor in the mathematics department (who, of course, believed that the effects of the different treatment combinations would be similar). The scores for the midterm examination are shown in Table 10.29. This was a multiple-choice exam, and the two versions of exam paper differed in terms of the numbers used in each problem and the order of presentation of the multiple-choice selections. Thus, the two treatment factors were:

A — Color; yellow or green, coded 1, 2.

B — Version; two levels, coded 1, 2.

Since the students were each assigned to one of three teaching assistants for the entire term, the experiment was divided into three blocks, and a random assignment of the four treatment combinations was made separately to the students in each block.

(a) Plot the data for each treatment combination in each block. Can you conclude anything from looking at the data plots?

(b) Fit a block–treatment model without block×treatment interaction. Using a computer package, calculate the analysis of variance table and state your conclusions.

(c) Check the assumptions on your model by plotting the standardized residuals.

Table 10.29 Data for the exam paper experiment

	Block I (Teaching Assistant 1)
A B	Response (order)
1 1	92 84 84 81 72 85 31
1 2	89 79 47 78 87 47 30 88 60 81 50
2 1	86 88 81 68 70 54 77 59 66 47 48 35
2 2	83 62 56 70 85 54 61 56 84
	Block II (Teaching Assistant 2)
A B	Response (order)
1 1	93 93 70 59 85 84 71 50 71 72 60 59 71 74 67
1 2	94 58 30 94 72 66 72 61 38
2 1	96 60 100 94 88 96 94 62 94 57 38
2 2	68 70 70 94 93 59 63 74 64 78 24 53 48 59
	Block III (Teaching Assistant 3)
A B	Response (order)
1 1	60 91 64 89 96 83 60 67 83 87 80
1 2	62 89 85 60 57 90 65 10 87 66
2 1	41 56 62 57 54 74 90 78 88 71 63 86
2 2	48 91 97 60 25 85 84 92 88

(d) If the same teaching assistant had been assigned to all three classes, should the experiment still have been designed as a block design? Discuss.

17. **Exercise experiment**

An experiment was run in 1995 by J. Cashy, D. Cui, T. Papa, M. Wishard, and C. Wong to examine how pulse rate changes due to different exercise intensities and programs on an athletic facility stationary bicycle. The subjects used in the experiment were a selection of graduate students at The Ohio State University. The experiment had two treatment factors. Factor C was the program setting on the bicycle (level 1 = "manual," level 2 = "hill"). Factor D was the intensity setting of each program (level 1 = "setting 3," level 2 = "setting 5"). A total of $n = 36$ subjects were recruited for the experiment. These were divided into blocks according to their normal exercise routine. Thus, $k = 12$ subjects who exercised 0–1 days per week constituted Block 1, those who exercised 2–4 days per week constituted Block 2 and 5–7 days per week constituted Block 3. Each subject was asked not to perform any strenuous exercise right before the experiment. Pulse rate was measured by means of a heart monitor strapped close to the subject's heart. An initial pulse rate reading was taken. Then the subject was asked to pedal at a constant rate of 80 rpm (monitored by the experimenter). Ten seconds after the end of the exercise program, the subject's pulse rate was measured a second time, and the response was the difference in the two readings. The data are shown in Table 10.30 together with the sex and the age of the subject.

(a) Using a block–treatment with no block×treatment interaction, check the assumptions on the error variables.

(b) Plot the standardized residuals against sex and age. Do you think that these variables should have been included in the model?

Table 10.30 Data for the exercise experiment (in heartbeats per minute) listed with order of observation (Ord)

		Block I—Infrequent exercise										
A B	y_{ijkl}	Age	Sex	Ord	y_{ijkl}	Age	Sex	Ord	y_{ijkl}	Age	Sex	Ord
1 1	55	25	0	6	36	34	0	8	41	25	0	10
1 2	74	26	0	17	64	25	0	20	68	25	0	18
2 1	36	26	1	11	26	31	1	14	52	25	0	23
2 2	72	24	1	4	51	24	1	13	44	24	0	16

		Block II—Medium exercise										
A B	y_{ijkl}	Age	Sex	Ord	y_{ijkl}	Age	Sex	Ord	y_{ijkl}	Age	Sex	Ord
1 1	17	26	0	2	51	29	0	21	45	31	0	29
1 2	72	24	1	7	88	26	1	27	53	24	0	30
2 1	45	26	1	1	34	30	1	28	26	24	1	35
2 2	50	23	0	12	47	23	0	15	53	29	1	22

		Block III—Frequent exercise										
A B	y_{ijkl}	Age	Sex	Ord	y_{ijkl}	Age	Sex	Ord	y_{ijkl}	Age	Sex	Ord
1 1	49	25	0	31	35	23	0	32	29	25	0	36
1 2	46	26	0	3	57	29	1	26	59	23	0	33
2 1	10	26	0	5	25	25	1	9	12	27	0	34
2 2	28	24	0	19	25	25	0	24	21	25	0	25

(c) Using whichever model you think is most appropriate, test any hypotheses that are of interest.

(d) What can you conclude from this experiment?

11 Incomplete Block Designs

11.1 Introduction
11.2 Design Issues
11.3 Analysis of General Incomplete Block Designs
11.4 Analysis of Balanced Incomplete Block Designs
11.5 Analysis of Group Divisible Designs
11.6 Analysis of Cyclic Designs
11.7 A Real Experiment—Plasma Experiment
11.8 Sample Sizes
11.9 Factorial Experiments
11.10 Using SAS Software
Exercises

11.1 Introduction

When an experiment involves a blocking factor, but practical considerations prevent the block size from being chosen to be a multiple of the number of treatments, then a complete block design (Chapter 10) cannot be used—an *incomplete block design* needs to be used instead. The most common incomplete block designs have block size smaller than the number of treatments, although larger block sizes can be used.

In this chapter we will discuss three of the more useful and efficient types of incomplete block designs, namely balanced incomplete block designs (Section 11.2.4), group divisible designs (Section 11.2.5), and cyclic designs (Section 11.2.6). We will develop the analysis of incomplete block designs, in general, in Section 11.3, and the particular analysis for balanced incomplete block designs in Section 11.4. Some formulas are given for confidence intervals for treatment contrasts in the group divisible design in Section 11.5, but these designs are more easily analyzed by computer. In Section 11.7, we describe and analyze an experiment that was designed as a cyclic group divisible design. Sample-size calculations are discussed in Section 11.8, and factorial experiments in incomplete block designs are considered in Section 11.9. Analysis by SAS of an incomplete block design is illustrated in Section 11.10.

11.2 Design Issues

11.2.1 Block Sizes

Block sizes are dictated by the availability of groups of similar experimental units. For example, in the Breathalyzer experiment examined in Section 10.3.1, page 297, the block size was chosen to be $k = 5$. This choice was made because the pilot experiment indicated that experimental conditions were fairly stable over a time span of five observations taken close together, and also because five observations could be taken by a single technician in a shift. In other experiments, the block size may be limited by the capacity of the experimental equipment, the availability of similar raw material, the length of time that a subject will agree to remain in the study, the number of observations that can be taken by an experimenter before fatigue becomes a problem, and so on.

Limiting the block size often means that the number of treatments is too large to allow every treatment to be observed in every block, and an incomplete block design must be used. The Breathalyzer experiment, for example, required the comparison of twelve different alcohol concentrations and three air-intake ports (two on one Breathalyzer machine and one on a second machine). Thus, in total there were $v = 36$ treatment combinations of which only five could be observed per block. Skill was then needed in selecting the best design that would still allow all treatment contrasts to be estimable.

11.2.2 Design Plans and Randomization

All the designs that we discuss in this chapter are equireplicate; that is, every treatment (or treatment combination) is observed r times in the experiment. Nonequireplicate designs are occasionally used in practice, and these can be analyzed by computer.

We use the symbol n_{hi} to denote the number of times that treatment i is observed in block h. In general, it is better to observe as many different treatments as possible in a block, since this tends to decrease the average variance of the contrast estimators. Therefore, when the block size is smaller than the number of treatments, each treatment should be observed either once or not at all in a block. Such block designs are called *binary*, and n_{hi} is either 0 or 1. For most purposes, the best binary designs are those in which pairs of treatments occur together in the same number, or nearly the same number, of blocks. These designs give rise to equal (or nearly equal) lengths of confidence intervals for pairwise comparisons of treatment effects.

There are two stages in designing an experiment with incomplete blocks. The first stage is to obtain as even a distribution as possible of treatment labels within the blocks. This results in an *experimental plan*. The plan in Table 11.1, for example, shows a design with $b = 8$

11.2 Design Issues

Table 11.1 An incomplete block design with $b = 8$, $k = 3$, $v = 8$, $r = 3$

Block				Block			
I	1	3	8	V	5	7	4
II	2	4	1	VI	6	8	5
III	3	5	2	VII	7	1	6
IV	4	6	3	VIII	8	2	7

blocks, labeled I, II, ..., VIII, each of size $k = 3$, which can be used for an experiment with $v = 8$ treatments each observed $r = 3$ times. The treatment labels are evenly distributed in the sense that no label appears more than once per block and pairs of labels appear together in a block either once or not at all (which is "as equal as possible").

The experimental plan is often called the "design," even though it is not ready for use until after the second stage, where the random assignments are made. There are two steps to the randomization procedure, as follows.

(i) Randomly assign the block labels in the plan to the levels of the blocking factor.

(ii) Randomly assign the experimental units in a block to those treatment labels allocated to that block.

The randomization procedure is illustrated in the following example.

Example 11.2.1 Metal alloy experiment

Suppose an experiment is to be run to compare $v = 7$ compositions of a metal alloy in terms of tensile strength. Further, suppose that only three observations can be taken per day, and that the experiment must be completed within seven days. It may be thought advisable to divide the experiment into blocks, with each day representing a block, since different technicians may work on the experiment on different days and the laboratory temperature may vary from day to day. Thus, an incomplete block design with $b = 7$ blocks of size $k = 3$ and with $v = 7$ treatment labels is needed. The plan shown in the first column of Table 11.2 is of the correct size. It is binary, with every treatment appearing 0 or 1 times per block and $r = 3$ times in total. Also, all pairs of treatments occur together in a block exactly once, so the treatment labels are evenly distributed over the blocks. Randomization now proceeds in two steps.

Step (i): The block labels need to be randomly assigned to days. Suppose we obtain the following pairs of random digits from a random number generator or from Table A.1 (starting at location 4, 2, 2, 4, 3) and associate them with the blocks as follows:

Random digits: 71 36 65 93 92 02 97
Block labels: I   II III IV V   VI VII

Then, sorting the random numbers into ascending order, the blocks are reordered and assigned to the seven days as follows:

Block labels: VI II III I   V   IV VII
Days: 1   2   3   4   5   6   7

The randomly ordered blocks are shown in the fourth column of Table 11.2.

Step (ii): Now we randomly assign time slots within each day to the treatment labels. Again, using pairs of random digits either from a random number generator or from where we left off in Table A.1, we associate the random digits with the treatment labels as follows:

Day:	Day 1			Day 2			Day 3		
Block:	(Block VI)			(Block II)			(Block III)		
Random digits:	50	29	03	65	34	30	74	56	88
Treatment labels:	6	7	2	2	3	5	3	4	6

Day:	Day 4			Day 5			Day 6		
Block:	(Block I)			(Block V)			(Block IV)		
Random digits:	33	05	75	83	98	13	27	55	67
Treatment labels:	1	2	4	5	6	1	4	5	7

Day:	Day 7		
Block:	(Block VII)		
Random digits:	71	94	88
Treatment labels:	7	1	3

Sorting the random numbers into ascending order for each day separately gives the treatment label order 2, 7, 6 for day 1, and 5, 3, 2 for day 2, and 4, 3, 6 for day 3, and so on. The design after step (ii) is shown in the last column in Table 11.2. □

11.2.3 Estimation of Contrasts (Optional)

The importance of selecting an experimental plan with an even distribution of treatment labels within the blocks is to try to ensure that all treatment contrasts are estimable and that pairwise comparison estimators have similar variances. The design in Table 11.3 does not have an even distribution of treatment labels. Some blocks contain all the even-numbered treatment labels, and the other blocks contain all the odd-numbered labels. The result is that every pairwise comparison between an even-numbered and an odd-numbered treatment is not estimable. The design is said to be *disconnected*.

Disconnectedness can be illustrated through a *connectivity graph* as follows. Draw a point for each treatment and then draw a line between every two treatments that occur together in any block of the design. The connectivity graph for the disconnected design in Table 11.3 is

Table 11.2 Randomization of an incomplete block design

Block Label	Unran- domized design	Day	Design After Step (i)	Design After Step (ii)
I	1 2 4	1	6 7 2	2 7 6
II	2 3 5	2	2 3 5	5 3 2
III	3 4 6	3	3 4 6	4 3 6
IV	4 5 7	4	1 2 4	2 1 4
V	5 6 1	5	5 6 1	1 5 6
VI	6 7 2	6	4 5 7	4 5 7
VII	7 1 3	7	7 1 3	7 3 1

11.2 Design Issues

Table 11.3 A disconnected incomplete block design with $b = 8$, $k = 3$, $v = 8$, $r = 3$

Block				Block			
I	1	3	5	V	5	7	1
II	2	4	6	VI	6	8	2
III	3	5	7	VII	7	1	3
IV	4	6	8	VIII	8	2	4

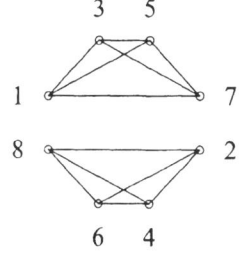 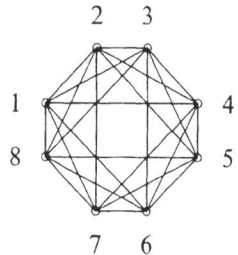

(a) Disconnected design   (b) Connected design

Figure 11.1 Connectivity graphs to check connectedness of designs

shown in Figure 11.1(a). Notice that the graph falls into two pieces. There is no line between any of the odd-labeled treatments and the even-labeled treatments.

A design is *connected* if every treatment can be reached from every other treatment via lines in the connectivity graph. The connectivity graph for the connected design in Table 11.1 is shown in Figure 11.1(b) and it can be verified that there is a path between every pair of treatments. For example, although treatments 1 and 5 never occur together in a block and so are not connected by a line, there is nevertheless a path from 1 to 4 to 5. All contrasts in the treatment effects are estimable in a design if and only if the design is connected. The connectivity graph therefore provides a simple means of checking estimability.

Although disconnected designs will be useful in Chapter 13 for single-replicate ($r = 1$) factorial experiments arranged in blocks, they need never be used for experiments with at least two observations per treatment. The three simplest, and often most efficient, types of design are the balanced incomplete block designs, group divisible designs, and cyclic designs. All balanced incomplete block designs are connected, as are almost all of the other two types of design.

11.2.4 Balanced Incomplete Block Designs

A *balanced incomplete block design* is a design with v treatment labels, each occurring r times, and with bk experimental units grouped into b blocks of size $k < v$ in such a way that the units within a block are alike and units in different blocks are substantially different. The plan of the design satisfies the following conditions:

(i) The design is binary (that is, each treatment label appears either once or not at all in a block).

(ii) Each pair of labels appears together in λ blocks, where λ is a fixed integer.

Block design randomization is carried out as illustrated in Example 11.2.2, and a third step is usually added to the randomization procedure, namely,

> Randomly assign the treatment labels in the plan to the actual levels of the treatment factor.

All balanced incomplete block designs have a number of desirable properties. For example, all treatment contrasts are estimable and all pairwise comparisons of treatment effects are estimated with the same variance so that their confidence intervals are all the same length. Balanced incomplete block designs also tend to give the shortest confidence intervals on the average for any large number of contrasts. For these reasons, the balanced incomplete block design is a popular choice among experimenters. The main drawback is that the design does not exist for many choices of v, k, b, and r.

An example of a balanced incomplete block design for $v = 7$ treatments and $b = 7$ blocks of size $k = 3$ is shown in Table 11.2. It can be seen that conditions (i) and (ii) are satisfied, with every pair of labels appearing together in exactly $\lambda = 1$ block. A second example of a balanced incomplete block design, prior to randomization, is shown in Table 11.4 for $v = 8$ treatments in $b = 14$ blocks of size $k = 4$. Again, conditions (i) and (ii) are satisfied, this time with $\lambda = 3$.

We can verify that the design in Table 11.1 (page 341) with $v = b = 8$, $r = k = 3$ is not a balanced incomplete block design. Label 2, for example, appears in one block with each of labels 1, 3, 4, 5, 7, and 8 but never with label 6. The following simple argument shows that no balanced incomplete block design can possibly exist for this size of experiment. In a balanced incomplete block design with $v = b = 8$, $r = k = 3$, label 2, for example, must appear in $r = 3$ blocks in the design, and in each block there are $k - 1 = 2$ other labels. So label 2 must appear in a block with a total of $r(k - 1) = 6$ other treatment labels. Consequently, if label 2 were to appear λ times with each of the other $v - 1 = 7$ labels, then 7λ would have to be equal to $r(k - 1) = 6$. This would require that $\lambda = 6/7 = r(k - 1)/(v - 1)$. Since λ is not an integer, a balanced incomplete block design cannot exist. For the balanced incomplete block design in Table 11.4, $\lambda = r(k - 1)/(v - 1) = 7(3)/(7) = 3$.

Table 11.4 A balanced incomplete block design with $v = 8$, $r = 7$, $b = 14$, $k = 4$, $\lambda = 3$

Block	Treatments				Block	Treatments			
I	1	2	3	4	VIII	2	3	5	8
II	1	2	5	6	IX	2	3	6	7
III	1	2	7	8	X	2	4	5	7
IV	1	3	5	7	XI	2	4	6	8
V	1	3	6	8	XII	3	4	5	6
VI	1	4	5	8	XIII	3	4	7	8
VII	1	4	6	7	XIV	5	6	7	8

There are three necessary conditions for the existence of a balanced incomplete block design, all of which are easy to check. These are

$$vr = bk,$$
$$r(k-1) = \lambda(v-1),$$
$$b \geq v.$$

The first condition is that satisfied by all block designs with equal replication and equal block sizes. The second condition is obtained by the argument above, and the third condition is called Fisher's inequality (see P. W. M. John, 1980, page 16). The three necessary conditions can be used to verify that a balanced incomplete block design of a given size possibly exists. They do not absolutely guarantee its existence. Lists of balanced incomplete block designs can be found in Cochran and Cox (1957, Chapter 11) and Fisher and Yates (1973).

11.2.5 Group Divisible Designs

A *group divisible design* is a design with $v = gl$ treatment labels (for some integers $g > 1$ and $l > 1$), each occurring r times, and bk experimental units grouped into b blocks of size $k < v$ in such a way that the units within a block are alike and units in different blocks are substantially different. The plan of the design satisfies the following conditions:

(i) The $v = gl$ treatment labels are divided into g groups of l labels—any two labels within a group are called *first associates* and any two labels in different groups are called *second associates*.

(ii) The design is binary (that is, each treatment label appears either once or not at all in a block).

(iii) Each pair of first associates appears together in λ_1 blocks.

(iv) Each pair of second associates appears together in λ_2 blocks.

Block design randomization is carried out as in Section 11.2.2.

Group divisible designs are often classified as one of several types of *partially balanced incomplete block designs with 2 associate classes*. It will be seen in Section 11.5 that the values of λ_1 and λ_2 govern the lengths of confidence intervals for treatment contrasts. Generally, it is preferable to have λ_1 and λ_2 as close as possible, which ensures that the confidence intervals of pairwise comparisons are of similar lengths. Group divisible designs with λ_1 and λ_2 differing by one are usually regarded as the best choice of incomplete block design when no balanced incomplete block design exists.

An example of a group divisible design (prior to randomization) is the experimental plan shown in Table 11.1 (page 341). It has the following $g = 4$ groups of $l = 2$ labels:

(1, 5), (2, 6), (3, 7), (4, 8).

Labels in the same group (first associates) never appear together in a block, so $\lambda_1 = 0$. Labels in different groups (second associates) appear together in one block, so $\lambda_2 = 1$.

Table 11.5 A group divisible design with
$v = 12, r = 3, b = 6, k = 6,$
$\lambda_1 = 3, \lambda_2 = 1$

Block	Treatments					
I	1	2	3	4	5	6
II	1	2	3	7	8	9
III	1	2	3	10	11	12
IV	4	5	6	7	8	9
V	4	5	6	10	11	12
VI	7	8	9	10	11	12

A second example is given in Table 11.5, and it has $g = 4$ groups of $l = 3$ labels:

$$(1, 2, 3), \quad (4, 5, 6), \quad (7, 8, 9), \quad (10, 11, 12),$$

and $\lambda_1 = 3, \lambda_2 = 1$.

There are four necessary conditions for the existence of a group divisible design with chosen values of $v, b, k, r, \lambda_1, \lambda_2$, namely,

$$vr = bk,$$
$$r(k-1) = \lambda_1(l-1) + \lambda_2 l(g-1),$$
$$r \geq \lambda_1,$$
$$rk \geq \lambda_2 v.$$

The first condition is satisfied by all block designs with the same parameters. The second condition can be argued as follows. Label 6 of the design in Table 11.5, for example, appears in $r = 3$ blocks, each time with $k - 1 = 5$ other labels. Also, since $\lambda_1 = 3$ and $\lambda_2 = 1$, label 6 must appear in $\lambda_1 = 3$ blocks with each of its $(l-1) = 2$ first associates and in $\lambda_2 = 1$ block with each of its $(g-1)l = 9$ second associates. So $\lambda_1(l-1) + \lambda_2(g-1)l = 15 = r(k-1)$. The third condition is true for any binary design, since a label must occur in at least as many blocks as it occurs with a first associate. The last condition is obtained from a matrix formulation of the analysis (see P. W. M. John, 1980, page 31).

All group divisible designs with $\lambda_2 = 0$ should be avoided, since not all of the treatment contrasts are estimable. (Optional note: It can be verified that the disconnected design of Table 11.3 is a group divisible design with groups $(1, 3, 5, 7)$ and $(2, 4, 6, 8)$ and with $\lambda_1 = 2$ and $\lambda_2 = 0$.) Lists of group divisible designs are given by Clatworthy (1973) and John and Turner (1977).

11.2.6 Cyclic Designs

A *cyclic design* is a design with v treatment labels, each occurring r times, and with bk experimental units grouped into $b = v$ blocks of size $k < v$ in such a way that the units within a block are alike and units in different blocks are substantially different. The experimental plan, using treatment labels $1, 2, \ldots, v$, can be obtained as follows:

(i) The first block, called the *initial block*, consists of a selection of k distinct treatment labels.

(ii) The second block is obtained from the initial block by *cycling the treatment labels*—that is, by replacing treatment label 1 with 2, 2 with 3, ..., $v-1$ with v, and v with 1. The third block is obtained from the second block by cycling the treatment labels once more, and so on until the vth block is reached.

Block design randomization is carried out as in Section 11.2.2.

The group divisible design in Table 11.1 is also a cyclic design and has initial block (1, 2, 4). The two cyclic designs in Table 11.6 both have block size $k = 4$ with initial block (1, 2, 3, 6), but one has $v = 7$ treatment labels and the other has $v = 6$. The first design is also a balanced incomplete block design with $\lambda = 2$. Although the second design does have pairs of treatments occurring together in either $\lambda_1 = 2$ or $\lambda_2 = 3$ blocks, which results in only two possible lengths of confidence intervals for pairwise comparisons, it is not a group divisible design (since the treatment labels cannot be divided into groups of first associates).

A cyclic design can have as many as $v/2$ different variances for the pairwise comparison estimators, yielding as many as $v/2$ different lengths of confidence intervals for pairwise comparisons of treatment effects. Again the best designs are usually regarded as those whose confidence intervals are close to the same lengths.

Some cyclic designs have duplicate blocks. These designs are useful when fewer than v blocks are required, since duplicate blocks can be ignored. Otherwise, designs with distinct blocks are usually better. Lists of cyclic designs are given by John, Wolock, and David (1972) and John (1987, Chapter 4).

Table 11.6 Cyclic designs with $k = 4$ generated by (1, 2, 3, 6) for $v = 7$ and $v = 6$

Design 1 $v = 7$		Design 2 $v = 6$	
Block	Treatments	Block	Treatments
1	1 2 3 6	1	1 2 3 6
2	2 3 4 7	2	2 3 4 1
3	3 4 5 1	3	3 4 5 2
4	4 5 6 2	4	4 5 6 3
5	5 6 7 3	5	5 6 1 4
6	6 7 1 4	6	6 1 2 5
7	7 1 2 5		

11.3 Analysis of General Incomplete Block Designs

11.3.1 Contrast Estimators and Multiple Comparisons

The standard block–treatment model for the observation on treatment i in block h in a binary incomplete block design is

$$Y_{hi} = \mu + \theta_h + \tau_i + \epsilon_{hi}, \qquad (11.3.1)$$

$$\epsilon_{hi} \sim N(0, \sigma^2),$$

ϵ_{hi}'s are mutually independent,

$h = 1, \ldots, b\,;\ i = 1, \ldots, v\,;\ (h, i)$ in the design.

The model, which assumes no block–treatment interaction, is almost identical to block–treatment model (10.4.1) for the randomized block design. The only difference is the phrase "(h, i) in the design," which means that the model is applicable only to those combinations of block h and treatment i that are actually observed. The phrase serves as a reminder that not all treatments are observed in each block.

For every experiment, the assumptions on the model should be checked. However, when all the treatments fail to appear in every block, it is difficult to check the assumption of no block–treatment interaction by plotting the data block by block, as was recommended for complete block designs in Section 10.7. Thus, it is preferable that a binary incomplete block design be used only when there are good reasons for believing that treatment differences are roughly the same in every block.

The least squares estimators for the treatment parameters in the model for an incomplete block design must include an adjustment for blocks, since some treatments may be observed in "better" blocks than others. This means that the least squares estimator for the pairwise comparison $\tau_p - \tau_i$ is *not* the *unadjusted estimator* $\overline{Y}_{.p} - \overline{Y}_{.i}$ as it would be for a randomized complete block design. For example, if metal alloys 2 and 7 were to be compared via the randomized balanced incomplete block design in Table 11.2, we see that alloy 2 is observed on days 1, 2, and 4, and alloy 7 is observed on days 1, 6, and 7. If we were to use $\overline{Y}_{.2} - \overline{Y}_{.7}$ to estimate $\tau_2 - \tau_7$, it would be biased, since

$$E[\overline{Y}_{.2} - \overline{Y}_{.7}] = E[\tfrac{1}{3}(Y_{12} + Y_{22} + Y_{42}) - \tfrac{1}{3}(Y_{17} + Y_{67} + Y_{77})]$$
$$= \tfrac{1}{3}(3\mu + \theta_1 + \theta_2 + \theta_4 + 3\tau_2) - \tfrac{1}{3}(3\mu + \theta_1 + \theta_6 + \theta_7 + 3\tau_7)$$
$$= (\tau_2 - \tau_7) + \frac{1}{3}(\theta_2 + \theta_4 - \theta_6 - \theta_7)$$
$$\neq (\tau_2 - \tau_7).$$

Now, if the experimental conditions were to change over the course of the experiment in such a way that observations on the first few days tended to be higher than observations on the last few days, then θ_2 and θ_4 would be larger than θ_6 and θ_7. If the two metals do not differ in their tensile strengths, then $\tau_2 = \tau_7$, but the above calculation shows that $\overline{Y}_{.2} - \overline{Y}_{.7}$ would nevertheless be expected to be large. This could cause the experimenter to conclude erroneously that alloy 2 was stronger than alloy 7. Thus, any estimator for $\tau_2 - \tau_7$ must contain an adjustment for the days on which the alloys were observed.

11.3 Analysis of General Incomplete Block Designs

The least squares estimators for the parameters τ_i in the block–treatment model (11.3.1) adjusted for blocks are complicated except in special cases. The general formula will be obtained in the optional Section 11.3.2 below and will be shown to be

$$r(k-1)\hat{\tau}_i - \sum_{p \neq i} \lambda_{pi} \hat{\tau}_p = kQ_i, \quad \text{for } i = 1, \ldots, v, \quad (11.3.2)$$

where λ_{pi} is the number of blocks containing both treatments p and i and Q_i is the ith *adjusted treatment total*; that is,

$$Q_i = T_i - \frac{1}{k}\sum_h n_{hi} B_h, \quad (11.3.3)$$

where T_i is the sum of all observations on treatment i; B_h is the sum of all observations in block h; n_{hi} is 1 if treatment i is observed in block h and zero otherwise. Then $\Sigma_h n_{hi} B_h$ represents the sum of all the observations in all blocks containing treatment p. Individual solutions to this equation are given in Sections 11.4 and 11.5 for balanced incomplete block designs and group divisible designs, respectively. For other incomplete block designs, the least squares estimates should be obtained from a computer package.

We could write the two quantities T_i and B_h as $y_{.i}$ and $y_{h.}$ in the usual way. The reason for changing notation is as a reminder that some of the y_{hi} are not actually observed. So, since n_{hi} is zero if treatment i is not observed in block h, and n_{hi} is one if it is observed, the quantities T_i and B_h are more accurately written as

$$T_i = \sum_h n_{hi} y_{hi} \quad \text{and} \quad B_h = \sum_i n_{hi} y_{hi}.$$

We will also use $G = \Sigma_h \Sigma_i n_{hi} y_{hi}$ to represent the "grand total" of all the observations.

The Bonferroni and Scheffé methods of multiple comparisons can be used for simultaneous confidence intervals of estimable contrasts in all incomplete block designs. The method of Tukey is applicable for balanced incomplete block designs, and it is believed to be conservative (true α level lower than stated) for other incomplete block designs, but this has not yet been proven. The methods of Dunnett and Hsu can be used in balanced incomplete block designs but not in other incomplete block designs without modification to our tables. For each method, the formula for a set of $100(1-\alpha)\%$ simultaneous intervals is

$$\sum c_i \tau_i \in \left(\sum c_i \hat{\tau}_i \pm w \sqrt{\widehat{\text{Var}}\left(\sum c_i \hat{\tau}_i\right)} \right)$$

exactly as in Section 4.4. The correct least squares estimate $\Sigma c_i \hat{\tau}_i$ and estimated variance $\widehat{\text{Var}}(\Sigma c_i \hat{\tau}_i)$ need to be calculated for the design being used.

Analysis of variance for incomplete block designs For an incomplete block design with block–treatment model (11.3.1), the sum of squares for error, ssE, is derived in the optional Section 11.3.2, and is shown to be equal to

$$ssE = \left(\sum_{h=1}^{b} \sum_{i=1}^{v} n_{hi} y_{hi}^2 - \frac{1}{bk} G^2 \right) - \left(\frac{1}{k} \sum_{h=1}^{b} B_h^2 - \frac{1}{bk} G^2 \right) - \sum_{i=1}^{v} Q_i \hat{\tau}_i. \quad (11.3.4)$$

The first term in parentheses in (11.3.4) is like a total sum of squares, so we can call it *sstot* as usual. The second term looks like a sum of squares for blocks, so we can label it $ss\theta$. The third term will turn out to be the quantity needed for testing equality of treatments. Since it is based on the adjusted treatment total, we call it ssT_{adj}. Thus we can write ssE as

$$ssE = sstot - ss\theta - ssT_{\text{adj}}.$$

When the null hypothesis $H_0^T : \{\tau_i \text{ all equal to } \tau\}$ is true, the reduced model obtained from the block–treatment model (11.3.1) looks like the one-way analysis of variance model in the block effects, that is,

$$Y_{hi} = \mu^* + \theta_h + \epsilon_{hi}$$

with $\mu^* = \mu + \tau$ and with k observations on each of the b blocks. Therefore, the sum of squares for error ssE_0 in the reduced model will look similar to the sum of squares for error in a completely randomized design, but with r_i replaced by k. Consequently, adapting (3.5.11), page 45, we have

$$ssE_0 = \sum_{h=1}^{b} \sum_{i=1}^{v} n_{hi} y_{hi}^2 - \sum_{h=1}^{b} B_h^2/k$$

$$= \left(\sum_{h=1}^{b} \sum_{i=1}^{v} n_{hi} y_{hi}^2 - \frac{1}{bk} G^2 \right) - \left(\frac{1}{k} \sum_{h=1}^{b} B_h^2 - \frac{1}{bk} G^2 \right)$$

$$= sstot - ss\theta.$$

So, the *sum of squares for treatments adjusted for blocks* is

$$ssT_{\text{adj}} = ssE_0 - ssE = \sum_{i=1}^{v} Q_i \hat{\tau}_i, \tag{11.3.5}$$

where $\hat{\tau}_i$ is the solution to (11.3.2) and Q_i is defined in (11.3.3). The number of degrees of freedom for treatments is $v - 1$, and the number of error degrees of freedom is

$$df = (bk - 1) - (b - 1) - (v - 1) = bk - b - v + 1, \tag{11.3.6}$$

where $bk = n$ is the total number of observations.

A test of $H_0^T :\{\text{all } \tau_i \text{ are equal}\}$ against $H_A^T :\{\text{at least two of the } \tau_i\text{'s differ}\}$ is given by the decision rule

$$\text{reject } H_0^T \quad \text{if} \quad \frac{msT_{\text{adj}}}{msE} > F_{v-1, bk-b-v+1, \alpha}$$

for some chosen significance level α, where $msT_{\text{adj}} = ssT_{\text{adj}}/(v-1)$, and $msE = ssE/(bk - b - v + 1)$, and where ssE and ssT_{adj} are given by (11.3.4) and (11.3.5). This test is most conveniently set out in an analysis of variance table, as in Table 11.7.

If evaluation of blocking for the purpose of planning future experiments is required, the quantity $ss\theta$ in (11.3.4) *is not the correct value* to use. It has not been adjusted for the fact that every block does not contain an observation on every treatment. In order to evaluate blocks, we need the *adjusted block sum of squares*. Some computer packages will give this value under the heading "adjusted" or "Type III" sum of squares. If the program does not

Table 11.7 Analysis of variance table for a binary incomplete block design with b blocks of size k, and v treatment labels appearing r times

Source of Variation	Degrees of Freedom	Sum of Squares	Mean Square	Ratio
Blocks (adj)	$b-1$	$ss\theta_{adj}$	$ms\theta_{adj}$	—
Blocks (unadj)	$b-1$	$ss\theta$	—	—
Treatments (adj)	$v-1$	ssT_{adj}	msT_{adj}	$\frac{msT_{adj}}{msE}$
Error	$bk-b-v+1$	ssE	msE	
Total	$bk-1$	$sstot$		

Formulae

$ss\theta = \sum_{h=1}^{b} B_h^2/k - G^2/(bk)$ $ssE = sstot - ss\theta - ssT_{adj}$

$ssT_{adj} = \sum_{i=1}^{v} Q_i \hat{\tau}_i$ $sstot = \sum_{h=1}^{b} \sum_{i=1}^{v} n_{hi} y_{hi}^2 - G^2/(bk)$

$Q_i = T_i - \sum_{h=1}^{b} n_{hi} B_h/k$. $ss\theta_{adj} = sstot - ssE - \left(\sum_{i=1}^{v} T_i^2/r - G^2/(bk) \right)$

automatically generate the adjusted value, it can be obtained from a "sequential sum of squares" by entering treatments in the model before blocks. The formula is similar to that for ssT_{adj} switching roles of blocks and treatments:

$$ss\theta_{adj} = sstot - ssE - ssT_{unadj}$$
$$= sstot - ssE - \left(\frac{1}{r} \sum_{i=1}^{v} T_i^2 - \frac{1}{vr} G^2 \right). \quad (11.3.7)$$

11.3.2 Least Squares Estimation (Optional)

A set of least squares estimates of the parameters in the block–treatment model (11.3.1) is obtained by minimizing the sum of squares of the errors (cf. Section 3.4.3). Since n_{hi} is equal to 1 if treatment i is observed in block h, and zero if it is not, then the sum of squares of the errors can be written for a binary design as

$$\sum_{h=1}^{b} \sum_{i=1}^{v} n_{hi} e_{hi}^2 = \sum_{h=1}^{b} \sum_{i=1}^{v} n_{hi} (y_{hi} - \mu - \theta_h - \tau_i)^2 .$$

To obtain a set of least squares estimates, we differentiate this expression with respect to each parameter $(\mu, \theta_1, \theta_2, \ldots, \theta_b, \tau_1, \tau_2, \ldots, \tau_v)$ in turn. If we set the derivatives equal to zero and note that $\sum_h n_{hi} = r$ and $\sum_i n_{hi} = k$, we obtain the following set of $1 + b + v$ normal equations:

$$G - rv\hat{\mu} - k \sum_{h=1}^{b} \hat{\theta}_h - r \sum_{i=1}^{v} \hat{\tau}_i = 0, \quad (11.3.8)$$

$$B_u - k\hat{\mu} - k\hat{\theta}_u - \sum_{i=1}^{v} n_{ui} \hat{\tau}_i = 0, \quad \text{for } u = 1, \ldots, b, \quad (11.3.9)$$

$$T_p - r\hat{\mu} - \sum_{h=1}^{b} n_{hp}\hat{\theta}_h - r\hat{\tau}_p = 0, \quad \text{for } p = 1, \ldots, v, \tag{11.3.10}$$

where

$$T_p = \sum_{h=1}^{b} n_{hp} y_{hp}, \quad B_u = \sum_{i=1}^{v} n_{ui} y_{ui}, \quad \text{and} \quad G = \sum_{h=1}^{b} \sum_{i=1}^{v} n_{hi} y_{hi}.$$

The sum of the b equations (11.3.9) gives equation (11.3.8), as does the sum of the v equations (11.3.10). Therefore, there are at most (and, in fact, exactly) $1 + v + b - 2$ distinct equations in $v + b + 1$ unknowns. We can choose any two extra distinct equations to obtain a set of least squares estimates of the parameters. If the two equations $\sum_i \hat{\tau}_i = 0$ and $\sum_h \hat{\theta}_h = 0$ are chosen, then (11.3.8) and (11.3.9) give the solutions

$$\hat{\mu} = \frac{1}{bk} G \quad \text{and} \quad \hat{\theta}_u = \frac{1}{k} B_u - \frac{1}{bk} G - \frac{1}{k} \sum_{i=1}^{v} n_{ui} \hat{\tau}_i. \tag{11.3.11}$$

Substituting these solutions into (11.3.10) gives

$$T_p - \frac{r}{bk} G - \sum_{h=1}^{b} n_{hp}\left(\frac{1}{k} B_h - \frac{1}{k}\sum_{i=1}^{v} n_{hi}\hat{\tau}_i - \frac{1}{bk} G\right) - r\hat{\tau}_p = 0.$$

Then, if we collect the terms involving the $\hat{\tau}_i$'s onto the left-hand side of the equation and interchange the order of summations, we have

$$\left(r - \frac{1}{k}\sum_{h=1}^{b} n_{hp}^2\right)\hat{\tau}_p - \frac{1}{k}\sum_{i\neq p}\sum_{h=1}^{b} n_{hp} n_{hi}\hat{\tau}_i = Q_p, \tag{11.3.12}$$

where

$$Q_p = T_p - \sum_{h=1}^{b} n_{hp} B_h / k.$$

Now, $\sum_h n_{hp}^2 = \sum_h n_{hp} = r$, and the quantity $\sum_{h=1}^{b} n_{hp} n_{hi}$ counts the number of blocks in which treatments i and p appear together. Writing this count as λ_{pi} and multiplying both sides of equation (11.3.12) by k, we obtain

$$r(k-1)\hat{\tau}_p - \sum_{i\neq p}\lambda_{pi}\hat{\tau}_i = kQ_p, \quad \text{for } p = 1, \ldots, v. \tag{11.3.13}$$

A least squares solution for each $\hat{\tau}_i$ requires a solution to (11.3.13), and this is not easy to obtain without a matrix formulation. However, we can obtain solutions in some special cases. We give an illustration below for the balanced incomplete block design and leave the group divisible design as an exercise (Exercise 13).

Solution for balanced incomplete block design For the balanced incomplete block design, all pairs of treatment labels appear together in λ blocks. Consequently, λ_{pi} in (11.3.13) is equal to a constant λ for all p and i, giving

$$r(k-1)\hat{\tau}_p - \lambda\sum_{i\neq p}\hat{\tau}_i = kQ_p, \quad \text{for } p = 1, \ldots, v.$$

In solving the normal equations, we used the extra equation $\sum_i \hat{\tau}_i = 0$, so we can write $\sum_{i \neq p} \hat{\tau}_i = -\hat{\tau}_p$, and obtain

$$(r(k-1) + \lambda)\hat{\tau}_p = kQ_p, \quad \text{for } p = 1, \ldots, v.$$

One of the necessary conditions given in Section 11.2.4 for the existence of a balanced incomplete block design is that $r(k-1) = \lambda(v-1)$ or, equivalently, $r(k-1) + \lambda = \lambda v$. Thus, a set of least squares estimators for the treatment parameters in a balanced incomplete block design is given by

$$\hat{\tau}_p = \frac{k}{\lambda v} Q_p, \quad \text{for } p = 1, \ldots, v.$$

Formula for ssE for any incomplete block design For any incomplete block design with r observations per treatment and with blocks of size k, the minimum sum of squares for error with the block–treatment model (11.3.1) is

$$ssE = \sum_{h=1}^{b}\sum_{i=1}^{v} n_{hi}\hat{e}_{hi}^2 = \sum_{h=1}^{b}\sum_{i=1}^{v} n_{hi}(y_{hi} - \hat{\mu} - \hat{\theta}_h - \hat{\tau}_i)^2,$$

where $\hat{\mu}, \hat{\theta}_h$, and $\hat{\tau}_i$ are given in (11.3.11) and the solution to (11.3.13). On multiplying out one copy of the squared factor, the previous equation becomes

$$ssE = \sum_h \sum_i n_{hi} y_{hi}(y_{hi} - \hat{\mu} - \hat{\theta}_h - \hat{\tau}_i) - \hat{\mu} \sum_h \sum_i n_{hi}(y_{hi} - \hat{\mu} - \hat{\theta}_h - \hat{\tau}_i)$$
$$- \sum_h \hat{\theta}_h \sum_i n_{hi}(y_{hi} - \hat{\mu} - \hat{\theta}_h - \hat{\tau}_i)$$
$$- \sum_i \hat{\tau}_i \sum_h n_{hi}(y_{hi} - \hat{\mu} - \hat{\theta}_h - \hat{\tau}_i),$$

which is equal to

$$ssE = \sum_h \sum_i n_{hi} y_{hi}(y_{hi} - \hat{\mu} - \hat{\theta}_h - \hat{\tau}_i) - \hat{\mu}(G - bk\hat{\mu} - k\sum_h \hat{\theta}_h - r\sum_i \hat{\tau}_i)$$
$$- \sum_h \hat{\theta}_h(B_h - k\hat{\mu} - k\hat{\theta}_h - \sum_i n_{hi}\hat{\tau}_i)$$
$$- \sum_i \hat{\tau}_i(T_i - r\hat{\mu} - \sum_h n_{hi}\hat{\theta}_h - r\hat{\tau}_i).$$

The last three terms are all zero by virtue of the normal equations, and so ssE becomes

$$ssE = \sum_i \sum_h n_{hi} y_{hi}^2 - G\hat{\mu} - \sum_h B_h \hat{\theta}_h - \sum_i T_i \hat{\tau}_i.$$

If we substitute the formulae for $\hat{\mu}$ and $\hat{\theta}_h$ from equation (11.3.11) into ssE above and collect the terms in $\hat{\tau}_i$ together, we obtain

$$ssE = \sum_h \sum_i n_{hi} y_{hi}^2 - \frac{1}{k}\sum_h B_h^2 - \sum_i Q_i \hat{\tau}_i,$$

with $Q_i = T_i - (1/k)\sum_h n_{hi} B_h$. Finally, if we add and subtract a term $G^2/(bk)$, the sum of squares for error for an incomplete block design can be written as

$$ssE = \left(\sum_{h=1}^{b}\sum_{i=1}^{v} n_{hi} y_{hi}^2 - \frac{1}{bk}G^2\right) - \left(\frac{1}{k}\sum_{h=1}^{b} B_h^2 - \frac{1}{bk}G^2\right) - \sum_{i=1}^{v} Q_i \hat{\tau}_i$$
$$= sstot - ss\theta - ssT_{\text{adj}},$$

where $\hat{\tau}_i$ is a least squares solution for τ_i obtained as a solution to (11.3.13).

11.4 Analysis of Balanced Incomplete Block Designs

11.4.1 Multiple Comparisons and Analysis of Variance

In optional Section 11.3.2, a set of least squares solutions for the treatment parameters in the block–treatment model (11.3.1) for the balanced incomplete block design were derived as

$$\hat{\tau}_i = \frac{k}{\lambda v} Q_i, \quad \text{for } i = 1, \ldots, v, \quad (11.4.14)$$

where Q_i is the adjusted treatment total, calculated as

$$Q_i = T_i - \frac{1}{k}\sum_{h=1}^{b} n_{hi} B_h,$$

and where λ is the number of times that every pair of treatments occurs together in a block, and $n_{hi} = 1$ if treatment i is observed in block h, and $n_{hi} = 0$ otherwise. Thus, the least squares estimator of contrast $\Sigma c_i \tau_i$ is

$$\sum c_i \hat{\tau}_i = \frac{k}{\lambda v} \sum c_i Q_i, \quad (11.4.15)$$

and it can be shown that

$$\text{Var}\left(\sum_i c_i \hat{\tau}_i\right) = \sum_i c_i^2 \left(\frac{k}{\lambda v}\right) \sigma^2. \quad (11.4.16)$$

The Bonferroni, Scheffé, Tukey, Dunnett, and Hsu methods of multiple comparisons can all be used for balanced incomplete block designs with error degrees of freedom $df = bk - b - v + 1$. The general formula for simultaneous $100(1-\alpha)\%$ confidence intervals for a set of contrasts $\Sigma c_i \tau_i$ is

$$\sum_i c_i \tau_i \in \left(\frac{k}{\lambda v} \sum_i c_i Q_i \pm w \sqrt{\Sigma c_i^2 \left(\frac{k}{\lambda v}\right) msE}\right), \quad (11.4.17)$$

where the critical coefficients for the five methods are, respectively,

$$w_B = t_{bk-b-v+1, \alpha/2m} \; ; \; w_S = \sqrt{(v-1)F_{v-1, bk-b-v+1, \alpha}} \; ;$$

$$w_T = q_{v, bk-b-v+1, \alpha}/\sqrt{2} \; ;$$

$$w_H = w_{D1} = t_{v-1, bk-b-v+1, \alpha}^{(0.5)} \; ; \; w_{D2} = |t|_{v-1, bk-b-v+1, \alpha}^{(0.5)}.$$

11.4 Analysis of Balanced Incomplete Block Designs

The analysis of variance table is given in Table 11.7, page 351. For testing m hypotheses of the general form $H_0 : \sum c_i \tau_i = 0$ against the corresponding alternative hypotheses $H_a : \sum c_i \tau_i \neq 0$, at overall significance level α, the decision rule using Bonferroni's method for preplanned contrasts is

$$\text{reject } H_0 \text{ if } \frac{ssc_{\text{adj}}}{msE} > F_{1, bk-b-v+1, \alpha/m}, \tag{11.4.18}$$

and the decision rule using Scheffé's method is

$$\text{reject } H_0 \text{ if } \frac{ssc_{\text{adj}}}{msE} > (v-1) F_{1, bk-b-v+1, \alpha}, \tag{11.4.19}$$

where

$$\frac{ssc_{\text{adj}}}{msE} = \frac{k(\sum c_i Q_i)^2}{\lambda v (\sum c_i^2) msE} = \frac{(\sum c_i \hat{\tau}_i)^2}{\left(\frac{k}{\lambda v}\right) \left(\sum c_i^2\right) msE}. \tag{11.4.20}$$

As for equireplicate completely randomized designs, two contrasts $\sum c_i \tau_i$ and $\sum d_i \tau_i$ are orthogonal in a balanced incomplete block design if $\sum c_i d_i = 0$. The adjusted treatment sum of squares can then be written as a sum of adjusted contrast sums of squares for a complete set of $(v-1)$ orthogonal contrasts. An example is given in the following section.

11.4.2 A Real Experiment—Detergent Experiment

An experiment to compare dishwashing detergent formulations was described by P. W. M. John in the journal *Technometrics* in 1961. The experiment involved three base detergents and an additive. Detergent I was observed with 3, 2, 1, and 0 parts of the additive, giving four treatment combinations, which we will code 1, 2, 3, and 4. Likewise, Detergent II was observed with 3, 2, 1, and 0 parts of the additive, giving four treatment combinations, which

Table 11.8 Design and number of plates washed for the detergent experiment

Block	Design			Plates Washed		
1	3	8	4	13	20	7
2	4	9	2	6	29	17
3	3	6	9	15	23	31
4	9	5	1	31	26	20
5	2	7	6	16	21	23
6	6	5	4	23	26	6
7	9	8	7	28	19	21
8	7	1	4	20	20	7
9	6	8	1	24	19	20
10	5	8	2	26	19	17
11	5	3	7	24	14	21
12	3	2	1	11	17	19

Source: John, P. W. M. (1961). Copyright © 1961 American Statistical Association. Reprinted with permission.

we will code 5, 6, 7, and 8. The standard detergent (Detergent III) with no additive served as a control treatment, which we will code as 9.

The experiment took place in a location where three sinks were available. Three people took part in the experiment and were instructed to wash plates at a common rate. An observation was the number of plates washed in a sink before the detergent foam disappeared. A block consisted of three observations, one per sink. The three people washing dishes rotated amongst the sinks after every five plates in order to reduce the person effect on the observation. The refilling of the three sinks with water and detergent constituted the beginning of a new block. The amount of soil on the plates was held constant. Differences between blocks were due to differences in the common washing rates, in water temperature, in experimenter fatigue, etc.

A design was required with blocks of size $k = 3$ and $v = 9$ treatment labels. A balanced incomplete block design was selected with $b = 12$ blocks giving $r = bk/v = 4$ observations per treatment and every pair of treatment labels occurring in $\lambda = r(k-1)/(v-1) = 1$ block. The design was randomized as in Section 11.2.2, but the randomization was not shown in the original article. We have shown a possible randomization in Table 11.8 together with the corresponding data. The positions within a block show the allocations of the three basins to treatments. The observations are plotted against treatment in Figure 11.2, ignoring the block from which the observation was collected.

Since each pair of treatments occurs together in only one block ($\lambda = 1$), a graphical approach for the evaluation of block–treatment interaction cannot be used. However, it appears from Figure 11.2 that block differences, block–treatment interaction effects, and random error variability must all be rather small compared with the large detergent differences.

Block–treatment model (11.3.1) for an incomplete block design was fitted to the data. The residual plots might lead us to question some of the error assumptions, but there are only 4 observations per treatment and these are all from different blocks, so it is difficult to make a proper assessment. We will proceed with the standard analysis for a balanced incomplete

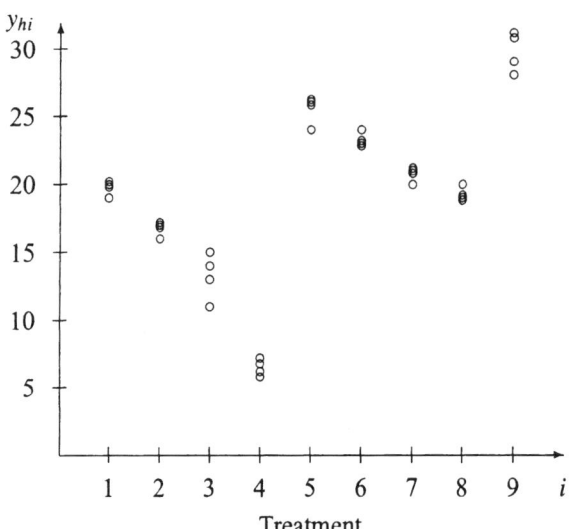

Figure 11.2
Data plot for the detergent experiment, ignoring block levels

11.4 Analysis of Balanced Incomplete Block Designs

block design, recognizing that the stated significance levels and confidence levels are only approximate.

Plotting the data adjusted for block effects In this detergent experiment, the treatment differences are fairly clear from the plot of the raw data in Figure 11.2. However, if the block effects had been substantial, such a plot of the raw data could have painted a muddled picture. In such cases, the picture can be substantially improved by adjusting each observation for the block effects before plotting. The observation y_{hi} is adjusted for the block effects as follows,

$$y_{hi}^* = y_{hi} - (\hat{\theta}_h - \widehat{\overline{\theta}_.}),$$

where $(\hat{\theta}_h - \widehat{\overline{\theta}_.})$ is the least squares estimator of $(\theta_h - \overline{\theta}_.)$. A SAS program that adjusts the observations for block effects and plots the adjusted observations is given in Table 11.20 in Section 11.10. It should be noted that since the variability due to block effects has been extracted, a plot of the adjusted observations will appear to exhibit less variability than really exists.

For this particular data set, the block differences are very small, so a plot of the adjusted data would provide information similar to that provided by the plot of the raw data in Figure 11.2. In Figure 11.3, the observations adjusted for blocks are plotted against "parts of additive" for each base detergent. It appears that the washing power decreases almost linearly as the amount of additive is decreased and also that the original detergent is superior to the two test detergents.

Analysis The analysis of variance table, given in Table 11.9, shows the treatment sum of squares and its decomposition into sums of squares for orthogonal contrasts. The eight orthogonal contrasts are the linear, quadratic, and cubic trends for each of detergents I and II (as the amount of additive increases), together with the "I vs. II" contrast that compares

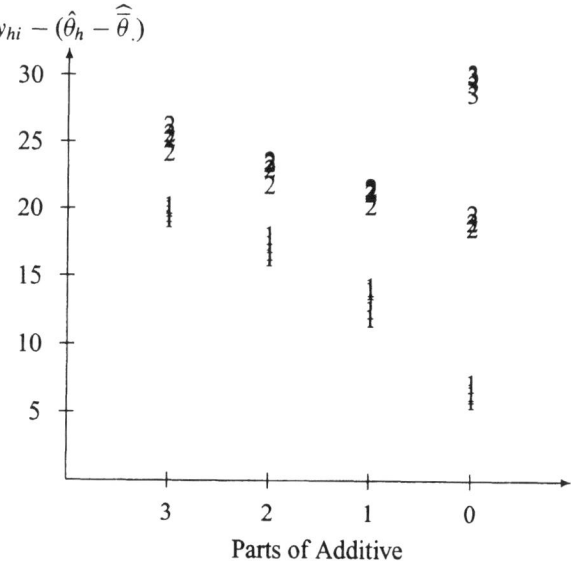

Figure 11.3 Plot of adjusted observations for the detergent experiment. Labels 1, 2, and 3 correspond to base detergents I, II, and III, respectively.

Table 11.9 Analysis of variance table for the detergent experiment

Source of Variation	Degrees of Freedom	Sum of Squares	Mean Square	Ratio	p-value
Blocks (adj)	11	10.06	0.91	–	–
Blocks (unadj)	11	412.75	–	–	–
Treatments(adj)	8	1086.81	135.85	164.85	0.0001
I linear	1	286.02	286.02	347.08	0.0001
I quadratic	1	12.68	12.68	15.38	0.0012
I cubic	1	0.22	0.22	0.27	0.6092
II linear	1	61.34	61.34	74.44	0.0001
II quadratic	1	0.15	0.15	0.18	0.6772
II cubic	1	0.03	0.03	0.04	0.8520
I vs II	1	381.34	381.34	462.75	0.0001
Control vs others	1	345.04	345.04	418.70	0.0001
Error	16	13.19	0.82		
Total	35	1512.75			

the effects of detergents I and II averaged over the levels of the additive, and the "control vs. others" contrast comparing the effect of the control detergent and the average effect of the other eight treatments. For example, the linear trend contrast for detergent I is $-3\tau_1 - \tau_2 + \tau_3 + 3\tau_4$, where the contrast coefficients are obtained from Table A.2. The contrast comparing detergents I and II is the difference of averages contrast

$$\frac{1}{4}(\tau_1 + \tau_2 + \tau_3 + \tau_4) - \frac{1}{4}(\tau_5 + \tau_6 + \tau_7 + \tau_8),$$

and the contrast comparing the control detergent with the others is the difference of averages contrast

$$\tau_9 - \frac{1}{8}(\tau_1 + \tau_2 + \tau_3 + \tau_4 + \tau_5 + \tau_6 + \tau_7 + \tau_8).$$

A set of simultaneous 99% confidence intervals for all treatment contrasts using Scheffé's method of multiple comparisons is given by (11.4.17); that is,

$$\sum_i c_i \tau_i \in \left(\frac{k}{\lambda v} \sum_i c_i Q_i \pm \sqrt{8 F_{8,16,.01}} \sqrt{\sum_i c_i^2 \left(\frac{k}{\lambda v}\right) msE} \right),$$

where $F_{8,16,.01} = 3.89$.

Using the data shown in Table 11.8, we have treatment totals

T_1	T_2	T_3	T_4	T_5	T_6	T_7	T_8	T_9
79	67	53	26	102	93	83	77	119

and block totals

B_1	B_2	B_3	B_4	B_5	B_6	B_7	B_8	B_9	B_{10}	B_{11}	B_{12}
40	52	69	77	60	55	68	47	63	62	59	47

Then, the first adjusted treatment total is

$$Q_1 = T_1 - \frac{1}{k}[B_4 + B_8 + B_9 + B_{12}]$$

11.4 Analysis of Balanced Incomplete Block Designs

The other adjusted treatment totals are calculated similarly, giving

Q_1	Q_2	Q_3	Q_4	Q_5	Q_6	Q_7	Q_8	Q_9
1.00	−6.67	−18.67	−38.67	17.67	10.67	5.00	−0.67	30.33

and since $k/(\lambda v) = 3/9$, the least squares estimate of the contrast $\Sigma c_i \tau_i$ is $\Sigma c_i \hat{\tau}_i = \Sigma c_i Q_i / 3$ ($i = 1, 2, \ldots, 9$). The least squares estimate for the control versus others contrast is then

$$\sum_{i=1}^{8} c_i \hat{\tau}_i = \hat{\tau}_9 - \frac{1}{8}\sum_{i=1}^{8} \hat{\tau}_i = \frac{1}{3}\left(Q_9 - \frac{1}{8}\sum_{i=1}^{8} Q_i\right) = 11.375,$$

with associated estimated variance

$$\widehat{\text{Var}}\left(\sum_{i=1}^{8} c_i \hat{\tau}_i\right) = \sum_{i=1}^{8} c_i^2 \left(\frac{k}{\lambda v}\right) msE = \left(1 + \frac{8}{64}\right)\left(\frac{3}{9}\right)(0.824) = 0.3075,$$

where $msE = 0.824$ is obtained from Table 11.9. Using Scheffé's method of multiple comparisons at overall level 99%, a confidence interval for the control versus others contrast is then

$$11.375 \pm \sqrt{8F_{8,16,0.01}}\sqrt{0.3075} = 11.375 \pm 3.093 = (8.282, 14.468),$$

showing that the control detergent washed between 8.3 and 14.5 more plates than the other detergents on average.

For each pairwise comparison $\tau_i - \tau_p$, we have $\sum c_i^2 = 2$, so

$$\widehat{\text{Var}}\left(\sum c_i \hat{\tau}_i\right) = \sum c_i^2 \left(\frac{k}{\lambda v}\right) msE = 2\left(\frac{3}{9}\right) msE = 0.547.$$

Hence, the minimum significant difference for treatment versus control contrasts, using the Scheffé method with overall significance level $\alpha = 0.01$, is

$$msd = \sqrt{8F_{8,16,.01}}\sqrt{0.547} = \sqrt{(31.12)(0.547)} \approx 4.125.$$

The treatment versus control least squares estimates are

$$\hat{\tau}_9 - \hat{\tau}_i = (Q_9 - Q_i)/3, \quad \text{for } i = 1, \ldots, 8.$$

Using the values of Q_i calculated above, we obtain

$$\hat{\tau}_9 - \hat{\tau}_1 = 9.78; \quad \hat{\tau}_9 - \hat{\tau}_2 = 12.33; \quad \hat{\tau}_9 - \hat{\tau}_3 = 16.33;$$
$$\hat{\tau}_9 - \hat{\tau}_4 = 23.00; \quad \hat{\tau}_9 - \hat{\tau}_5 = 4.22; \quad \hat{\tau}_9 - \hat{\tau}_6 = 6.55;$$
$$\hat{\tau}_9 - \hat{\tau}_7 = 8.44; \quad \hat{\tau}_9 - \hat{\tau}_8 = 10.33.$$

Each contrast estimate exceeds $msd = 4.125$ in magnitude, so we conclude that the control detergent on average washes more dishes than each of the 8 experimental detergents (although the interval that compares detergent 9 with that of detergent 5 only just excludes zero).

The sum of squares for treatments adjusted for blocks is

$$ssT_{\text{adj}} = \sum \hat{\tau}_i Q_i = \frac{k}{\lambda v} \sum Q_i^2 = 1086.81.$$

Since

$$\frac{msT_{\text{adj}}}{msE} = \frac{1086.81/8}{0.824} = 164.85 > F_{8,16,0.01} = 3.89,$$

we reject the hypothesis of no treatment differences.

The eight orthogonal contrasts can be tested simultaneously using the method of Scheffé. For example, the confidence interval for the "control versus others" contrast calculated above as part of a 99% simultaneous set of intervals does not contain zero, so the hypothesis that the control treatment does not differ from the others would be rejected. The overall significance level for all such tests would be $\alpha = 0.01$. Equivalently, the contrasts can be tested by the Scheffé method using the decision rule (11.4.19), page 355; that is,

$$\text{reject } H_0 : \sum c_i \tau_i = 0 \text{ if } \frac{ssc_{\text{adj}}}{msE} > 8F_{8,df,.01} = 31.12,$$

where

$$\frac{ssc_{\text{adj}}}{msE} = \frac{\left(\sum c_i \hat{\tau}_i\right)^2}{\widehat{\text{Var}}\left(\sum c_i \hat{\tau}_i\right)} = \frac{k\left(\sum c_i Q_i\right)^2}{\lambda v \left(\Sigma c_i^2\right) msE}.$$

The ratios ssc_{adj}/msE are provided in Table 11.9. Comparing their values with 31.12, we see that the linear trends are significantly different from zero for each of the base detergents I and II, as are the comparison of detergents I and II on average and the comparison of the control detergent with the average effects of the other 8 treatments. From significance of the linear trend contrasts, coupled with the direction of the trends, one can conclude that detergents I and II get better as the level of additive is increased.

We cannot use the unadjusted block sum of squares to evaluate the usefulness of blocking. We would need to calculate the adjusted block sum of squares as (11.3.7), page 351; that is,

$$ss\theta_{\text{adj}} = sstot - ssE - \left(\frac{1}{r}\sum_{i=1}^{v} T_i^2 - \frac{1}{vr}G^2\right)$$

$$= 1512.75 - 13.19 - \left(\frac{1}{4}(79^2 + 67^2 + \cdots + 119^2) - \frac{1}{36}699^2\right)$$

$$= 1512.75 - 13.19 - 1489.50 = 10.06.$$

The adjusted block mean square is $ms\theta_{\text{adj}} = 10.06/11 = 0.91$, which is not much larger than the error mean square, $msE = 0.82$, so the blocking did not help with increasing the power of the hypothesis tests. Nevertheless, it was natural to design this experiment as a block design, and the creation of blocks was a wise precaution against changing experimental conditions.

11.5 Analysis of Group Divisible Designs

11.5.1 Multiple Comparisons and Analysis of Variance

Group divisible designs were described in Section 11.2.5, page 345, and illustrations were shown in Tables 11.1 and 11.5. The v treatment labels are divided into g groups of l labels each. Treatment labels in the same group are called *first associates*, and those in different groups are called *second associates*. Pairs of treatment labels that are first associates occur together in λ_1 blocks and those that are second associates in λ_2 blocks.

The least squares estimators for the treatment parameters τ_i adjusted for blocks are obtained from the general formula (11.3.2), page 349, with $\lambda_{pi} = \lambda_1$ if treatments p and i are first associates, and with $\lambda_{pi} = \lambda_2$ if they are second associates. The reader is asked in Exercise 13 to show that for a group divisible design, $\hat{\tau}_i$ is equal to

$$\hat{\tau}_i = \left[\frac{k}{(r(k-1) + \lambda_1)v\lambda_2} \right] \times \left[(v\lambda_2 + (\lambda_1 - \lambda_2))Q_i + (\lambda_1 - \lambda_2)\sum_{(1)} Q_p \right],$$

(11.5.21)

where $Q_i = T_i - (1/k)\sum_h n_{hi} B_h$ is the adjusted treatment total as in (11.3.3), page 349, and where $\sum_{(1)} Q_p$ denotes the sum of the Q_p corresponding to the treatment labels that are the first associates of treatment label i.

In general, the variance of the least squares estimator $\Sigma c_i \hat{\tau}_i$ of an estimable contrast $\Sigma c_i \tau_i$ is

$$\text{Var}\left(\sum_{i=1}^{v} c_i \hat{\tau}_i\right) = \sum_{i=1}^{v} c_i^2 \, \text{Var}(\hat{\tau}_i) + 2 \sum_{i=1}^{v-1} \sum_{p=i+1}^{v} c_i c_p \, \text{Cov}(\hat{\tau}_i, \hat{\tau}_p),$$

where

$$\text{Var}(\hat{\tau}_i) = \frac{k[v\lambda_2 + (\lambda_1 - \lambda_2)]}{v\lambda_2[v\lambda_2 + l(\lambda_1 - \lambda_2)]} \sigma^2$$

and

$$\text{Cov}(\hat{\tau}_i, \hat{\tau}_p) = \begin{cases} \dfrac{k(\lambda_1 - \lambda_2)\sigma^2}{v\lambda_2[v\lambda_2 + l(\lambda_1 - \lambda_2)]}, & \text{if } i \text{ and } p \text{ are first associates,} \\ 0, & \text{if } i \text{ and } p \text{ are second associates.} \end{cases}$$

The variance of the least squares estimator of the pairwise comparison $\tau_i - \tau_p$ is then

$$\text{Var}(\hat{\tau}_i - \hat{\tau}_p) = \begin{cases} \dfrac{2k\sigma^2}{[v\lambda_2 + l(\lambda_1 - \lambda_2)]}, & \text{if } i \text{ and } p \text{ are first associates,} \\ \dfrac{2k[v\lambda_2 + (\lambda_1 - \lambda_2)]\sigma^2}{v\lambda_2[v\lambda_2 + l(\lambda_1 - \lambda_2)]}, & \text{if } i \text{ and } p \text{ are second associates.} \end{cases}$$

In order to obtain the variances for the pairwise comparisons as close together as possible, it is clear that λ_1 and λ_2 need to be as close as possible.

The Bonferroni and Scheffé methods of multiple comparisons can be used for group divisible designs. The Tukey method is believed to be conservative (α-level lower than stated). The Dunnett and Hsu methods are not available using our tables, since the critical values can be used only for designs in which $\text{Cov}(\hat{\tau}_i, \hat{\tau}_p)$ are equal for all i and p. The analysis of variance table for the group divisible design is that given in Table 11.7, page 351, with $\hat{\tau}_i$ as in (11.5.21) above. An example of an experiment designed as a group divisible design is discussed in Section 11.7. SAS programs are illustrated in Section 11.10 that can be used to analyze any group divisible design.

11.6 Analysis of Cyclic Designs

Cyclic designs were described in Section 11.2.6, page 346, and illustrated in Tables 11.1 and 11.6. They are incomplete block designs that may or may not possess the properties of balanced incomplete block designs or group divisible designs. When they do possess these properties, they can be analyzed as indicated in Sections 11.4 and 11.5. We recommend analysis by computer, since in general, the least squares estimators $\hat{\tau}_i$ have no simple form. The Bonferroni and Scheffé methods of multiple comparisons can be used for cyclic designs.

In Section 11.7 we reproduce the checklist and analysis of an experiment that was designed as a cyclic group divisible design, and in Section 11.10 we illustrate SAS computer programs that can be used to analyze any cyclic incomplete block design.

11.7 A Real Experiment—Plasma Experiment

The plasma experiment was run by Ernesto Barrios, Jin Feng, and Richard Kibombo in 1992 in the Engineering Research Center at the University of Wisconsin. The following checklist has been extracted verbatim from the experimenters' report, and our comments are in parentheses. The design used was a cyclic group divisible design. Notice that the experimenters moved step (e) of the checklist forward. They had made a list of all potential sources of variation, but they needed a pilot experiment to help determine which sources they could control and which they could not.

CHECKLIST

(a) **Define the objectives of the experiment.**
In physics, plasma is an ionized gas with essentially equal densities of positive and negative charges. It has long been known that plasma can effect desirable changes in the surface properties of materials.
The purpose of this experiment is to study the effects of different plasma treatments of certain plastic pipet tips on the capillary action of the pipets. Capillary action concerns the movement of a liquid up the pipet—a small tube. Before a plasma treatment, the capillarity conduct of the tips is too narrow to permit water to move up. Changes in capillary action effected by plasma treatment can be measured by suspending the tip of a vertical pipet into a bed of water and measuring the height of the column of water in the tube.

(e) **Run a pilot experiment.**
At this stage we decided to make a test run to become familiar with the process of setting up and running the experiment, to determine the appropriate treatment factor levels, and to help identify the major sources of variation that could be controlled, and to identify other variables that might affect the response but which could not be controlled.

(b) **Identify all sources of variation.**
From the test run, we determined that pressure and voltage could not both be effectively controlled. More generally, it would be difficult to vary all of the variables initially listed

11.7 A Real Experiment—Plasma Experiment

(gas flow rate, type of gas, pressure, voltage, presence or absence of a ground shield, and exposure time of the pipet tips to the ionized gas).

Also, the following factors were potential sources of variation.

- Experimenters. Despite the fact that all of the experimenters were to play certain roles during each run of the experiment, it was noted that most of the variation due to the personnel could be attributed to the person who actually connects the pipet tips to the gas tube in the ionization chamber and takes the readings of the final response using vernier calipers.

- Room conditions. It was thought that variations in both room temperature and atmospheric pressure could have an effect on response.

- Water purity. If the water used to measure the capillarity has a substantial amount of impurities, especially mineral salts, then the response may be greatly affected, either because of variability in cohesion and adhesion forces of different types of substances, or because of a reaction between the impurities (salts) and the pipet tips.

- Materials. Variability in the quality of both the pipet tips and the gases used is likely to introduce some variation in the response. Within an enclosed room such as a laboratory, the composition of air may vary significantly over time.

Taking into account the results of the pilot run, and given the fact that we are all amateurs in the field, the following decisions were made.

(i) Treatment factors and their levels.

Scale down the variables of interest to three by keeping both the pressure and voltage constant at 100 mm Torres and 5 volts, respectively, and by keeping the ground shield on. Distilled water will be used to control for impurities in the water. Pipet tips from a single package will be used, so the pipets are more likely to be from the same batch and hence more likely to be homogeneous.

No attempt will be made to control for variation in the composition or purity of the gases used. Thus, the only factors that made up the various treatment combinations were gas flow rate, type of gas, and exposure time.

Set the lower and upper levels of each factor far apart in order to make any (linear) effect more noticeable. Also, we decided to include only 6 of the 8 possible treatment combinations. These were:

	Factors and Levels		
Treatment	Type of Gas	Exposure Time (sec)	Gas Flow Rate (cc/sec)
1	Argon	180	10
2	Air	180	10
3	Argon	180	30
4	Argon	60	30
5	Air	60	30
6	Air	60	10

(ii) Experimental units.

The experimental units are the (combinations of) pipets and time order of observations.

(iii) Blocking factors, noise factors, and covariates.

The two blocking factors are "experimenter"—namely, who connects the pipet and measures the resulting water column height—and "day."

(No covariates or noise factors were included.)

(c) **Specify a rule by which to assign the experimental units to the treatments.**

(There were 3 experimenters collecting observations each day, and a total of only 9 observations could be taken per day within the time available.)

The design will be an incomplete block design with blocks of size three, and three blocks of data will be collected on each of two days. We will use the cyclic design for $v = 6 = b$ and $k = 3 = r$ generated by the treatment labels 1, 4, 5. The labels in the design will be randomly assigned to the six treatment combinations, and the treatments within each block will be randomly ordered.

(The selected cyclic design is shown in Table 11.10 in nonrandomized order so that the cyclic nature of the design can be seen more easily. The design also happens to be a group divisible design (see Exercise 14). The smallest balanced incomplete block design with $v = 6$ and $k = 3$ has $r = 5$ and $b = 10$ and would require more observations.)

(d) **Specify the measurements to be made, the experimental procedure, and the anticipated difficulties.**

The height of the water column will be measured for each pipet. In order to make the measurements as uniform as possible, a device has been constructed, consisting of a rectangular sheet of plexiglass with a small hole in which to place the pipet tip. Placing this pipet holder on a water vessel suspends about 2 mm of the tip of the pipet into the water. After 60 seconds, a mark will be made on the pipet indicating the water level reached. The distance of the mark from the tip of the pipet will be measured using a vernier caliper with tenth of a millimeter precision.

The experimental procedure for each observation is as follows: Place a pipet on the tube through which the plasma will flow, screw a glass tube, turn on the pump and wait 40 seconds, open the Baratron, open the gas, turn a controller to auto, set the flow to a specified level, turn the pressure controller to auto and set the level, set the voltage, time the treatment, turn off flow and shut off the gas, set the pressure to open, wait until the pressure is less than 20, turn off the Baratron, turn off the pump, unscrew the glass tube and pull out a cone, (wearing a glove) take out the pipet, place the pipet in water (using the device for this purpose), and mark the height of the water column, then go on to the next observation.

Table 11.10 Design and data for the plasma experiment

Block	Day	Experimenter	Response (Treatment)		
1	1	Feng	0.459 (4)	0.458 (5)	0.482 (1)
2	1	Barrios	0.465 (5)	0.467 (6)	0.464 (2)
3	1	Kibombo	0.472 (6)	0.495 (1)	0.473 (3)
4	2	Feng	0.325 (1)	0.296 (2)	0.283 (4)
5	2	Barrios	0.390 (2)	0.248 (3)	0.410 (5)
6	2	Kibombo	0.239 (3)	0.350 (4)	0.384 (6)

11.7 A Real Experiment—Plasma Experiment

Anticipated difficulties: Differences in the way people would mark or measure the water column heights would cause variation. Running the experiment as consistently as possible.

(f) **Specify the model.**
(The standard block–treatment model (11.3.1) for an incomplete block design was specified.)

(g) **Outline the analysis.**
An analysis of variance test for equality of the treatment effects will be performed. Then confidence intervals for all pairwise comparisons will be obtained, with a simultaneous 95% confidence level using the method of Scheffé. (Since Tukey's method gives tighter confidence intervals, the calculations below have been modified from the experimenters' article.) Model assumptions will be evaluated.

(h) **Calculate the number of observations to be taken.**
(This was not discussed.)

(i) **Review the above decisions. Revise if necessary.**
(No revisions were made at this stage.)

Results of the experiment During the experiment, an unexpected event occurred. A little tube through which the gas passes was broken, allowing for some leaking of gas. We realized this after our first day's runs and tried to fix this problem the next day, using tape, as a new tube was unavailable. As can be seen from the results, given in Table 11.10, the responses from the last nine runs, corresponding to the second day, were consistently smaller than those from the first nine runs. This underscores the advantage of using time as a blocking factor.

Data Analysis The analysis of variance is given in Table 11.11. It was obtained via a SAS computer program similar to the one in Table 11.16 in Section 11.10. The adjusted block mean square, $ms\theta_{adj} = 0.0161$, is twelve times larger than the error mean square, so blocking was certainly worthwhile.

The ratio $msT_{adj}/msE = 2.99$ does not exceed the critical value $F_{5,7,.05} = 3.97$ for testing equality of the treatment effects at the 5% significance level (equivalently, the p-value is greater than 0.05). Based on this result, examination of any individual treatment contrasts may seem unwarranted, as the results will not be statistically significant.

Table 11.11 Analysis of variance table for the plasma experiment

Source of Variation	Degrees of Freedom	Sum of Squares	Mean Square	Ratio	p-value
Blocks (adj)	5	0.0805	0.0161	–	–
Blocks	5	0.0992	–	–	–
Treatments (adj)	5	0.0196	0.0039	2.99	0.0932
Error	7	0.0092	0.0013		
Total	17	0.1279			

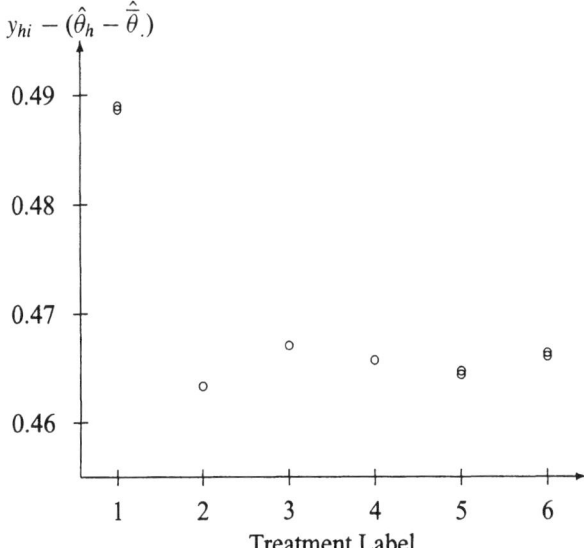

Figure 11.4
Plasma data adjusted for block effects—day one only

However, the broken tube discovered after the first day of runs is an important consideration in this experiment. It is quite possible that the treatments are not the same on day one as on day two. For example, the broken tube may change the gas flow rate or the type of gas to which the pipet is exposed. So, one must ask the question, "Is there anything to be salvaged from this experiment?"

First of all, as has already been noted, the results of the first day were all higher than the results of the second day. Since the objective is to increase capillarity, the results of the first day are thus consistently better. If an explanation can be found for this difference, and if the effect of this difference between days can be repeated, then perhaps quite a lot can be learned "by accident" from this experiment!

If the broken tube has in fact changed the treatments, and if the breakage occurred after the first day's runs, then it might be most useful to analyze the data for one or both days separately. The design for each day is no longer a cyclic design or a group divisible design. Nevertheless, the designs for each day are still connected incomplete block designs and can still be analyzed by computer (see Section 11.10).

Table 11.12 Analysis of variance for the plasma experiment—day one only

Source	Degrees of Freedom	Sum of Squares	Mean Square	Ratio	p-value
Blocks (adj)	2	0.0001213	0.0000607	364.00	—
Block	2	0.0004029	—	—	—
Treatments (adj)	5	0.0007112	0.0001422	853.40	0.026
Error	1	0.0000002	0.0000002		
Total	8	0.0011142			

11.7 A Real Experiment—Plasma Experiment

Table 11.13 Pairwise comparisons for the plasma experiment using the Tukey method and confidence level 95%—day one only

i, p	$\hat{\tau}_i - \hat{\tau}_p$	$\sqrt{\widehat{\text{Var}}(\hat{\tau}_i - \hat{\tau}_p)}$	msd	Significant
1, 2	0.025500	0.000646	0.0184	yes
1, 3	0.021833	0.000553	0.0158	yes
1, 4	0.023167	0.000553	0.0158	yes
1, 5	0.024333	0.000471	0.0135	yes
1, 6	0.022667	0.000471	0.0135	yes
2, 3	−0.003667	0.000746	0.0213	
2, 4	−0.002333	0.000746	0.0213	
2, 5	−0.001167	0.000553	0.0158	
2, 6	−0.002833	0.000553	0.0158	
3, 4	0.001333	0.000746	0.0213	
3, 5	0.002500	0.000646	0.0184	
3, 6	0.000833	0.000553	0.0158	
4, 5	0.001167	0.000553	0.0158	
4, 6	−0.000500	0.000646	0.0184	
5, 6	−0.001667	0.000471	0.0135	

If a test of the null hypothesis H_0^T of equal treatment effects is conducted separately for each day's data, it can be verified that H_0^T would not be rejected at the 5% significance level for the data collected on day two but would be rejected for the data of day one.

The analysis of variance for day one is shown in Table 11.12. The test ratio is 853.40—which is larger than $F_{5,1,.05} = 230$. The mean square for blocks adjusted for treatments is 0.0000607, which is 364 times larger than msE, so blocking was helpful for the observations collected on day one. With only one degree of freedom for error, use of residuals to check model assumptions is of little value. Figure 11.4 shows the day-one observations adjusted for block effects, $y_{hi} - (\hat{\theta}_h - \hat{\bar{\theta}}_.)$ plotted against treatment. It appears that treatment 1 (Argon at 10 cc per second for 180 seconds) is very different from the other treatments. Clearly, the effect of such a break in the tube should be investigated rather carefully via further experimentation.

Table 11.13 contains information for applying Tukey's method of multiple comparisons to the day-one data, using a simultaneous 95% confidence level. The least squares estimates were obtained using a SAS computer program (see Table 11.19 in Section 11.10). The minimum significant difference is smaller than $\hat{\tau}_i - \hat{\tau}_p$ for all pairwise comparisons with $i = 1$, but for none of the others. Consequently, the only confidence intervals that do not contain zero are those involving treatment 1. We conclude that based on the first day's data, treatment 1 is significantly better than each of the other 5 treatments. However, without further experimentation, this difference cannot be explained.

11.8 Sample Sizes

Given the number of treatments v and the block size k, how many blocks b are required to achieve confidence intervals of a specified length or a hypothesis test of specified power? Since for most purposes the balanced incomplete block design is the best incomplete block design when it is available, we start by calculating b and the treatment replication $r = bk/v$ for this design. Then if a balanced incomplete block design cannot be found with b and r close to the calculated values, a group divisible, cyclic, or other incomplete block design can be considered. Since balanced incomplete block designs are the most efficient, other incomplete block designs would generally require b and r to be a little larger.

Example 11.8.1 Sample size to achieve confidence interval length

Suppose Tukey's method for all pairwise comparisons will be used to analyze an experiment with $v = 5$ treatments and block size $k = 3$. It is thought unlikely that msE will be larger than 2.0 units^2. Suppose that the experimenters want the length of simultaneous 95% confidence intervals for pairwise comparisons to be at most 3.0 units (that is, a minimum significant difference of at most 1.5). A balanced incomplete block design will ensure that the interval lengths will all be the same.

Using the facts that $\lambda = r(k-1)/(v-1)$ for a balanced incomplete block design and $b = vr/k$ for a block design, the error degrees of freedom can be written as

$$df = bk - b - v + 1 = vr - vr/k - v + 1 \tag{11.8.22}$$
$$= [vr(k-1) - k(v-1)]/k = [10r - 12]/3.$$

The minimum significant difference for a confidence interval for any pairwise treatment comparison, using Tukey's method with an overall 95% confidence level, is given by (11.4.17), page 354; that is,

$$msd = (q_{v,df,.05}/\sqrt{2})\sqrt{2\left[\frac{k(v-1)}{rv(k-1)}\right]msE} = q_{5,df,.05}\sqrt{\frac{12}{5r}},$$

where $df = [10r/3] - 4$. For the minimum significant difference to be at most 1.5 units, it is necessary that

$$r \geq 1.0667 q_{5,df,.05}^2.$$

Trial and error shows that around 17–18 observations per treatment would be needed to satisfy the inequality; that is, 85–90 observations in total, which would require 28–30 blocks of size 3. A balanced incomplete block design exists with $v = 5, k = 3, b = 10, r = 6$ (all possible combinations of five treatments taken three at a time as blocks). Repeating this entire design three times would give a balanced incomplete block design with $r = 18$, which will give confidence intervals of length about

$$2q_{5,56,.05}\sqrt{12/(5 \times 18)} \approx 2.92 < 3.0. \qquad \square$$

Example 11.8.2 Sample size to achieve specified power

Suppose a test of the null hypothesis $H_0 : \{\tau_i \text{ all equal}\}$ is required to detect a difference in the treatment effects of $\Delta = 1$ unit with probability 0.95, using significance level $\alpha = 0.05$ for a balanced incomplete block design with $v = 5, k = 3$

The least squares estimator $\hat{\tau}_i - \hat{\tau}_p$ of a pairwise comparison contrast $\tau_i - \tau_p$ for a balanced incomplete block design has variance given by (11.4.16), page 354, with $\Sigma c_i^2 = 2$; that is,

$$\text{Var}(\Sigma c_i \hat{\tau}_i) = 2 \frac{k}{\lambda v} \sigma^2 = 2 \left[\frac{k(v-1)}{rv(k-1)} \right] \sigma^2 . \tag{11.8.23}$$

The number r of observations needed per treatment is calculated via a formula similar to (6.6.49), page 168, with $a = v$ and with $2\sigma^2/b$ replaced by the variance (11.8.23); that is,

$$r = \frac{2v\sigma^2\phi^2}{\Delta^2} \left[\frac{k(v-1)}{v(k-1)} \right] .$$

Suppose that σ^2 is believed to be at most 1.0 unit^2; then $r = 12\phi^2$. The power tables in Appendix A.7 can be used to find ϕ^2. The numerator degrees of freedom are $\nu_1 = v - 1$ and the denominator degrees of freedom ν_2 are the error degrees of freedom (11.8.22). So for our example, $\nu_1 = 4$ and $\nu_2 = (10r - 12)/3$. Trial and error shows that about $r = 48$ observations per treatment are needed to satisfy the equality. A balanced incomplete block design exists with $v = 5, k = 3, b = 10, r = 6$ (all possible selections of three treatments taken as blocks). Repeating the entire design eight times would give a balanced incomplete block design with $r = 48$, as required. □

11.9 Factorial Experiments

11.9.1 Factorial Structure

Any of the incomplete block designs that we have discussed can be used for a factorial experiment by taking the treatment labels to represent treatment combinations. The incomplete block designs that are the most suitable for factorial experiments allow the adjusted treatment sum of squares ssT_{adj} to be written as a sum of the adjusted sums of squares for main effects and interactions. Thus, for an experiment with two factors A and B, for example, we would like to have

$$ssT_{\text{adj}} = ssA_{\text{adj}} + ssB_{\text{adj}} + ssAB_{\text{adj}} .$$

Such block designs are said to have *factorial structure*.

One benefit of this property is that the computations for, and interpretation of, the analysis of variance are simplified. A design with factorial structure requires only that main-effect and interaction contrast estimates be adjusted for block effects. In designs without factorial structure, the contrast estimates have to be adjusted not only for blocks but also for contrasts in all the other main effects and interactions. Although a computer program, such as the SAS system, can handle this adjustment, uncorrelated estimates are much easier to interpret and are, therefore, preferred.

Table 11.14 Design and data for the step experiment

Block	\multicolumn{6}{c}{Treatment Combination}					
	11	12	13	21	22	23
1		75	87	84	93	99
2	93	84	96	90	108	
3	99	93	96		123	129
4	99	108	99	99		120
5	99		111	90	129	141
6	129	135		120	147	153

All balanced incomplete block designs have factorial structure, and the features are illustrated in the following example.

Example 11.9.1 Step experiment

An experiment was run by S. Guerlain, B. Busam, D. Huland, P. Taige, and M. Pavol in 1993 to investigate the effects on heart rate due to the use of a step machine. The experimenters were interested in checking the theoretical model that says that heart rate should be a function of body mass, step height, and step frequency. The experiment involved the two treatment factors "step height" (factor C) and "step frequency" (factor D). Levels of "step height" were 5.75 and 11.5 inches, coded 1 and 2. "Step frequency" had three equally spaced levels, 14, 21, and 28 steps per minute, coded 1, 2, 3. The response variable was pulse rate in beats per minute.

The experiment used $b = 6$ subjects as blocks, and each subject was measured under $k = 5$ of the $v = 6$ combinations of step height and step frequency. The design was a balanced incomplete block design with blocks corresponding to different combinations of subject, run timer, and pulse measurer. All pairs of treatment combinations appeared together in $\lambda = 4$ blocks. The data are shown in Table 11.14.

Writing the treatment combinations as two-digit codes, the block–treatment model (11.3.1), page 348, becomes

$$Y_{hij} = \mu + \theta_h + \tau_{ij} + \epsilon_{hij},$$

and the least squares estimates of the treatment parameters adjusted for subject are given by (11.4.14), page 354, with two-digit codes; that is,

$$\hat{\tau}_{ij} = \frac{k}{\lambda v} Q_{ij} = \frac{k}{\lambda v}\left[T_{ij} - \frac{1}{k}\sum_{h=1}^{b} n_{hij} B_h\right],$$

where T_{ij} is the total of the $r = 5$ observations on step height i, step frequency j, B_h is the total of the $k = 5$ observations on the hth subject; and n_{hij} is 1 if treatment combination ij is observed for subject h and is zero otherwise. We obtain

$\hat{\tau}_{11}$	$\hat{\tau}_{12}$	$\hat{\tau}_{13}$	$\hat{\tau}_{21}$	$\hat{\tau}_{22}$	$\hat{\tau}_{23}$
−8.125	−7.625	−4.125	−11.375	12.375	18.875

Using the p-values in Table 11.15, the experimenters rejected the hypothesis of negligible interaction. A plot of the data (not shown) suggests that heart rate increases linearly as the step frequency is increased, but that the linear trend is not the same for the two step heights. The experimenters wanted to examine the average behavior of the two factors, so despite this interaction, they decided to examine the main effects. In Exercise 10, the reader is asked to examine the linear trends at each step height separately.

The adjusted sums of squares for the main effects of C (step height) and D (step frequency) and for their interaction are shown in Table 11.15. For a balanced incomplete block design, these sums of squares can be obtained by hand by using the values of $\hat{\tau}_{ij}$ in place of $y_{ij.}$, and $k/(\lambda v)$ in place of r in the formulae (6.4.21)–(6.4.26), page 154, or, equivalently, in Rule 4, page 202, which leads to

$$ssC_{adj} = \frac{\lambda v}{k}\left[\frac{1}{d}\sum_{i=1}^{v}\hat{\tau}_{i.}^2 - \frac{1}{cd}\hat{\tau}_{..}^2\right] = \left(\frac{\lambda v}{k}\right)\frac{1}{d}\sum_{i=1}^{v}\hat{\tau}_{i.}^2$$

$$= \left(\frac{24}{5}\right)\frac{1}{3}(-19.875^2 + 19.875^2) = 1264.05.$$

For simplicity of notation, we now drop the subscript "adj." However, all estimates and sums of squares are adjusted for block effects. The experimenters were interested in examining the linear and quadratic trend contrasts for step frequency, that is,

$$D_L = -\overline{\tau}_{.1} + \overline{\tau}_{.3} = -\frac{1}{2}(\tau_{11} + \tau_{21}) + \frac{1}{2}(\tau_{13} + \tau_{23}),$$

$$D_Q = -\overline{\tau}_{.1} + 2\overline{\tau}_{.2} - \overline{\tau}_{.3} = -\frac{1}{2}(\tau_{11} + \tau_{21}) + \frac{2}{2}(\tau_{12} + \tau_{22}) - \frac{1}{2}(\tau_{13} + \tau_{23}).$$

The least squares estimate for a contrast $\sum c_{ij}\tau_{ij}$ and the associated variance are given by (11.4.15) and (11.4.16), page 354; that is,

$$\sum\sum c_{ij}\hat{\tau}_{ij} = \frac{k}{\lambda v}\sum\sum c_{ij}Q_{ij}$$

and

$$\text{Var}\left(\sum\sum c_{ij}\hat{\tau}_{ij}\right) = \sum\sum c_{ij}^2\left(\frac{k}{\lambda v}\right)\sigma^2.$$

Table 11.15 Analysis of variance for the step experiment

Source of Variation	Degrees of Freedom	Sum of Squares	Mean Square	Ratio	p-value
Subject (Block) (adj)	5	6685.05	1337.01	–	–
Subject (Block) (unadj)	5	7400.40	–	–	–
Height (C) (adj)	1	1264.05	1264.05	28.63	0.0001
Frequency (D) (adj)	2	1488.90	744.45	16.86	0.0001
Ht×Freq (CD) (adj)	2	990.90	495.45	11.22	0.0006
Error	19	838.95	44.16		
Total	29	11983.20			

Using these formulae, we find that the least squares estimates of the linear and quadratic trend contrasts for step frequency (adjusted for subjects) are

$$\hat{D}_L = 17.125 \quad \text{and} \quad \hat{D}_Q = -7.125.$$

The linear trend is positive, suggesting that the average pulse rate increases as the step frequency increases, and the quadratic trend is negative, suggesting that the increase in pulse rate is greater from 14 to 21 steps per minute than it is from 21 to 28 steps per minute. The null hypotheses $H_0^L : \{D_L = 0\}$ and $H_0^Q : \{D_Q = 0\}$ should be tested to check whether the perceived trends are significantly different from zero. The variances of the contrast estimators are

$$\text{Var}(\hat{D}_L) = \sum\sum c_{ij}^2 \left(\frac{k}{\lambda v}\right)\sigma^2 = \left(\frac{4}{4}\right)\left(\frac{5}{24}\right)\sigma^2 = 0.2083\sigma^2,$$

$$\text{Var}(\hat{D}_Q) = \left(\frac{12}{4}\right)\left(\frac{5}{24}\right)\sigma^2 = 0.625\sigma^2.$$

The contrast sum of squares for testing the null hypothesis $H_0^L : \{D_L = 0\}$ is obtained from (11.4.20), page 355, as

$$ss(D_L) = \frac{(\hat{D}_L)^2}{\sum\sum c_{ij}^2\left(\frac{k}{\lambda v}\right)} = \frac{17.125^2}{0.2083} = 1407.675,$$

and the contrast sum of squares for testing $H_0^Q : \{D_Q = 0\}$ is

$$ss(D_Q) = \frac{(\hat{D}_Q)^2}{\sum\sum c_{ij}^2\left(\frac{k}{\lambda v}\right)} = \frac{(-7.125)^2}{0.625} = 81.225.$$

The linear and quadratic contrasts are orthogonal in a balanced incomplete block design even after adjusting for blocks, and we can now verify that indeed, $ssD = ss(D_L) + ss(D_Q)$.

To test the null hypotheses H_0^L and H_0^Q against their respective alternative hypotheses that the null hypothesis is false, we compare each of $ss(D_L)/msE = 31.88$ and $ss(D_Q)/msE = 1.84$ with $2F_{2,19,.01} = 7.04$ for Scheffé's method and an overall level of $\alpha = 0.01$. We conclude that the quadratic trend is negligible, but there is a nonnegligible linear trend in the heart rate as the stepping frequency increases (averaged over step height). □

11.10 Using SAS Software

11.10.1 Analysis of Variance and Estimation of Contrasts

In this section, sample programs are given to illustrate the analysis of incomplete block designs using the SAS software. The programs shown are for the detergent experiment of Section 11.4.2 and the plasma experiment of Section 11.7, but similar programs can be used to analyze the data collected in any incomplete block design.

Table 11.16 contains the first sample program. The data are entered into a data set called ONE, using the variables BLOCK, TRTMT, and Y for the block, treatment, and response value, respectively. PROC PLOT is used to plot the observations against treatments, analogous to

11.10 Using SAS Software

Table 11.16 SAS program for analysis of a balanced incomplete block design—detergent experiment

```
DATA ONE;
  INPUT BLOCK TRTMT Y;
  LINES;
    1  3 13
    1  8 20
    :  :  :
   12  1 19
;
PROC PLOT;
  PLOT Y*TRTMT=BLOCK / VPOS=19 HPOS=50;
;
PROC GLM;
  CLASS BLOCK TRTMT;
  MODEL Y = BLOCK TRTMT;
  OUTPUT OUT=RESIDS PREDICTED=PREDY RESIDUALS=Z;
;
* contrast sums of squares for 8 orthogonal contrasts;
  CONTRAST 'I linear'         TRTMT -3 -1  1  3  0  0  0  0  0;
  CONTRAST 'I QUADRATIC'      TRTMT  1 -1 -1  1  0  0  0  0  0;
  CONTRAST 'I cubic'          TRTMT -1  3 -3  1  0  0  0  0  0;
  CONTRAST 'II linear'        TRTMT  0  0  0  0 -3 -1  1  3  0;
  CONTRAST 'II quadratic'     TRTMT  0  0  0  0  1 -1 -1  1  0;
  CONTRAST 'II cubic'         TRTMT  0  0  0  0 -1  3 -3  1  0;
  CONTRAST 'I vs II'          TRTMT  1  1  1  1 -1 -1 -1 -1  0;
  CONTRAST 'others vs control' TRTMT 1  1  1  1  1  1  1  1 -8;
;
* estimation of treatment versus control contrasts via LSMEANS;
  LSMEANS TRTMT / PDIFF=CONTROL('9') CL ADJUST=DUNNETT;
;
* estimation of treatment versus control contrasts via ESTIMATE;
  ESTIMATE 'Det 9-1' TRTMT -1  0  0  0  0  0  0  0  1;
  ESTIMATE 'Det 9-2' TRTMT  0 -1  0  0  0  0  0  0  1;
```

Figure 11.2, page 356, but using block labels as the plotting legend (the plot is not shown here). PROC GLM is used to fit the block–treatment model (11.3.1), generate the analysis of variance table, and save the predicted values and residuals in the output data set RESIDS. Residuals can be standardized and plotted as in Chapter 6.

The output from PROC GLM is reproduced in Table 11.17. Since BLOCK has been entered before TRTMT in the model statement, the sum of squares for treatments adjusted for blocks is listed under Type I (or sequential) sums of squares as well as under the Type III sums of squares. The adjusted block sum of squares is listed under the Type III sums of squares. In order to use the sequential or Type I sums of squares, one would need to rerun the program with TRTMT entered before BLOCK in the model statement. In Table 11.16, the sums of squares corresponding to 8 orthogonal treatment contrasts are requested via the

Table 11.17 Partial output from PROC GLM for analysis of an incomplete block design—detergent experiment

```
                      The SAS System
                General Linear Models Procedure
Dependent Variable: Y
                          Sum of         Mean
Source            DF     Squares       Square    F Value    Pr > F
Model             19    1499.5648      78.9245     95.77    0.0001
Error             16      13.1852       0.8241
Corrected Total   35    1512.7500

Source            DF    Type I SS   Mean Square   F Value    Pr > F
BLOCK             11     412.7500       37.5227     45.53    0.0001
TRTMT              8    1086.8148      135.8519    164.85    0.0001

Source            DF   Type III SS  Mean Square   F Value    Pr > F
BLOCK             11      10.0648       0.9150      1.11    0.4127
TRTMT              8    1086.8148     135.8519    164.85    0.0001

Contrast          DF   Contrast SS  Mean Square   F Value    Pr > F
I linear           1     286.01667    286.01667    347.08    0.0001
I quadratic        1      12.67593     12.67593     15.38    0.0012
I cubic            1       0.22407      0.22407      0.27    0.6092
II linear          1      61.34074     61.34074     74.44    0.0001
II quadratic       1       0.14815      0.14815      0.18    0.6772
II cubic           1       0.02963      0.02963      0.04    0.8520
I vs II            1     381.33796    381.33796    462.75    0.0001
others vs control  1     345.04167    345.04167    418.70    0.0001
```

CONTRAST statements, and it can be verified from Table 11.17 that the contrast sums of squares add to the treatment sum of squares.

Simultaneous confidence intervals for pairwise comparisons can be obtained via the ESTIMATE statements or via LSMEANS with options as discussed in Section 6.8, page 175. One such set of options is

LSMEANS TRTMT / PDIFF=CONTROL('9') CL ADJUST=DUNNETT;

The PDIFF=CONTROL('9') option specifies that level 9 is the control, as was the case in the detergent experiment. If the designation "('9')" had been omitted, then the lowest level would have been taken to be the control treatment by default. Partial output is shown in

The first section of the table shows the usual output from the ESTIMATE statement. The middle section provides least squares estimates of the parameters $\hat{\tau}_i + \hat{\mu} + \overline{\beta}_.$, which for the balanced incomplete block design are $(k/(\lambda v))Q_i + G/(bk)$. Results are also given in the middle section for simultaneous tests for whether or not each of the treatment-versus-control comparisons $\tau_i - \tau_9$ is zero, or equivalently whether $\tau_i = \tau_9$, using Dunnett's method. The

Table 11.18 Partial output from ESTIMATE and LSMEANS for an incomplete block design—detergent experiment with detergent 9 as the control treatment

```
                        The SAS System
                 General Linear Models Procedure

                              T for H0:    Pr > |T|   Std Error of
Parameter     Estimate     Parameter=0                Estimate
Det 9-1       9.7777778       13.19        0.0001     0.74120356
Det 9-2      12.3333333       16.64        0.0001     0.74120356

                      Least Squares Means
           Adjustment for multiple comparisons: Dunnett-Hsu

           TRTMT            Y         Pr > |T| H0:
                          LSMEAN      LSMEAN=CONTROL
             1          19.7500000       0.0001
             2          17.1944444       0.0001
             3          13.1944444       0.0001
             4           6.5277778       0.0001
             5          25.3055556       0.0002
             6          22.9722222       0.0001
             7          21.0833333       0.0001
             8          19.1944444       0.0001
             9          29.5277778

           Adjustment for multiple comparisons: Dunnett-Hsu
              Least Squares Means for effect TRTMT
           95% Confidence Limits for LSMEAN(i)-LSMEAN(j)

                     Simultaneous                    Simultaneous
                        Lower       Difference          Upper
                      Confidence     Between         Confidence
            i     j     Limit         Means             Limit
            1     9   -11.981915    -9.777778        -7.573641
            2     9   -14.537470   -12.333333       -10.129196
            3     9   -18.537470   -16.333333       -14.129196
            4     9   -25.204137   -23.000000       -20.795863
            5     9    -6.426359    -4.222222        -2.018085
            6     9    -8.759692    -6.555556        -4.351419
            7     9   -10.648581    -8.444444        -6.240308
            8     9   -12.537470   -10.333333        -8.129196
```

third section of the output gives simultaneous 95% confidence intervals for the treatment-versus-control comparisons using Dunnett's method. A word of warning is in order here. If the treatments had been labeled anything other than 1, 2, ..., 9, at this point SAS would have relabeled them lexicographically. For example, if the control treatment had been labeled as

Table 11.19 Partial output from PROC GLM and LSMEANS in a SAS program for analyzing an incomplete block design—plasma experiment

```
                        The SAS System
                  General Linear Models Procedure
Dependent Variable: Y
                             Sum of          Mean
Source            DF        Squares        Square    F Value   Pr > F
Model              7      0.00111406    0.00015915     954.90   0.0249
Error              1      0.00000017    0.00000017
Corrected Total    8      0.00111422

Source            DF       Type I SS   Mean Square   F Value   Pr > F
BLOCK              2      0.00040289    0.00020144   1208.67   0.0203
TRTMT              5      0.00071117    0.00014223    853.40   0.0260

Source            DF     Type III SS   Mean Square   F Value   Pr > F
BLOCK              2      0.00012133    0.00006067    364.00   0.0370
TRTMT              5      0.00071117    0.00014223    853.40   0.0260

                      Least Squares Means
         Adjustment for multiple comparisons: Tukey-Kramer
              Least Squares Means for effect TRTMT
            95% Confidence Limits for LSMEAN(i)-LSMEAN(j)

                  Simultaneous                  Simultaneous
                     Lower        Difference        Upper
                  Confidence       Between       Confidence
      i    j        Limit           Means          Limit
      1    2       0.007057        0.025500       0.043943
      1    3       0.006039        0.021833       0.037627
      1    4       0.007373        0.023167       0.038961
      1    5       0.010864        0.024333       0.037802
      1    6       0.009198        0.022667       0.036136
      2    3      -0.024963       -0.003667       0.017630
      :    :         :               :              :
```

0 and the test treatments as $1, \ldots, 8$, SAS would have relabeled the control as treatment 1 and the test treatments as $2, \ldots, 9$.

Table 11.19 shows partial output for the first day's data from the plasma experiment (Section 11.7), which was a nonstandard incomplete block design. The Type I and Type III sums of squares are shown, together with partial output from the LSMEANS statement

```
LSMEANS TRTMT / PDIFF = ALL CL ADJUST=TUKEY;
```

which was used to compile Table 11.13 on page 367.

Table 11.20 SAS program to plot data adjusted for block effects—plasma experiment, day one only.

```
* This program requires 3 runs, adding more information in each run;
DATA ONE;
  INPUT BLOCK TRTMT Y;
  LINES;
    1  4  0.459
    1  5  0.467
    :  :   :
    3  3  0.473
  ;
* Get block effect estimates;
PROC GLM;
  CLASS BLOCK TRTMT;
  MODEL Y = BLOCK TRTMT / SOLUTION;
PROC SORT;  BY BLOCK;
  ;
* Add the following code for the second run;
* values BHAT are solutions for block parameters from first run;
DATA TWO;
  INPUT BLOCK BHAT;
  LINES;
      1   -.0126666667
      2   -.0053333333
      3    0.0000000000
PROC MEANS MEAN; * print average of BHAT values;
  VAR BHAT;
  ;
* Add the following code for the third run;
* The number -0.006 below is average BHAT calculated in second run;
DATA THREE;
  MERGE ONE TWO;
  BY BLOCK;
  Y_ADJ=Y - (BHAT-(-0.006));
PROC PLOT;
  PLOT Y_ADJ*TRTMT / VPOS=19 HPOS=50;
```

11.10.2 Plots

Table 11.20 contains a sample SAS program illustrating how to plot the data adjusted for blocks against the treatment labels, using the day one data of the plasma experiment, (Table 11.10, page 364). The program, as written, must be run in three passes. In successive passes, information generated by earlier passes must be added as input in later parts of the program. First, the data are entered into a data set called ONE. Since the block effect estimates are needed to adjust the observations, the option SOLUTION is included in the MODEL statement of PROC GLM. This causes a (nonunique) solution to the normal equations for $\hat{\mu}$, $\hat{\tau}_i$, and $\hat{\theta}_h$ to be printed. The solutions will all be labeled "B" for "biased," meaning that the corresponding parameters are not estimable (see Table 10.17, page 324, for example).

The least squares solutions $\hat{\theta}_h$ are then entered into the data set TWO in the second run of the program. PROC MEANS is used to compute and print the average value $\hat{\bar{\theta}}_.$. Finally, in the third run of the program, the block-effect estimates and their average value are used to adjust the data values. The adjusted values are then plotted against treatment. The SAS plot is not shown here, but it is similar to the plot in Figure 11.4 (page 366).

Exercises

1. **Connectedness and estimability**
   (a) For each of the three block designs in Table 11.21, draw the connectivity graph for the design, and determine whether the design is connected.
   (b) If the design is connected, determine whether or not it is a balanced incomplete block design.
   (c) For designs II and III, determine graphically whether or not $\tau_1 - \tau_5$ and $\tau_1 - \tau_6$ are estimable.
   (d) For design III, use expected values to show that $\tau_1 - \tau_8$ is estimable.

2. **Connectedness**
   (a) Determine whether or not the cyclic design with initial block (1, 3, 5) is a connected design if $v = 8$ or $v = 9$.
   (b) Determine whether or not the cyclic design with initial block (1, 4, 7) is a connected design if $v = 8$ or $v = 9$.

3. **Randomization**
   Conduct a block design randomization for design II in Table 11.21.

4. **Cyclic designs**
   Determine whether or not the cyclic design obtained from each initial block below is a balanced incomplete block design or a group divisible design or neither.

Table 11.21 Three incomplete block designs

Design I		Design II		Design III	
Block	Treatments	Block	Treatments	Block	Treatments
1	1 2	1	1 2 3	1	1 2 6
2	1 3	2	4 5 6	2	3 4 5
3	1 4	3	7 8 9	3	2 6 8
4	2 3	4	1 4 7	4	4 5 7
5	2 4	5	2 5 8	5	1 6 8
6	3 4	6	3 6 9	6	3 5 7
		7	1 5 9	7	1 2 8
		8	2 6 7	8	3 4 7
		9	3 4 8		

(a) Initial block: 1, 3, 4; $v = 7$.

(b) Initial block: 1, 2, 4, 8; $v = 8$.

(c) Initial block: 1, 2, 4; $v = 5$.

5. **Balanced incomplete block design**

   Consider an experiment to compare 7 treatments in blocks of size 5. Taking all possible combinations of five treatments from seven gives a balanced incomplete block design with $r = 15$.

   (a) How many blocks does the design have?

   (b) Show that r must be a multiple of five for a balanced incomplete block design with $v = 7$ treatments and blocks of size $k = 5$ to exist.

   (c) Show that the smallest balanced incomplete block design has $r = 15$ observations per treatment.

6. **Sample sizes**

   Consider an experiment to compare 7 treatments in blocks of size 5, with an anticipated error variance of at most 30 square units.

   (a) Assuming that a balanced incomplete block design will be used, how many observations would be needed for the minimum significant difference to be about 50 units for a pairwise comparison using Tukey's method and a 95% simultaneous confidence level?

   (b) Repeat part (a) for a minimum significant difference of 25 units.

   (c) Repeat part (a) using Dunnett's method for treatment versus control comparisons.

7. **Least squares estimator**

   For the balanced incomplete block design in Table 11.8,

   (a) Show that the least squares estimator $\hat{\tau}_4 - \hat{\tau}_6$ is an unbiased estimator of $\tau_4 - \tau_6$ under the block–treatment model (11.3.1) (that is, show that $E[\hat{\tau}_4 - \hat{\tau}_6] = \tau_4 - \tau_6$).

   (b) Verify that the variance of $\hat{\tau}_4 - \hat{\tau}_6$ is $2k\sigma^2/(\lambda v)$.

8. **Isomer experiment**

   The following experiment was described by Kuehl (1994) and run by J. Berry and A. Deutschman at the University of Arizona to obtain specific information about the effect of pressure on percent conversion of methyl glucoside to monovinyl isomers. The conversion is achieved by addition of acetylene to methyl glucoside in the presence of a base under high pressure. Five pressures were examined in the experiment, but only three could be examined at any one time under identical experimental conditions. A balanced incomplete block design was used. The data and design are shown in Table 11.22.

   (a) Was blocking worthwhile in this experiment?

   (b) Write down a model for this experiment and test the hypothesis of no difference in the effects of the pressures on the percent conversion of methyl glucoside, against the alternative hypothesis that at least two pressures differ. Use a significance level of 0.01.

Table 11.22 Percent conversion of methyl glucoside to monovinyl isomers

Pressure (psi)	Block									
	I	II	III	IV	V	VI	VII	VIII	IX	X
250	16	19				20	13	21	24	
325	18		26		19		13		10	24
400			39	21		33	34	30		31
475	32	46		35	47	31				37
550		45	61	55	48			52	50	

Source: From *Statistical Principles of Research Design and Analysis* by R. O. Kuehl. Copyright © 1994 Brooks/Cole Publishing Company, Pacific Grove, CA 93950, a division of International Thomson Publishing Inc. By Permission of the publisher.

(c) What does " a significance level of 0.01" in part (b) mean?

(d) Give a formula for a 95% set of simultaneous confidence intervals for pairwise comparisons among the pressures in this experiment. Calculate, by hand, the interval comparing pressures 5 and 4 (that is, $\tau_5 - \tau_4$) for illustration. Compare your answer with that obtained from your computer output.

9. **Balanced incomplete block design**
An experiment is to be run to compare the effects of four different formulations of a drug to relieve an allergy. In a pilot experiment, four subjects are to be used, and each is to be given a sequence of three of the four drugs. The measurements are the number of minutes that the subject appears to be free of allergy symptoms. Suppose the following design is selected for the experiment.

Block	Levels of Treatment Factor		
I	0	1	2
II	0	1	3
III	0	2	3
IV	1	2	3

(a) Check that this design is a balanced incomplete block design. (Show what you are checking.)

(b) What would you randomize when using this design?

(c) The typical model for a balanced incomplete block design is the block–treatment model (11.3.1), page 348. Do you think this is a reasonable model for the experiment described? Why or why not?

The experiment was run as described, and the model indicated in (c) was used to analyze it. Some information for the analysis is shown in Table 11.23. (Even if you criticized the model, answer the rest of the questions.)

(d) Show that $Q_2 = 81.667$.

(e) Give a confidence interval for $\tau_3 - \tau_2$ assuming that it is part of a set of 95% Tukey confidence intervals.

(f) Test the hypothesis that there is no difference between the drugs.

Table 11.23 Partial information for analysis of the drug experiment

Block totals	Treatment totals	Q_i	LSMEANS
$B_1 = 417$	$T_1 = 385$	$Q_1 = -79.333$	134.41
$B_2 = 507$	$T_2 = 582$	$Q_2 = 81.667$	195.79
$B_3 = 469$	$T_3 = 329$	$Q_3 = -158.667$	104.67
$B_4 = 577$	$T_4 = 674$	$Q_4 = 156.333$	222.79
$\bar{y}_{..} = 164.1667$	$msE = 3.683$	$\Sigma Q_i^2 = 62{,}578.335$	

10. **Step experiment, continued**

    The step experiment was described in Example 11.9.1 and the data are shown in Table 11.14, page 370.

    (a) Prepare a plot of the treatment averages and examine the linear trends in the heart rate due to step frequency at each level of step height.

    (b) Fit a block–treatment model to the data with $v = 6$ treatments representing the six treatment combinations.

    (c) Estimate the linear trends in the heart rate due to step frequency at each level of step height separately, and calculate confidence intervals for these.

    (d) Write down a contrast that compares the linear trends in part (c) and test the hypothesis that the linear trends are the same against the alternative hypothesis that they are different.

11. **Beef experiment**

    Cochran and Cox (1957) describe an experiment that was run to compare the effects of cold storage on the tenderness of beef roasts. Six periods of storage (0, 1, 2, 4, 9, and 18 days) were tested and coded 1–6. It was believed that roasts from similar positions on the two sides of the animal would be similar, and therefore the experiment was run in $b = 15$ blocks of size $k = 2$. The response y_{hi} from treatment i in block h is the tenderness score. The maximum score is 40, indicating very tender beef. The design and responses are shown in Table 11.24.

    (a) What is the value of λ for this balanced incomplete block design?

    (b) What benefit do you think the experimenters expected to gain by using a block design instead of a completely randomized design?

    (c) Calculate the least squares estimate of $\tau_6 - \tau_1$ and its corresponding variance.

    (d) Calculate a confidence interval for $\tau_6 - \tau_1$ as though it were part of a set of 95% confidence intervals using Tukey's method of multiple comparisons.

    (e) Calculate a confidence interval for the difference of averages contrast

    $$\frac{1}{3}(\tau_4 + \tau_5 + \tau_6) - \frac{1}{3}(\tau_1 + \tau_2 + \tau_3),$$

    assuming that you want the overall level of this interval together with the intervals in part (d) to be at least 94%. What does your interval tell you about storage time and tenderness of beef?

Table 11.24 Design and data for the beef experiment

Block	Treatment					
	1	2	3	4	5	6
I	7	17				
II			26	25		
III					33	29
IV	17		27			
V		23			27	
VI				29		30
VII	10			25		
VIII		26				37
IX			24		26	
X	25				40	
XI		25		34		
XII			34			32
XIII	11					27
XIV		24	21			
XV					26	32

Source: *Experimental Designs*, Second Edition, by W. G. Cochran and G. M. Cox, Copyright © 1957, John Wiley & Sons, New York. Adapted by permission of John Wiley & Sons, Inc.

12. **Balanced incomplete block design**
    (a) Explain why you might choose to use a block design rather than a completely randomized design.
    (b) For a balanced incomplete block design, why is it incorrect to estimate the difference in the effects of treatments i and p as $\bar{y}_{i.} - \bar{y}_{p.}$? What is the correct least squares estimate?
    (c) Give a formula for a set of 95% confidence intervals for the pairwise differences in the effects of the treatments.
    (d) The information in Table 11.25 is taken from a computer analysis of a balanced incomplete block design with treatment factor "temperature" having $v = 7$ levels of increasing dosage and with blocking factor "block" having $b = 7$ levels. Blocks are of size $k = 3$. What can you deduce?

13. **Group divisible design least squares estimates**
    Show that equation (11.3.2) on page 349 gives the solution (11.5.21), page 361, for the group divisible design.

14. **Plasma experiment, continued**
    In the plasma experiment of Section 11.7, the experimenters used the cyclic design for six treatments generated by 1, 4, 5.
    (a) Show that the cyclic design is also a group divisible design by determining the groups and the values of λ_1 and λ_2.

Table 11.25 Partial analysis of a balanced incomplete block design

```
                        The SAS System

                           Sum of        Mean
Source              DF    Squares        Square        Ratio    p-value
Model               12   385.56000000    32.13000000   20.60    0.0001
BLOCK (unadjusted)   6   163.67809524    27.27968254   17.49    0.0003
TEMPR   (adjusted)   6   221.88190476    36.98031746   23.71    0.0001
Error                8    12.47809524     1.55976190
Corrected Total     20   398.03809524


Contrast            DF   Contrast SS     Mean Square   F Value  p-value
LIN TEMPR            1   176.88137755    176.88137755  113.40   0.0001


Parameter                Estimate
LIN TEMPR                46.0714286

          Least Squares Means
          TEMPR     LSMEAN
            1       20.4904762
            2       20.6476190
            3       18.9476190
            4       24.9190476
            5       27.8047619
            6       27.1904762
            7       28.5333333
```

(b) The design for day one of the experiment consisted of the following three blocks of size three: (1, 4, 5); (2, 5, 6); (3, 6, 1). Show that this design is connected by drawing the connectivity graph of the design.

15. **Vitamin D experiment**

    M. N. Das and G. A. Kulkarni (*Biometrics*, 1966) presented a set of data that they modified from an original study by K. H. Coward and E. W. Kassner (*Biochemical Journal*, 1941). The study concerned the potency of different doses and preparations of vitamin D. The experimental units were rats, and the rats within each block were from the same litter. We present the modified data of Das and Kulkarni in Table 11.26. There are six treatments. Treatments 1, 2, and 3 are the standard preparation of three doses of vitamin D equally spaced in ascending order on the logarithmic scale. Treatments 4, 5, and 6 are the test preparation of three doses of vitamin D equally spaced in descending order on the logarithmic scale.

    (a) Show that the design consists of a group divisible design with three blocks repeated six times. What are the groups in the group divisible design? What are the values of λ_1 and λ_2 for the entire design with $b = 18$ blocks?

Table 11.26 Responses for the vitamin D experiment

Block	\multicolumn{6}{c}{Treatment}					
	1	2	3	4	5	6
I	2	8			9	7
II	6		9	3		8
III		6	12	4	6	
IV	9	11			14	13
V	10		17	8		10
VI		7	5	6	9	
VII	4	10			11	13
VIII	11		9	3		15
IX		9	14	5	8	
X	4	7			10	10
XI	12		9	15		15
XII		8	11	7	8	
XIII	4	4			5	9
XIV	7		8	3		9
XV		15	10	6	8	
XVI	2	4			6	6
XVII	4		13	5		12
XVIII		10	13	4	18	

Source: Das, M. N. and Kulkarni, G. A. (1966).
Copyright © 1966 International Biometric Society.
The data were adapted from Coward, K. H. and
Kassner, E. W. (1941). Reprinted with permission of
The Biochemical Society and Portland Press.

(b) Calculate the variance for a pairwise comparison between two treatments that are first associates, and also for two treatments that are second associates.

(c) Which treatments are first associates and which treatments are second associates? Why do you think the authors chose this particular design? [Hint: Use the information in part (b).]

(d) Write down the contrast coefficient lists for the following contrasts:

   (i) comparison of the standard preparation with the test preparation, averaged over doses (that is, a difference-of-averages contrast comparing the average effect of treatments 1, 2, and 3 with the average effects of treatments 4, 5, and 6);

   (ii) the comparison of the two preparations at each dose separately;

   (iii) the linear effect of log(dose) on the response for each preparation;

   (iv) the quadratic effect of log(dose) on the response for each preparation.

(e) Prepare an analysis of variance table and test the hypotheses that each contrast in (d) is negligible. Use an overall significance level of not more than 0.1. State your conclusions.

16. **Air rifle experiment**

    This is a dangerous experiment that should not be copied! It requires proper facilities and expert safety supervision.

    An experiment was run in 1995 by C.-Y. Li, D. Ranson, T. Schneider, T. Walsh, and P.-J. Zhang to examine the accuracy of an air rifle shooting at a target. The two treatment factors of interest were the projectile type (factor A at levels 1 and 2) and the number of pumps of the rifle (factor B, 2, 6, and 10 pumps, coded 1, 2, 3). The paper covering the target had to be changed after every four observations, and since there were $v = 6$ treatment combinations, an incomplete block design was selected.

    Two copies of the following incomplete block design (called a generalized cyclic design) were used:

Block	Treatment Combination			
I	11	21	13	22
II	12	22	11	23
III	13	23	12	21
IV	21	11	23	12
V	22	12	21	13
VI	23	13	22	11

    The total of 12 blocks were randomly ordered, as were the treatment combinations within each block. The data, shown in Table 11.27, are distances from the center of the target measured in millimeters.

    (a) Write down a suitable model for this experiment.

    (b) Check that the assumptions on your model are satisfied.

    (c) The experimenters expected to see a difference in accuracy of the two projectile types. Using a computer package, analyze the data and determine whether or not this was the case.

    (d) For each projectile type separately, examine the linear and quadratic trends in the effects of the number of pumps. State your conclusions.

Table 11.27 Data for the rifle experiment

Block	Treatment Combination (Response)			
I	22 (2.24)	23 (6.02)	12 (11.40)	11 (26.91)
II	13 (7.07)	22 (8.49)	21 (19.72)	11 (24.21)
III	12 (10.63)	23 (6.32)	21 (9.06)	13 (29.15)
IV	11 (11.05)	22 (6.32)	23 (7.21)	13 (23.02)
V	23 (6.71)	22 (15.65)	12 (11.40)	11 (23.02)
VI	21 (17.89)	12 (8.60)	22 (10.20)	13 (10.05)
VII	11 (18.38)	13 (11.18)	23 (11.31)	22 (11.70)
VIII	22 (1.00)	11 (30.87)	21 (20.10)	13 (17.03)
IX	11 (18.03)	21 (8.25)	23 (6.08)	12 (19.24)
X	21 (15.81)	13 (2.24)	12 (17.09)	22 (7.28)
XI	23 (8.60)	21 (15.13)	12 (14.42)	11 (25.32)
XII	13 (8.49)	12 (14.32)	23 (11.66)	21 (17.72)

12 Designs with Two Blocking Factors

12.1 Introduction
12.2 Design Issues
12.3 Model for a Row–Column Design
12.4 Analysis of Row–Column Designs (Optional)
12.5 Analysis of Latin Square Designs
12.6 Analysis of Youden Designs
12.7 Analysis of Cyclic and Other Row–Column Designs
12.8 Checking the Assumptions on the Model
12.9 Factorial Experiments in Row–Column Designs
12.10 Using SAS Software
Exercises

12.1 Introduction

In Chapters 10 and 11, we discussed designs for experiments involving a single system of blocks. As we saw in the randomized complete block design of Table 10.2, page 300, a block label can represent a combination of levels of several factors. The design in Table 10.2 was presented as having six blocks—the six block labels being the six combinations of levels of the factors "run of the oven" and "shelf." When a block design is used in this way, the $b-1$ degrees of freedom for the block effects include not only those degrees of freedom for the effects of the two factors, but also for their interaction.

In this chapter, we look at designs for experiments that involve two blocking factors that *do not* interact. When the blocking factor interactions can be omitted from the model, fewer observations are needed, and the experiment can be designed with only one observation per combination of levels of the blocking factors. The plan of the design is written as an array (that is, a table) with the levels of one blocking factor providing the row headings and those of the other providing the column headings. These designs are often called *row–column designs*, and the two sets of blocks are called row blocks and column blocks or, more simply, rows and columns. Exactly one treatment label is allocated to each cell of the table.

In Section 12.2.2, Latin square designs are described. These are the two-dimensional counterparts of randomized complete block designs in that row blocks are complete blocks and column blocks are also complete blocks. Latin square designs require that the numbers of levels of both blocking factors be the same as (or a multiple of) the number of treatments. Section 12.2.3 concerns Youden designs, in which the column blocks form a randomized block design and the row blocks form a balanced incomplete block design (or vice versa). Cyclic and other row–column designs are discussed in Section 12.2.4.

The randomization procedure needed for row–column designs is given in Section 12.2.1, and the standard row–column–treatment model for row–column designs is given in Section 12.3. Analysis of variance and confidence intervals for row–column designs are derived in the optional section, Section 12.4. The analysis simplifies for both Latin square designs and Youden designs, and their simplified versions are discussed in Sections 12.5 and 12.6, respectively. All row–column designs can be analyzed using statistical software (see Section 12.10).

Some experiments are designed with more than two blocking factors. Block designs with a single system of blocks can still be used where each block represents some combination of the levels of the three or more blocking factors, but designs with more than two blocking factors and one observation per combination of their levels are not discussed in this book.

12.2 Design Issues

12.2.1 Selection and Randomization of Row–Column Designs

In most row–column designs all treatment contrasts are estimable. This is not as easy to verify as it was for incomplete block designs (see optional Section 11.2.3), since it cannot be deduced from the row blocks and column blocks separately. However, it can be shown that all Latin square designs and all Youden designs, which are introduced in Sections 12.2.2 and 12.2.3, allow estimability of all treatment contrasts. One way to check that a miscellaneous row–column design is suitable for a planned experiment is to enter a set of hypothetical data for the design into a computer package and see whether the required contrasts are estimable.

Once the numbers of rows, columns, and treatments have been determined, there are two stages to designing an experiment with two blocking factors and one observation at each combination of their levels. The first stage is to choose an experimental plan, such as that in Table 12.1. It has two sets of blocks corresponding to two blocking factors—one with 7 levels represented by the $b = 7$ rows, and the other with 4 levels represented by the $c = 4$ columns. There are $v = 7$ treatment labels, each appearing $r = 4$ times in the design. The second stage is the random assignment of the labels in the design to the levels of the treatment factors and the blocking factors in the experiment, as follows:

Table 12.1 A row–column experimental plan with $b = 7$, $c = 4$, $v = 7$, $r = 4$

		Column Blocks			
		I	II	III	IV
	I	1	3	6	7
	II	2	4	7	1
Row	III	3	5	1	2
Blocks	IV	4	6	2	3
	V	5	7	3	4
	VI	6	1	4	5
	VII	7	2	5	6

(i) The row-block labels in the design are randomly assigned to the levels of the first blocking factor.

(ii) The column-block labels in the design are randomly assigned to the levels of the second blocking factor.

(iii) The treatment labels in the design are randomly assigned to the levels of the treatment factor.

Since there is only one experimental unit in each cell, there is no need for random assignment of experimental units to treatment labels within a cell.

12.2.2 Latin Square Designs

A $v \times v$ *Latin square* is an arrangement of v Latin letters into a $v \times v$ array (a table with v rows and v columns) in such a way that each letter occurs once in each row and once in each column. For example, the following 3×3 array is a 3×3 Latin square:

A B C
B C A
C A B

A *Latin square design* is a design with v treatment labels and v^2 experimental units arranged in v row blocks and v column blocks, where experimental units within each row block are alike, units within each column block are alike, and units not in the same row block nor column block are substantially different. The experimental plan of the design is a $v \times v$ Latin square. Randomization of row block, column block, and treatment labels in the plan is carried out as in Section 12.2.1.

If we look only at the row blocks of a Latin square design, ignoring the column blocks, we have a randomized complete block design, and if we look at the column blocks alone, ignoring the row blocks, we also have a randomized complete block design. Each level of the treatment factor is observed $r = v$ times—once in each row block and once in each column block.

The 3×3 Latin square shown above is a "standard, cyclic Latin square." A Latin square is a *standard Latin square* if the letters in the first row and in the first column are in alphabetical

Table 12.2 Latin squares with $b = c = v = 4$

Square 1	Square 2	Square 3	Square 4
A B C D	A B C D	A B C D	A B C D
B C D A	B A D C	B A D C	B D A C
C D A B	C D A B	C D B A	C A D B
D A B C	D C B A	D C A B	D C B A

order, and it is *cyclic* if the letters in each row can be obtained from those in the previous row by cycling the letters in alphabetical order (cycling back to letter A after the vth letter).

There is only one standard 3×3 Latin square, but there are four standard 4×4 Latin squares, and these are shown in Table 12.2. The first square is the cyclic standard Latin square. A standard cyclic Latin square exists for any number of treatments.

An example of a 6×6 Latin square design was shown in Table 2.5 (page 20). It was a design that was considered for the cotton-spinning experiment. The row blocks represented the different machines with their attendant operators, and the column blocks represented the different days over which the experiment was to be run. The treatment labels were the combinations of degrees of twist and flyer used on the cotton-spinning machines. After careful consideration, the experimenters decided not to use this design, since it required the same six machines to be available over the same six days, and this could not be guaranteed in the factory setting.

Latin square designs are often used in experiments where subjects are allocated a sequence of treatments over time and where the time effect is thought to have a major effect on the response. For example, in an experiment to compare the effects of v drugs, the rows of the Latin square might correspond to v subjects to whom the drugs are given, and the columns might correspond to v time periods, with each subject receiving one drug during each time period. An experiment of this type, in which each subject receives a sequence of treatments, is called a *crossover experiment*.

In a crossover experiment, it is possible that a treatment will continue to have some effect on the response when a subsequent treatment is administered to the same subject. Such an effect is called a *carryover effect* or *residual effect* and must be accounted for in the design and analysis of the experiment. This is outside the scope of this book, but for further information, see Ratkowsky, Evans, and Alldredge (1993). We will consider experiments in which either there is no carryover effect or in which the gap between the administration of one treatment and the next is sufficient to allow the carryover effect to diminish (and, hopefully, to disappear). The "gap" is called a *washout period*.

In order that every possible Latin square have the same chance of being selected as the design for an experiment, a standard square should be selected from the list of all possible standard squares and the randomization procedure of Section 12.2.1 performed. Thus for an experiment with $b = 4$ levels of the row blocking factor, $c = 4$ levels of the column blocking factor, and $v = 4$ levels of the treatment factor, one of the four standard squares of Table 12.2 would be selected at random. For larger squares, we have not provided a complete list of all standard squares, since there are so many. A complete list for $v = 5$ and 6, and selected squares for $v = 7$–12 can be found in the statistical tables of Fisher and Yates. For the larger values of v, it is permissible to select the standard cyclic Latin square and perform the randomization procedure on this, rather than selecting from the set of standard squares.

Replication of Latin squares A design based on a single Latin square has $r = v$ observations on each treatment, which may not be adequate. One way to use multiple Latin squares is to piece together a number, s say, of $v \times v$ Latin squares. We will call such a design an *s-replicate Latin square*. Use of an s-replicate Latin square requires the column (or row) blocks to be of size vs. For example, stacking two 3×3 Latin squares, one above the other, we can obtain two possible 2-replicate Latin squares as follows.

Plan 1	Plan 2
A B C	A B C
B C A	B C A
C A B	C A B
A B C	A C B
B C A	B A C
C A B	C B A

For either plan, the number of observations per treatment is now $r = 6 = 2v$ rather than only $r = 3 = v$. Plan 2 is probably preferable, because the row blocks consist of each possible ordering of the three treatments (and this will remain true even after randomization).

Suppose we were to use Plan 2 for an experiment to compare the efficacy of $v = 3$ drugs, where there are two blocking factors, say "subjects"—the people to whom the drugs will be administered, and "time period"—the time during which each subject receives a single drug and a response is measured. If rows correspond to subjects, then $b = 6$ subjects would be required over $c = 3$ time periods, but if columns correspond to subjects and rows to time periods, then each of $c = 3$ subjects would stay in the study for $b = 6$ periods. In practice, in drug studies, subjects are rarely used for more than 4 time periods, since the drop-out rate tends to be high after this length of time.

An alternative way to obtain $r = vs$ observations per treatment is to use s separate Latin squares. For example, taking Square 1 and Square 2 from Table 12.2 and using them separately gives $r = 8$ observations per treatment and uses $b = 8$ row blocks and $c = 8$ column blocks all of size 4. On the other hand, stacking them to obtain a 2-replicate Latin square gives $r = 8$ observations per treatment with $b = 8$ row blocks of size 4 and $c = 4$ column blocks of size 8. The latter arrangement would allow column–treatment interactions to be measured, but also requires that the column-block sizes be large.

12.2.3 Youden Designs

A $v \times c$ *Youden square* is an arrangement of v Latin letters into a $v \times c$ array (with $c < v$) in such a way that each letter occurs once in each column and at most once in each row. In addition, all pairs of treatments occur together in the same number of rows. Notice that a Youden square is not, in fact, square! We have defined a Youden square to have more rows than columns, but the array could be turned so that there are more columns than rows. The following is a 4×3 Youden square with $v = 4$ treatment labels.

Plan 3
A B C
B C D
C D A
D A B

A *Youden design* is a design with v treatment labels and with vc experimental units arranged in $b = v$ row blocks and $c < v$ column blocks where experimental units within each row block are alike, experimental units within each column block are alike, and experimental units not in the same row block or column block are substantially different. Also, the column blocks form a randomized complete block design, and the row blocks form a balanced incomplete block design. Each level of the treatment factor is observed $r = c$ times—once in each column and at most once in each row. The experimental plan is a $v \times c$ Youden square, and randomization of row, column, and treatment labels is carried out as in Section 12.2.1.

A Youden design would be selected for an experiment when the block size for one of the blocking factors cannot be chosen to be as large as the number of treatments (cf. incomplete block designs, Section 11.2.1). Plan 3 shown above is a *cyclic Youden design* prior to randomization. The row blocks form a cyclic balanced incomplete block design with every pair of treatment labels occurring in $\lambda = 2$ rows. Any cyclic balanced incomplete block design (with the full set of v blocks) can be used as a cyclic Youden design. The design in Table 12.1 is also a cyclic Youden design, having $b = v = 7$ row blocks and $c = 4$ column blocks and every pair of treatments occurring in $\lambda = 2$ row blocks. Youden designs with $c = v - 1$ column blocks can be obtained by deleting any column from a $v \times v$ Latin square.

Replication of Youden squares We can obtain $r = cs$ observations per treatment either by using s Youden squares separately (with $b = vs$ row blocks and cs column blocks) or by stacking s Youden squares one above another (giving $b = vs$ row blocks and c column blocks). The latter requires large column-block sizes but allows the estimation of column×treatment interactions. In either case, the row blocks still form a balanced incomplete block design, and the column blocks still form a complete block design.

Suppose $v = 7$ drugs are to be compared and that a number of subjects are each to be given a sequence of 4 of the drugs over $c = 4$ time periods, with washout periods in between to avoid carryover effects. The experimental plan in Table 12.1, which uses 7 subjects, would be suitable for the experiment if $r = 4$ observations per drug is deemed adequate. Otherwise, several copies of this design could be stacked. If, for example, $s = 3$ copies of the design are pieced together, the resulting 3-replicate Youden design would require 21 subjects, 4 time periods, and would have $r = cs = 12$ observations per treatment, with $b = 21$, $c = 4$, $v = 7$, and $\lambda = 6$. The design would be randomized before use as described in Section 12.2.1.

12.2.4 Cyclic and Other Row–Column Designs

Any arrangement of v treatment labels into b rows and c columns can be used as a row–column design for an experiment with two blocking factors. As with incomplete block

12.2 Design Issues

designs, some row–column designs are better than others. The better designs (in terms of average length of confidence intervals for pairwise comparisons) have every pair of treatment labels occurring the same, or nearly the same, number of times in the row blocks and also in the column blocks. This is satisfied by Latin square designs, Youden designs, and also some cyclic designs.

A *cyclic row–column design with complete column blocks* is a row–column design in which the row blocks form a cyclic block design and the column blocks are complete blocks. For example, the experimental plan in Table 12.1 is a cyclic row–column design. The class of cyclic row–column designs is very large, and a design with $b = vs$ rows can always be found when a Youden design does not exist. For example, consider an experiment for comparing $v = 8$ treatments when there are two blocking factors having $b = 8$ and $c = 3$ levels, respectively. There does not exist a Youden design for $b = v = 8$ and $c = 3$, because there does not exist a balanced incomplete block design for 8 treatments in 8 blocks of size $c = 3$. (Notice that $\lambda = r(c-1)/(v-1) = 6/7$ is not an integer.) A cyclic design that may be suitable for the experiment is given as Plan 4 in Table 12.3. The design is obtained by cycling the labels in the first row block. Each treatment occurs $r = c = 3$ times. Treatment pairs (1, 5), (2, 6), (3, 7), and (4, 8) never occur together in a row block, but all other pairs of treatments occur in exactly one row block, so this should be a reasonably good design. The design should be randomized via the procedure in Section 12.2.1 before it is used.

Plan 5 of Table 12.3, which is neither a Youden nor a cyclic design, could also be used, but we might guess that it is not quite as good for pairwise comparisons. Treatment 1, for example, appears in one block with each of treatments 4 and 8, in two blocks with 3 and 5, and not at all with treatments 2, 6, or 7. Thus, the treatments are not quite so evenly distributed in the row blocks. Nevertheless, this design was used for the exercise bike experiment described below, and will be analyzed using the SAS software in Section 12.10.

Example 12.2.1 Exercise bicycle experiment

Yuedong Wang, Dong Xiang, and Yili Lu conducted an experiment in 1992 at the University of Wisconsin to investigate the effects of exercise on pulse rate. The exercise was performed on a stationary bicycle that included both foot pedals and hand bars. The experiment involved three treatment factors each at two levels. These were "time duration of exercise" with levels

Table 12.3 Incomplete-row complete-column designs with $v = b = 8$ and $r = c = 3$

	Plan 4				Plan 5		
	I	II	III		I	II	III
I	1	2	4	I	1	2	7
II	2	3	5	II	2	4	1
III	3	4	6	III	3	7	6
IV	4	5	7	IV	4	8	2
V	5	6	8	V	5	6	8
VI	6	7	1	VI	6	3	5
VII	7	8	2	VII	7	1	3
VIII	8	1	3	VIII	8	5	4

1 and 3 minutes, coded as 1 and 2, "exercise speed" with levels 40 and 60 rpm, coded as 1 and 2, and "pedal type" with levels foot pedal and hand bars, coded as 1 and 2.

The three experimenters served as the $b = 3$ subjects in the experiment, since they were interested in the effects of the different exercises on themselves rather than on a large population. Had the latter been the case, then the subjects would have been selected at random from the population of interest.

Each subject was assigned a different exercise on each of 8 different days. To minimize any carryover effect, there was a training period prior to the experiment, and there was at least one day of rest between observations. The subject's pulse was taken immediately after completion of the exercise, and the response variable was the time (in seconds) for 50 heart beats.

Plan 5 of Table 12.3 was used. The design, after randomization, and the corresponding data are shown in Table 12.4. The rows were randomly assigned to days in the order V, III, VI, VIII, VII, IV, I, II. The columns were randomly assigned to subjects in the order I = Lu, II=Wang, III=Xiang. The treatment labels were randomly assigned to treatment combinations in the order

1	5	2	4	7	6	8	3
111	112	121	122	211	212	221	222

12.3 Model for a Row–Column Design

For a row–column design with b rows and c columns, the row–column–treatment model for an observation on treatment i in row-block h and column-block q is

$$Y_{hqi} = \mu + \theta_h + \phi_q + \tau_i + \epsilon_{hqi}, \quad (12.3.1)$$

$$\epsilon_{hqi} \sim N(0, \sigma^2),$$

ϵ_{hqi}'s are mutually independent,

$h = 1, \ldots, b; \quad q = 1, \ldots, c; \quad i = 1, \ldots, v; \quad (h, q, i)$ in the design.

Table 12.4 Row–column design showing treatments and data for the exercise bicycle experiment

Day	Subject		
	Lu	Wang	Xiang
1	112 45	212 25	221 18
2	222 27	211 20	212 32
3	212 40	222 23	112 28
4	221 17	112 32	122 24
5	211 30	111 36	222 20
6	122 29	221 13	121 20
7	111 34	121 18	211 25
8	121 21	122 22	111 34

The term "(h, q, i) in the design" means that the model is valid for whichever treatment i is observed in the cell defined by the hth row block and qth column block. The model includes the usual assumptions that the error terms ϵ_{hqi} are independent and normally distributed with constant variance. It also includes the assumptions of no interaction between the row and column blocking factors or between these and the treatment factors

If $b > v$ or $c > v$, or both, there may be sufficient degrees of freedom to be able to estimate one or both block-treatment interactions. For example, if $c = v = 3$, and $b = 6$, there are 18 observations, giving 17 total number of degrees of freedom. Of these, $2+5+2 = 9$ are needed to measure treatments, row blocks, and column blocks. Of the remaining 8 degrees of freedom, 4 could be used to measure column×treatment interaction, in which case the model would be of the form

$$Y_{hqi} = \mu + \theta_h + \phi_q + \tau_i + (\phi\tau)_{qi} + \epsilon_{hqi} . \tag{12.3.2}$$

In the next section, we derive the least squares parameter estimators and the analysis for row–column designs in general. This material is optional, since the specific analyses are given for Latin square designs and Youden designs in Sections 12.5 and 12.6, respectively, and analysis of more general designs is illustrated via the SAS computer package in Section 12.10.

12.4 Analysis of Row–Column Designs (Optional)

12.4.1 Least Squares Estimation (Optional)

In this section, we derive least squares parameter estimates and the error sum of squares for row–column designs with b row blocks, c column blocks, and r observations on each of v treatments.

The sum of squares of the errors for the row–column–treatment model (12.3.1) is

$$\sum_{h=1}^{b}\sum_{q=1}^{c}\sum_{i=1}^{v} n_{hqi} e_{hqi}^2 = \sum_{h=1}^{b}\sum_{q=1}^{c}\sum_{i=1}^{v} n_{hqi}\left(y_{hqi} - \mu - \theta_h - \phi_q - \tau_i\right)^2 ,$$

where $n_{hqi} = 1$ if treatment label i is allocated to the combination of row block h and column block q and zero otherwise. Notice that

$$\sum_{h=1}^{b}\sum_{q=1}^{c} n_{hqi} = r \quad \text{and} \quad \sum_{i=1}^{c} n_{hqi} = 1 .$$

So,

$$\sum_{h=1}^{b}\sum_{i=1}^{v} n_{hqi} = b, \quad \sum_{q=1}^{c}\sum_{i=1}^{v} n_{hqi} = c, \quad \sum_{h=1}^{b}\sum_{q=1}^{c}\sum_{i=1}^{v} n_{hqi} = bc = rv .$$

Also,

$$\sum_{q=1}^{c} n_{hqi} = n_{h.i} \quad \text{and} \quad \sum_{h=1}^{b} n_{hqi} = n_{.qi} ,$$

where $n_{h.i}$ and $n_{.qi}$ are respectively the number of times that treatment label i appears in the hth row block and the qth column block.

We define T_i, B_h, C_q, and G to be respectively the sum of the observations on the ith treatment, the hth row block, the qth column block, and the grand total of all the observations. We use these symbols, rather than $y_{..i}$, $y_{h..}$, $y_{.q.}$, and $y_{...}$, since only bc of the bcv possible y_{hqi} are observed. Thus,

$$T_i = \sum_{h=1}^{b}\sum_{q=1}^{c} n_{hqi} y_{hqi}, \quad B_h = \sum_{q=1}^{c}\sum_{i=1}^{v} n_{hqi} y_{hqi},$$

$$C_q = \sum_{h=1}^{b}\sum_{i=1}^{v} n_{hqi} y_{hqi}, \quad G = \sum_{h=1}^{b}\sum_{q=1}^{c}\sum_{i=1}^{v} n_{hqi} y_{hqi}.$$

To obtain least squares estimates for the treatment contrasts, we first differentiate the sum of squares of the errors with respect to each of the parameters in the model in turn and set the derivatives equal to zero. This gives the following $1 + b + c + v$ normal equations:

$$G - bc\hat{\mu} - c\sum_{h}\hat{\theta}_h - b\sum_{q}\hat{\phi}_q - r\sum_{i}\hat{\tau}_i = 0, \qquad (12.4.3)$$

$$B_h - c\hat{\mu} - c\hat{\theta}_h - \sum_{q}\hat{\phi}_q - \sum_{i} n_{h.i}\hat{\tau}_i = 0, \quad \text{for } h = 1, \ldots, b, \qquad (12.4.4)$$

$$C_q - b\hat{\mu} - \sum_{h}\hat{\theta}_h - b\hat{\phi}_q - \sum_{i} n_{.qi}\hat{\tau}_i = 0, \quad \text{for } q = 1, \ldots, c, \qquad (12.4.5)$$

$$T_p - r\hat{\mu} - \sum_{h} n_{h.p}\hat{\theta}_h - \sum_{q} n_{.qp}\hat{\phi}_q - r\hat{\tau}_p = 0, \quad \text{for } p = 1, \ldots, v. \qquad (12.4.6)$$

The sum of the v equations (12.4.6) gives equation (12.4.3), as do the sum of the b equations (12.4.4) and the sum of the c equations (12.4.5). Therefore, there are at most (and, in fact, exactly) $1 + b + c + v - 3$ distinct equations, so an extra three equations (distinct from the normal equations) need to be added to the set in order to obtain a solution. If we choose $\Sigma_h \hat{\theta}_h = 0$, $\Sigma_q \hat{\phi}_q = 0$, and $\Sigma_i \hat{\tau}_i = 0$, then equation (12.4.3) becomes

$$\hat{\mu} = \frac{1}{bc} G, \qquad (12.4.7)$$

and (12.4.4) and (12.4.5) become

$$\hat{\theta}_h = \frac{1}{c} B_h - \frac{1}{bc} G - \frac{1}{c}\sum_{i} n_{h.i}\hat{\tau}_i, \quad \text{for } h = 1, \ldots, b, \qquad (12.4.8)$$

$$\hat{\phi}_q = \frac{1}{b} C_q - \frac{1}{bc} G - \frac{1}{b}\sum_{i} n_{.qi}\hat{\tau}_i, \quad \text{for } q = 1, \ldots, c. \qquad (12.4.9)$$

If we now substitute these expressions for $\hat{\mu}$, $\hat{\theta}_h$, and $\hat{\phi}_q$ into (12.4.6), we can obtain an equation involving only the $\hat{\tau}_p$'s. After rearranging the terms in this equation and writing $\lambda_{pi} = \Sigma_h n_{h.i} n_{h.p}$ for the number of row blocks in which treatment labels i and p appear together, $\delta_{pi} = \Sigma_q n_{.qi} n_{.qp}$ for the number of column blocks in which treatment labels i and

p appear together, and $\Sigma_h n_{h.p} = \Sigma_q n_{.qp} = r$, we obtain

$$r\hat{\tau}_p - \frac{1}{c}\sum_i \lambda_{pi}\hat{\tau}_i - \frac{1}{b}\sum_i \delta_{pi}\hat{\tau}_i = Q_p, \text{ for } p = 1, \ldots, v, \quad (12.4.10)$$

where

$$Q_p = T_p - \frac{1}{c}\sum_h n_{h.p} B_h - \frac{1}{b}\sum_q n_{.qp} C_q + \frac{r}{bc} G.$$

12.4.2 Solution for Complete Column Blocks (Optional)

The three types of design that we looked at in Section 12.2 all have binary row blocks ($c < v$) and complete column blocks ($b = vs$). For these designs, every treatment appears s times in every column block, so $n_{.qp} = s$ for all q and p, and $\delta_{pi} = cs^2$ for all p and i, so

$$\frac{1}{b}\sum_q n_{.qp} C_q = \frac{s}{b}\sum_q C_q = \frac{r}{bc} G,$$

since $\sum C_q = G$ and $r = cs$. Also, $\Sigma_i \delta_{pi}\hat{\tau}_i = cs^2 \Sigma_i \hat{\tau}_i$, which is equal to zero because of the extra equation $\Sigma_i \hat{\tau}_i = 0$ that was added to the normal equations. We can use these facts to simplify (12.4.10), and we obtain

$$r\hat{\tau}_p - \frac{1}{c}\sum_i \lambda_{pi}\hat{\tau}_i = Q_p, \quad (12.4.11)$$

where Q_p reduces to

$$Q_p = T_p - \frac{1}{c}\sum_h n_{h.p} B_h. \quad (12.4.12)$$

Since $\lambda_{pp} = r$, we can write

$$r(c-1)\hat{\tau}_p - \sum_{i \neq s} \lambda_{pi}\hat{\tau}_i = cQ_p, \text{ for } p = 1, \ldots, v, \quad (12.4.13)$$

which is identical to equation (11.3.13) for incomplete block designs with block size $k = c$.

Solution for Latin square designs (optional) For an s-replicate Latin square design obtained by stacking s Latin squares one above the other, not only are the column blocks complete blocks, but so are the row blocks. Now, $b = r = vs$, $c = v$, $\lambda_{pi} = vs$ for all i and p, and $n_{h.p} = 1$ for all p and h. If we feed these values into (12.4.11) and (12.4.12) and use the extra equation $\Sigma_i \hat{\tau}_i = 0$, we obtain

$$\hat{\tau}_p = \frac{1}{vs} Q_p = \frac{1}{vs} T_p - \frac{1}{vs^2} G, \text{ for all } p = 1, \ldots, v.$$

The least squares estimate of a treatment contrast $\Sigma d_i \tau_i$ (with $\Sigma d_i = 0$) is then

$$\sum_i d_i \hat{\tau}_i = \frac{1}{vs} \sum_i d_i T_i, \quad (12.4.14)$$

where T_i is the sum of the observations on the ith treatment, and the corresponding variance is

$$\text{Var}\left(\sum d_i \hat{\tau}_i\right) = \frac{1}{vs}\sum d_i^2 \sigma^2 .$$

Solution for Youden designs (optional) For an s-replicate Youden design with $b = vs$ rows and $c < v$ columns, the row blocks form a balanced incomplete block design with row-block size c. Thus, $\lambda_{pi} = \lambda$ is constant for all $p \neq i$, and, as for any balanced incomplete block design with blocks of size $k = c$, we have $r(c-1) = \lambda(v-1)$. If we use these values, together with the extra equation $\Sigma_i \hat{\tau}_i = 0$, then (12.4.13) reduces to

$$\hat{\tau}_p = \frac{c}{\lambda v} Q_p, \quad \text{for all } p = 1, \ldots, v,$$

where Q_p is defined in (12.4.12). The least squares estimate of a treatment contrast $\Sigma d_i \tau_i$ is then

$$\sum_i d_i \hat{\tau}_i = \frac{c}{\lambda v} \sum_i d_i Q_i , \qquad (12.4.15)$$

which is the same as equation (11.4.15) for a balanced incomplete block design with blocks of size $k = c$. The corresponding variance can be shown to be

$$\text{Var}\left(\sum d_i \hat{\tau}_i\right) = \left(\frac{c}{\lambda v}\right) \sum d_i^2 \sigma^2 . \qquad (12.4.16)$$

12.4.3 Formula for *ssE* (Optional)

The minimum sum of squares for error, when a row–column–treatment model is used for a row–column design, is of the form

$$ssE = \sum_h \sum_q \sum_i n_{hqi} \hat{e}_{hqi}^2 = \sum_h \sum_q \sum_i n_{hqi}(y_{hqi} - \hat{\mu} - \hat{\theta}_h - \hat{\phi}_q - \hat{\tau}_i)^2 ,$$

where $\hat{\mu}, \hat{\theta}_h, \hat{\phi}_q, \hat{\tau}_i$ are a set of least squares solutions to the normal equations. If we follow the trick of Section 11.3.2 and multiply out one copy of the squared factor and use the normal equations (12.4.3)–(12.4.6) to set some of the terms equal to zero, we then obtain

$$ssE = \sum_h \sum_q \sum_i n_{hqi} y_{hqi}^2 - G\hat{\mu} - \sum_h B_h \hat{\theta}_h - \sum_q C_q \hat{\phi}_q - \sum_i T_i \hat{\tau}_i .$$

Then using (12.4.7)–(12.4.9), the right-hand side becomes

$$\sum_h \sum_q \sum_i n_{hqi} y_{hqi}^2 - \frac{1}{bc}G^2 - \left[\frac{1}{c}\sum_h B_h^2 - \frac{1}{bc}G^2 - \frac{1}{c}\sum_h \sum_i n_{h.i} B_h \hat{\tau}_i\right]$$

$$- \left[\frac{1}{b}\sum_q C_q^2 - \frac{1}{bc}G^2 - \frac{1}{b}\sum_q \sum_i n_{.qi} C_q \hat{\tau}_i\right] - \sum_i T_i \hat{\tau}_i . \qquad (12.4.17)$$

12.4 Analysis of Row–Column Designs (Optional)

Gathering together the terms involving $\hat{\tau}_i$, and noting that $\frac{r}{bc} G \sum_i \hat{\tau}_i = 0$ due to the extra equation $\sum_i \hat{\tau}_i = 0$ added to the normal equations, we have

$$ssE = \left(\sum_h \sum_q \sum_i n_{hqi} y_{hqi}^2 - \frac{1}{bc} G^2\right) - \left(\frac{1}{c} \sum_h B_h^2 - \frac{1}{bc} G^2\right)$$
$$- \left(\frac{1}{b} \sum_q C_q^2 - \frac{1}{bc} G^2\right) - \sum_i Q_i \hat{\tau}_i, \qquad (12.4.18)$$

where

$$Q_i = T_i - \frac{1}{c} \sum_h n_{h.i} B_h - \frac{1}{b} \sum_q n_{.qi} C_q + \frac{r}{bc} G$$

and where $\hat{\tau}_i$ is obtained as a solution of (12.4.10).

For designs with complete column blocks, we know that $n_{.qi} = s$, $r = cs$, and $\frac{1}{b} \sum_q n_{.qi} C_q = \frac{r}{bc} G$. So Q_i reduces to

$$Q_i = T_i - \frac{1}{c} \sum_h n_{h.i} B_h,$$

as in (12.4.12). We usually write (12.4.18) as

$$ssE = sstot - ss\theta - ss\phi - ssT_{\text{adj}},$$

where $ss\theta$ and $ss\phi$ are the unadjusted sums of squares for rows and columns.

It can be shown that SSE/σ^2 has a chi-squared distribution with degrees of freedom

$$df = (bc - 1) - (b - 1) - (c - 1) - (v - 1) = bc - b - c - v + 2,$$

and $E[MSE] = E[SSE/(bc - b - c - v + 2)] = \sigma^2$.

12.4.4 Analysis of Variance for a Row–Column Design (Optional)

Under the null hypothesis

$$H_0^T : \{\tau_i \text{ all equal to } \tau\},$$

the row–column–treatment model (12.3.1) becomes a two-way main-effects model in the effects of the two blocking factors; that is, the reduced model is

$$Y_{hqi} = (\mu + \tau) + \theta_h + \phi_q + \epsilon_{hqi}.$$

The sum of squares $ss\theta$ for row blocks, the sum of squares $ss\phi$ for column blocks, and the sum of squares ssE_0 for error will look similar (apart from the order of the subscripts) to the sums of squares for treatment factors A and B, and the sum of squares for error, respectively, for a two-way main-effects model that was studied in Chapter 6. Consequently, for the reduced

Table 12.5 Analysis of variance table for a connected row–column design with complete column blocks and no interactions

Source of Variation	Degrees of Freedom	Sum of Squares	Mean Square	Ratio
Rows (unadj)	$b-1$	$ss\theta$	—	—
Columns (unadj)	$c-1$	$ss\phi$	—	—
Treatments(adj)	$v-1$	ssT_{adj}	msT_{adj}	msT_{adj}/msE
Error	$bc-b-c-v+2$	ssE	msE	
Total	$bc-1$	$sstot$		

Formulae

$ssT_{adj} = \sum_{i=1}^{v} Q_i \hat{\tau}_i$ $\qquad Q_i = T_i - \frac{1}{c}\sum_{h=1}^{b} n_{h.i} B_h$

$sstot = \sum_{h=1}^{b}\sum_{q=1}^{c}\sum_{i=1}^{v} n_{hqi} y_{hqi}^2 - \frac{1}{bc}G^2$ $\qquad ss\theta = \frac{1}{c}\sum_{h=1}^{b} B_h^2 - \frac{1}{bc}G^2$

$ssE = sstot - ss\theta - ss\phi - ssT_{adj}$ $\qquad ss\phi = \frac{1}{b}\sum_{q=1}^{c} C_q^2 - \frac{1}{bc}G^2$

$T_i = \sum_{h=1}^{b}\sum_{q=1}^{c} n_{hqi} y_{hqi}$ $\qquad B_h = \sum_{q=1}^{c}\sum_{i=1}^{v} n_{hqi} y_{hqi}$

$C_q = \sum_{h=1}^{b}\sum_{i=1}^{v} n_{hqi} y_{hqi}$ $\qquad G = \sum_{h=1}^{b}\sum_{q=1}^{c}\sum_{i=1}^{v} n_{hqi} y_{hqi}$

model, we have

$$ss\theta = \frac{1}{c}\sum_{h=1}^{b} B_h^2 - \frac{1}{bc}G^2,$$

$$ss\phi = \frac{1}{b}\sum_{q=1}^{\phi} C_q^2 - \frac{1}{bc}G^2,$$

$$sstot = \sum_h\sum_q\sum_i n_{hqi} y_{hqi}^2 - \frac{1}{bc}G^2,$$

and

$$ssE_0 = sstot - ss\theta - ss\phi,$$

where B_h is the total of the observations in the hth row block, C_q is the total of the observations in the qth column block, and G is the grand total of all the observations.

The sum of squares for error for the row–column–treatment model (12.3.1) was calculated in optional Section 12.4.3 as

$$ssE = sstot - ss\theta - ss\phi - ssT_{adj},$$

where $ssT_{adj} = \sum_{i=1}^{v} Q_i \hat{\tau}_i$ with $Q_i = \sum_i T_i - \frac{1}{c}\sum_h n_{h.i} B_h - \frac{1}{b}\sum_q n_{.qi} C_q$, which reduces to $Q_i = \sum_i T_i - \frac{1}{c}\sum_h n_{h.i} B_h$ for designs with complete column blocks. The sum of squares for treatments adjusted for row-block and column-block effects is obtained as the difference in the error sum of squares between the reduced and the full models; that is,

$$ssE_0 - ssE = ssT_{adj}. \qquad (12.4.19)$$

It can be shown that if H_0^T is true, then SST_{adj}/σ^2 has a χ^2 distribution with $v-1$ degrees of freedom, and also, SST_{adj} is independent of SSE. Therefore, MST_{adj}/MSE has an F-distribution with $v-1$ and $bc-b-c-v+2$ degrees of freedom.

A test of the null hypothesis H_0^T against the general alternative hypothesis H_A^T :{at least two of the τ_i's differ} is given by the decision rule

$$\text{reject } H_0^T \text{ if } msT_{\text{adj}}/msE > F_{v-1,bc-b-c-v+2,\alpha} \tag{12.4.20}$$

for some chosen significance level α.

The test (12.4.20) is most conveniently set out in an analysis of variance table. The table for row–column designs with complete column blocks is shown in Table 12.5 for future reference. If column blocks are not complete, then Q_i needs to be modified accordingly.

When the number of row blocks is a multiple of v, and $c \geq 3$, there are sufficient degrees of freedom to be able to measure a column-treatment interaction, if one is thought to exist. The model would be modified as in (12.3.2), and the analysis of variance table would contain an extra row measuring this source of variation. The interaction must also be adjusted for row blocks, and the analysis is best done by computer.

12.4.5 Confidence Intervals and Multiple Comparisons

The multiple-comparison methods of Bonferroni and Scheffé can be applied for any row–column design. Simultaneous $100(1 - \alpha)\%$ confidence intervals for treatment contrasts $\Sigma_i d_i \hat{\tau}_i$ are of the form

$$\left(\sum_i d_i \hat{\tau}_i \pm w \sqrt{\widehat{\text{Var}}(\sum_i d_i \hat{\tau}_i)} \right), \tag{12.4.21}$$

where the critical coefficient w is

$$w_B = t_{bc-b-c-v+2,\alpha/2m} \text{ or } w_S = \sqrt{(v-1)F_{v-1,bc-b-c-v+2,\alpha}},$$

for the Bonferroni or Scheffé methods, respectively.

12.5 Analysis of Latin Square Designs

12.5.1 Analysis of Variance for Latin Square Designs

Table 12.5, page 400, shows the analysis of variance table for the row–column–treatment model (12.3.1) for any row–column design with complete column blocks. The notation T_i, B_h, C_q, and G refers to the sum of the observations on the ith level of the treatment factor, the hth level of the row-blocking factor, the qth level of the column-blocking factor, and the grand total of all the observations, respectively. The constant n_{hqi} is equal to 1 if treatment i is observed in the combination of row block h and column block q, and n_{hqi} is equal to zero otherwise. The term $\Sigma_h \Sigma_q \Sigma_i n_{hqi} y_{hqi}^2$ represents the sum of squares of all the observations. For an s-replicate Latin square design, we have $b = vs$ and $c = v$. The adjusted treatment sum of squares is $\Sigma Q_i \hat{\tau}_i$, where we showed in optional Section 12.4.2 that

$$\hat{\tau}_i = \frac{1}{vs} Q_i$$

and
$$Q_i = T_i - \frac{1}{c}\sum_h n_{h.i} B_h = T_i - \frac{1}{v}G.$$

Thus, for a Latin square design, the treatment sum of squares can be rewritten as
$$ssT_{adj} = \sum_{i=1}^{v} Q_i \hat{t}_i = \frac{1}{vs}\sum_i \left(T_i - \frac{1}{v}G\right)^2,$$

and we see that the "adjusted" treatment sum of squares for a Latin square design is actually *not* adjusted for row-block effects nor for column-block effects. This is because every treatment is observed the same number of times in every row block and in every column block even though only a few of the row–column–treatment combinations are actually observed. The computational formula is obtained by expanding the terms in parentheses,
$$ssT = \frac{1}{vs}\sum_i T_i^2 - \frac{1}{v^2 s}G^2.$$

The error sum of squares and degrees of freedom are obtained by subtraction, as shown in the analysis of variance table, Table 12.5. For an s-replicate Latin square design with $b = vs$ rows and $c = v$ columns, we have error degrees of freedom equal to
$$\begin{aligned} df &= bc - b - c - v + 2 \\ &= v^2 s - vs - 2v + 2 \\ &= (vs - 2)(v - 1). \end{aligned} \tag{12.5.22}$$

The test for equality of treatment effects compares the ratio of the treatment and error mean squares with the corresponding value from the F-distribution, in the usual way (see Example 12.5.1). We will not utilize a test for the hypothesis of negligible row-block or of negligible column-block effects. However, we conclude that the current experiment has benefitted from the use of the row (or column) blocking factor if the row (or column) mean square exceeds the mean square for error.

Example 12.5.1 Dairy cow experiment

Cochran and Cox (1957, page 135) described an experiment that studied the effects of three diets on the milk production of dairy cows. The $v = 3$ diets (levels of a treatment factor) consisted of roughage (level 1), limited grain (level 2), and full grain (level 3). A crossover experiment was used, with each of the $b = 6$ cows being fed each diet in turn, each for a six-week period. The response variable was the milk yield for each cow during each of the $c = 3$ periods. Data were collected using a 2-replicate Latin square with $v = 3$ treatment labels, $b = vs = 6$ rows, $c = 3$ columns, and with $v = 3$ treatments each observed $r = vs = 6$ times. The data are reproduced in Table 12.6.

Provided that each measurement on milk yield has been taken after a cow has been on a new diet for a sufficiently long period of time to allow the effect of the previous diet to "wash out" of the system, we need not be concerned with carryover effects. For this experiment, a model without carryover effects describes the data fairly well.

12.5 Analysis of Latin Square Designs

Table 12.6 Unrandomized design (and data in parentheses) for the dairy cow experiment

	Period		
Cow	1	2	3
1	1 (38)	2 (25)	3 (15)
2	2 (109)	3 (86)	1 (39)
3	3 (124)	1 (72)	2 (27)
4	1 (86)	3 (76)	2 (46)
5	2 (75)	1 (35)	3 (34)
6	3 (101)	2 (63)	1 (1)

Source: *Experimental Designs*, Second Edition, by W. G. Cochran and G. M. Cox, Copyright © 1957, John Wiley & Sons, New York. Adapted by permission of John Wiley & Sons, Inc.

Table 12.7 Analysis of variance table for the dairy cow experiment

Source of Variation	Degrees of Freedom	Sum of Squares	Mean Square	Ratio	p-value
Cow (row)	5	5781.11	1156.22	–	–
Period (column)	2	11480.11	5740.06	–	–
Diets (treatment)	2	2276.78	1138.39	11.05	0.0050
Error	8	824.44	103.06		
Total	17	20362.44			

Table 12.7 shows the analysis of variance table. To test the null hypothesis H_0^T that all three diets have the same effect, the decision rule is

reject H_0^T if $msT/msE > F_{2,8,.01} = 8.65$,

with a probability of $\alpha = 0.01$ of making a Type I error. Since $msT/msE = 1138.39/103.06 = 11.05 > 8.65$, we reject the null hypothesis at the $\alpha = 0.01$ significance level and conclude that the diets do not all have the same effect.

To evaluate whether or not blocking was worthwhile, observe that the mean squares for periods and cows are both considerably larger than the error mean square. Thus, inclusion of both blocking factors in the experiment was beneficial in reducing the experimental error. □

12.5.2 Confidence Intervals for Latin Square Designs

All treatment contrasts are estimable in Latin square designs. Using the row–column–treatment model (12.3.1), the least squares estimate of a treatment contrast $\Sigma d_i \tau_i$ was shown in the optional Section 12.4.2 to be

$$\sum d_i \hat{\tau}_i = \frac{1}{vs} \sum d_i T_i \,, \qquad (12.5.23)$$

where T_i is the total of all the observations on the ith treatment, and the corresponding variance is

$$\text{Var}\left(\sum d_i \hat{\tau}_i\right) = \sum d_i^2 \left(\frac{\sigma^2}{vs}\right). \qquad (12.5.24)$$

The multiple-comparison methods of Bonferroni, Scheffé, Tukey, Dunnett, and Hsu can be used for Latin square designs. Formulae for confidence intervals for treatment contrasts $\Sigma d_i \tau_i$ are of the form

$$\left(\frac{1}{vs}\Sigma_i d_i T_i \pm w\sqrt{msE\left(\frac{\Sigma d_i^2}{vs}\right)}\right), \qquad (12.5.25)$$

where the appropriate critical coefficients w for the five methods are,

$$w_B = t_{(vs-2)(v-1),\alpha/2m} \; ; \; w_S = \sqrt{(v-1)F_{v-1,(vs-2)(v-1),\alpha}} \; ;$$

$$w_T = q_{v,(vs-2)(v-1),\alpha}/\sqrt{2} \; ;$$

$$w_H = w_{D1} = t_{v-1,(vs-2)(v-1),\alpha}^{(0.5)} \; ; \; w_{D2} = |t|_{v-1,(vs-2)(v-1),\alpha}^{(0.5)}.$$

Example 12.5.2 Dairy cow experiment, continued

The dairy cow experiment was described in Example 12.5.1, page 402. Tukey's method of all pairwise comparisons can be applied to determine which diets, if any, are significantly different from any of the others. The treatment sample means, which can be computed from the data in Table 12.6, are

$$\frac{1}{vs}T_1 = 45.167, \quad \frac{1}{vs}T_2 = 57.500, \quad \frac{1}{vs}T_3 = 72.667,$$

so that the least squares estimates of the pairwise differences are

$$\hat{\tau}_2 - \hat{\tau}_1 = 57.500 - 45.167 = 12.333,$$
$$\hat{\tau}_3 - \hat{\tau}_1 = 72.667 - 45.167 = 27.500,$$
$$\hat{\tau}_3 - \hat{\tau}_2 = 72.667 - 57.500 = 15.167.$$

The value of the error mean square is $msE = 103.06$ from the analysis of variance table, Table 12.7. The error degrees of freedom are given by (12.5.22) as

$$df = (vs - 2)(v - 1) = (2 \times 3 - 1)(3 - 1) = 8.$$

Using a simultaneous confidence level of 95%, we obtain the critical coefficient for Tukey's method as $w_T = q_{3,8,.05}/\sqrt{2} = 4.04/\sqrt{2}$. Hence, using (12.5.25), the minimum significant difference for pairwise differences is

$$msd = (4.04/\sqrt{2})\sqrt{msE(2/6)} = 16.743.$$

The simultaneous 95% confidence intervals are therefore

$$\tau_3 - \tau_1 \in (12.333 \pm 16.743) = (10.58, 44.24),$$
$$\tau_2 - \tau_1 \in (27.500 \pm 16.743) = (-4.41, 29.08),$$

12.5 Analysis of Latin Square Designs 405

$$\tau_2 - \tau_3 \in (15.167 \pm 16.743) = (-31.91, 1.58).$$

From these intervals, we can deduce that at overall significance level $\alpha = 0.05$, the full grain diet (level 3) results in a mean yield of 10.58–44.24 units higher than the roughage diet (level 1), but the limited grain diet (level 2) is not significantly different from either of the other two. □

12.5.3 How Many Observations?

The formula for determining the sample size needed to achieve a power $\pi(\Delta)$ of detecting a difference Δ in the treatment effects for given v, α, and σ^2, using a Latin square design, is the same as that for a randomized complete block design but with b replaced by v (since $r = vs$ rather than $r = bs$). Thus, we need to find s to satisfy

$$s \geq \frac{2\sigma^2\phi^2}{\Delta^2}. \tag{12.5.26}$$

Alternatively, the confidence interval formula (12.5.25) can be used to calculate the sample sizes needed for achieving confidence intervals of a desired width (see Example 12.5.3).

Example 12.5.3 Sample-size calculation for a Latin square design

Consider an experiment to compare $v = 3$ computer keyboard designs with respect to the time taken to type an article of given length. Typists and time periods are the two blocking factors, with each of $b = 3s$ typists using each of the keyboards in a sequence of $c = 3$ time periods. An s-replicate Latin square design will be used, with $r = 3s$ observations per keyboard layout (treatment).

With 3 treatments, there are 3 pairwise comparisons. Using Tukey's method of multiple comparisons and a simultaneous confidence level of 95%, suppose that the experimenters want confidence intervals with half-width (minimum significant difference) of 10 minutes or less. The error standard deviation is expected to be at most 15 minutes (a variance of at most 225 minutes²). The error degrees of freedom, as given in (12.5.22), are

$$df = (vs - 2)(v - 1) = 2(3s - 2) = 6s - 4.$$

Then, using the confidence interval formula (12.5.25) for Tukey's method of pairwise comparisons, the minimum significant difference is

$$msd \approx \frac{q_{3,6s-4,.05}}{\sqrt{2}}\sqrt{\frac{225 \times 2}{vs}} = q_{3,6s-4,.05}\sqrt{\frac{225}{3s}}.$$

To obtain $msd \leq 10$, we require $0.75 q_{3,6s-4,.05}^2 \leq s$. Sample size is then computed by trial and error as follows:

s	$6s-4$	$q_{3,6s-4,.05}$	$0.75 q_{3,6s-4,.05}^2$	Required
	∞	3.31	8.22	$s \geq 9$
9	50	3.42	8.77	$s = 9$

So, $s = 9$, and a 9-replicate Latin square would be needed, giving $msd \approx 3.42\sqrt{\frac{225}{27}} = 9.87$ minutes and requiring $b = vs = 27$ typists and $r = vs = 27$ observations per keyboard layout. □

12.6 Analysis of Youden Designs

12.6.1 Analysis of Variance for Youden Designs

The s-replicate Youden design with v treatments has complete column blocks of size $b = vs$ and row blocks forming a balanced incomplete block design with blocks of size c. Each treatment is observed $r = cs$ times. For the row–column–treatment model (12.3.1) with no interactions, a solution to the normal equations was derived in optional Section 12.4.2 as

$$\hat{\tau}_i = \frac{c}{\lambda v} Q_i, \quad \text{where} \quad Q_i = T_i - \frac{1}{c}\sum_h n_{h.i} B_h, \tag{12.6.27}$$

for $i = 1, \ldots, v$, where $\lambda = r(c-1)/(v-1)$, analogous to the solution for a balanced incomplete block design.

As in the analysis of a balanced incomplete block design, the estimators of the treatment effects in a Youden design are not independent of the estimators of the row-block effects. As a result, the treatment-effect estimators must be adjusted for row blocks. However, since the column blocks form a randomized complete block design, no adjustment is needed for these. Table 12.5, page 400, shows the analysis of variance table for a row–column design with complete column blocks and no interactions. The adjusted treatment sum of squares is given in the table as

$$ssT_{\text{adj}} = \sum_{i=1}^{v} Q_i \hat{\tau}_i = \frac{c}{\lambda v}\sum_{i=1}^{v} Q_i^2, \quad \text{where} \quad Q_i = T_i - \frac{1}{c}\sum_{h=1}^{b} n_{h.i} B_h,$$

exactly as for a balanced incomplete block design with blocks of size $k = c$ (see Table 11.7, page 351).

From Table 12.5, the number of error degrees of freedom for an s-replicate Youden square is

$$\begin{aligned} df &= bc - b - c - v + 2 \\ &= vsc - vs - c - v + 2 \\ &= (vs - 1)(c - 1) - (v - 1). \end{aligned} \tag{12.6.28}$$

To test the null hypothesis H_0^T of no treatment differences, the decision rule is

$$\text{reject } H_0^T \quad \text{if} \quad \frac{msT_{\text{adj}}}{msE} > F_{v-1, df, \alpha}.$$

12.6.2 Confidence Intervals for Youden Designs

All treatment contrasts $\Sigma d_i \tau_i$ are estimable in all Youden designs. The least squares estimate of the contrast $\Sigma d_i \tau_i$ is

$$\sum d_i \hat{\tau}_i = \frac{c}{\lambda v} \sum_i d_i Q_i = \left(\frac{v-1}{vs(c-1)}\right) \sum_i d_i Q_i ,$$

since $\lambda = r(c-1)/(v-1)$. The variance of the corresponding estimator is

$$\text{Var}\left(\sum d_i \hat{\tau}_i\right) = \left(\frac{c}{\lambda v}\right) \sum d_i^2 \sigma^2 = \left(\frac{v-1}{vs(c-1)}\right) \sum d_i^2 \sigma^2 . \quad (12.6.29)$$

Confidence interval formulae for treatment contrasts $\Sigma d_i \tau_i$ are the same as those for a balanced incomplete block design with block size $k = c$; that is,

$$\left(\frac{v-1}{vs(c-1)}\right) \sum_i d_i Q_i \pm w \sqrt{msE \left(\frac{v-1}{vs(c-1)}\right) \sum d_i^2} , \quad (12.6.30)$$

where w, as usual, is the critical coefficient for the Bonferroni, Scheffé, Tukey, Dunnett, or Hsu method, given by

$$w_B = t_{(vs-1)(c-1)-(v-1),\alpha/2m} \; ; \quad w_S = \sqrt{(v-1)F_{v-1,(vs-1)(c-1)-(v-1),\alpha}} \; ;$$

$$w_T = q_{v,(vs-1)(c-1)-(v-1),\alpha}/\sqrt{2} \; ;$$

$$w_H = w_{D1} = t^{(0.5)}_{v-1,(vs-1)(c-1)-(v-1),\alpha} \; ; \quad w_{D2} = |t|^{(0.5)}_{v-1,(vs-1)(c-1)-(v-1),\alpha} .$$

12.6.3 How Many Observations?

The methods of sample-size calculation for an s-replicate Youden square are analogous to those for computing samples sizes for a balanced incomplete block design. To calculate the number of observations $r = cs$ per treatment required to achieve a power $\pi(\Delta)$ of detecting a difference Δ in treatment effects for given v, α, and σ^2, the power tables in Appendix A.7 can be used in the same way as for a balanced incomplete block design (Section 11.8), with $b = c$ and $r = cs$. Thus, we need to find s satisfying

$$cs \geq \frac{2v\sigma^2 \phi^2}{\Delta^2} \left[\frac{c(v-1)}{v(c-1)}\right] ,$$

which reduces to

$$s \geq \frac{2\sigma^2 \phi^2 (v-1)}{\Delta^2 (c-1)} .$$

Alternatively, the confidence interval formula (12.6.30) can be used to calculate the sample sizes needed for achieving confidence intervals of a desired width (see Example 12.6.3).

Example 12.6.1

Suppose an experiment is run to compare six paint additive formulations (levels 2–7 of the treatment factor) with a standard "control" formulation (level 1) with respect to the drying time (in minutes). The paint is sprayed through $c = 4$ different nozzles so that $c = 4$ paints can be sprayed simultaneously. A total of $b = 7s$ panels will each be painted with strips of 4 of the 7 paints. The error variability is expected to be at most 25 minutes² (standard deviation at most 5 minutes), and simultaneous 90% confidence intervals with half-width of at most 3.5 minutes are required for the six treatment-versus-control contrasts.

A Youden square with 7 rows, 4 columns, and 7 treatments is shown in Table 12.1. Suppose that s copies of this basic square are to be used, giving an s-replicate Youden design with the same number of observations on the experimental and control treatments. The number of error degrees of freedom, given in (12.6.28), is then

$$df = (vs - 1)(c - 1) - (v - 1) = (7s - 1)(4 - 1) - (7 - 1) = 21s - 9.$$

Using Dunnett's method of multiple comparisons for treatment versus control, the minimum significant difference is

$$msd = w_{D2}\sqrt{msE \times \left(\frac{(v-1)}{vs(c-1)}\right) \times 2},$$

so we require

$$msd \approx w_{D2}\sqrt{\frac{(25)(6)(2)}{(7)s(3)}} \leq 3.5,$$

that is,

$$w_{D2}^2 \leq 0.8575s,$$

where $w_{D2} = |t|_{6,21s-9,.1}^{(.5)}$. Using Table A.10 for the values of $w_{D2} = |t|_{6,21s-9,.1}^{(.5)}$, we have

s	$21s-9$	$w_{D2} = |t|_{6,21s-9,.1}^{(.5)}$	w_{D2}^2	$0.8575s$	Action
20	411	2.30	5.29	17.15	Decrease s
6	117	2.32	5.38	5.15	Increase s
7	138	2.32	5.38	6.00	Decrease s

So we see that $s = 7$ Youden squares would be adequate, requiring $b = vs = 49$ panels and giving $r = sc = 28$ observations on each of the $v = 7$ paint formulations, both experimental and control. If 49 panels are not available for the experiment, then the experimenter would have to be satisfied with wider confidence intervals. □

12.7 Analysis of Cyclic and Other Row–Column Designs

In general, it is not easy to obtain least squares solutions for the treatment parameters $\hat{\tau}_i$ in row–column designs other than Latin square and Youden designs. We will illustrate the analysis of a more complicated row–column design via the SAS computer package in Section 12.10.

12.8 Checking the Assumptions on the Model

The error assumptions on the row–column–treatment model (12.3.1) and on the model (12.3.2), which includes interaction, can be checked by plotting the standardized residuals against the run order, the predicted values $\hat{y}_{ijk}$, the levels of the row blocking factor, the levels of the column blocking factor, the levels of the treatment factor, and the normal scores.

The data collected from a row–column design can be examined by plotting the adjusted observations against the treatment labels. The observations are adjusted by subtracting the treatment-adjusted row-block and column-block effect estimators:

$$y^*_{hqi} = y_{hqi} - (\hat{\theta}_h - \hat{\bar{\theta}}_.) - (\hat{\phi}_q - \hat{\bar{\phi}}_.). \tag{12.8.31}$$

For the s-replicate Latin square design with $b = vs$ row blocks and $c = v$ column blocks, and the row–column–treatment model (12.3.1), the row-block and column-block effect estimators are independent of treatment effects, and (12.8.31) becomes

$$y^*_{hqi} = y_{hqi} - \left(\frac{1}{vs}B_h - \frac{1}{v^2s}G\right) - \left(\frac{1}{v}C_q - \frac{1}{v^2s}G\right).$$

For other designs, the adjusted observations can be calculated by computer, as shown in Section 12.10. Since the variability due to the blocking factors has been extracted from the adjusted observations, the data plots will exhibit less variability than really exists.

Example 12.8.1 Dairy cow experiment

For the dairy cow experiment, which was run as a Latin square design, the adjusted observations are plotted against treatment labels in Figure 12.1. The plot shows how milk yield tends to increase as the quality of the diet improves from roughage (level 1) to limited grain (level 2) to full grain (level 3). In Example 12.5.1, simultaneous confidence intervals (with an overall 95% confidence level) showed a significant difference in diets 1 and 3, but were unable to distinguish between diets 1 and 2 and between diets 2 and 3. □

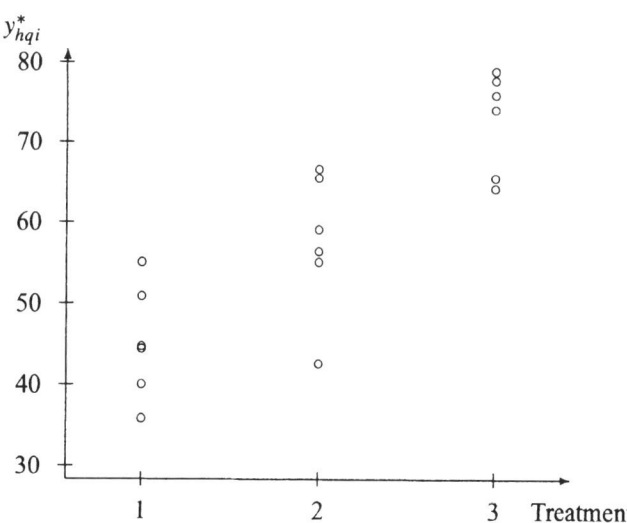

Figure 12.1 Adjusted data for the dairy cow experiment

The row–column–treatment model (12.3.1) assumes that the two blocking factors do not interact with each other or with the treatment factor. If interactions are present, the error variance estimate will be inflated, decreasing the powers of hypothesis tests and widening confidence intervals for treatment contrasts. When the column blocks are complete blocks, the column–treatment interaction can be checked by plotting the row-adjusted observations (see Chapter 10) against treatments, with plot labels being the column-block labels. Interaction is indicated by nonparallel lines. The row–column interaction can be investigated in the same way using the treatment-adjusted observations and plotting against row labels. The row–treatment interactions can only be investigated if the row blocks are also complete blocks as in Latin square designs.

12.9 Factorial Experiments in Row–Column Designs

The exercise bike experiment of Example 12.2.4, page 393, was a factorial experiment with three treatment factors having two levels each. These were "time duration of exercise" (1 and 3 minutes, coded as 1 and 2), "exercise speed" (40 and 60 rpm, coded as 1 and 2), and "pedal type" (foot pedal and hand bars, coded as 1 and 2). The data were shown in Table 12.4.

If we use the row–column–treatment model with three-digit labels for the treatment combinations, then we can write the model as

$$Y_{hqijk} = \mu + \theta_h + \phi_q + \tau_{ijk} + \epsilon_{hqijk},$$

where θ_h is effect of day h, ϕ_q is the effect of subject q, and τ_{ijk} is the effect of treatment combination ijk. If we now rewrite τ_{ijk} in terms of its constituent main effects and interactions, we obtain the following form of the row–column–treatment model:

$$\begin{aligned} Y_{hqijk} &= \mu + \theta_h + \phi_q + \alpha_i + \beta_j + \gamma_k \\ &\quad + (\alpha\beta)_{ij} + (\alpha\gamma)_{jk} + (\beta\gamma)_{jk} + (\alpha\beta\gamma)_{ijk} + \epsilon_{hqijk}, \\ h &= 1, \ldots, 8; \ q = 1, 2, 3; \ i = 1, 2; \ j = 1, 2; \ k = 1, 2; \end{aligned}$$

where α_i is the effect of the ith duration, β_j is the effect of the jth speed, and γ_k is the effect of the kth pedal type, and the other terms represent interactions between the treatment factors. The analysis of this experiment in terms of both forms of the model by the computer program SAS is shown in Section 12.10. If some of the treatment interactions are thought to be negligible, they can be dropped from the latter model. We note that there are 7 degrees of freedom for error, so it would be possible to measure, say, the day×duration interaction.

12.10 Using SAS Software

In this section, a sample program is given to illustrate the analysis of row–column designs using the SAS software. The program uses the data of the exercise bicycle experiment, which was described in Example 12.2.4, page 393. The design is a cyclic row–column design with $v = 8$ treatment labels representing the eight treatment combinations shown in Table 12.4,

12.10 Using SAS Software

Table 12.8 SAS program for analysis of a row–column design—Exercise bicycle experiment

```
DATA BIKE;
  INPUT DAY SUBJECT PULSE DURAT$ SPEED$ PEDAL$;
  TRTMT = trim(DURAT)||trim(SPEED)||trim(PEDAL);
  LINES;
    1 1  45  1 1 2
    1 2  25  2 1 2
    1 3  18  2 2 1
    2 1  27  2 2 2
    : :   :  : : :
    8 3  34  1 1 1
  ;
PROC PRINT;
*    row-column-treatment model;
PROC GLM;
  CLASSES DAY SUBJECT TRTMT;
  MODEL PULSE = DAY SUBJECT TRTMT / SOLUTION;
  OUTPUT OUT=RESIDS PREDICTED=PRED RESIDUAL=Z;
  ESTIMATE 'DURATION DIFF'  TRTMT -1 -1 -1 -1  1  1  1  1 / DIVISOR=4;
  ESTIMATE 'SPEED DIFF'     TRTMT -1 -1  1  1 -1 -1  1  1 / DIVISOR=4;
  ESTIMATE 'PEDAL DIFF'     TRTMT -1  1 -1  1 -1  1 -1  1 / DIVISOR=4;
* Standardize residuals and compute normal scores;
PROC STANDARD STD=1.0; VAR Z;
PROC RANK NORMAL=BLOM; VAR Z; RANKS NSCORE;
* Generate residual plots;
PROC PLOT;
  PLOT Z*PRED Z*TRTMT Z*DAY Z*SUBJECT / VREF=0 VPOS=19 HPOS=50;
  PLOT Z*NSCORE / VREF=0 HREF=0 VPOS=19 HPOS=50;
```

and with $b = 8$ row blocks representing days and $c = 3$ column blocks representing subject. The treatment combinations were combinations of the levels of the three treatment factors "time duration of exercise," "exercise speed," and "pedal type."

A SAS program for analyzing this experiment is shown in Table 12.8. The data are entered into a data set called BIKE, using DAY and SUBJECT as the two blocking factors, and using DURAT, SPEED, and PEDAL as the three treatment factors, and PULSE for the response variable "pulse rate." The combinations of the levels of the three treatment factors are recoded as levels of a factor TRTMT.

The MODEL statement in the first GLM procedure causes generation of the analysis of variance table shown in Table 12.9. The blocking factors are entered into the model before the treatment factor, so the treatment sum of squares is adjusted for block effects whether one looks at the Type I or Type III sums of squares. The row and column effects are independent of each other, since there is one observation at each combination of levels of the row and column blocking factors. Consequently, the Type I sums of squares reproduce the analysis of variance table, Table 12.5, with unadjusted block effects and adjusted treatment effects. In this particular experiment, the column blocks (subjects) are complete blocks, and so the

Table 12.9 Analysis of variance for a row–column design—Exercise bicycle experiment

```
                          The SAS System
                    General Linear Models Procedure

Dependent Variable: PULSE
                             Sum of         Mean
Source              DF      Squares       Square    F Value    Pr > F
Model               16    1257.6786      78.6049       3.28    0.0588
Error                7     167.9464      23.9923
Corrected Total     23    1425.6250

Source              DF    Type I SS   Mean Square   F Value   Pr > F
DAY                  7    202.29167      28.89881      1.20   0.4062
SUBJECT              2    201.00000     100.50000      4.19   0.0636
TRTMT                7    854.38690     122.05527      5.09   0.0238

Source              DF   Type III SS  Mean Square   F Value   Pr > F
DAY                  7     23.72024       3.38861      0.14   0.9903
SUBJECT              2    201.00000     100.50000      4.19   0.0636
TRTMT                7    854.38690     122.05527      5.09   0.0238

                                        T for H0:   Pr > |T|   Std Error of
Parameter              Estimate       Parameter=0               Estimate
INTERCEPT           19.86607143 B          3.55      0.0093     5.59247154
DAY         1        3.99107143 B          0.73      0.4868     5.43709471
            2        1.56250000 B          0.29      0.7821     5.43709471
            3        2.57142857 B          0.46      0.6574     5.55403486
            4        0.80357143 B          0.16      0.8754     4.94173873
            5        2.51785714 B          0.51      0.6261     4.94173873
            6        1.63392857 B          0.37      0.7208     4.39085009
            7        0.20535714 B          0.05      0.9640     4.39085009
            8        0.00000000 B           .          .           .
SUBJECT     1        5.25000000 B          2.14      0.0693     2.44909917
            2       -1.50000000 B         -0.61      0.5596     2.44909917
            3        0.00000000 B           .          .           .
TRTMT     111       12.64285714 B          2.56      0.0376     4.94173873
            :           :                  :          :           :
          221       -7.25892857 B         -1.34      0.2236     5.43709471
          222        0.00000000 B           .          .           .

NOTE: The X'X matrix has been found to be singular and a generalized
      inverse was used to solve the normal equations.  Estimates
      followed by the letter 'B' are biased, and are not unique
      estimators of the parameters.
```

12.10 Using SAS Software

Table 12.10 Output from the ESTIMATE statement—Exercise bicycle experiment

```
                        The SAS System
    Dependent Variable: PULSE
                                T for H0:    Pr > |T|   Std Error of
    Parameter        Estimate   Parameter=0             Estimate
    DURATION DIFF   -5.3437500     -1.97      0.0901    2.71854736
    SPEED DIFF     -10.3214286     -4.18      0.0042    2.47086937
    PEDAL DIFF       4.3080357      1.58      0.1571    2.71854736
```

Table 12.11 Analysis of variance for a row–column design—Exercise bicycle experiment

```
                          The SAS System

    Source            DF   Type III SS   Mean Square   F Value   Pr > F
    DAY                7      23.72024      3.38861      0.14    0.9903
    SUBJECT            2     201.00000    100.50000      4.19    0.0636
    DURAT              1      92.70245     92.70245      3.86    0.0901
    SPEED              1     418.65163    418.65163     17.45    0.0042
    PEDAL              1      60.25006     60.25006      2.51    0.1571
    DURAT*SPEED        1       7.28581      7.28581      0.30    0.5987
    DURAT*PEDAL        1      21.66949     21.66949      0.90    0.3736
    SPEED*PEDAL        1      11.57667     11.57667      0.48    0.5097
    DURAT*SPEED*PEDAL  1       7.71429      7.71429      0.32    0.5884
```

treatment sum of squares is actually only adjusted for row-block (day) effects. The Type III sums of squares show the row-block (day) effects adjusted for the treatment effects.

The ESTIMATE statements request SAS to calculate the information needed to calculate three confidence intervals. The three selected contrasts are the main-effect contrasts comparing the effect on pulse rate of the two levels of each treatment factor (averaging over any interaction that might be present). The output, shown in Table 12.10, gives the least squares estimates of the contrasts and their associated standard errors. From this information, confidence intervals can be calculated in the usual way. The contrast estimates are adjusted for the incomplete row blocks.

12.10.1 Factorial Model

To write the row–column–treatment model in terms of factorial treatment combinations, the model statement in Table 12.8 is replaced by

```
MODEL PULSE  =   DAY SUBJECT DURAT SPEED PEDAL DURAT*SPEED
                 DURAT*PEDAL SPEED*PEDAL DURAT*SPEED*PEDAL;
```

The Type III sums of squares are shown in Table 12.11, and we can see that the sums of squares for the main effects and interactions of the three treatment factors do not add to the

Table 12.12 SAS code for calculating and plotting the adjusted observations—Exercise bicycle experiment

```
* Add the following code for the second run of the program;
DATA BIKE3; * input subject effect estimates from first run;
  INPUT SUBJECT SHAT @@;
  LINES;
    1  5.25    2 -1.50    3  0.00
PROC MEANS MEAN; * print average of the subject effect estimates;
  VAR SHAT;
DATA BIKE4; * input day effect estimates from first run;
  INPUT DAY DHAT @@;
  LINES;
    1 3.9911   2 1.5625   3 2.5714   4 0.8036
    5 2.5179   6 1.6334   7 0.2054   8 0.0000
PROC MEANS MEAN; * print average of the day effect estimates;
  VAR DHAT;
;
* Add the following code for the third run;
* Adjust data for subject and day effects, then plot adjusted data;
DATA BIKE5; SET BIKE;
  IF SUBJECT=1 THEN YADJ=PULSE-(5.25-1.25);
    ELSE IF SUBJECT=2 THEN YADJ=PULSE-(-1.50-1.25);
      ELSE IF SUBJECT=3 THEN YADJ=PULSE-(0.00-1.25);
  IF DAY=1 THEN YADJ=YADJ-(3.9911-1.660);
    ELSE IF DAY=2 THEN YADJ=YADJ-(1.5625-1.6607);
      ELSE IF DAY=3 THEN YADJ=YADJ-(2.5714-1.6607);
        ELSE IF DAY=4 THEN YADJ=YADJ-(0.8036-1.6607);
          ELSE IF DAY=5 THEN YADJ=YADJ-(2.5179-1.6607);
            ELSE IF DAY=6 THEN YADJ=YADJ-(1.6334-1.6607);
              ELSE IF DAY=7 THEN YADJ=YADJ-(0.2054-1.6607);
                ELSE IF DAY=8 THEN YADJ=YADJ-(0.0000-1.6607);
PROC PLOT;
  PLOT YADJ*TRTMT / VPOS=19 HPOS=50;
```

Type III treatment sum of squares in Table 12.9 due to the individual adjustments for block effects and for the other treatment factor effects.

The ESTIMATE statements for the factorial model become

```
ESTIMATE 'DURATION DIFF'  DURAT -1 1;
ESTIMATE 'SPEED DIFF'     SPEED -1 1;
ESTIMATE 'FOOT/HAND DIFF' PEDAL -1 1;
```

and give output identical to that of Table 12.10.

12.10.2 Plots

The statements needed for calculating the standardized residuals and normal scores and for generating the residual plots are also shown in Table 12.8. These are as discussed in Section 5.8. PROC PLOT is used to generate the usual residual plots (not shown here).

The SOLUTION option in the MODEL statement causes a set of least squares solutions to the normal equations to be printed. These are subsequently used to calculate the adjusted data values needed for examining the data.

For obtaining the adjusted observations, the statements in Table 12.12 would be added to the program in Table 12.8 for a second and third run of the program. The data are adjusted using (12.8.31); that is,

$$y^*_{hqi} = y_{hqi} - (\hat{\theta}_h - \bar{\hat{\theta}}_.) - (\hat{\phi}_q - \bar{\hat{\phi}}_.),$$

where $\hat{\theta}_h$ and $\hat{\phi}_q$ are obtained from the output of the SOLUTION option shown in Table 12.9. The second run of the program takes as input the values of $\hat{\theta}_h$ and $\hat{\phi}_q$ from the first run of the program and calculates $\bar{\hat{\theta}}_. = \Sigma \hat{\theta}_h/3 = 1.25$ and $\bar{\hat{\phi}}_. = \Sigma \hat{\phi}_q/8 = 1.66$, which are then copied by hand into the statements for the third run of the program. The symbols @@ allow the input to be entered with more than one record per line.

In the third run of the program, the data set BIKE5 is created as a copy of the data set BIKE. This is done to create the new variable YADJ, which contains the values of PULSE, adjusted for the row and column effects. The PLOT procedure is used to plot the adjusted observations by treatment. The SAS plot is analogous to that in Figure 12.1 (page 409) and is not shown here.

Exercises

1. **Randomization**

    (a) Randomize the design in Table 12.1 (page 389) so that it can be used for an experiment with seven subjects, each being assigned a sequence of four out of a possible seven antihistamines over four time periods.

    (b) Discuss whether or not one could avoid a carryover effect from becoming a problem in this type of experiment.

2. **Latin squares**

    (a) Show that there is only one standard 3×3 Latin square. (Hint: Given the letters in the first row and the first column, show that there is only one way to complete the Latin square.)

    (b) Show that there are exactly four standard 4×4 Latin squares.

3. **Sample sizes**

    Consider an experiment to compare 4 degrees of twist in a cotton-spinning experiment with respect to the number of breaks per 100 pounds. A replicated Latin square design is to be used, with time periods and machines being the row and column blocking factors.

(a) Determine the number s of Latin squares and the number r of observations per degree of twist to include in the experiment if each interval in a simultaneous set of 95% confidence intervals for all pairwise comparisons is to have a minimum significant difference (half-width) of 5 breaks per 100 pounds. The error standard deviation is thought to be at most 6 breaks per 100 pounds. Investigate both the Tukey and the Bonferroni methods.

(b) Discuss how the resulting design would be randomized.

4. **Youden design randomization**

    (a) Find a Youden square (plan of treatment labels in rows and columns) for 5 treatments in 5 rows and 4 columns.

    (b) Randomize the design found in part (a), assigning the rows to 5 different drying temperatures, the columns to 4 different paint nozzles, and the treatment labels to 5 different paint formulations.

5. **Row–column design randomization**

    Consider an experiment to compare 5 protocols with respect to a resting metabolism rate measurement. A row–column design is to be used, blocking on subjects and time periods. Since subjects prefer to stay in a study for a short length of time, only 3 time periods will be used, with each subject assigned a different protocol in each of the 3 time periods. For 10 subjects, the following experimental plan with 10 rows and 3 columns could be used:

	Column				Column		
Row	I	II	III	Row	I	II	III
I	1	2	3	VI	1	2	4
II	2	3	4	VII	2	3	5
III	3	4	5	VIII	3	4	1
IV	4	5	1	IX	4	5	2
V	5	1	2	X	5	1	3

    (a) Does this experimental plan have treatments evenly distributed across rows and columns? Explain what you mean by "evenly distributed."

    (b) Determine the number of replicates s of this experimental plan, and the corresponding number of observations r per protocol to include in the experiment if each interval in a set of simultaneous 95% confidence intervals for all pairwise comparisons is to have a minimum significant difference (half-width) of 150 units. The error standard deviation is thought to be at most 250 units. Investigate both the Tukey and the Bonferroni methods.

    (c) Discuss how the resulting design would be randomized.

6. **Video game experiment**

    Professor Robert Wardrop, of the University of Wisconsin, conducted an experiment in 1991 to evaluate in which of five sound modes he best played a certain video game. The first three sound modes corresponded to three different types of background music,

Table 12.13 Latin square design showing treatments and data for the video game experiment

		\multicolumn{5}{c}{Day}
		1	2	3	4	5
Time Order	1	1 94	3 100	4 98	2 101	5 112
	2	3 103	2 111	1 51	5 110	4 90
	3	4 114	1 75	5 94	3 85	2 107
	4	5 100	4 74	2 70	1 93	3 106
	5	2 106	5 95	3 81	4 90	1 73

as well as game sounds expected to enhance play. The fourth mode had game sounds but no background music. The fifth mode had no music or game sounds. Denote these sound modes by the treatment factor levels 1–5, respectively.

The experimenter observed that the game required no warmup, that boredom and fatigue would be a factor after 4 to 6 games, and that his performance varied considerably on a day-to-day basis. Hence, he used a Latin square design, with the two blocking factors being "day" and "time order of the game." The response measured was the game score, with higher scores being better. The design and resulting data are given in Table 12.13.

(a) Write down a possible model for these data and check the model assumptions. If the assumptions appear to be approximately satisfied, then answer parts (b)–(f).

(b) Plot the adjusted data and discuss the plot.

(c) Complete an analysis of variance table.

(d) Evaluate whether blocking was effective.

(e) Construct simultaneous 95% confidence intervals for all pairwise comparisons, as well as the "music versus no music" contrast

$$\tfrac{1}{3}(\tau_1 + \tau_2 + \tau_3) - \tfrac{1}{2}(\tau_4 + \tau_5)$$

and the "game sound versus no game sound" contrast

$$\tfrac{1}{4}(\tau_1 + \tau_2 + \tau_3 + \tau_4) - \tau_5 \,.$$

(f) What are your conclusions from this experiment? Which sound mode(s) should Professor Wardrop use?

7. **Video game experiment, continued**
Suppose that in the video game experiment of Exercise 6, Professor Wardop had run out of time and that only the first four days of data had been collected. The design would then have been a Youden design. Repeat parts (c), (e), and (f) of Exercise 6. Do your conclusions remain the same? Is this what you expected? Why or why not?

8. **Air freshener experiment**
A. Cunningham and N. O'Connor (1968, *British Journal of Marketing* 2, 147–151) conducted a two-replicate Latin square design to compare the effects of four price-and-display treatments on the sales of a brand of air fresheners. Treatments 1–3 corresponded to high, middle, and low prices, respectively, and each had an extra display. Treatment

Table 12.14 Latin square design and data for the air freshener sales experiment

Week	Store							
	1	2	3	4	5	6	7	8
1	2 31	1 23	3 12	4 3	1 10	3 30	2 23	4 14
2	1 19	4 16	2 14	3 4	2 21	4 25	3 17	1 14
3	4 15	3 30	1 12	2 6	3 12	1 47	4 5	2 3
4	3 16	2 27	4 5	1 11	4 12	2 38	1 13	3 6

Source: Cunningham, A. and O'Connor, N. (1968).

4 corresponded to the middle price and no extra display. The response variable was the unit sales for a one-week period. The experiment involved two blocking factors defined by stores ($c = 8$ levels) and one-week periods ($b = 4$ levels). The design and data are given in Table 12.14.

(a) Factors such as product location and shelf stocking could affect sales. Discuss how these factors might be controlled in such an experiment.

(b) Check the model assumptions.

(c) Plot the adjusted data and comment on the results.

(d) Complete an analysis of variance table.

(e) Evaluate whether blocking was effective.

(f) Test for equality of treatment effects using a 5% significance level.

(g) Construct simultaneous 95% confidence intervals for all pairwise comparisons of the treatments. What would you recommend for the sales of air fresheners if the results of this experiment are still valid today?

9. **Air freshener experiment, continued**

Suppose that the air freshener experiment of Exercise 8 had to be stopped after only 3 weeks. The resulting design would then be a replicated Youden design. Repeat parts (d)–(g) of Exercise 8. Do your conclusions remain the same? Is this what you expected? Why or why not?

10. **Quantity perception experiment**

An experiment was run in 1996 by M. Gbenado, A. Veress, L. Heimenz, J. Monroe, and S. Yu to investigate the effect of color on the perception of quantity. Subjects were selected at random from the student population. A number of small candies of a specific color were tipped onto a flat try. A subject was allowed to view the tray for 3 seconds and then asked to make a guess as to the number of candies on the tray.

The treatment factors of interest were "actual number of candies on the tray" and "color." The selected levels were 17, 29, and 41 for the treatment factor "number" and yellow, orange, brown for the factor "color." The experimenters decided that each subject should view all nine treatment combinations, and they based their design on 9×9 Latin squares.

We consider only part of the original study, constituting a 2-replicate Latin square, to save space. The data are shown in Table 12.15. Subjects represent the row blocks, and

Table 12.15 2-replicate Latin square design and data in parentheses for the quantity perception experiment

Subj	\multicolumn{9}{c}{Time Order}								
	1	2	3	4	5	6	7	8	9
1	23 (4)	22 (−3)	11 (0)	12 (−3)	32 (−1)	13 (−3)	31 (−6)	33 (−9)	21 (−1)
2	12 (2)	31 (16)	32 (21)	21 (9)	22 (4)	33 (16)	23 (9)	11 (2)	13 (2)
3	21 (4)	23 (−1)	12 (7)	13 (−5)	33 (−1)	11 (−13)	32 (−9)	31 (−19)	22 (−16)
4	32 (21)	12 (4)	22 (3)	31 (11)	13 (0)	21 (4)	11 (0)	23 (4)	33 (11)
5	31 (7)	11 (−2)	21 (2)	33 (3)	12 (−3)	23 (3)	13 (−4)	22 (−5)	32 (−7)
6	11 (3)	33 (7)	31 (14)	23 (11)	21 (12)	32 (17)	22 (10)	13 (5)	12 (0)
7	22 (11)	21 (14)	13 (0)	11 (1)	31 (16)	12 (1)	33 (13)	32 (14)	23 (7)
8	13 (7)	32 (16)	33 (16)	22 (4)	23 (4)	31 (16)	21 (14)	12 (7)	11 (2)
9	33 (21)	13 (2)	23 (10)	32 (24)	11 (6)	22 (13)	12 (2)	21 (8)	31 (20)
10	33 (16)	31 (20)	22 (6)	21 (6)	11 (7)	23 (6)	12 (2)	13 (3)	32 (14)
11	12 (2)	22 (4)	32 (11)	13 (2)	21 (9)	33 (1)	23 (4)	31 (−4)	11 (7)
12	13 (−4)	23 (−11)	33 (−4)	11 (−3)	22 (4)	31 (11)	21 (−1)	32 (1)	12 (−3)
13	21 (4)	12 (−1)	11 (2)	32 (11)	31 (11)	13 (−3)	33 (1)	22 (−1)	23 (11)
14	22 (2)	13 (−7)	12 (−9)	33 (8)	32 (−2)	11 (−6)	31 (−9)	23 (4)	21 (2)
15	31 (21)	32 (21)	23 (14)	22 (14)	12 (4)	21 (16)	13 (0)	11 (5)	33 (11)
16	11 (2)	21 (9)	31 (21)	12 (6)	23 (9)	32 (18)	22 (9)	33 (16)	13 (2)
17	32 (6)	33 (6)	21 (−1)	23 (−1)	13 (2)	22 (4)	11 (−3)	12 (−1)	31 (6)
18	23 (4)	11 (2)	13 (2)	31 (11)	33 (6)	12 (2)	32 (6)	21 (−1)	22 (4)

time order represents the column blocks. The treatment combinations have been coded as follows:

$1 = (17, \text{yellow})$ $2 = (17, \text{orange})$ $3 = (17, \text{brown})$
$4 = (29, \text{yellow})$ $5 = (29, \text{orange})$ $6 = (29, \text{brown})$
$7 = (41, \text{yellow})$ $8 = (41, \text{orange})$ $9 = (41, \text{brown})$

(a) The experiment was conducted in a busy hallway in the Ohio Union at The Ohio State University. Subjects were recruited from the population of noncolorblind students walking past the table. Recruited subjects were not allowed to view the experiment in progress with previous subjects, but they were paid for their participation in candies. Discuss whether or not the subjects in this study are likely to be representative of some larger population of subjects. Are the conclusions of the study likely to be relevant to people in general?

(b) Fit a model that includes the effects of the two blocking factors "subject" and "time order," the treatment effect, and the treatment × time interaction. Check whether the residuals are approximately normally distributed and whether they have approximately the same variance for each treatment. Do you prefer to use the original response variable "guessed number" or the transformed response "square root of guessed number" or "(true number − guessed number)/(true number)" or some other transformation?

(c) Present an analysis of variance table and test any hypotheses that you think are of interest. State your conclusions.

(d) Rewrite the treatment parameter in your model in terms of main effects and interactions of the two treatment factors. Redo your analysis of variance table. What can you conclude from the experiment?

(e) If you were to plan a followup experiment, what would you wish to study? Write up a checklist for such an experiment.

13 Confounded Two-Level Factorial Experiments

13.1 Introduction
13.2 Single replicate factorial experiments
13.3 Confounding Using Contrasts
13.4 Confounding Using Equations
13.5 A Real Experiment—Mangold Experiment
13.6 Plans for Confounded 2^p Experiments
13.7 Multireplicate Designs
13.8 Complete Confounding: Repeated Single-Replicate Designs
13.9 Partial Confounding
13.10 Comparing the Multireplicate Designs
13.11 Using SAS Software
Exercises

13.1 Introduction

In Chapters 6 and 7 we discussed factorial experiments arranged as completely randomized designs, and in Chapters 10 and 11 we looked at factorial experiments arranged as block designs. Factorial experiments that involve several treatment factors tend to be large. Even a modest experiment with four factors having 2, 2, 3, and 3 levels has a total of 36 treatment combinations. Since experimenters generally are working to a restricted budget and since observations cost time and money, many factorial experiments are single-replicate experiments (one observation per treatment combination). In this chapter we consider single-replicate experiments arranged in blocks where every treatment factor has two levels. This will be extended in Chapter 14 to cover treatment factors with more than two levels.

In Section 13.2.1 we discuss alternative codings of treatment combinations and, in Sections 13.2.2 and 13.2.3, the general problem of confounding and its implications for analysis.

Methods of designing single-replicate experiments so that information is lost on as few lower-order treatment contrasts as possible are the main focus of Sections 13.3 and 13.4. Section 13.5 contains an example.

In Sections 13.7–13.10 we return to the subject of multi-replicate factorial experiments in blocks and compare the traditional incomplete block designs with the multiple use of single-replicate confounded designs. Analysis of confounded factorial experiments by the computer package SAS is considered briefly in Section 13.11.

13.2 Single replicate factorial experiments

13.2.1 Coding and notation

A factorial experiment that involves two treatment factors each having two levels is known as a 2×2, or 2^2, experiment. Similarly, an experiment with two factors each having 3 levels is known as a 3×3, or 3^2, experiment. A $2^4 \times 3^2$ experiment has six treatment factors, the first four having two levels each, and the last two having three levels each. Other factorial experiments are described in a similar manner. A factorial experiment is called y if all factors have the same number of levels. Otherwise, it is called *asymmetric*. In this chapter we will deal only with symmetric 2^p experiments. Other situations will be discussed in Chapter 14.

The levels of a two-level treatment factor are often referred to as the "low" and "high" levels, and in Chapters 6 and 7 we coded these as 1 and 2. The codings 0 and 1, or -1 and $+1$, are also commonly used. A 2^2 experiment then has four treatment combinations coded as (11, 12, 21, 22) or as (00, 01, 10, 11) or as ($-1-1$, $-1+1$, $+1-1$, $+1+1$). A fourth standard coding for the treatment combinations is ((1), b, a, ab), where the letter a or b appears if the corresponding factor A or B is at its high level, and is absent if the corresponding factor is at its low level. The symbol (1) means that both factors are at their low level.

Coding is a matter of personal choice. Although we have coded the levels as 1 and 2 until now, the other three codings are more usual in talking about single-replicate factorial experiments. We will code the levels as 0 and 1 throughout this and the next three chapters.

13.2.2 Confounding

A factorial experiment with v treatment combinations uses $v - 1$ degrees of freedom to measure all of the main effects and interactions. In a single-replicate experiment, there are only v observations and $v - 1$ total degrees of freedom. Thus, the experiment is not large enough to allow measurement of all of the factorial effects and also estimation of the error variance. Three ways around this problem for statistical inference in completely randomized designs were discussed in Sections 6.7 and 7.5.

The problem is worse when the experiment is to be run as a block design. If there are b blocks in the design, $b - 1$ of the total degrees of freedom are used to measure the block differences, leaving only $(v - 1) - (b - 1) = v - b$ degrees of freedom available for measuring the treatment contrasts and the error variance. The result of this is that $b - 1$ of

Table 13.1 Outline analysis of variance table for single-replicate factorial experiments constructed by the methods of this chapter

Source of Variation	Degrees of Freedom	Sum of Squares
Blocks	$b-1$	$ss\theta = \frac{1}{k}\Sigma B_h^2 - \frac{1}{v}G^2$
⋮	⋮	⋮
$\Sigma c_i \tau_i$ (at most $v-b$ of these)	1	$ssc = \frac{(\Sigma\Sigma c_i y_{hi})^2}{\Sigma c_i^2}$
⋮	⋮	⋮
Error	df (by subtraction)	ssE (by subtraction)
Total	$v-1$	$sstot = \Sigma\Sigma y_{hi}^2 - \frac{1}{v}G^2$

the treatment contrasts can no longer be measured. They cannot be distinguished from block contrasts and are said to be *confounded* with blocks. Such a design is useful only when at most $v - b$ treatment contrasts are to be measured.

Care is required in designing this type of experiment. If v treatment combinations are arbitrarily divided into b blocks of size v/b, the important treatment contrasts will not necessarily be estimable. The estimable contrasts are those that are orthogonal to the confounded contrasts. This means that the experiment should be designed in such a way that the confounded contrasts belong only to interactions that are expected to be negligible. Fortunately, in some cases this is not difficult to achieve, and we will examine these cases in this chapter. For a 2^p experiment, we will restrict attention to designs with $b = 2^s$ blocks of size $k = 2^{p-s}$.

13.2.3 Analysis

The standard block–treatment model for a single-replicate factorial experiment arranged as an incomplete block design has the same form as model (11.3.1) used for incomplete block designs in Chapter 11; that is,

$$Y_{hi} = \mu + \theta_h + \tau_i + \epsilon_{hi},\qquad(13.2.1)$$

$$\epsilon_{hi} \sim N(0, \sigma^2),$$

ϵ_{hi}'s are mutually independent,

$h = 1, \ldots, b; \quad i = 1, \ldots, v; \quad (h, i)$ is in the design.

As usual, Y_{hi} is the random variable representing the observation on treatment combination i in block h (if it appears in the design), ϵ_{hi} is the corresponding error random variable, μ is a constant, τ_i is the effect of the ith treatment combination, and θ_h is the effect of the hth block. The block×treatment interaction is assumed to be negligible.

Analysis of all single-replicate designs described in this chapter is straight-forward. Because of the way in which the designs will be constructed, contrasts in the important main effects and interactions will be completely orthogonal to block contrasts. As a consequence, these contrasts will have no adjustment for blocks, and their estimates and sums of squares can be calculated in exactly the same way as for completely randomized designs (see Chapters 6 and 7). An outline of an analysis of variance table is shown in Table 13.1. The maximum number of degrees of freedom available for estimating main effects and interactions is $v - b$ if no estimate of the error variance is required; otherwise, it is $v - b - 1$. The unadjusted sum of squares for blocks $ss\theta$ can be calculated either as the total of all the confounded contrast sums of squares or by the usual formula, which was given in Table 11.7 as

$$ss\theta = \frac{1}{k}\Sigma_h B_h^2 - \frac{1}{v}G^2, \qquad (13.2.2)$$

where $B_h = y_{h.}$ is the total of the observations in the hth block and $G = y_{..}$ is the grand total of all the observations.

13.3 Confounding Using Contrasts

13.3.1 Contrasts

Treatment contrasts for factorial experiments were discussed in Sections 6.3, 7.2.4, and 7.3. When there are two factors, A and B, each having two levels, there are $v = 4$ treatment combinations in total, and it is possible to find a set of three orthogonal contrasts, one for the main effects of each of A and B and one for their interaction. The coefficient lists $[c_{00}, c_{01}, c_{10}, c_{11}]$ for these contrasts are

For A: $[-1, -1, 1, 1]$,

For B: $[-1, 1, -1, 1]$,

For AB: $[1, -1, -1, 1]$.

Each coefficient for the AB interaction is the product of the corresponding coefficients for the main effects of A and B. Similarly, for three factors, the coefficient lists for three of the seven contrasts are

For A: $[-1, -1, -1, -1, 1, 1, 1, 1]$,

For AB: $[1, 1, -1, -1, -1, -1, 1, 1]$,

For ABC: $[-1, 1, 1, -1, 1, -1, -1, 1]$,

again with interaction coefficients obtained as the product of corresponding main-effect coefficients.

Such contrasts in a 2^p experiment all have the same variance, since $\Sigma c_i^2 = v = 2^p$ for all contrasts and $\mathrm{Var}(\Sigma c_i \hat{\tau}_i) = \Sigma c_i^2 \sigma^2 = v\sigma^2$. The main effect of A is often measured by the A contrast divided by $v/2$, so that it compares the average of all treatment combinations at the high level of A with the average of the treatment combinations at the low level. If

Table 13.2 Contrasts for a 2^3 experiment

	A	B	C	AB	AC	BC	ABC
000	−1	−1	−1	1	1	1	−1
001	−1	−1	1	1	−1	−1	1
010	−1	1	−1	−1	1	−1	1
011	−1	1	1	−1	−1	1	−1
100	1	−1	−1	−1	−1	1	1
101	1	−1	1	−1	1	−1	−1
110	1	1	−1	1	−1	−1	−1
111	1	1	1	1	1	1	1

the interaction contrasts are also divided by $v/2$, the contrast estimators all have variance $\sum c_i^2 \sigma^2 = 4\sigma^2/v$, and the AB interaction, for example, then compares the average response when factors A and B are at the same level with the average response when they are at different levels. We shall use either $v/2$ or 1 for the divisor for all contrasts in 2^p experiments both here and in Chapter 15.

The full set of factorial treatment contrasts (without divisors) for a 2^3 experiment is shown in Table 13.2 written as columns. The row headings are the treatment combinations (in lexicographical order) whose observations are to be multiplied by the contrast coefficients when estimating the contrast.

The contrasts in Table 13.2 are orthogonal. This can be verified by multiplying together corresponding digits in any two columns and showing that the sum of the products is zero. A table of orthogonal contrasts, such as Table 13.2, is sometimes called an *orthogonal array*.

13.3.2 Experiments in Two Blocks

We start with an example. Suppose that a single-replicate 2^3 experiment is to be run in two blocks of size four. Suppose also that the experimenter knows that one of the factors, say factor A, does not interact with either of the other two factors. This means that the interactions AB, AC, and ABC may be assumed to be negligible and that the contrasts labeled A, B, C, and BC in Table 13.2 are the only contrasts to be measured.

Since there will be $b = 2$ blocks, it follows that $b - 1 = 1$ degree of freedom will be used to measure block differences and one treatment contrast will be confounded with blocks. Without too much difficulty, we can ensure that the confounded contrast is one of the negligible contrasts. For example, we can confound the negligible ABC contrast by placing in one block those treatment combinations corresponding to -1 in the ABC contrast, and placing in the second block those treatment combinations corresponding to $+1$ in the same contrast. Referring to Table 13.2, we can see that the design in Table 13.3 results. The ABC contrast is now identical to a block contrast that compares Block I with Block II, and

Table 13.3 2^3 experiment in 2 blocks of 4, confounding ABC

Block I	000	011	101	110
Block II	001	010	100	111

Table 13.4 Data for the field experiment (*ABCD* is confounded)

Block I		Block II	
TC	Response	TC	Response
0000	58	0001	55
0011	51	0010	45
0101	44	0100	42
0110	50	0111	36
1001	43	1000	53
1010	50	1011	55
1100	41	1101	41
1111	44	1110	48

Source: *Experimental Designs*, Second Edition, by W. G. Cochran and G. M. Cox, Copyright © 1957, John Wiley & Sons, New York. Adapted by permission of John Wiley & Sons, Inc.

consequently, the contrast is confounded with blocks. The other two negligible contrasts, AB and AC, provide two degrees of freedom to estimate σ^2. Since all the nonnegligible factorial contrasts are orthogonal to AB, AC, and ABC, they can be measured as though there were no blocks present. Block design randomization (see Section 11.2.2) needs to be carried out before the design in Table 13.3 can be used in practice.

A similar method of confounding can be used for any 2^p experiment in $b = 2$ blocks of size $k = 2^{p-1}$. All factorial contrasts except for the one confounded contrast can be estimated.

Example 13.3.1 Field experiment

The data shown in Table 13.4 form part of the results of a field experiment on the yield of beans using various types of fertilization. The experiment was conducted at Rothamsted Experimental Station in 1936 and was reported by W. G. Cochran and G. M. Cox in their book *Experimental Designs*. There were four treatment factors each at two levels. Factor A was the amount of dung (0 or 10 tons) spread per acre, factors B, C, and D were the amounts of nitrochalk (0 and 45 lb), superphosphate (0 and 67 lb), and muriate of potash (0 and 112 lb), respectively, per acre. The experimental area was divided into two possibly dissimilar blocks of land, each of which was subdivided into eight plots (experimental units). Since this was a single-replicate experiment with $2^4 = 16$ treatment combinations (TC) divided into $b = 2$ blocks of size $k = 8$, one treatment contrast had to be confounded. The experimenters chose to confound the $ABCD$ contrast, since the four-factor interaction was of least interest. The $ABCD$ contrast is shown below, and it can be verified that the treatment combinations corresponding to contrast coefficient -1 appear in Block I of Table 13.4, while those corresponding to coefficient $+1$ appear in Block II.

All the other factorial contrasts are orthogonal to the $ABCD$ contrast, so they can all be estimated without adjusting for the block effects. We take as examples the B and BC contrasts shown below.

13.3 Confounding Using Contrasts

TC	0000	0001	0010	0011	0100	0101	0110	0111
y_{hijkl}	58	55	45	51	42	44	50	36
B	−1	−1	−1	−1	1	1	1	1
BC	1	1	−1	−1	−1	−1	1	1
ABCD	1	−1	−1	1	−1	1	1	−1

TC	1000	1001	1010	1011	1100	1101	1110	1111
y_{hijkl}	53	43	50	55	41	41	48	44
B	−1	−1	−1	−1	1	1	1	1
BC	1	1	−1	−1	−1	−1	1	1
ABCD	−1	1	1	−1	1	−1	−1	1

Using rule 10 of Section 7.3, the least squares estimate of the B contrast is $\bar{y}_{.1..} - \bar{y}_{.0..} = \frac{1}{8}\Sigma c_{ijkl} y_{hijkl}$, where the sum is to be taken over the four subscripts, and the contrast coefficients c_{ijkl} are given in standard order above. Multiplying the contrast coefficients by the data values, we obtain

For B: $\quad \frac{1}{8}\sum c_{ijkl} y_{hijkl} = -8.00$.

Similarly, if we divide the BC contrast shown above by the same divisor $v/2$, we obtain the contrast estimate

For BC: $\quad \frac{1}{8}\sum c_{ijkl} y_{hijkl} = \frac{1}{8}(y_{..00.} - y_{..01.} - y_{..10.} + y_{..11.}) = 2.25$.

Using (6.7.54), the sum of squares for testing the hypothesis that the main effect of B is negligible is

$$ssB = \frac{(\frac{1}{8}\sum c_{ijkl} y_{hijkl})^2}{\sum(\frac{1}{8}c_{ijkl})^2} = \frac{(-8.00)^2}{\frac{16}{64}} = 256.0.$$

Similarly, the sum of squares for testing the hypothesis that the interaction between B and C is negligible is

$$ss(BC) = \frac{(2.25)^2}{\frac{16}{64}} = 20.25.$$

An alternative way to calculate the sums of squares is to use the method of Section 7.3. Following the rules in that section, we obtain

$$ssB = acd \sum_j \bar{y}_{..j..}^2 - abcd\bar{y}_{.....}^2$$
$$= 8\left[(51.25)^2 + (43.25)^2\right] - 16(47.25)^2 = 256.0$$

and

$$ss(BC) = ad \sum_j \sum_k \bar{y}_{..jk.}^2 - acd \sum_j \bar{y}_{..j..}^2 - abd \sum_k \bar{y}_{...k.}^2 + abcd\bar{y}_{.....}^2$$
$$= 4\left[(52.25)^2 + (50.25)^2 + (42.00)^2 + (44.50)^2\right]$$
$$\quad - 8\left[(51.25)^2 + (43.25)^2\right] - 8\left[(47.125)^2 + (47.375)^2\right]$$
$$\quad + 16(47.25)^2 = 20.25.$$

The complete analysis of variance table is shown in Table 13.5. The important contrasts can be identified using one of the methods of Section 7.5. A normal probability plot of the 14

Table 13.5 Analysis of variance for the field experiment

Source of Variation	Degrees of Freedom	Sum of Squares	Contrast Estimate (divisor v/2)
Block (ABCD)	1	2.25	
A	1	2.25	−0.75
B	1	256.00	−8.00
C	1	0.25	0.25
D	1	20.25	−2.25
AB	1	6.25	1.25
AC	1	81.00	4.50
AD	1	0.00	0.00
BC	1	20.25	2.25
BD	1	12.25	−1.75
CD	1	1.00	0.50
ABC	1	16.00	−2.00
ABD	1	16.00	2.00
ACD	1	20.25	2.25
BCD	1	121.00	−5.50
Total	15	575.00	

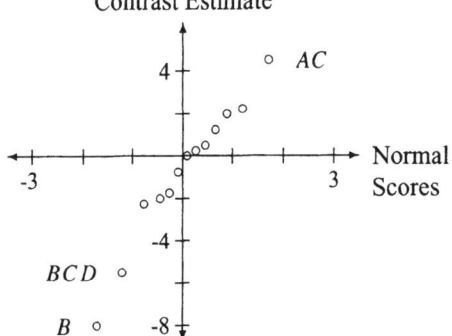

Figure 13.1 Normal probability plot for the contrasts of the field experiment

contrast estimates is shown in Figure 13.1. Note that we have not included the confounded $ABCD$ contrast in the normal probability plot. Although it is not the case here, the block effect is usually expected to be large and may draw attention away from the important treatment contrasts.

The important contrasts appear to be those of B, BCD, and AC. We notice that the contrast estimate for B is negative, suggesting that the addition of nitrochalk decreased the yield of beans when averaged over the levels of A, C, and D. The interaction plot for BCD is shown in Figure 13.2 and that for AC in Figure 13.3. We see from Figure 13.2 that the BCD interaction can be characterized by the fact that the CD interaction changes as B changes from its low level to its high level. If the objective of the experiment is to increase yield, then comparison of the two plots in Figure 13.2 suggests that B should be set at its low level, unless the high level of C and the low level of D are used. This tends to agree

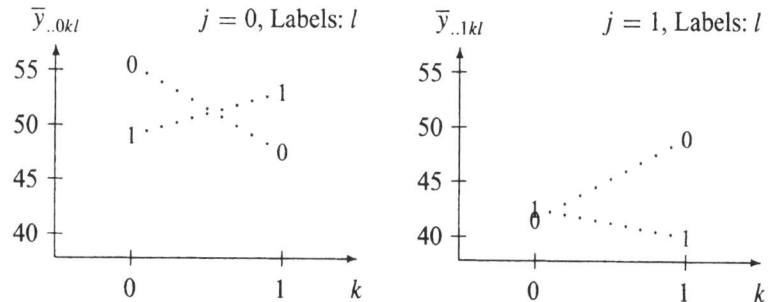

Figure 13.2 BCD interaction plot for the field experiment

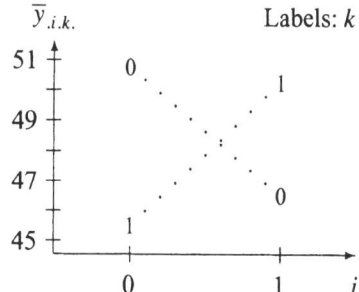

Figure 13.3 AC interaction plot for the field experiment

with the earlier observation that the contrast estimate for B is negative, suggesting that the low level is better. If B is set at its low level, the left-hand graph of Figure 13.2 suggests that both C and D should be at their low levels. The AC interaction plot in Figure 13.3 shows that either both C and A should be at their low levels or both C and A should be at their high levels. The contrast estimators for A and D are both negative, suggesting that on average the low level is better, although the difference in yield is minor. More importantly, the low levels in this experiment are cheaper. Therefore, all the evidence points towards not adding any fertilizer ingredients in the quantities studied in the experiment. A followup experiment could be run with the same four factors but with an increased "high" level of C and lower "high" levels of A, B, and D. Since it is possible that the response is quadratic for each of the factors, a 3^4 experiment could be run.

Suppose the experimenters had known ahead of time that factors A and D do not interact, so that interactions AD, ABD and ACD could have been assumed negligible; then the corresponding terms would have been omitted from the model. There would then have been 3 degrees of freedom for estimating the error variance. The error sum of squares would have been the total of the sums of squares for AD, ABD, and ACD listed in Table 13.5, so that

$$\hat{\sigma}^2 = msE = \tfrac{1}{3}(ss(AD) + ss(ABD) + ss(ACD)) = \tfrac{1}{3}(36.25) = 12.0833.$$

The analysis of variance table would then have been as shown in Table 13.6, and we see that at an overall significance level of at most $\alpha = 11(0.005) = 0.055$, none of the contrasts would have been judged as significantly different from zero, since $F_{1,3,0.005} = 55.55$.

Table 13.6 Analysis of variance for the field experiment

Source of Variation	Degrees of Freedom	Sum of Squares	Mean Square	Ratio	p-value
Block	1	2.25	2.25	–	–
A	1	2.25	2.25	0.186	0.6952
B	1	256.00	256.00	21.186	0.0193
C	1	0.25	0.25	0.021	0.8947
D	1	20.25	20.25	1.676	0.2861
AB	1	6.25	6.25	0.517	0.5240
AC	1	81.00	81.00	6.703	0.0811
BC	1	20.25	20.25	1.676	0.2861
BD	1	12.25	12.25	1.014	0.3882
CD	1	1.00	1.00	0.083	0.7923
ABC	1	16.00	16.00	1.324	0.3332
BCD	1	121.00	121.00	10.014	0.0507
Error	3	36.25	12.0833		
Total	15	575.00			

Confidence intervals could be calculated for each contrast at an overall confidence level of at least 94.5%, using the Bonferroni method (formula (4.4.21)) with error degrees of freedom $df = 3$ and with $r_i = 8$ being the number of observations averaged over to obtain the estimate. For example, a confidence interval for the difference in the high and low levels of B is

$$\beta_1^* - \beta_0^* \in \left(\tfrac{1}{8} \Sigma c_{ijkl} y_{hijkl} \pm t_{3, 0.0025} \sqrt{msE \, (16/64)} \right)$$
$$= (-8 \pm 7.4532 \times 1.7381) = (-20.954, 4.954).$$

We remind the reader that if both of the above analyses are done, that is, if the interactions AD, ABD, and ACD are dropped from the model after examining the normal probability plot, then it is no longer meaningful to talk about the significance levels of the tests or the confidence levels of the intervals (see Section 6.5.6). □

13.3.3 Experiments in Four Blocks

We can extend the method of confounding that we used for two blocks to obtain $b = 4 = 2^2$ blocks. We then need to use two contrasts to divide up the treatment combinations. For example, suppose that in a 2^4 experiment, all interactions except for the two-factor interactions are thought to be negligible. We can select one of the negligible interactions to produce two blocks of size 8 and then select a second interaction to subdivide each of these two blocks into two smaller blocks, giving a total of 4 blocks of size 4. Now, $b-1 = 3$ degrees of freedom are used to measure blocks, which means that a third treatment contrast must also be confounded. Since we require this third contrast to be among the negligible contrasts, care must be taken as to which pair of contrasts is initially selected for confounding. The choice of $ABCD$ and ABC, for example, is a very poor choice even if these high-order interactions may be thought to be negligible. We can see this by examining the design and the third confounded contrast as follows.

13.3 Confounding Using Contrasts

Table 13.7 2^4 experiment in 4 blocks of 4, confounding $ABCD$, ABC, D

TC	0000	0001	0010	0011	0100	0101	0110	0111
ABCD	1	−1	−1	1	−1	1	1	−1
ABC	−1	−1	1	1	1	1	−1	−1
TC	1000	1001	1010	1011	1100	1101	1110	1111
ABCD	−1	1	1	−1	1	−1	−1	1
ABC	1	1	−1	−1	−1	−1	1	1

Block	Contrast Coefficients (ABCD, ABC)	Treatment Combinations			
I	(−1, −1)	0000	0110	1010	1100
II	(−1, 1)	0011	0101	1001	1111
III	(1, −1)	0001	0111	1011	1101
IV	(1, 1)	0010	0100	1000	1110

The two contrasts and the corresponding treatment combinations (TC) are shown in Table 13.7. The treatment combinations are divided into 2 blocks of size 8 according to the coefficients in the $ABCD$ contrast. Each of these blocks is then subdivided into 2 blocks of size 4 according to the coefficients in the ABC contrast. The $b=4$ blocks of size $k=4$ are therefore determined by the pairs of coefficients $(ABCD, ABC) = (-1, -1)$ or $(-1, 1)$ or $(1, -1)$ or $(1, 1)$ in the two contrasts. The resulting design is shown in Table 13.7 (prior to randomization).

Examination of the blocks in the design shows that all of the treatment combinations in Block I and Block IV have the fourth digit equal to 0, and all of those in Blocks II and III have the fourth digit equal to 1. This means that the high and low levels of factor D cannot be compared within the same block, and therefore the contrast for the main effect of D must be the third contrast confounded with blocks.

We could have predicted this outcome, since if corresponding coefficients of the two contrasts $ABCD$ and ABC shown in Table 13.7 are multiplied together, the coefficients of the D contrast results. Notice that in symbols, we can write

$$(ABCD)(ABC) = A^2 B^2 C^2 D = D,$$

where any letter with exponent 2 is ignored. The squared coefficients of any 2^p factorial contrast are all $+1$, so multiplying the D contrast by C^2, say, is the same as multiplying the contrast coefficients by $+1$ and $C^2 D = D$. Multiplication of the contrast names in this way gives a quick, easy method of checking which third contrast is confounded without writing out the contrasts and without writing out the design.

The above design is not suitable for the stated experiment. Suppose that the contrasts ABD and BCD were selected for confounding instead. The third confounded contrast would then be $(ABD)(BCD) = AB^2CD^2 = AC$. This, too, is not suitable since all two-factor interactions were to have been measured. Unfortunately, there is no choice that will meet the specifications of this particular experiment. The number of blocks and the block sizes are too small to measure everything that is required. At least one two-factor interaction would have to be sacrificed, or a larger experiment must be run.

Example 13.3.2 2^5 experiment in 4 blocks of 8

In Section 13.5 we will describe a 2^5 experiment that was run in $b = 4$ blocks of size $k = 8$. In designing such an experiment, one needs to select two contrasts for confounding and to check that the third confounded contrast is acceptable. As in the discussion above, selecting the 5-factor interaction for confounding will generally be a poor choice, since no matter which other interaction is selected for the second confounded interaction, a 2-factor interaction or main effect will be among the confounded contrasts. For example,

$$(ABCD)(ABCDE) = E \text{ and } (ABC)(ABCDE) = DE.$$

If an experimenter knew ahead of time that factors D and E do not interact, then the second choice might be acceptable. In general, though, most experimenters would prefer not to confound low-order interactions. So, a selection of a 3-factor interaction and a 4-factor interaction with as few letters in common as possible will generally be the best choice. For example,

$$(ABCD)(CDE) = ABE.$$

There are many selections of this type, and the experimenter would wish to avoid confounding any 3-factor interaction that might be of some interest. The selection made in Section 13.5 is ABD, BCE, and their product $ACDE$. If the treatment combinations are written out in standard order together with the contrast coefficients for ABD and BCE, it can be verified that the pairs of contrast coefficients give the four blocks shown in Table 13.11 (page 438). □

13.3.4 Experiments in Eight Blocks

If an experiment is required in $b = 2^3$ blocks, then three contrasts must be selected for confounding. A single-replicate design in eight blocks confounds $b - 1 = 7$ treatment contrasts in total, including the three contrasts initially selected and all products of these. For example, suppose that a 2^6 experiment is required in $b = 2^3 = 8$ blocks of 8, and that the two-factor interactions are of interest together with the four three-factor interactions ACE, ACD, ADE, and CDE (these are the four 3-factor interactions that do not contain B or F). A suitable choice might be to confound the interactions BCD, ABE, and ADF. The other four confounded contrasts would be

$$(BCD)(ABE) = ACDE,$$
$$(BCD)(ADF) = ABCF,$$
$$(ABE)(ADF) = BDEF,$$
$$(BCD)(ABE)(ADF) = CEF.$$

The list of seven confounded contrasts is called the *confounding scheme* for the design. The reader is invited to write out the three selected contrasts and verify that the design in

Table 13.10 (page 436) results. The fact that $ACDE$, $ABCF$, $BDEF$, and ABE are also confounded can be verified by showing that the coefficients of each of these four contrasts are constant for all treatment combinations within each block.

Note that the same design will be obtained for any initial selection of three of the above seven confounded contrasts, *provided that no selected contrast is a product of the other two*. For example, suppose that $ABCF$, ABE, and CEF are initially selected. The selected contrasts $ABCF$ and ABE divide the treatment combinations into four blocks, but CEF does not subdivide these blocks further, since it is the third contrast automatically confounded in the four blocks, i.e., $(ABCF)(ABE) = CEF$. A different third contrast needs to be chosen, and any of the remaining four contrasts will do. Three selected contrasts satisfying the requirement that no selected contrast be a product of the other two is called a set of three *independent* contrasts.

13.3.5 Experiments in More Than Eight Blocks

The same ideas can be used for 2^p experiments in 2^s blocks of size $k = 2^{p-s}$ by selecting s independent contrasts to subdivide the treatment combinations into blocks—s contrasts are independent if none can be obtained as the product of two or more of the other $s-1$ contrasts. Although the multiplication of contrast names is a convenient method to determine the list of confounded contrasts, it becomes harder to use the contrasts themselves for constructing the design as the number of factors increases. In the next section, we present a method of constructing block designs for single-replicate factorial experiments that avoids writing out the contrasts.

13.4 Confounding Using Equations

13.4.1 Experiments in Two Blocks

The design in Table 13.8 (given previously in Table 13.3) was constructed by allocating the treatment combinations to blocks in such a way that the contrast that compares Block I with Block II is identical to the ABC contrast. The ABC contrast is confounded with blocks and cannot be estimated, but all of the other factorial contrasts are estimable because they are orthogonal to the confounded contrast.

If the treatment combinations in the two blocks of the design are examined closely, an interesting property becomes apparent. All the treatment combinations in the first block have an even number of 1's, and all those in the second block have an odd number of 1's. We could, in fact, have predicted this property. The ABC contrast is the product of the A, B, and C contrasts (see Section 13.3.1), and so the only way to achieve a coefficient -1

Table 13.8 2^3 experiment in 2 blocks of 4, confounding ABC

Block I	000	011	101	110
Block II	001	010	100	111

in the ABC contrast is for there to be an odd number of -1's among the corresponding coefficients in the A, B, and C contrasts. This means that there must be an odd number of 0's and an even number of 1's in the corresponding treatment combination. Thus, if we want to confound ABC without writing out the contrasts, we can simply allocate a treatment combination $a_1 a_2 a_3$ to Block I if it has an even number of 1's among its digits, and to Block II if it has an odd number of 1's. Equivalently, we allocate $a_1 a_2 a_3$ to Block I if $a_1 + a_2 + a_3$ is an even number and to Block II if $a_1 + a_2 + a_3$ is an odd number. Instead of writing "$a_1 + a_2 + a_3$ is an even number," we write "$a_1 + a_2 + a_3 = 0 \pmod 2$" and instead of writing "$a_1 + a_2 + a_3$ is an odd number," we write "$a_1 + a_2 + a_3 = 1 \pmod 2$". Working mod 2, or modulo 2, means that we subtract 2 repeatedly from the number until we reach either 0 or 1, or equivalently, we divide by 2 and take the remainder which is either 0 or 1. For example, $5 = 1 \pmod 2$, but $8 = 0 \pmod 2$. We call the pair of equations "$a_1 + a_2 + a_3 = 0 \pmod 2$" and "$a_1 + a_2 + a_3 = 1 \pmod 2$" the *confounding equations*. The design of Tables 13.3 and 13.8 is constructed using the following rule:

Block I: Treatment combinations with $a_1 + a_2 + a_3 = 0 \pmod 2$,

Block II: Treatment combinations with $a_1 + a_2 + a_3 = 1 \pmod 2$.

If the AC contrast were to be confounded in the 2^3 experiment instead of the ABC contrast, only the first and third digits of each treatment combination would be used to allocate it to a block. This is because only the first and third digits of the treatment combination govern the coefficients in the AC contrast. We would use the pair of confounding equations $a_1 + a_3 = 0 \pmod 2$ and $a_1 + a_3 = 1 \pmod 2$, and the design would be constructed using the rule

Block I: Treatment combinations with $a_1 + a_3 = 0 \pmod 2$,

Block II: Treatment combinations with $a_1 + a_3 = 1 \pmod 2$,

giving the design of Table 13.9. It can be verified that AC is indeed confounded with blocks in this design, since the coefficients of the AC contrast (shown in Table 13.2) are all equal to $+1$ for the treatment combinations in Block I and -1 for those in Block II. All other contrasts are estimable because they are orthogonal to the confounded AC contrast.

We may now generalize to 2^p experiments in 2 blocks of size 2^{p-1}. If the interaction $A^{z_1} B^{z_2} C^{z_3} \cdots P^{z_p}$ is to be confounded with blocks, where $z_i = 1$ if the factor is present in the interaction and $z_i = 0$ if it is not, the blocks of the design are

Block I: Treatment combinations with
$$z_1 a_1 + z_2 a_2 + z_3 a_3 + \cdots + z_p a_p = 0 \pmod 2 ,$$

Block II: Treatment combinations with
$$z_1 a_1 + z_2 a_2 + z_3 a_3 + \cdots + z_p a_p = 1 \pmod 2 .$$

Table 13.9 2^3 experiments in 2 blocks of 4, confounding AC

Block I	000	010	101	111
Block II	001	011	100	110

13.4.2 Experiments in More Than Two Blocks

We obtain designs with 4, 8, 16, ... blocks by using more than one pair of confounding equations. We label the pairs of confounding equations as L_1, L_2, etc. For example, the (unsatisfactory) design

Block	Treatment Combinations			
I	0000	0110	1010	1100
II	0011	0101	1001	1111
III	0001	0111	1011	1101
IV	0010	0100	1000	1110

from Table 13.7 for a 2^4 experiment in 4 blocks of size 4 was produced by confounding the $ABCD$ and ABC contrasts. Using the $ABCD$ contrast to produce two blocks is equivalent to using the pair of confounding equations $L_1 = a_1 + a_2 + a_3 + a_4 = 0$ and $L_1 = 1$ (mod 2), and using the ABC contrast is equivalent to using the pair of confounding equations $L_2 = a_1 + a_2 + a_3 = 0$ and $L_2 = 1$ (mod 2). Thus there are four possible values for the pair (L_1, L_2), and it can be verified that the blocks of the design satisfy

Block I: $L_1 = a_1 + a_2 + a_3 + a_4 = 0$; $L_2 = a_1 + a_2 + a_3 = 0$ (mod 2);
Block II: $L_1 = a_1 + a_2 + a_3 + a_4 = 0$; $L_2 = a_1 + a_2 + a_3 = 1$ (mod 2);
Block III: $L_1 = a_1 + a_2 + a_3 + a_4 = 1$; $L_2 = a_1 + a_2 + a_3 = 0$ (mod 2);
Block IV: $L_1 = a_1 + a_2 + a_3 + a_4 = 1$; $L_2 = a_1 + a_2 + a_3 = 1$ (mod 2).

We already know from Section 13.3.3 that a third contrast, namely contrast D, is confounded in this design. If we add the two confounding equations used to create each block, we have, for Block I,

$$\begin{aligned} L_1 &= a_1 + a_2 + a_3 + a_4 = 0 \pmod{2} \\ +L_2 &= a_1 + a_2 + a_3 = 0 \pmod{2} \\ \hline L_1 + L_2 &= 2a_1 + 2a_2 + 2a_3 + a_4 = 0 \pmod{2} \end{aligned}$$

If all coefficients are reduced modulo 2, the sum gives $a_4 = 0$ (mod 2), which indicates that the D contrast is also confounded.

As has been demonstrated, there is a correspondence between the contrasts, the contrast names, and the confounding equations. The contrast names are the most convenient for checking the total list of confounded contrasts, and the equations are the most convenient for constructing the design.

Design construction can be done in several ways. One way is to examine each of the 2^p treatment combinations and allocate them to blocks according to their values obtained in the left sides of the confounding equations. Another way is to identify the treatment combinations that make the confounding equations equal to zero. These will form Block I of the design. The other blocks of the design are obtained by adding (modulo 2) to the treatment combinations in Block I any treatment combination that has not yet appeared in a block. This is illustrated in the following example.

Table 13.10 2^6 experiment in 8 blocks of 8, confounding ABE, ADF, BCD, CEF, $ABCF$, $ACDE$, $BDEF$

Block	L_1, L_2, L_3	Treatment Combinations			
I	0,0,0	000000	001101	010111	011010
		100011	101110	110100	111001
II	0,0,1	000001	001100	010110	011011
		100010	101111	110101	111000
III	0,1,0	000010	001111	010101	011000
		100001	101100	110110	111011
IV	0,1,1	000011	001110	010100	011001
		100000	101101	110111	111010
V	1,0,0	000101	001000	010010	011111
		100110	101011	110001	111100
VI	1,0,1	000100	001001	010011	011110
		100111	101010	110000	111101
VII	1,1,0	000111	001010	010000	011101
		100100	101001	110011	111110
VIII	1,1,1	000110	001011	010001	011100
		100101	101000	110010	111111

Example 13.4.1 2^6 experiment in 8 blocks of 8

A confounding scheme was found in Section 13.3.4 for a 2^6 experiment in 8 blocks of 8 by selecting contrasts BCD, ABE, and ADF for confounding. It was shown that contrasts $ACDE$, $ABCF$, $BDEF$, and CEF were also confounded. Writing out the equations for the three selected contrasts, we have

$$L_1 = a_2 + a_3 + a_4 = 0 \text{ or } 1 \pmod 2,$$
$$L_2 = a_1 + a_2 + a_5 = 0 \text{ or } 1 \pmod 2,$$
$$L_3 = a_1 + a_2 + a_3 + a_6 = 0 \text{ or } 1 \pmod 2.$$

The equations corresponding to the other four confounded contrasts are obtained by setting $L_1 + L_2$, $L_1 + L_3$, $L_2 + L_3$, and $L_1 + L_2 + L_3$ equal to 0 or 1 (mod 2).

The design is shown in Table 13.10. It can be constructed systematically as follows. The first treatment combination 000000 gives 0 for each of L_1, L_2, L_3 and is allocated to Block I. The second treatment combination 000001 gives values 0, 0, 1 for the three L_i and is allocated to Block II, and so on.

Alternatively, one can look for the eight treatment combinations that give zero for each of L_1, L_2, and L_3 and construct Block I first. Solving $L_1 = 0$, $L_2 = 0$, and $L_3 = 0$ each for the last a_i gives $a_4 = a_2 + a_3 \pmod 2$, $a_5 = a_1 + a_2 \pmod 2$ and $a_6 = a_1 + a_4 \pmod 2$, respectively. For each of the eight combinations a_1, a_2, a_3 of factors A, B, and C, the corresponding values of a_4, a_5, and a_6 can thus be computed to obtain one of the eight treatment combinations of Block I. For example, if $a_1 = a_2 = a_3 = 1$, then $a_4 = a_2 + a_3 = 1 + 1 = 0 \pmod 2$, $a_5 = a_1 + a_2 = 1 + 1 = 0 \pmod 2$, and $a_6 = a_1 + a_4 = 1 + 0 = 1 \pmod 2$, so the treatment combination 111001 is in Block I.

Each of the other blocks is obtained by adding a new treatment combination to those in Block I—"new" meaning not yet in a block. For example, the treatment combination 000001 is not in Block I. If 000001 is added modulo 2 to the eight treatment combinations in Block I, then Block II results. As an illustration, it is added to 111001 of Block I by adding corresponding digits and reducing modulo 2; that is,

$$000001 + 111001 = 111002 = 111000 \pmod 2,$$

so 111000 is also in Block II.

A third block can be obtained by taking any treatment combination not in the first two blocks, and adding it to each treatment combination in Block I. Proceeding in this fashion, blocks are constructed until each treatment combination has been allocated to some block. □

13.5 A Real Experiment—Mangold Experiment

O. Kempthorne in his book *Design and Analysis of Experiments* describes an experiment run at Rothamsted Agricultural Station to investigate the effects of five different fertilizers on the growth of mangold roots. The five factors were Sulphate of ammonia (factor A at levels 0 or 0.6 cwt per acre), Superphosphate (factor B at levels 0 or 0.5 cwt per acre), Muriate of potash (factor C at levels 0 or 1.0 cwt per acre), Agricultural salt (factor D at levels 0 or 5 cwt per acre), and Dung (factor E at levels 0 or 10 tons per acre). The experimental area was divided into $b = 4$ blocks of size $k = 8$. All 3-, 4-, and 5-factor interactions were expected to be negligible. The two three-factor interactions ABD, BCE, and their product $ACDE$ were selected for confounding.

The division of the 32 treatment combinations into the four blocks was then determined from the confounding equations corresponding to ABD and BCE; that is,

$$L_1 = a_1 + a_2 \quad\quad + a_4 \quad\quad = 0 \text{ or } 1 \pmod 2,$$
$$L_2 = \quad\quad a_2 + a_3 \quad\quad + a_5 = 0 \text{ or } 1 \pmod 2.$$

The blocks can be formed by systematically working through all 32 treatment combinations and assigning them to the blocks according to the values of L_1 and L_2 and the rule

Block I: $L_1 = 0 \pmod 2$ and $L_2 = 0 \pmod 2$,
Block II: $L_1 = 0 \pmod 2$ and $L_2 = 1 \pmod 2$,
Block III: $L_1 = 1 \pmod 2$ and $L_2 = 0 \pmod 2$,
Block IV: $L_1 = 1 \pmod 2$ and $L_2 = 1 \pmod 2$.

Alternatively, we can notice that $L_1 = 0$ gives $a_4 = a_1 + a_2$ and $L_2 = 0$ gives $a_5 = a_2 + a_3 \pmod 2$. Computing these for each combination of levels $a_1 a_2 a_3$ of the first three factors gives Block I as

Block I: 00000   10010   00101   10111
           11100   01110   11001   01011

Any treatment combination that is not in Block I can be added to the treatment combinations in Block I to obtain a second block. So, if we add 00001, for example, we obtain

Table 13.11 Yields (in pounds) of mangold roots for the mangold experiment

Block	Treatment Combinations (Yield)			
I	01101 (844)	00011 (1104)	10100 (1156)	11111 (1508)
	11010 (1312)	00110 (1000)	10001 (1176)	01000 (888)
II	00001 (1248)	01111 (1100)	00100 (784)	10011 (1376)
	11101 (1356)	10110 (1376)	11000 (1008)	01010 (964)
III	00101 (896)	11001 (1284)	01011 (996)	01110 (860)
	10010 (1184)	11100 (984)	00000 (740)	10111 (1468)
IV	10101 (1328)	11110 (1292)	00111 (1008)	11011 (1324)
	01001 (1008)	01100 (692)	00010 (780)	10000 (1108)

Source: *Design and Analysis of Experiments*, by O. Kempthorne, Copyright © 1976, John Wiley & Sons, Inc. Reprinted by permission of John Wiley & Sons, Inc.

Block II: 00001 10011 00100 10110
           11101 01111 11000 01010

Any treatment combination not in Blocks I and II can now be added to the treatment combinations in Block I to obtain a third block. Since 00010, for example, has not appeared in Blocks I or II, we can add it to the treatment combinations in Block I to obtain Block III. Block IV can then be obtained by adding a treatment combination, say 00011, that has not appeared in the previous three blocks.

Block III: 00010 10000 00111 10101
            11110 01100 11011 01001
Block IV: 00011 10001 00110 10100
            11111 01101 11010 01000

The order of the treatment combinations was randomized within each block, and the order of the blocks was randomized. The final design, together with the resulting yields (in pounds) of mangold roots, is shown in Table 13.11.

The contrasts for the main effects and interactions are obtained as usual. The contrast for comparing the high and low levels of superphosphate (B), for example, is $\sum_i c_i \tau_i$, where τ_i is the effect of the ith treatment combination, and the coefficient c_i is the ith element of

$$\frac{1}{16}[-1, -1, -1, -1, -1, -1, -1, -1, 1, 1, 1, 1, 1, 1, 1, 1,$$
$$-1, -1, -1, -1, -1, -1, -1, -1, 1, 1, 1, 1, 1, 1, 1, 1].$$

13.5 A Real Experiment—Mangold Experiment

Table 13.12 Analysis of variance for the mangold experiment (ABD, BCE, ACDE are confounded)

Source of Variation	Degrees of Freedom	Sum of Squares	Mean Square	Ratio	Contrast Estimate	p-values
Block	3	52832	17610.67	–	–	–
A	1	887112	887112.00	130.63	333.0	0.0001
B	1	3042	3042.00	0.45	−19.5	0.5150
C	1	722	722.00	0.11	9.5	0.7496
D	1	144722	144722.00	21.31	134.5	0.0005
E	1	262088	262088.00	38.59	181.0	0.0001
AB	1	338	338.00	0.05	6.5	0.8269
AC	1	48050	48050.00	7.08	77.5	0.0196
AD	1	16562	16562.00	2.44	45.5	0.1424
AE	1	288	288.00	0.04	−5.0	0.8400
BC	1	6272	6272.00	0.92	−28.0	0.3541
BD	1	5832	5832.00	0.86	27.0	0.3710
BE	1	98	98.00	0.01	−3.5	0.9062
CD	1	30752	30752.00	4.53	62.0	0.0530
CE	1	882	882.00	0.13	−10.5	0.7244
DE	1	13778	13778.00	2.03	−41.5	0.1779
Error	13	88286	6791.23			
Total	31	1561656				

Equivalently, $c_i = -1/16$ if the ith treatment combination has B at the low level, and $c_i = 1/16$ otherwise. There is only one observation on each treatment combination, so a least squares estimate of τ_i is just y_i (the observation on the ith treatment combination). The least squares estimate of the contrast for comparing the high and low levels of B is

$$\sum_i c_i \hat{\tau}_i = -312/16 = -19.5 \,.$$

The contrast estimates for each of the main effects and also the interactions are shown in Table 13.12.

To test the hypothesis that adding superphosphate has no effect on the yield of mangold roots, we test the null hypothesis $H_0^B : \sum_i c_i \tau_i = 0$ against the alternative hypothesis $H_A^B : \sum_i c_i \tau_i \neq 0$ using (4.4.24); that is,

$$\text{reject } H_0^B \text{ if } \frac{ssB}{msE} = \frac{(\sum_i c_i \hat{\tau}_i)^2}{(\sum_i c_i^2)\, msE} = \frac{(\sum_i c_i y_i)^2}{(32/16^2)\, msE} > F_{1,13,\alpha/m} \,.$$

If the 3-, 4-, and 5-factor interactions are assumed to be negligible, the total of their contrast sums of squares, apart from ABD, BCE, and $ACDE$, which are confounded with blocks, forms the error sum of squares with 13 degrees of freedom, ($10 - 2 = 8$ from the 3-factor interactions, $5 - 1 = 4$ from the 4-factor interactions, and 1 from the 5-factor interaction). There are $m = 15$ contrasts of interest, so if each individual contrast is tested at significance level $\alpha/m = 0.005$, the overall significance level would be at most $\alpha = 0.075$. Since

$$\frac{(\sum_i c_i y_i)^2}{(32/16^2)\, msE} = \frac{8(-19.5)^2}{msE} = \frac{3042.00}{6791.23} < F_{1,13,0.005} = t^2_{13,0.0025} = 11.37 \,,$$

we fail to reject the null hypothesis H_0^B and conclude that there is no evidence to suggest a difference in yield due to the high and low levels of superphosphate. The ratio for each of the other hypothesis tests is given in Table 13.12, and we see that the only hypotheses that would be rejected at overall significance level 0.075 are the hypotheses of no effect of A or D or E.

The block sum of squares is the sum of the ABD, BCE, and $ACDE$ contrast sums of squares, or alternatively, it can be calculated as

$$ss\theta = \frac{1}{8}\sum_h B_h^2 - \frac{1}{32}G^2 = 52832.0$$

as in (13.2.2), where $B_h = y_{h....}$ and $G = y_{.....}$. The block mean square is over twice the size of the error mean square, indicating that the creation of blocks in this experiment was useful for reducing the error variability, assuming that the effects confounded with blocks are negligible.

Since none of the interactions appear to be significantly different from zero, the main effects can be investigated. The contrast estimates of the significant main effects are all positive, suggesting that the high levels of A, D, and E increase the yield of mangold roots significantly. Confidence intervals for all the main-effect contrasts can be obtained via Bonferroni's method using formula (4.4.21), but with error degrees of freedom $df = 13$. If we select the confidence level to match the α level of the hypothesis tests, we obtain a set of simultaneous confidence intervals with overall confidence level at least 92.5% as follows:

For A: $\left(\bar{y}_{.1....} - \bar{y}_{.0....} \pm t_{13,0.0025}\sqrt{32\hat{\sigma}^2/16^2}\right)$
$= (333.0 \pm 3.372\sqrt{6791.23/8})$
$= (333.0 \pm 98.25) = (234.75,\ 431.25)$;

For B: $(-19.5 \pm 98.25) = (-117.75,\ 78.75)$;

For C: $(9.5 \pm 98.25) = (-88.75,\ 107.75)$;

For D: $(134.5 \pm 98.25) = (36.25,\ 232.75)$;

For E: $(181.0 \pm 98.25) = (82.75,\ 279.25)$.

At a somewhat higher overall significance level, the hypothesis of no interaction between A and C would have been rejected. The AC interaction plot is shown in Figure 13.4. This plot agrees with the earlier observation that A should be set at its high level. Unless the cost is high, the plot suggests that C should also be set at its high level. Factor B does not seem to affect the yield much and can be set at its low (zero) level. As stated earlier, D and E should be at their high levels.

After an experiment of this type, it is good policy to run a *confirmatory experiment* verifying that the selected levels are, indeed, a good combination. Here the recommendation is to use treatment combination 10111. Certainly, in the main experiment, this treatment combination gave the highest yield in Block III, but it cannot easily be compared with the other observations because of the large block differences. A few more observations on this particular treatment combination would help to verify that it is consistently a good choice.

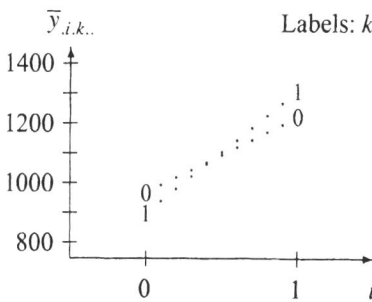

Figure 13.4 *AC* interaction plot for the mangold experiment

13.6 Plans for Confounded 2^p Experiments

Suggestions for confounding schemes useful for constructing designs for 2^p experiments in $b = 2^s$ blocks of size $k = 2^{p-s}$ are given in Table 13.28 at the end of this chapter. These have been chosen to allow all main effects to be estimable and as many two-factor interactions as possible. The first block of each design can be obtained from the given relations on the factor levels, and the other blocks can be obtained by adding (modulo 2) new treatment combinations to those in Block I. This process was illustrated in Example 13.4.2, page 436.

If one or more of the listed confounded contrasts is an important contrast in the experiment being designed, the factors should be relabeled. For example, suppose that a design is required for a 2^6 experiment in 8 blocks of 8. The design listed in Table 13.28 confounds BCD, ABE, $ACDE$, ADF, $ABCF$, $BDEF$, and CEF (this is also the design of Table 13.10). Suppose that the experimenter wished to estimate the contrasts ABC and BCD. The design, as listed, confounds BCD. However, if the experimenter were to switch the labels B and E of the actual treatment factors, then the important contrasts would be called ACE and CDE, both of which can be estimated in the listed design. Equivalently, the experimenter could switch the labels B and E of the listed design, so as to confound CDE, ABE, $ABCD$, ADF, $ACEF$, $BDEF$, and BCF, leaving ABC and BCD unconfounded.

13.7 Multireplicate Designs

When the experiment is large enough that each treatment combination can be observed $r > 1$ times, we have a choice of several different ways to design the experiment. If the block size can be chosen to be as large as the number of treatment combinations, then a randomized block design (Chapter 10) can be used.

If an incomplete block design is required, we could choose to use one of the standard incomplete block designs described in Chapter 11, such as a balanced incomplete block design. Alternatively, we could take a single-replicate design, confounding one or more interaction contrasts, and repeat the design r times. Alternatively, again, we could piece together r different single-replicate designs, confounding different contrasts in each. The confounding of a contrast in some but not all replicates is called *partial confounding*.

The choice between these three types of incomplete block designs involves tradeoffs concerning what is to be estimated and how accurately. The analysis of a standard incomplete block design of Chapter 11 requires that *every* contrast estimate be adjusted for blocks. Thus, while all contrasts are estimable, their least squares estimators have higher variances than they would if they were unadjusted for blocks—there is some *loss of information* on each contrast. If a balanced incomplete block design is used, then there is the same loss of information on every treatment contrast.

A repeated single-replicate experiment, on the other hand, loses information completely on the confounded contrasts—they are not estimable and are said to be *completely confounded*—but all contrasts orthogonal to these are unconfounded and can be estimated with no block adjustment and so with no loss of information. The estimator of any unconfounded contrast is the same as it would be in a complete block design, and with the same variance formula. The use of smaller blocks should, however, result in a smaller error variance σ^2.

The designs with partial confounding fall between these two extremes, allowing all contrasts to be estimable but with different levels of adjustment and loss of information. Complete and partial confounding are illustrated in Sections 13.8–13.9 and are compared via an example in Section 13.10.

13.8 Complete Confounding: Repeated Single-Replicate Designs

If the number of treatment combinations v is divisible by the block size k, the v/k blocks of a single-replicate design can be repeated r times to give an incomplete block design with $b = rv/k$ blocks. The contrasts that are confounded in the single-replicate design are also confounded in the r-replicate design—they cannot be estimated and are said to be *completely confounded*. For all other contrasts, the estimators and corresponding variance formulae are as in a complete block design—no block adjustments are needed.

13.8.1 A Real Experiment—Decontamination Experiment

An experiment was described by M. K. Barnett and F. C. Mead, Jr. in the journal *Applied Statistics* in 1956 to explore the effect of four factors on the efficiency of a decontamination process for the removal of radioactive isotopes from liquid waste. The four treatment factors were:

A: The amount of aluminum sulphate added to the liquid waste (two levels, 0.4 g and 2.5 g per liter, coded 0, 1).

B: The amount of barium chloride added to the liquid waste (two levels, 0.4 g and 2.5 g per liter, coded 0, 1).

C: The amount of carbon added to the liquid waste (two levels, 0.08 g and 0.4 g per liter, coded 0, 1).

Table 13.13 Repeated single-replicate design for a 2^4 experiment in 2×2 blocks of size 8, confounding $ABCD$. Data for the decontamination alpha-particle experiment are shown in parentheses.

Block	Treatment Combinations (Response)							
I	1010 (183)	1111 (350)	0110 (188)	0000 (881)	1100 (225)	0101 (298)	0011 (1039)	1001 (466)
II	0010 (650)	0001 (1180)	0111 (238)	1000 (191)	1101 (420)	0100 (289)	1110 (135)	1011 (781)
III	0101 (273)	1001 (890)	1100 (370)	0000 (834)	1010 (193)	1111 (389)	0110 (163)	0011 (1146)
IV	0001 (1193)	1110 (156)	1000 (257)	0100 (178)	0111 (254)	1011 (775)	0010 (494)	1101 (429)

Source: Barnett, M. K. and Mead, F. C. Jr. (1956). Copyright © 1956 Blackwell Publishers. Reprinted with permission.

D: Final pH of liquid waste (two levels, 6 and 10, coded 0, 1) achieved by adding sodium hydroxide or hydrochloric acid.

The experimental units were portions of a typical laboratory waste of pH 8.3 and in which the principal radioactivity was attributable to salts of radium, thorium, and actinium. The measurements taken after the experimental decontamination process were the counts per minute per milliliter of alpha and beta particles. Here, we will only reproduce the data for the alpha particles (see Table 13.13).

Only 8 of the 16 treatment combinations could be examined per day. Four days were available for the experiment, allowing each treatment combination to be measured twice during the course of the experiment. The experimenters anticipated day-to-day variations in the observations and decided to run a block design with $b = 4$ blocks of size $k = 8$. In fact, an unforeseen change of operators became necessary at the end of the first day. Since a block design had been used, any shift in the observations due to the operator change was absorbed into the block differences and did not affect the measurements on the treatment combinations.

The experimenters first selected a single-replicate design in $b^* = 2$ blocks of size $k = 8$ that confounded the 4-factor interaction contrast $ABCD$ (which they thought unlikely to exist in this experiment). By repeating this single-replicate design twice, they obtained a design with $r = 2$ observations per treatment combination and with $b = 2b^* = 4$ blocks in which all contrasts except for $ABCD$ could be measured without adjustments for blocks. (The single-replicate design selected is that listed in Table 13.28.) The treatment combinations were randomly ordered within each block, and the final design is shown in Table 13.13.

The chosen model included all main effects and all 2-factor and 3-factor interactions. The treatment–block interaction was assumed to be negligible. Thus, the model was

$$Y_{hijkl} = \mu + \theta_h + \alpha_i + \beta_j + \gamma_k + \delta_l$$
$$+ (\alpha\beta)_{ij} + (\alpha\gamma)_{ik} + (\alpha\delta)_{il} + (\beta\gamma)_{jk} + (\beta\delta)_{jl} + (\gamma\delta)_{kl}$$
$$+ (\alpha\beta\gamma)_{ijk} + (\alpha\beta\delta)_{ijl} + (\alpha\gamma\delta)_{ikl} + (\beta\gamma\delta)_{jkl} + \epsilon_{hijkl},$$

$\epsilon_{hijkl} \sim N(0, \sigma^2)$ and mutually independent,

$h = 1, 2, 3, 4; \quad i = 0, 1; \quad j = 0, 1; \quad k = 0, 1; \quad l = 0, 1;$

(h, i, j, k, l) in the design.

The analysis of variance table is shown in Table 13.14. There are $b = 4$ blocks in the design, but $ABCD$ is the only treatment contrast confounded. The sum of squares for $ABCD$ is included in the block sum of squares. The sums of squares for the other interactions and main effects can be obtained from rules 4 or 12 in Section 7.3. For example, the sum of squares for testing the hypothesis of negligible AC interaction is

$$ss(AC) = \sum_i \sum_k \frac{y_{.i.k.}^2}{8} - \sum_i \frac{y_{.i...}^2}{16} - \sum_k \frac{y_{...k.}^2}{16} + \frac{y_{.....}^2}{32}$$

$$= \frac{1}{8}(5126^2 + 4172^2 + 3248^2 + 2962^2) - \frac{1}{16}(9298^2 + 6210^2)$$

$$- \frac{1}{16}(8374^2 + 7134^2) + \frac{1}{32}(15508^2)$$

$$= 7{,}875{,}551 - 7{,}813{,}556.5 - 7{,}563{,}614.5 + 7{,}515{,}564.5$$

$$= 13{,}944.5.$$

Alternatively, we can calculate the sum of squares for the AC contrast as follows. The contrast coefficients, which are 1 if the levels of factors A and C are both high or both low, and -1 otherwise, are

$[1, \ 1, \ -1, \ -1, \ 1, \ 1, \ -1, \ -1, \ -1, \ -1, \ 1, \ 1, \ -1, \ -1, \ 1, \ 1 \ ].$

Then, as in Section 4.3.3, we have

$$ss(AC) = \frac{\left(\sum_i \sum_j \sum_k \sum_l c_{ijkl} \overline{y}_{.ijkl}\right)^2}{\left(\sum_i \sum_j \sum_k \sum_l c_{ijkl}^2/r\right)} = \frac{(334)^2}{16/2} = 13{,}944.5.$$

The error sum of squares is obtained by subtracting the sums of squares for the main effects and interactions from the total sum of squares, where the latter is

$$sstot = \sum_h \sum_i \sum_j \sum_k \sum_l y_{hijkl}^2 - \frac{G^2}{32}.$$

Similarly, the error degrees of freedom are obtained by subtraction.

There are fourteen hypothesis tests to be done. If each is done at a significance level 0.01, the overall level is at most $\alpha = 0.14$. At this level $F_{1,14,.01} = 8.86$, and the significant effects are the AB and BD interactions (and the main effects of A, B, and D averaged over the levels of the other factors). The hypotheses of negligible C main effect and ABC interaction would be rejected at a slightly higher significance level.

Table 13.14 Analysis of variance for the decontamination experiment

Source of Variation	Degrees of Freedom	Sum of Squares	Ratio	p-value
Blocks	3	28262.500	–	–
A	1	297992.000	40.25	0.0001
B	1	1444150.125	195.07	0.0001
C	1	48050.000	6.49	0.0232
D	1	700336.125	94.60	0.0001
AB	1	570846.125	77.11	0.0001
AC	1	13944.500	1.88	0.1915
AD	1	22366.125	3.02	0.1041
BC	1	15.125	0.00	0.9646
BD	1	252050.000	34.05	0.0001
CD	1	24531.125	3.31	0.0902
ABC	1	38226.125	5.16	0.0394
ABD	1	144.500	0.02	0.8909
ACD	1	66.125	0.01	0.9260
BCD	1	5618.000	0.76	0.3984
Error	14	103645.000		
Total	31	3550243.500		

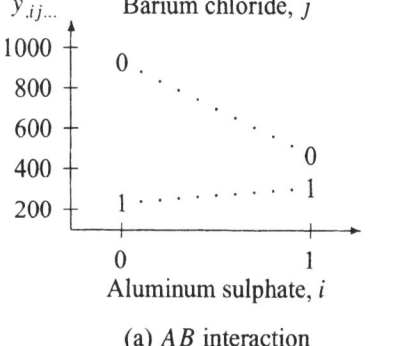
(a) AB interaction

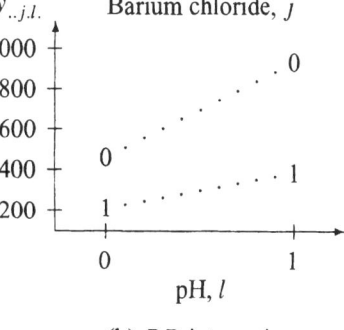
(b) BD interaction

Figure 13.5 Interaction plots for the decontamination experiment

Interaction plots of the two most important interactions (AB and BD) are shown in Figures 13.5(a) and 13.5(b). Figure 13.5(a) suggests that B should be set at its high level to achieve a lower radioactivity. The interaction is caused by the fact that the benefit of setting B at its high level is more marked when A is at its low level than at its high level. If B is at its high level, there is a slight preference for A to be at its low level to achieve a lower radioactivity. On the other hand, the system is more stable when A is at its high level, meaning that the radioactivity is not so sensitive to the level of B. So the choice for the setting of A is not completely obvious. Figure 13.5(b) shows a similar picture. Again B should be at its high level with a preference for D at its low level (which also produces the more stable system). The main effect of C is not significant, but the data suggest that it should be set at its high level unless this increases the cost substantially.

Thus, the results of the analysis of variance table suggest that suitable treatment combinations for the decontamination process are 0110 or 1110. These are the same two treatment combinations that would be selected from a perusal of the data in Table 13.13. The benefit of the investigation of main effects and interactions is that it suggests directions for future experiments (raising the level of B further while keeping D fairly low and perhaps raising A a little). It also allows the chemical analysts to better understand the nature of the chemical system (see the original article of Barnett and Mead). Finally, it helps to ensure that the treatment combination that appears to be the best is not just the result of error variability or a spurious observation (outlier).

13.9 Partial Confounding

Partial confounding is the term applied to a design that is the combination of the blocks from different single-replicate designs with different confounding schemes. A contrast that is confounded in some replicates but not in others is said to be *partially confounded* with blocks. A partially confounded contrast can be estimated using only the data of those replicates in which it is unconfounded. Thus, the variance of the contrast estimator is inversely proportional to the number of replicates in which it is estimable.

For example, the design in Table 13.15 for a 2^3 experiment in 8 blocks of size 4 has four observations on each treatment combination. It is made up of four single-replicate designs; the first confounds the contrast from the ABC interaction, while the second confounds the AB contrast, the third confounds AC, and the fourth BC. This means that the ABC contrast is estimable from the second, third, and fourth single-replicate designs, but not the first, and the AB contrast is estimable from the first, third, and fourth single-replicate designs, but not the second. Similarly, the AC and BC contrasts are estimable from three of the four replicates. The main-effect contrasts A, B, and C are estimable from all four replicates. Consequently, all factorial treatment contrasts can be estimated, but the variance associated with each partially confounded contrast will be larger than that associated with each of the unconfounded contrasts by a factor of four-thirds. The benefit of using a partially confounded design instead of a repeated single-replicate design is that each treatment contrast is estimable, yet all totally unconfounded contrasts are still estimated with the maximum possible precision.

Example 13.9.1 Coil experiment

C. Hicks, in his textbook *Fundamental Concepts in the Design of Experiments*, describes an experiment to examine the variability of outside diameters of coils of wire. There were three treatment factors of interest, as follows:

A: Two winding machines, coded 0, 1.

B: Two wire stocks, coded 0, 1.

C: Two positions on the coil, coded 0, 1.

13.9 Partial Confounding

Table 13.15 Design and data (in parentheses) for the coil experiment

Confounded	Block	Treatment Combination (Response)			
ABC	I	000 (2208)	110 (2133)	101 (2459)	011 (3096)
	II	100 (2196)	010 (2086)	001 (3356)	111 (2776)
AB	III	000 (2004)	110 (2112)	001 (3073)	111 (2631)
	IV	100 (2179)	010 (2073)	101 (3474)	011 (3360)
AC	V	001 (2839)	100 (2189)	011 (3522)	110 (2095)
	VI	000 (1916)	101 (2979)	010 (2151)	111 (2500)
BC	VII	100 (2056)	000 (2010)	011 (3209)	111 (3066)
	VIII	010 (1878)	110 (2156)	001 (3423)	101 (2524)

Source: From *Fundamental Concepts in the Design of Experiments, Fourth Edition*, by Charles R. Hicks. Copyright © 1964, 1973, 1982, 1993 by Oxford University Press, Inc. Used by permission of Oxford University Press, Inc.

Only four of the $v = 8$ treatment combinations could be measured at any one time. Consequently, the experiment was divided into blocks of size $k = 4$. A total of $n = 32$ observations could be taken, and a block design of $b = 8$ blocks of size $k = 4$ was needed.

It is easily verified that no balanced incomplete block design exists of this size (since $r = 4$ and $\lambda = 12/7$). A cyclic design could have been used, but cyclic designs do not have orthogonal factorial structure in general. A partially confounded design was selected, consisting of four single-replicate designs each confounding a different interaction (ABC, AB, AC, and BC). The design and responses are shown in Table 13.15.

We use the standard model for 3 treatment factors and one block factor with no treatment block interaction.

$$\begin{aligned} Y_{hijk} &= \mu + \theta_h + \tau_{ijk} + \epsilon_{hijk} \\ &= \mu + \theta_h + \alpha_i + \beta_j + \gamma_k + (\alpha\beta)_{ij} + (\alpha\gamma)_{ik} \\ &\quad + (\beta\gamma)_{jk} + (\alpha\beta\gamma)_{ijk} + \epsilon_{hijk}, \end{aligned}$$

$\epsilon_{hijk} \sim N(0, \sigma^2)$ and mutually independent,

$h = 1, \ldots, 8; \quad i = 0, 1; \quad j = 0, 1; \quad k = 0, 1;$

(h, i, j, k) in the design.

The contrast for comparing the first two levels of factor B (averaged over the levels of A and C), for example, is estimated using all of the data. Since the B contrast is orthogonal to blocks, no block adjustments are needed. Consequently, the contrast

$$\beta_1^* - \beta_0^*, \quad \text{where } \beta_j^* = \beta_j + \overline{(\alpha\beta)}_{.j} + \overline{(\beta\gamma)}_{j.} + \overline{(\alpha\beta\gamma)}_{.j.},$$

has least squares estimate

$$\hat{\beta}_1^* - \hat{\beta}_0^* = \frac{y_{..1.}}{16} - \frac{y_{..0.}}{16} = 2552.75 - 2555.31 = -2.56.$$

Equivalently, in terms of the treatment combinations the contrast is $\sum_i \sum_j \sum_k c_{ijk} \tau_{ijk}$, where the coefficients c_{ijk} in standard order are given by

$$\tfrac{1}{4}[-1, \ -1, \ 1, \ 1, \ -1, \ -1, \ 1, \ 1\,].$$

with least squares estimate $\sum_i \sum_j \sum_k c_{ijk}(y_{.ijk}/4) = -2.56$.

The test of the hypothesis $H_0 : \{\beta_1^* - \beta_0^* = 0\}$ against the alternative hypothesis $H_A : \{\beta_1^* - \beta_0^* \neq 0\}$ is similar to (4.3.15) (page 76); that is,

$$\text{reject } H_0 \text{ if } \frac{ssc}{msE} = \frac{(\overline{y}_{..1.} - \overline{y}_{..0.})^2}{(2/16)msE} > F_{1,df,\alpha/m},$$

where df is the error degrees of freedom obtained from the analysis of variance table, which is shown in Table 13.16, m is the number of hypotheses to be tested, and $ssc = 8(-2.56)^2 = 52.53$. To test the equivalent hypothesis $H_0^B : \{\beta_0^* = \beta_1^*\}$ using the rules of Chapter 7, we obtain

$$ssB = acr \sum_j \overline{y}_{.j.}^2 - abcr\overline{y}_{....}^2 = 52.53.$$

The sum of squares for each of the other main effects can be calculated in a similar fashion. The sum of squares for the interactions can be calculated similarly, except that only three of the four replicates are used. For example, the BC interaction contrast can be estimated only from the first three replicates (that is, from 24 observations, not 32). Thus, the contrast $\tfrac{1}{2}[(\beta\gamma)_{00}^* - (\beta\gamma)_{01}^* - (\beta\gamma)_{10}^* + (\beta\gamma)_{11}^*]$ (where $(\beta\gamma)_{jk}^* = (\beta\gamma)_{jk} + \overline{(\alpha\beta\gamma)}_{.jk}$) has least squares estimate

$$\frac{1}{2}\left[\frac{y_{..00}}{6} - \frac{y_{..01}}{6} - \frac{y_{..10}}{6} + \frac{y_{..11}}{6}\right]$$
$$= \frac{1}{2}[2080.86 - 3029.76 - 2099.38 + 3006.11]$$
$$= -21.085,$$

and only six observations y_{hijk} are used in the calculation of $y_{..jk}$ for each combination of B and C. In terms of the treatment combinations, the contrast coefficients c_{ijk} are

$$\tfrac{1}{4}[1, \ -1, \ -1, \ 1, \ 1, \ -1, \ -1, \ 1\,],$$

and the contrast estimate is

$$\sum\sum\sum c_{ijk}(y_{.ijk}/3) = -21.085,$$

where again, only the three observations from the first three single-replicate designs are used in the calculation of each $y_{.ijk}$.

To test the hypothesis $H_0^{BC} : [(\beta\gamma)_{00}^* - (\beta\gamma)_{01}^* - (\beta\gamma)_{10}^* + (\beta\gamma)_{11}^* = 0]$, we reject H_0^{BC} if

$$\frac{(-21.083)^2}{(4/24)msE} = 2667.04 > F_{1,17,\alpha/m}.$$

The complete analysis of variance table is shown in Table 13.16. The adjusted mean square for blocks, which is 43,344.88, is smaller than the mean square for error. Thus, blocking was not helpful in this experiment. □

Table 13.16 Analysis of variance for the coil experiment

Source of Variation	Degrees of Freedom	Sum of Squares	Mean Square	Ratio	p-value
Blocks (adj)	7	303,414.14	43,344.88	—	
Blocks (unadj)	7	439,777.72	—	—	
A	1	224,282.53	224,282.53	3.97	0.0627
B	1	52.53	52.53	0.00	0.9760
C	1	6,886,688.28	6,886,688.28	121.79	0.0001
AB (adj)	1	737.04	737.04	0.01	0.9104
AC (adj)	1	416,066.67	416,066.67	7.36	0.0148
BC (adj)	1	2,667.04	2,667.04	0.05	0.8307
ABC (adj)	1	70,742.04	70,742.04	1.25	0.2789
Error	17	961,283.11	56,546.07		
Total	31	9,002,296.97			

Table 13.17 A balanced incomplete block design with 8 treatment labels and 14 blocks of size 4

						Blocks							
I	II	III	IV	V	VI	VII	VIII	IX	X	XI	XII	XIII	XIV
1	5	1	3	1	2	1	2	1	3	1	2	1	2
2	6	2	4	3	4	4	3	2	4	3	4	4	3
3	7	7	5	6	5	6	5	5	7	5	6	5	6
4	8	8	6	8	7	7	8	6	8	7	8	8	7

13.10 Comparing the Multireplicate Designs

For some block sizes, we have a choice of possible designs for a multireplicate factorial experiment. For example, suppose that a design is required with blocks of size $k = 4$ for a factorial experiment involving 3 factors, each having two levels (so $v = 8$). Practical considerations dictate that at most $b = 14$ blocks can be used. The contrasts of interest are all of the main-effect and interaction contrasts with coefficients ± 1 (for simplicity) as listed in Table 13.2 (page 425). Three possible ways of designing the experiment with 14 blocks are as follows.

Design possibility 1 The first possibility is to use a balanced incomplete block design, since one exists with $\lambda = 3$ and $r = 7$, $b = 14$, $v = 8$, $k = 4$. The design (before randomization) is given in Table 13.17, where the blocks are shown as columns. The eight treatment labels of the design are randomly assigned to the $v = 8$ treatment combinations $(000, 001, \ldots, 111)$ to obtain a balanced incomplete block design suitable for a factorial experiment.

If we let the effect of treatment combination ijk be denoted by τ_{ijk}, the least squares estimator of a contrast $\Sigma\Sigma\Sigma c_{ijk}\tau_{ijk}$ is $\Sigma\Sigma\Sigma c_{ijk}\widehat{\tau}_{ijk} = \frac{k}{\lambda v}\Sigma\Sigma\Sigma c_{ijk}Q_{ijk}$, where $Q_{ijk} = T_{ijk} - \frac{1}{k}\sum_h n_{hijk}B_h$, and T_{ijk} is the total of the observations on treatment combination ijk, B_h is the total of the observations in the hth block, and n_{hijk} is 1 if treatment combination

ijk is in block h, and otherwise n_{hijk} is 0. The variance of the least squares estimator is

$$\text{Var}\left(\sum_i\sum_j\sum_k c_{ijk}\hat{\tau}_{ijk}\right) = \frac{k}{\lambda v}\sum_i\sum_j\sum_k c_{ijk}^2\,\sigma^2 = \frac{8\sigma^2}{6}, \qquad (13.10.3)$$

and this is the same for all main-effect and interaction contrasts for a 2^3 experiment, (since all of the c_{ijk}'s are $+1$ or -1).

Design possibility 2 For the same experiment, suppose that the 3-factor interaction ABC is expected to be negligible and is of no interest. Then the balanced incomplete block design discussed above is not ideal, because ABC contrast is measured with the same precision as the main-effect and 2-factor interaction contrasts. Suppose, instead, we decide to confound the ABC contrast in each of seven replicates. Using the equations

$$a_1 + a_2 + a_3 = 0 \text{ or } 1 \text{ (mod 2)},$$

the following single-replicate design in two blocks would be obtained:

Block I: 000   011   101   110
Block II: 001   010   100   111

The design in $b = 14$ blocks is obtained by repeating these two blocks $r = 7$ times. Since ABC is confounded in every replicate, it is not estimable—it cannot be measured. ABC is said to be *completely confounded*. All other orthogonal contrasts (including the main-effect and 2-factor interaction contrasts) are unconfounded, so can be estimated without adjusting for blocks.

Let $\Sigma\Sigma\Sigma c_{ijk}\tau_{ijk}$ be a contrast measuring a 2-factor interaction or a main effect. Then its least squares estimator is $\Sigma\Sigma\Sigma c_{ijk}\overline{Y}_{.ijk}$, where the average is taken over the 7 replicates or repeated pairs of blocks. The corresponding variance is

$$\text{Var}\left(\Sigma\Sigma\Sigma c_{ijk}\hat{\tau}_{ijk}\right) = \text{Var}\left(\Sigma\Sigma\Sigma c_{ijk}\overline{Y}_{.ijk}\right) = \frac{8\sigma^2}{7}. \qquad (13.10.4)$$

Comparing (13.10.4) with (13.10.3), we see that the effect of losing all of the information on the ABC contrast is to reduce the variance of all other factorial contrasts from $8\sigma^2/6$ to $8\sigma^2/7$.

Design possibility 3 Instead of repeating the same design seven times as was done above, we could try to spread the loss of information due to confounding across several of the interaction contrasts by using partial confounding. Suppose that we take four copies of the two blocks that confound ABC, together with one pair of blocks that confounds AB, one pair that confounds AC, and one pair that confounds BC. This seven-replicate design is shown in Table 13.18.

The ABC contrast is confounded in replicates I–IV, but can be estimated (without block adjustments) from replicates V–VII (that is, from Blocks IX–XIV, or three pairs of blocks). The variance of the ABC contrast estimator is then $8\sigma^2/3$, compared with $8\sigma^2/6$ in the balanced incomplete block design—it is completely confounded in the second design.

Each 2-factor interaction contrast can be estimated without block adjustments from the six replicates, or pairs of blocks, in which it is not confounded. The variances of their least-

Table 13.18 Partial confounding of ABC, AB, AC, BC contrasts in a design with 14 blocks of size 4 for 3 factors at two levels each

Replicates	Confounded	Blocks	Treatment Combinations
I–IV	ABC	I, III, V, VII	000 011 101 110
		II, IV, VI, VIII	001 010 100 111
V	AB	IX	000 001 110 111
		X	100 101 110 011
VI	AC	XI	000 010 101 111
		XII	001 011 100 110
VII	BC	XIII	000 011 100 111
		XIV	001 010 101 110

Table 13.19 Variances of contrast estimators for three design possibilities for $v = 8, r = 7, b = 14, k = 4$

Contrast	Design 1 BIBD	Design 2 Complete Confounding of ABC	Design 3 Partial Confounding of ABC, AB, AC, BC
ABC	$\frac{8\sigma^2}{6}$	not estimable	$\frac{8\sigma^2}{3}$
AB, AC, BC	$\frac{8\sigma^2}{6}$	$\frac{8\sigma^2}{7}$	$\frac{8\sigma^2}{6}$
A, B, C	$\frac{8\sigma^2}{6}$	$\frac{8\sigma^2}{7}$	$\frac{8\sigma^2}{7}$

squares estimators are then $8\sigma^2/6$, the same as in the balanced incomplete block design, but worse than the value $8\sigma^2/7$ for the design with ABC completely confounded.

The main effects can be estimated from all seven replicates. The variances of their contrast estimators are all $8\sigma^2/7$, the same as for the second design, but better than the value of $8\sigma^2/6$ for the balanced incomplete block design.

Summary A summary of the variances of the least squares estimators of the factorial contrasts is given in Table 13.19. No one design is the best for all seven factorial effects. The choice of design would depend upon the importance of estimating the ABC contrast relative to the 2-factor and main-effect contrasts.

We have not exhausted all the possible designs that can be obtained by partial confounding. For example, one could confound the two-factor interactions in two pairs of blocks and the three-factor interaction in one pair of blocks, or alternatively, one could confound each of the seven factorial contrasts in turn in one pair of blocks. (This latter option gives design possibility 1.) In every case, the smaller contrast variances will coincide with the contrasts that are confounded less often, the contrast variance being $8\sigma^2/r$ if the contrast is unconfounded in r replicates.

13.11 Using SAS Software

Analyzing factorial experiments with confounding using the SAS software is straightforward for the types of designs discussed in this chapter. The SAS statements required for the analysis are the same as those outlined in Chapters 6, 7, and 10. Using the GLM procedure, the blocking factor is listed in the MODEL statement first, so that the Type I sums of squares for factorial effects are appropriately adjusted for blocks.

Any effect that is completely confounded—including effects confounded in a single replicate design—should *not* be included in the MODEL statement. If included after the blocking factor, a completely confounded effect would show zero degrees of freedom under the Type I and Type III sums of squares. The corresponding degree of freedom would already be accounted for under block effects. Partially confounded effects, however, should be included in the model statement as illustrated in the following example.

Example 13.11.1 Partial confounding—Coil experiment, continued

Table 13.20 contains a SAS program for analysis of the coil experiment data. Corresponding output is given in Table 13.21. The coil experiment was a four-replicate 2^3 experiment with partial confounding—each of the four interaction effects was confounded in one of the four

Table 13.20 SAS program for analysis of an experiment with partial confounding—the coil experiment

```
DATA COIL;
 INPUT BLOCK A B C Y;
 LINES;
   1  0 0 0  2208
   1  1 1 0  2133
   1  1 0 1  2459
   1  0 1 1  3096
   2  1 0 0  2196
   :  : : :    :
   8  1 1 0  2156
   8  0 0 1  3423
   8  1 0 1  2524
 ;
PROC GLM;
  CLASS BLOCK A B C;
  MODEL Y = BLOCK A B C A*B A*C B*C A*B*C;
  ESTIMATE 'A'  A -1 1;
  ESTIMATE 'B'  B -1 1;
  ESTIMATE 'C'  C -1 1;
  ESTIMATE 'AB' A*B  1 -1 -1  1 / DIVISOR=2;
  ESTIMATE 'AC' A*C  1 -1 -1  1 / DIVISOR=2;
  ESTIMATE 'BC' B*C  1 -1 -1  1 / DIVISOR=2;
  ESTIMATE 'ABC' A*B*C -1 1 1 -1 1 -1 -1 1 / DIVISOR=4;
```

13.11 Using SAS Software

Table 13.21 SAS program partial output illustrating partial confounding—the coil experiment

```
                       The SAS System
                 General Linear Models Procedure

Dependent Variable: Y
                              Sum of        Mean
Source              DF       Squares      Square    F Value    Pr > F
Model               14      8041013.9     574358.1    10.16    0.0001
Error               17       961283.1      56546.1
Corrected Total     31      9002297.0

Source              DF    Type I SS    Mean Square   F Value   Pr > F
BLOCK                7     439777.7       62825.4      1.11    0.4003
A                    1     224282.5      224282.5      3.97    0.0627
B                    1         52.5          52.5      0.00    0.9760
C                    1    6886688.3     6886688.3    121.79    0.0001
A*B                  1        737.0         737.0      0.01    0.9104
A*C                  1     416066.7      416066.7      7.36    0.0148
B*C                  1       2667.0        2667.0      0.05    0.8307
A*B*C                1      70742.0       70742.0      1.25    0.2789

Source              DF   Type III SS   Mean Square   F Value   Pr > F
BLOCK                7     303414.1       43344.9      0.77    0.6225

                               T for H0:    Pr > |T|   Std Error of
Parameter      Estimate      Parameter=0                Estimate
A             -167.437500       -1.99       0.0627     84.0729338
B               -2.562500       -0.03       0.9760     84.0729338
C              927.812500       11.04       0.0001     84.0729338
AB              11.083333        0.11       0.9104     97.0790619
AC            -263.333333       -2.71       0.0148     97.0790619
BC             -21.083333       -0.22       0.8307     97.0790619
ABC           -108.583333       -1.12       0.2789     97.0790619
```

replicates. In the GLM procedure, the blocking factor is entered into the MODEL statement first. As a result, the Type I sum of squares for blocks is unadjusted for treatment effects, whereas the Type I sums of squares for each treatment interaction effect is adjusted for block effects. The Type III sum of squares for blocks is adjusted for treatment effects, so can be used to assess the usefulness of blocking in this experiment.

The divisors used in the ESTIMATE statements of the GLM procedure cause use of divisor $v/2 = 4$ for the contrast coefficients c_{ijk} for each contrast. Thus, all contrasts would have been estimated with the same variance had there been no partial confounding. Because each interaction contrast is confounded in one of the four replicates, the variance of each interaction contrast estimator is larger than each main effect contrast estimator by a factor of four-thirds.   □

Exercises

1. Construct a single-replicate 2^3 design confounding AB with blocks. In other words, list the treatment combinations block by block.

2. Construct a single replicate 2^5 design confounding ABC and CDE. Determine the other effect that is confounded.

3. **Projectile experiment**

   N. L. Johnson and F. C. Leone, in their 1977 book *Statistics and Experimental Design in Engineering and the Physical Sciences*, described a single-replicate 2^4 experiment concerning the performance of a new rifle under test. Under study were the effects on projectile velocity of the factors charge weight (A), projectile weight (B), propellant web (C), and weapon (D), where two rifles were used. The design included two blocks each of size eight, corresponding to the two days on which data were collected, confounding $ABCD$. The coded velocity data are given in Table 13.22.

   (a) Fit a model including block effects, treatment main effects, and 2-factor interactions. Use residual plots to check the standard model assumptions.

   (b) Conduct the analysis of variance, and discuss the results.

   (c) Construct simultaneous confidence intervals for any interesting treatment contrasts using an appropriate method of multiple comparisons.

   (d) Reanalyze the data using the Voss–Wang method, including all estimable treatment effects in the analysis.

4. **Field experiment, continued**

   (a) For the field experiment of Example 13.3.2, verify that the sum of squares and the contrast estimate for BD are as shown in Table 13.5, page 428.

   (b) Draw the BD interaction plot. Does this plot also suggest that B and D should be at their low levels?

Table 13.22 Projectile experiment data, confounding $ABCD$

	Day 1			Day 2	
Run	TC	y_{1ijkl}	Run	TC	y_{2ijkl}
1	0000	97	13	0001	75
7	0011	26	11	0010	39
5	0101	53	9	0100	68
3	0110	15	15	0111	−16
6	1001	145	10	1000	151
4	1010	100	16	1011	97
2	1100	150	14	1101	141
8	1111	54	12	1110	66

Source: Johnson, N. L. and Leone, F. C. (1977). Copyright © 1977 Johnson and Leone. Reprinted with permission.

(c) Suppose that the experimenters had expected all of the 3-factor interactions to be negligible and had omitted the corresponding terms from the model (instead of those involving AD). Reanalyze the experiment accordingly. What would have been concluded?

(d) Apply the Voss–Wang method to analyze the data of the field experiment. Relevant information is given in Table 13.5, page 428.

5. Suggest a confounding scheme for a 2^6 experiment in 8 blocks of 8, assuming that all 2-factor interactions are to be estimated, as are the 3-factor interactions involving both A and F. List all effects confounded. List the treatment combinations in the design block by block.

6. Suggest a confounding scheme for a 2^8 experiment in 16 blocks of 16, assuming that all 2-factor and 3-factor interactions are to be estimated. List all effects confounded. List the treatment combinations in Block I and in two other blocks.

7. **Mangold experiment, continued**

   (a) For the mangold experiment of Section 13.5, verify that the sum of squares and the contrast estimate for CD are as shown in Table 13.12, page 439.

   (b) Draw the CD interaction plot. Does this plot agree with the factor levels suggested in Section 13.5 for increasing the yield?

   (c) Check that the assumption of normality of the error variables is satisfied. Also check that the variances of the errors appear to be equal for each level of the four factors.

   (d) Draw a normal probability plot of all of the contrast estimates (including the higher-order interactions). Does it appear that the experimenters made the correct assumptions of negligible higher-order interactions?

8. **Decontamination experiment—Beta particles**

   An experiment was described by M. K. Barnett and F. C. Mead, Jr. in the journal *Applied Statistics* in 1956 to explore the effect of four factors on the efficiency of a decontamination process for the removal of radioactive isotopes from liquid waste. The measurements taken after the decontamination process were the counts per minute per milliliter of alpha and beta particles. Data for the alpha particles and further description of the experiment were given in Section 13.8.1. We consider here part of the data for the beta particles, shown in Table 13.23. The four treatment factors were:

   A: 0.4 g and 2.5 g per liter of aluminum sulphate (coded 0, 1);

   B: 0.4 g and 2.5 g per liter of barium chloride (coded 0, 1);

   C: 0.08 g and 0.4 g per liter of carbon (coded 0, 1);

   D: Final pH of liquid waste (6 and 10, coded 0, 1).

   The experimenters selected a design in $b = 2$ blocks of $k = 8$ that confounded the four-factor interaction contrast $ABCD$. The fitted model included all main-effects and all 2-factor and 3-factor interactions. The treatment–block interaction was assumed to be negligible.

Table 13.23 Randomized design for a 2^4 experiment in 2 blocks of size 8, confounding $ABCD$. Data for the decontamination beta-particle experiment are shown in parentheses.

Block	Treatment Combinations (Response)							
I	1010	1111	0110	0000	1100	0101	0011	1001
	(716)	(686)	(498)	(1437)	(527)	(579)	(1433)	(906)
II	0010	0001	0111	1000	1101	0100	1110	1011
	(1024)	(1364)	(475)	(574)	(664)	(579)	(507)	(1130)

Source: Barnett, M. K. and Mead, F. C. Jr. (1956). Copyright © 1956 Blackwell Publishers. Reprinted with permission.

(a) Use a normal probability plot to identify the important contrasts.

(b) Use the method of Voss–Wang to check your selection in part (a).

(c) Draw an interaction plot of any interaction that appears to be nonnegligible by either analysis.

(d) Looking at the results of your analysis, which settings of the factors would you recommend for reducing the beta particle counts?

(e) Suppose that the experimenters had believed before the experiment that the three-factor interactions were all negligible. What would the analysis of variance table have looked like? Would your recommendations have been any different?

9. **Penicillin experiment**

An experiment is described in Example 9.2 of the book *Design and Analysis of Industrial Experiments* edited by O. L. Davies that investigates the effects of various factors on the yield of penicillin in surface culture experiments. The five factors of interest were added to the nutrient medium, which was inoculated with a spore suspension of *P. Chrysogenum*. The spores rise to the surface, causing the growth of mycelium accompanied by the formation of penicillin. The factors and their levels were corn steep liquor (factor A, 2% and 3% strength), lactose (factor B, 2% and 3% strength), precursor (factor C, 0% and 0.05%), sodium nitrate (factor D, 0% and 0.3%), and glucose (factor E, 0% and 0.5%). Only 16 of the 32 treatment combinations could be carried out at one time, and the experimenters decided to observe 16 treatment combinations in one week and the remaining 16 in the following week. Large week-to-week variations were known to exist, and therefore the experiment was designed as a block design with two blocks, confounding the 5-factor interaction $ABCDE$. The observed yields of penicillin are shown in Table 13.24. Prior to the experiment, it was believed that all that all 3- and 4-factor interactions would be negligible, and also that the CE interaction would be important.

(a) Analyze the data, assuming that all 3- and 4-factor interactions are negligible. Do not forget to check the assumptions on the model.

(b) The experimenters decided to use logarithms of the data. Does your assumption check in part (a) confirm that this should be done? If so, reanalyze the data and state your conclusions.

Table 13.24 Data for the penicillin experiment

Block I		Block II	
Treatment Combination	Yield	Treatment Combination	Yield
00000	142	00001	106
00011	101	00010	148
00101	113	00100	185
00110	200	00111	130
01001	88	01000	129
01010	146	01011	140
01100	200	01101	166
01111	145	01110	215
10001	106	10000	114
10010	108	10011	114
10100	162	10101	88
10111	83	10110	164
11000	109	11001	98
11011	72	11010	195
11101	79	11100	172
11110	118	11111	110

Source: Davies, O. L. (1963). Reprinted by permission of Addison Wesley Longman Ltd.

(c) Using the logarithms of the data, draw a normal probability plot of the contrast estimates without using any knowledge that the higher-order interactions are likely to be negligible. Do your conclusions remain the same? Which analysis do you prefer? Why?

10. **Peas experiment**

    The following experiment was run at Biggelswade, in England, and reported by F. Yates in his 1935 paper *Complex Experiments*. The three treatment factors were the standard fertilizers, nitrogen, phosphate, and potassium (factors N, P, and K) each at two levels. The experimental area was divided into $b = 6$ blocks of $1/70$ of an acre. Each block was large enough for four plots on which a certain variety of pea was sown, and the fertilizer combinations shown in Table 13.25 were added. The design consists of three identical single-replicate designs each of which confounds the 3-factor interaction NPK. Each block has been separately randomized.

    (a) Estimate the treatment contrasts for all main effects and interactions.

    (b) Calculate the analysis of variance table for this experiment and test all relevant hypotheses. State the overall significance level.

    (c) Draw interaction plots for any important interactions. Give a set of 95% confidence intervals for the main-effect contrasts, if appropriate.

    (d) State your overall recommendations about the fertilizers in this experiment. Would you recommend a followup experiment? If so, what would you investigate?

Exercises

Table 13.25 Data for the peas experiment

Block	Treatment Combinations (Yield)		Block	Treatment Combinations (Yield)	
I	011 (49.5)	000 (46.8)	II	100 (62.0)	001 (45.5)
	110 (62.8)	101 (57.0)		111 (48.8)	010 (44.2)
III	100 (59.8)	001 (55.5)	IV	110 (52.0)	101 (49.8)
	111 (58.5)	010 (56.0)		000 (51.5)	011 (48.8)
V	010 (62.8)	100 (69.5)	VI	101 (57.2)	011 (53.2)
	111 (55.8)	001 (55.0)		110 (59.0)	000 (56.0)

Source: Yates, F. (1935). Copyright © 1935 Blackwell Publishers. Reprinted with permission. (Reprinted in *Experimental Design* (1970), Charles Griffin and Company, Ltd., London. Copyright © 1970 Edward Arnold/Hodder & Stoughton Educational. Reprinted with permission.)

Table 13.26 Data for the field experiment, by block and treatment combination (TC)

Block I		Block II		Block III		Block IV	
TC	y_{1ijkl}	TC	y_{2ijkl}	TC	y_{3ijkl}	TC	y_{4ijkl}
0000	58	0001	55	0000	57	0001	50
0011	51	0010	45	0011	56	0010	39
0101	44	0100	42	0101	43	0100	47
0110	50	0111	36	0110	39	0111	43
1001	43	1000	53	1001	52	1000	42
1010	50	1011	55	1010	52	1011	44
1100	41	1101	41	1100	42	1101	34
1111	44	1110	48	1111	54	1110	52

Source: *Experimental Designs*, Second Edition, by W. G. Cochran and G. M. Cox, Copyright © 1957, John Wiley & Sons, New York. Adapted by permission of John Wiley & Sons, Inc.

11. **Field experiment, continued**

    The field experiment was described in Example 13.3.2. There were four treatment factors (A, B, C, and D) at two levels each, and the $v = 16$ treatment combinations were observed twice. Each of the $r = 2$ sets of treatment combinations were divided into blocks of size 8. The first two blocks, which confounded the $ABCD$ interaction, were shown in Table 13.4. The complete design, which is shown in Table 13.26, consisted of two such single-replicate designs.

    (a) Calculate the analysis of variance table for this experiment. Now that $r = 2$, there is an estimate for error variability. Test any hypotheses of interest. Are the results similar to those obtained from the first two blocks only?

    (b) Draw any interaction plots of interest. If the yield is to be increased, what recommendations would you make about the levels of the factors?

Exercises   459

Table 13.27 Data for the catalytic reaction experiment

Run	Block	TC	y_{hijk}	Run	Block	TC	y_{hijk}
1	1	011	89.5	9	3	010	86.2
2	1	101	84.2	10	3	100	81.8
3	1	110	85.2	11	3	001	90.4
4	1	000	89.9	12	3	111	83.6
5	2	010	85.1	13	4	110	75.3
6	2	111	83.5	14	4	000	84.6
7	2	001	90.8	15	4	011	86.7
8	2	100	81.8	16	4	101	82.2

Source: Reprinted with permission from Bainbridge, J. R. (1951). Copyright 1951 *American Chemical Society*.

12. Construct a four-replicate 2^3 design in eight blocks of size four, partially confounding each interaction effect. Compare the variance of each interaction contrast with that of each main effect, using divisor $v/2 = 4$ for each contrast.

13. **Catalytic reaction experiment**

    J. R. Bainbridge, in his 1951 article in the journal *Industrial and Engineering Chemistry*, described a factorial experiment conducted at a small plant carrying out a catalytic gaseous synthesis reaction to remove the product as a liquid solution. A 2-replicate 2^3 experiment was conducted to study the effects of converter reaction temperature (factor A), throughput rate through the converter (factor B), and the concentration of the active ingredient in the makeup gas (factor C) on each of several response variables, including the strength of the product solution (y_{hijk}). The design was composed of four blocks of size four, with the ABC interaction completely confounded. The design and data are provided in Table 13.27, including the run order. (The observations in Table 13.27 are "uncoded," each value being 80 plus one-tenth the coded value given by Bainbridge.)

    (a) Based on the run order, discuss how the design was probably randomized.

    (b) Fit an appropriate model, and use residual plots to check the standard model assumptions.

    (c) Conduct the analysis of variance, and discuss the results.

    (d) Using a simultaneous confidence level of 95% for all six factorial effects, construct confidence intervals for those effects found to be significant in the analysis of variance.

14. **Catalytic reaction experiment, continued**

    In the experiment described in Exercise 13, the covariate "makeup gas purity" was measured. The covariate values were 17, 12, 10, 10, 13, 14, 10, 16, 12, 13, 13, 11, 16, 11, 12, and 11, corresponding to runs 1–16, respectively. Repeat Exercise 13, but for an analysis of covariance.

Table 13.28 Confounding schemes for 2^p experiments in $b = 2^s$ blocks of size $k = 2^{p-s}$. For each design, s independent generators are underlined, and s corresponding equations are given. To obtain Block I of a design, list all k combinations of the first a_i's shown, then use the equations modulo 2 to complete each treatment combination.

2^p	b	k	Confounded Contrasts	Block I
2^3	2	4	<u>ABC</u>	a_1, a_2 $a_3 = a_1 + a_2$
2^4	2	8	<u>ABCD</u>	a_1, a_2, a_3 $a_4 = a_1 + a_2 + a_3$
2^4	4	4	<u>AC</u>, <u>ABD</u>, BCD	a_1, a_2 $a_3 = a_1$ $a_4 = a_1 + a_2$
2^5	2	16	<u>ABCDE</u>	a_1, a_2, a_3, a_4 $a_5 = a_1 + a_2 + a_3 + a_4$
2^5	4	8	<u>ABCD</u>, <u>ABE</u>, CDE	a_1, a_2, a_3 $a_4 = a_1 + a_2 + a_3$ $a_5 = a_1 + a_2$
2^5	8	4	<u>AC</u>, <u>BD</u>, ABCD, <u>ABE</u>, BCE, ADE, CDE	a_1, a_2 $a_3 = a_1$ $a_4 = a_2$ $a_5 = a_1 + a_2$
2^6	2	32	<u>ABCDEF</u>	a_1, a_2, a_3, a_4, a_5 $a_6 = a_1 + a_2 + a_3 + a_4 + a_5$
2^6	4	16	<u>ABCD</u>, <u>CDEF</u>, ABEF	a_1, a_2, a_3, a_5 $a_4 = a_1 + a_2 + a_3$ $a_6 = a_3 + a_4 + a_5$
2^6	8	8	<u>BCD</u>, <u>ABE</u>, ACDE, <u>ABCF</u>, ADF, CEF, BDEF	a_1, a_2, a_3 $a_4 = a_2 + a_3$ $a_5 = a_1 + a_2$ $a_6 = a_1 + a_2 + a_3$
2^7	4	32	<u>ABCDE</u>, <u>ABFG</u>, CDEFG	a_1, a_2, a_3, a_4, a_6 $a_5 = a_1 + a_2 + a_3 + a_4$ $a_7 = a_1 + a_2 + a_6$
2^7	8	16	<u>ABCD</u>, <u>CDEF</u>, ABEF, <u>ACEG</u>, BDEG, ADFG, BCFG	a_1, a_2, a_3, a_5 $a_4 = a_1 + a_2 + a_3$ $a_6 = a_3 + a_4 + a_5$ $a_7 = a_1 + a_3 + a_5$
2^7	16	8	<u>ABC</u>, <u>CDE</u>, ABDE, <u>BDF</u>, ACDF, BCEF, AEF, <u>ADG</u>, BCDG, ACEG, BEG, ABFG, CFG, ABCDEFG, DEFG	a_1, a_2, a_4 $a_3 = a_1 + a_2$ $a_5 = a_3 + a_4$ $a_6 = a_2 + a_4$ $a_7 = a_1 + a_4$

14. Confounding in General Factorial Experiments

14.1 Introduction
14.2 Confounding with Factors at Three Levels
14.3 Designing Using Pseudofactors
14.4 Designing Confounded Asymmetrical Experiments
14.5 Using SAS Software
Exercises

14.1 Introduction

In Chapter 13, incomplete block designs for 2^p factorial experiments were obtained by confounding one or more interaction contrasts with block contrasts. In this chapter, we extend the idea of confounding to encompass experiments in which not all factors have two levels. We will code the levels of an m-level factor as $0, 1, \ldots, m-1$.

In Section 14.2 we consider single-replicate 3^p experiments arranged in $b = 3^s$ blocks of size $k = 3^{p-s}$. The techniques used in designing these types of experiment can be adapted for m^p experiments in m^s blocks of size m^{p-s} where m is a prime number.

Pseudofactors are introduced in Section 14.3 to facilitate confounding in symmetrical 4^p experiments and asymmetrical $2^p 4^q$ experiments. Then, in Section 14.4 we consider asymmetrical experiments involving factors or pseudofactors at both two and three levels, allowing us to look at more complicated situations where the treatment factors have a mixture of 2, 3, 4, and 6 levels.

Analysis of a two-replicate 3^3 experiment with partial confounding is illustrated using the SAS software in Section 14.5.

Table 14.1 Sets of orthogonal contrasts measuring the interaction in a 3^2 experiment

TC	$A_L B_L$	$A_L B_Q$	$A_Q B_L$	$A_Q B_Q$	$(AB;$	$A^2B^2)$	$(AB^2;$	$A^2B)$
00	1	−1	−1	1	−1	1	−1	1
01	0	2	0	−2	0	−2	1	1
02	−1	−1	1	1	1	1	0	−2
10	0	0	2	−2	0	−2	0	−2
11	0	0	0	4	1	1	−1	1
12	0	0	−2	−2	−1	1	1	1
20	−1	1	−1	1	1	1	1	1
21	0	−2	0	−2	−1	1	0	−2
22	1	1	1	1	0	−2	−1	1

14.2 Confounding with Factors at Three Levels

14.2.1 Contrasts

In a factorial experiment where all treatment factors have 3 levels, each main effect has 2 degrees of freedom associated with it, each two-factor interaction has $2 \times 2 = 4$ degrees of freedom, etc. (see Section 7.3). Therefore, we can find 2 orthogonal contrasts to measure each main effect, 4 orthogonal contrasts to measure each two-factor interaction, and so on.

In a 3^2 experiment, for example, two orthogonal contrasts measuring the main effect of each of factors A and B are the linear and quadratic trend contrasts. Similarly, four orthogonal trend contrasts $A_L B_L$, $A_L B_Q$, $A_Q B_L$, and $A_Q B_Q$ measuring the interaction are reproduced in Table 14.1 (see Section 6.3). A different set of four orthogonal contrasts, labeled in pairs as $(AB; A^2B^2)$ and $(AB^2; A^2B)$, is also shown in Table 14.1. Although this second set of contrasts is less useful than the set of trend contrasts in measuring details of the interaction, it will prove extremely useful for confounding purposes. The reader is asked to verify that *any* contrasts that measure the main effects of A and B are orthogonal to all the contrasts in Table 14.1 measuring the interaction.

Notice that the pair of contrasts labeled $(AB; A^2B^2)$ are two orthogonal contrasts that compare the three groups of treatment combinations (00, 12, 21) and (01, 10, 22) and (02, 11, 20). Any linear combination of these two contrasts is also a contrast between these three groups of treatment combinations. We have illustrated these groups of treatment combinations in the left-hand side of Table 14.2, where treatment combinations with the same superscript are in the same group. Notice that each group contains one treatment combination from each row and each column, making sure that each level of each factor is represented once in each group.

Similarly, the pair of contrasts labeled $(AB^2; A^2B)$ comprise two orthogonal contrasts that compare the three groups of treatment combinations (00, 11, 22) and (01, 12, 20) and (02, 10, 21). Any linear combination of these two contrasts is also a contrast between these three groups of treatment combinations. The groups are illustrated in the right-hand side of Table 14.2 and also have the property that each group contains one treatment combination from each row and each column.

14.2 Confounding with Factors at Three Levels   463

Table 14.2 Groups of treatment combinations corresponding to orthogonal interaction contrasts in a 3^2 experiment

$(AB;\ A^2B^2)$			$(AB^2;\ A^2B)$		
00*	01†	02+	00*	01†	02+
10†	11+	12*	10+	11*	12†
20+	21*	22†	20†	21+	22*

The reason for the labeling $(AB;\ A^2B^2)$ and $(AB^2;\ A^2B)$ is to match the contrasts with the equation method of confounding in Section 14.2.3. The contrast names themselves have little meaning, except to acknowledge that each contrast belongs to the AB interaction and, as will be seen, each pair corresponds to a set of equations that partitions the treatment combinations into three groups.

Many texts list only one of the two labels in each pair, since each is the square of the other. For example, when the exponents are reduced modulo 3, then $A^2B = (AB^2)^2$. The convention is then to list AB^2 rather than A^2B, for example, since the leading exponent is one. However, we will list both labels to aid in identifying a complete set of confounded contrasts in designs with more than three blocks.

An alternative labeling for the contrasts that can be found in a number of texts and articles is $AB(J)$ for the pair of contrasts $(AB;\ A^2B^2)$, and $AB(I)$ for the pair of contrasts $(AB^2;\ A^2B)$.

14.2.2 Confounding Using Contrasts

In this section we consider the division of treatment combinations into blocks by deliberately confounding negligible contrasts, as in Section 13.3.2 for 2^p experiments. For 3^p experiments, we look at designs with 3^s blocks of size 3^{p-s}, starting with 3 blocks of size 3^{p-1}. For a design with $b = 3$ blocks, two degrees of freedom are used to measure the block differences. Therefore, in a single-replicate design, we must confound a pair of treatment contrasts.

As a simple example, we start with an experiment with two factors A and B in which the interaction is known to be negligible. We will attempt to use two of the interaction contrasts shown in Table 14.1 to divide the treatment combinations into 3 blocks. A pair of trend contrasts, such as $A_L B_Q$ and $A_Q B_Q$ cannot be used to give blocks of equal size, since the values of the coefficients do not fall into 3 groups of 3. However, the pair of contrasts labeled $(AB;\ A^2B^2)$ have three pairs of coefficients $(-1,\ 1)$, $(0,\ -2)$, and $(1,\ 1)$ each of which appear three times. If we use these as a guide to dividing the treatment combinations into blocks, we obtain the design in Table 14.3.

Any contrast that is orthogonal to the two confounded contrasts can be estimated without requiring block adjustments. Estimable contrasts include all contrasts measuring the main effects of A and B and the remaining two interaction contrasts labeled $(A^2B;\ AB^2)$ and linear combinations of these. The trend contrasts in Table 14.1 are not orthogonal to any of the AB, A^2B^2, AB^2, A^2B contrasts, so they do not fall into either the confounded or the

Table 14.3 3^2 experiment in 3 blocks of 3, confounding $(AB; A^2B^2)$

Block	Contrast Coefficients	Treatment Combinations		
I	(−1, 1)	00	12	21
II	(0, −2)	01	10	22
III	(1, 1)	02	11	20

Table 14.4 3^2 experiment in 3 blocks of 3, confounding $(A^2B; AB^2)$

Block	Contrast Coefficients	Treatment Combinations		
I	(−1, 1)	00	11	22
II	(1, 1)	01	12	20
III	(0, −2)	02	10	21

estimable category. They are *partly confounded*. In general, interaction trend contrasts can be estimated completely only when no contrasts from the interaction are confounded.

In the present example, the interaction has four degrees of freedom. Two are used to measure blocks. The other two correspond to two estimable contrasts, which are negligible and provide two degrees of freedom to measure σ^2.

If the contrasts labeled $(A^2B; AB^2)$ in Table 14.1 were used instead of the contrasts labeled $(AB; A^2B^2)$ to provide three blocks, the design of Table 14.4 would result. This has the same properties as the design in Table 14.3 in that all main-effect contrasts are estimable and there are two estimable contrasts $(AB; A^2B^2)$ remaining in the interaction that provide an estimate of σ^2. Neither design is better than the other, and a choice can be made at random. Block design randomization should be carried out before the design is used.

As we saw in 2^p experiments, there is a correspondence between the contrasts used for confounding, the contrast names, and the equation method of confounding. In the next section we show how to obtain the design of Table 14.3 by the equation method. This is then the only method that we shall use for more complicated designs.

14.2.3 Confounding Using Equations

3^p experiments in three blocks The design in Table 14.3, which was obtained by confounding the two interaction contrasts labeled $(AB; A^2B^2)$ in Table 14.1, can be obtained by an equation method similar to that of Section 13.4. Notice that in Block I the digits of the three treatment combinations add to 0 or 3. In Block II they add to 1 or 4, and in Block III they add to 2. Now that both factors have three levels, we work modulo 3, which means that we subtract 3 from the sum of the digits until we obtain one of 0, 1, or 2, or equivalently, we take the remainder on division by 3. Writing the treatment combinations as a_1a_2, the blocks can be defined by the confounding equations

Block I: Treatment combinations with $L = a_1 + a_2 = 0 \pmod 3$,
Block II: Treatment combinations with $L = a_1 + a_2 = 1 \pmod 3$,
Block III: Treatment combinations with $L = a_1 + a_2 = 2 \pmod 3$.

Equivalently, the same three blocks can be obtained if the equations are multiplied by 2; that is,

Block I: Treatment combinations with $2L = 2a_1 + 2a_2 = 0 \pmod 3$,
Block II: Treatment combinations with $2L = 2a_1 + 2a_2 = 2 \pmod 3$,
Block III: Treatment combinations with $2L = 2a_1 + 2a_2 = 1 \pmod 3$.

Thus, if the contrasts labeled $(AB; A^2B^2)$ in Table 14.1 are confounded with blocks, the treatment combinations in the three blocks satisfy

$$L = a_1 + a_2 = 0, 1, \text{ or } 2 \pmod 3,$$

and also

$$2L = 2a_1 + 2a_2 = 0, 2, \text{ or } 1 \pmod 3.$$

Similarly, if the contrasts labeled $(AB^2; A^2B)$ are to be confounded, the equations

$$L = a_1 + 2a_2 = 0, 1, \text{ or } 2 \pmod 3$$

or, multiplying by 2,

$$2L = 2a_1 + a_2 = 0, 2, \text{ or } 1 \pmod 3$$

will produce the design in Table 14.4. Notice that the coefficients in the confounding equations correspond to the exponents in the contrast names. A set of equations defines the same set of blocks when it is multiplied by 2. Therefore, the confounded contrast names always come in pairs—one name being the square of the other—$(AB^2)^2 = A^2B^4 = A^2B$, reducing exponents (mod 3).

In general, in a 3^p experiment, if the equations

$$L = z_1 a_1 + z_2 a_2 + \cdots + z_p a_p = 0, 1, \text{ or } 2 \pmod 3$$

are used to produce three blocks, two contrasts will be confounded that can be labeled $(A^{z_1} B^{z_2} \cdots P^{z_p}; A^{2z_1} B^{2z_2} \cdots P^{2z_p})$, where z_i is 1 or 2 if the factor is present in the interaction, and 0 if it is not, and where the exponent is reduced modulo 3. For example, in a 3^5 experiment, the equations

$$L = a_1 + 2a_2 + a_4 = 0, 1, \text{ or } 2 \pmod 3$$

will give 3 blocks of size 3^4 confounding AB^2D and $A^2B^4D^2 = A^2BD^2$, which represent two contrasts from the three-factor interaction ABD. It is rarely of importance to identify exactly what the contrasts look like (they are any pair of orthogonal contrasts between the groups of treatment combinations in the three blocks). What is important is the knowledge that the confounded contrasts belong to a particular interaction, and therefore that all other main-effect and interaction contrasts are estimable.

3^p *experiments in nine blocks* The equation method of confounding can be used to produce $b = 9 = 3^2$ blocks of size 3^{p-2} in a 3^p experiment by selecting two pairs of contrasts to be confounded. If the pair $(A^{z_1} B^{z_2} \cdots P^{z_p}; A^{2z_1} B^{2z_2} \cdots P^{2z_p})$ is chosen for confounding together with the pair $(A^{y_1} B^{y_2} \cdots P^{y_p}; A^{2y_1} B^{2y_2} \cdots P^{2y_p})$, the $b = 9$ blocks are produced from the the nine combinations of values of the two linear functions

$$L_1 = z_1 a_1 + z_2 a_2 + \cdots + z_p a_p \pmod{3},$$

$$L_2 = y_1 a_1 + y_2 a_2 + \cdots + y_p a_p \pmod{3},$$

that is, from the nine possible pairs of values of the two equations

$$L_1 = z_1 a_1 + z_2 a_2 + \cdots + z_p a_p = 0, 1, \text{ or } 2 \pmod{3},$$

$$L_2 = y_1 a_1 + y_2 a_2 + \cdots + y_p a_p = 0, 1, \text{ or } 2 \pmod{3}.$$

The $b - 1 = 8$ confounded contrasts are the four originally chosen, together with all possible products. This is most conveniently set out as a table. The selected pairs of contrasts are written in the first row and first column. The table is then filled out by multiplication, and the exponents are reduced modulo 3, as follows:

	$A^{y_1} B^{y_2} \cdots P^{y_p}$	$A^{2y_1} B^{2y_2} \cdots P^{2y_p}$
$A^{z_1} B^{z_2} \cdots P^{z_p}$	$A^{z_1+y_1} B^{z_2+y_2} \cdots P^{z_p+y_p}$	$A^{z_1+2y_1} B^{z_2+2y_2} \cdots P^{z_p+2y_p}$
$A^{2z_1} B^{2z_2} \cdots P^{2z_p}$	$A^{2z_1+y_1} B^{2z_2+y_2} \cdots P^{2z_p+y_p}$	$A^{2z_1+2y_1} B^{2z_2+2y_2} \cdots P^{2z_p+2y_p}$

If $b = 3^s$ blocks of size 3^{p-s} are required, then s independent pairs of contrast names need to be chosen for confounding. All possible products determine the entire set of $b - 1 = 3^s - 1$ confounded contrasts.

Example 14.2.1 3^4 experiment in 9 blocks of size 9

Suppose that a 3^4 experiment, with factors A, B, C, D, is to be run in $b = 9$ blocks of size 9. The only interactions thought to be important are the 2-factor interactions. The 2-factor interactions should therefore not be confounded. Now $b = 3^2$ blocks are required, so 2 pairs of contrasts should be chosen for confounding. The $ABCD$ interaction has 16 degrees of freedom, so we can find 16 orthogonal contrasts and label them in pairs as

$(ABCD; A^2 B^2 C^2 D^2)$, $(AB^2 CD; A^2 BC^2 D^2)$,
$(ABCD^2; A^2 B^2 C^2 D)$, $(AB^2 CD^2; A^2 BC^2 D)$,
$(ABC^2 D; A^2 B^2 CD^2)$, $(AB^2 C^2 D; A^2 BCD^2)$,
$(ABC^2 D^2; A^2 B^2 CD)$, $(AB^2 C^2 D^2; A^2 BCD)$.

Selecting two pairs of contrasts from the 4-factor interaction for confounding contrasts is not a good choice. For example, if $(ABCD^2; A^2 B^2 C^2 D)$ and $(ABCD; A^2 B^2 C^2 D^2)$ were chosen, the set of eight confounded degrees of freedom would be

$ABCD$	$ABCD^2$	$A^2 B^2 C^2 D$
	$A^2 B^2 C^2$	D^2
$A^2 B^2 C^2 D^2$	D	ABC

and we can see that the two orthogonal contrasts in the main-effect D would also be confounded. All possible selections of two pairs of contrasts from the $ABCD$ interaction will confound either a main effect or a two-factor interaction. However, the three-factor interactions are also thought to be negligible, so one possible choice is to confound $(ABD; A^2B^2D^2)$ together with $(BCD^2; B^2C^2D)$. This gives the following set of eight confounded degrees of freedom.

	BCD^2	B^2C^2D
ABD	AB^2C	AC^2D^2
$A^2B^2D^2$	A^2CD	A^2BC^2

Thus, each 3-factor interaction (which has 8 degrees of freedom) has two orthogonal contrasts confounded with blocks and six estimable contrasts, which are assumed to be negligible. This means that there are 24 degrees of freedom from the 3-factor interactions and a further 16 degrees of freedom from the $ABCD$ interaction available for estimating σ^2. The design is obtained by using the linear functions L_1 and L_2, corresponding to the selected confounded contrasts ABD and BCD^2 as follows. For each treatment combination, compute the values of L_1 and L_2 modulo 3:

$$L_1 = a_1 + a_2 \quad\quad + a_4 = 0, 1, \text{ or } 2 \pmod 3.$$
$$L_2 = \quad\quad a_2 + a_3 + 2a_4 = 0, 1, \text{ or } 2 \pmod 3.$$

The design is given in Table 14.5, and it can be verified that the nine blocks are obtained from the nine possible pairs of values of L_1 and L_2. □

14.2.4 A Real Experiment—Dye Experiment

An experiment is described in the book *Design and Analysis of Industrial Experiments*, edited by O. L. Davies, that investigates three reactants (the base material and two inorganic materials, called here M and N) in the manufacture of a cotton dyestuff. The three factors of interest in the experiment were the concentration of M in the free water in the reaction mixture (factor A at three equally spaced levels), the volume of free water in the reaction mixture (factor B at three equally spaced levels), and the concentration of N in the free water in the reaction mixture (factor C at three equally spaced levels).

Although it was possible to control the conditions in the laboratory fairly accurately, the experimenters divided the treatment combinations into blocks of size 9. This was done as a

Table 14.5 3^4 experiment in 3^2 blocks of 9; confounding $(ABD; A^2B^2D^2)$, $(BCD^2; B^2C^2D)$, $(AB^2C; A^2BC^2)$, and $(AC^2D^2; A^2CD)$

Block	L_1, L_2	Treatment Combinations
I	0,0	0000 0112 0221 1022 1101 1210 2011 2120 2202
II	1,2	0001 0110 0222 1020 1102 1211 2012 2121 2200
III	2,1	0002 0111 0220 1021 1100 1212 2010 2122 2201
IV	0,1	0010 0122 0201 1002 1111 1220 2021 2100 2212
V	1,0	0011 0120 0212 1000 1112 1221 2022 2101 2210
VI	2,2	0012 0121 0210 1001 1110 1222 2020 2102 2211
VII	1,1	0100 0212 0021 1122 1201 1010 2111 2220 2002
VIII	2,0	0101 0210 0022 1120 1202 1011 2112 2221 2000
IX	0,2	0102 0211 0020 1121 1200 1012 2110 2222 2001

Table 14.6 Data for dye experiment

Block I		Block II		Block III	
TC	Volume	TC	Volume	TC	Volume
000	74	020	69	010	13
021	130	011	46	001	112
012	56	002	71	022	125
110	110	100	211	120	199
101	166	121	220	111	218
122	227	112	216	102	201
220	195	210	147	200	74
211	146	201	47	221	198
202	90	222	164	212	102

Source: Davies, O. L. (1963). Reprinted by permission of Addison Wesley Longman Ltd.

safeguard against time trends, because the time required to complete the investigation was reasonably long. The experiment involved $r = 2$ replications of each treatment combination, but here we will analyze only the first replicate.

The observations were the volumes of dyestuff resulting from the chemical reactions and are shown in Table 14.6. Looking at the treatment combinations (TC) listed in Block I, we can see that they all satisfy the confounding equation $a_1 + 2a_2 + 2a_3 = 0 \pmod 3$. Consequently, the experimenters have confounded two contrasts from the 3-factor interaction, which we can label as $(AB^2C^2; A^2BC)$. Since there are only three blocks, these are the only two contrasts confounded. If the 3-factor interaction can be assumed to be negligible, the remaining six degrees of freedom can be used to measure the error variability. The analysis of variance table is shown in Table 14.7. The sum of squares for testing that the main effect of A (averaged over the levels of the other factors) can be calculated either by using the formulae of Chapter 7 or by adding together the sums of squares for two orthogonal contrasts. For example, rule 4 of Section 7.3 gives

$$ssA = 9 \sum_{i=1}^{3} \bar{y}_{i..}^2 - 27 \bar{y}_{...}^2$$
$$= 9(5980.44 + 38{,}590.42 + 16{,}698.38) - 27(18{,}045.44)$$
$$= 64{,}196.222.$$

Two orthogonal contrasts for A are the linear and quadratic contrasts. From Table A.2, the coefficients for the (nonnormalized) linear contrast are $(-1, 0, 1)$, and those for the quadratic contrast are $(1, -2, 1)$. The least squares estimates for these two contrasts are

$$\hat{A}_L = (-\bar{y}_{.0..} + \bar{y}_{.2..}) = 51.889$$

and

$$\hat{A}_Q = (\bar{y}_{.0..} - 2\bar{y}_{.1..} + \bar{y}_{.2..}) = -186.333.$$

14.2 Confounding with Factors at Three Levels

To normalize the contrasts, one would divide $\hat{A}_L$ by $\sqrt{\Sigma c_i^2/(rbc)} = \sqrt{2/9}$ and divide $\hat{A}_Q$ by $\sqrt{\Sigma c_i^2/(rbc)} = \sqrt{6/9}$.

The sum of squares for testing the hypothesis that the linear contrast for A is negligible is the square of the normalized contrast estimate,

$$ss(A_L) = \frac{(-\bar{y}_{.0..} + \bar{y}_{.2..})^2}{2/9} = \frac{(51.889)^2}{2/9} = 12{,}116.06;$$

the sum of squares for testing the hypothesis that the quadratic contrast for A is negligible is

$$ss(A_Q) = \frac{(\bar{y}_{.0..} - 2\bar{y}_{.1..} + \bar{y}_{.2..})^2}{6/9} = \frac{(-186.333)^2}{6/9} = 52{,}080.17;$$

and we see that

$$ss(A_L) + ss(A_Q) = 12{,}116.06 + 52{,}080.17 = 64{,}196.23 = ssA.$$

The other sums of squares in Table 14.7 can be obtained in a similar way.

For testing the hypotheses that the three main effects and the three 2-factor interactions are negligible at individual significance levels $\alpha^* = 0.01$ (an overall level significance level of $\alpha \leq 0.06$), we would compare the ratios in the analysis of variance table (Table 14.7) with the critical values from the F-distribution ($F_{2,6,0.01} = 10.9$ for the main effects and $F_{4,6,0.01} = 9.15$ for the 2-factor interactions), and we would reject only the hypothesis that the main effect of A is negligible. Plots for the average response due to A and B are shown in Figure 14.1. We can see from the plot of the A average responses that as the levels of the concentration of inorganic material M in the free water increase, the volumes of dyestuff first increase and then begin to decrease. We might expect to see both a significant linear trend and a significant quadratic trend. Testing the two hypotheses that the linear trend in A is negligible and the quadratic trend in A is negligible, each at level 0.005 (to give an overall significance level of $\alpha^* \leq 0.01$, we have

$$ss(A_L)/msE = 10.04 < F_{1,6,0.005} = 18.6$$

and

$$ss(A_Q)/msE = 43.16 > F_{1,6,0.005} = 18.6,$$

and we conclude that there is a quadratic trend in the levels of A, and that the turning point is towards the center of the range of levels investigated (otherwise, the linear trend would also have been significantly different from zero). Since the objective of the experiment was to boost the volume of dyestuff produced, the results of the experiment suggest that further investigation around the second concentration of inorganic material M might be wise. Although the hypothesis of no effect of B was not rejected, Figure 14.1 suggests that further experimentation with higher volumes of free water in the reaction mixture is worth consideration.

The above method of testing these two hypotheses uses Bonferroni's method of combining significance levels. An alternative method is to use Scheffé's method of multiple

Table 14.7 Analysis of variance for the dye experiment

Source of Variation	Degrees of Freedom	Sum of Squares	Mean Square	Ratio	p-values
Block	2	182.00			
A	2	64,196.22	32,098.11	26.60	0.0010
A_L	1	12,116.06	12,116.06	10.04	0.0194
A_Q	1	52,080.17	52,080.17	43.16	0.0006
B	2	16,857.56	8,428.78	6.98	0.0271
B_L	1	12,853.39	12,853.39	10.65	0.0172
B_Q	1	4,004.17	4,004.17	3.32	0.1184
C	2	2,334.89	1,167.44	0.97	0.4324
C_L	1	1,422.22	1,422.22	1.18	0.3193
C_Q	1	912.67	912.67	0.76	0.4179
AB	4	12,512.89	3,128.22	2.59	0.1428
AC	4	4,044.89	1,011.22	0.84	0.5481
BC	4	2,698.89	674.72	0.56	0.7015
Error	6	7,240.67	1,206.78		
Total	26	110,068.00			

comparisons and test the two hypotheses simultaneously at level 0.01. Since

$$ss(A_L)/msE = 10.04 < 2F_{2,6,0.01} = 21.8$$

and

$$ss(A_Q)/msE = 43.16 > 2F_{2,6,0.01} = 21.8,$$

we arrive at the same conclusion. The more powerful method here is the first since

$$F_{1,6,0.005} < 2F_{2,6,0.01}.$$

14.2.5 Plans for Confounded 3^p Experiments

At the end of the chapter we give a table (Table 14.21) of suggested confounding schemes for 3^p experiments in blocks of size 3, 9, or 27. As illustrated in Section 13.6, if the design in the table confounds important contrasts in the experiment, then a relabeling of treatment factors should be attempted.

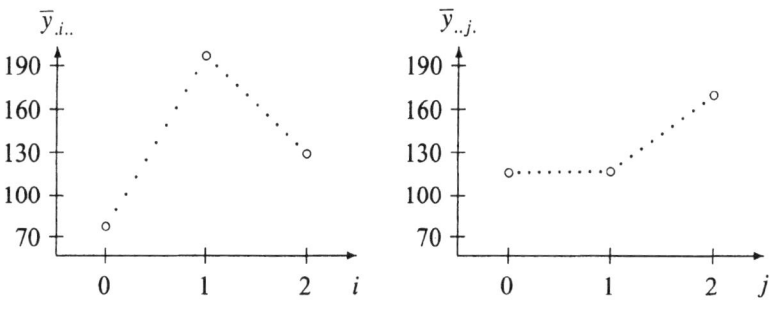

Figure 14.1 A and B main-effect plots for the dye experiment

14.3 Designing Using Pseudofactors

14.3.1 Confounding in 4^p Experiments

A treatment factor F with four levels coded 0, 1, 2, 3 can be represented by two factors F_1 and F_2 each having two levels coded 0, 1. The levels of F_1 and F_2 taken together correspond to the levels of the original factor F. One possible correspondence is given below:

F	F_1	F_2
0	0	0
1	0	1
2	1	0
3	1	1

The factors F_1 and F_2 are called *pseudofactors*. All factors in a 4^p experiment can be represented by pseudofactors. Thus, a 4^p experiment in 4^s blocks of size 4^{p-s} can be represented as a 2^{2p} experiment in 2^{2s} blocks of size $2^{2(p-s)}$. The techniques of confounding in a 2^{2p} experiment as discussed in Section 13.3 can therefore be used. The only difference is that an interaction of pseudofactors of the form $F_1G_1G_2$, say, does not represent 3-factor interaction. It represents one of nine orthogonal contrasts measuring the two-factor interaction, FG. Similarly, F_1F_2 does not represent a contrast in a two-factor interaction. It represents one of three orthogonal contrasts measuring the main effect of factor F.

Example 14.3.1 4^2 experiment in 4 blocks of size 4

Consider a 4^2 experiment with two factors F and G to be run in 4 blocks of size 4. The main effects are to be estimated, but the interaction is thought to be negligible. If F and G are represented by pseudofactors F_1, F_2, G_1, G_2 having two levels each, we can consult Table 13.28 hoping to find a suitable 2^4 experiment in 4 blocks of size 4.

In Table 13.28, we find a design that confounds AC, ABD, and BCD. If we make the correspondence $F_1 = A$, $F_2 = B$, $G_1 = C$, $G_2 = D$, then the design confounds F_1G_1, $F_1F_2G_2$, $F_2G_1G_2$, all three of which belong to the interaction of F and G. All main-effect contrasts of F and G are orthogonal to all interaction contrasts and can therefore be estimated without adjustment for blocks. The design is shown in Table 14.8, with blocks corresponding to combinations of values of $L_1 = a_1 + a_3$ (mod 2) and $L_2 = a_1 + a_2 + a_4$ (mod 2).

If we make a different correspondence, say $F_1 = A$, $F_2 = D$, $G_1 = B$, $G_2 = C$, then a slightly different design is obtained, this time confounding F_1G_2, $F_1F_2G_1$, and $F_2G_1G_2$, which again belong to the interaction of F and G. There is no particular reason to prefer one design over the other. However, a third correspondence, $F_1 = A$, $F_2 = C$, $G_1 = B$, $G_2 = D$, would not be good, since it confounds F_1F_2, $F_1G_1G_2$, $F_2G_1G_2$, and this includes one degree of freedom F_1F_2 from the main effect of F. □

Since two-level pseudofactors are being used, block sizes need only be a power of two, not necessarily a power of four.

Table 14.8 4^2 experiment in 4 blocks of 4, confounding three degrees of freedom (F_1G_1, $F_1F_2G_2$, $F_2G_1G_2$) from FG

Block	L_1, L_2	Pseudofactors F_1, F_2, G_1, G_2	Factors F, G
I	0,0	0000 0101 1011 1110	00 11 23 32
II	0,1	0001 0100 1010 1111	01 10 22 33
III	1,0	0010 0111 1001 1100	02 13 21 30
IV	1,1	0011 0110 1000 1101	03 12 20 31

14.3.2   Confounding in $2^p \times 4^q$ Experiments

Since factors with 4 levels can be written in terms of pseudofactors having 2 levels each, a $2^p \times 4^q$ experiment can be written in terms of pseudofactors as a $2^{(p+2q)}$ experiment, and no new techniques are needed.

Example 14.3.2 $2^3 \times 4$ experiment in 4 blocks of size 8

Suppose that a $2^3 \times 4$ experiment with factors F, G, H, and J is to be run in 4 blocks of size 8. This could be designed using pseudofactors by selecting a design for a 2^5 experiment in 4 blocks from Table 13.28. A design is shown that confounds ABE, CDE, and $ABCD$. If we let the combination of levels of A and C represent the levels of the 4-level factor $J_1 J_2 = J$ with the representation $00 = 0$, $01 = 1$, $10 = 2$, $11 = 3$, and let the levels of B, D, and E respectively represent the levels of F, G, and H, we obtain the design of Table 14.9, that confounds one contrast from each of the 3-factor interactions FHJ, GHJ, and FGJ. All main effects and 2-factor interactions can be estimated. There are 10 degrees of freedom available for estimating σ^2. These come from the two unconfounded degrees of freedom from each of FHJ, GHJ, and FGJ and the one degree of freedom from FGH and the three from $FGHJ$. □

14.4   Designing Confounded Asymmetrical Experiments

A factorial experiment is called an asymmetric experiment when the treatment factors do not all have the same number of levels. For example, $2^2 \times 4^2$, $2^5 \times 3$, $2^2 \times 3^2 \times 4^2$, and 3×6 experiments are all asymmetric experiments. We have already discussed the design of asymmetric $2^p \times 4^q$ experiments in Section 14.3.2. We used pseudofactors for the factors with four levels, thus allowing the symmetric designs for 2^{p+2q} experiments to be used.

Table 14.9 $2^3 \times 4$ experiment in 4 blocks of 8, using pseudofactors and confounding one degree of freedom from interactions FGJ, GHJ, and FHJ

Blocks	Treatment Combinations (Factors F, G, H, J)							
I	0000	1002	0101	1103	0013	1011	0112	1110
II	0010	1012	0111	1113	0003	1001	0102	1100
III	0100	1102	0001	1003	0113	1110	0012	1010
IV	0110	1112	0011	1013	0103	1101	0002	1000

14.4 Designing Confounded Asymmetrical Experiments

We can use this idea only when the numbers of levels of all factors are powers of the same prime number. For all of the other examples mentioned above, the use of pseudofactors would transform the experiment into a $2^p \times 3^q$ experiment. Consequently, we concentrate on this type of situation in this section.

Since 2 and 3 are relatively prime, the only type of design that can be constructed using the equation method will confound contrasts within the two symmetric parts of the experiment. Consequently, to obtain a design for a $2^p \times 3^q$ experiment in $2^s \times 3^t$ blocks of size $2^{p-s} \times 3^{q-t}$, we combine a design for a 2^p experiment in 2^s blocks with a design for a 3^q experiment in 3^t blocks, using the idea of a *crossed array* as illustrated in the following example. The total number of blocks created in the combined design is always the product of the numbers of blocks in the original two designs. Likewise, the block sizes in the combined design are products of the block sizes in the original two designs. The confounded contrasts in the combined design are those confounded in the separate designs together with those indicated by all possible products of contrast names.

Example 14.4.1 $2^2 \times 3^2$ experiment in 6 blocks of size 6

Suppose that a $2^2 \times 3^2$ experiment is to be run in 6 blocks of size 6. We label the two 2-level factors as A and B and the two 3-level factors as C and D. Since the design must confound within the two symmetric parts of the experiment, one contrast from A, B, or AB must be confounded to divide the 2^2 treatment combinations into two blocks, and one pair of contrasts from C, D, or CD must be confounded to divide the 3^2 treatment combinations into three blocks. The confounded contrasts in the combined design are those confounded in the separate designs together with their products.

For example, we could combine the two designs in Table 14.10. The design labeled d_1 is for a 2^2 experiment in two blocks of size 2 confounding AB, with treatment combinations (TC) grouped into blocks determined by the two values of $L_1 = a_1 + a_2 \pmod 2$. The design labeled d_2 is for a 3^2 experiment in three blocks of size 3 confounding the pair of contrasts $(CD^2; C^2D)$, with blocks determined by the three values of $L_2 = a_3 + 2a_4 \pmod 3$. The combined array in Table 14.11 divides the treatment combinations into blocks according to the six combinations of values of L_1 and L_2.

A quick way to obtain the crossed array is as follows. Each of the 2 blocks of d_1 is combined with each of the 3 blocks of d_2 to obtain the 2×3 blocks of the combined design. For example, to combine the first blocks of d_1 and d_2, each of the combinations 00 and 11 in block I$_1$ of d_1 is combined with each of the combinations 00, 11, and 22 in block I$_2$ of d_2 to give the treatment combinations 0000, 1100, 0011, 1111, 0022, 1122 in the first block

Table 14.10 Design d_1 for a 2^2 experiment confounding AB and design d_2 for a 3^2 experiment confounding $(CD^2; C^2D)$

Design	L_1	Block	TC	Design	L_2	Block	TC
d_1	0	I$_1$	00 11	d_2	0	I$_2$	00 11 22
	1	II$_1$	01 10		1	II$_2$	02 10 21
					2	III$_2$	01 12 20

Table 14.11 $2^2 \times 3^2$ experiment in 6 blocks of 6, confounding AB, $(CD^2; C^2D)$, $(ABCD^2; ABC^2D)$

L_1, L_2			Blocks	Treatment Combinations
0,0	I_1	I_2	I	0000 0011 0022 1100 1111 1122
0,1	I_1	II_2	II	0002 0010 0021 1102 1110 1121
0,2	I_1	III_2	III	0001 0012 0020 1101 1112 1120
1,0	II_1	I_2	IV	0100 0111 0122 1000 1011 1022
1,1	II_1	II_2	V	0102 0110 0121 1002 1010 1021
1,2	II_1	III_2	VI	0101 0112 0120 1001 1012 1020

of the combined design. The other blocks of the combined design are obtained in a similar way.

The $b - 1 = 5$ confounded contrasts are those corresponding to the original confounding schemes, namely the contrast AB and the pair of contrasts represented by $(CD^2; C^2D)$, together with the pair of contrasts represented by the products of these labels—namely $(ABCD^2; ABC^2D)$. □

Example 14.4.2 $4 \times 6 \times 3$ experiment in 6 blocks of size 12

Suppose that a $4 \times 6 \times 3$ experiment with factors F, G, H is to be run in 6 blocks of size 12. If we use the pseudofactor labels F_1, F_2, G_1, G_2, and H, then the factors F_1, F_2, and G_1 are in the 2^3 pseudofactor experiment and G_2 and H are in the 3^2 pseudofactor experiment. In the 2^3 experiment, we confound $F_1F_2G_1$ to give the two blocks of the design d_1 of Table 14.12, and in the 3^2 experiment, we confound the pair of contrasts $(G_2H; G_2^2H^2)$ to give the two blocks of the design d_2. Combining each treatment combination in design d_1 with those in d_2 gives the design in Table 14.13. The $b - 1 = 5$ confounded degrees of freedom correspond to the original three confounded contrasts together with their products, that is, $F_1F_2G_1$, $(G_2H; G_2^2H^2)$, and $(F_1F_2G_1G_2H; F_1F_2G_1G_2^2H^2)$.

Translating back to the original factors, we can see that one degree of freedom from the interaction FG is confounded, together with two degrees of freedom from each of GH and FGH. This means that all contrasts from the three main effects and also from the interaction FH can be estimated.

If we take the mapping of pseudofactor levels to factor levels as follows, then the design of Table 14.13 is as shown in Table 14.14.

F	F_1	F_2		G	G_1	G_2
0	0	0		0	0	0
1	0	1		1	0	1
2	1	0		2	0	2
3	1	1		3	1	0
				4	1	1
				5	1	2

□

Table 14.12 Design d_1 for a 2^3 experiment confounding $F_1F_2G_1$ and design d_2 for a 3^2 experiment confounding $G_2H; G_2^2H^2$

Design	Block	Treatment Combinations	Design	Block	Treatment Combinations
d_1	I_1	000 011 101 110	d_2	I_2	00 12 21
	II_1	001 010 100 111		II_2	01 10 22
				III_2	02 11 20

Table 14.13 $3 \times 4 \times 6$ experiment in 6 blocks of size 12, confounding one degree of freedom from FG and two degrees of freedom from each of GH and FGH

		Block	Pseudofactor Combinations					
I_1	I_2	I	00000	00012	00021	01100	01112	01121
			10100	10112	10121	11000	11012	11021
I_1	II_2	II	00001	00010	00022	01101	01110	01122
			10101	10110	10122	11001	11010	11022
I_1	III_2	III	00002	00011	00020	01102	01111	01123
			10102	10111	10120	11002	11011	11020
II_1	I_2	IV	00100	00112	00121	01000	01012	01021
			10000	10012	10021	11100	11112	11121
II_1	II_2	V	00101	00110	00122	01001	01010	01022
			10001	10010	10022	11101	11110	11122
II_1	III_2	VI	00102	00111	00120	01002	01011	01023
			10002	10011	10020	11102	11111	11120

Table 14.14 $3 \times 4 \times 6$ experiment in 6 blocks of size 12, confounding one degree of freedom from FG and two degrees of freedom from each of GH and FGH.

Block	Treatment Combinations
I	000 012 021 130 142 151 230 242 251 300 312 321
II	001 010 022 131 140 152 231 240 252 301 310 322
III	002 011 020 132 141 153 232 241 250 302 311 320
IV	030 042 051 100 112 121 200 212 221 330 342 351
V	031 040 052 101 110 122 201 210 222 331 340 352
VI	032 041 050 102 111 123 202 211 220 332 341 350

14.5 Using SAS Software

In this section we illustrate the use of the SAS software in analyzing a two-replicate factorial experiment with partial confounding. This we do via an example. The analysis is straightforward, as was illustrated in the previous chapter. Along with the correct analysis, we also fit an incorrect model—one without block effects—to illustrate the effect of partial confounding on the analysis.

Table 14.15 Data for the dye experiment

Block 1		Block 2		Block 3	
TC	Volume	TC	Volume	TC	Volume
000	74	020	69	010	13
021	130	011	46	001	112
012	56	002	71	022	125
110	110	100	211	120	199
101	166	121	220	111	218
122	227	112	216	102	201
220	195	210	147	200	74
211	146	201	47	221	198
202	90	222	164	212	102
Block 4		Block 5		Block 6	
TC	Volume	TC	Volume	TC	Volume
000	85	010	12	020	115
011	52	021	107	001	148
022	70	002	75	012	47
120	164	100	184	110	145
101	288	111	204	121	142
112	239	122	265	102	216
210	104	220	183	200	75
221	165	201	65	211	124
202	60	212	70	222	114

Source: Davies, O. L. (1963) Reprinted by permission of Addison Wesley Longman Ltd.

Example 14.5.1 Dye experiment, continued

The dye experiment was described in Section 14.2.4, where part of the data was analyzed as though it came from a single-replicate confounded experiment. In fact, in the original experiment, the design was a partially confounded design made up of two single-replicate 3^3 designs with different confounding schemes. The three factors of interest in the experiment were the concentration of inorganic material M in the free water in the reaction mixture (factor A at three equally spaced levels), the volume of free water in the reaction mixture (factor B at three equally spaced levels), and the concentration of inorganic material N in the free water in the reaction mixture (factor C at three equally spaced levels). The observations were the volumes of dyestuff resulting from the chemical reactions and are shown in Table 14.15 together with the design (prior to randomization). The contrasts $(AB^2C^2; A^2BC)$ are confounded in the first set of three blocks and estimable in the second set, whereas the contrasts $(ABC^2; A^2B^2C)$ are confounded in the second set of three blocks and estimable in the first set.

Since no contrast is completely confounded, no terms need be omitted from the model. Table 14.16 shows the SAS input statements for analyzing this experiment with partial confounding. The statements are exactly as they would be for a replicated experiment with three factors and no confounding. A second run of PROC GLM with no block parameter in the model is included for illustration purposes to show the effect of the partial confounding.

Table 14.16 SAS program for the dye experiment

```
******** to input the data;
DATA DYE;
INPUT BLK A B C Y;
  LINES;
  1 0 0 0  74
  1 0 2 1 130
  1 0 1 2  56
  1 1 1 0 110
  : : : :   :
  6 2 2 2 114
*** analysis of variance -- correct, with block effect;
PROC GLM;
  CLASSES  BLK A B C ;
  MODEL Y = BLK A B C A*B A*C B*C A*B*C ;
*** analysis of variance -- without block effect, for comparison;
PROC GLM;
  CLASSES  A B C;
  MODEL Y =  A B C A*B A*C B*C A*B*C;
```

All contrasts from the main-effects and 2-factor interactions are orthogonal to the block contrasts and can be estimated without adjustment for blocks. Consequently, the sums of squares for these terms are the same whether or not the block parameter is in the model. This can be verified by comparing the Type III sums of squares for the two runs of PROC GLM shown in Tables 14.17 and 14.18. Inclusion of the block parameter in the model changes the sum of squares for the three-factor interaction, since the three-factor interaction is partially confounded with blocks. The degrees of freedom for the three-factor interaction remain at 8, as all 8 orthogonal contrasts can be estimated from some portion of the data.

The analysis of variance table (Table 14.17) provides *no* evidence that certain contrasts are partially confounded. However, partially confounded contrasts are estimated with larger variance due to the adjustment for blocks. As a result, for the corresponding effects, confidence intervals are wider and tests are less powerful. □

Exercises

1. Suggest a confounding scheme for a 3^5 experiment in 9 blocks of size 27 if all 2-factor interactions and the 3-factor interaction ABE are to estimated.

2. Suggest a confounding scheme for a 3^5 experiment in 27 blocks of size 9 if all 2-factor interactions and the 3-factor interaction ABE are to estimated.

Table 14.17 Correct analysis of variance for the dye experiment

```
The SAS System
General Linear Models Procedure
Dependent Variable: Y
```

Source	DF	Sum of Squares	Mean Square	F Value	Pr > F
Model	31	221034.796296	7130.154719	8.25	0.0001
Error	22	19010.851852	864.129630		
Corrected Total	53	240045.648148			

Source	DF	Type III SS	Mean Square	F Value	Pr > F
BLK	5	2027.648148	405.529630	0.47	0.7950
A	2	140999.703704	70499.851852	81.58	0.0001
B	2	19447.814815	9723.907408	11.25	0.0004
C	2	4934.481481	2467.240741	2.86	0.0790
A*B	4	27922.629630	6980.657407	8.08	0.0004
A*C	4	13043.629630	3260.907407	3.77	0.0175
B*C	4	2913.185185	728.296296	0.84	0.5130
A*B*C	8	10794.814815	1349.351852	1.56	0.1935

Table 14.18 Incorrect analysis of variance, omitting the blocking factor to show the effect of partial confounding

```
The SAS System
General Linear Models Procedure
Dependent Variable: Y
```

Source	DF	Sum of Squares	Mean Square	F Value	Pr > F
Model	26	219007.148148	8423.351852	10.81	0.0001
Error	27	21038.500000	779.203704		
Corrected Total	53	240045.648148			

Source	DF	Type III SS	Mean Square	F Value	Pr > F
A	2	140999.703704	70499.851852	90.48	0.0001
B	2	19447.814815	9723.907407	12.48	0.0001
C	2	4934.481481	2467.240741	3.17	0.0582
A*B	4	27922.629630	6980.657407	8.96	0.0001
A*C	4	13043.629630	3260.907407	4.18	0.0092
B*C	4	2913.185185	728.296296	0.93	0.4588
A*B*C	8	9745.703704	1218.212963	1.56	0.1826

3. **Dye experiment, continued**

   (a) For the dye experiment of Section 14.2.4, check that the variances of the errors appear to be equal for the different levels of the three factors. Check also that the assumption of normality of the error variables is reasonable.

(b) Calculate the normalized contrast estimate for Linear $A \times$ Linear B, using the method outlined in Section 14.2.4.

(c) Compute the sum of squares for testing the hypothesis that the Linear $A \times$ Linear B contrast is negligible, using the method outlined in Section 14.2.4.

(d) Test the hypothesis that the Linear $A \times$ Linear B contrast is negligible, using an individual significance level of 0.01.

(e) Draw an interaction plot for AC and verify that the interaction appears to be negligible.

(f) Assuming that the contrasts were preplanned, calculate confidence intervals for the pairwise differences in yields due to the three different levels of each of A, B and C. State your overall confidence level.

4. **Dye experiment, continued**

   The experimenters who ran the dye experiment were interested in the linear and quadratic components of the main effects and interactions. Analyze the experiment accordingly. What information have you gathered about the levels of the factors if high yield is of importance?

5. A set of hypothetical data is given in Table 14.19 for a partially confounded 3^2 experiment in 6 blocks of 3. The design is made up of two single-replicate designs: The first confounds the contrasts $(AB; A^2B^2)$ from the interaction, while the second confounds the contrasts $(AB^2; A^2B)$.

   (a) By hand, write out the estimates of the linear and quadratic contrasts for the main effects and their associated variances.

   (b) Using the contrast estimates in part (a), calculate the sums of squares for A and B.

   (c) Calculate the least squares estimates of a pair of orthogonal contrasts for $(AB^2; A^2B)$ from the first replicate and the estimates of a pair of orthogonal contrasts for $(AB; A^2B^2)$ from the second replicate. Using these contrast estimates, calculate the sum of squares for the AB interaction (adjusted for blocks).

   (d) Prepare an analysis of variance table. Test any hypotheses that you think are of interest and state your conclusions about the two factors and their interaction.

   (e) Check your analysis in part (d) using a computer program.

Table 14.19 Partially confounded 3^2 experiment in $b = 6$ blocks of $k = 3$. Hypothetical data are shown in parentheses with corresponding treatment combinations.

Replicate	Block	Treatment Combinations (Response)		
1	I	00 (53)	12 (59)	21 (80)
Confounds $(AB; A^2B^2)$	II	01 (66)	10 (71)	22 (78)
	III	02 (69)	11 (91)	20 (92)
2	IV	00 (46)	11 (62)	22 (58)
Confounds $(AB^2; A^2B)$	V	01 (65)	12 (61)	20 (76)
	VI	02 (34)	10 (50)	21 (66)

Table 14.20 Data for the sugar beet experiment

Block I		Block II		Block III	
Levels of N, P, K	Yield	Levels of N, P, K	Yield	Levels of N, P, K	Yield
211	2575	121	2599	202	2189
120	2472	220	2517	020	2093
200	2411	022	2411	210	2354
002	2403	110	2252	111	2268
010	2220	212	2381	001	1926
021	2252	201	2067	122	2152
101	2295	102	2021	221	2349
112	2362	011	1953	012	2025
222	2434	000	1989	100	2106

Source: Yates, F. (1935). Copyright © 1935 Blackwell Publishers. Reprinted with permission. (Reprinted in *Experimental Design* (1970), Charles Griffin and Company, Ltd., London. Copyright 1970 Edward Arnold/Hodder & Stoughton Educational. Reprinted with permission.)

6. **Sugar beet experiment**

    F. Yates, in a 1935 paper published in a supplement to the *Journal of the Royal Statistical Society*, describes an agricultural experiment on the yield of sugar beet. The three factors of interest were three standard fertilizers, nitrogen, phosphate, and potassium (factors N, P, and K) each at three equally spaced levels. The experimental field was divided into $b = 3$ blocks and each block subdivided into $k = 9$ 0.1 acre plots. The experiment was designed so that the contrasts (NP^2K; N^2PK^2) were confounded with blocks. The randomized design and yields of sugar beet are shown in Table 14.20.

    (a) Prepare an analysis of variance table for the data, assuming that the three-factor interaction is negligible.

    (b) Investigate the linear and quadratic trends of the main effects and the two factor interactions. Yates assumed in his analysis that the only important contrast for each two factor interaction was the linear×linear contrast. Is this assumption supported by your analysis?

    (c) Draw any plots that help to illustrate the important features of the analysis.

7. **Example 14.3.2, continued**

    In Example 14.3.2, we showed one way of associating design factors F, G, H, and J of a $2^3 \times 4$ factorial experiment to the 2-level pseudofactors A–E of a specific design from Table 13.28. There are 10 different ways to make this association (since there are 10 ways of selecting two of A–E to represent J_1 and J_2).

    (a) Investigate the confounding schemes for each of the ten possible associations. Specifically, for each association, determine the number of contrasts confounded for each effect, and compare the results.

    (b) State under which circumstances you would recommend each design.

Exercises 481

8. Consider a $2^2 \times 3^2$ design confounding AB, $(CD^2; C^2D)$, and $(ABCD^2; ABC^2D)$.
   (a) Give the design—namely, list the treatment combinations block by block.
   (b) Describe how to randomize the design.
   (c) Give a set of five orthogonal treatment contrasts that are confounded with blocks.

9. Suggest a confounding scheme for a $2^3 \times 3^3$ experiment in 12 blocks of size 18. Under what circumstances would the design be useful? Write out two blocks of the design.

10. Suggest a confounding scheme for a $2^2 \times 3^2 \times 4$ experiment in 12 blocks of size 12. Under what circumstances would the design be useful? Write out two blocks of the design.

11. Suggest a confounding scheme for a $2^2 \times 3^2 \times 6$ experiment in 9 blocks of size 24. Under what circumstances would the design be useful? Explain how to find the blocks of the design.

12. Suggest a confounding scheme for a $2^2 \times 3^2 \times 6$ experiment in 12 blocks of size 18. Under what circumstances would the design be useful? Write out two blocks of the design.

Table 14.21 Confounding schemes for 3^p experiments in $b = 3^s$ blocks of size $k = 3^{p-s}$. For each design, s independent generators are underlined, and s corresponding equations are given. To obtain Block I of a design, list all k combinations of the first a_i's shown, then use the equations modulo 3 to complete each treatment combination.

3^p	b	k	Confounded Contrasts	Block I
3^2	3	3	($\underline{AB}$; A^2B^2)	a_1
				$a_2 = 2a_1$
3^3	3	9	($\underline{ABC^2}$; A^2B^2C)	a_1, a_2
				$a_3 = a_1 + a_2$
3^3	9	3	($\underline{AB^2}$; A^2B), ($\underline{AC}$; A^2C^2),	a_1
			(BC; B^2C^2), (ABC^2; A^2B^2C)	$a_2 = a_1$
				$a_3 = 2a_1$
3^4	3	27	($\underline{ABCD^2}$; $A^2B^2C^2D$)	a_1, a_2, a_3
				$a_4 = a_1 + a_2 + a_3$
3^4	9	9	($\underline{AB^2C}$; A^2BC^2), ($\underline{ABD}$; $A^2B^2D^2$),	a_1, a_2
			(AC^2D^2; A^2CD), (BCD^2; B^2C^2D)	$a_3 = 2a_1 + a_2$
				$a_4 = 2a_1 + 2a_2$
3^5	9	27	($\underline{ABCD^2}$; $A^2B^2C^2D$), ($\underline{AB^2E^2}$; A^2BE)	a_1, a_2, a_3
			(AC^2DE; $A^2CD^2E^2$),	$a_4 = a_1 + a_2 + a_3$
			(BC^2DE^2; B^2CD^2E)	$a_5 = a_1 + 2a_2$

15. Fractional Factorial Experiments

15.1 Introduction
15.2 Fractions from Block Designs; Factors with 2 Levels
15.3 Fractions from Block Designs; Factors with 3 Levels
15.4 Fractions from Block Designs; Other Experiments
15.5 Blocked Fractional Factorial Experiments
15.6 Fractions from Orthogonal Arrays
15.7 Design for the Control of Noise Variability
15.8 Using SAS Software
Exercises

15.1 Introduction

Fractional factorial experiments are used frequently in industry, especially in various stages of product development and in process and quality improvement. In a *fractional factorial experiment*, only a fraction of the treatment combinations are observed. This has the advantage of saving time and money in running the experiment, but has the disadvantage that each main-effect and interaction contrast will be confounded with one or more other main-effect and interaction contrasts and cannot be estimated separately. Two factorial contrasts that are confounded are referred to as being *aliased*. The term "confounded" is generally reserved for the indistinguishability of a treatment contrast and a block contrast.

We look at two methods of obtaining fractional factorial designs that can be analyzed in a straightforward manner. The first method, described in Sections 15.2–15.4, is to select one block from one of the single-replicate designs in Chapter 13. Blocked fractions are discussed in Section 15.5. The second method, described in Section 15.6, which is popular in industry, uses the concept of an *orthogonal array*. Throughout this chapter, we revisit some of the ideas introduced in Section 7.6 to reduce the sensitivity to noise variables of a product or manufacturing process.

The use of SAS in analyzing fractional factorial experiments is explored in Section 15.8.

15.2 Fractions from Block Designs; Factors with 2 Levels

15.2.1 Half-Fractions of 2^p Experiments; 2^{p-1} Experiments

We start with a very small example to illustrate the ideas. Suppose that an experiment is to be run with three treatment factors A, B, and C, each having two levels. There are no blocking factors, so the experiment will be run as a completely randomized design. However, only four observations can be taken.

We can obtain a design with just 4 of the 8 total treatment combinations by selecting at random one of the blocks of a single-replicate design with two blocks of size 4. For illustration, consider the block design that confounds the ABC contrast, given in Table 15.1. Suppose we select the second block, which is (001, 010, 100, 111) and is defined by the equation

$$a_1 + a_2 + a_3 = 1 \pmod 2.$$

The four treatment combinations constitute a $\frac{1}{2}$-*fraction* or $\frac{1}{2}$-*replicate* of a 2^3 experiment, called a 2^{3-1} design, and the ABC contrast is called the *defining contrast* for the fraction. We write

$$I = ABC,$$

which is called the *defining relation* for the fraction.

With only $n = 4$ observations, there are only $n - 1 = 3$ total degrees of freedom. This means that it is not possible to estimate each of the six remaining contrasts (A, B, C, AB, AC, BC) even if no estimate of σ^2 is required. If we look at the contrasts for a 2^3 experiment (shown in Table 13.2, page 425) and cross out the rows corresponding to the unobserved treatment combinations, we are left with the contrast coefficients in Table 15.2.

Table 15.1 2^3 experiment in 2 blocks of 4, confounding ABC

Block I	000	011	101	110
Block II	001	010	100	111

Table 15.2 Contrasts for the $\frac{1}{2}$-fraction (001, 010, 100, 111) of a 2^3 experiment

	A	B	C	AB	AC	BC	ABC
001	−1	−1	1	1	−1	−1	1
010	−1	1	−1	−1	1	−1	1
100	1	−1	−1	−1	−1	1	1
111	1	1	1	1	1	1	1

Table 15.2 shows several interesting features. First, the column corresponding to ABC is not a contrast. The coefficients are the coefficients that one would use in obtaining the sum of the four observations, which is a multiple of the mean. So, ABC is confounded with the mean, and the ABC contrast cannot be measured. The notation $I = ABC$ of the defining relation indicates the equivalence of ABC and the sum of the observations, since I corresponds to a list of coefficients all equal to $+1$.

Secondly, we see from Table 15.2 that the contrast coefficients for A and BC are identical, the contrasts for B and AC are identical, and the contrasts for C and AB are identical. The main effect of A and the interaction BC are said to be *aliased*, as are B and AC, and C and AB. We write

$$A = BC, \quad B = AC, \quad C = AB.$$

Thus, there are three estimable contrasts in the fraction, but each measures more than one factorial effect. For example, using the cell-means model

$$Y_{ijk} = \mu + \tau_{ijk} + \epsilon_{ijk},$$

the "$A = BC$" contrast with divisor $v/2$ (where v is the number of treatment combinations in the fraction) is obtained by multiplying the τ_{ijk}'s by the coefficients in Table 15.2, that is,

$$\tfrac{1}{2}[-\tau_{001} - \tau_{010} + \tau_{100} + \tau_{111}].$$

This is an estimable contrast with least squares estimate

$$\tfrac{1}{2}[-y_{001} - y_{010} + y_{100} + y_{111}].$$

This "$A = BC$" contrast is equal to

$$\frac{1}{4}[-\tau_{000} - \tau_{001} - \tau_{010} - \tau_{011} + \tau_{100} + \tau_{101} + \tau_{110} + \tau_{111}]$$
$$+ \frac{1}{4}[+\tau_{000} - \tau_{001} - \tau_{010} + \tau_{011} + \tau_{100} - \tau_{101} - \tau_{110} + \tau_{111}]$$
$$= [-\overline{\tau}_{0..} + \overline{\tau}_{1..}] + \frac{1}{2}[\overline{\tau}_{00.} - \overline{\tau}_{01.} - \overline{\tau}_{10.} + \overline{\tau}_{11.}].$$

Thus, what is being estimated is the sum of the A contrast and the BC contrast, and we could refer to the "$A = BC$" contrast as the $A + BC$ contrast. However, for simplicity, we often refer to it as the A contrast, remembering the role of BC from the list of aliased contrasts.

If in a hypothesis test or normal probability plot the main effect of factor A is nonsignificant, then there are two possibilities. One is that neither A nor its alias, BC, is significantly different from zero. The alternative is that the two contrasts have equal and opposite effects and cancel each other out. Since the former is much more likely, this is the assumption that is usually made. If the main effect of A appears to be significant, then it is not clear whether the observed effect is due to the main effect of A or to the BC interaction or the combination of both effects. Because of the aliasing problem, fractional factorial experiments are most often run as *screening experiments*. The word "screening" means that the experimenter is trying to determine which of a large number of factors affect the response.

Table 15.3 Aliasing scheme for the $\frac{1}{2}$-fraction (001, 010, 100, 111) of a 2^3 experiment

I	$=$	ABC
A	$=$	BC
B	$=$	AC
C	$=$	AB

The list of aliased contrasts is called the *aliasing scheme* for the design. We generally write this as in Table 15.3, with the first row showing the defining relation and the following rows listing the aliased contrasts. The number of rows in the aliasing scheme is the same as the number of observations in the design.

It can be verified that if Block I of the single replicate design in Table 15.1 were to be used as the $\frac{1}{2}$-fraction instead of Block II, exactly the same aliasing scheme would result, except that in each pair of aliased contrasts, the coefficients would differ from each other in sign. The estimable A contrast in the three-way model is of the form

$$\frac{1}{2}[-\tau_{000} - \tau_{011} + \tau_{101} + \tau_{110}]$$

$$= \frac{1}{4}[-\tau_{000} - \tau_{001} - \tau_{010} - \tau_{011} + \tau_{100} + \tau_{101} + \tau_{110} + \tau_{111}]$$

$$- \frac{1}{4}[+\tau_{000} - \tau_{001} - \tau_{010} + \tau_{011} + \tau_{100} - \tau_{101} - \tau_{110} + \tau_{111}]$$

$$= [-\overline{\tau}_{0..} + \overline{\tau}_{1..}] - \frac{1}{2}[\overline{\tau}_{00.} - \overline{\tau}_{01.} - \overline{\tau}_{10.} + \overline{\tau}_{11.}],$$

which we can refer to as the $A - BC$ contrast. This difference in signs can be highlighted by including the information in the aliasing scheme, that is, $I = -ABC$, $A = -BC$, $B = -AC$, and $C = -AB$. In order to disentangle the effects of A and BC, both blocks of the design would need to be run.

The observant reader will have noticed that the entire aliasing scheme in Table 15.3 can be deduced from the defining relation without writing out the contrasts. Using the contrast names, we can multiply the defining relation by A to obtain

$$A \times I = A \times ABC.$$

Treating I as a multiplicative identity so that $A \times I = A$, and reducing superscripts modulo 2 so that $A^2 BC = BC$, we obtain $A = BC$. The other two rows of the scheme can be obtained in a similar fashion. From now on, we will avoid writing out the contrasts and use the contrast names and the defining relation to obtain the aliasing scheme.

Half-fractions of 2^p experiments are called 2^{p-1} experiments, since they have $\frac{1}{2}2^p$ treatment combinations. They are almost always obtained by selecting one block from a block design that confounds the highest-order interaction. Thus for $p = 4$, for example, the fraction satisfies either

$$a_1 + a_2 + a_3 + a_4 = 0 \pmod{2}$$

and has defining relation $I = ABCD$, or it satisfies

$$a_1 + a_2 + a_3 + a_4 = 1 \pmod{2}$$

and has defining relation $I = -ABCD$. For $p = 5$, the fraction satisfies either

$$a_1 + a_2 + a_3 + a_4 + a_5 = 0 \pmod{2}$$

and has defining relation $I = -ABCDE$, or it satisfies

$$a_1 + a_2 + a_3 + a_4 + a_5 = 1 \pmod{2}$$

and has defining relation $I = ABCDE$. Notice that the sign of the contrast in the defining relation is positive if the equation contains an even number of a_i's and also is set equal to 0 (mod 2). It is also positive if the equation contains an odd number of a_i's and also is set equal to 1 (mod 2). Otherwise, the sign is negative. This always holds, even for the more complicated fractions of 2^p experiments discussed in the following sections.

For most purposes, we do not need to know whether the contrasts listed in the defining relation differ in sign. Consequently, unless they are needed, we shall usually ignore the signs in the aliasing scheme.

15.2.2 Resolution and Notation

The defining relation is a list of contrasts such as AB, ABC, etc. that are aliased with the mean. The contrasts in the defining relation are called *words*. The *resolution* of a design is the number of letters in the shortest word in the defining relation.

The design in Table 15.3 is a Resolution III design, since the only word in the defining relation is ABC, which has three letters. In all Resolution III designs, main-effect contrasts are aliased with 2-factor interaction contrasts. In a Resolution IV design, the defining relation contains words with 4 or more letters. Main effects then are aliased with 3-factor interactions and 2-factor interactions aliased with other 2-factor interactions. In a Resolution V design, such as that in Table 15.5, main effects are aliased with 4-factor interactions, and 2-factor interactions are aliased with 3-factor interactions. This is summarized in Table 15.4.

Since the main-effects and low-order interactions are usually the most important factorial effects to be measured, it is generally beneficial to select a design with as high resolution as can be found. The designs in Table 15.55 (page 542) all satisfy this requirement.

A $1/2^q$ fraction of a 2^p experiment is often referred to as a 2^{p-q} *fractional factorial experiment*. The resolution number is sometimes added as a subscript. A resolution III design, for example, can be written as a 2^{p-q}_{III} design.

15.2.3 A Real Experiment—Soup Experiment

L. B. Hare, Manager of Statistical Services at Thomas J. Lipton, Inc., described an experiment on a dry soup mix filling process in the 1988 issue of the *Journal of Quality Technology*.

Table 15.4 Resolution numbers of fractional factorial experiments

Resolution	Main effects aliased with:	2-factor interactions aliased with:
III	2-factor interactions or higher	Main effects and interactions
IV	3-factor interactions or higher	2-factor interactions or higher
V	4-factor interactions or higher	3-factor interactions or higher

Table 15.5 Design, data (measure of weight variability), and main-effect contrasts for the soup experiment. Defining relation $I = ABCDE$.

Levels of A, B, C, D, E.	y_{ijklm}	A	B	C	D	E
01011	0.78	−1	1	−1	1	1
11111	1.10	1	1	1	1	1
10000	1.70	1	−1	−1	−1	−1
11100	1.28	1	1	1	−1	−1
00001	0.97	−1	−1	−1	−1	1
01101	1.47	−1	1	1	−1	1
00010	1.85	−1	−1	−1	1	−1
10110	2.10	1	−1	1	1	−1
00111	0.76	−1	−1	1	1	1
10011	0.62	1	−1	−1	1	1
01110	1.09	−1	1	1	1	−1
01000	1.13	−1	1	−1	−1	−1
11001	1.25	1	1	−1	−1	1
10101	0.98	1	−1	1	−1	1
11010	1.36	1	1	−1	1	−1
00100	1.18	−1	−1	1	−1	−1

Source: Hare, L. B. (1988). Copyright © 1997 American Society for Quality. Reprinted with Permission.

The company was concerned about keeping the weight of the mix as uniform as possible. They found that most of the variability was due to the uneven flow of the "intermix," which is a mixture of vegetable oil, salt, and other ingredients, during the mixing process. The researchers prepared a list of five treatment factors that they thought might be influential in controlling the mixing process. The factors were:

A: the number of mixer ports through which the vegetable oil was added (two levels, 1 and 3);

B: temperature of mixer jacket (two levels; ambient temperature, presence of cooling water);

C: mixing time (two levels; 60 seconds and 80 seconds);

D: batch weight (two levels; 1500 lb and 2000 lb);

E: delay between mixing and packaging (two levels; 1 day and 7 days).

This was a screening experiment, since the researchers had little idea of which factors were going to turn out to be important in affecting the variability of the soup mix weight. They decided to run a $\frac{1}{2}$-fraction to investigate the five factors, and follow up the experiment with a more detailed study of the important factors later. They chose a Resolution V design with defining relation $I = ABCDE$, which allowed them to include all main effects and first-order interactions in the model. The corresponding block design, which confounds $ABCDE$, is listed in Table 13.28 (page 460). The experimenters chose the second block, as it contained the treatment combination that represented the normal operating conditions prior to the experiment. These were 00010, that is, one port, presence of cooling water, 60 seconds mix, 2000 lb batch weight, and one day delay before packaging.

The experiment was designed so that it could be run with very little disruption to the daily production routine. Sets of 5 samples were taken every 15 minutes during the production run for each treatment combination and weighed. The response variable was a measure of variation based on these weights (see the original article for the description).

The randomized design and the responses obtained are shown in Table 15.5. Also shown in the table are the contrasts for the main effects. As in Chapter 13, the contrast has coefficient -1 when the corresponding factor is at its low level and coefficient $+1$ when it is at its high level. The contrasts for the interactions are the products of the main-effect contrasts.

The experimenters assumed that all 3-, 4-, and 5-factor interactions would be negligible and included all the main effects and 2-factor interactions in the model. Since there were $16 - 1 = 15$ degrees of freedom in total and 15 contrasts to estimate (5 main effects and 10 two-factor interactions), there were no degrees of freedom available to estimate the error variability. The experimenters calculated all the contrast estimates and prepared a normal probability plot (Section 7.5) to find the important contrasts. For example, the contrast estimate for the main effect of E is

$$\hat{E} = (0.78 + 1.10 - 1.70 - 1.28 + 0.97 + 1.47 - 1.85 - 2.10 + 0.76$$
$$+ 0.62 - 1.09 - 1.13 + 1.25 + 0.98 - 1.36 - 1.18)/8 = -0.47.$$

The factorial contrast estimates are shown in Table 15.6, and the normal probability plot is shown in Figure 15.1. It can be seen that the most important contrasts appear to be BE, DE, and E.

Interaction plots of the two interactions BE and DE are shown in Figures 15.2(a) and 15.2(b). The response is a measure of weight variability, and the experimenters wanted to reduce this as much as possible. The estimate of the E contrast is negative, indicating that the low level of E (one-day delay before packaging) is more variable that the high level (seven-day delay). The two interaction plots also indicate that a seven-day delay before packaging would be beneficial using the ambient temperature and the large batch size

Table 15.6 Contrast estimates (with divisor $v/2 = 8$) for the soup experiment

Contrasts	A	B	C	D	E	AB	AC	
Estimates	0.145	−0.088	0.038	−0.038	−0.470	−0.015	0.095	
Contrasts	AD	AE	BC	BD	BE	CD	CE	DE
Estimates	0.030	−0.153	0.068	−0.163	0.405	0.073	0.135	−0.315

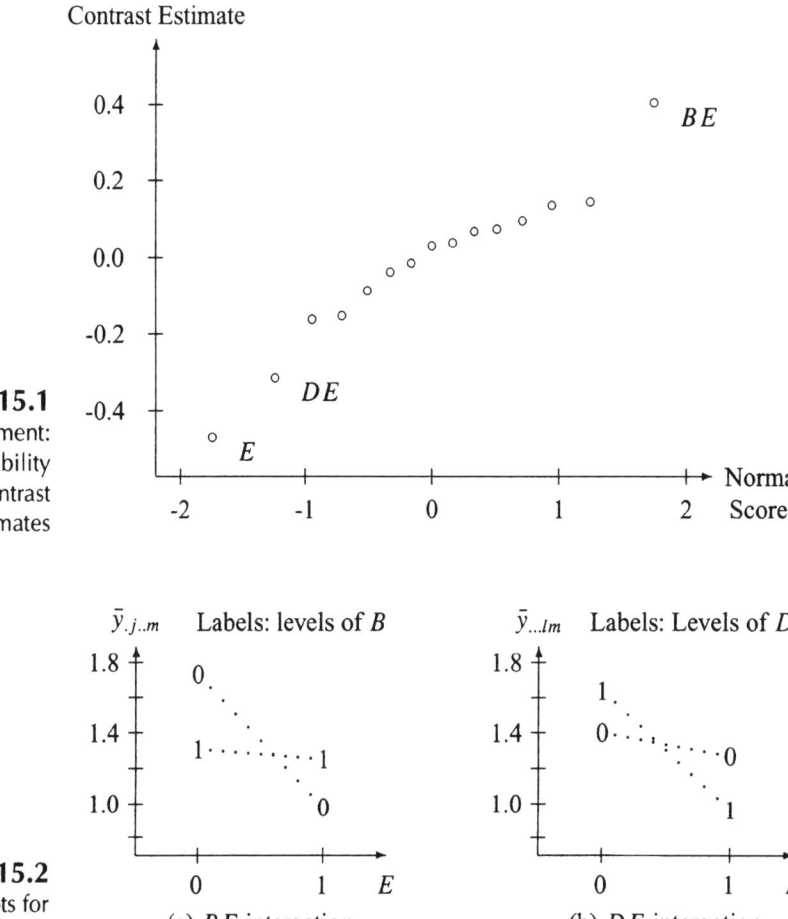

Figure 15.1 Soup experiment: normal probability plot of contrast estimates

Figure 15.2 Interaction plots for the soup experiment

(2000 lb). The interaction plots also indicate that if a seven-day delay is not feasible, then it is better to use cooling water and a small batch size. The production management of the company agreed to a seven-day delay, and the researchers decided to investigate these three factors (B, D, and E) in more detail in a followup experiment, together with factor C, whose interaction with E was the next largest effect.

15.2.4 Quarter-Fractions of 2^p Experiments; 2^{p-2} Experiments

We can obtain a $\frac{1}{4}$-fraction of a 2^p experiment by selecting at random one of the blocks from a single-replicate confounded design with 4 blocks of size 2^{p-2}. The defining relation is then the set of three contrasts that were confounded to obtain the block design.

For example, suppose a 2^5 experiment was to be run as a completely randomized design, but only eight observations could be taken. Table 13.28 (page 460) lists a design in 4 blocks of 8 that confounds the three interactions ABD, CDE, and $ABCE$. Suppose that we select one block for the $\frac{1}{4}$-fraction, specifically the block that satisfies

$$a_1 + a_2 + a_4 = 1 \pmod 2 \quad \text{and} \quad a_3 + a_4 + a_5 = 0 \pmod 2.$$

The treatment combinations in this fraction are

00011 00110 01000 01101 10000 10101 11011 11110,

and the defining relation for the fraction is

$$I = ABD = CDE = ABCE.$$

(If we work out the contrast coefficients for this fraction, we find that the coefficients for ABD are all $+1$, while those for CDE and $ABCE$ are all -1. Thus, if the signs of the contrasts were taken into account, the defining relation would be $I = ABD = -CDE = -ABCE$.) The other seven rows of the aliasing scheme are obtained by multiplying the defining relation by each of the contrast names in turn. The resulting aliasing scheme (ignoring signs) is shown in Table 15.7.

Only one factorial effect from each row of the aliasing scheme (and none from the defining relation) can be entered into the model for analyzing the experiment. So, for example, we could include all main effects and the two 2-factor interactions AC and AE in the model.

If the D effect, for example, is insignificant, the interactions AB, CE, and $ABCDE$ are also regarded as insignificant. But if the analysis shows that D has a significant effect on the response, it is unknown whether the effect is due to the main effect of D, or to AB, or to CE, or to $ABCDE$ (although this latter effect is the least likely), or to some combination of all four. The design is useful for screening when it is believed that most main effects and interactions will be negligible but one or two factors will possibly have an important effect on the response.

This design is clearly ideal if all of the interactions are negligible, or if all interactions except exactly one of AC, BE, AE, and BC are thought to be negligible. In the first case, two degrees of freedom are available to estimate σ^2. In the second case, all of the main effects and the one interaction can be measured, and one degree of freedom remains to estimate σ^2. If all main effects and, say, the CD interaction were required to be estimated, then a different block design should be chosen. For example, a suitable design could be obtained by interchanging A and D in the list of confounded contrasts. In other words, the

Table 15.7 Aliasing scheme (ignoring signs) for a $\frac{1}{4}$-fraction of a 2^5 experiment with the defining relation $I = ABD = CDE = ABCE$

I	=	ABD	=	CDE	=	$ABCE$
A	=	BD	=	$ACDE$	=	BCE
B	=	AD	=	$BCDE$	=	ACE
C	=	$ABCD$	=	DE	=	ABE
D	=	AB	=	CE	=	$ABCDE$
E	=	$ABDE$	=	CD	=	ABC
AC	=	BCD	=	ADE	=	BE
AE	=	BDE	=	ACD	=	BC

design obtained by confounding ABD, ACE, and $BCDE$ will give a $\frac{1}{4}$–fraction in which CD is not aliased with main effects.

A list of useful $\frac{1}{4}$–fractions is given in Table 15.55 at the end of the chapter.

Example 15.2.1 Sludge experiment

S. R. Schmidt and R. G. Launsby, in their textbook *Understanding Industrial Designed Experiments*, include an article by J. Brickell and K. Knox on the operation of a biological treatment system (known as an activated sludge system) used in wastewater treatment plants. The details of the system are given in the article. The response, Y, is the removal of "biochemical oxygen demand," which is related to the quality of water. The water quality increases as more biochemical oxygen demand is removed, so the response Y is to be maximized. The experiment described in the article investigates the effect of five factors on Y:

A: Reactor biomass concentration (3000 and 6000 mg/l),

B: Clarifier biomass concentration (8000 and 12000 mg/l),

C: Waste sludge flow rate (78.5 and 940 m^3/d),

D: Biological growth rate constant (0.040 and 0.075 d^{-1})

E: Fraction of food to biomass (0.4 and 0.8 kg/kg).

Since this experiment was an illustration of what could be run in a water treatment plant, it was necessary to keep the number of observations small, and a $\frac{1}{4}$–fraction was selected with defining relation $I = ABD = CDE = ABCE$. This gives an aliasing scheme that is similar to that shown in Table 15.7 but with D and E interchanged.

The experimenters selected the fraction whose treatment combinations, written as $a_1a_2a_3a_4a_5$, satisfied

$$a_1 + a_2 \quad + a_4 \quad\quad = 1 \ (\text{mod } 2)$$
$$a_3 + a_4 + a_5 = 1 \ (\text{mod } 2)$$

The design, prior to randomization, is shown in Table 15.8 together with the responses obtained.

The experimenters included all main effects and the 2–factor interactions AC and BC in their model. The contrast estimates (with divisors $v/2 = 4$) are listed in Table 15.9, and a normal probability plot of the seven contrast estimates is shown in Figure 15.3. There are too few contrast estimates to be able to draw good conclusions from the normal probability plot. Nevertheless, the most important effect appears to be the main effect of C and, perhaps to a lesser extent, E. Now, C is aliased with DE, and E is aliased with CD. A followup experiment investigating the effects of C, D, and E would certainly be advisable.

If we try to draw conclusions from the results of the present experiment, and if we are willing to assume that the main effects are the dominant effects in any alias sets, it would seem advisable to set C and possibly B at their high levels in order to maximize the response, and to set E and possibly D at their low levels. On the other hand, if we assume

Table 15.8 $\frac{1}{4}$-fraction of a 2^5 experiment and data from the sludge experiment.

Levels of A, B, C, D, E	y_{ijklm}	Contrasts						
		A	B	C	D	E	AC	BC
00010	195	−1	−1	−1	1	−1	1	1
00111	496	−1	−1	1	1	1	−1	−1
01001	87	−1	1	−1	−1	1	1	−1
01100	1371	−1	1	1	−1	−1	−1	1
10001	102	1	−1	−1	−1	1	−1	1
10100	1001	1	−1	1	−1	−1	1	−1
11010	354	1	1	−1	1	−1	−1	−1
11111	775	1	1	1	1	1	1	1

Source: Brickell, J. and Knox, K. (1992). Copyright © 1992 Air Academy Press. Reprinted with permission.

Table 15.9 Contrast estimates (with divisor $v/2 = 4$) for the sludge experiment

Contrast	A	B	C	D	E	AC	BC
Estimate	20.75	198.25	726.25	−185.25	−365.25	−66.25	126.25

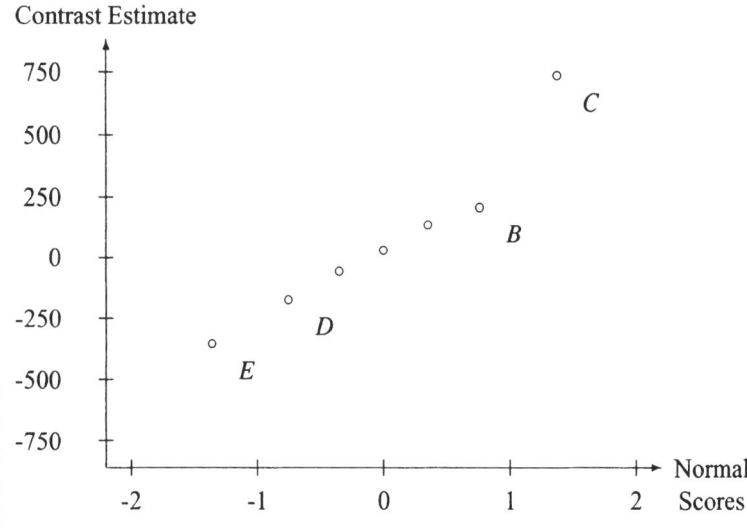

Figure 15.3 Normal probability plot of contrast estimates for the sludge experiment

that the interactions in the alias sets might be important, we would examine the DE and CD interaction plots (see Figure 15.4). The DE plot suggests that D and E should both be at their low levels, and the CD plot suggests that C should be at its high level and D at its low level. Since the recommendations from the interaction plots agree with those from the main effect comparisons, we would feel comfortable in recommending that the process be set at the cheaper of 01100 or 11100. Notice that the first of these was included among the experimental runs (and happened to give rise to the largest observed yield), whereas the

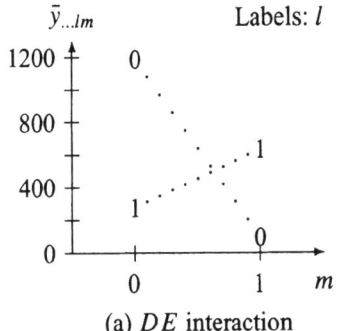
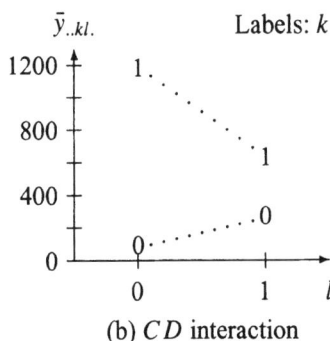

Figure 15.4
Interaction plots for the sludge experiment

(a) DE interaction (b) CD interaction

second was not included. In either case, it would be advisable to rerun the experiment at the chosen setting to confirm the results.

The authors of the article point out that other considerations, such as cost, come into play before any system can be changed. In an actual water treatment plant, it would be expensive to change the levels of D (biological growth rate constant) and E (fraction of food to biomass) from their current settings. Also, increasing the waste sludge flow rate (C) increases cost. A followup experiment could not only verify that the above recommendations were correct, but also could examine intermediate values of C. □

15.2.5 Smaller Fractions of 2^p Experiments

Smaller fractions of a 2^p experiment can be obtained in exactly the same way as the $\frac{1}{2}$–fractions and $\frac{1}{4}$–fractions of the preceding subsections. For a $1/2^s$ fraction, the first step is to find a design with 2^s blocks of size 2^{p-s} that confounds negligible interactions. One block is selected at random. The aliasing scheme is then checked to ensure that as few important contrasts as possible are aliased with each other. If the aliasing scheme is not suitable, then an attempt is made to obtain a better design by interchanging letters in the confounding scheme, or by investigating different confounding schemes.

A list of useful $\frac{1}{8}$–fractions and $\frac{1}{16}$–fractions of 2^p experiments is given in Table 15.55 at the end of the chapter.

Example 15.2.2 Welding experiment

An experiment was discussed by A. K. Shahani in *The Statistician* in 1970 which involved a (1/1024) fraction of a 2^{21} experiment (that is, a 2^{21-16} experiment). The experiment, which required only $v = 2^5 = 32$ observations, was designed by Dr. Shahani for Bristol Aerojet Ltd and concerned the "pull strength" of welds resulting from a certain welding process. The company wished to discover which settings of the 21 factors would give welds with pull strength exceeding a given size. Of the 21 factors, only a few were expected to have important effects on the pull strength, and this allowed the use of such a highly fractionated design.

15.2 Fractions from Block Designs; Factors with 2 Levels   495

All 21 factors were easy to manipulate, and the engineers selected two reasonable settings for each factor (coded 0 and 1). For some of the factors, the two levels chosen were at equal distances on each side of the current operating conditions. For others, such as factors A, D, and W, the low levels were at the current operating conditions and could not be lowered further. If we label the factors $A, B, \ldots, W$ omitting I and O, the contrasts selected for confounding were as follows:

ABV	*ACW*	*ADT*	*AES*
BCU	*ABEN*	*ACDQ*	*ACEP*
ADEM	*BCER*	*BDEL*	*CDEK*
ABCEH	*ABDEJ*	*ACDEG*	*BCDEF*

The defining relation consists of these 16 contrasts together with all their possible products. Since the shortest word in the defining relation is of length 3, the design is Resolution III. Although each main effect is aliased with several two-factor interactions, the main effects are not aliased with each other.

The 32 treatment combinations and their responses are shown in Table 15.10. (We have corrected typing errors that occurred in the original paper in the two treatment combinations in the second row of our table). The responses are in coded units, details of which were not given in the original paper.

This experiment has too many factors to be able to analyze it easily by hand. The main-effect contrast estimates (with divisors $v/2 = 16$), obtained from the SAS computer package, are shown in Table 15.11. Under the current operating conditions, it was known that the error standard deviation σ was about 60 units. The experimenters were willing to assume that this would not change appreciably under different operating conditions and therefore

Table 15.10 Treatment combinations and responses for the 2^{21-16} welding experiment

Treatment Combination	Response	Treatment Combination	Response
000001111000000011111	430	100001000001111000100	422
010000100010100111001	336	110000011011011100010	380
001000001100011111010	438	101000110101100100001	96
011001010110111011100	394	111001101111000000111	319
000100010111001010111	334	100100101110110001100	202
010101001101101110001	322	110101110100010101010	238
001101100011010110010	184	101101011010101101001	188
011100111001110010100	348	111100000000001001111	−234
000010000111110101111	384	100010111110001110100	338
010011011101010001001	404	110011100100101010010	370
001011110011101001010	542	101011001010010010001	114
011010101001001101100	316	111010010000110110111	432
000111101000111100111	256	100111010001000111100	206
010110110010011000001	82	110110001011100011010	106
001110011100000000010	528	101110100101011011001	110
011111000110000100100	35	111111111111111111111	370

Source: Shahani, A. K. (1970). Copyright © 1970 Blackwell Publishers. Reprinted with permission.

calculated the standard error of a main effect contrast $\Sigma_i c_i \tau_i$ to be

$$\sqrt{\text{Var}(\Sigma_i c_i \hat{\tau}_i)} = \sigma\sqrt{\Sigma c_i^2} = 60\sqrt{32/16^2} = 21.21.$$

Without assuming that the coded responses follow a normal distribution, the experimenters then deemed any contrast whose estimated absolute value turned out to be several times larger than 21.21 to be important.

The contrast estimates whose absolute values exceed 63.63 are those for the main effects of A, D, H, J, N, V, and W. The estimates for main effects of M, K, and S are all around 2 standard errors, with those for C and F a little smaller.

Since there are 32 observations, a total of 31 orthogonal contrasts can be measured. Thus there are 10 sets of confounded interaction contrasts that can be measured in addition to the 21 main-effect contrasts. The identification of these contrast sets requires writing out the entire aliasing scheme—a daunting task! A proper analysis of the main effects also requires knowledge about which interactions are aliased with which main effects. A followup experiment to separate out the most likely aliased effects would be needed.

Assuming, temporarily, that the process can be improved by considering the main effects only, the contrast estimates (high level minus low level) suggest that factors A, D, and W (whose contrast estimates are negative) should be set at their low levels and factors H, J, K, N, V, M, and S (whose contrast estimates are positive) should be set at their high levels. As mentioned above, A, D, and W were already set at the lowest possible values in the original process, and therefore further experimentation with these factors is unnecessary. The other seven factors were discussed by the engineers and new (higher) settings selected for these, resulting in an improved process that met the pull strength requirements. The author of the article commented that the research and development department should give consideration to a further experiment involving these seven factors in which main effects and two-factor interactions could all be measured. He suggested the use of a 2^{7-1} experiment, which would require 64 observations. Fewer observations would require aliasing some of the 2-factor interactions (see Table 15.55). □

15.3 Fractions from Block Designs; Factors with 3 Levels

15.3.1 One-Third Fractions of 3^p Experiments; 3^{p-1} Experiments

To obtain a fraction of a 3^p experiment, we use the same idea that we used for 2^p experiments. We select one block at random from a confounded single-replicate design with 3^s blocks

Table 15.11 Contrast estimates for the welding experiment (with divisor 16)

Contrast	A	B	C	D	E	F	G
Estimate	−104.8	−34.6	−39.4	−152.5	12.3	37.4	5.3
Contrast	H	J	K	L	M	N	P
Estimate	101.9	70.5	48.4	−23.4	43.5	100.1	32.9
Contrast	Q	R	S	T	U	V	W
Estimate	16.6	3.0	42.1	−8.1	7.1	72.8	−69.0

of size 3^{p-s} with a suitable confounding scheme. For example, suppose a $\frac{1}{3}$-fraction of a 3^4 experiment is required (that is, a total of $3^{4-1} = 27$ observations). The highest-order interaction in a 3^4 experiment that can be confounded is the 4-factor interaction. Therefore, the maximum number of letters in a word in the defining relation of the fraction is also four. For a Resolution IV design, when 3- and 4-factor interactions are negligible, the main-effect contrasts can be estimated, but the 2-factor interactions will be aliased. This is the best design available, and unless a larger budget can be obtained to allow more observations, some aliasing among the low-order interactions will have to be tolerated.

Suppose the selected single-replicate confounded design is that which confounds the pair of contrasts $(ABCD; A^2B^2C^2D^2)$ from the 4-factor interaction. The block design is constructed using the equations

$$a_1 + a_2 + a_3 + a_4 = 0, 1, \text{ or } 2 \pmod{3}$$

as in Section 14.2.3, and one block is selected at random for the $\frac{1}{3}$-fraction. Since there are 27 treatment combinations to be observed, the aliasing scheme has 27 rows. Seven rows from the aliasing scheme are given in Table 15.12. The remaining rows contain main effects or 2-factor interaction contrasts (such as AB^2 or A^2B) that are aliased only with higher-order interactions. The 27 rows of the aliasing scheme include one for effects aliased with the mean plus 13 pairs of rows, such as rows including AB and A^2B^2, which represent the same contrasts. The two rows containing AB and A^2B^2 indicate, for example, that the contrasts $(AB; A^2B^2)$, $(CD; C^2D^2)$, and $(ABC^2D^2; A^2B^2CD)$ are aliased. Use of this design is illustrated in Example 15.3.1.

Example 15.3.1   Refinery experiment

P. W. M. John (1971) describes an experiment of Vance (1962) to find a set of operating conditions to optimize the quality of lube oil treated at a refinery. There were four factors of interest, called here A, B, C, and D, and three equally spaced levels were selected for each of these so that quadratic trends could be measured.

Since this was a preliminary experiment, a $\frac{1}{3}$-fraction of Resolution IV was thought to be adequate. The experimenters used a design with defining relation $I = ABCD = A^2B^2C^2D^2$. Part of the aliasing scheme is shown in Table 15.12. We see that two degrees of freedom $(AB; A^2B^2)$ from AB are aliased with two degrees of freedom $(CD; C^2D^2)$ from CD and two degrees of freedom from the four-factor interaction. The other two degrees of freedom from each of these interactions are aliased with higher-order interactions. (For example, the pair $(AB^2; A^2B)$ is aliased with the pairs $(A^2CD; AC^2D^2)$ and $(BC^2D^2; B^2CD)$). A similar confounding occurs with AC and BD and also with AD and BC.

Table 15.12 Seven rows from the aliasing scheme for a $\frac{1}{3}$-fraction of a 3^4 experiment with the defining relation $I = ABCD = A^2B^2C^2D^2$

I	=	$ABCD$	=	$A^2B^2C^2D^2$				
AB	=	A^2B^2CD	=	C^2D^2	A^2B^2	=	CD	=	ABC^2D^2
AC	=	A^2BC^2D	=	B^2D^2	A^2C^2	=	BD	=	AB^2CD^2
AD	=	A^2BCD^2	=	B^2C^2	A^2D^2	=	BC	=	AB^2C^2D

Table 15.13 $\frac{1}{3}$-fraction of a 3^4 experiment and data from the refinery experiment

Treatment Combination	y_{ijkl}	A_L	A_Q	B_L	B_Q	C_L	C_Q	D_L	D_Q
0000	4.2	−1	1	−1	1	−1	1	−1	1
0012	5.9	−1	1	−1	1	0	−2	1	1
0021	8.2	−1	1	−1	1	1	1	0	−2
0102	13.1	−1	1	0	−2	−1	1	1	1
0111	16.4	−1	1	0	−2	0	−2	0	−2
0120	30.7	−1	1	0	−2	1	1	−1	1
0201	9.5	−1	1	1	1	−1	1	0	−2
0210	22.2	−1	1	1	1	0	−2	−1	1
0222	31.0	−1	1	1	1	1	1	1	1
1002	7.7	0	−2	−1	1	−1	1	1	1
1011	16.5	0	−2	−1	1	0	−2	0	−2
1020	14.3	0	−2	−1	1	1	1	−1	1
1101	11.0	0	−2	0	−2	−1	1	0	−2
1110	29.0	0	−2	0	−2	0	−2	−1	1
1122	55.0	0	−2	0	−2	1	1	1	1
1200	8.5	0	−2	1	1	−1	1	−1	1
1212	37.4	0	−2	1	1	0	−2	1	1
1221	66.3	0	−2	1	1	1	1	0	−2
2001	11.4	1	1	−1	1	−1	1	0	−2
2010	21.1	1	1	−1	1	0	−2	−1	1
2022	57.9	1	1	−1	1	1	1	1	1
2100	13.5	1	1	0	−2	−1	1	−1	1
2112	51.6	1	1	0	−2	0	−2	1	1
2121	76.5	1	1	0	−2	1	1	0	−2
2202	31.0	1	1	1	1	−1	1	1	1
2211	74.5	1	1	1	1	0	−2	0	−2
2220	85.1	1	1	1	1	1	1	−1	1

Source: John, P. W. M. (1971). Copyright © 1971 P. W. M. John. Reprinted with permission.

The treatment combinations are obtained from the equation

$$a_1 + a_2 + a_3 + a_4 = 0 \pmod 3$$

and are shown in Table 15.13, prior to randomization, together with the data collected. Also shown are the linear and quadratic contrast coefficients for the main effects. The objective of the experiment was to select factor levels that would increase the response (a measure of quality).

The analysis of variance is complicated by the aliasing of pairs of degrees of freedom for two-factor interactions. We have not tried to separate these but have listed the contributions of the pairs of interactions on the same line of the analysis of variance table shown in Table 15.14. Without information concerning negligible interactions, we are unable to obtain an estimate for the error variance. The most important interactions appear to be the AC, BD, BC, and AD interactions, and the corresponding interaction plots are shown in Figure 15.5. In each case, the plots indicate that in order to increase the response, factors A, B, and

15.3 Fractions from Block Designs; Factors with 3 Levels

Table 15.14 Analysis of variance for the refinery experiment

Source of Variation	Degrees of Freedom	Sum of Squares	Mean Square
A	2	4496.29	2248.14
A_L	1	4399.22	4399.22
A_Q	1	97.07	97.07
B	2	2768.69	1384.35
B_L	1	2647.49	2647.49
B_Q	1	121.20	121.20
C	2	5519.79	2759.89
C_L	1	5516.00	5516.00
C_Q	1	3.79	3.79
D	2	283.37	141.68
D_L	1	213.56	213.56
D_Q	1	69.81	69.81
AB, CD	6	339.00	56.50
AC, BD	6	1384.24	230.71
AD, BC	6	753.38	125.56
Total	26	15,544.66	

C should all be set at their high levels, cost permitting, and factor D should be set at its middle level. They also indicate that since the lines are not too far from parallel, it would be reasonable to examine the main-effect contrasts.

Normalized linear and quadratic main-effect contrast estimates are obtained as

$$\frac{1}{d}\sum_i\sum_j\sum_k\sum_l c_{ijkl}\hat{\tau}_{ijkl} = \frac{1}{d}\sum_i\sum_j\sum_k\sum_l c_{ijkl}y_{ijkl},$$

where the c_{ijkl}'s are the contrast coefficients in Table 15.13 and the divisor d is the square root of the sum of squares of the coefficients (that is, $\sqrt{18}$ for the linear contrasts and $\sqrt{54}$ for the quadratic contrasts). These estimates are listed in Table 15.15, and a normal probability plot of the estimates is shown in Figure 15.6. Eight estimates are too few to make a good judgment, but the most important effects appear to be the linear trends in C, A, and B (in that order). All of these contrast estimates are positive, suggesting that the high levels should be selected in order to increase the response. This agrees with the conclusions from the interaction plots. Note that we could have examined interactions more closely by including individual interaction contrast estimates in the normal probability plot. We have not done this because of the complicated confounding of the interactions.

Since we have no estimate for error, we are unable to test any hypotheses. However, had the experimenters believed, prior to the experiment, that some or all of the interactions were

Table 15.15 Normalized contrast estimates for the refinery experiment

	$\hat{A}_L$	$\hat{A}_Q$	$\hat{B}_L$	$\hat{B}_Q$	$\hat{C}_L$	$\hat{C}_Q$	$\hat{D}_L$	$\hat{D}_Q$
Estimate	66.33	9.85	51.45	−11.01	74.27	−1.95	14.61	−8.36

negligible, then tests would have been done for the remaining interactions and the main effects. The sums of squares for testing the linear and quadratic main-effect contrasts are merely the squares of the corresponding normalized contrast estimates in Table 15.15. For

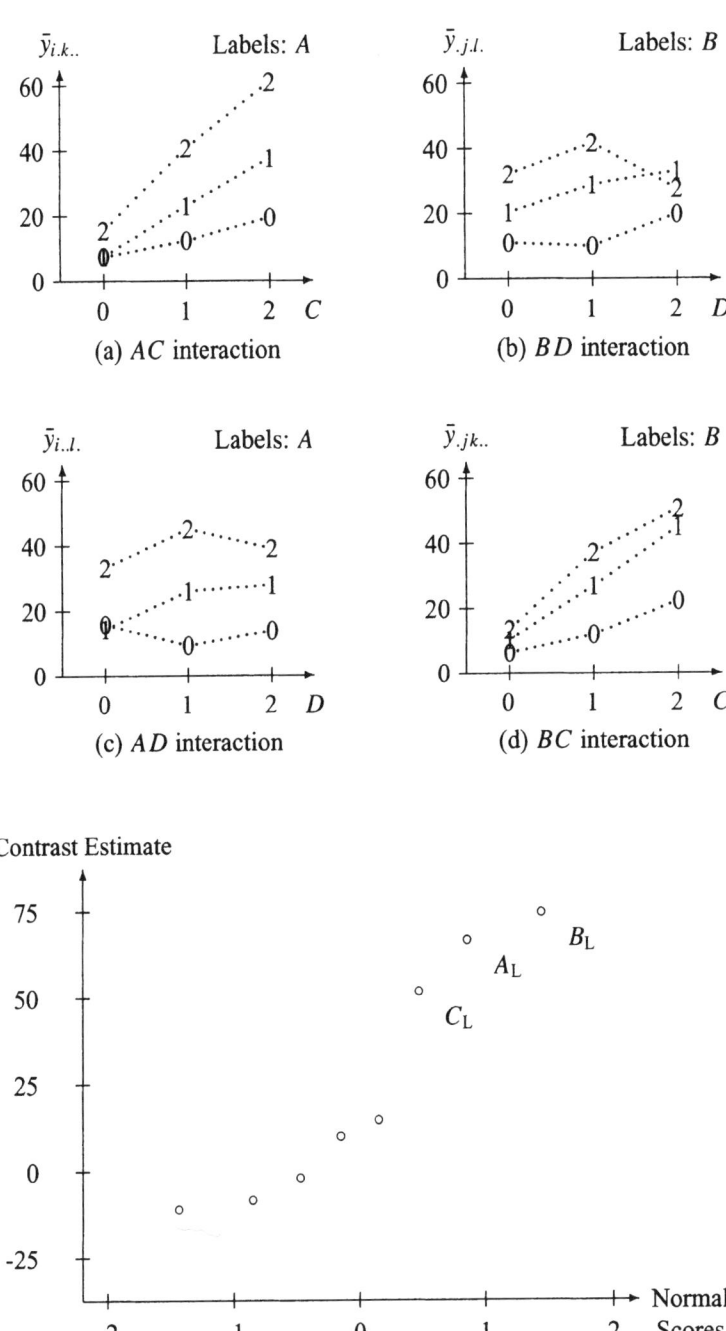

Figure 15.5 Interaction plots for the refinery experiment

Figure 15.6 Normal probability plot for the refinery experiment contrast estimates

example, the sums of squares for testing the hypothesis that the linear trend of factor A is negligible, against the alternative hypothesis that it is not negligible, is

$$ss(A_L) = 66.327^2 = 4399.22.$$

The sums of squares for the main effects of factors A, B, C, and D can be obtained either by adding their respective linear and quadratic contrast sums of squares, or by using the rules of Chapter 7 with $r = 1/3$ (since this is a one-third fraction). For example,

$$ssA = ss(A_L) + ss(A_Q) = 66.327^2 + 9.852^2 = 4399.22 + 97.07 = 4496.29,$$

or

$$ssA = 9\Sigma \bar{y}_{i...}^2 - 27\bar{y}_{....}^2 = 28766.30 - 24270.01 = 4496.29. \qquad \square$$

15.3.2 One-Ninth Fractions of 3^p Experiments; 3^{p-2} Experiments

As an example of a $\frac{1}{9}$-fraction, we take the sixth block of the 3^4 single-replicate confounded design shown in Table 14.5. The list of confounded interactions in the block design provides the defining relation for the 3^{4-2} fractional factorial design, namely

$$\begin{aligned}
I &= AB^2C = A^2BC^2 \\
= ABD &= A^2CD = B^2C^2D \\
= A^2B^2D^2 &= BCD^2 = AC^2D^2.
\end{aligned}$$

The confounded contrasts are $(AB^2C; A^2BC^2)$, $(ABD; A^2B^2D^2)$, $(AC^2D^2; A^2CD)$, and $(BCD^2; B^2C^2D)$.

This design has Resolution III (since the shortest word has 3 letters), and main-effect contrasts will be aliased with 2-factor interaction contrasts. The nine observations provide 8 degrees of freedom, which is sufficient to estimate the four main effects (with two degrees of freedom each). Therefore, the design would be useful if all two-factor interactions were believed to be negligible. Since there are no degrees of freedom available for estimating σ^2, a normal probability plot of normalized contrast estimates would be drawn as in Figure 15.6.

15.4 Fractions from Block Designs; Other Experiments

15.4.1 $2^p \times 4^q$ Experiments

The simplest way to design a fractional factorial experiment when all factors have four levels, or when some factors have two levels and the others have four levels, is to use pseudofactors. For example, suppose we require a design for a $2^3 \times 4$ experiment with eight observations. A design in four blocks of size 8 is shown in Table 14.9 (page 472). The confounded contrasts are FGJ_1J_2, GHJ_2, and FHJ_1, where J_1 and J_2 are the two 2-level pseudofactors making up the 4-level factor J. Suppose Block I is selected from the design to give a $\frac{1}{4}$-fraction of a $2^3 \times 4$ experiment, then the defining relation is

$$I = FGJ_1J_2 = GHJ_2 = FHJ_1$$

Table 15.16 Aliasing scheme for a $\frac{1}{4}$-fraction of a $2^3 \times 4$ experiment

I	$=$	FGJ_1J_2	$= GHJ_2$	$=$	FHJ_1
F	$=$	GJ_1J_2	$= FGHJ_2$	$=$	HJ_1
G	$=$	FJ_1J_2	$= HJ_2$	$=$	$FGHJ_1$
H	$=$	$FGHJ_1J_2$	$= GJ_2$	$=$	FJ_1
J_1	$=$	FGJ_2	$= GHJ_1J_2$	$=$	FH
J_2	$=$	FGJ_1	$= GH$	$=$	FHJ_1J_2
J_1J_2	$=$	FG	$= GHJ_1$	$=$	FHJ_2
FJ_2	$=$	GJ_1	$= FGH$	$=$	HJ_1J_2

and the design is Resolution III. The aliasing scheme, shown in Table 15.16, indicates that the F contrast, for example, is aliased with one contrast from each of the GJ, $FGHJ$ and HJ interactions. There are 3 contrasts (J_1, J_2 and J_1J_2) for the 4-level factor J. The aliasing scheme shows that J is aliased with the FGJ, GHJ, FH, GH, FHJ, FG contrasts.

An experiment involving pseudofactors will be illustrated in Example 15.5 in Section 15.5.

15.4.2 $2^p \times 3^q$ Experiments

Suppose that a $\frac{1}{6}$-fraction is required from a $2^3 \times 3^3$ experiment giving a total of 36 observations. We select a $\frac{1}{2}$-fraction of a 2^3 experiment confounding, say, ABC, and a $\frac{1}{3}$-fraction from a 3^3 experiment confounding, say, $(DE^2F; D^2EF^2)$. Combining the treatment combinations as in a combined array (see Example 14.4, page 473, for illustration) gives a design in $b = 6$ blocks of 36 that confounds the five contrasts ABC, $(DE^2F; D^2EF^2)$, and $(ABCDE^2F; ABCD^2EF^2)$. If one block is selected, we have a Resolution III design with defining relation

$$I \quad = DE^2F \quad = D^2EF^2$$
$$= ABC = ABCDE^2F = ABCD^2EF^2,$$

and the contrasts ABC, $(DE^2F; D^2EF^2)$, and $(ABCDE^2F; ABCD^2EF^2)$ are aliased with the mean.

The aliasing scheme (which has 36 rows) includes the following rows:

$$A = BC = ADE^2F = AD^2EF^2 = BCDE^2F = BCD^2EF^2,$$
$$B = AC = BDE^2F = BD^2EF^2 = ACDE^2F = ACD^2EF^2,$$
$$C = AB = CDE^2F = CD^2EF^2 = ABDE^2F = ABD^2EF^2.$$

Thus, the 2-level factors A, B, and C are aliased with 2-factor interactions between the 2-level factors plus some higher-order interactions. For example, the A contrast is aliased with the contrasts BC, $(ADE^2F; AD^2EF^2)$, and $(BCDE^2F; BCD^2EF^2)$.

A similar aliasing happens for the 3-level factors. For example,

$$D = ABCD = D^2E^2F = EF^2 = ABCD^2E^2F = ABCEF^2,$$
$$D^2 = ABCD^2 = E^2F \quad = DEF^2 = ABCE^2F \quad = ABCDEF^2,$$

so the pairs of contrasts $(D; D^2)$, $(ABCD; ABCD^2)$, $(DEF^2; D^2E^2F)$, $(EF^2; E^2F)$, and $(ABCEF^2; ABCE^2F)$ are aliased with one another.

Finally, there is aliasing of interactions involving both 2- and 3-level factors, for example

$$AD = BCD = AD^2E^2F = AEF^2 = BCD^2E^2F = BCEF^2,$$
$$AD^2 = BCD^2 = AE^2F = ADEF^2 = BCE^2F = BCDEF^2,$$

so the pairs of contrasts $(AD; AD^2)$, $(BCD; BCD^2)$, $(ADEF^2; AD^2E^2F)$, $(AEF^2; AE^2F)$, and $(BCEF^2; BCE^2F)$ are aliased with one another.

The design would be useful mainly when most of the interactions were expected to be negligible.

15.5 Blocked Fractional Factorial Experiments

If experimental conditions are not constant over the entire experiment, it may be necessary to arrange a fractional factorial experiment in blocks. For example, consider the soup experiment in Section 15.2.3 (page 487), for which the experimenters used the resolution V 2^{5-1} fraction with defining relation $I = ABCDE$. Suppose the experimenters had decided that the experimental conditions could be kept fairly stable over the course of 8 observations but not 16. The treatment combinations would then have been divided into two blocks of size 8. If the fraction is divided into $b = 2$ blocks, then $b - 1 = 1$ contrast *and its alias* must be confounded. If CDE, for example, is selected for confounding, then the aliased pair of contrasts $CDE = AB$ is confounded with blocks, and neither of these contrasts can be measured. Rather than confound a 2-factor interaction, an alternative might be to select the Resolution IV design with defining relation $I = ABDE$ and to confound the aliased pair of contrasts $BCE = ACD$. Then, all two-factor interactions can be estimated, although six of them will be in aliased pairs. The choice between these two designs is the choice of losing information on one 2-factor interaction completely while aliasing the others with high-order interactions, or aliasing three pairs of 2-factor interactions.

Table 15.55, page 542, gives a list of confounding schemes that are useful in conjunction with the fractional factorial experiments also listed in the table.

Example 15.5.1 Flour experiment

M. G. Tuck, S. M. Lewis, and J. I. L. Cottrell describe a series of four experiments in the 1993 issue of *Applied Statistics* that were carried out at Spillers Milling Ltd. in order to reduce the variability of their bread products. Here, we discuss the third experiment in the series. In this experiment, four flour formulations were investigated (four levels of factor A). Four noise factors each at two levels were also investigated. These were amount of yeast (factor N, low or high), proof time (factor S, short or long), degree of mixing and moulding (factor Q, "undermixing, little water, heavy pressure" or "overmixing, much water, little pressure"), and dough time delay (factor T, short or long). Thus, this was a 4×2^4 experiment, with 64 treatment combinations. The time allowed for only 32 treatment combinations to

Table 15.17 A blocked $\frac{1}{2}$-fraction of a 4×2^4 experiment and experiment.

Block (Day)	Treatment Combination	Av. Specific Volume	A_1	A_2	A_1A_2	N	S	Q	T
I	000011	436	−1	−1	1	−1	−1	1	1
	000110	507	−1	−1	1	−1	1	1	−1
	001001	434	−1	−1	1	1	−1	−1	1
	001100	508	−1	−1	1	1	1	−1	−1
	010010	436	−1	1	−1	−1	−1	1	−1
	010111	508	−1	1	−1	−1	1	1	1
	011000	404	−1	1	−1	1	−1	−1	−1
	011101	510	−1	1	−1	1	1	−1	1
	100010	440	1	−1	−1	−1	−1	1	−1
	100111	517	1	−1	−1	−1	1	1	1
	101000	442	1	−1	−1	1	−1	−1	−1
	101101	501	1	−1	−1	1	1	−1	1
	110011	458	1	1	1	−1	−1	1	1
	110110	536	1	1	1	−1	1	1	−1
	111001	464	1	1	1	1	−1	−1	1
	111100	532	1	1	1	1	1	−1	−1
II	000000	567	−1	−1	1	−1	−1	−1	−1
	000101	549	−1	−1	1	−1	1	−1	1
	001010	391	−1	−1	1	1	−1	1	−1
	001111	418	−1	−1	1	1	1	1	1
	010001	458	−1	1	−1	−1	−1	−1	1
	010100	499	−1	1	−1	−1	1	−1	−1
	011011	381	−1	1	−1	1	−1	1	1
	011110	451	−1	1	−1	1	1	1	−1
	100001	499	1	−1	−1	−1	−1	−1	1
	100100	483	1	−1	−1	−1	1	−1	−1
	101011	368	1	−1	−1	1	−1	1	1
	101110	456	1	−1	−1	1	1	1	−1
	110000	475	1	1	1	−1	−1	−1	−1
	110101	597	1	1	1	−1	1	−1	1
	111010	414	1	1	1	1	−1	1	−1
	111111	452	1	1	1	1	1	1	1

Source: Tuck, M. G., Lewis, S. M., and Cottrell, J. I. L. (1993). Copyright © 1993 Blackwell Publishers. Reprinted with permission.

be observed in total, and these were divided into two blocks of size 16, representing the number of observations that could be taken per day.

Since the purpose of the experiment was to find the flour that was least variable under the different levels of the noise variables, the interactions of A with the noise variables were of primary interest. The 4-level factor A was written in terms of two pseudofactors A_1 and A_2, with the level correspondence $0 = 00$, $1 = 01$, $2 = 10$, $3 = 11$. The researchers selected the first block of the $\frac{1}{2}$-fraction with defining relation $I = A_1 A_2 NSQT$. The aliased pair of contrasts selected for confounding were $NQ = A_1 A_2 ST$. The treatment combinations in each block are shown, prior to randomization, in Table 15.17, together with the main-

15.5 Blocked Fractional Factorial Experiments

Table 15.18 Contrast estimates (with divisor 16) for the flour experiment

$\widehat{A_1}$	$\widehat{A_2}$	$\widehat{A_1 A_2}$	$\widehat{N}$	$\widehat{S}$	$\widehat{Q}$	$\widehat{T}$
11.06	3.69	24.06	−52.44	59.81	−47.06	0.56
$\widehat{A_1 N}$	$\widehat{A_2 N}$	$\widehat{A_1 A_2 N}$	$\widehat{A_1 S}$	$\widehat{A_2 S}$	$\widehat{A_1 A_2 S}$	
5.44	7.56	−11.56	4.44	14.56	−2.31	
$\widehat{A_1 Q}$	$\widehat{A_2 Q}$	$\widehat{A_1 A_2 Q}$	$\widehat{A_1 T}$	$\widehat{A_2 T}$	$\widehat{A_1 A_2 T}$	
3.06	9.19	−17.19	9.19	9.56	−15.81	

effect contrast coefficients. All of the contrasts, apart from the confounded contrast NQ and its alias $A_1 A_2 ST$, are orthogonal to the block contrast, and the estimates of their aliased pairs can be calculated without block adjustments. The response variable for each treatment combination was the average specific volume of three loaves, and the observed values are listed in Table 15.17.

Multiplying the responses by the A_1 contrast coefficients and dividing by 16, we obtain the estimate of the difference in the effect of the high and low levels of A_1. Translating back to the original levels of factor A, this contrast compares the average of the third and fourth flours with the average of the first and second flours. The contrast estimate is $(7634 - 7457)/16 = 11.0625$.

The contrast for the interaction of the noise variable N with the pseudofactor A_1 is obtained by multiplying the A_1 and N contrast coefficients in Table 15.17 and dividing by $v/2 = 16$ to obtain the same standard error as the main-effect contrasts. Thus the contrast has coefficients

[1, 1,−1,−1, 1, 1,−1,−1,−1,−1, 1, 1,−1,−1, 1, 1,
  1, 1,−1,−1, 1, 1,−1,−1,−1,−1, 1, 1,−1,−1, 1, 1]

with divisor 16, and the contrast estimate is $(7589 - 7502)/16 = 5.4375$. All of the main-effect and 2-factor interaction contrast estimates are shown in Table 15.18, and the analysis of variance table is shown in Table 15.19.

Selecting individual significance levels of $\alpha^* = 0.01$ for each hypothesis test, (for an overall Type I error probability of at most $\alpha = 0.09$), we compare the F-ratios in Table 15.19 with either $F_{1,11,0.01} = 9.65$ or $F_{3,11,0.01} = 6.22$ as appropriate. The interactions of the flour formulations with the noise variables are not significantly different from zero, but the noise variables N, S, Q themselves do have a large effect on the specific volume. Although the flours are not significantly different, the contrast $A_1 A_2$ appears to be the most important of the three flour contrasts investigated. This contrast compares the average of flours 1 and 4 with the average of flours 2 and 3. The first pair give the higher average specific volume. Before the experiment took place, the experimenters had expected flour 3 (coded 11) to be the best. The difference of averages contrast, which compares flour 3 with the average of

Table 15.19 Analysis of variance for the flour experiment

Source of Variation	Degrees of Freedom	Sum of Squares	Mean Square	Ratio	p-value
Block	1	957.03	–		
A	3	5719.84	1906.62	3.06	0.0737
A_1	1	979.03	979.03	1.57	0.2363
A_2	1	108.78	108.78	0.17	0.6843
$A_1 A_2$	1	4632.03	4632.03	7.42	0.0198
N	1	21997.53	21997.53	35.26	0.0001
S	1	28620.28	28620.28	45.88	0.0001
Q	1	17719.03	17719.03	28.40	0.0002
T	1	2.53	2.53	0.00	0.9504
AN	3	1763.59	587.86	0.94	0.4533
AS	3	1896.84	632.28	1.01	0.4235
AQ	3	3113.59	1037.87	1.66	0.2318
AT	3	3407.09	1135.69	1.82	0.2017
Error	11	6862.34	623.85		
Total	31	92059.72			

the other three flours, has least squares estimate

$$\bar{y}_{11...} - \frac{1}{3}(\bar{y}_{00...} + \bar{y}_{01...} + \bar{y}_{10...})$$
$$= 491.000 - \frac{1}{3}(476.250 + 455.875 + 463.250)$$
$$= 25.875.$$

A preplanned 95% confidence interval for this contrast is given by

$$\bar{y}_{11...} - \tfrac{1}{3}(\bar{y}_{00...} + \bar{y}_{01...} + \bar{y}_{10...}) \pm t_{11,0.025} \sqrt{mse\left(8\left(\tfrac{1}{8}\right)^2 + 24\left(\tfrac{1}{24}\right)^2\right)}$$
$$= 25.875 \pm 2.201\sqrt{(623.849)(0.1667)}$$
$$= 25.875 \pm 22.443$$
$$= (3.432, 48.318).$$

At the 95% confidence level, it does appear that flour 3 (coded 11) has specific volume between 3.4 and 48.3 larger than the average of the other three flours. (We can draw this conclusion only because the contrast was preplanned. Otherwise, we would have needed to use Scheffé's method of multiple comparisons with $t_{11,0.025}$ replaced by $\sqrt{3F_{3,11,0.05}} = 3.24$, and the interval would have included zero). □

15.6 Fractions from Orthogonal Arrays

15.6.1 2^p Orthogonal Arrays

The simplest type of orthogonal array is that shown in Table 15.20, consisting of a set of $2^p - 1$ orthogonal contrasts. The first column has the first half of its 8 entries equal to

15.6 Fractions from Orthogonal Arrays

Table 15.20 Contrasts for a 2^3 experiment

	A	B	C	AB	AC	BC	ABC
000	−1	−1	−1	1	1	1	−1
001	−1	−1	1	1	−1	−1	1
010	−1	1	−1	−1	1	−1	1
011	−1	1	1	−1	−1	1	−1
100	1	−1	−1	−1	−1	1	1
101	1	−1	1	−1	1	−1	−1
110	1	1	−1	1	−1	−1	−1
111	1	1	1	1	1	1	1

−1 and the second half equal to +1. The second column has the first quarter of its entries equal to −1, the second quarter equal to +1, the third quarter equal to −1 again and the fourth quarter equal to +1 again. The third column is divided into eighths with alternating −1's and +1's. If the columns had been longer, the next column would have been divided into sixteenths, and so on. These are the "independent" columns. The fourth, fifth and sixth columns of Table 15.20 are the products of the first three columns in pairs, and the last column is the triple product of the first three columns. The result is a table with $p = 8$ rows and $p − 1 = 7$ columns in which any pair of columns are orthogonal.

The independent columns of an orthogonal array define the treatment combinations for a 2^3 design. As usual, a contrast coefficient of −1 in a column corresponds to level 0 in the corresponding factor, and a contrast coefficient of +1 in a column corresponds to level 1. If all eight treatment combinations of Table 15.20 are used in the experiment, and if there are only three factors of interest, then the orthogonal array defines a full factorial experiment, and no aliasing of contrasts occurs.

Now suppose that only four observations can be taken in a 2^3 experiment. Instead of proceeding as in Section 15.2.1 and choosing a defining relation, we could first construct an orthogonal array with $n = 4$ rows and $n − 1 = 3$ columns. One is shown in Table 15.21, where the first column has the first half of its entries −1, and the second half +1, the second column is divided into quarters, and the third column is the product of the first two. Since we have 3 factors, suppose we label the columns in order as A, B, C. The three columns then show the parts of the A, B, and C contrasts corresponding to a $\frac{1}{2}$-fraction. However, the third column is also the product of the first two columns, so it not only represents the C contrast but also the interaction between A and B. Consequently, C is aliased with AB. Similarly, the first column is the product of the last two columns, so A is aliased with BC. Similarly, again, B is aliased with AC. The defining relation must be $I = ABC$ in order to produce this aliasing scheme.

The coefficients in the contrasts tell us which treatment combinations are represented, and the design is "Design d_1" shown in Table 15.22. Notice that this is the same design that would have been produced from the equation $a_1 + a_2 + a_3 = 1 \pmod{2}$. We could obtain the $\frac{1}{2}$-fraction corresponding to $a_1 + a_2 + a_3 = 0 \pmod 2$, by multiplying any one of the columns by −1 (see Design d_2 in Table 15.22, where the second column has been multiplied by −1).

Table 15.21 An orthogonal array for four observations

−1	−1	1
−1	1	−1
1	−1	−1
1	1	1

Table 15.22 $\frac{1}{2}$-fractions of a 2^3 experiment obtained from orthogonal arrays

	Design d_1				Design d_2		
TC	A	B	C	TC	A	B	C
001	−1	−1	1	011	−1	1	1
010	−1	1	−1	000	−1	−1	−1
100	1	−1	−1	110	1	1	−1
111	1	1	1	101	1	−1	1

Thus, we have arrived back at the same type of design that we studied in Section 15.2, and this will often (but not always) be the case. The main difference in procedure is that when we start with an orthogonal array, we are starting with an unlabeled list of contrasts which can be labeled in any way we please. The labeling then determines the defining relation and the design.

Any columns in an orthogonal array can be multiplied by −1 and we still obtain an orthogonal array, although the treatment combinations may not be identical, or they may be identical but in a different order (try multiplying the B and C columns for the designs in Table 15.22 by −1 and see whether the same design results).

Now we return to the orthogonal array of Table 15.20, which is reproduced in Table 15.23 with column headings indicating which columns are products of which other columns. For example, column 7 is the product of columns 1, 2, and 3. We consider using this array for a 2^5 experiment instead of a 2^3 experiment. Since there are only 8 rows, we will be looking for a $\frac{1}{4}$-replicate (that is, a 2^{5-2} fractional factorial experiment).

Table 15.23 An orthogonal array for 8 observations

			Columns			
1	2	3	12	13	23	123
−1	−1	−1	1	1	1	−1
−1	−1	1	1	−1	−1	1
−1	1	−1	−1	1	−1	1
−1	1	1	−1	−1	1	−1
1	−1	−1	−1	−1	1	1
1	−1	1	−1	1	−1	−1
1	1	−1	1	−1	−1	−1
1	1	1	1	1	1	1

Suppose that we label the first 5 columns as A, B, C, D, E. Since the product of the first two columns gives column 4 and the product of the first and third columns gives column 5, aliasing would occur between D and AB, and between E and AC. Consequently, ABD and ACE must be in the defining relation, together with their product, so we have

$$I = ABD = ACE = BCDE.$$

The rest of the aliasing scheme can be written out also, and we would see that A is aliased with BD and CE, that B is aliased with AD, and that C is aliased with AE. The sixth column, which is the product of columns 2 and 3, and also of columns 4 and 5, can be labeled BC or DE, and these two interactions are aliased. The seventh column is $ABC = CD = BE = ADE$. The eight treatment combinations are deduced from the -1's and $+1$'s in the first five columns; that is,

00011, 00110, 01001, 01100, 10000, 10101, 11010, 11111.

Different sets of treatment combinations corresponding to the same defining relation (but with different signs in the aliasing scheme) can be obtained by multiplying one or more columns of Table 15.23 by -1.

There is nothing special about labeling the first five columns of Table 15.23 as A, B, C, D, E. Any five columns could have been chosen. Different choices may lead to different aliasing schemes, and sometimes these aliasing schemes may not be equally good. Table 15.57 (page 544) lists orthogonal arrays for various-sized experiments. Some useful column labelings for various fractional factorial experiments are suggested in the table.

The standard notation, used by industrial statisticians and engineers, for an orthogonal array is the letter L with subscript equal to the number of runs. Sometimes, the largest Resolution III design that can be used with the array is added in brackets. The orthogonal array in Table 15.21 provides a Resolution III design with 4 observations for 3 or fewer two-level factors and would be written as $L_4(2^3)$. Similarly, the design of Table 15.23 would be written as $L_8(2^7)$. Occasionally, an orthogonal array for 2^p experiments will be written using factor levels rather than contrast coefficients. The orthogonality could then be checked by ensuring that in every pair of columns, all possible pairs of factor levels (00, 01, 10, and 11) appear the same number of times (see, for example, the factor levels shown together with designs (a) and (b) of Table 15.22).

Example 15.6.1 Wafer experiment

R. Kackar and A. Shoemaker (*AT&T Technical Journal*, 1986) describe an experiment they helped to run at AT&T to try to reduce the variability of the thickness of an "epitaxial layer" deposited onto silicon wafers during the manufacture of integrated circuit devices.

The wafers were mounted on a seven-sided "susceptor" with two wafers (one above the other) on each side. The susceptor rotated inside a heated bell jar as chemical vapors were introduced via a nozzle near the top of the jar. The chemicals were deposited on the wafers, and the bell jar was cooled when the thickness of the deposited layer was close to the target of 14.5 micrometers.

Table 15.24 Treatment factors and their levels for the wafer experiment

Factors	Prior Level	Experimental levels	
		Low (0)	High (1)
A (rotation method)	oscillating	continuous	oscillating
B (wafer batch)		668G4	678D4
C (deposition temperature)	1215°C	1210°C	1220°C
D (deposition time)	low	high	low
E (arsenic flow rate)	57%	55%	59%
F (acid etch temp.)	1200°C	1180°C	1215°C
G (acid flow rate)	12%	10%	14%
H (nozzle position)	4	2	6

Table 15.25 An orthogonal array for 16 observations: An $L_{16}(2^{15})$

\	\	\	\	\	\	\	Columns	\	\	\	\	\	\	\
1	2	12	3	13	23	123	4	14	24	124	34	134	234	1234
−1	−1	1	−1	1	1	−1	−1	1	1	−1	1	−1	−1	1
−1	−1	1	−1	1	1	−1	1	−1	−1	1	−1	1	1	−1
−1	−1	1	1	−1	−1	1	−1	1	1	−1	−1	1	1	−1
−1	−1	1	1	−1	−1	1	1	−1	−1	1	1	−1	−1	1
−1	1	−1	−1	1	−1	1	−1	1	−1	1	1	−1	1	−1
−1	1	−1	−1	1	−1	1	1	−1	1	−1	−1	1	−1	1
−1	1	−1	1	−1	1	−1	−1	1	−1	1	−1	1	−1	1
−1	1	−1	1	−1	1	−1	1	−1	1	−1	1	−1	1	−1
1	−1	−1	−1	−1	1	1	−1	−1	1	1	1	1	−1	−1
1	−1	−1	−1	−1	1	1	1	1	−1	−1	−1	−1	1	1
1	−1	−1	1	1	−1	−1	−1	1	1	−1	−1	1	1
1	−1	−1	1	1	−1	−1	1	1	−1	−1	1	1	−1	−1
1	1	1	−1	−1	−1	−1	−1	−1	−1	−1	1	1	1	1
1	1	1	−1	−1	−1	−1	1	1	1	1	−1	−1	−1	−1
1	1	1	1	1	1	1	−1	−1	−1	−1	−1	−1	−1	−1
1	1	1	1	1	1	1	1	1	1	1	1	1	1	1
A	B		C				D	E			F		G	H

The engineers identified eight factors that might affect the variability of the thickness of the epitaxial layer. These are shown in Table 15.24 together with the operating factor levels prior to the experiment and the levels selected for the experiment.

The experimenters decided to take 16 observations. The 16 treatment combinations were selected via the orthogonal array $L_{16}(2^{15})$ shown in Table 15.25. The orthogonal array is constructed as described earlier in this section. The labels in the row headings of the table identify which columns are products of which other columns. The assignment of factors to columns chosen by the experimenters is indicated in the foot of the table. The experiment is a 2^{8-4} experiment, and the defining relation is generated by 4 confounded interactions. Notice, from the heading and the foot of Table 15.25, that D must be aliased with ABC, F must be aliased with ABE, G with ACE, and H with BCE. Thus, the defining relation includes

$ABCD, ABEF, ACEG, BCEH$, and all their possible products (a total of $2^4 = 16$ terms in the defining relation):

$$\begin{aligned}
I &= ABCD = ABEF = CDEF \\
&= ACEG = BDEG = BCFG = ADFG \\
&= BCEH = ADEH = ACFH = BDFH \\
&= ABGH = CDGH = EFGH = ABCDEFGH \ .
\end{aligned}$$

This is a Resolution IV design, and there is considerable aliasing between 2-factor interactions. For example, the contrast listed in column 12 of Table 15.25 not only measures the 2-factor interaction AB, but also measures its aliased 2-factor interactions CD, EF, GH (and some higher-order interactions).

There were 70 measurements taken for each treatment combination (5 measurements on each of the 2 wafers on the 7 sides of the receptor). From these, two different response variables were calculated—the average of the 70 measurements (which we denote by x) and the log sample variance of the 70 measurements (which we denote by v). The treatment combinations, corresponding to the orthogonal array in Table 15.25, together with the two response variables, are shown in Table 15.26.

The experimenters first analyzed the log variance response. The contrast estimates (high level minus low level) for this response variable are shown in Table 15.27. The contrast estimates for factors A and H are considerably larger than those for the other factors. Consequently, A and H should be investigated for reducing variability in the response. Since the log variance response is to be reduced, and the contrast estimate for A is positive

Table 15.26 Treatment combinations and response variables for the wafer experiment

Treatment Combination $ABCDEFGH$	Average Response $x_{ijklmnpq}$	Log Variance Response $v_{ijklmnpq}$
0 0 0 0 0 0 0 0	14.821	−0.4425
0 0 0 0 1 1 1 1	14.888	−1.1989
0 0 1 1 0 0 1 1	14.037	−1.4307
0 0 1 1 1 1 0 0	13.880	−0.6505
0 1 0 1 0 1 0 1	14.165	−1.4230
0 1 0 1 1 0 1 0	13.860	−0.4969
0 1 1 0 0 1 1 0	14.757	−0.3267
0 1 1 0 1 0 0 1	14.921	−0.6270
0 0 0 1 0 1 1 0	13.972	−0.3467
0 0 0 1 1 0 0 1	14.032	−0.8563
0 0 1 0 0 1 0 1	14.843	−0.4369
0 0 1 0 1 0 1 0	14.415	−0.3131
0 1 0 0 0 0 1 1	14.878	−0.6154
0 1 0 0 1 1 0 0	14.932	−0.2292
0 1 1 1 0 0 0 0	13.907	−0.1190
0 1 1 1 1 1 1 1	13.914	−0.8625

Source: Kackar, R. N. and Shoemaker, A. C. (1986). Copyright © 1986 AT&T. All rights reserved. Reprinted from the AT&T Technical Journal with permission.

Table 15.27 Contrast estimates for log sample variance response variable

Contrast	A	B	C	D	E	F	G	H
Estimate	0.352	0.122	0.105	−0.249	−0.012	−0.072	−0.101	−0.566

Table 15.28 Contrast estimates for the mean response variable

Contrast	A	B	C	D	E	F	G	H
Estimate	−0.055	0.056	−0.109	−0.836	−0.067	0.060	−0.098	0.142

while that for H is negative, we would want to set A at its low level (continuous rotation) and H at its high level (position 6). All other factors can be set at their current operating conditions.

The second requirement of the experimenters was to achieve an average thickness of 14.5 micrometers. Contrast estimates for the average response are shown in Table 15.28. Not surprisingly, factor D, deposition time, has by far the largest effect on the mean response, and the experimenters were able to adjust this factor in order to meet the target.

As with any good experiment, the experimenters wished to confirm their results. Their confirmation experiment investigated two treatment combinations. The first treatment combination consisted of the prior operating levels of factors A and C–H, with factor B at level 1 as shown in Table 15.24, and the second treatment combination was the same except that the levels of A and H were changed as discussed above. The confirmation experiment showed that the variance of the thickness had been reduced by a factor 2.5—quite a remarkable result! □

15.6.2 Saturated Designs

In the context of fractional factorial experiments, a saturated design is one that uses only $n = p + 1$ treatment combinations to estimate the main effects of p factors independently (assuming that all interactions are negligible). Saturated designs are used in the early stages of experimentation to try to screen out unimportant factors from among a large number of possible factors.

The two designs in Table 15.22 (page 508) are saturated designs for 3 factors, and the design of Table 15.23 is a saturated design for 7 factors. In each case, if the interactions are negligible, the main effects can be independently estimated because their contrasts are orthogonal. The designs are called *saturated* because the main-effect contrasts take up all the available degrees of freedom, leaving none to estimate σ^2. They are also known as *orthogonal main-effect plans*, since the main-effect contrasts are orthogonal but there are no degrees of freedom available for estimating interaction contrasts.

We obtained the orthogonal arrays of Tables 15.22 and 15.23 by selecting k independent columns, for $n = 2^k$, and then multiplying these columns together. The designs obtained via this method are all equivalent to designs that would be obtained by randomly selecting a block from a single-replicate confounded design. R. L. Plackett and J. P. Burman provided an alternative set of saturated designs for values of n that are divisible by 4 in a paper in *Biometrika* in 1946. Except for $n = 8$, these designs are quite different from the designs considered so far, and we cannot easily write down an aliasing scheme for them. Nevertheless, the main-effect contrasts are all orthogonal.

Plackett and Burman's method of construction of the designs is quick and easy. They listed the first row of each orthogonal array, called the *generator*. The entire orthogonal array is obtained from the generator by cycling it to the right, and then adding a row of -1's. For example, for $n = 12$, if we take the generator

$$1\ \ 1\ \ 1\ -1\ \ 1\ \ 1\ -1\ \ 1\ -1\ -1\ -1,$$

this gives us the first row of the orthogonal array. Cycling this to the right, and wrapping the end round to the beginning, we get the second row of the array as

$$-1\ \ 1\ \ 1\ \ 1\ -1\ \ 1\ \ 1\ -1\ \ 1\ -1\ -1\ .$$

The cycling procedure gives 11 rows. The 12th row of -1's is then added, giving the entire array as shown in Table 15.29. Interestingly, no column is the product of any two or more other columns, and this is the reason that the aliasing scheme is not straightforward. The designs should be used only when all interactions are expected to be negligible. Generators for other cyclically generated orthogonal main-effect plans are listed in Table 15.58 (page 544). These were obtained by a computer program described by Dean and Draper (1998).

15.6.3 $2^p \times 4^q$ Orthogonal Arrays

The orthogonal arrays of Section 15.6.1 can be used when one or more factors have 4 levels. Each 4-level factor requires 3 independent columns to represent 3 orthogonal contrasts. For example, the orthogonal array in Table 15.23 could be used for a $2^3 \times 4$ experiment as follows. The first three columns (which are independent) could be labeled A, B, and C. If the 4th column is labeled D_1, then D_1 is aliased with AB. If the 7th column is labeled D_2, then D_2 is aliased with ABC. The product of the coefficients in the 4th and 7th columns gives the coefficients in the 3rd column, so $D_1 D_2$, the remaining contrast for the 4-level

Table 15.29 A saturated orthogonal main-effect plan for 11 factors and 12 observations

1	1	1	−1	1	1	−1	1	−1	−1	−1
−1	1	1	1	−1	1	1	−1	1	−1	−1
−1	−1	1	1	1	−1	1	1	−1	1	−1
−1	−1	−1	1	1	1	−1	1	1	−1	1
1	−1	−1	−1	1	1	1	−1	1	1	−1
−1	1	−1	−1	−1	1	1	1	−1	1	1
1	−1	1	−1	−1	−1	1	1	1	−1	1
1	1	−1	1	−1	−1	−1	1	1	1	−1
−1	1	1	−1	1	−1	−1	−1	1	1	1
1	−1	1	1	−1	1	−1	−1	−1	1	1
1	1	−1	1	1	−1	1	−1	−1	−1	1
−1	−1	−1	−1	−1	−1	−1	−1	−1	−1	−1

factor, is aliased with C, and the defining relation is

$$I = ABD_1 = ABCD_2 = CD_1D_2.$$

This is a Resolution II design, which should be avoided if possible, since it confounds two main effects (C and D). A better design is to assign D_2 to the 5th column, where it is aliased with AC. The product of the 4th and 5th columns gives the 6th column, so that D_1D_2 is aliased with BC. The defining relation is

$$I = ABD_1 = ACD_2 = BCD_1D_2,$$

which is Resolution III. The 7th column of Table 15.23 corresponds to the ABC contrast, which is aliased with AD_1D_2, CD_1, and BD_2 contrasts. The complete aliasing scheme is

$$\begin{aligned}
I &= ABD_1 &&= ACD_2 &&= BCD_1D_2 \\
A &= BD_1 &&= CD_2 &&= ABCD_1D_2 \\
B &= AD_1 &&= ABCD_2 &&= CD_1D_2 \\
C &= ABCD_1 &&= AD_2 &&= BD_1D_2 \\
D_1 &= AB &&= ACD_1D_2 &&= BCD_2 \\
D_2 &= ABD_1D_2 &&= AC &&= BCD_1 \\
D_1D_2 &= ABD_2 &&= ACD_1 &&= BC \\
ABC &= CD_1 &&= BD_2 &&= AD_1D_2
\end{aligned}$$

The design would be useful for an experiment where all interactions were expected to be negligible, in which case one degree of freedom would be available to estimate σ^2.

15.6.4 3^p Orthogonal Arrays

The orthogonal arrays for 2^p experiments introduced in Section 15.6.1 have the property that any pair of columns in the array are orthogonal (that is, the sum of the products of corresponding coefficients is zero). An examination of the arrays in Tables 15.21–15.29 reveals that this orthogonality arises because every pair of coefficients $(-1, -1)$, $(-1, 1)$, $(1, -1)$ and $(1, 1)$ occurs equally often in every pair of columns. We could rewrite the array to contain the factor labels 0, 1 instead of the contrast coefficients $-1, 1$, and every pair of labels would occur the same number of times in every pair of columns. This is the way that orthogonal arrays are defined for 3^p experiments.

An orthogonal array with 9 treatment combinations is shown as columns 1–4 in Table 15.30 for four factors, each having 3 levels. If any pair of columns is selected, it can be verified that each of the nine pairs of levels (0, 0), (0, 1), (0, 2), (1, 0), (1, 1), (1, 2), (2, 0), (2, 1), (2, 2) occurs once. The first column consists of three copies of each of 0, 1, and 2. The second column consists of 0, 1, and 2, in order, repeated three times. The third column is obtained from the sum of the coefficients in the first two columns reduced modulo 3 (thereby ensuring that any factor assigned to the 3rd column will be aliased with the interaction between the first two factors). The fourth column is obtained from twice the sum of columns 2 and 3 (ensuring that any factor assigned to the fourth column will be aliased

Table 15.30 A 3^p orthogonal array for 9 observations

Columns				Contrasts							
1	2	3	4								
0	0	0	0	−1	1	−1	1	−1	1	−1	1
0	1	1	1	−1	1	0	−2	0	−2	0	−2
0	2	2	2	−1	1	1	1	1	1	1	1
1	0	1	2	0	−2	−1	1	0	−2	1	1
1	1	2	0	0	−2	0	−2	1	1	−1	1
1	2	0	1	0	−2	1	1	−1	1	0	−2
2	0	2	1	1	1	−1	1	1	1	0	−2
2	1	0	2	1	1	0	−2	−1	1	1	1
2	2	1	0	1	1	1	1	0	−2	−1	1
A	B	C	D	A_L	A_Q	B_L	B_Q	C_L	C_Q	D_L	D_Q

with the interaction of factors assigned to columns 2 and 3 and with the interaction of the first two factors). It is not possible to find more than four orthogonal columns with only 9 observations.

In Table 15.30, a pair of orthogonal contrasts is given corresponding to each of the four columns in the orthogonal array. It can be verified that this set of 8 contrasts is orthogonal. As for 2^p experiments, a 3^p orthogonal array with n rows can have at most $n − 1$ orthogonal columns of contrast coefficients, and therefore can accommodate at most $(n − 1)/2$ 3-level factors. An experiment is discussed in Section 15.7.1 that uses part of the orthogonal array for 3-level factors and 27 observations listed in Table 15.59.

15.7 Design for the Control of Noise Variability

Experiments that involve both design and noise factors are often known colloquially as *Taguchi experiments*. There are two different types of designs for such an experiment— "product arrays" and "mixed arrays." The *product arrays* are composed of two fractional factorial experiments, one for the design factors and one for the noise factors, and every combination of design factors is observed in conjunction with every combination of noise factors. (Product arrays were introduced in Section 7.6.) *Mixed arrays*, on the other hand, are ordinary fractional factorial designs in which the difference between the design and noise factors is ignored at the design stage except to ensure that the design-by-noise interactions are estimable.

Product arrays are usually observed in the following way. The order of the design combinations is randomized. For each design combination in turn, observations are taken across all of the noise combinations in a random order. This produces what is called a "split-plot design, with noise combinations on the split plots and design combinations on the whole plots." (Split-plot designs will be discussed in Chapter 19—familiarity with them is not needed here.) Such designs are usually analyzed by calculating, for each design combination, the average and log sample variance of the responses obtained under the different noise combinations. The average response and the log variance response are then taken as two separate sets of data values for the design factor combinations. The objective of the experiment is to find out which factors most affect the log variance response, and which factors most affect the average response. Design combinations are then sought that give a low sample variance across the noise combinations and that give a response close to the target value. Finally, confirmatory observations are taken.

Product arrays can also be randomized in the same way as mixed arrays, as follows. For a mixed array, all of the combinations of both design and noise factors are observed in a random order. This allows a different analysis where the experiment is analyzed without averaging the data across the noise combinations. Instead, the design-by-noise interactions are studied with the view of identifying those levels of the design factors that are least affected by changing the levels of the noise factors, and identifying those design factors that most affect the mean. The use of a (blocked) mixed array was illustrated in the flour experiment of Example 15.5. The use of a product array was illustrated in the wafer experiment of Example 15.6.1 for 2-level factors and will be illustrated for the inclinometer experiment with 3-level factors in Section 15.7.1. The computer analysis of the latter experiment is discussed in Section 15.8.2.

15.7.1 A Real Experiment—Inclinometer Experiment

A collaborative study involving statisticians and mechanical engineers was described by S. Lewis, B. Hodgson, R. New, and C. Sexton in the 1989 *Proceedings of the Institute of Mechanical Engineers International Conference on Engineering Design*. The experiment sought to improve the performance of an inclinometer, which is an instrument that records the angle of tilt of an object such as a crane jib. The design of the inclinometer is described in the article as follows.

> The basic design of the product is composed in four parts: a bob-weight and flexure, a flanged flywheel and a copper-plated disc (PCB). All are attached to a shaft supported in low-friction bearings. When the object to which the flywheel is attached is tilted, the bob-weight assembly moves to stay perpendicular to the earth, causing the PCB to rotate relative to the casing.
>
> The main performance difficulty of the inclinometer is that it does not immediately register the true angle of tilt. Spurious swing of the disc is produced by movement of the object.

The purpose of the experiment was to vary the relative sizes of the parts of the inclinometer to find a combination of factors that would reduce the swing. The engineers identified 7 factors that could be altered and that might affect the swing. Three levels were selected for each factor so that linear and quadratic trends could be investigated. The levels of the first six factors were selected to be equally spaced. The factors were:

A: Flexure length (30.00, 31.25, 32.5)

B: Flexure thickness (0.05, 0.275, 0.5)

C: Flexure width (4.0, 5.0, 6.0)

15.7 Design for the Control of Noise Variability

D: Flange thickness (1.0, 3.5, 6.0)

E: Flange width (6.0, 10.5, 15.0)

F: Bob-weight length (12.0, 20.0, 28.0)

G: Copper plating thickness (0.0175, 0.035, 0.07)

All measurements are in millimeters, and the levels of all factors are coded 0, 1, and 2. A $\frac{1}{9}$ fraction would have been possible except for the fact that there were other considerations that needed to be taken into account. Under designled experimental conditions, it was possible to produce the factor levels exactly as specified, but in mass production variability naturally creeps in. The experimenters decided to build the production variability into the experiment as noise factors as follows (measured in mm, except where stated),

H: Flexure length (-0.25, +0.25)

P: Flexure thickness (-0.005, +0.005)

J: Flange thickness (-0.025, +0.025)

K: Flange width (-0.025, +0.025)

L: Copper plating thickness (-0.005, +0.005)

M: Tolerance on bob weight mass (-9.0, +9.0 $\times (1/100)g$)

N: Maximum horizontal amplitude of vibration (5, 25)

The two levels of each noise factor were coded as 0 and 1. Thus, the entire experiment was a $3^7 \times 2^7$ factorial experiment, where the 3-level factors were the design factors and the 2-level factors were the noise factors (see Section 7.6). The treatment combination 0000000 of the design factors in conjunction with the combination 0000000 of the noise factors would have flexure length (A) of $(30.00 - 0.25)$ mm $= 29.75$ mm, flexure thickness of $(0.050 - 0.005)$ mm $= 0.045$ mm, and so on.

The objective of the experiment was to select the combinations of the design factors that gave the least amount of swing. In terms of producing a product of consistently high quality, it was also important that the variability of the amount of swing also remain low across the different noise combinations.

The experimenters selected a product array formed from a $(\frac{1}{3})^4$ fraction of the 3^7 design-treatment combinations and a $(\frac{1}{2})^4$ fraction of the 2^7 noise combinations. This gave a total of $27 \times 8 = 216$ observations. For the $(\frac{1}{3})^4$ fraction of the 3^7 factorial experiment, seven columns of the orthogonal array $L_{27}(3^{13})$ were selected. These are indicated in Table 15.59 (page 545). For the $(\frac{1}{2})^4$ fraction of the 2^7 factorial experiment, the orthogonal array $L_8(2^7)$ is shown in Table 15.23 (page 508), with the noise factors assigned to the columns in the order $H, P, K, -J, -L, -M, N$, where the minus signs indicate that the column was multiplied by -1 (thus reversing the high and low levels).

The maximum absolute angle of swing was ascertained for each of the selected combinations of design- and noise-factor levels. The data are reproduced in Table 15.31. The

Table 15.31 Maximum angle of swing for the inclinometer experiment. Combinations of design factors A–G are in rows, and combinations of noise factors H–N are in columns.

	H	0	0	0	0	1	1	1	1		
	P	0	0	1	1	0	0	1	1		
Noise	J	0	0	1	1	1	1	0	0		
Factors:	K	0	1	0	1	0	1	0	1		
	L	0	1	0	1	1	0	1	0		
	M	0	1	1	0	0	1	1	0		
	N	0	1	1	0	1	0	0	1		

Design Factors ABCDEFG									Mean	$\ln(s^2)$
0 0 0 0 0 0 0	0.62	3.54	3.56	0.62	3.09	0.71	0.73	3.20	2.01	0.73
0 0 1 1 1 1 1	0.59	3.11	3.11	0.59	2.98	0.63	0.64	3.02	1.83	0.53
0 0 2 2 2 2 2	0.59	3.01	3.02	0.59	2.97	0.61	0.62	3.00	1.80	0.50
0 1 0 1 1 2 2	0.51	2.65	2.65	0.50	2.53	0.53	0.54	2.56	1.56	0.21
0 1 1 2 2 0 0	0.18	0.96	0.96	0.18	0.89	0.19	0.20	0.90	0.56	−1.85
0 1 2 0 0 1 1	1.88	9.58	9.55	1.85	9.30	1.92	1.94	9.48	5.69	2.80
0 2 0 2 2 1 1	0.19	1.03	1.03	0.19	0.97	0.21	0.21	0.93	0.60	−1.72
0 2 1 0 0 2 2	1.85	9.46	9.42	1.82	9.19	1.90	1.92	9.35	5.61	2.77
0 2 2 1 1 0 0	0.52	2.73	2.72	0.51	2.61	0.55	0.56	2.64	1.61	0.27
1 0 0 1 2 1 2	0.29	1.56	1.56	0.29	1.45	0.31	0.32	1.47	0.91	−0.87
1 0 1 2 0 2 0	0.95	4.98	4.93	0.94	4.79	0.99	1.00	4.82	2.93	1.48
1 0 2 0 1 0 1	1.16	6.09	6.09	1.13	5.70	1.21	1.26	5.93	3.57	1.87
1 1 0 2 0 0 1	0.26	1.45	1.45	0.25	1.30	0.29	0.30	1.30	0.83	−1.05
1 1 1 0 1 1 2	1.15	5.99	5.92	1.13	5.69	1.19	1.22	5.84	3.51	1.84
1 1 2 1 2 2 0	0.85	4.31	4.30	0.84	4.23	0.86	0.88	4.28	2.57	1.21
1 2 0 0 1 2 0	1.10	5.74	5.67	1.07	5.43	1.14	1.17	5.57	3.36	1.75
1 2 1 1 2 0 1	0.29	1.55	1.55	0.28	1.45	0.31	0.32	1.47	0.90	−0.88
1 2 2 2 0 1 2	0.91	4.64	4.66	0.90	4.56	0.94	0.95	4.57	2.77	1.35
2 0 0 2 1 2 1	0.39	2.05	2.06	0.39	1.96	0.41	0.42	1.97	1.21	−0.30
2 0 1 0 2 0 2	0.67	3.61	3.57	0.65	3.27	0.72	0.74	3.41	2.08	0.79
2 0 2 1 0 1 0	1.42	7.31	7.38	1.41	7.14	1.48	1.51	7.24	4.36	2.27
2 1 0 0 2 1 0	0.69	3.66	3.60	0.67	3.37	0.73	0.74	3.47	2.12	0.82
2 1 1 1 0 2 1	1.18	6.04	6.06	1.17	5.90	1.21	1.23	5.95	3.59	1.88
2 1 2 2 1 0 2	0.37	1.95	1.95	0.37	1.87	0.39	0.40	1.88	1.15	−0.40
2 2 0 1 0 0 2	0.39	2.15	2.16	0.38	1.94	0.44	0.44	1.96	1.23	−0.25
2 2 1 2 1 1 0	0.44	2.29	2.29	0.43	2.21	0.46	0.47	2.22	1.35	−0.07
2 2 2 0 2 2 1	1.84	9.35	9.19	1.79	9.06	1.85	1.89	9.28	5.53	2.74

Source: Lewis, S. M., Hodgson, B. A., New, R. E., and Sexton, C. J. (1989). Copyright © 1989 Mechanical Engineering Publications. Reprinted with permission.

noise-factor combinations label the columns, and the design-factor combinations label the rows. The last two columns of the table show the average and log sample variance of the observations for the design combinations calculated across the noise combinations.

Consider first using the log sample variance $\ln(s^2)$ of the observations as the response variable. The analysis of variance table is shown in Table 15.32. We have included the

15.7 Design for the Control of Noise Variability

Table 15.32 Analysis of variance for $\ln(s^2)$ response for the inclinometer experiment

Source of Variation	Degrees of Freedom	Sum of Squares	Mean Square	Ratio	p-value
A	2	0.6316	0.3158		
Linear A	1	0.5798	0.5798	22.73	0.0005
Quadratic A	1	0.0519	0.0519	2.03	0.1794
B	2	0.1358	0.0679		
Linear B	1	0.0581	0.0581	2.28	0.1571
Quadratic B	1	0.0777	0.0777	3.05	0.1064
C	2	9.8448	4.9224		
Linear C	1	9.8241	9.8241	385.18	0.0001
Quadratic C	1	0.0207	0.0207	0.81	0.3852
D	2	18.8987	9.4493		
Linear D	1	18.3769	18.3769	720.53	0.0001
Quadratic D	1	0.5217	0.5217	20.46	0.0007
E	2	7.0366	3.5183		
Linear E	1	7.0044	7.0044	274.63	0.0001
Quadratic E	1	0.0322	0.0322	1.26	0.2829
F	2	9.5150	4.7575		
Linear F	1	9.4043	9.4043	368.73	0.0001
Quadratic F	1	0.1106	0.1106	4.34	0.0593
G	2	0.0354	0.0177	0.69	0.5184
Error	12	0.3061	0.0255		
Total	26	46.4039			

Table 15.33 Contrast estimates (log var response)

Lin A	Lin C	Lin D	Quad D	Lin E	Lin F
0.359	1.478	−2.021	0.590	−1.248	1.446

information needed for testing the hypotheses of negligible linear and quadratic trends in each of the factors except for G. The levels of G are not equally spaced, and therefore the correct trend contrast coefficients are not those shown in Table A.2.

If we test the hypotheses of negligible contrasts for each trend contrast shown in Table 15.32 at individual significance levels $\alpha^* = 0.01$ and test the hypothesis of no effect of factor G at level $\alpha^* = 0.01$ (for an overall level of at most $\alpha = 0.13$), we reject the hypotheses of negligible linear trends in factors A, C, D, E, F and of a negligible quadratic trend in factor D. Factors B and G show very little effect on log variance response, so these factors (flexure thickness and copper plating thickness) cannot be employed to achieve less variability in the swing in the inclinometer. The contrast estimates for the nonnegligible contrasts are shown in Table 15.33. From the signs on the contrast estimates we see that in order to reduce the variability of the swing, factors A, C, and F should be set at their low levels, while factors D and E should be set at their high levels.

In order to reduce the size of the swing, we need to use as response variable the average swing for each design combination (averaged over the noise combinations). These are listed

Table 15.34 Analysis of variance for average response for the inclinometer experiment

Source of Variation	Degrees of Freedom	Sum of Squares	Mean Square	Ratio	p-value
A	2	0.1288	0.0644		
Linear A	1	0.1023	0.1023	0.53	0.4813
Quadratic A	1	0.0264	0.0264	0.14	0.7183
B	2	0.2899	0.1449		
Linear B	1	0.2850	0.2850	1.47	0.2486
Quadratic B	1	0.0049	0.0049	0.03	0.8768
C	2	12.9528	6.4764		
Linear C	1	12.8863	12.8863	66.48	0.0001
Quadratic C	1	0.0665	0.0665	0.34	0.5689
D	2	24.6042	12.3021		
Linear D	1	22.9193	22.9193	118.23	0.0001
Quadratic D	1	1.6850	1.6850	8.69	0.0122
E	2	9.0561	4.5280		
Linear E	1	7.9385	7.9385	40.95	0.0001
Quadratic E	1	1.1177	1.1177	5.77	0.0334
F	2	11.5710	5.7855		
Linear F	1	11.2476	11.2476	58.02	0.0001
Quadratic F	1	0.3234	0.3234	1.67	0.2208
G	2	0.6725	0.3362	1.73	0.2179
Error	12	2.3262	0.1938		
Total	26	61.6014			

Table 15.35 Contrast estimates (average response)

Lin C	Lin D	Quad D	Lin E	Quad E	Lin F
1.692	−2.257	1.060	−1.328	0.863	1.581

in Table 15.31. The analysis of variance (shown in Table 15.34) identifies the linear trends of factors C, D, E, and F as having large effects on the swing. The contrast estimates are shown in Table 15.35. The signs of the estimates suggest that factors D and E should be set at their high levels and factors C and F at their low levels. Since this agrees with the conclusions of the analysis of variability, it is possibile to reduce the size and the variability of the swing simultaneously.

Plots of the least squares estimates of the effect of the levels of factor D for both log sample variance and average response are shown in Figure 15.7. The conclusions of the experiment are that the dimensions of the flexure and bob-weight (A, C, F) should be decreased, while the dimensions of the flange (D, E) should be increased. The experimenters comment in the article that the results match what would be expected by engineering principles.

The SAS commands for analyzing this experiment are discussed in Example 15.8.2, Section 15.8. The analysis of the same experiment using a mixed array approach, rather than a product array approach, are discussed in Example 15.8.2.

15.8 Using SAS Software

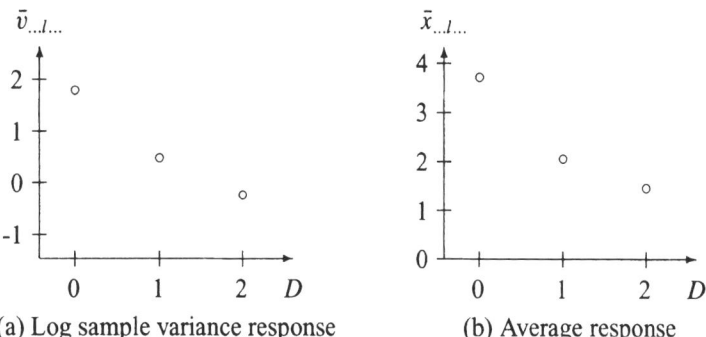

Figure 15.7
Plots of the effect of the levels of factor D for the inclinometer experiment, where $x_{ijklmnp}$ denotes average response and $v_{ijklmnp}$ denotes the log sample variance $\ln(s^2)$ for corresponding design factor combinations

(a) Log sample variance response   (b) Average response

Table 15.36 SAS program for the sludge experiment—cell-means model

```
DATA SLG;
 INPUT  A  B  C  D  E  Y;
 TC=((((((A*10)+B)*10)+C)*10)+D)*10)+E;
 LINES;
 0 0 0 1 0   195
 0 0 1 1 1   496
 0 1 0 0 1    87
 0 1 1 0 0  1371
 1 0 0 0 1   102
 1 0 1 0 0  1001
 1 1 0 1 0   354
 1 1 1 1 1   775
 ;
PROC PRINT;
 ;
PROC GLM;
 CLASSES TC;
 MODEL Y= TC;
 ESTIMATE 'A'  TC -1 -1 -1 -1  1  1  1  1 / DIVISOR=4;
 ESTIMATE 'B'  TC -1 -1  1  1 -1 -1  1  1 / DIVISOR=4;
 ESTIMATE 'C'  TC -1  1 -1  1 -1  1 -1  1 / DIVISOR=4;
 ESTIMATE 'D'  TC  1  1 -1 -1 -1 -1  1  1 / DIVISOR=4;
 ESTIMATE 'E'  TC -1  1  1 -1  1 -1 -1  1 / DIVISOR=4;
 ESTIMATE 'AC' TC  1 -1  1 -1 -1  1 -1  1 / DIVISOR=4;
 ESTIMATE 'BC' TC  1 -1 -1  1  1 -1 -1  1 / DIVISOR=4;
```

15.8 Using SAS Software

15.8.1 Fractional Factorials

The analysis of a fractional factorial experiment by computer is identical to that of a single-replicate factorial experiment (Section 13.11) except that only one main effect or interaction

Table 15.37 Output from SAS program for the sludge experiment—cell-means model

```
                          The SAS System
                  General Linear Models Procedure
Dependent Variable: Y
                            Sum of        Mean
Source              DF     Squares       Square    F Value    Pr > F
Model                7    1510451.9     215778.8       .          .
Error                0        .             .
Corrected Total      7    1510451.9

Source              DF   Type III SS  Mean Square  F Value    Pr > F
TC                   7    1510451.9     215778.8       .          .

                              T for H0:    Pr > |T|    Std Error of
Parameter     Estimate       Parameter=0                 Estimate
A             20.750000       99999.99       0.0            0
B            198.250000       99999.99       0.0            0
C            726.250000       99999.99       0.0            0
D           -185.250000       99999.99       0.0            0
E           -365.250000       99999.99       0.0            0
AC           -66.250000       99999.99       0.0            0
BC           126.250000       99999.99       0.0            0
```

Table 15.38 SAS program for the sludge experiment—five-way model

```
PROC GLM;
 CLASSES A B C D E;
 MODEL Y= A B C D E A*C B*C;
 ESTIMATE 'A' A -1 1;
 ESTIMATE 'B' B -1 1;
 ESTIMATE 'C' C -1 1;
 ESTIMATE 'D' D -1 1;
 ESTIMATE 'E' E -1 1;
 ESTIMATE 'AC' A*C 1 -1 -1 1 / DIVISOR=2;
 ESTIMATE 'BC' B*C 1 -1 -1 1 / DIVISOR=2;
```

can be entered into the model from each line of the aliasing scheme (and none from the defining relation). If two aliased effects are entered into the model, the Type I sum of squares and degrees of freedom will be recorded as zero for the second effect entered. The Type III sum of squares and degrees of freedom will be recorded as zero for both.

In Table 15.36 we show a straightforward program for analyzing the sludge experiment of Example 15.2.4. The cell-means model in terms of the treatment combinations TC is used. (Variables A–E are created for later use.) The main effect contrasts are obtained using the contrast coefficients listed in Table 15.8 (page 493). Using PROC GLM, the analysis of

15.8 Using SAS Software

Table 15.39 Output from the SAS program for the sludge experiment—five-way model

```
                          The SAS System
                   General Linear Models Procedure
Dependent Variable: Y
Source              DF    Type III SS    Mean Square    F Value    Pr > F
A                    1          861.1          861.1        .         .
B                    1        78606.1        78606.1        .         .
C                    1      1054878.1      1054878.1        .         .
D                    1        68635.1        68635.1        .         .
E                    1       266815.1       266815.1        .         .
A*C                  1         8778.1         8778.1        .         .
B*C                  1        31878.1        31878.1        .         .
```

variance is generated in the usual way by the MODEL statement, while the contrast estimates are obtained by the ESTIMATE statements. The output is shown in Table 15.37. The main effects are each aliased with 2-factor (and 3-factor) interactions (see page 493). The 2-factor interactions AC and BC are aliased with BE and AE, respectively. Since there is only one observation on each of the observed treatment combinations, the cell-means model leaves no degrees of freedom for error—this is why the p-values and values of test statistics and standard errors are either missing or meaningless. The inclusion of the DIVISOR=4 options in the ESTIMATE statements ensures that all the contrasts listed in Table 15.8 will be divided by $v/2 = 4$ and give the same estimates as those in Table 15.9 (page 493).

In Table 15.38, we show the SAS program for the equivalent model written in terms of main-effect and interaction parameters. The ESTIMATE statements for the main effects need no divisors, as they are automatically divided by 4 (the number of observations on the high and low levels). However, the ESTIMATE statements for the interaction contrasts include the option DIVISOR=2, to increase the actual divisor by a factor of 2. Without this option, the interaction estimates would be calculated with divisor 2 (the number of observations on each combination of levels of the two factors). The main-effect and interaction sums of squares are shown in Table 15.39. The output from the ESTIMATE statements is identical to that obtained from the cell-means model.

Again, there are no degrees of freedom for error, since a term has been included in the model from every row of the aliasing scheme. If all 2-factor interactions can be assumed to be negligible, then AC and BC would be omitted from the model, leaving 2 degrees of freedom for error.

Table 15.40 shows what happens when two aliased terms are entered into the model. The defining relation for the $\frac{1}{4}$–fraction was stated in Example 15.2.4 to be $I = ABD = CDE = ABCE$. Consequently, A is aliased with BD. Adding BD into the model subsequent to A gives Type I sum of squares and degrees of freedom for BD equal to zero. This is because BD adds no more information if A is already in the model. The Type III sums of squares are zero for both A and BD, since each is added into the model assuming that the other is already in the model.

Table 15.40 Output from SAS program for the sludge experiment—five-way model, with too many terms

```
                          The SAS System
                   General Linear Models Procedure
Dependent Variable: Y

Source              DF    Type I SS    Mean Square    F Value    Pr > F
A                    1        861.1         861.1         .         .
B                    1      78606.1       78606.1         .         .
C                    1    1054878.1     1054878.1         .         .
D                    1      68635.1       68635.1         .         .
E                    1     266815.1      266815.1         .         .
A*C                  1       8778.1        8778.1         .         .
B*C                  1      31878.1       31878.1         .         .
B*D                  0          0.0             .         .         .

Source              DF   Type III SS   Mean Square    F Value    Pr > F
A                    0          0.0             .         .         .
B                    1      78606.1       78606.1         .         .
C                    1    1054878.1     1054878.1         .         .
D                    1      68635.1       68635.1         .         .
E                    1     266815.1      266815.1         .         .
A*C                  1       8778.1        8778.1         .         .
B*C                  1      31878.1       31878.1         .         .
B*D                  0          0.0             .         .         .
```

In Exercise 14, information is given on how to use the SAS software for analysis of a 3_{IV}^{4-1} design, using the data of the refinery experiment of Example 15.3.1.

15.8.2 Design for the Control of Noise Variability

We now turn to the analysis of experiments involving design and noise factors, often known as Taguchi experiments. These were reviewed in Section 15.7. There are two approaches to the analysis. The first approach involves the analysis of the mean and variance of the response observed for each design-treatment combination, calculated over the levels of the noise factors. The second approach involves the study of design-by-noise interactions. The first approach requires every noise combination to be observed with every design combination (that is, a product array), and the second approach requires randomization of all observed combinations of noise and design factors taken together.

Example 15.8.1 Inclinometer experiment—product-array approach

The inclinometer experiment was described in Section 15.7.1, and the data are shown in Table 15.31 (page 518). The SAS program in Table 15.41 reads in the data corresponding to each combination of levels of the seven design factors (A–G) without identifying the levels

Table 15.41 SAS program for the product array analysis of the inclinometer experiment.

```
DATA INCLP;
  INPUT A B C D E F G  Y1 Y2 Y3 Y4 Y5 Y6 Y7 Y8;
  SUM = (Y1 + Y2 + Y3 + Y4 + Y5 + Y6 + Y7 + Y8);
  AVY = SUM/8;
  VAR = ((Y1*Y1 + Y2*Y2 + Y3*Y3 + Y4*Y4 + Y5*Y5 + Y6*Y6
                          + Y7*Y7 + Y8*Y8) - SUM*SUM/8)/7;
  LNVAR = LOG(VAR);
  LINES;
  0 0 0 0 0 0 0 0.62 3.54 3.56 0.62 3.09 0.71 0.73 3.20
  0 0 1 1 1 1 1 0.59 3.11 3.11 0.59 2.98 0.63 0.64 3.02
  : : : : : : :  :    :    :    :    :    :    :    :
  2 2 2 0 2 2 1 1.84 9.35 9.19 1.79 9.06 1.85 1.89 9.28
;
PROC PRINT;
;
* Analysis of the log sample variance;
PROC GLM;
  CLASS A B C D E F G;
  MODEL LNVAR = A B C D E F G;
  ESTIMATE 'Lin A'  A -1  0  1;
  ESTIMATE 'Quad A' A  1 -2  1;
      :          :  :  :  :
  ESTIMATE 'Quad F' F  1 -2  1;
  CONTRAST 'Lin A'  A -1  0  1;
  CONTRAST 'Quad A' A  1 -2  1;
      :          :  :  :  :
  CONTRAST 'Quad F' F  1 -2  1;
;
* Analysis of the sample mean;
PROC GLM;
  CLASS A B C D E F G;
  MODEL AVY = A B C D E F G;
  ESTIMATE 'Lin A'  A -1  0  1;
  ESTIMATE 'Quad A' A  1 -2  1;
      :          :  :  :  :
  ESTIMATE 'Quad F' F  1 -2  1;
  CONTRAST 'Lin A'  A -1  0  1;
  CONTRAST 'Quad A' A  1 -2  1;
      :          :  :  :  :
  CONTRAST 'Quad F' F  1 -2  1;
```

of the noise factors. The average AVY and the log sample variance LNVAR of the observations for each design-treatment combination is computed and added to the data set. Since only 27 (i.e., 3^3) of the 3^7 design combinations are observed, we have a 3^{7-4} fractional factorial experiment with two possible response variables.

Table 15.42 SAS program for the mixed-array analysis of the inclinometer experiment

```
DATA INCLM;
  INPUT A B C D E F G  H P J K L M N   Y;
  LINES;
       0 0 0 0 0 0 0  0 0 0 0 0 0 0  0.62
       0 0 0 0 0 0 0  0 0 0 1 1 1 1  3.54
       0 0 0 0 0 0 0  0 1 1 0 0 1 1  3.56
       0 0 0 0 0 0 0  0 1 1 1 1 0 0  0.62
       0 0 0 0 0 0 0  1 0 1 0 1 0 1  3.09
       0 0 0 0 0 0 0  1 0 1 1 0 1 0  0.71
       0 0 0 0 0 0 0  1 1 0 0 1 1 0  0.73
       0 0 0 0 0 0 0  1 1 0 1 0 0 1  3.20
       0 0 1 1 1 1 1  0 0 0 0 0 0 0  0.59
       0 0 1 1 1 1 1  0 0 0 1 1 1 1  3.11
       : : : : : : :  : : : : : : :   :
       2 2 2 0 2 2 1  1 1 0 0 1 1 0  1.89
       2 2 2 0 2 2 1  1 1 0 1 0 0 1  9.28
;
PROC GLM;
  CLASS A B C D E F G  H P J K L M N;
  MODEL Y = A B C D E F G  H P J K L M N
    A*H B*H C*H D*H E*H F*H G*H  A*P B*P C*P D*P E*P F*P G*P
    A*J B*J C*J D*J E*J F*J G*J  A*K B*K C*K D*K E*K F*K G*K
    A*L B*L C*L D*L E*L F*L G*L  A*M B*M C*M D*M E*M F*M G*M
    A*N B*N C*N D*N E*N F*N G*N;
  CONTRAST 'Lin B'  B -1  0  1;
  CONTRAST 'Quad B' B  1 -2  1;
  ESTIMATE 'G2-G0'  G -1  0  1;
  ESTIMATE 'G2-G1'  G  0 -1  1;
  ESTIMATE 'G1-G0'  G -1  1  0;
  ESTIMATE 'Lin B'  B -1  0  1;
  ESTIMATE 'Quad B' B  1 -2  1;
```

Two analyses are requested in Table 15.41. The first uses the response variable LNVAR. An analysis of variance table is requested via the PROC GLM statement for the model that includes main effects but no interactions. The least squares means for the levels of the design variables are requested via the LSMEANS statement, and these can be used to prepare plots such as those shown in Figure 15.7 (page 521). Linear and quadratic trends in each design factor can be tested via ESTIMATE or CONTRAST statements, only two of which are shown in the program. The final section of the program in Table 15.41 uses the response variable AVY. The output is similar to that shown in Tables 15.32–15.35, pages 519–520. □

Example 15.8.2 Inclinometer experiment—mixed-array approach

In the mixed-array approach, it is necessary to input the levels of the noise factors as well as those of the design factors. There are 216 observations in total—a sufficient number to allow

15.8 Using SAS Software

Table 15.43 Output from the SAS program for the inclinometer experiment—mixed-array analysis

```
                         The SAS System
                 General Linear Models Procedure
Dependent Variable: Y
                           Sum of         Mean
Source             DF     Squares       Square    F Value    Pr > F
Model             119   1245.7533      10.4685      37.32    0.0001
Error              96     26.9252       0.2805
Corrected Total   215   1272.6785
```

Source	DF	Type I SS	Mean Square	F Value	Pr > F
A	2	1.03058	0.51529	1.84	0.1648
B	2	2.31901	1.15951	4.13	0.0190
C	2	103.62218	51.81109	184.73	0.0001
D	2	196.83383	98.41692	350.90	0.0001
E	2	72.44887	36.22444	129.16	0.0001
F	2	92.56799	46.28399	165.02	0.0001
G	2	5.37960	2.68980	9.59	0.0002
:	:	:	:	:	:
M	1	0.48356	0.48356	1.72	0.1923
N	1	559.82920	559.82920	1996.03	0.0001
:	:	:	:	:	:
B*N	2	1.02817	0.51409	1.83	0.1655
C*N	2	46.05211	23.02606	82.10	0.0001
D*N	2	87.46056	43.73028	155.92	0.0001
E*N	2	32.21601	16.10800	57.43	0.0001
F*N	2	41.17091	20.58545	73.40	0.0001
G*N	2	2.40028	1.20014	4.28	0.0166

Contrast	DF	Contrast SS	Mean Square	F Value	Pr > F
Lin B	1	2.2801000	2.2801000	8.13	0.0053
Quad B	1	0.0389120	0.0389120	0.14	0.7104

Parameter	Estimate	T for H0: Parameter=0	Pr > |T|	Std Error of Estimate
G2-G0	-0.02583333	-0.29	0.7704	0.08826590
G2-G1	-0.34694444	-3.93	0.0002	0.08826590
G1-G0	0.32111111	3.64	0.0004	0.08826590
Lin B	0.25166667	2.85	0.0053	0.08826590
Quad B	0.05694444	0.37	0.7104	0.15288103

estimation of the main effects of the seven design 3-level factors (2 degrees of freedom each) and the seven noise 2-level factors (1 degree of freedom each), all of the 49 design-by-noise interactions (2 degrees of freedom each). A few nonaliased design-by-design interactions (4 degrees of freedom each) are also estimable.

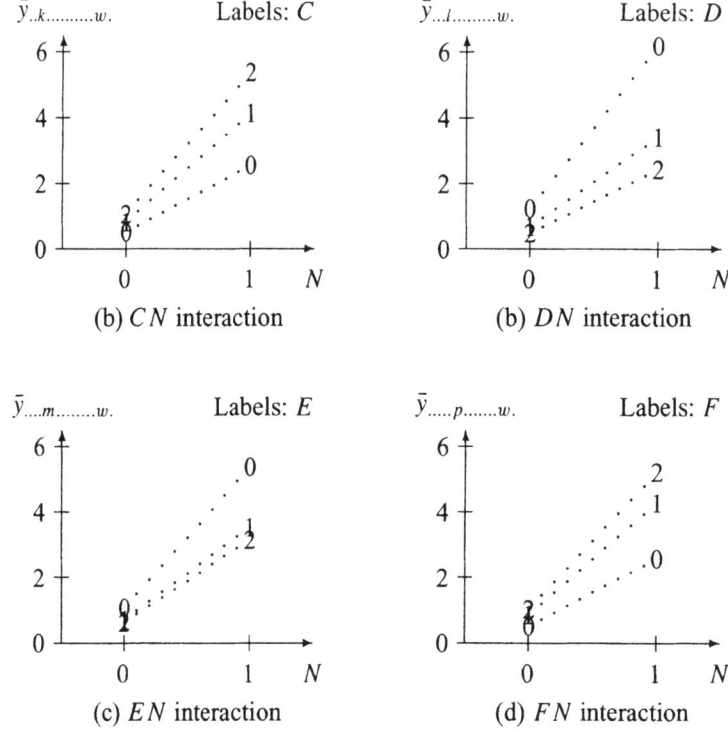

Figure 15.8
Interaction plots for the inclinometer experiment

The model statement shown in Table 15.42 includes only the main effects of the design and noise factors and the design-by-noise interactions. Most of the design-by-noise interactions are negligible, with p-values over 0.9. Since there are 63 terms in the model, we show in Table 15.43 only that part of the SAS analysis of variance for main effects and for interactions whose associated p-values are less than 0.2. For an overall significance level of at most $\alpha = 0.126$, we would test each individual hypothesis at level $\alpha^* = 0.002$. Looking at the p-values in Table 15.43, we note that the only design-by-noise interactions that appear to be significantly different from zero are the interactions of noise factor N with design factors C, D, E, and F. Interaction plots for these four interactions are shown in Figure 15.8. We can see that in order to reduce the dependency of factors C, D, E, and F on noise factor N, factors C and F would be set at their low levels and factors D and E at their high levels. These are also the levels that reduce the total amount of swing.

We can also see from the analysis of variance table, Table 15.43, that factor G (copper plating thickness) affects the amount of swing, although its interaction with noise factor N (maximum horizontal amplitude of vibration) is smaller. From the ESTIMATE statements in the SAS program, the least squares estimates for the three contrasts in G (averaged over the levels of the other factors) are

$$\widehat{G2-G0} = -0.026, \quad \widehat{G2-G1} = -0.347, \quad \widehat{G1-G0} = 0.321,$$

indicating that the smallest amount of swing is given by level 2, closely followed by level 0. Thus, G could be set at either the low or high level. Although we do not have enough evidence to conclude that factor B (flexure thickness) affects the swing significantly, it may still be of interest to examine the direction of any trend that may be present (the p-value for the test of no effect of factor B is 0.019). The least squares estimate for the linear contrast in the levels of B is 0.251, indicating that the low level possibly gives the smallest amount of swing.

Thus the final recommendation is to set C and F low, and to set D and E high, with G at either the high or low level. Factor A can be set at the cheapest level. Factor B could be set at the low level or to the cheapest level. The product-array analysis in Section 15.7.1 gave the same conclusions for the highly significant factors C, D, E, and F, and any followup experiment would want to concentrate on these four factors together with the noise factor N. □

Exercises

1. **Decontamination experiment, continued**

   Suppose that only the first block of the data (beta particles) had been obtained in the decontamination experiment described in Exercise 8 of Chapter 13 (page 455). The design would then have been a $\frac{1}{2}$-fraction of a 2^4 experiment with defining relation $I = ABCD$. The half fraction is shown in Table 15.44. Analyze the data and compare your conclusions with those of the full experiment in Exercise 8 of Chapter 13. Explain the circumstances under which a half fraction would be preferred to a single-replicate factorial experiment.

2. **Mangold experiment, continued**

   The mangold experiment in Section 13.5, page 437, was a single replicate confounded design for a 2^5 experiment in $b = 4$ blocks of size 8. The five factors were Sulphate of Ammonia (factor A at levels 0 or 0.6 cwt per acre), Superphosphate (factor B at levels 0 or 0.5 cwt per acre), Muriate of Potash (factor C at levels 0 or 1.0 cwt per acre), Agricultural Salt (factor D at levels 0 or 5 cwt per acre), and Dung (factor E at levels 0 or 10 tons per acre). All of the 3-, 4-, and 5-factor interactions were expected to be negligible. The two three-factor interactions ABD, BCE and their product $ACDE$ were selected for confounding.

   Suppose that the data from only the third block had been available, so that we have a $\frac{1}{4}$-fraction. The data are reproduced in Table 15.45.

   (a) Write down the aliasing scheme for this fractional factorial experiment.

Table 15.44 Block I of the decontamination experiment

Treatment Combinations (Response)							
1010	1111	0110	0000	1100	0101	0011	1001
(716)	(686)	(498)	(1437)	(527)	(579)	(1433)	(906)

Source: Barnett, M. K. and Mead, F. C. Jr. (1956). Copyright © 1956 Blackwell Publishers. Reprinted with permission.

Table 15.45 Yields (in pounds) of mangold roots for Block III of the mangold experiment

Treatment Combinations (Yield)							
00101	11001	01011	01110	10010	11100	00000	10111
(896)	(1284)	(996)	(860)	(1184)	(984)	(740)	(1468)

Source: *Design and Analysis of Experiments*, by O. Kempthorne, Copyright © 1976, John Wiley & Sons, Inc. Reprinted by permission of John Wiley & Sons, Inc.

(b) Analyze the data. What conclusions can you draw?

(c) Comparing your conclusions with those of Section 13.5, what extra information do you gain by running the single-replicate design instead of the fraction?

(d) When would you recommend that an experimeter consider using a fractional factorial design rather than a single-replicate design?

3. **Dye experiment, continued**

   The dye experiment was discussed in Section 14.2.4 (page 467). There were three factors: the concentration of inorganic material M in the free water in the reaction mixture (factor A at three equally spaced levels), the volume of free water in the reaction mixture (factor B at three equally spaced levels), and the concentration of inorganic material N in the free water in the reaction mixture (factor C at three equally spaced levels). The data for the original experiment were given in Table 14.6 (page 468) and the first replicate is reproduced in Table 15.46. The design for the first replicate was a single-replicate design that confounded (AB^2C^2; A^2BC). Analyze the data of Block II as though it had come from a $\frac{1}{3}$-fraction. State your conclusions.

4. **Sugar beet experiment, continued**

   The sugar beet experiment described in Exercise 6 of Chapter 14 concerned the effects of three standard fertilizers, nitrogen, phosphate, and potassium (factors N, P, and K), each at three equally spaced levels, on sugar beet yield. The experiment was run as a single-replicate confounding the contrasts (NP^2K; N^2PK^2). Suppose the only data available were those of Block III, reproduced in Table 15.47.

   (a) If the only data available were those from Block III, write out the aliasing scheme for the design.

   (b) Analyze the data from Block III as though they came from a $\frac{1}{3}$-fraction. State your conclusions.

Table 15.46 Volume of dyestuff for Block I of the dye experiment

Treatment Combinations (Yield)							
000	021	012	110	101	122	220	211	202
(74)	(130)	(56)	(110)	(166)	(227)	(195)	(146)	(90)

Source: Davies, O. L. (1963). Reprinted by permission of Addison Wesley Longman Ltd.

Table 15.47 Yields of sugar beet for Block III of the sugar-beet experiment

Treatment Combinations (Yield)								
202	020	210	111	001	122	221	012	100
(2198)	(2093)	(2354)	(2268)	(1926)	(2152)	(2349)	(2025)	(2106)

Source: Yates, F. (1935). Copyright © 1935 Blackwell Publishers. Reprinted with permission. (Reprinted in *Experimental Design* (1970), Charles Griffin and Company, Ltd., London. Copyright 1970 Edward Arnold/Hodder & Stoughton Educational. Reprinted with permission.)

5. **Flour experiment number 3, continued**

   Suppose that the data from Block II of the 4×2^4 experiment in Table 15.17 (page 504) had been lost, so that only Block I remained. This would then constitute a $\frac{1}{4}$-fraction.

   (a) Write out the aliasing scheme for the design. What is the resolution number. Is this a good design?

   (b) Bearing in mind the purpose of the experiment, can you find a better $\frac{1}{4}$-fraction? If so, write out the design and its aliasing scheme.

   (c) Analyze the data from Block I of Table 15.17. What can you conclude?

6. **Handwheel experiment**

   E. N. Corlett and G. Gregory describe an experiment in the 1960 issue of *Applied Statistics* that was concerned with finding the design of a machine tool handwheel that would maximize the accuracy on the part of the operator in the setting of the machine tool handwheel. The apparatus consisted of an optical dividing head with a dial mounted onto a mandrel to which was connected the handwheel spindle. The spindle was provided with an adjustable friction brake. The operator first offset the dial by 15 degrees and then moved the handwheel so that a line on the dial was brought "into coincidence with a fixed line on the dividing head, making the final adjustment by means of a series of taps by hand on the handwheel rim."

   Seven factors, each at two levels (coded 0 and 1) were investigated as follows.

   A: Handwheel diameter (5.5 in., 10 in.)

   B: Dial diameter (4 in., 8 in.)

   C: Thickness of the dial line (0.008 in., 0.064 in.)

   D: Friction of the spindle (7.5 lb.-in., 45 lb.-in.)

   E: Level of operator's elbow relative to height of handwheel
   (Level with center of spindle, 6 in. above spindle center)

   F: Previous experience of operator (Practiced, Nonpracticed)

   G: Knowledge of accuracy of previous setting (Feedback, No feedback)

   The response variable was $\ln(s^2)$, where s^2 was the sample variance of 25 repeated observations for a particular treatment combination. It was estimated that each set of 25 repeated observations would take about 15 minutes to complete, including setup time. In a morning or afternoon session of four hours, therefore, sixteen observations could

Table 15.48 Log variance of observations for the handwheel experiment

Block I ABCDEFG	ln(s^2)	Block II ABCDEFG	ln(s^2)
0 0 0 0 0 0 0	0.7044	1 1 0 0 0 0 0	0.0561
1 0 1 0 0 0 0	0.5907	0 1 1 0 0 0 0	0.3615
0 0 1 0 0 0 1	−0.0297	1 1 1 0 0 0 1	−0.1158
1 0 0 0 0 0 1	0.3914	0 1 0 0 0 0 1	−0.1952
0 1 0 1 0 0 0	0.0792	1 0 0 1 0 0 0	0.4585
1 1 1 1 0 0 0	0.3228	0 0 1 1 0 0 0	0.2531
0 1 1 1 0 0 1	−0.1599	1 0 1 1 0 0 1	0.2727
1 1 0 1 0 0 1	−0.0996	0 0 0 1 0 0 1	0.6861
0 0 0 1 1 0 0	0.5878	1 1 0 1 1 0 0	−0.0074
1 0 1 1 1 0 0	0.3577	0 1 1 1 1 0 0	−0.2328
0 0 1 1 1 0 1	0.1847	1 1 1 1 1 0 1	−0.1046
1 0 0 1 1 0 1	0.5706	0 1 0 1 1 0 1	−0.2069
0 1 0 0 1 0 0	−0.1805	1 0 0 0 1 0 0	0.8051
1 1 1 0 1 0 0	−0.3224	0 0 1 0 1 0 0	0.4634
0 1 1 0 1 0 1	−0.1433	1 0 1 0 1 0 1	0.2904
1 1 0 0 1 0 1	0.1354	0 0 0 0 1 0 1	0.4692

Block III ABCDEFG	ln(s^2)	Block IV ABCDEFG	ln(s^2)
0 1 0 0 0 1 0	−0.6760	1 0 0 0 0 1 0	0.5457
1 1 1 0 0 1 0	−0.3824	0 0 1 0 0 1 0	0.0846
0 1 1 0 0 1 1	−0.2996	1 0 1 0 0 1 1	0.4453
1 1 0 0 0 1 1	−0.4539	0 0 0 0 0 1 1	0.2361
0 0 0 1 0 1 0	0.2970	1 1 0 1 0 1 0	−0.5069
1 0 1 1 0 1 0	0.1646	0 1 1 1 0 1 0	−0.3299
0 0 1 1 0 1 1	0.3878	1 1 1 1 0 1 1	−0.3245
1 0 0 1 0 1 1	0.2168	0 1 0 1 0 1 1	−0.3233
0 1 0 1 1 1 0	0.0148	1 0 0 1 1 1 0	0.4199
1 1 1 1 1 1 0	−0.4898	0 0 1 1 1 1 0	0.2957
0 1 1 1 1 1 1	0.1308	1 0 1 1 1 1 1	0.2278
1 1 0 1 1 1 1	−0.1829	0 0 0 1 1 1 1	0.4269
0 0 0 0 1 1 0	0.0182	1 1 0 0 1 1 0	−0.4798
1 0 1 0 1 1 0	0.2070	0 1 1 0 1 1 0	−0.0669
0 0 1 0 1 1 1	0.1101	1 1 1 0 1 1 1	−0.0584
1 0 0 0 1 1 1	0.2642	0 1 0 0 1 1 1	−0.6856

Source: Corlett, E. N. and Gregory, G. (1960). Copyright © 1960 Blackwell Publishers. Reprinted with permission.

be taken. The experiment was to last over two days, which meant that a 2^{7-1} fractional factorial experiment was required, divided into 4 blocks of 16.

The highest-order interaction was selected for the defining relation of the fraction, that is, $I = ABCDEFG$. Only two operators were used for the experiment, one for each level of practice. The difference between these operators was not of interest, only the interaction of the level of practice with the other factors. Rather unusually, then, the main effect of F was selected as one of the contrasts for confounding. All the 2-factor interactions and most of the 3-factor interactions were thought to be of interest. Unlikely 3-factor interactions included ACG and BDE, which were also chosen for confounding with blocks. The complete set of confounded contrasts was F, ACG, $ACFG$ together with its set of aliases $ABCDEG$, $BDEF$, BDE. All other main-effect, 2-factor, and 3-factor interaction contrasts could be estimated.

The data obtained from the experiment are shown in Table 15.48.

(a) Write out the aliasing scheme for the design.

(b) Using a computer package, estimate the (estimable) main-effect and interaction contrasts.

(c) Prepare a normal probability plot of the contrast estimates and identify the most important main effects and interactions.

(d) The authors of the article point out that if the responses are normally distributed and n is large (where n is the number of repeated observations, 25 in this experiment), then the response variable $\ln(s^2)$ has approximately constant variance equal to $2/(n-1)$. Calculate the standard error for each of the contrasts estimated in part (c). Using Bonferroni's method with an individual significance level of 0.001 for each test (giving an overall level of at most 0.06), which main effects and interactions are significantly different from zero? Do these results agree with the results from part (c)? Discuss why or why not.

(e) Draw interaction plots of the important interactions and discuss recommended settings for the six factors A, B, C, D, E, and G for the practiced and nonpracticed operators individually.

(f) Would you recommend further experimentation? If so, which factors and which settings would you recommend? Can you suggest a suitable design?

7. **Paint experiment**

   (a) Suppose that you need to design an experiment involving 6 factors (A, B, C, D, E, F) at 2 levels each (64 treatment combinations) and that only 8 observations can be taken. You decide to sacrifice information on the ABF, $ACDF$, and $ABCE$ contrasts. Write out the defining relation and the two rows of the aliasing scheme showing the aliasing of A and the aliasing of AC.

   (b) Explain what aliasing means.

   (c) An experiment was run in Germany by S. Eibl, U. Kess, and F. Pukelsheim (*Journal of Quality Technology*, 1992) on the thickness of a paint coating. Prior to the experiment, the thickness achieved was around 2 mm, much higher than the target 0.8 mm. They selected the following six factors, each at two levels:

   A: Belt speed

   B: Tube width

   C: Pump pressure

   D: Paint viscosity

   E: Tube height

Table 15.49 Paint thickness for the paint experiment

A	B	C	D	E	F	y_{i1}	y_{i2}	y_{i3}	y_{i4}	$\bar{y}_{i.}$	$\ln(s_i^2)$
1	0	1	0	0	0	1.09	1.12	0.83	0.88	0.9800	−3.8444
0	0	1	0	1	1	1.62	1.49	1.48	1.59	1.5450	−5.3050
1	1	0	0	0	1	0.88	1.29	1.04	1.31	1.1300	−3.1497
0	1	0	0	1	0	1.83	1.65	1.71	1.76	1.7375	−5.1456
0	0	0	1	0	1	1.46	1.51	1.59	1.40	1.4900	−5.0411
1	0	0	1	1	0	0.74	0.98	0.79	0.83	0.8350	−4.5375
0	1	1	1	0	0	2.05	2.17	2.36	2.12	2.1750	−4.0380
1	1	1	1	1	1	1.51	1.46	1.42	1.40	1.4475	−6.0498

Source: Eibl, S., Kess, U., and Pukelsheim, F. (1992). Copyright © 1997 American Society for Quality. Reprinted with Permission.

F: Heating temperature

They used the $\frac{1}{8}$-fraction with the aliasing scheme in part (a), and they decided to ignore all interactions for this first experiment. Since they wanted to monitor the variation of the thickness, they took four observations on each of the 8 treatment combinations in the fraction. The data are shown in Table 15.49.

Calculate the analysis of variance table and contrast estimates using response variable LNVAR (the log variance). What do you conclude?

(d) Calculate the analysis of variance table and also contrast estimates of interest, using the 32 observations separately (without combining them into an average). Remembering that the goal is to reduce the thickness, what conclusions would you draw from this particular experiment?

(e) The experimenters decided to run a followup experiment with at most 16 observations. You can use any of the original 6 factors and you can change the levels from their original settings. The ultimate goal is to achieve a coating of 0.8 mm. Suggest a followup experiment.

8. **Laser printer experiment**

The laser printer experiment was run by H.-P. Chu, M. Lagus, and P. Weiss at the University of Wisconsin as a class project. They identified options that could be set for drawing a picture on a laser printer using the MacDraw Pro package, and they wanted to find out which of the options would speed up or slow down the printing of a reasonably detailed picture. The 13 factors and their levels (low, high) are listed below.

A: Orientation (vertical, horizontal)

B: Scale (100%, 120%)

C: Complexity (simple, complex)

D: Print (color/grey-scale, black-white)

E: Font substitution (no, yes)

F: Text smoothing (no, yes)

G: Graphics smoothing (no, yes)

H: Faster bitmap printing (no, yes)

Table 15.50 Times of printing (seconds) for the laser printer experiment

Block	A B C D E F G H J K L M N	Repl. I	Repl. II
II	0 0 0 0 1 1 1 1 1 0 0 0 0	1511	1506
I	1 0 0 0 0 0 1 1 1 1 1 1 0	375	370
I	0 1 0 0 1 1 0 0 1 1 1 0 1	392	392
II	1 1 0 0 0 0 0 0 1 0 0 1 1	523	520
I	0 0 1 0 0 1 0 1 0 1 0 1 1	481	479
II	1 0 1 0 1 0 0 1 0 0 1 0 1	2238	2230
II	0 1 1 0 0 1 1 0 0 0 1 1 0	550	551
I	1 1 1 0 1 0 1 0 0 1 0 0 0	546	542
I	0 0 0 1 1 0 1 0 0 0 1 1 1	369	367
II	1 0 0 1 0 1 1 0 0 1 0 0 1	2242	2243
II	0 1 0 1 1 0 0 1 0 1 0 1 0	584	592
I	1 1 0 1 0 1 0 1 0 0 1 0 0	446	447
II	0 0 1 1 0 0 0 0 1 1 1 0 0	1776	1782
I	1 0 1 1 1 1 0 0 1 0 0 1 0	1053	1059
I	0 1 1 1 0 0 1 1 1 0 0 0 1	462	463
II	1 1 1 1 1 1 1 1 1 1 1 1 1	582	582

J: Flip vertical (no, yes)

K: Invert image (no, yes)

L: Precision bitmap (no, yes)

M: Font (Palatino, Times Italic)

N: Flip horizontal (no, yes)

The factor "complexity" referred to the image being printed. At the high level, more detail was added to the picture. The other factors were set via the print option of the drawing package. Two different laser printers and operators were used with the two printer/operator combinations constituting two blocks. The response variable was speed of printing, and this was measured by the operator, one of whom used a stop-watch and the other a wrist-watch.

They used a 2^{13-9} fractional factorial, obtained by confounding the following 9 interactions: ACE, ADF, AGJ, AHK, ALN, BCG, BDH, BMN, and CDN. The interaction AM (together with its aliases) was confounded with blocks. Thus, the final design had 16 treatment combinations divided into 2 blocks of size 8, as shown in Table 15.50. The defining relation for the design contains 2^9 terms, and so does each row of the aliasing scheme. The parts of the aliasing scheme involving main effects and

2-factor interactions are

A	$= CE$	$= DF$	$= GJ$	$= HK$	$= LN$	
B	$= CG$	$= DH$	$= EJ$	$= FK$	$= MN$	
C	$= AE$	$= BG$	$= DN$	$= FL$	$= HM$	
D	$= AF$	$= BH$	$= CN$	$= EL$	$= GM$	
E	$= AC$	$= BJ$	$= DL$	$= FN$	$= KM$	
F	$= AD$	$= BK$	$= CL$	$= EN$	$= JM$	
G	$= AJ$	$= BC$	$= DM$	$= HN$	$= KL$	
H	$= AK$	$= BD$	$= CM$	$= GN$	$= JL$	
J	$= AG$	$= BE$	$= FM$	$= HL$	$= KN$	
K	$= AH$	$= BF$	$= EM$	$= GL$	$= JN$	
L	$= AN$	$= CF$	$= DE$	$= GK$	$= HJ$	
M	$= BN$	$= CH$	$= DG$	$= EK$	$= FJ$	
N	$= AL$	$= BM$	$= CD$	$= EF$	$= GH$	$= JK$
AB	$= CJ$	$= CK$	$= EG$	$= FH$	$= LM$	
Block	$= AM$	$= BL$	$= CK$	$= DJ$	$= EH$	$= FG$

(a) The first replicate of data shown in Table 15.50 was analyzed by the experimenters. Duplicate their analysis by estimating the 13 main-effect contrasts and the AB interaction contrast using only the Replicate I data. Prepare a normal probability plot of the contrast estimates. What can you conclude?

(b) The experimenters later decided to obtain an estimate of error, and they repeated the entire 16 runs in two more blocks. The second set of data is listed as Replicate II in Table 15.50. Analyze all four blocks of data. What can you now conclude? Discuss whether statistical significance and practical significance always coincide.

(c) What does the analysis in part (b) tell you about the dangers of relying heavily on normal probability plots to pick out important interactions?

(d) This is a highly fractionated experiment, which has resulted in a lot of confounding. Suppose that you were going to run an experiment somewhat similar to this with 32 observations, how would you design it?

9. **Flour experiment number 1**

The flour experiment was introduced in Example 15.5.1. In Table 15.51, we show part of the design for the first experiment in the series. Six ingredients, A, B, C, D, E, F, added to the flour were to be investigated in the experiment. In addition, there were three noise factors: Factor P (which was a combination of factors N and S in Example 15.5.1) had two levels ("high yeast with long proof time" or "low yeast with short proof time"), Factor Q, (as in Example 15.5.1, two levels "undermixing, little water, heavy pressure" or "overmixing, much water, little pressure"), and Factor R (two levels, underbake or overbake).

Exercises 537

Table 15.51 Specific volume for part of experiment 1 of the flour experiment

	Noise Combinations			
Design Combinations	Day 1 (111)	Day 2 (101)	Day 3 (000)	Day 4 (011)
000000	519	446	337	415
000011	503	468	343	418
001101	567	471	355	424
001110	552	489	361	425
010101	534	466	356	431
010110	549	461	354	427
011000	560	480	345	437
011011	535	477	363	418
100100	558	483	376	418
100111	551	472	349	426
101001	576	487	358	434
101010	569	494	357	444
110001	562	474	358	404
110010	569	494	348	400
111100	568	478	367	463
111111	551	500	373	462

Source: Tuck, M. G., Lewis, S. M., and Cottrell, J. I. L. (1993). Copyright © 1993 Blackwell Publishers. Reprinted with permission.

A crossed array was selected. The noise array was a $\frac{1}{2}$-fraction with defining relation $I = PQR$. Each of the four noise combinations was run on a single day, so that the experiment ran over four days. The design array was a $\frac{1}{4}$-fraction with defining relation $I = ABCD = BCEF = ADEF$, and this was run on each day. Thus the noise contrasts are confounded with days and cannot be analyzed. However, the object of the experiment was to examine the average yield (specific volume, ml/100 g) and the variance of the yield for the design factors across the noise factors.

(a) Calculate the average yield and the log variance of the yield for each design-treatment combination.

(b) Analyze the two sets of data separately. What recommendations would you make if the objective is to reduce the variability and increase the specific volume?

10. **Injection molding experiment**

S. R. Schmidt and R. G. Launsby in their book *Understanding Industrial Designed Experiments* describe an experiment on the effect of six factors on the shrinkage of a part produced by injection molding. The six factors were injection velocity (factor A), cooling time (factor B), barrel zone temperature (factor C), mold temperature (factor D), hold pressure (factor E), and back pressure (factor F). Each factor had two levels coded 0 and 1.

Table 15.52 Lengths and widths of parts after shrinkage in the injection molding experiment

Treatment Combinations A B C D E F	Length (Deviation from 14.5)×10⁴					Width (Deviation from 9.35)×10⁴				
0 0 0 0 0 0	0	5	0	0	5	75	60	70	85	90
0 0 0 1 1 1	75	90	70	65	65	50	40	40	40	45
0 1 1 0 0 1	45	50	45	45	45	45	45	45	50	40
0 1 1 1 1 0	100	105	105	110	105	130	130	125	135	135
1 0 1 0 1 0	105	110	105	120	100	55	60	60	55	60
1 0 1 1 0 1	45	55	65	50	50	80	65	50	40	45
1 1 0 0 1 1	150	140	155	50	145	100	80	85	90	85
1 1 0 1 0 0	55	65	55	55	60	65	60	65	65	60

Source: Schmidt, S. R. and Launsby, R. G. (1992). Copyright © 1992 Air Academy Press. Reprinted with permission.

There were two responses of interest, the length and width of the part after shrinkage. The purpose of the experiment was to find settings of the six variables that would enable the parts to be "on target," that is, a post-shrinkage length of 14.5 and width of 9.38. The orthogonal array in Table 15.23 was selected with columns 1–6 labeled D, C, B, A, E, F. Columns 5 and 6 were multiplied by -1. One degree of freedom (corresponding to column 7) is available to measure σ^2 or one of the two-factor interactions. Five parts were measured at each treatment combination, and the lengths and widths are recorded in Table 15.52.

(a) Write down the defining relation for the $\frac{1}{8}$-fraction and the aliasing scheme. The investigators assumed that all the interactions were negligible. If they had not done so, which interactions could be measured?

(b) For the length data, calculate the average response and the standard deviation of the response for each treatment combination.

(c) Can you recommend which factors should be investigated more thoroughly in order to find a setting that would give the required length and also factors that could be set to reduce the variability?

(d) Repeat parts (a) and (b) for the width data.

(e) Can you make any overall recommendation?

(f) Write down the assumptions on the model that would need to be true in order to interpret the analysis of variance. Are these assumptions likely to be valid for this experiment?

11. **Spectrometer experiment, continued**

Read the details of the spectrometer experiment in Chapter 7, page 236. You will need to have access to your solutions to that exercise to answer this question.

Suppose that you are consultant for a different company and that they wish to run a similar experiment, with the same five factors, but with a total of 64 observations. To keep things simple, you might recommend that factors A and C be examined at 2 levels each rather than 3 levels in your first experiment (even though you may suspect that some of the factors have quadratic trends). Thus, you have a 2^5 experiment. List 5 interactions that you are particularly interested in studying. You should use information from your answer to (a) and (b) in choosing the interactions. Design a factorial experiment in 4 blocks of size 8. State exactly how you chose your design. Write out at least three of the treatment combinations in two of the blocks and explain how you obtained them.

12. **Design of industrial experiment**

    Suppose that you are asked to design an experiment for 6 treatment factors each having two levels. Only 64 observations can be taken in total, and these should be divided into 8 blocks of size 8. Suppose that you decide to confound the interaction contrasts ABD, DEF, and $ACDF$.

    (a) Can all the other interaction contrasts be estimated?

    (b) What does the statement "ABD is confounded" mean?

    (c) How would you obtain the 8 blocks? Write out two blocks as an example.

    (d) Suppose that the budget is cut before the experiment can take place, and only 8 observations can be taken in total. How would you decide which 8 observations to take? What can be estimated?

    (e) Suppose that you were fairly sure that all interactions involving 4 factors or more were negligible and that D does not interact with any of the other factors. Suppose that the analysis of variance table obtained from the results of the experiment is as in Table 15.53. What would you investigate in a followup experiment? Give your reasons.

13. Suppose that you wish to run an experiment with four treatment factors (A, B, C, D) each having three levels. The only likely interactions are AB, AC, and ABC. The experiment needs to be run in blocks of size at most $k = 9$.

    (a) Design a 3^4 experiment in $b = 3^2$ blocks of size $k = 3^2$ confounding ABD and AB^2CD. What else is confounded? Are you happy with this design? Why or why not?

    (b) Show how you would obtain the nine blocks and show one of the blocks as an illustration.

Table 15.53 Analysis of variance for the industrial experiment

Source of Variation	Degrees of Freedom	Sum of Squares	Mean Square	Ratio	p-value
A	1	262.205	262.205	54.57	0.0857
B	1	11.045	11.045	2.30	0.3712
C	1	981.245	981.245	204.21	0.0445
D	1	5.120	5.120	1.07	0.4899
E	1	1568.000	1568.000	326.33	0.0352
F	1	8.820	8.820	1.84	0.4048
Error	1	4.805	4.805		
Total	7	2841.240			

(c) Write out the degrees of freedom column for the analysis of variance table. (Read the question again before you do this.)

(d) Suppose that the blocks are randomly ordered. After the first block is run, the budget for the experiment is cut, so only nine observations are available. The design is now a 3^{4-2} fractional factorial design. Write down the defining relation for the design. Is this design going to be useful in examining the main effects and the interactions of interest? Why or why not?

14. **Refinery experiment, continued**

The refinery experiment was discussed in Example 15.3.1, with a corresponding analysis of variance table given in Table 15.14 on page 499. A SAS program for generating similar information is shown in Table 15.54.

(a) Run the SAS program shown in Table 15.54. (The rest of the data are shown in Table 15.13, page 498.)

(b) Examine the output generated by PROC PRINT. Discuss how the variables AB and AB2 are related to the four degrees of freedom for the AB interaction.

(c) Using the output generated by the SAS program, verify that the information in Table 15.14 (page 499) is correct.

Table 15.54 SAS program for analysis of the refinery experiment

```
DATA REFINERY;
  INPUT A B C D Y;
  AB=MOD(A+B,3); AB2=MOD(A+2*B,3);
  AC=MOD(A+C,3); AC2=MOD(A+2*C,3);
  AD=MOD(A+D,3); AD2=MOD(A+2*D,3);
  BC=MOD(B+C,3); BC2=MOD(B+2*C,3);
  BD=MOD(B+D,3); BD2=MOD(B+2*D,3);
  CD=MOD(C+D,3); CD2=MOD(C+2*D,3);
  LINES;
  0 0 0 0   4.2
  0 0 1 2   5.9
  : : : :   :
  2 2 2 0  85.1
;
PROC PRINT;
PROC GLM;
  CLASS A B C D AB AB2 AC AC2 AD AD2 BC BC2 BD BD2 CD CD2;
  MODEL Y = A B C D AB AB2 AC AC2 AD AD2 BC BC2 BD BD2 CD CD2;
  CONTRAST 'A LIN'  A -1  0  1;
  CONTRAST 'A QUAD' A  1 -2  1;
  CONTRAST 'B LIN'  B -1  0  1;
  CONTRAST 'B QUAD' B  1 -2  1;
  CONTRAST 'C LIN'  C -1  0  1;
  CONTRAST 'C QUAD' C  1 -2  1;
  CONTRAST 'D LIN'  D -1  0  1;
  CONTRAST 'D QUAD' D  1 -2  1;
PROC GLM;  CLASS A B C D;
  MODEL Y = A B C D A*B A*C A*D B*C B*D C*D;
```

Table 15.55 2^{p-s} fractions of 2^p experiments. For each defining relation, s independent generators are underlined, and s corresponding equations are given. To obtain the $v = 2^{p-s}$ treatment combinations in the fraction, list all v combinations of levels a_i of the $p - s$ factors not determined by the equations, then use the equations modulo 2 to complete each treatment combination. For two blocks, confound the effect in parentheses and its aliases.

2^{p-s}	v	Defining Relation	Equations
2^{5-2}_{III}	8	$I = \underline{ABCD} = \underline{ABE} = CDE$ (AC)	$a_4 = a_1 + a_2 + a_3$ $a_5 = a_1 + a_2$
2^{6-2}_{IV}	16	$I = \underline{ABCD} = \underline{CDEF} = ABEF$ (ACE)	$a_4 = a_1 + a_2 + a_3$ $a_6 = a_3 + a_4 + a_5$
2^{7-2}_{IV}	32	$I = \underline{ABCDE} = \underline{ABFG} = CDEFG$ (AEF)	$a_5 = a_1 + a_2 + a_3 + a_4$ $a_7 = a_1 + a_2 + a_6$
2^{8-2}_{V}	64	$I = \underline{ABCDE} = \underline{DEFGH} = ABCFGH$ (CEF)	$a_5 = a_1 + a_2 + a_3 + a_4$ $a_8 = a_4 + a_5 + a_6 + a_7$
2^{6-3}_{III}	8	$I = \underline{BCD} = \underline{ABE} = ACDE$ $= \underline{ABCF} = ADF = CEF$ $= BDEF$ (AC)	$a_4 = a_2 + a_3$ $a_5 = a_1 + a_2$ $a_6 = a_1 + a_2 + a_3$
2^{7-3}_{IV}	16	$I = \underline{ABCD} = \underline{CDEF} = ABEF$ $= \underline{ACEG} = BDEG = ADFG$ $= BCFG$ (ACF)	$a_4 = a_1 + a_2 + a_3$ $a_6 = a_3 + a_4 + a_5$ $a_7 = a_1 + a_3 + a_5$
2^{8-3}_{IV}	32	$I = \underline{ABCD} = \underline{CDEF} = ABEF$ $= \underline{ACEGH} = BDEGH = ADFGH$ $= BCFGH$ (ABG)	$a_4 = a_1 + a_2 + a_3$ $a_6 = a_3 + a_4 + a_5$ $a_8 = a_1 + a_3 + a_5 + a_7$
2^{9-3}_{IV}	64	$I = \underline{CDEF} = \underline{ACEGH} = ADFGH$ $= \underline{ABCDJ} = ABEFJ = BDEGHJ$ $= BCFGHJ$ (ACF)	$a_4 = a_1 + a_2 + a_3$ $a_8 = a_1 + a_3 + a_5 + a_7$ $a_9 = a_1 + a_2 + a_3 + a_4$
2^{7-4}_{III}	8	$I = \underline{ABCD} = \underline{BCE} = ADE$ $= \underline{ACF} = BDF = ABEF$ $= CDEF = \underline{ABG} = CDG$ $= ACEG = BDEG = BCFG$ $= ADFG = EFG = ABCDEFG$	$a_4 = a_1 + a_2 + a_3$ $a_5 = a_2 + a_3$ $a_6 = a_1 + a_3$ $a_7 = a_1 + a_2$
2^{8-4}_{IV}	16	$I = \underline{ABCD} = \underline{CDEF} = ABEF$ $= \underline{ADFG} = BCFG = ACEG$ $= BDEG = \underline{ABGH} = CDGH$ $= ABCDEFGH = EFGH = BDFH$ $= ACFH = BCEH = ADEH$ (ADH)	$a_4 = a_1 + a_2 + a_3$ $a_6 = a_3 + a_4 + a_5$ $a_7 = a_1 + a_4 + a_6$ $a_8 = a_1 + a_2 + a_7$
2^{9-4}_{IV}	32	$I = \underline{ABCD} = \underline{CDEF} = ABEF$ $= \underline{ADFGH} = BCFGH = ACEGH$ $= BDEGH = \underline{ABGJ} = CDGJ$ $= ABCDEFGH = EFGJ = BDFHJ$ $= ACFHJ = BCEHJ = ADEHJ$ (ABH)	$a_4 = a_1 + a_2 + a_3$ $a_6 = a_3 + a_4 + a_5$ $a_7 = a_1 + a_4 + a_5 + a_6$ $a_8 = a_1 + a_2 + a_7$

Table 15.56 3^{p-s} fractions of 3^p experiments. For each defining relation, s independent generators are underlined, and s corresponding equations are given. To obtain the $v = 3^{p-s}$ treatment combinations in the fraction, list all v combinations of levels a_i of the $p - s$ factors not determined by the equations, then use the equations modulo 3 to complete each treatment combination.

3^{p-s}	v	Defining Relation	Equations
3^{4-2}_{III}	9	$I = \underline{AB^2C} = A^2BC^2$	$a_3 = 2a_1 + a_2$
		$= \underline{ABD} = A^2CD = B^2C^2D$	$a_4 = 2a_2 + 2a_3$
		$= A^2B^2D^2 = BCD^2 = AC^2D^2$	
3^{5-2}_{III}	27	$I = \underline{ABC^2D^2} = A^2B^2CD$	$a_4 = a_1 + a_2 + 2a_3$
		$= \underline{ADE^2} = A^2BC^2E^2 = B^2CD^2E^2$	$a_5 = a_1 + a_4$
		$= A^2D^2E = BC^2DE = AB^2CE$	
3^{6-2}_{IV}	81	$I = \underline{ABC^2D^2} = A^2B^2CD$	$a_4 = a_1 + a_2 + 2a_3$
		$= \underline{ACEF} = A^2BD^2EF = B^2C^2DEF$	$a_6 = 2a_1 + 2a_3 + 2a_5$
		$= A^2C^2E^2F^2 = BCD^2E^2F^2 = AB^2DE^2F^2$	
3^{6-3}_{III}	27	$I = \underline{ABC^2D^2} = A^2B^2CD$	$a_4 = a_1 + a_2 + 2a_3$
		$= \underline{BDE^2} = AB^2C^2E^2 = A^2CD^2E^2$	
		$= B^2D^2E = AC^2DE = A^2BCE$	$a_5 = a_2 + a_4$
		$= \underline{CDF^2} = ABF^2 = A^2B^2C^2D^2F^2$	$a_6 = a_3 + a_4$
		$= BCD^2E^2F^2 = AB^2DE^2F^2 = A^2C^2E^2F^2$	
		$= B^2CEF^2 = AD^2EF^2 = A^2BC^2DEF^2$	
		$= C^2D^2F = ABCDF = A^2B^2F$	
		$= BC^2E^2F = AB^2CD^2E^2F = A^2DE^2F$	
		$= B^2C^2DEF = ACEF = A^2BD^2EF$	
3^{7-3}_{IV}	81	$I = \underline{ABC^2D^2} = A^2B^2CD$	$a_4 = a_1 + a_2 + 2a_3$
		$= \underline{ACEF} = A^2BD^2EF = B^2C^2DEF$	$a_6 = 2a_1 + 2a_3 + 2a_5$
		$= A^2C^2E^2F^2 = BCD^2E^2F^2 = AB^2DE^2F^2$	
		$= \underline{BC^2E^2FG} = AB^2CD^2E^2FG = A^2DE^2FG$	$a_7 = 2a_2 + a_3$
		$= ABF^2G = A^2B^2C^2D^2F^2G = CDF^2G$	$\quad + a_5 + 2a_6$
		$= A^2BCEG = B^2D^2EG = AC^2DEG$	
		$= B^2CEF^2G^2 = AD^2EF^2G^2 = A^2BC^2DEF^2G^2$	
		$= AB^2C^2E^2G^2 = A^2CD^2E^2G^2 = BDE^2G^2$	
		$= AB^2FG^2 = C^2D^2FG^2 = ABCDFG^2$	

Table 15.57 Orthogonal arrays with 2^p observations and useful column labelings. Assign factors in alphabetical order.

No. of factors	8 observations—Design of Table 15.23
	Columns
	1 2 3 12 13 23 123
3–6	A B C E F G D

No. of factors	16 observations—Design of Table 15.25
	Columns
	1 2 12 3 13 23 123 4 14 24 124 34 134 234 1234
4–5	A B C D E
6–15	A B K C L M D E N P F Q G H J

No. of factors	32 observations
	Columns
	1 2 12 3 13 23 123 4 14 24 124 34 134 234 1234
4–22	A B C M D N P Q G
	Columns
	5 15 25 125 35 135 235 1235 45 145 245 1245 345 1345 2345 12345
4–22	E R S T H U V J W K L F

Table 15.58 Generators for cyclically generated orthogonal main-effect plans. These are saturated designs for factors each at two levels, for n observations with n divisible by 4 but not a power of 2. To generate a design, systematically cycle the generator to the right to obtain $n - 1$ rows; then include a final row of -1's.

n	Generator
12	1 1 1 –1 1 1 –1 1 –1 –1 –1
20	1 1 1 1 –1 1 –1 1 –1 –1 –1 –1 1 1 –1 1 1 –1 –1
24	1 1 1 1 1 –1 1 –1 1 1 –1 –1 1 1 –1 –1 1 –1 1 –1 –1 –1 –1

Table 15.59 A 3^p orthogonal array for 27 observations: An $L_{27}(3^{11})$

						Columns						
1	2	3	4	5	6	7	8	9	10	11	12	13
0	0	0	0	0	0	0	0	0	0	0	0	0
0	0	0	0	1	1	1	1	1	1	1	1	1
0	0	0	0	2	2	2	2	2	2	2	2	2
0	1	1	1	0	0	1	1	1	2	2	0	2
0	1	1	1	1	1	2	2	2	0	0	1	0
0	1	1	1	2	2	0	0	0	1	1	2	1
0	2	2	2	0	0	2	2	2	1	1	0	1
0	2	2	2	1	1	0	0	0	2	2	1	2
0	2	2	2	2	2	1	1	1	0	0	2	0
1	0	1	2	0	1	0	1	2	1	2	2	0
1	0	1	2	1	2	1	2	0	2	0	0	1
1	0	1	2	2	0	2	0	1	0	1	1	2
1	1	2	0	0	1	1	2	0	0	1	2	2
1	1	2	0	1	2	2	0	1	1	2	0	0
1	1	2	0	2	0	0	1	2	2	0	1	1
1	2	0	1	0	1	2	0	1	2	0	2	1
1	2	0	1	1	2	0	1	2	0	1	0	2
1	2	0	1	2	0	1	2	0	1	2	1	0
2	0	2	1	0	2	0	2	1	2	1	1	0
2	0	2	1	1	0	1	0	2	0	2	2	1
2	0	2	1	2	1	2	1	0	1	0	0	2
2	1	0	2	0	2	1	0	2	1	0	1	2
2	1	0	2	1	0	2	1	0	2	1	2	0
2	1	0	2	2	1	0	2	1	0	2	0	1
2	2	1	0	0	2	2	1	0	0	2	1	1
2	2	1	0	1	0	0	2	1	1	0	2	2
2	2	1	0	2	1	1	0	2	2	1	0	0
A	B		C			D	E	F	G			

16 Response Surface Methodology

16.1 Introduction
16.2 First-Order Designs and Analysis
16.3 Second-Order Designs and Analysis
16.4 Properties of Second-Order Designs: CCDs
16.5 A Real Experiment: Flour Production Experiment, Continued
16.6 Box–Behnken Designs
16.7 Using SAS Software
Exercises

16.1 Introduction

Response surface methodology was developed by Box and Wilson in 1951 to aid the improvement of manufacturing processes in the chemical industry. The purpose was to optimize chemical reactions to obtain, for example, high yield and purity at low cost. This was accomplished through the use of sequential experimentation involving factors such as temperature, pressure, duration of reaction, and proportion of reactants. The same methodology can be used to model or optimize any response that is affected by the levels of one or more quantitative factors. The models are generalizations of the polynomial regression models studied in Chapter 8.

The general scenario is as follows. The response is a quantitative continuous variable (e.g., yield, purity, cost), and the mean response is a smooth but unknown function of the levels of p factors (e.g., temperature, pressure), and the levels are real-valued and accurately controllable. The mean response, when plotted as a function of the treatment combinations, is a surface in $p + 1$ dimensions, called the *response surface*. For example, Figure 16.1 shows a response surface for two factors A and B. The levels at which the observations were taken are marked on the plot.

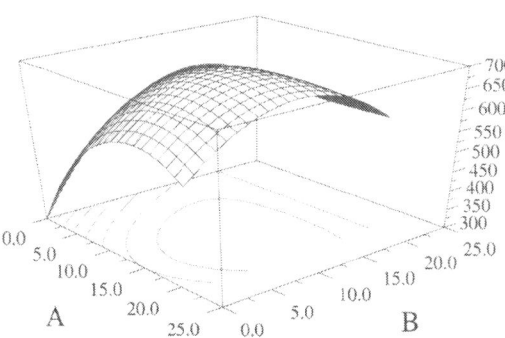

Figure 16.1
Hypothetical response surface for two factors

We will denote the levels of A by values of x_1 or x_A and the levels of B by values of x_2 or x_B. We will denote a treatment combination by $\mathbf{x} = (x_1\ x_2 \ldots x_p)$ or by $\mathbf{x} = (x_A\ x_B \ldots x_P)$ and the mean response at $\mathbf{x}$ by $\eta_\mathbf{x} = E[Y_\mathbf{x}]$. The general response surface model is of the form

$$Y_\mathbf{x} = \eta_\mathbf{x} + \epsilon_\mathbf{x},$$

where $\epsilon_\mathbf{x}$ is a random error variable.

The objective of obtaining a response surface is twofold:

(i) to locate a feasible treatment combination $\mathbf{x}$ for which the mean response is maximized (or minimized, or equal to a specific target value); and

(ii) to estimate the response surface in the vicinity of this good location or region, in order to better understand the "local" effects of the factors on the mean response.

In general, throughout the chapter we will think about maximizing the response, but we show via an example that exactly the same techniques can be used for minimizing a response. The techniques can easily be adapted when the goal is to have the response close to a target value.

One possible approach to achieving the objective involves collecting observations at each location on a grid of treatment combinations spanning the entire experimental region of interest (as illustrated in Figure 16.1). However, the number of observations required by such a comprehensive approach can be very large, and it grows very quickly as the number of factors under study increases. Also, somewhat sophisticated modeling techniques would generally be needed to obtain an adequate fit of a model over the entire region. Instead, it is generally more efficient to conduct a sequence of small "local" experiments with which to search out the location of the peak mean response and then to study its vicinity.

Seeking out the peak is analogous to climbing an unfamiliar mountain under conditions of limited visibility—the mountain is the response surface, and your location on the mountainside is a treatment combination, say $\mathbf{x}_a$. Standing at position $\mathbf{x}_a$, you look around and can see enough to determine in which direction to go to continue a steep ascent. Then you climb in the determined direction as long as it continues to take you up, not looking about lest you lose footing. Then you stop and look around again to determine whether you are at the top of the mountain or in which direction you need to continue your ascent. Of course, when you reach a peak, due to the limited visibility, you may not be sure that you have actually reached the highest peak.

How does one do this experimentally? Looking around with limited visibility is equivalent to analyzing the data of a *local experiment*, consisting of observations on treatment combinations $\mathbf{x}$ close to your current position, $\mathbf{x}_a$. The local terrain is assessed by fitting a local model. Collecting observations in sufficiently close proximity to one another generally allows the local response surface to be well approximated by a rather simple polynomial regression model. When still far from the peak, a first-order model is often adequate. The fitted first-order model is a plane, from which the direction or *path of steepest ascent* is easily determined. Then observations are collected along this path as long as the response continues to increase. When the response stops increasing, another local experiment can be conducted to determine a new path of steepest ascent. This process can be iterated until the first-order model no longer adequately describes the local true surface. For example, close to the peak, the true surface generally exhibits greater curvature, and a first-order regression model becomes inadequate, exhibiting lack of fit. A larger number of observations is needed to fit a higher-order model with which to locate and study the vicinity of the peak. Typically, a second-order model is suitable.

A flow chart describing the steps in this process is shown in Figure 16.2. While a surface is difficult to envisage in more than three dimensions, the process can work well for any number of factors. How well it works depends on several decisions requiring judgment on the part of the experimenter. The first part of this chapter (Section 16.2) looks at the left-hand portion of the flow chart and investigates first-order designs and first-order models, including lack of fit and the path of steepest ascent. Section 16.3 addresses the right-hand portion of the flow chart, which becomes relevant when the vicinity of the peak is reached. Second-order designs and models are described. More details about second-order designs are given in Section 16.4, and an experiment conducted in the flour milling industry is described in Section 16.5. The collection of observations as a block design is discussed in Section 16.6. Section 16.7 describes the use of the SAS software.

16.2 First-Order Designs and Analysis

16.2.1 Models

Before the peak of the response surface is reached, a small local experiment is conducted to assess the local terrain. If the local experiment is not in the vicinity of the peak, then a first-order regression model often provides an adequate approximation to the local response

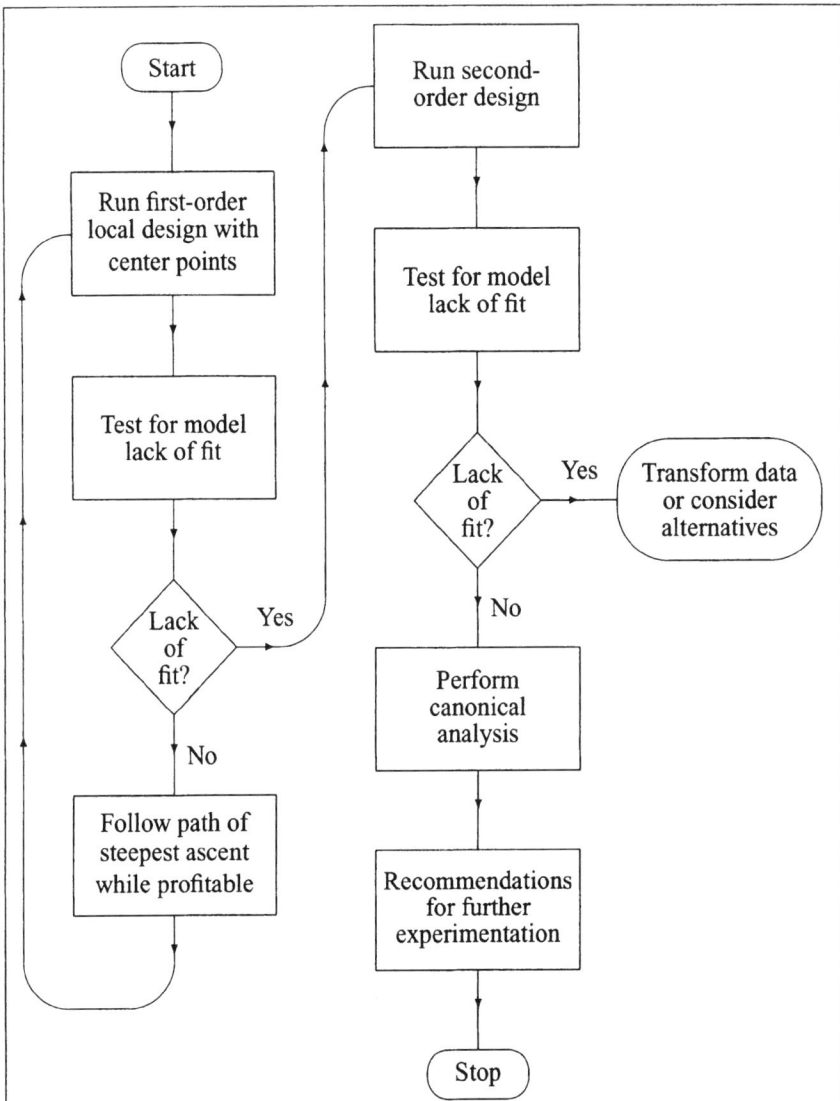

Figure 16.2
Flow chart for response surface methods

surface. For p factors, the standard *first-order model* is a first-order polynomial regression model:

$$Y_{\mathbf{x},t} = \beta_0 + \beta_1 x_1 + \cdots + \beta_p x_p + \epsilon_{\mathbf{x},t}, \tag{16.2.1}$$

where $Y_{\mathbf{x},t}$ denotes the tth observation at treatment combination $\mathbf{x} = (x_1 \ldots x_p)$, and the error variables $\epsilon_{\mathbf{x},t}$ are assumed to be independent with $N(0, \sigma^2)$ distributions. The parameter β_i is a measure of the local *linear effect* of the ith factor ($i = 1, \ldots, p$).

We often code the levels of each factor in each local experiment so that zero represents the *midrange* of the levels of the factor (the average of the highest and lowest levels included in the experiment) and $+1$ and -1 represent the highest and lowest levels of the factor, respectively. For the ith factor, such coded levels z_i are obtained as

$$z_i = (x_i - m_i)/h_i, \qquad (16.2.2)$$

where m_i denotes the midrange of the values of x_i of the ith factor, and h_i denotes the *half-range*—half of the range. So, in terms of coded levels, the center of the design corresponds to the point $\mathbf{z}_0 = (0\,0\ldots 0)$.

The first-order model (16.2.1) can be rewritten in terms of the coded factor levels as follows:

$$Y_{\mathbf{z},t} = \gamma_0 + \gamma_1 z_1 + \cdots + \gamma_p z_p + \epsilon_{\mathbf{z},t}. \qquad (16.2.3)$$

The parameters in models (16.2.1) and (16.2.3) are related, since

$$\beta_0 = \gamma_0 - \sum_i m_i \gamma_i / h_i \quad \text{and} \quad \beta_i = \gamma_i / h_i \ (i = 1, 2, \ldots, p).$$

A design for estimating the parameters of a first-order model is called a *first-order design*. A first-order design should (i) allow for efficient estimation of each linear effect β_i or γ_i, (ii) allow a test for lack of fit of the first-order model, and (iii) be expandable to a good second-order design.

As long as there is no significant model lack of fit but there are significant linear effects, the fitted first-order model can be used to estimate the path of steepest ascent. If there is significant lack of fit of the first-order model, then additional observations may be collected to augment the first-order design so that a second-order polynomial regression model can be fitted to the data.

If there is no significant model lack of fit and also no significant linear effects, then more data may be needed to increase precision of the parameter estimators. Alternatively, the experimenters may need to change the factors under study or increase the range of levels.

16.2.2 Standard First-Order Designs

Throughout the rest of Section 16.2, we consider designs which we refer to as *standard first-order designs*. These designs consist of n_f "factorial points" and n_0 "center points." The *factorial points* consist of the treatment combinations of a 2^p factorial experiment run as a completely randomized design as in Chapter 7, or a 2^{p-s} fractional factorial design of Resolution III or higher. The *center points* are observations collected at the center of the local region under study; that is, at $z_0 = (0\,0\ldots 0)$. These are needed to provide error degrees of freedom and to provide adequate power for a test for model lack of fit.

Standard first-order designs are *orthogonal*, which means that

(i) for each factor, the sum of the coded levels used in the design is zero, ($\sum z_i = 0$, summing over observations), so half of the n_f factorial points in the design have each factor at its high level and the other half have each factor at its low level; and

(ii) for each pair of factors, the sum of cross products of the coded levels in the design is zero ($\sum z_i z_j = 0$, summing over observations).

The factorial portion of the design is chosen to be at least Resolution III so that the linear effects can be estimated. Higher resolution allows model lack of fit due to two-factor interaction effects to be tested. The 2^{p-s} orthogonal fractional factorial designs and the Plackett–Burman designs of Chapter 15 are the most efficient designs for estimation of the linear effects.

16.2.3 Least Squares Estimation

The method of least squares (as shown in optional Section 8.3) is used to fit a first-order model to the data. Denote the fitted model by

$$\hat{y}_\mathbf{x} = \hat{\beta}_0 + \hat{\beta}_1 x_1 + \cdots + \hat{\beta}_p x_p \tag{16.2.4}$$

or, in coded form,

$$\hat{y}_\mathbf{z} = \hat{\gamma}_0 + \hat{\gamma}_1 z_1 + \cdots + \hat{\gamma}_p z_p . \tag{16.2.5}$$

If a standard first-order design is used, with the extreme levels of each factor coded as $+1$ and -1, then the least squares estimator $\hat{\gamma}_i$ of the linear effect γ_i of the ith factor is

$$\hat{\gamma}_i = (\overline{Y}_{z_i(+1)} - \overline{Y}_{z_i(-1)})/2 , \tag{16.2.6}$$

where $\overline{Y}_{z_i(+1)}$ and $\overline{Y}_{z_i(-1)}$ denote the averages of the observations at the high and the low level of the ith factor, respectively. The parameter $2\gamma_i$ denotes the change in the mean response between the high and low levels of the ith factor. This is the same as the main-effect contrast for the ith factor. The least squares estimator of β_i in the uncoded model is $\hat{\beta}_i = \hat{\gamma}_i/h_i$, where h_i is the half-range of the uncoded levels of the ith factor.

Example 16.2.1 Paint experiment

Several experiments were run in Germany by S. Eibl, U. Kess, and F. Pukelsheim (1992) on the thickness of a paint coating. Prior to the experiments, the paint thickness achieved was around 2 mm, much higher than the target 0.8 mm. In order to study how to decrease the mean thickness, they selected the following six factors, each at two levels:

A: belt speed   B: tube width   C: pump pressure
D: paint viscosity   E: tube height   F: heating temperature

The first experiment that was conducted consisted of a 2_{III}^{6-3} fractional factorial design with defining relation generated by BCD, ADE, and ABF; that is,

$$I = BCD = ADE = ABCE = ABF = ACDF = BDEF = CEF .$$

The experimenters decided to ignore all interactions for this first experiment. Since they wanted to monitor the variation of the paint thickness, they took four observations on each of the 8 treatment combinations in the fraction. The data are reproduced in Table 16.1, with factor levels coded as -1 and 1.

16.2 First-Order Designs and Analysis

Table 16.1 Paint thickness for the paint experiment

z_A	z_B	z_C	z_D	z_E	z_F	y_{z1}	y_{z2}	y_{z3}	y_{z4}	$\bar{y}_{z.}$	s_z^2
1	−1	1	−1	−1	−1	1.09	1.12	0.83	0.88	0.9800	0.021400
−1	−1	1	−1	1	1	1.62	1.49	1.48	1.59	1.5450	0.004967
1	1	−1	−1	−1	1	0.88	1.29	1.04	1.31	1.1300	0.042867
−1	1	−1	−1	1	−1	1.83	1.65	1.71	1.76	1.7375	0.005825
−1	−1	−1	1	−1	1	1.46	1.51	1.59	1.40	1.4900	0.006467
1	−1	−1	1	1	−1	0.74	0.98	0.79	0.83	0.8350	0.010700
−1	1	1	1	−1	−1	2.05	2.17	2.36	2.12	2.1750	0.017633
1	1	1	1	1	1	1.51	1.46	1.42	1.40	1.4475	0.002358

Source: Eibl, S., Kess, U., and Pukelsheim, F. (1992). Copyright © 1997 American Society for Quality. Reprinted with Permission.

Using $z_A, \ldots, z_F$ rather than $z_1, \ldots, z_6$ to denote the factor levels, the fitted first-order model for the mean response is

$$\hat{y}_z = \hat{\gamma}_0 + \hat{\gamma}_A z_A + \cdots + \hat{\gamma}_F z_F$$
$$= 1.42 - 0.32 z_A + 0.21 z_B + 0.12 z_C + 0.07 z_D - 0.03 z_E - 0.01 z_F,$$

where, for example, the parameter estimate $\hat{\gamma}_D$ is calculated as

$$\hat{\gamma}_D = (\bar{y}_{z_D(+1)} - \bar{y}_{z_D(-1)})/2 = (1.493125 - 1.348125)/2 = 0.0725 \approx 0.07,$$

where $\bar{y}_{z_D(+1)}$ is the average of the observations with D at its high level and $\bar{y}_{z_D(-1)}$ is the average of the observations with D at its low level. □

16.2.4 Checking Model Assumptions

Before progressing with the analysis of the fitted model, the model assumptions should be checked. We shall discuss tests for model lack of fit in Section 16.2.6.

If there is no model lack of fit, then the error assumptions may be checked using residual plots. If the observations were collected sequentially in a known run order, then the residuals are plotted against run order to check for independence of observations. Residuals are plotted against predicted values to assess equality of error variances. Normality is checked by plotting residuals versus normal scores.

16.2.5 Analysis of Variance

Suppose that a standard first-order design has been used, with the extreme levels of each factor coded as −1 and +1. Under the first-order model, it follows from equation (16.2.6) that

$$\text{Var}(\hat{\gamma}_i) = \left(\frac{\sigma^2}{n_f/2} + \frac{\sigma^2}{n_f/2} \right) / 4 = \sigma^2/n_f,$$

for any $i = A, B, C, \ldots$. The sum of squares for testing that the main-effect contrast γ_A is zero (that is, $H_0^A : \gamma_A = 0$ against $H_A^A : \gamma_A \neq 0$) is

$$ssA = \hat{\gamma}_A^2/(1/n_f) = n_f \hat{\gamma}_A^2,$$

Table 16.2 Analysis of variance for the first-order model

Source of Variation	Degrees of Freedom	Sum of Squares	Mean Square	Ratio	Expected Mean Square
A	1	ssA	msA	msA/msE	$\sigma^2 + n_f \gamma_A^2$
B	1	ssB	msB	msB/msE	$\sigma^2 + n_f \gamma_B^2$
⋮	⋮	⋮	⋮	⋮	⋮
Error	$n - p - 1$	ssE	msE		σ^2
Total	$n - 1$	sstot			
Computational Formulae					
$ssi = n_f \hat{\gamma}_i^2 = n_f(\bar{y}_{z_i(+1)} - \bar{y}_{z_i(-1)})^2/4$, for $i = A, B, C, \ldots$					
ssE by subtraction			$sstot = \sum_z \sum_t y_{zt}^2 - n\bar{y}_{..}^2$		

and since there is only one degree of freedom for the A contrast, $msA = ssA$. For the first-order model and a standard first-order design, we have

$$E[MSA] = n_f E[\hat{\gamma}_A^2] = n_f \text{Var}(\hat{\gamma}_A) + n_f (E[\hat{\gamma}_A])^2 = \sigma^2 + n_f \gamma_A^2.$$

It can also be shown that $msE = ssE/(n - p - 1)$ is an unbiased estimate of σ^2, where ssE is obtained by subtraction in the analysis of variance table. Consequently, the decision rule for testing H_0^A against H_A^A is

reject H_0^A if $msA/msE > F_{1, n-p-1, \alpha}$.

Similar formulae hold for each main effect. The analysis of variance for the first-order model and a standard first-order design for p factors are shown in outline in Table 16.2.

Example 16.2.2 Paint experiment, continued

The paint experiment was introduced in Example 16.2.1, and the data were given in Table 16.1. The purpose of the experiment was to study the effects of six factors on paint thickness. The experimental design consisted of four observations on each of the treatment combinations of a 2_{III}^{6-3} design, which is an orthogonal factorial design with $n_f = 32$ factorial points and no center points. The corresponding analysis of variance is shown in Table 16.3. The linear effect of each of factors A, B, C, and D is significantly different from zero, but factors E and F appear to have little effect on the response. □

16.2.6 Tests for Lack of Fit

A first-order design allows the experimenter to determine when the first-order model is no longer adequate, provided that there are more design points than first-order model parameters, and the design includes replication at one or more points. There is said to be model *lack of fit* when the model does not adequately represent the mean response as a function of the factor levels. Lack of fit of the first-order model occurs when the local response surface is no longer a plane.

16.2 First-Order Designs and Analysis

Table 16.3 Analysis of variance for the paint experiment

Source of Variation	Degrees of Freedom	Sum of Squares	Mean Square	Ratio	p-value	Expected Mean Square
A	1	3.2640	3.2640	242.07	0.0001	$\sigma^2 + 32\gamma_A^2$
B	1	1.3448	1.3448	99.73	0.0001	$\sigma^2 + 32\gamma_B^2$
C	1	0.4560	0.4560	33.82	0.0001	$\sigma^2 + 32\gamma_C^2$
D	1	0.1540	0.1540	11.42	0.0024	$\sigma^2 + 32\gamma_D^2$
E	1	0.0221	0.0221	1.64	0.2127	$\sigma^2 + 32\gamma_E^2$
F	1	0.0066	0.0066	0.49	0.4902	$\sigma^2 + 32\gamma_F^2$
Error	25	0.3371	0.0135			σ^2
Total	31	5.5846				

Generic test Let n_d denote the number of *distinct* coded treatment combinations **z**. For each treatment combination for which there is replication, the sample variance s_z^2 of the n_z observations at that treatment combination provides an unbiased estimate of the error variance σ^2. These sample variances can be pooled together to obtain a *sum of squares for pure error*

$$ssPE = \sum_{\mathbf{z}} (n_{\mathbf{z}} - 1)s_{\mathbf{z}}^2 \tag{16.2.7}$$

with $n - n_d$ degrees of freedom, giving a *mean square for pure error*

$$msPE = ssPE/(n - n_d).$$

The error sum of squares ssE is obtained from fitting the first-order model (Table 16.2), and the difference

$$ssLOF = ssE - ssPE \tag{16.2.8}$$

is called the *sum of squares for lack of fit*. The corresponding mean square is

$$msLOF = ssLOF/(n - p - 1).$$

Then the ratio

$$msLOF/msPE$$

is used to test the null hypothesis of no model lack of fit. The null hypothesis is rejected at level α if this ratio exceeds $F_{n_d-p-1,n-n_d,\alpha}$. This lack-of-fit test is summarized in Table 16.4.

Example 16.2.3 Paint experiment, continued

The paint experiment was introduced in Example 16.2.1. The analysis of variance for the first-order model is shown in Table 16.3, giving $ssE = 0.3371$ with 25 degrees of freedom. There were $n_z = 4$ observations at each of eight factorial points, and the corresponding eight sample variances, each with three degrees of freedom, were given in Table 16.1, page 553.

Table 16.4 Generic lack-of-fit test for the first-order model

Source of Variation	Degrees of Freedom	Sum of Squares	Mean Square	Ratio	Expected Mean Square
Lack of fit	$n_d - p - 1$	ssLOF	msLOF	msLOF/msPE	$\sigma^2 + \theta^2$
Pure Error	$n - n_d$	ssPE	msPE		σ^2
Error	$n - p - 1$	ssE			
Computational Formulae					
ssE from Table 16.2, $ssPE = \sum_z (n_z - 1)s_z^2$, ssLOF by subtraction, n_d distinct design points, n observations total, θ depends on the nature of model lack of fit					

Table 16.5 Generic lack-of-fit test for the paint experiment

Source of Variation	Degrees of Freedom	Sum of Squares	Mean Square	Ratio	p-value
Lack of fit	1	0.0004	0.0004	0.03	0.8594
Pure error	24	0.3367	0.0140		
Error	25	0.3371	0.0135		

These eight sample variances can be pooled together to obtain

$$ssPE = \sum_z (4 - 1)s_z^2 = 0.3367$$

based on $n - n_d = 32 - 8 = 24$ degrees of freedom. The sum of squares for lack of fit is

$$ssLOF = ssE - ssPE = 0.3371 - 0.3367 = 0.0004,$$

and the test is summarized in Table 16.5. Since the *p*-value is large, there is no evidence of lack of fit of the first-order model. □

Test for second-order lack of fit If the generic test indicates lack of fit of the first-order model, this provides no insight into why the model is not fitting well. To understand the nature of the lack of fit, it can be helpful to consider what the mean square for lack of fit measures in terms of higher-order models. If the first-order model is inadequate, the next possibility is that a second-order model would provide an adequate approximation to the local response surface. If so, then lack of fit of the first-order model is attributable to the presence of either two-factor interactions or to quadratic effects or to both.

If the only lack of fit is due to two-factor interaction effects, this corresponds to a twisting of the response surface. Such lack of fit can be tested if the first-order design allows estimation of two-factor interactions in addition to providing error degrees of freedom. In the paint experiment, for example, it is possible to estimate the *AC* interaction effect, in addition to the six main effects, provided that all other interaction effects are known to be negligible.

If the center of the experimental design is near the peak of the response surface, then one would expect quadratic effects, or curvature, to be present and a higher mean response near the design center than at the factorial points. Multiple center points $\mathbf{z}_0 = (0, \ldots, 0)$ are usually included in a first-order design, because comparison of the mean response at the center of the design region with the mean response at the factorial points provides an effective test for lack of fit due to quadratic effects.

So, to assess second-order lack of fit we fit a second-order polynomial regression model under the alternative hypothesis. With respect to the coded factor levels, the standard *second-order model* for p factors is

$$Y_{z,t} = \gamma_0 + \sum_i \gamma_i z_i + \sum_i \gamma_{ii} z_i^2 + \sum_{i<j} \gamma_{ij} z_i z_j + \epsilon_{z,t},$$

where the parameter γ_i represents the linear effect of the ith factor, γ_{ii} represents the quadratic effect of the ith factor, and γ_{ij} represents the cross product effect between the ith and jth factors.

If the factorial portion of the standard first-order design is either a complete factorial design or a fraction of resolution V or higher, then all two-factor interaction parameters γ_{ij} in the second-order model are estimable (assuming higher-order interactions to be negligible). For testing for second-order lack of fit, we add the sums of squares for these two-factor interactions to obtain a pooled interaction sum of squares, ssI. If the factorial portion of the design is a fraction of resolution less than V, then not all two-factor interactions are estimable, and only the sums of squares of those two-factor interactions which are not aliased with main effects may be pooled—one sum of squares from each alias set.

The quadratic-effect parameters are not individually estimable from a standard first-order design. They are aliased with one another, and only their sum can be estimated. It can be shown that

$$E[\overline{Y}_f - \overline{Y}_0] = \sum_{i=1}^{p} \gamma_{ii},$$

Table 16.6 Lack-of-fit test for the first-order model, given the data of a standard first-order design, with p factors $A, B, \ldots$ and m alias sets for interaction effects clear of main effects

Source of Variation	Degrees of Freedom	Sum of Squares	Mean Square	Ratio	Expected Mean Square
Interaction	m	$ssI = ssAB + \cdots$	msI	$\frac{msI}{msPE}$	$\sigma^2 + \frac{n_f}{m}\theta_1$
Quadratic	1	ssQ	msQ	$\frac{msQ}{msPE}$	$\sigma^2 + \frac{n_0 n_f}{n}\theta_2^2$
Higher-order	$n_d - p - m - 2$	ssH			
Pure Error	$n - n_d$	$ssPE$	$msPE$		σ^2
Error	$n - p - 1$	ssE			

Computational Formulae

$ssAB = n_f \hat{\gamma}_{AB}^2 = n_f(\overline{y}_{z_A z_B(+1)} - \overline{y}_{z_A z_B(-1)})^2/4$ ssE from Table 16.2
$ssQ = (n_0 n_f/n)^2(\overline{y}_f - \overline{y}_0)^2$ $\theta_1 = \gamma_{AB}^2 + \cdots$
$ssPE = \sum_z (n_z - 1)s_z^2$ $\theta_2 = \gamma_{AA} + \gamma_{BB} + \cdots$
$ssH = (ssE - ssPE) - ssI - ssQ$

with

$$\text{Var}(\overline{Y}_f - \overline{Y}_0) = \left(\frac{1}{n_0} + \frac{1}{n_f}\right)\sigma^2 = \frac{n}{n_f n_0}\sigma^2,$$

where $\overline{Y}_f$ and $\overline{Y}_0$ denote the average of the n_f factorial points and the average of the n_0 center points, respectively. It follows that the corresponding sum of squares for testing whether or not the sum of the quadratic parameters is zero is

$$ssQ = \frac{n_f n_0}{n}(\overline{Y}_f - \overline{Y}_0)^2,$$

with one degree of freedom. The expected mean square is

$$E[MSQ] = \sigma^2 + \frac{n_f n_0}{n}\left(\sum_{i=1}^{p}\gamma_{ii}\right)^2.$$

In the generic test for lack of fit of the first-order model, ssI and ssQ are part of $ssLOF$. Thus, we can write

$$ssLOF = ssI + ssQ + ssH,$$

where ssH is the sum of squares for lack of fit due to a higher-order model. Then lack of fit due specifically to interaction terms and quadratic terms can be investigated separately. The tests are summarized in Table 16.6 for a standard first-order design.

For all tests for lack of fit, an adequate number of pure error degrees of freedom are needed for the test power to be reasonably high. Since $\text{Var}(\overline{Y}_f - \overline{Y}_0) > \sigma^2/n_0$, the test for lack of fit due to quadratic effects will have low power if there are few center points. Typically, 3–6 center points would be used.

Example 16.2.4 Acid copper pattern plating experiment

G. K. K. Poon (1995) conducted a sequence of fractional factorial and response surface experiments each involving as many as seven factors to minimize the coating thickness variation of an acid copper-plating process. In the final experiments, conducted in the vicinity of minimum thickness variation, response surface methods were utilized to study the effects of anode-cathode separation (factor A) and cathodic current density (factor B) on the standard deviation of coating thickness. One experiment used the factorial points of a single replicate 2^2 design, augmented by two center points. The response was the standard deviation (in μm) of copper-plating thickness. The coded and uncoded factor levels, together with the resulting data, are given in Table 16.7.

The midrange of levels of factor A is $(11.5 + 9.5)/2 = 10.5$, and the half-range is $(11.5 - 9.5)/2.0 = 1.0$. So the coded levels are given by

$$z_A = x_A - 10.5.$$

The midrange and half-range of the factor B levels are $(41+31)/2 = 36$ and $(41-31)/2 = 5$, respectively, so the coded levels of factor B are

$$z_B = (x_B - 36)/5.$$

16.2 First-Order Designs and Analysis

Table 16.7 Data for the acid copper pattern plating experiment

Anode–cathode Separation (in.)		Current Density (asf)		Standard Deviation (μm)
Coded	Uncoded	Coded	Uncoded	
−1	9.5	−1	31	5.60
−1	9.5	1	41	6.45
1	11.5	−1	31	4.84
1	11.5	1	41	5.19
0	10.5	0	36	4.32
0	10.5	0	36	4.25

Source: Poon, G. K. K. (1995). Reprinted with permission.

Table 16.8 Analysis of variance and lack-of-fit test for the acid copper pattern plating experiment

Source of Variation	Degrees of Freedom	Sum of Squares	Mean Square	Ratio	p-value	Expected Mean Square
A	1	1.0201	1.0201	1.46	0.3137	$\sigma^2 + n_f \gamma_A^2$
B	1	0.3600	0.3600	0.51	0.5250	$\sigma^2 + n_f \gamma_B^2$
Error	3	2.0986	0.6995			
Total	5	3.4787				
Interaction AB	1	0.0625	0.0625	25.51	0.1244	$\sigma^2 + n_f \gamma_{AB}^2$
Quadratic	1	2.0336	2.0336	830.05	0.0221	$\sigma^2 + \frac{n_0 n_f}{n} \theta^2$
Pure Error	1	0.0025	0.0025			
Error	3	2.0986	0.6995			
where $\theta = \gamma_{AA} + \gamma_{BB}$						

Table 16.8 shows the analysis of variance, including tests for lack of fit, due to a second-order model. The analyses are identical for coded and uncoded factor levels. There are significant quadratic effects—an indication that quadratic terms for either or both of factors A and B are needed to adequately model the response surface. The first-order design is inadequate, then, because not all parameters in the second-order model are estimable. The solution is to collect some additional observations, as will be illustrated in Example 16.3.2. □

16.2.7 Path of Steepest Ascent

If there are significant linear effects and there is no significant lack of fit of the first-order model, then the *path of steepest ascent* may be followed to climb towards the maximum of the response surface.

Given the fitted first-order regression model (16.2.5), the path of steepest ascent from the current position $\mathbf{z}_a$ is determined as follows. If $\hat{\gamma}_i$ is positive, increase z_i to increase predicted mean response $\hat{y}_\mathbf{z}$. If $\hat{\gamma}_i$ is negative, decrease z_i to increase $\hat{y}_\mathbf{z}$. To follow the path of steepest ascent up the fitted response surface, change each z_i in proportion to the magnitude of $\hat{\gamma}_i$.

Example 16.2.5   Paint experiment, continued

The paint experiment was introduced in Example 16.2.3, page 552. The experimenters conducted an experiment to study how to decrease the thickness of a paint coating to the target 0.8 mm. Four observations were taken at each treatment combination of a 2_{III}^{6-3} design. The resulting data were shown in Table 16.1, page 553.

The target thickness is approximately achieved at the experimental design point $\mathbf{z} =$ (1, −1, −1, 1, 1, −1), so perhaps no further analysis or experimentation is needed. Nevertheless, we will use these data to illustrate how to find the minimum of a response surface.

Since a minimum is required, we need to identify the path of steepest *descent*. The fitted first-order model is obtained from Example 16.2.3 as

$$\hat{y}_z = \hat{\gamma}_0 + \hat{\gamma}_A z_A + \cdots + \hat{\gamma}_F z_F$$
$$= 1.42 - 0.32 z_A + 0.21 z_B + 0.12 z_C + 0.07 z_D - 0.03 z_E - 0.01 z_F.$$

The analysis of variance conducted in Example 16.2.5 suggests that only factors A, B, C, and D significantly affect the response. So, these four factors should be adjusted in an attempt to reduce paint thickness.

Based on the signs of the parameter estimates in the fitted model, we ought to be able to effect a reduction in mean response if we increase the level of factor A and decrease the level of any of factors B, C, and D. To follow the path of steepest descent, we change the levels of these factors each in proportion to the magnitude of its corresponding parameter estimate, $\hat{\gamma}_i$. So, if we increase z_A by $0.32u$ units for some real number u, then we decrease z_B by $0.21u$ units, decrease z_C by $0.12u$ units, and decrease z_D by $0.07u$ units.

Observations along the path of steepest descent moving away from the center of the current design, $\mathbf{z}_0 = (0, 0, 0, 0, 0, 0)$, consist of treatment combinations $(0.32u, -0.21u, -0.12u, -0.07u, 0, 0)$ corresponding to increasing values of u, such as $u = 3, 6, 9, \ldots$. The suggested values of u start at $u = 3$. This value is large enough to move the level of factor A near to the edge of the region of the current local experiment, but other values of u could also have been chosen. Observations may then be collected along this path setting u equal to, say, multiples of 3 until the target thickness is achieved, or until the response stops decreasing before reaching the target level. In the latter case, at the point of lowest response along the path another first-order design could be run to determine a new path of steepest descent. □

In the previous example, the effects of factors E and F were not found to be significantly different from zero, so their levels were not changed in following the estimated path of steepest descent. There are a variety of reasons why the effect of a factor may be negligible. The obvious reason is that response is independent of the factor. However, it could also be that the levels used for the factor may be near the optimal value, so the response surface may be relatively flat with respect to small changes in the level of that factor. Alternatively, the levels of the factor may simply be too close together to give rise to a detectable change in the mean response. In subsequent experiments, the levels of such factors can be chosen farther apart to guard against the last scenario.

16.3 Second-Order Designs and Analysis

16.3.1 Models and Designs

Second-order designs and analysis are used when the test for lack of fit of the first-order model indicates that the vicinity of the maximum (or minimum) of the response surface has been reached and a second-order model should be fitted. For p factors, the standard second-order model is

$$Y_{\mathbf{x},t} = \beta_0 + \sum_{i=1}^{p} \beta_i x_i + \sum_{i=1}^{p} \beta_{ii} x_i^2 + \sum_{i<j} \beta_{ij} x_i x_j + \epsilon_{\mathbf{x},t}, \tag{16.3.9}$$

where $Y_{\mathbf{x},t}$ denotes the tth response observed for treatment combination $\mathbf{x} = (x_1 \, x_2 \ldots x_p)$. The random-error variables $\epsilon_{\mathbf{x},t}$ are assumed to be independent with $N(0, \sigma^2)$ distributions. The parameter β_i represents the linear effect of the ith factor. The parameter β_{ii} represents the quadratic effect of the ith factor, and β_{ij} represents the cross product effect, or interaction effect, between the ith and jth factors.

With respect to the coded factor levels $z_i = (x_i - m_i)/h_i$, the second-order model is

$$Y_{\mathbf{z},t} = \gamma_0 + \sum_{i=1}^{p} \gamma_i z_i + \sum_{i=1}^{p} \gamma_{ii} z_i^2 + \sum_{i<j} \gamma_{ij} z_i z_j + \epsilon_{\mathbf{z},t}. \tag{16.3.10}$$

Experimental designs used to fit a second-order model are referred to as *second-order designs*. A second-order design should (i) allow for efficient estimation of the response surface, in the sense of having $\text{Var}(\widehat{Y}_{\mathbf{z}})$ be small; (ii) allow a test for lack of fit of the second-order model; and (iii) allow for efficient estimation of all model parameters. Second-order designs must have at least $(p+1)(p+2)/2$ distinct design points; otherwise, not all of the $(p+1)(p+2)/2$ parameters in the second-order model can be estimated. We will consider only such designs in this chapter. Observations at even more points are needed, plus some replication, in order to be able to conduct a generic test for model lack of fit. Other properties of second-order designs that are sometimes desirable include rotatability, orthogonal blocking, and orthogonality—these will be discussed in Sections 16.4.1–16.4.2.

The method of least squares is used to fit the second-order model to the data. This method is exactly as discussed in optional Section 8.3, with each second-order term z_i^2 or $z_i z_j$ being treated as a single regressor. In terms of the uncoded and coded factor levels, the fitted models are, respectively,

$$\hat{y}_{\mathbf{x}} = \hat{\beta}_0 + \sum_i \hat{\beta}_i x_i + \sum_i \hat{\beta}_{ii} x_i^2 + \sum_{i<j} \hat{\beta}_{ij} x_i x_j \tag{16.3.11}$$

and

$$\hat{y}_{\mathbf{z}} = \hat{\gamma}_0 + \sum_i \hat{\gamma}_i z_i + \sum_i \hat{\gamma}_{ii} z_i^2 + \sum_{i<j} \hat{\gamma}_{ij} z_i z_j, \tag{16.3.12}$$

where hats on the parameters denote the least squares estimates. Although it is possible to obtain explicit formulae for the least squares estimates for any specific design, the formulae for the quadratic parameter estimates $\hat{\gamma}_{ii}$ are complicated. Consequently, we rely on statistical computer software to obtain the least squares estimates (see Section 16.7 for the use of the SAS software).

As long as there is no significant lack of fit, the fitted second-order model can be used to study the local response surface. Generally, there will be a unique treatment combination $\mathbf{x}_s = (x_{s1}\, x_{s2} \ldots x_{sp})$, called the *stationary point*, at which $\hat{y}$ is maximized, minimized, or is at a saddle point. The surface close to a *saddle point* is reminiscent of a horse saddle—rising up from front to back but sloping down from side to side. A saddle point yields neither a maximum nor a minimum for the fitted model. Instead, these will be found at the edge of the local design region.

If there is significant lack of fit of the second-order model, a higher-order model could be used, or a more local experiment could be run.

16.3.2 Central Composite Designs

Central composite designs were first described by Box and Wilson in 1951, and they are nowadays the most popular second-order designs. Each design consists of a standard first-order design with n_f orthogonal factorial points and n_0 center points, augmented by n_a "axial points."

We follow the convention of coding the factor levels so the factorial points have coded levels ± 1 for each factor. However, it should be noted that some software packages, including SAS, will recode the levels in a central composite design before doing the analysis. Under our convention, *axial points* are points located at a specified distance α from the design center in each direction on each axis defined by the coded factor levels. On the z_i-axis, for example, two axial points are obtained by setting $z_i = \pm \alpha$, with $z_j = 0$ for all $j \neq i$. Thus, if there are p factors, there are $2p$ distinct axial points. Axial points are also commonly referred to as *star points*. Figure 16.3 shows central composite designs for $p = 2$ and $p = 3$ factors.

A central composite design is easily built up from a standard first-order design by the addition of axial points, and possibly some extra factorial and center points. If the factorial portion of the design is a complete factorial or a fractional factorial of resolution V or

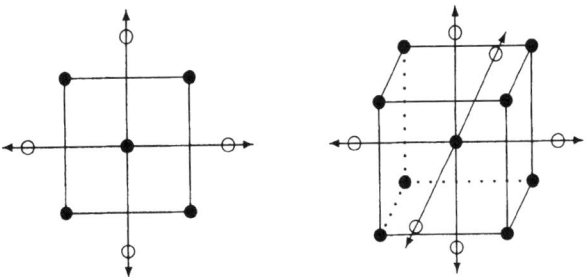

Figure 16.3 Central composite designs for $p = 2$ and $p = 3$ factors

16.3 Second-Order Designs and Analysis

more, all model parameters are estimable. Otherwise, some aliasing will occur, and some terms will need to be omitted from the second-order model. A design should include enough replication, often at the center points, to allow for a test for model lack of fit. The axial points are located at a distance α from the center of the design, where the choice of α depends on the properties required of the design. A popular choice is $\alpha = (n_f)^{1/4}$ (see Section 16.4.1).

Example 16.3.1 Acid copper pattern plating experiment, continued

In Example 16.2.4, page 558, a standard first-order design was used to study the effects of anode–cathode separation (factor A) and cathodic current density (factor B) on the standard deviation of a copper-plating thickness. The first-order design involved the $n_f = 4$ factorial points of a single-replicate 2^2 design, augmented by $n_0 = 2$ center points. There was significant lack of fit of the first-order model, so additional observations needed to be taken in order to fit a second-order model. The experimenters augmented the first-order design with four axial points, using $\alpha = (n_f)^{1/4} = \sqrt{2}$, giving the central composite design and data shown in Table 16.9.

The second-order model is fitted by a computer regression package. In terms of the uncoded factor levels, the fitted model is given by

$$\hat{y}_{\mathbf{x}} = \hat{\beta}_0 + \hat{\beta}_A x_A + \hat{\beta}_B x_B + \hat{\beta}_{AA} x_A^2 + \hat{\beta}_{BB} x_B^2 + \hat{\beta}_{AB} x_A x_B$$
$$= 79.6898 - 7.8187 x_A - 1.8126 x_B$$
$$+ 0.3926 x_A^2 + 0.0294 x_B^2 - 0.0250 x_A x_B ,$$

and, in terms of the coded factor levels, $z_A = (x_A - 10.5)$, $z_B = (x_B - 36)/5$, the fitted model is

$$\hat{y}_{\mathbf{z}} = \hat{\gamma}_0 + \hat{\gamma}_A z_A + \hat{\gamma}_B z_B + \hat{\gamma}_{AA} z_A^2 + \hat{\gamma}_{BB} z_B^2 + \hat{\gamma}_{AB} z_A z_B$$
$$= 4.2939 - 0.4741 z_A + 0.2134 z_B$$
$$+ 0.3926 z_A^2 + 0.7353 z_B^2 - 0.1250 z_A z_B .$$

Table 16.9 Data for the acid copper pattern plating experiment—Central composite design

Anode–cathode Separation (in.)		Current Density (asf)		Standard Deviation (μm)
Coded	Uncoded	Coded	Uncoded	
-1	9.5	-1	31	5.60
-1	9.5	1	41	6.45
1	11.5	-1	31	4.84
1	11.5	1	41	5.19
0	10.5	0	36	4.32
0	10.5	0	36	4.25
$-\sqrt{2}$	9.0	0	36	5.76
$\sqrt{2}$	12.0	0	36	4.42
0	10.5	$-\sqrt{2}$	29	5.46
0	10.5	$\sqrt{2}$	43	5.81

Source: Poon, G. K. K. (1995). Reprinted with permission.

Figure 16.4
Response surface contour plot and response surface plot of fitted second-order model for the acid copper pattern plating experiment

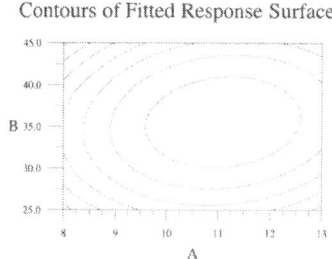
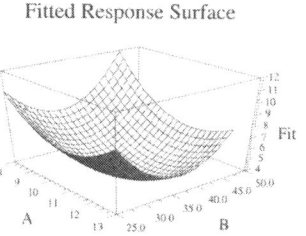

Figure 16.4 shows both a contour plot and a surface plot of the fitted model for uncoded factor levels. The stationary point is in the center of the ellipses. Clearly, the stationary point provides a minimum. The exact location of the stationary point will be determined in the next section. □

16.3.3 Generic Test for Lack of Fit of the Second-Order Model

If the second-order design includes n_d distinct treatment combinations, with n_d larger than $(p+2)(p+1)/2$, and replication at one or more of these, then a generic test for lack of fit of the second-order model can be conducted, just as for the first-order model (Section 16.2.6). The sum of squares for pure error, $ssPE$, and the sum of squares for lack of fit, $ssLOF$, are calculated as in (16.2.7) and (16.2.8). The error sum of squares ssE and the error degrees of freedom are obtained from the analysis of variance table of the second-order model. The test proceeds exactly as in Table 16.4 except that the error degrees of freedom are $n-[(p+2)(p+1)/2]$ and the degrees of freedom for lack of fit are then $n_d-[(p+2)(p+1)/2]$. The test will be illustrated for the acid copper-plating experiment in Example 16.3.2 in the next subsection.

16.3.4 Analysis of Variance for a Second-Order Model

Table 16.10 shows an outline analysis of variance table for a second-order design and second-order model, assuming that all parameters are estimable. The degrees of freedom associated

Table 16.10 Analysis of variance for a second-order design and model

Source of Variation	Degrees of Freedom	Sum of Squares (Type I)	Mean Square (Type I)	Ratio	Expected Mean Square
L	p	ssL	msL	$\frac{msL}{msE}$	$\sigma^2 + \sum_i a_i \beta_i^2$
$Q\|L$	p	$ss(Q\|L)$	$ms(Q\|L)$	$\frac{ms(Q\|L)}{msE}$	$\sigma^2 + \sum_i a_{ii} \beta_{ii}^2$
$I\|L, Q$	$\frac{1}{2}p(p-1)$	$ss(I\|L, Q)$	$ms(I\|L, Q)$	$\frac{ms(I\|L,Q)}{msE}$	$\sigma^2 + \sum_{i<j} a_{ij} \beta_{ij}^2$
Error	df	ssE	msE		σ^2
Total	$n-1$	$sstot$			
Formula:	$df = n - \frac{1}{2}(p+2)(p+1)$				

with the linear effects have been added (pooled) together, as have those of the quadratic effects and those of the interaction (cross product) effects. Sequential, or Type I, sums of squares are listed for each of these pooled sources of variation. These include the sum of squares for all linear terms, $ss(L)$; the sum of squares for adding all quadratic terms to the model, given that all linear terms are already included, $ss(Q|L)$; and the sum of squares for adding all interaction terms to the model, given that all linear and quadratic terms are already in the model, $ss(I|L, Q)$. Using these sequential sums of squares, the analysis of variance is the same whether factor levels are coded or not. The coefficients a_i, a_{ii}, and a_{ij} listed in the expected mean squares are positive and depend on the design and the model. If coded factor levels are used, the expected mean squares would involve the parameters γ instead of the parameters β, but would have the same form.

If a central composite design is used and factor levels are coded in the usual way, the linear, quadratic and interaction sums of squares are independent of one another, so the corresponding sums of squares are the same, no matter in which order the terms are fitted. Also, the individual linear and interaction (cross product) parameters are estimated independently of one another and of the quadratic effects. The quadratic parameters are estimated independently of each other only if α and the number of center points n_0 are chosen to satisfy certain restrictions (see Section 16.4.2).

Example 16.3.2 Acid copper pattern plating experiment, continued

The data for the central composite design of the acid copper pattern plating experiment were shown in Table 16.9, page 563. The analysis of variance for these data is given in Table 16.11. The same analysis is obtained whether the model is fitted using coded or uncoded factor levels. The table shows the decomposition of the linear sum of squares with respect to the individual linear effects. Each of the two quadratic effects is shown adjusted for the other quadratic effect. If we test each hypothesis at individual level 0.01, the linear effect of factor A is significantly different from zero, as is the adjusted quadratic effect of each factor. Consequently, the model should include these three terms. We would also include the linear effect of B, since the higher-order (quadratic) term is included. The AB-interaction effect, or cross product effect, is not significantly different from zero.

Table 16.11 Analysis of variance for the acid copper pattern plating experiment

Source of Variation	Degrees of Freedom	Sum of Squares	Mean Square	Ratio	p-value
Linear	2	2.2713			
A_L	1	1.9107	1.9107	52.97	0.0019
B_L	1	0.3606	0.3606	10.00	0.0341
Quadratic	2	2.4117			
$A_Q|B_Q$	1	0.8509	0.8509	23.59	0.0083
$B_Q|A_Q$	1	2.3407	2.3407	64.89	0.0013
Interaction	1	0.0625	0.0625	1.73	0.2584
Error	4	0.1443	0.0361		
Total	9	4.8898			

Before settling on a final model, we should check the lack of fit of the second-order model. The only replication consisted of two center-point observations with values 5.46 and 5.81. The sample variance of these two observations is $s_0^2 = 0.0613$, so $ssPE = 0.0613$ with one degree of freedom. From the analysis of variance table, Table 16.11, we see that $ssE = 0.1443$ with 4 degrees of freedom. So,

$$ssLOF = ssE - ssPE = 0.1443 - 0.0613 = 0.0830$$

with $4 - 1 = 3$ degrees of freedom for lack of fit. There is significant lack of fit of the second-order model if

$$msLOF/msPE > F_{3,1,\alpha},$$

for appropriate significance level α. Here,

$$msLOF/msPE = (0.0830/3)/(0.0613/1) = 0.4519,$$

which is less than 1.0, so there is no significant lack of fit of the second-order model, and the model fitted in Example 16.3.2, page 563, should be a reasonable approximation to the true surface in the local region under study ($9.5 \le x_A \le 11.5$; $31 \le x_B \le 41$). □

16.3.5 Canonical Analysis of a Second-Order Model

After fitting a second-order model, we need to determine the location of the stationary point. We think of each treatment combination $\mathbf{x}$ as a point in p-dimensional space, $\mathbf{x} = (x_1, x_2, \ldots, x_p)$, so the stationary point that we are trying to find is the point $\mathbf{x}_s = (x_{s1}, x_{s2}, \ldots, x_{sp})$. We then change to a new coordinate system of points in two steps. First we set $\mathbf{v} = \mathbf{x} - \mathbf{x}_s$, so that $v_i = x_i - x_{si}$ for $i = 1, 2, \ldots, p$. This moves the coordinate system so that the stationary point is at the origin with respect to the v_i-axes, so the stationary point is now $\mathbf{v}_s = (0, 0, \ldots, 0)$. The other points $\mathbf{v} = \mathbf{x} - \mathbf{x}_s$ measure position relative to the stationary point $\mathbf{x}_s$. This eliminates all linear terms from the model. As the second step, the v_i-axes are rotated to obtain the w_i-axes, with the rotation chosen to eliminate the cross product terms from the model. We show how to do this in the following subsections.

In terms of each of these coordinate systems, the fitted model has the following equivalent representations:

$$\hat{y}_\mathbf{x} = \hat{\beta}_0 + \sum_{i=1}^p \hat{\beta}_i x_i + \sum_{i=1}^p \hat{\beta}_{ii} x_i^2 + \sum_{i<j} \hat{\beta}_{ij} x_i x_j,$$

$$\hat{y}_\mathbf{v} = \hat{y}_{\mathbf{v}_s} + \sum_{i=1}^p \hat{\beta}_{ii} v_i^2 + \sum_{i<j} \hat{\beta}_{ij} v_i v_j,$$

$$\hat{y}_\mathbf{w} = \hat{y}_{\mathbf{w}_s} + \sum_{i=1}^p \hat{\lambda}_{ii} w_i^2,$$

where $\hat{y}_{\mathbf{v}_s}$ and $\hat{y}_{\mathbf{w}_s}$ are equal and each denotes the predicted response at the stationary point.

The last equation is said to be in *canonical form*, and in this form, we can immediately tell whether the stationary point is a maximum, a minimum, or a saddle point. If all of the

$\hat{\lambda}_{ii}$'s are negative, then the fitted model is concave down and has a maximum at the stationary point. If all of the $\hat{\lambda}_{ii}$'s are positive, then the fitted model is concave up and has a minimum at the stationary point. If some of the $\hat{\lambda}_{ii}$'s are positive and some are negative, the stationary point is a saddle point. The $\hat{\lambda}_{ii}$ are called the *canonical coefficients*.

If a specific $\hat{\lambda}_{ii}$ is relatively large in magnitude, then $\hat{y}_w$ will change rapidly for changes away from the stationary point $\mathbf{w}_s = (0, 0, \ldots, 0)$ in the w_i direction. Thus, if the stationary point is a saddle point and if $\hat{\lambda}_{\ell\ell}$ is the largest positive $\hat{\lambda}_{ii}$, movement in either direction away from the stationary point along the w_ℓ-axis provides a path of steepest ascent. On the other hand, if a specific $\hat{\lambda}_{ii}$ is relatively small in magnitude, then $\hat{y}_w$ is relatively unaffected by changes away from the stationary point along the w_i-axis.

Canonical analysis for $p = 2$ factors We consider first the fitted second-order model

$$\hat{y}_\mathbf{x} = \hat{\beta}_0 + \hat{\beta}_A x_A + \hat{\beta}_B x_B + \hat{\beta}_{AA} x_A^2 + \hat{\beta}_{BB} x_B^2 + \hat{\beta}_{AB} x_A x_B$$

for $p = 2$ factors A and B. The results below are special cases of the general formulae in the following optional subsection for $p \geq 2$ factors.

The canonical coefficients $\hat{\lambda}_{11}$ and $\hat{\lambda}_{22}$ are

$$\frac{1}{2}\left[\hat{\beta}_{AA} + \hat{\beta}_{BB} \pm \sqrt{(\hat{\beta}_{AA} - \hat{\beta}_{BB})^2 + \hat{\beta}_{AB}^2}\right], \qquad (16.3.13)$$

and the stationary point is $\mathbf{x}_s = (x_{sA}, x_{sB})$, where

$$x_{sA} = \frac{1}{4D}(\hat{\beta}_B \hat{\beta}_{AB} - 2\hat{\beta}_A \hat{\beta}_{BB}), \qquad (16.3.14)$$

$$x_{sB} = \frac{1}{4D}(\hat{\beta}_A \hat{\beta}_{AB} - 2\hat{\beta}_B \hat{\beta}_{AA}),$$

and

$$D = \hat{\lambda}_{11}\hat{\lambda}_{22} = \hat{\beta}_{AA}\hat{\beta}_{BB} - \frac{1}{4}\hat{\beta}_{AB}^2.$$

The stationary point $\mathbf{x}_s$ maximizes $\hat{y}_\mathbf{x}$ if both canonical coefficients are negative, minimizes it if both are positive, and yields a saddle point of the fitted surface if one is positive and the other negative.

The w_1 *canonical axis* consists of all points (x_A, x_B) of the form

$$(x_{sA}, x_{sB}) + u\left(\frac{\hat{\beta}_{AB}}{2}, \hat{\lambda}_{11} - \hat{\beta}_{AA}\right) \qquad (16.3.15)$$

for real numbers u, and the canonical axis w_2 consists of all points (x_A, x_B) of the form

$$(x_{sA}, x_{sB}) + q\left(\frac{\hat{\beta}_{AB}}{2}, \hat{\lambda}_{22} - \hat{\beta}_{AA}\right) \qquad (16.3.16)$$

for real numbers q.

The canonical analysis can also be done with respect to the coded factor levels. The SAS software does this, for example.

Example 16.3.3 Acid copper pattern plating experiment, continued

In Example 16.3.1, page 563, a second-order model was fitted to data collected from a central composite design. The experiment was run in order to study the effects of anode-cathode separation (factor A) and cathodic current density (factor B) on the standard deviation of copper-plating thickness. In terms of the uncoded factor levels, the fitted model was

$$\hat{y}_\mathbf{x} = \hat{\beta}_0 + \hat{\beta}_A x_A + \hat{\beta}_B x_B + \hat{\beta}_{AA} x_A^2 + \hat{\beta}_{BB} x_B^2 + \hat{\beta}_{AB} x_A x_B$$
$$= 79.6898 - 7.8187 x_A - 1.8126 x_B$$
$$+ 0.3926 x_A^2 + 0.0294 x_B^2 - 0.0250 x_A x_B \,.$$

Now,

$$\hat{\beta}_{AA} + \hat{\beta}_{BB} = 0.3926 + 0.0294 = 0.4220 \,,$$

and

$$\sqrt{(\hat{\beta}_{AA} - \hat{\beta}_{BB})^2 + \hat{\beta}_{AB}^2} = \sqrt{(0.3926 - 0.0294)^2 + (-0.0250)^2} = 0.3640 \,.$$

Then, from equation (16.3.13), the canonical coefficients are

$$\hat{\lambda}_{11} = \tfrac{1}{2}(0.4220 + 0.3640) = 0.3930 \,,$$
$$\hat{\lambda}_{22} = \tfrac{1}{2}(0.4220 - 0.3640) = 0.0290 \,.$$

Since both canonical coefficients are positive, the stationary point minimizes the estimated standard deviation of coating thickness.

From equation (16.3.14), the stationary point is $\mathbf{x}_s = (x_{sA}, x_{sB})$, where

$$x_{sA} = \frac{1}{4(0.0114)}[(-1.8126)(-0.0250) - 2(-7.8187)(0.0294)] = 11.0758 \,,$$

$$x_{sB} = \frac{1}{4(0.0114)}[(-7.8187)(-0.0250) - 2(-1.8126)(0.3926)] = 35.4983 \,.$$

Using these values of x_A and x_B in the fitted model, we obtain the predicted minimum response as $\hat{y}_{\mathbf{x}_s} = 4.144$.

Now, $\hat{\lambda}_{11}$ is much larger than $\hat{\lambda}_{22}$, so the surface will rise more rapidly as we move away from $\mathbf{x}_s$ in the w_1 direction than in the w_2 direction. From equation (16.3.15), the w_1 canonical axis consists of all points (x_A, x_B) of the form

$$(x_{sA}, x_{sB}) + u \left(\frac{\hat{\beta}_{AB}}{2}, \hat{\lambda}_{11} - \hat{\beta}_{AA} \right) = (11.0758, 35.4983) + u(-0.0125, 0.0004)$$

for real numbers u. If we divide u by $\sqrt{(-0.0125)^2 + (0.0004)^2} = 0.01251$ and multiply the point $(-0.0125, 0.0004)$ by the same constant, we are then looking at steps of size 1.0 along the w_1-axis; that is,

$$(11.0758, 35.4983) + u(-0.9995, 0.0320) \,.$$

This is not very different from

$$(11.0758, 35.4983) + u(-1.0, 0.0),$$

so the w_1-axis has not been rotated very far from the A-axis (or x_1-axis). We can verify this from Figure 16.4 on page 564, which shows the surface contours with axes almost parallel to the A and B axes. This means that the level of A must be controlled more precisely than the level of B in order to maintain a minimum response. The same conclusion can be reached by examining the fitted equation, since the coefficients of x_A and x_A^2 are considerably larger than those of x_B, x_B^2, and $x_A x_B$. □

Canonical analysis for $p \geq 2$ factors (optional) This subsection requires the knowledge of matrices and vectors. Consider the fitted second-order model

$$\hat{y}_\mathbf{x} = \hat{\beta}_0 + \sum_i \hat{\beta}_i x_i + \sum_i \hat{\beta}_{ii} x_i^2 + \sum_{i<j} \hat{\beta}_{ij} x_i x_j$$

for p factors. Let $\mathbf{b}$ denote the $p \times 1$ vector of linear parameter estimates, with ith entry $\hat{\beta}_i$. Let $\mathbf{B}$ denote the $p \times p$ matrix with ith diagonal element $\hat{\beta}_{ii}$ and with off-diagonal (ij)th entry $\hat{\beta}_{ij}/2$. Then the least squares fitted model can be written in matrix terms as

$$\hat{y}_\mathbf{x} = \hat{\beta}_0 + \mathbf{x}'\mathbf{b} + \mathbf{x}'\mathbf{B}\mathbf{x}.$$

Furthermore, the stationary point is

$$\mathbf{x}_s = -\frac{1}{2}\mathbf{B}^{-1}\mathbf{b}$$

with corresponding predicted mean response

$$\hat{y}_{\mathbf{x}_s} = \hat{\beta}_0 - \mathbf{x}_s'\mathbf{B}\mathbf{x}_s = \hat{\beta}_0 + \frac{1}{2}\mathbf{x}_s'\mathbf{b}.$$

The canonical coefficients are the eigenvalues of the matrix $\mathbf{B}$. The eigenvectors of $\mathbf{B}$ determine the canonical axes. The use of SAS in obtaining these will be illustrated in Example 16.7.2, and the interpretation of the resulting formulae will also be discussed.

16.4 Properties of Second-Order Designs: CCDs

In this section we discuss some desirable properties—rotatability, orthogonal blocking, and orthogonality—of second-order designs. The discussion here focuses on central composite designs (CCDs) because their properties can be controlled by judicious choice of the number of center points n_0 and the distance α of the axial points from the design center. In addition to rotatability, orthogonal blocking, and orthogonality, a design should include enough center points (say 3–6) to provide a reasonably sensitive test for lack of fit.

16.4.1 Rotatability

A design is *rotatable* if the variance $\text{Var}(\widehat{Y}_\mathbf{z})$ of the predicted response is the same for all coded points $\mathbf{z} = (z_1, z_2, \ldots, z_p)$ at any given distance $d = (\sum_i z_i^2)^{1/2}$ from the design center, $\mathbf{z}_0 = (0, 0, \ldots, 0)$. Thus, there is the same amount of information about the response surface at the same distance d in any direction from the design center. This is a reasonable requirement of a design, since data are generally collected without knowing in which direction from the design center the stationary point of the fitted surface will be located.

Rotatable Central Composite Designs Suppose we take a central composite design for p factors, with one observation at each axial point located a distance α from the design center, and with one observation at each of the n_f factorial points. It can be shown that such a central composite design is rotatable if

$$\alpha = (n_f)^{1/4}, \tag{16.4.17}$$

and if each axial point is observed r_a times, then the requirement for rotatability becomes

$$\alpha = (n_f/r_a)^{1/4}.$$

The details can be found in the articles by Box and Hunter (1957) and Draper (1982).

Example 16.4.1 Acid copper pattern plating experiment, continued

In Example 16.3.2, page 563, a central composite design was used for $p = 2$ factors. The design involved one observation at each $n_f = 4$ factorial points and $n_a = 4$ axial points, plus two center points. If the model, in terms of coded factor levels, is fitted using $\alpha = (n_f)^{1/4} = \sqrt{2}$, the design is rotatable with respect to the coded factor levels. For example, it can be verified that the estimate of the variance is

$$\widehat{\text{Var}}(\widehat{Y}_\mathbf{z}) = 0.0182$$

at each point $\mathbf{z} = (z_1, z_2)$ at distance $\sqrt{2}$ from the design center. This includes each factorial point and each axial point. For comparison, $\widehat{\text{Var}}(\widehat{Y}) = 0.0145$ at the center point and $\widehat{\text{Var}}(\widehat{Y}) = 0.0100$ at the points $(-1, 0)$, $(1, 0)$, $(0, -1)$, and $(0, 1)$, which are each a distance 1.0 from the design center. □

16.4.2 Orthogonality

The second-order model (16.3.10) includes $(p+1)(p+2)/2$ parameters, including the intercept γ_0. A second-order design is *orthogonal* if the sums of squares, $ss(\gamma_i|\gamma_0)$ ($i = 1, 2, \ldots, p$), $ss(\gamma_{ii}|\gamma_0)$ ($i = 1, 2, \ldots, p$), and $ss(\gamma_{ij}|\gamma_0)$ ($1 \le i < j \le p$), each adjusted for the intercept γ_0, are independent. In the analysis of variance of an orthogonal design, the sums of squares associated with these $(p+2)(p+1)/2 - 1$ parameters are independent, and do not depend on the order in which the parameters are entered into the model. Orthogonality is advantageous if the experimenter is interested in evaluating which of the linear, quadratic, and cross product effects are significantly different from zero.

Orthogonal central composite designs Suppose we take a central composite design with one observation at each of the n_f factorial points and $2p$ axial points, and with n_0 observations at the center. As shown by Khuri and Cornell, 1987, page 119, the design is orthogonal if

$$(n_f + 2\alpha^2)^2 = n_f n,$$

where n is the total number of observations; that is, $n = n_f + 2p + n_0$. So, a central composite design with n_f factorial points and $2p$ axial points can be made orthogonal by appropriate choice of α or n_0. For example, if the number of center points is fixed at n_0, then n is fixed, and a central composite design is orthogonal if

$$\alpha = \left(\frac{\sqrt{n_f n} - n_f}{2}\right)^{1/2}. \tag{16.4.18}$$

If a central composite design is to be rotatable and n_0 is not fixed, then we would choose $\alpha = (n_f)^{1/4}$, and the design would also be orthogonal if the number of center points was chosen to be

$$n_0 = 4\sqrt{n_f} + 4 - 2p. \tag{16.4.19}$$

This may not be achievable, since n_0 must be an integer. Rounding (16.4.19) to the nearest integer gives a rotatable design that is nearly orthogonal.

Example 16.4.2 Flour production experiment

In Section 16.5, we will consider the last of four experiments described by M. G. Tuck, S. M. Lewis, and J. I. L. Cottrell (1993) to develop robust bread flours. This experiment was run using a central composite design for three factors, with one observation at each of $n_f = 8$ factorial points and $2p = 6$ axial points. From equation (16.4.19), since $\sqrt{n_f} = \sqrt{8}$ is not an integer, the design with $n_f = 8$ cannot be both orthogonal and rotatable. The experimenters used only $n_0 = 2$ center points, giving $n = 16$ observations in total. From equation (16.4.18), the design is orthogonal if

$$\alpha = \left(\frac{\sqrt{(8)(16)} - 8}{2}\right)^{1/2} = 1.2872.$$

This value of α was used by the experimenters. □

16.4.3 Orthogonal Blocking

If a second-order design is conducted as a block design, then the second-order model (16.3.10) is modified to include additive block effects. For p factors, the model is

$$Y_{h,\mathbf{z},t} = \gamma_0 + \theta_h + \sum_{i=1}^{p} \gamma_i z_i + \sum_{i=1}^{p} \gamma_{ii} z_i^2 + \sum_{i<j} \gamma_{ij} z_i z_j + \epsilon_{h,\mathbf{z},t}, \tag{16.4.20}$$

where $Y_{h,\mathbf{z},t}$ denotes the tth observation at coded treatment combination $\mathbf{z} = (z_1 z_2 \ldots z_p)$ in block h, and the error variables $\epsilon_{h,\mathbf{z},t}$ are independent with $N(0, \sigma^2)$ distributions. The parameter θ_h denotes the effect of the hth block, and the other parameters are defined as in the second-order model (16.3.10).

A design is said to have *orthogonal blocking* if the least squares estimates of the linear, quadratic, and cross product effect parameters are the same under model (16.4.20), which includes block effects as under the model (16.3.10) without block effects; that is, the linear, quadratic, and cross product effects are estimated independently of the block effects. The primary advantage of orthogonal blocking as compared with nonorthogonal blocking is that an orthogonally blocked design gives the smallest values of $\text{Var}(\widehat{Y})$, $\text{Var}(\hat{\gamma}_i)$, $\text{Var}(\hat{\gamma}_{ii})$, and $\text{Var}(\hat{\gamma}_{ij})$. A second advantage is that a rotatable design conducted with orthogonal blocking is still rotatable.

Given a design in b blocks with orthogonal blocking, the analysis under the block design model (16.4.20) is almost the same as it would be under model (16.3.10) for the design with no blocking. However, the sum of squares for blocks is extracted from the sum of squares for error, and there are $b - 1$ degrees of freedom for blocks giving $b - 1$ fewer degrees of freedom for error. The sum of squares for blocks is

$$ss\theta = \sum_{h=1}^{b} k_h (\bar{y}_{h..} - \bar{y}_{...})^2 = \sum_{h=1}^{b} y_{h..}^2 / k_h - y_{...}^2 / n,$$

where $y_{h..}$ is the sum of the observations in the hth block, k_h is the size of the hth block, and $y_{...}$ is the sum of all n observations in the design.

In their 1957 article, Box and Hunter developed the following general conditions under which a second-order design can be blocked orthogonally.

(1) Each block must be a first-order orthogonal design: that is, (i) for each block and each factor i, the sum of coded levels of the factor, $\sum z_i$, is zero; and (ii) for each block and each pair of factors i and j, the sum of cross products, $\sum z_i z_j$, is zero. (Each sum is over all the observations in the block.)

(2) For each block and each factor i, the sum of squares $\sum z_i^2$ of the coded levels of the ith factor in the block must be proportional to the number of observations in the block.

Orthogonal blocking of central composite designs For a central composite design, we first divide the observations into two blocks: an *axial-points block* consisting of the n_a axial points plus n_{0a} center points, and a *factorial-points block* consisting of the n_f factorial points plus n_{0f} center points. This division into blocks is natural if, for example, a first-order design results in lack of fit, so that axial and additional center points are added at a later date to build up to a second-order design. Each of the blocks is a first-order orthogonal design, meeting condition (1) for orthogonal blocking. Concerning condition (2), the sum of squares $\sum z_i^2$ of the coded levels of each factor is $2\alpha^2$ in the axial block and n_f in the factorial block. So, condition (2) requires that

$$\frac{2\alpha^2}{n_f} = \frac{n_a + n_{0a}}{n_f + n_{0f}}.$$

Solving for α, a central composite design has orthogonal blocking if

$$\alpha = \left(\frac{n_f(n_a + n_{0a})}{2(n_f + n_{0f})} \right)^{1/2}. \tag{16.4.21}$$

The design is also rotatable if $\alpha = (n_f)^{1/4}$, in which case we require

$$n_{0f} = (\sqrt{n_f}/2)(n_a + n_{0a}) - n_f. \tag{16.4.22}$$

If the numbers of center points, n_{0a} and n_{0f}, in the blocks can be chosen to satisfy this equation, then the design will be rotatable and can be orthogonally blocked. When this is not possible, it is preferable to maintain orthogonal blocking but to relax rotatability. To accomplish this, the numbers n_{0a} and n_{0f} can be chosen such that equation (16.4.22) is approximately satisfied, and then α can be computed from equation (16.4.21).

It is sometimes possible to block a central composite design orthogonally in more than two blocks. The axial block cannot be further subdivided, but the factorial block can sometimes be divided into 2^m factorial blocks while maintaining orthogonal blocking if the number of factorial center points n_{0f} is divisible by 2^m so the factorial center points can be equally divided among the 2^m factorial blocks. This is done by confounding interaction effects between three or more factors. Box and Hunter (1957, page 233) provide a table of blocking arrangements for rotatable and near-rotatable central composite designs. Notice that if center points are spread across b blocks, then they provide $b-1$ fewer pure error degrees of freedom than they would in a design that is not blocked.

Example 16.4.3 PAH recovery experiment

I. J. Barnabas, J. R. Dean, W. R. Tomlinson, and S. P. Owen (1995) used a central composite design to study the effects of four factors—pressure, temperature, extraction time, and methanol content—on the total recovery of polycyclic aromatic hydrocarbons (PAHs) when extracted from soil. The design was composed of $n_f = 2^4 = 16$ factorial points and $n_a = 2p = 8$ axial points. Taking $\alpha = 16^{1/4} = 2$ would give a rotatable design. From equation (16.4.22),

$$n_{0f} = (\sqrt{16}/2)(8 + n_{0a}) - 16 = 2n_{0a},$$

so use of twice as many factorial center points as axial center points would give a rotatable design that could be orthogonally blocked.

The experimenters chose to use $n_{0a} = 2$ axial center points and $n_{0f} = 4$ factorial center points. This gave an axial block of size 10 and a factorial block of size 20. They then subdivided the factorial block into two blocks each of size 10 by confounding the four-factor interaction and including two of the four factorial center points in each factorial block. The resulting design was rotatable with orthogonal blocking. Analysis of the design is discussed in Section 16.7 using the SAS computer program. The design itself is shown in Table 16.19, page 582, where the first ten observations comprise the first factorial block, the second ten the second factorial block, and the final ten the axial block. □

16.5 A Real Experiment: Flour Production Experiment, Continued

M. G. Tuck, S. M. Lewis and J. I. L. Cottrell (1993) described a series of four related experiments, involving quality improvement in the milling industry. The collective purpose of the experiments was to develop a bread flour that would give high loaf volume despite fluctuations in the bread-making process. We consider here their fourth experiment.

Bread flour consists of wheat plus a small number of minor ingredients. Their fourth experiment was concerned with the effects of three such ingredients (labeled design factors B, C, and D) on loaf volume. An orthogonal central composite design, involving eight factorial points, six axial points, and two center points, was used. For the axial points, the value $\alpha = 1.2872$ was used to make the design orthogonal (see Example 16.4.2).

When a product consists of a mixture of ingredients, and the total volume of the mixture is held constant, the fractions associated with the ingredients in the mixture necessarily sum to one. This has implications for the model and data analysis. However, in this experiment, the minor ingredients constituted such a small portion of the mixture that the total volume did not need to remain fixed, and standard response surface methods could be used to study the design factors.

There were a number of sources of variation in the production process that constituted noise factors. The production factors were paired in order to keep the experiment small. So, noise factor G represented oven bake and proof time, noise factor J represented yeast and water level, and noise factor K represented degree of mixing and moulding pressure. Each of these composite factors had two levels, "high" and "low." A 2_{III}^{3-1} fraction in the composite noise factors was used, with defining relation

$$I = GJK.$$

The experimental design used was a product design. It included $16 \times 4 = 64$ observations—each of the 16 design factor combinations of the central composite design was observed with each of the four noise factor combinations of the noise array. Also, the noise factors were difficult to change, so each noise factor combination constituted a different block, and in each block the design factor treatment combinations $(z_B z_C z_D)$ were randomly ordered. Observations were collected over two days using half-days as blocks, with the four blocks collected in the order $(z_G z_J z_K) = 111, 100, 010, 001$. As a result, noise factor effects are also confounded with changes in conditions from half-day to half-day. For each observation, three loaves were baked from a single dough, then the average specific volume of the three loaves recorded. The resulting data y_{hz} are shown in Table 16.12.

For each of the 16 treatment combinations $\mathbf{z}$ of the central composite design in turn, the sample mean $\bar{y}_z$ and the log sample variance ($\times 100$) were computed from the observations y_{hz} in the four blocks ($h = 1, 2, 3, 4$). The effects of the design factors on these two response variables were studied separately by fitting second-order response surface regression models to each set of 16 responses.

The analysis of variance for fitting the second-order model to the response $\bar{y}_z$ is shown in Table 16.13. Because the design is orthogonal, the effects can be assessed for significance independently of their order of entry into the model. The only effects that are significantly different from zero at an individual significance level of 0.01 are the main effects of each of the three factors. The overall significance level for the nine tests is at most 0.09. If the corresponding first-order model is fitted to $\bar{y}_z$, we obtain

$$\hat{\bar{y}}_z = 475.50 + 4.42 z_B + 10.24 z_C + 6.87 z_D.$$

The coefficients of z_B, z_C, and z_D are all positive. Thus, increasing the level of design factors B, C, and D has a positive effect on the mean loaf specific volume.

16.5 A Real Experiment: Flour Production Experiment, Continued

Table 16.12 Flour production experiment: Average specific volume y_{hz} of loaves on half-day h; $\alpha = 1.2872$

z_B	z_C	z_D	y_{1z}	y_{2z}	y_{3z}	y_{4z}	$\bar{y}_{.z}$	$100 \log_{10}(s_z)$
-1	-1	-1	586	399	418	404	451.75	195.36
-1	-1	1	615	411	435	421	470.50	198.60
-1	1	-1	611	422	431	439	475.75	195.63
-1	1	1	639	436	444	454	493.25	198.88
1	-1	-1	603	422	400	430	463.75	197.17
1	-1	1	622	411	425	436	473.50	199.79
1	1	-1	634	471	436	425	491.50	198.68
1	1	1	673	433	423	462	497.75	207.19
α	0	0	618	414	419	477	482.00	197.80
$-\alpha$	0	0	586	421	420	455	470.50	189.60
0	α	0	621	426	427	458	483.00	196.94
0	$-\alpha$	0	629	412	412	426	469.75	202.68
0	0	α	631	411	433	453	482.00	200.35
0	0	$-\alpha$	587	413	419	430	462.25	192.15
0	0	0	604	432	416	438	472.50	194.53
0	0	0	602	425	407	439	468.25	195.48

Source: Tuck, M. G., Lewis, S. M., and Cottrell, J. I. L. (1993). Copyright © 1993 Blackwell Publishers. Reprinted with permission.

Table 16.13 Flour production experiment: Analysis of variance for $\bar{y}_{.z}$

Source of Variation	Degrees of Freedom	Sum of Squares	Mean Square	Ratio	p-value
z_B	1	221.4366	221.4366	9.77	0.0204
z_C	1	1185.3603	1185.3603	52.30	0.0004
z_D	1	533.2415	533.2415	23.53	0.0029
z_B^2	1	48.0081	48.0081	2.12	0.1958
z_C^2	1	50.4906	50.4906	2.23	0.1862
z_D^2	1	1.1997	1.1997	0.05	0.8257
$z_B z_C$	1	3.4453	3.4453	0.15	0.7101
$z_B z_D$	1	51.2578	51.2578	2.26	0.1833
$z_C z_D$	1	2.8203	2.8203	0.12	0.7363
Error	6	135.9897	22.6650		
Total	15	2233.2500			

The analysis of variance for the response $100 \log_{10}(s_z)$ is shown in Table 16.14. No effects can be regarded as significantly different from zero at an individual 0.01 significance level. However, in this setting it would not be particularly bad to make a Type I error, and if we raise the individual significance level we would select the linear effects of factors B and D and the quadratic effect of C as being the important effects. If the corresponding model is fitted and the linear effect of C is also included, we obtain

$$\widehat{\log_{10}(s_z)} = 195.19 + 0.18 z_C + 3.35 z_C^2 + 2.20 z_B + 2.49 z_D$$
$$\approx 195.19 + 3.35(z_C + 0.027)^2 + 2.20 z_B + 2.49 z_D.$$

Table 16.14 Flour production experiment: Analysis of variance for $100\log_{10}(s_z)$

Source of Variation	Degrees of Freedom	Sum of Squares	Mean Square	Ratio	p-value
z_B	1	54.9174	54.9174	8.24	0.0284
z_C	1	0.3730	0.3730	0.06	0.8208
z_D	1	70.1514	70.1514	10.53	0.0176
z_B^2	1	0.6409	0.6409	0.10	0.7669
z_C^2	1	61.4625	61.4625	9.23	0.0229
z_D^2	1	7.8587	7.8587	1.18	0.3191
$z_B z_C$	1	8.7175	8.7175	1.31	0.2963
$z_B z_D$	1	2.6931	2.6931	0.40	0.5484
$z_C z_D$	1	4.3269	4.3269	0.65	0.4511
Error	6	39.9752	6.6625		
Total	15	251.1164			

Taking the two fitted models, we see that not only does the mean response increase as the levels of factors B, C, and D are increased, but so does the variability. The minimum variability with respect to factor C is achieved at $z_C = -0.027$. However, the amount of factor C in the loaf cannot be negative, and so the minimum variability is achieved when the amount of factor C, as well as factors B and D, is zero.

The end result was that the experimenters set $z_C = 0$ to achieve low variability and adjusted the level of factor B (which has the slightly smaller effect on the variance, and may have been less costly than factor D) to raise mean response to the desired level.

16.6 Box–Behnken Designs

A central composite design has five levels for each factor, $\pm 1, \pm \alpha, 0$. For a given experiment, circumstances may dictate the use of fewer levels, but at least three levels per factor are needed for quadratic terms to be estimable in the second-order model. Use of 3^p factorial designs or regular 3^{p-s} fractional factorial designs might be considered. These tend to be large, however, and the smaller ones tend to be of resolution III or IV so that two-factor interactions are confounded with main effects or other two-factor interactions. For fitting a second-order response model a different type of design, called a *Box–Behnken design*, is often preferred, since interaction parameter estimates are not completely confounded, and in many cases, these designs are considerably smaller than 3^{p-s} fractional factorial designs.

A Box–Behnken design for p factors is constructed by a composition of an incomplete block design for p treatments in b blocks of size k and a 2^k factorial design having factor levels coded $+1$ and -1. The method of composition is illustrated in Example 16.6. In addition to the points generated by the composition, center points must be added to the design for all model parameters to be estimable.

A list of Box–Behnken designs can be found in the article of Box and Behnken (1960). The designs have p factors with each factor observed at 3 levels, for $p = 3$–7, 9–12, and 16. The designs for $p = 4$ and 7 are rotatable, and the others are nearly rotatable. The designs

16.6 Box–Behnken Designs

for $p = 4$–$7, 9, 10, 12$, or 16 allow orthogonal blocking. All of the designs possess a high degree of orthogonality, the only correlation being between the estimators of the intercept and the quadratic terms.

Example 16.6.1 Construction of a Box–Behnken Design: $p = 4$

Suppose we require a second-order design for $p = 4$ factors, each observed at three levels, and with a total of 27 observations. As illustrated by Box and Behnken (1960), a Box–Behnken design can be constructed from a composition of a balanced incomplete block design in $b = 6$ blocks of size $k = 2$ and a 2^2 factorial design as follows. The balanced incomplete block design, shown below left, consists of all possible combinations of four treatment labels taken two at a time. Shown to its right are the $v = 4$ treatment combinations of a 2^2 design, with factor levels coded $+1$ and -1. These two designs are composed as follows. In each of the six blocks of the incomplete block design, the treatment labels are replaced by the symbol ± 1 and the blank "–" by 0 to give the Box–Behnken design represented in condensed form (and without center points) below right.

$$\begin{bmatrix} 1 & 2 & - & - \\ - & - & 3 & 4 \\ 1 & - & 3 & - \\ - & 2 & - & 4 \\ 1 & - & - & 4 \\ - & 2 & 3 & - \end{bmatrix} \text{ with } \begin{bmatrix} -1 & -1 \\ -1 & 1 \\ 1 & -1 \\ 1 & 1 \end{bmatrix} \text{ gives } \begin{bmatrix} \pm 1 & \pm 1 & 0 & 0 \\ 0 & 0 & \pm 1 & \pm 1 \\ \pm 1 & 0 & \pm 1 & 0 \\ 0 & \pm 1 & 0 & \pm 1 \\ \pm 1 & 0 & 0 & \pm 1 \\ 0 & \pm 1 & \pm 1 & 0 \end{bmatrix}$$

The same design, but expanded out and augmented with three center points, is shown in Table 16.15. The first ± 1 in each row of the condensed design is replaced by the first column of levels of the 2^2 design, the second ± 1 is replaced by the second column of levels, and each 0 is replaced by a column of $v = 4$ zeros. With the addition of three center points, this gives the design with 27 treatment combinations shown as the 27 rows of Table 16.15. Although this design has the same number of treatment combinations as a 3_{IV}^{4-1} design, it does not have complete confounding of the two-factor interactions in pairs. □

Table 16.15 Box–Behnken design: $p = 4$ factors, $n = 27$ treatment combinations

$$\begin{bmatrix} -1 & -1 & 0 & 0 \\ -1 & 1 & 0 & 0 \\ 1 & -1 & 0 & 0 \\ 1 & 1 & 0 & 0 \\ 0 & 0 & -1 & -1 \\ 0 & 0 & -1 & 1 \\ 0 & 0 & 1 & -1 \\ 0 & 0 & 1 & 1 \\ 0 & 0 & 0 & 0 \end{bmatrix} \begin{bmatrix} -1 & 0 & -1 & 0 \\ -1 & 0 & 1 & 0 \\ 1 & 0 & -1 & 0 \\ 1 & 0 & 1 & 0 \\ 0 & -1 & 0 & -1 \\ 0 & -1 & 0 & 1 \\ 0 & 1 & 0 & -1 \\ 0 & 1 & 0 & 1 \\ 0 & 0 & 0 & 0 \end{bmatrix} \begin{bmatrix} -1 & 0 & 0 & -1 \\ -1 & 0 & 0 & 1 \\ 1 & 0 & 0 & -1 \\ 1 & 0 & 0 & 1 \\ 0 & -1 & -1 & 0 \\ 0 & -1 & 1 & 0 \\ 0 & 1 & -1 & 0 \\ 0 & 1 & 1 & 0 \\ 0 & 0 & 0 & 0 \end{bmatrix}$$

In general, the composition of an incomplete block design for p treatments in b blocks of size k with a factorial design with $v = 2^k$ treatment combinations yields a Box–Behnken design for p factors with bv treatment combinations. The ith of the k treatment labels in each block is replaced by the ith of the k columns of the factorial design, and each "–" is replaced by a column of v zeros.

In general, if the incomplete block design is a balanced incomplete block design with $r = 3\lambda$, as in Example 16.6, then the resulting Box–Behnken design is rotatable—otherwise not. If there does not exist a balanced incomplete block design with $r = 3\lambda$, then one can either use a balanced incomplete block design with $r \neq 3\lambda$ or use a partially balanced incomplete block design. If a partially balanced incomplete block design is used, each pair of treatment labels must occur together in at least one block for all second-order model parameters to be estimable.

Orthogonal blocking Many Box–Behnken designs can be blocked orthogonally. The requirements for orthogonal blocking of a second-order design were given in Section 16.4.3, and these imply that a Box–Behnken design can be blocked orthogonally under either of two circumstances.

First, if the blocks of the incomplete block design in the composition can be partitioned into equireplicate sets, then the same partition of observations in the Box–Behnken design provides orthogonal blocking as long as the same number of center points is included in each block. Such is the case for the design of Example 16.6, since each pair of blocks in the balanced incomplete block design includes every treatment label exactly once. For the resulting Box–Behnken design in Table 16.15, each bracketed set of nine treatment combinations is a corresponding block with one center point included.

The second situation that allows orthogonal blocking occurs when interactions involving three or more factors can be confounded in the generating factorial design. An example follows:

Example 16.6.2 Example of orthogonal blocking

For $p = 4$ factors, the balanced incomplete block design with blocks consisting of the four combinations of three treatment labels can be combined with the 2^3 factorial design as follows.

$$\begin{bmatrix} 1 & 2 & 3 & - \\ 1 & 2 & - & 4 \\ 1 & - & 3 & 4 \\ - & 2 & 3 & 4 \end{bmatrix} \text{ with } \begin{bmatrix} -1 & -1 & -1 \\ -1 & -1 & 1 \\ -1 & 1 & -1 \\ -1 & 1 & 1 \\ 1 & -1 & -1 \\ 1 & -1 & 1 \\ 1 & 1 & -1 \\ 1 & 1 & 1 \end{bmatrix} \text{ gives } \begin{bmatrix} \pm 1 & \pm 1 & \pm 1 & 0 \\ \pm 1 & \pm 1 & 0 & \pm 1 \\ \pm 1 & 0 & \pm 1 & \pm 1 \\ 0 & \pm 1 & \pm 1 & \pm 1 \end{bmatrix}$$

where the ith occurrence of ± 1 in any row of the combined design is replaced by the ith column of the factorial design, and each 0 in the combined design is replaced by a column of eight 0's. The resulting 32-run Box–Behnken design can be partitioned into two blocks of size 16 by confounding the three-factor interaction in the generating factorial design. Thus, treatment combinations in the combined design are divided into two blocks, the division depending on whether they include an even or odd number of factors at level "-1." An equal number of center points must be added to each block. □

16.7 Using SAS Software

In this section we illustrate the analysis of a standard first-order design and a central composite design using the SAS procedures GLM and RSREG, respectively.

16.7.1 Analysis of a Standard First-Order Design

The acid copper pattern plating experiment of Poon (1995) was introduced in Example 16.2.6 (page 558). This small experiment involved four factorial points and two center points. A SAS program using the GLM procedure for the analysis of this standard first-order design is shown in Table 16.16. After reading the data and coding the factor levels, there are two calls of PROC GLM. Neither of these calls includes a CLASS statement, since the goal is to fit a model to the levels of the quantitative factors and not to compare the effects of their levels.

Table 16.16 SAS program for first-order response surface regression.

```
* Enter data of the first-order design and code levels;
DATA COPPER;
  INPUT XA XB S;
  ZA = (XA - 10.5);
  ZB = (XB - 36)/5;
  LINES;
    9.5  31  5.60
    9.5  41  6.45
   11.5  31  4.84
   11.5  41  5.19
   10.5  36  4.32
   10.5  36  4.25
;
* Analysis of the first-order design;
PROC GLM;
  MODEL S = ZA ZB;
;
* Add model terms to test for lack of fit;
PROC GLM;
  MODEL S = ZA ZB ZA*ZB ZA*ZA;
```

Table 16.17 SAS output from the first call of PROC GLM: Analysis of variance and parameter estimates for a first-order design

```
                        The SAS System
                  General Linear Models Procedure

Dependent Variable: S
                              Sum of          Mean
Source              DF       Squares        Square    F Value     Pr > F
Model                2      1.3801000      0.6900500     0.99      0.4686
Error                3      2.0985833      0.6995278
Corrected Total      5      3.4786833


Source              DF    Type III SS    Mean Square    F Value    Pr > F
ZA                   1      1.0201000      1.0201000      1.46      0.3137
ZB                   1      0.3600000      0.3600000      0.51      0.5250


                                    T for H0:      Pr > |T|     Std Error of
Parameter         Estimate         Parameter=0                    Estimate
INTERCEPT        5.108333333          14.96          0.0006        0.34144980
ZA              -0.505000000          -1.21          0.3137        0.41818889
ZB               0.300000000           0.72          0.5250        0.41818889
```

Table 16.18 SAS output from the second call of PROC GLM: Test for lack of fit of the first-order model

```
                        The SAS System
                  General Linear Models Procedure

Dependent Variable: S
                              Sum of          Mean
Source              DF       Squares        Square    F Value     Pr > F
Model                4      3.4762333      0.8690583    354.72     0.0398
Error                1      0.0024500      0.0024500
Corrected Total      5      3.4786833


Source              DF    Type III SS    Mean Square    F Value    Pr > F
ZA                   1      1.0201000      1.0201000     416.37     0.0312
ZB                   1      0.3600000      0.3600000     146.94     0.0524
ZA*ZB                1      0.0625000      0.0625000      25.51     0.1244
ZA*ZA                1      2.0336333      2.0336333     830.05     0.0221
```

(See page 582.)

In the first call, the first-order model (16.2.3) is fitted, generating the output shown in Table 16.17. Neither main effect is significantly different from zero, indicating either that the experimental region is in the vicinity of the peak, or that neither factor affects the response.

Table 16.19 SAS program for response surface regression (PAH recovery experiment)

```
DATA PAH;
  INPUT RUN B1 B2 PRES TEMP    ET  MC     Y;
  LINES;
        1   1  0  250  55  47.5  15  391.8
        2   1  0  150  85  47.5  15  413.6
        3   1  0  250  55  22.5   5   68.7
        4   1  0  250  85  47.5   5  143.0
        5   1  0  150  85  22.5   5  104.0
        6   1  0  150  55  22.5  15  309.1
        7   1  0  200  70  35.0  10  400.6
        8   1  0  250  85  22.5  15  402.5
        9   1  0  150  55  47.5   5   77.7
       10   1  0  200  70  35.0  10  426.5
       11   0  1  250  85  47.5  15  457.5
       12   0  1  150  55  22.5   5   56.9
       13   0  1  250  85  22.5   5   94.1
       14   0  1  250  55  22.5  15  409.7
       15   0  1  150  55  47.5  15  410.9
       16   0  1  150  85  22.5  15  375.8
       17   0  1  150  85  47.5   5  110.5
       18   0  1  200  70  35.0  10  387.8
       19   0  1  250  55  47.5   5  103.0
       20   0  1  200  70  35.0  10  399.1
       21  -1 -1  200  70  35.0  10  416.9
       22  -1 -1  200  40  35.0  10  359.8
       23  -1 -1  200  70  10.0  10  276.1
       24  -1 -1  200  70  60.0  10  462.3
       25  -1 -1  100  70  35.0  10  311.5
       26  -1 -1  200  70  35.0  10  346.5
       27  -1 -1  200  70  35.0   0   46.8
       28  -1 -1  200  70  35.0  20  418.7
       29  -1 -1  200 100  35.0  10  413.9
       30  -1 -1  300  70  35.0  10  429.4
;
* Sort by independent variables for lack of fit test;
PROC SORT;
  BY B1 B2 PRES TEMP ET MC;
;
* Response surface regression;
PROC RSREG;
  MODEL Y = B1 B2 PRES TEMP ET MC / COVAR=2 LACKFIT;
```

Source: The data in the program are reprinted with permission from Barnabas, I. J., Dean, J. R., Tomlinson, W. R., and Owen, S. P. (1995). Copyright © 1995 American Chemical Society.
(See page 582.)

Table 16.20 SAS output from PROC RSREG: Coding of factor levels

```
              The SAS System
Coding Coefficients for the Independent Variables

     Factor    Subtracted off    Divided by
     PRES      200.000000        100.000000
     TEMP       70.000000         30.000000
     ET         35.000000         25.000000
     MC         10.000000         10.000000
```

(See page 584.)

In the second call of PROC GLM, the interaction term and one quadratic term are added to the model to test for lack of fit of the first-order model—the model would contain too many parameters if both quadratic terms were added. Some of the resulting output is shown in Table 16.18. At an overall level of 0.1 for the four tests (each being done at individual level $\alpha^* = 0.025$), the quadratic term ZA*ZA is significantly different from zero, indicating the presence of significant curvature. This fact caused the experimenters to add axial points to the first-order design to obtain a central composite design (see Example 16.3.2).

16.7.2 Analysis of a Second-Order Design

The SAS procedure RSREG is used to fit a second-order response surface regression model. This is illustrated in Table 16.19 in the context of the PAH recovery experiment that was introduced in Example 16.4.3, page 573. A rotatable central composite design with orthogonal blocking was used to study the effects of four factors—pressure (PRES), temperature (TEMP), extraction time (ET), and methanol content (MC)—on the total recovery of polycyclic aromatic hydrocarbons (Y) when extracted from soil.

The SAS program shown in Table 16.19 reads the run number, the levels of the block indicator variables, the uncoded levels of the four factors, and the data into data set ONE. Until now, we have always declared a block variable to be a classification variable via the CLASSES statement and listed its levels as $1, 2, \ldots, b$. However, PROC RSREG does not recognize classification variables, and if a single block factor were included in the model, it would be interpreted as a quantitative variable possessing one degree of freedom. We have included in the model the pair of covariates (B1, B2), for which we have selected the three coded pairs of levels $(1, 0), (0, 1)$ and $(-1, -1)$. The three pairs of levels distinguish the three blocks and provide two block degrees of freedom.

Only the factor *names* need be listed in the MODEL statement in RSREG, as all quadratic and cross product terms in the factors are automatically included in the model. To avoid treatment–block interactions from being included, B1 and B2 are declared to be covariates. This is done via the option COVAR=2, which indicates that the first two listed independent variables are covariates and should not be included in any interactions.

A generic test for model lack of fit can optionally be requested if the SAS data set has been sorted by the independent variables in the model to cluster replicated observations.

Table 16.21 SAS output from PROC RSREG: Analysis of variance, lack-of-fit test, and parameter estimates for uncoded factor levels

```
                          The SAS System
                    Response Surface for Variable Y

                 Response Mean          300.823333
                 Root MSE                58.911606
                 R-Square                 0.9280
                 Coef. of Variation      19.5835

              Degrees
                of      Type I Sum
Regression   Freedom    of Squares    R-Square   F-Ratio   Prob > F
Covariates      2          33884       0.0540     4.882     0.0262
Linear          4         447761       0.7141    32.254     0.0000
Quadratic       4          99227       0.1583     7.148     0.0029
Crossproduct    6        1007.770000   0.0016     0.0484    0.9993
Total Regress  16         581880       0.9280    10.479     0.0001

              Degrees
                of       Sum of
Residual     Freedom    Squares      Mean Square   F-Ratio   Prob > F
Lack of Fit    10         42240      4224.017550    4.404     0.1246
Pure Error      3       2877.330000   959.110000
Total Error    13         45118      3470.577346

              Degrees
                of       Parameter     Standard     T for H0:
Parameter    Freedom     Estimate      Error        Parameter=0
INTERCEPT       1       -1238.272778   552.190337    -2.242
PRES            1           3.998267     2.492765     1.604
TEMP            1          12.755444     8.744702     1.459
ET              1          12.320000     9.182487     1.342
MC              1          65.649667    21.967351     2.989
PRES*PRES       1          -0.008863     0.004499    -1.970
TEMP*PRES       1          -0.002117     0.019637    -0.108
TEMP*TEMP       1          -0.080250     0.049994    -1.605
ET*PRES         1          -0.004660     0.023565    -0.198
ET*TEMP         1           0.003067     0.078549     0.0390
ET*ET           1          -0.143800     0.071991    -1.997
MC*PRES         1           0.023100     0.058912     0.392
MC*TEMP         1          -0.014500     0.196372    -0.0738
MC*ET           1           0.066200     0.235646     0.281
MC*MC           1          -2.263250     0.449945    -5.030
B1              1         -27.073333    15.210911    -1.780
B2              1         -20.293333    15.210911    -1.334
```

(See page 584.)

Table 16.22 SAS output from PROC RSREG: Parameter estimates for coded factor levels, and further analysis of variance

```
                        The SAS System

                                        Parameter
                                        Estimate
                                        from Coded
        Parameter       Prob > |T|        Data
        INTERCEPT         0.0430        396.233333
        PRES              0.1327         37.300000
        TEMP              0.1684         31.783333
        ET                0.2027         54.966667
        MC                0.0105        263.066667
        PRES*PRES         0.0706        -88.625000
        TEMP*PRES         0.9158         -6.350000
        TEMP*TEMP         0.1325        -72.225000
        ET*PRES           0.8463        -11.650000
        ET*TEMP           0.9695          2.300000
        ET*ET             0.0671        -89.875000
        MC*PRES           0.7013         23.100000
        MC*TEMP           0.9423         -4.350000
        MC*ET             0.7832         16.550000
        MC*MC             0.0002       -226.325000
        B1                0.0985        -27.073333
        B2                0.2051        -20.293333
```

Factor	Degrees of Freedom	Sum of Squares	Mean Square	F-Ratio	Prob > F
PRES	5	22522	4504.412929	1.298	0.3235
TEMP	5	15068	3013.620690	0.868	0.5279
ET	5	32390	6478.018262	1.867	0.1690
MC	5	503862	100772	29.036	0.0000

(See page 585.)

PROC SORT is used to sort the data, and a test for lack of fit is requested via the option LACKFIT in the model statement of PROC RSREG.

PROC RSREG codes the levels of each factor so that $+1$ and -1 represent the extreme levels of each factor. For example, the axial points of a central composite design would typically be coded ± 1 by SAS rather than the conventional $\pm \alpha$. Table 16.20 shows how SAS codes the factor levels, and Table 16.21 shows the resulting analysis of variance table. The analysis of variance table includes Type I sums of squares for covariates, linear terms, quadratic terms, and cross product terms, adding the terms to the model in that order. These Type I sums of squares are the same, whether coded or uncoded factor levels are specified in the model statement. Information is included for the lack of fit test, the least squares parameter estimates, and the corresponding t-tests. Observe that for this experiment, cross

16.7 Using SAS Software

Table 16.23 SAS output from PROC RSREG: Canonical analysis

```
                      The SAS System
              Canonical Analysis of Response Surface
                      (based on coded data)

                              Critical Value
              Factor        Coded         Uncoded
              PRES          0.259473      225.947276
              TEMP          0.195926       75.877786
              ET            0.347209       43.680235
              MC            0.605224       16.052237

          Predicted value at stationary point    493.335662

                              Eigenvectors
  Eigenvalues     PRES         TEMP         ET           MC
  -71.256548    -0.233360    0.964432    0.121729    -0.024413
  -84.167652     0.712899    0.255316   -0.652941     0.016008
  -93.760753     0.655835    0.067269    0.744877     0.102535
 -227.865046    -0.084838    0.012632   -0.063313     0.994301

                   Stationary point is a maximum.
```

product terms are not significantly different from zero, and there is no significant lack of fit of the model.

Type III sums of squares are also provided for each factor (see Table 16.22), pooling together the sums of squares for all terms—linear, quadratic and interaction—involving the factor. This information can be used for assessing whether any single factor can be removed from the model. These Type III sums of squares are also the same using either the coded or uncoded factor levels. The Type III sums of squares indicate that the factor methanol content (MC) is needed in the model, but perhaps not the other factors.

In Table 16.23, the canonical analysis is shown, including the stationary point (*Critical Value*) in terms of both coded and uncoded factor levels; the predicted value at the stationary point; the canonical coefficients (*Eigenvalues*); classification of the stationary point as a maximum, minimum, or saddle point; and the direction of each canonical axis (*Eigenvectors*). The canonical coefficients and axes are with respect to the coded factor levels.

For this experiment, all eigenvalues (canonical coefficients) are negative, so the stationary point is a maximum. The eigenvectors are each scaled to be of length one. The last eigenvalue, with value -227.865046, is the largest in magnitude. For the corresponding eigenvector, the primary component is that of MC with value 0.994301. So, the fitted model has greatest curvature at the stationary point when moving in either direction determined by this fourth eigenvector, which is nearly parallel to the MC-axis.

Exercises

1. **Paint experiment, continued**

   The paint experiment of S. Eibl, U. Kess, and F. Pukelsheim (1992) was discussed in Example 16.2.1 (page 552), where the first-order model was fitted to the data. For the fitted first-order model, do the following.

   (a) Plot the residuals versus run order, and use the plot to check the independence assumption. (The order of the observations was not randomized in this experiment. Rather, the observations were collected in the order they are shown row by row in Table 16.1, page 553.)

   (b) Plot the residuals versus the predicted values, and use the plot to check the assumption of equal variance.

   (c) Plot the residuals versus their normal scores, and use the plot to check the normality assumption.

   (d) Verify that the design is orthogonal.

2. **Paint followup experiment**

   The data of the second paint experiment described by S. Eibl, U. Kess, and F. Pukelsheim (1992) are given in Table 16.24. This experiment involves factors A–D, as these had significant effects in the first experiment (Example 16.2.1). The factors are

   A: belt speed B: tube width
   C: pump pressure D: paint viscosity

   All four factors are at lower levels than in the first experiment. Lowering the levels of factors B–D was indicated by the analysis of the first experiment. Lowering the level of factor A was based on a conjecture of the experimenters.

   (a) The experiment consists of two replicates of a half-fraction. Find the defining relation for the half-fraction.

Table 16.24 Paint thickness y_{zt} for the paint followup experiment

z_A	z_B	z_C	z_D	y_{z1}	y_{z2}
−1.5	0	−2	0	1.71	1.61
0.5	0	−2	0	0.91	1.30
−1.5	−2	0	0	1.71	1.60
0.5	−2	0	0	1.15	1.29
−1.5	0	0	−2	1.33	1.06
0.5	0	0	−2	1.74	1.98
−1.5	−2	−2	−2	0.64	0.78
0.5	−2	−2	−2	1.51	1.18

Source: Eibl, S., Kess, U., and Pukelsheim, F. (1992). Copyright © 1997 American Society for Quality. Reprinted with Permission.

Table 16.25 Purified lecithin yield and phosphatidylcholine enrichment (PCE), given extraction time (z_1), solvent volume (z_2), ethanol concentration (z_3), and temperature (z_4); fractionation experiment

Run	z_1	z_2	z_3	z_4	Yield	PCE
1	1	1	1	1	27.6	43.8
2	−1	−1	1	1	16.6	27.2
3	1	−1	−1	1	15.4	23.6
4	−1	1	−1	1	17.4	26.2
5	1	−1	1	−1	17.0	27.8
6	−1	1	1	−1	19.0	30.2
7	1	1	−1	−1	17.4	25.2
8	−1	−1	−1	−1	12.6	18.8
9	1	−1	1	1	18.6	28.8
10	−1	1	1	1	22.4	36.8
11	1	1	−1	1	21.4	33.4
12	−1	−1	−1	1	14.0	21.0
13	1	1	1	−1	24.0	38.0
14	−1	−1	1	−1	15.6	23.6
15	1	−1	−1	−1	13.0	20.2
16	−1	1	−1	−1	14.4	22.6
17	0	0	0	0	22.6	
18	$\sqrt{2}$	0	0	0	23.4	
19	$-\sqrt{2}$	0	0	0	20.6	
20	0	$\sqrt{2}$	0	0	22.6	
21	0	$-\sqrt{2}$	0	0	13.4	
22	0	0	$\sqrt{2}$	0	20.6	
23	0	0	$-\sqrt{2}$	0	15.6	
24	0	0	0	$\sqrt{2}$	21.0	
25	0	0	0	$-\sqrt{2}$	17.6	

Source: Sosada, M. (1993). Copyright © 1993 American Oil Chemists Society. Reprinted with permission.

(b) Fit the first-order model, recoding the factor levels as ±1.

(c) Test for lack of fit of the first-order model.

(d) What would you recommend the experimenters do next?

3. **Fractionation experiment**

M. Sosada (1993) studied the effects of extraction time, solvent volume, ethanol concentration, and temperature on the yield and phosphatidylcholine enrichment (PCE) of deoiled rapeseed lecithin when fractionated with ethanol.

Initially, a single-replicate 2^4 experiment was conducted, augmented by three center points. The results for the 16 factorial points are shown as the first 16 runs in Table 16.25.

(a) Fit the first-order model for the response variable "PCE" and conduct the corresponding analysis of variance.

(b) The design also included $n_0 = 3$ center-point observations of PCE. The sample variance of these three observations was $s_0^2 = 1.120$. Test the first-order model for lack of fit, using a 5% level of significance. (Hint: Since the factorial points include no replication, $msPE = s_0^2$, and ssE based on all 19 runs is equal to ssE from the factorial portion of the design plus $(n_0 - 1)s_0^2$.)

(c) Based on the results of parts (a) and (b), what subsequent experimentation would you recommend?

4. **Fractionation experiment, continued**

   The fractionation experiment was described in Exercise 3. Consider here the analysis of "Yield" based on the initial first-order design, shown as the first 16 runs in Table 16.25.

   (a) Fit the first-order model for the response variable "Yield" and conduct the corresponding analysis of variance.

   (b) At the design center point, three additional observations were collected, for which the sample variance was $s_0^2 = 0.090$. Test the first-order model for lack of fit, using a 5% level of significance. (Hint: Since the factorial points include no replication, $msPE = s_0^2$, and ssE based on all 19 runs is equal to ssE from the factorial portion of the design plus $(n_0 - 1)s_0^2$.)

   (c) Based on the results of parts (a) and (b), what subsequent experimentation would you recommend?

5. **Fractionation experiment, continued**

   The fractionation experiment was described in Exercise 3, and analysis of the first-order model for "Yield" was considered in Exercise 4. Based on the analysis of the first-order design, the experimenter chose to augment the 16 factorial points of the first-order design into a 25-run central composite design, the yields from which are shown in Table 16.25.

   (a) Determine whether the central composite design used is rotatable or orthogonal.

   (b) Fit the second-order response surface model and determine which effects are significantly different from zero.

   (c) Conduct a canonical analysis and discuss the results with respect to the following items. What is the nature of the critical point? Noting that the objective is to increase yield, in what direction should one move in subsequent experimentation?

6. **Film viscosity experiment**

   B. Cuq, C. Aymard, J.-L. Cuq, and S. Guilbert (1995, *Journal of Food Science*) used a central composite design to study the effects of protein concentration (g/100 g solution), pH, and temperature (°C), denoted by P, H, and T, respectively, on the apparent viscosity Y (mPa) of film-forming solution, in the development of edible packaging films based on fish myofibrillar proteins. The data are shown in Table 16.26.

   (a) Is this central composite design rotatable or orthogonal?

   (b) Fit the second-order model to the data using the coded factor levels, and check the model assumptions. Would you recommend that a transformation of the data be taken?

Exercises 589

Table 16.26 Apparent viscosity y_{zt} of film-forming solution, for combinations of levels of protein concentration (g/100 g solution), pH, and temperature (°C)

Design point	P		H		T		y
	z_P	x_P	z_H	x_H	z_T	x_T	
1	−1	1.25	−1	2.75	−1	20	50
2	1	2.75	−1	2.75	−1	20	48
3	−1	1.25	1	3.25	−1	20	16700
4	1	2.75	1	3.25	−1	20	560
5	−1	1.25	−1	2.75	1	40	320
6	1	2.75	−1	2.75	1	40	18
7	−1	1.25	1	3.25	1	40	19000
8	1	2.75	1	3.25	1	40	5000
9	−2	0.50	0	3.00	0	30	12700
10	2	3.50	0	3.00	0	30	182
11	0	2.00	−2	2.50	0	30	14
12	0	2.00	2	3.50	0	30	27800
13	0	2.00	0	3.00	−2	10	133
14	0	2.00	0	3.00	2	50	4300
15	0	2.00	0	3.00	0	30	57
16	0	2.00	0	3.00	0	30	70
17	0	2.00	0	3.00	0	30	58
18	0	2.00	0	3.00	0	30	56

Source: Cuq, B., Aymard, C., Cuq, J.-L., and Guilbert, S. (1995). Copyright © 1995 Inst. of Food Technologists. Reprinted with permission.

(c) Fit the second-order model to the natural log of the data, ln(y), using the coded factor levels.

(d) Conduct the test for lack of fit of the second-order model for ln(y).

(e) Check the model assumptions for ln(y).

(f) Conduct the canonical analysis for ln(y).

(g) Conduct the analysis of variance for ln(y).

(h) Compute the coefficient of multiple determination R^2 for the second-order model for ln(y).

(i) Assess the results of the experiment, based on the model for ln(y).

7. **Flour production experiment, continued**

   Consider again the bread-baking experiment of Section 16.5. The data were given in Table 16.12 (page 575), along with the statistics $\bar{y}_{.z}$ and $100 \log_{10}(s_z)$ computed for the observations at each design-factor combination z.

   (a) Plot $\log_{10}(s_z)$ versus $\log_{10}(\bar{y}_{.z})$, and use the methods of Section 5.6.2 to determine an appropriate variance-stabilizing transformation for these data. (Use of $\log_{10}$ is equivalent to use of ln for choosing a transformation.)

(b) Repeat the first analysis of variance of Section 16.5, for which the response variable was $\bar{y}_{.z}$, after applying the transformation determined in part (a) to the observations y_{hz}. Compare your conclusions with those reached in Section 16.5.

(c) Repeat the second analysis of variance of Section 16.5, for which the response variable was $100 \log_{10}(s_z)$, after applying the transformation determined in part (a) to the observations y_{hz}. Compare your conclusions to those reached in Section 16.5.

8. **Central composite design**

Consider using a central composite design for three factors, to include eight factorial points and six axial points.

(a) Determine the value of α to make the design rotatable.

(b) Investigate how α and the number of center points should be chosen to make the design both rotatable and orthogonal, if possible. If this is not possible, how can the design be made rotatable and nearly orthogonal?

(c) Investigate whether the design can be rotatable with orthogonal blocking. If not, then investigate whether orthogonal blocking is possible. If so, how many blocks could be used? Investigate whether orthogonal blocking and near rotatability is possible.

9. **Central composite design**

Repeat Exercise 8 for a central composite design for four factors, to include 16 factorial points and eight axial points.

10. **Resin impurity experiment**

An experiment was conducted using a design close to a central composite design to study the effects of drying time (hours) and temperature (°C) on the content y (ppm) of undesirable compounds in a resin. The data are shown in Table 16.27.

Table 16.27 Resin impurity content y_{zt} (ppm)

Design point	Time	Temp.	$y_{x,t}$
1	7.0	232.4	18.5
2	3.0	220.0	22.5
3	11.0	220.0	17.2
4	1.3	190.0	42.2
5	7.0	190.0	28.6
6	7.0	190.0	19.8
7	7.0	190.0	23.6
8	7.0	190.0	24.1
9	7.0	190.0	24.2
10	12.7	190.0	19.1
11	3.0	160.0	54.1
12	11.0	160.0	33.8
13	7.0	147.6	55.4

(a) Determine the coded levels of time and temperature, as well as the values of n_f, n_a, n_0. What values of α for each factor were selected by the experimenters for the axial points? Why is the design not quite a central composite design?

(b) Fit the second-order model, using coded factor levels.

(c) Test for model lack of fit.

(d) Check the equal variance and normality assumptions of the model using residual plots.

(e) Conduct the canonical analysis.

(f) Conduct the analysis of variance.

(g) Summarize the results.

11. **Resin moisture experiment**

A Box–Behnken design was used to determine whether specific drying conditions for a process could yield a resin that is sufficiently devoid of moisture and low-molecular-weight components. The three factors T, H, and P under study were temperature (150, 185, 220°C), relative humidity (0, 50, 100%), and air pressure (1, 5, 9 torr). The response variable y was a measure of product degradation (ppm). The design and data are shown in Table 16.28.

(a) Fit the second-order model, using coded factor levels.

(b) Test for model lack of fit.

(c) Check the equal-variance and normality model assumptions using residual plots.

(d) Conduct the canonical analysis.

Table 16.28 Resin degradation (ppm) for the resin moisture experiment

Design point	T		H		P		y
	z_T	x_T	z_H	x_H	z_P	x_P	
1	−1	150	−1	0	0	4	83
2	−1	150	0	50	−1	1	103
3	−1	150	0	50	1	9	94
4	−1	150	1	100	0	4	98
5	0	185	−1	0	−1	1	51
6	0	185	−1	0	1	9	48
7	0	185	1	100	−1	1	106
8	0	185	1	100	1	9	108
9	1	220	−1	0	0	4	36
10	1	220	0	50	−1	1	153
11	1	220	0	50	1	9	107
12	1	220	1	100	0	4	87
13	0	185	0	50	0	4	80
14	0	185	0	50	0	4	81
15	0	185	0	50	0	4	77
16	0	185	0	50	0	4	80
17	0	185	0	50	0	4	82

(e) Conduct the analysis of variance.

(f) Summarize the results.

12. **Box–Behnken design**
    (a) Construct a Box–Behnken design for three factors based on the balanced incomplete block design for three treatments in three blocks of size two and the 2^2 factorial design.
    (b) Determine whether the design constructed in part (a) is rotatable.
    (c) For the design constructed in part (a), determine whether orthogonal blocking is possible.

13. **Box–Behnken design.**
    (a) Construct a Box–Behnken design for five factors based on the balanced incomplete block design for five treatments in 10 blocks of size two.
    (b) Determine whether the design constructed in part (a) is rotatable.
    (c) For the design constructed in part (a), determine whether orthogonal blocking is possible.

17 Random Effects and Variance Components

17.1 Introduction
17.2 Some Examples
17.3 One Random Effect
17.4 Sample Sizes for an Experiment with One Random Effect
17.5 Checking Assumptions on the Model
17.6 Two or More Random Effects
17.7 Mixed Models
17.8 Rules for Analysis of Random and Mixed Models
17.9 Block Designs and Random Blocking Factors
17.10 Using SAS Software
Exercises

17.1 Introduction

Until now, we have looked only at treatment factors whose levels have been specifically chosen. We have tested hypotheses about, and calculated confidence intervals for, comparisons in the effects of these particular treatment factor levels. These treatment effects are known as *fixed effects*, since we represent them in the model as unknown constants (parameters). Models that contain only fixed effects are called *fixed-effects models*.

As mentioned in step (f) of the checklist in Section 2.2, page 7, there are occasions when we are interested in a large population of possible levels of a treatment factor, and the levels that are actually used in the experiment are a random sample from this population. The effects of the levels used in the experiment are then represented as random variables whose distributions are the distributions of values in the population. Such treatment-factor effects are called *random effects*, and the corresponding models are called *random-effects models*. We are not interested in just the levels that happen to be in the experiment. Rather, we are concerned with the variability of the effects of all the levels in the population. Consequently, random effects are handled somewhat differently from fixed effects. Some examples of experiments involving random effects are given in Section 17.2.

In Section 17.3 we look at experiments with a single random effect. The selection of sample sizes and model assumption checking are discussed in Sections 17.4 and 17.5. These ideas are extended to experiments with two or more random effects in Section 17.6.

An experiment may involve both random and fixed effects, and the corresponding model is then known as a *mixed model*. Such experiments are discussed in Section 17.7. Block effects may also be random effects, and these are discussed in Section 17.9. Rules for obtaining confidence intervals and testing hypotheses for random effects are given in Section 17.8. The use of SAS is considered in Section 17.10.

17.2 Some Examples

Before running an experiment, the checklist (Section 2.2, page 7) should be completed as usual. In the case of a random effect, the treatment factor will have an extremely large number of levels, only a very small proportion of which can be observed in the experiment. Throughout this chapter, we will assume that the total possible number of levels of each treatment factor will be at least 100 times larger than the numbers of levels that can be observed. Typically, the population of possible levels *will* meet this requirement, and for the purposes of writing down a model, we may regard the population as infinite. Otherwise, one needs to make a correction for a "finite population" in all of the formulae, and this is beyond the scope of this book. Some examples of "infinite" populations are given in Example 17.2.

Example 17.2.1 Infinite populations

Suppose that a manufacturer of canned tomato soup wishes to reduce the variability in the thickness of the soup. Suppose that the most likely causes of the variability are the quality of the cornflour (cornstarch) received from the supplier and the actions of the machine operators. Let us consider two different scenarios:

Scenario 1: The machine operators are highly skilled and have been with the company for a long time. Thus, the most likely cause of variability is the quality of the cornflour delivered to the company. The treatment factor is "cornflour," and its possible levels are all the possible batches of cornflour that the supplier could deliver. Theoretically, this is an infinite population of batches. We are interested not only in the batches of cornflour that have currently been delivered, but also in all those that might be delivered in the future. If we assume that the batches delivered to the company are a random sample from all batches that could be delivered, and if we take a random sample of delivered batches to be observed in the experiment, then the effect of the cornflour on the thickness is a random effect and can be modeled by a random variable.

Scenario 2: It is known that the quality of the cornflour is extremely consistent, so the most likely cause of variability is due to the different actions of the machine operators. The company is large and machine operators change quite frequently. Consequently, those that

17.2 Some Examples

are available to take part in the experiment are only a small sample of all operators employed by the company at present or that might be employed in the future. If we can assume that the operators available for the experiment are representative of the population, then we can assume that they are similar to a random sample from a very large population of possible operators, present and future. Since we would like to know about the variability of the entire population, we model the effect of the operators as random variables, and call them random effects. □

In the absence of any blocking factors, a completely randomized design would be used. The levels of the random treatment factor are first selected at random from the population of all possible levels, and then the experimental units are randomly assigned to these selected levels as usual. At step (h) of the checklist, we need to calculate the number of levels v of the treatment factor to be observed in the experiment in addition to r, the number of observations on each level. Since this calculation uses the formulas for confidence intervals and hypothesis tests, we will postpone the discussion to Section 17.4. As a general rule, if the variability of the treatment effects is much greater than the error (measurement) variability, then v should be large and r small; and vice versa.

Example 17.2.2 Clean wool experiment

The clean wool experiment was reported by J. M. Cameron, of the National Bureau of Standards, in the 1952 volume of *Biometrics*. The following checklist has been compiled from the information given in the article.

(a) **Define the objectives of the experiment.**

Raw wool contains varying amounts of grease, dirt, and foreign material which must be removed before manufacturing begins. The purchase price and customs levy of a shipment are based on the actual amount of wool present, i.e., on the amount of wool present after thorough cleaning—the "clean content." The clean content is expressed as the percentage the weight of the clean wool is of the original weight of the raw wool.

The experiment was run in order to estimate the variability in "clean content" of bales of wool in a shipment.

(b) **Identify all sources of variation.**

(i) Treatment factors and their levels.

The treatment factor was "wool bale" and its levels were the entire population of bales in a particular shipment. Seven bales were observed in the experiment, and these were selected at random from the shipment. The shipment was large enough to allow the bales used in the experiment to be regarded as a random sample from an infinite population of bales. The treatment factor "wool bale" was therefore regarded as a random effect.

(ii) Experimental units.

The experimental units were time slots, so that allocation of these to the levels of the treatment factor determined the order in which the wool bales were observed.

(iii) Blocking factors, noise factors, and covariates.

No nuisance factors were identified as major sources of variation.

(c) Choose a rule by which to assign the experimental units to the levels of the treatment factors.

A completely randomized design was selected.

(d) Specify the measurements to be made, the experimental procedure, and the anticipated difficulties.

A machine was used to bore through a bale of wool and extract a core of wool. Several cores were taken from each of the seven selected bales so that several observations on clean content could be made on each bale. Each core of wool was weighed and then cleaned by scouring, removing burrs, etc. After cleaning, the wool was reweighed and the clean content calculated as the ratio of the clean wool to the initial weight, times 100%.

An anticipated difficulty was that the scouring process, which works well with large amounts of wool, proves difficult with a small core of wool, so that the experimental error observed in the experiment may be larger than would normally be observed in routine production.

The observations on the clean content of the seven bales are shown in Table 17.1. Model selection for this experiment and its analysis via SAS are discussed in Section 17.10.

17.3 One Random Effect

17.3.1 The Random-Effects One-Way Model

For a completely randomized design, with v randomly selected levels of a treatment factor T, the random-effects one-way model is

$$Y_{it} = \mu + T_i + \epsilon_{it}, \qquad (17.3.1)$$

$$\epsilon_{it} \sim N(0, \sigma^2), \quad T_i \sim N(0, \sigma_T^2),$$

ϵ_{it}'s and T_i's are all mutually independent,

$$t = 1, \ldots, r_i, \quad i = 1, \ldots, v.$$

Compare this with the fixed-effects one-way analysis of variance model (3.3.1), page 36. The form of the model and the error assumptions are exactly the same. The only difference is

Table 17.1 Data for the clean wool experiment

	\multicolumn{7}{c}{Bale}						
	1	2	3	4	5	6	7
Clean Content	52.33	56.99	54.64	54.90	59.89	57.76	60.27
	56.26	58.69	57.48	60.08	57.76	59.68	60.30
	62.86	58.20	59.29	58.72	60.26	59.58	61.09
	50.46	57.35	57.51	55.61	57.53	58.08	61.45

Source: Cameron, J. M. (1951). Copyright © 1951 International Biometric Society. Reprinted with permission.

17.3 One Random Effect

in the modeling of the treatment effect. Since the ith level of the treatment factor T observed in the experiment has been randomly selected from the "infinite" population, its observed effect is an observation of a random variable T_i. The distribution of T_i is the distribution of treatment effects in the whole population. We have assumed in (17.3.1) that the population of effects follows a normal distribution with variance σ_T^2, and this assumption will need to be checked along with the error assumptions. The mean of the treatment-effect population has been absorbed into the constant μ, so the distribution of T_i is listed as $N(0, \sigma_T^2)$. The variance σ_T^2 is the parameter of interest, since if the effects of all of the treatment-factor levels are the same, then σ_T^2 is zero. If the effects of the levels are quite different, then σ_T^2 is quite large.

Our final assumption is one of independence. If the treatment-factor levels are selected at random, then the assumption of independence of $T_1, T_2, \ldots, T_v$ is reasonable. However, if, as in Example 17.2 Scenario 2, the levels are a "convenient sample," then this assumption should be investigated carefully. Independence of the T_i and ϵ_{it} requires that the treatment factor not affect any source of variation that has been absorbed into the error variable.

In a random-effects model, the expected value of the response is μ, since

$$E[Y_{it}] = E[\mu] + E[T_i] + E[\epsilon_{it}] = \mu .$$

The variance of Y_{it} is

$$\text{Var}(Y_{it}) = \text{Var}(\mu + T_i + \epsilon_{it}) = \text{Var}(T_i) + \text{Var}(\epsilon_{it}) + 2\text{Cov}(T_i, \epsilon_{it}) = \sigma_T^2 + \sigma^2 ,$$

since T_i and ϵ_{it} are mutually independent and so have zero covariance. Therefore, the distribution of Y_{it} is

$$Y_{it} \sim N(\mu, \; \sigma_T^2 + \sigma^2). \tag{17.3.2}$$

The two components σ_T^2 and σ^2 of the variance of Y_{it} are known as *variance components*. Observations on the same treatment are correlated, with

$$\text{Cov}(Y_{it}, Y_{is}) = \text{Cov}(\mu + T_i + \epsilon_{it}, \mu + T_i + \epsilon_{is}) = \text{Var}(T_i) = \sigma_T^2 .$$

17.3.2 Estimation of σ^2

In order to be able to test hypotheses about σ_T^2 or to calculate confidence intervals, we need an unbiased estimate of σ^2. The random-effects one-way model (17.3.1) is very similar to the fixed effects one-way analysis of variance model (3.3.1), page 36, so a natural question is whether the fixed-effects mean square for error MSE provides an unbiased estimator for σ^2 in the random-effects model also. The answer, happily, is "yes," and we can check it by calculating $E[MSE]$ for the random-effects model, as shown below.

From (3.4.7), page 43, the fixed-effects sum of squares for error is

$$SSE = \sum_{i=1}^{v} \sum_{t=1}^{r_i} Y_{it}^2 - \sum_{i=1}^{v} r_i \overline{Y}_{i.}^2 .$$

Remember that the variance of a random variable X is calculated as $\text{Var}(X) = E[X^2] - (E[X])^2$. So, we have

$$E[Y_{it}^2] = \text{Var}(Y_{it}) + (E[Y_{it}])^2 = (\sigma_T^2 + \sigma^2) + \mu^2.$$

Now,

$$\overline{Y}_{i.} = \mu + T_i + \frac{1}{r_i}\sum_{t=1}^{r_i} \epsilon_{it},$$

so

$$\text{Var}(\overline{Y}_{i.}) = \sigma_T^2 + \frac{\sigma^2}{r_i} \quad \text{and} \quad E[\overline{Y}_{i.}] = \mu. \tag{17.3.3}$$

Consequently,

$$E[\overline{Y}_{i.}^2] = \left(\sigma_T^2 + \frac{\sigma^2}{r_i}\right) + \mu^2.$$

Thus,

$$E[SSE] = \sum_{i=1}^{v}\sum_{t=1}^{r_i}(\sigma_T^2 + \sigma^2 + \mu^2) - \sum_{i=1}^{v} r_i\left(\sigma_T^2 + \frac{\sigma^2}{r_i} + \mu^2\right)$$

$$= n\sigma^2 - v\sigma^2 \quad \left(\text{where } n = \sum_{i=1}^{v} r_i\right)$$

$$= (n-v)\sigma^2,$$

giving

$$E[MSE] = E[SSE/(n-v)] = \sigma^2.$$

So MSE is an unbiased estimator for σ^2, and the observed value of the mean square for error, msE, is an unbiased estimate for σ^2 in the random-effects one-way model, as well as in the fixed-effects one-way model.

Confidence bounds for σ^2 can be computed as under fixed-effects models (Section 3.4.6), that is,

$$\sigma^2 \leq \frac{ssE}{\chi^2_{n-v,1-\alpha}}, \tag{17.3.4}$$

where $\chi^2_{n-v,1-\alpha}$ is the percentile of the chi-squared distribution with $n-v$ degrees of freedom and with probability of $1-\alpha$ in the right-hand tail.

17.3.3 Estimation of σ_T^2

Since the fixed-effects mean square for error msE provides an unbiased estimate of σ^2, the next question that is natural to ask is whether the fixed-effects mean square for treatments

17.3 One Random Effect

msT provides an unbiased estimate for σ_T^2. The answer is "not quite," but we can certainly use it to find an estimate. Now $msT = ssT/(v-1)$, and ssT is given in (3.5.12), page 46, as

$$ssT = \sum_{i=1}^{v} r_i \bar{y}_{i.}^2 - n\bar{y}_{..}^2$$

with corresponding random variable

$$SST = \sum_{i=1}^{v} r_i \bar{Y}_{i.}^2 - n\bar{Y}_{..}^2 \,.$$

Using the same type of calculation as in Section 17.3.2 above, we have

$$\bar{Y}_{..} = \mu + \frac{1}{n}\sum_i r_i T_i + \frac{1}{n}\sum_{i=1}^{v}\sum_{t=1}^{r_i} \epsilon_{it} \,.$$

So

$$E[\bar{Y}_{..}] = \mu \quad \text{and} \quad \text{Var}(\bar{Y}_{..}) = \frac{\sum r_i^2}{n^2}\sigma_T^2 + \frac{n}{n^2}\sigma^2 \,.$$

Also, from (17.3.3),

$$E[\bar{Y}_{i.}] = \mu \quad \text{and} \quad \text{Var}(\bar{Y}_{i.}) = \sigma_T^2 + \frac{\sigma^2}{r_i} \,.$$

Therefore,

$$E[SST] = \sum_{i=1}^{v} r_i \left(\sigma_T^2 + \frac{\sigma^2}{r_i} + \mu^2\right) - n\left(\frac{\sum r_i^2}{n^2}\sigma_T^2 + \frac{\sigma^2}{n} + \mu^2\right)$$

$$= \left(n - \frac{\sum r_i^2}{n}\right)\sigma_T^2 + (v-1)\sigma^2 \,.$$

Since $MST = SST/(v-1)$, we have

$$E[MST] = c\sigma_T^2 + \sigma^2, \quad \text{where} \quad c = \frac{n^2 - \sum r_i^2}{n(v-1)} \,.$$

Notice that if all r_i are equal to r, then $n = vr$ and $c = r$.

We see that MST is an unbiased estimator of $c\sigma_T^2 + \sigma^2$, not σ_T^2. Nevertheless, we can easily find an unbiased estimator of σ_T^2, since

$$E\left[\frac{MST - MSE}{c}\right] = \sigma_T^2 \,. \tag{17.3.5}$$

It is, unfortunately, possible for the observed value of this estimator to be negative even though σ_T^2 cannot be negative. This will occur when msE happens to be greater than msT, and this is most likely when σ_T^2 is close to zero. If msE is considerably greater than msT, then the model should be questioned, as it is unlikely to be a good description of the data.

Table 17.2 Melting times for three randomly selected flavors of ice cream. Order of observation in parentheses.

Flavor	Time in seconds (order of observation)					
1	924 (1)	876 (2)	1150 (5)	1053 (7)	1041 (10)	1037 (12)
	1125 (15)	1075 (16)	1066 (20)	977 (22)	886 (25)	
2	891 (3)	982 (4)	1041 (8)	1135 (13)	1019 (14)	1093 (18)
	994 (27)	960 (30)	889 (31)	967 (32)	838 (33)	
3	817 (6)	1032 (9)	844 (11)	841 (17)	785 (19)	823 (21)
	846 (23)	840 (24)	848 (26)	848 (28)	832 (29)	

Example 17.3.1 Ice cream experiment

The following experiment was run by Sue Hubbard in 1986 to determine whether or not different flavors of ice cream melt at different speeds. A random sample of three flavors was selected from a large population of flavors offered to the customer by a single manufacturer in May 1986. It is not obvious that the selected flavors are representative of all possible ice cream flavors, since some may include an ingredient that inhibits melting. The theoretical population is therefore the population of all flavors that could be made with ingredients similar to those flavors available.

The three flavors of ice cream were stored in the same freezer in similar-sized containers. For each observation, one teaspoonful of ice cream was taken from the freezer, transferred to a plate, and the melting time at room temperature was observed to the nearest second. Eleven observations were taken on each flavor. These are shown, together with their order of observation, in Table 17.2 and plotted in Figure 17.1.

Now,

$$ssE = \sum\sum y_{it}^2 - 11 \sum \bar{y}_{i.}^2$$

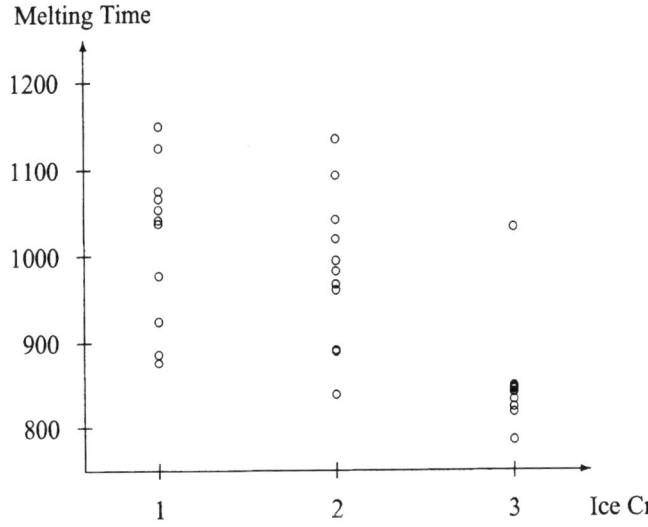

Figure 17.1 Plot of data for the ice cream experiment

$$= 30{,}206{,}485 - 30{,}003{,}028.8181$$
$$= 203{,}456.1819\,.$$

So an unbiased estimate of σ^2 is

$$msE = ssE/(33-3) = 6781.8727 \text{ seconds}^2\,.$$

Similarly,

$$ssT = 11\sum \bar{y}_{i.}^2 - 33\bar{y}_{..}^2$$
$$= 30{,}003{,}028.8181 - 29{,}830{,}018.9393$$
$$= 173{,}009.8787\,.$$

So $msT = ssT/(3-1) = 86{,}504.9394$ seconds^2, and an unbiased estimate of σ_T^2 is given by

$$\frac{msT - msE}{c} = \frac{86{,}504.9394 - 6781.8727}{11}$$
$$= 7247.5515 \text{ seconds}^2\,. \qquad \square$$

17.3.4 Testing Equality of Treatment Effects

When the treatment factor is random, we are interested in the variability of the treatment effects in the entire population of levels, not just those in the experiment. Since the variance of the effects in the population is σ_T^2, the null hypothesis of interest is of the form

$$H_0^T : \sigma_T^2 = 0\,,$$

and the alternative hypothesis is

$$H_A^T : \sigma_T^2 > 0.$$

It would be very convenient if we could use the same hypothesis-testing rule as we used for testing equality of treatment effects in the fixed-effects model. The fixed-effects decision rule was to reject the hypothesis of no difference in the treatments if $msT/msE > F_{v-1, n-v, \alpha}$, (see (3.5.17), page 47). Let us examine the ratio msT/msE in the random-effects one-way model (17.3.1). In Section 17.3.2 we showed that

$$E[MSE] = \sigma^2\,,$$

and in Section 17.3.3 we showed that

$$E[MST] = c\sigma_T^2 + \sigma^2\,,$$

where $c = (n^2 - \sum r_i^2)/n(v-1)$, and if all r_i are equal to r, then $c = r$.

So, if $H_0^T : \sigma_T^2 = 0$ is true, then the expected value of the numerator of the ratio MST/MSE is equal to σ^2, the same as the expected value of the denominator. Then, if H_0^T is true, the ratio should be in the region of 1.0. But if σ_T^2 is large, the expected value of the numerator is larger than the denominator, and the ratio should be large and positive. This situation is similar to that for the fixed-effects case. The only remaining question is whether MST/MSE has an F distribution with $v-1$ and $n-v$ degrees of freedom when H_0^T is true.

It can be shown that
$$SST/(c\sigma_T^2 + \sigma^2) \sim \chi_{v-1}^2 \qquad (17.3.6)$$
and
$$SSE/\sigma^2 \sim \chi_{n-v}^2$$
and that SST and SSE are independent. Consequently, we have
$$\frac{SST/((c\sigma_T^2 + \sigma^2)(v-1))}{SSE/(\sigma^2(n-v))} = \frac{MST/(c\sigma_T^2+\sigma^2)}{MSE/\sigma^2} \sim \frac{\chi_{v-1}^2/(v-1)}{\chi_{n-v}^2/(n-v)} \sim F_{v-1,n-v}, \qquad (17.3.7)$$
and when $\sigma_T^2 = 0$, then
$$\frac{MST}{MSE} \sim F_{v-1,n-v}.$$

Thus, to test $H_0 : \sigma_T^2 = 0$ against $H_A : \sigma_T^2 > 0$, our decision rule is to
$$\text{reject } H_0^T \text{ if } \frac{msT}{msE} > F_{v-1,n-v,\alpha} \qquad (17.3.8)$$
for some chosen value of the significance level α. The test can be set out in an analysis of variance table in the usual way; see Table 17.3. We have included the expected mean squares in the table for easy reference.

Rather than testing whether or not the variance of the population of treatment effects is zero, it may be of more interest to test whether the variance is less than or equal to some proportion of the error variance, that is,
$$H_0^{\gamma T} : \sigma_T^2 \leq \gamma \sigma^2 \text{ and } H_A^{\gamma T} : \sigma_T^2 > \gamma \sigma^2,$$
for some constant γ. From (17.3.7), we see that if $H_0^{\gamma T}$ is true with $\sigma_T^2 = \gamma \sigma^2$, then
$$\frac{MST/(\sigma^2(c\sigma_T^2/\sigma^2 + 1))}{MSE/\sigma^2} = \frac{MST}{MSE(c\gamma + 1)} \sim F_{v-1,n-v}.$$
So, our decision rule (17.3.8) needs only the minor modification of including the constant $(c\gamma + 1)$, that is,
$$\text{reject } H_0^{\gamma T} \text{ if } \frac{msT}{msE} > (c\gamma + 1) F_{v-1,n-v,\alpha}. \qquad (17.3.9)$$

Table 17.3 Analysis of variance table for the random-effects one-way model

Source of Variation	Degrees of Freedom	Sum of Squares	Mean Square	Ratio	Expected Mean Square
Treatments	$v-1$	ssT	$\frac{ssT}{v-1}$	$\frac{msT}{msE}$	$c\sigma_T^2 + \sigma^2$
Error	$n-v$	ssE	$\frac{ssE}{n-v}$		σ^2
Total	$n-1$	$sstot$			
Computational Formulae					
$ssT = \sum_i r_i \bar{y}_{i.}^2 - n\bar{y}_{..}^2$			$ssE = \sum_i \sum_t y_{it}^2 - \sum_i r_i \bar{y}_{i.}^2$		
$sstot = \sum_i \sum_t y_{it}^2 - n\bar{y}_{..}^2$			$c = \frac{n^2 - \sum_i r_i^2}{n(v-1)}$		

17.3 One Random Effect

Table 17.4 Analysis of variance table for the ice cream experiment

Source of Variation	Degrees of Freedom	Sum of Squares	Mean Square	Ratio	p-value
Flavor	2	173009.8788	86504.9394	12.76	0.0001
Error	30	203456.1818	6781.8727		
Total	32	376466.0606			

If we choose $\gamma = 0$, then the decision rule (17.3.9) reduces to rule (17.3.8) for testing the null hypothesis $H_0^T : \sigma_T^2 = 0$ against its alternative hypothesis $H_A^T : \sigma_T^2 > 0$.

Example 17.3.2   Ice cream experiment, continued

The analysis of variance table for the ice cream experiment of Example 17.3.3 is shown in Table 17.4. If we test the null hypothesis that the variance of melting times in the population of ice creams is negligible against the alternative hypothesis that it is not (that is, $H_0^T : \sigma_T^2 = 0$ versus $H_A^T : \sigma_T^2 > 0$) with a Type I error probability of $\alpha = 0.05$, we would reject H_0^T, since

$$msT/msE = 12.76 > F_{2,30,0.05} = 3.32,$$

or equivalently, the p-value is less than 0.05.

In such an experiment there will clearly be considerable error variability in the data due to fluctuations of room temperature and the difficulty of determining the exact time at which the ice cream has melted completely. Variability in the melting time of different flavors is unlikely to be of interest to the experimenter unless it is larger than the error variability. Suppose, therefore, instead of testing the hypothesis H_0^T against H_A^T, we test the null hypothesis $H_0^{\gamma T} : \{\sigma_T^2 \leq \sigma^2\}$ against $H_A^{\gamma T} : \{\sigma_T^2 > \sigma^2\}$. Since there are $r = 11$ observations on each ice cream, the constant is $c = 11$, and the hypothesis-testing rule (17.3.9) with $\gamma = 1.0$ becomes

$$\text{reject } H_0^{\gamma T} \text{ if } \frac{msT}{msE} > (11+1) F_{2,30,\alpha},$$

that is,

$$\text{reject } H_0^{\gamma T} \text{ if } 12.76 > 12 \, F_{2,30,\alpha}.$$

It can be seen from the table in Appendix A.6 that for any practical choice of α, there is not sufficient evidence to reject the null hypothesis. Thus, although the variation in the melting times of the different flavors is significant, so apparently $\sigma_T^2 > 0$, sufficient evidence has not been gathered to be able to claim that the variation is significantly larger than the error variation in the data.   □

17.3.5   Confidence Intervals for Variance Components

We showed in Section 17.3.1, page 596, that the response variable Y_{it} in a random-effects one-way model (17.3.1) has a normal distribution with variance $\sigma^2 + \sigma_T^2$, where σ^2 is the variance of the error variables and σ_T^2 is the variance of the treatment effects in the population. In order to assess the variability of the treatment-effect population, we may wish to calculate a confidence interval for σ_T^2 or, alternatively, for σ_T^2/σ^2 if we want to assess the treatment variability relative to the error variability. Since the latter is the easier calculation, we investigate this first.

Confidence intervals for σ_T^2/σ^2 From (17.3.7), page 602, we know that

$$\frac{MST}{MSE(c\sigma_T^2/\sigma^2 + 1)} \sim F_{v-1,n-v}, \qquad (17.3.10)$$

where $c = (n^2 - \Sigma r_i^2)/(n(v-1))$, and if the r_i are all equal to r, then $c = r$. From this, we can write down an interval in which MST/MSE lies with probability $1 - \alpha$; that is,

$$P\left(F_{v-1,n-v,1-\alpha/2} \leq \frac{MST}{MSE(c\sigma_T^2/\sigma^2 + 1)} \leq F_{v-1,n-v,\alpha/2}\right) = 1 - \alpha.$$

If we rearrange the left-hand inequality, we find that

$$c\sigma_T^2/\sigma^2 \leq \frac{MST}{MSE\, F_{v-1,n-v,1-\alpha/2}} - 1,$$

and similarly for the right-hand inequality,

$$c\sigma_T^2/\sigma^2 \geq \frac{MST}{MSE\, F_{v-1,n-v,\alpha/2}} - 1.$$

So, replacing the random variables by their observed values, we obtain a $100(1-\alpha)\%$ confidence interval for σ_T^2/σ^2 as

$$\frac{1}{c}\left[\frac{msT}{msE\, F_{v-1,n-v,\alpha/2}} - 1\right] \leq \frac{\sigma_T^2}{\sigma^2} \leq \frac{1}{c}\left[\frac{msT}{msE\, F_{v-1,n-v,1-\alpha/2}} - 1\right]. \qquad (17.3.11)$$

A drawback of this interval is that if msT is not much larger than msE (or perhaps smaller), then it is possible for the left-hand end of the interval to be negative even though σ_T^2/σ^2 can never be negative. Although we could replace a negative lower bound by zero, we will not do so, since it can result in a short interval, giving the misleading impression that the experiment was more accurate than it actually was.

For calculation of the interval, remember that $F_{v-1,n-v,\alpha/2}$ denotes the percentile of the $F_{v-1,n-v}$ distribution corresponding to a probability of $\alpha/2$ in the right-hand tail. Also, $F_{v-1,n-v,1-\alpha/2}$ denotes the percentile corresponding to a probability of $\alpha/2$ in the left-hand tail, that is, $1 - \alpha/2$ in the right-hand tail. Since $F_{v-1,n-v,1-\alpha/2}$ is not tabulated in Appendix A.6, it is important to note that

$$F_{v-1,n-v,1-\alpha/2} = (F_{n-v,v-1,\alpha/2})^{-1}. \qquad (17.3.12)$$

Example 17.3.3 Ice cream experiment, continued

In the ice cream experiment of Examples 17.3.3 and 17.3.4, pages 600 and 603, the variance σ_T^2 in the melting times (in seconds) of the population of different flavors of ice cream is

17.3 One Random Effect

substantially greater than zero but not substantially greater than the error variance σ^2. A confidence interval for σ_T^2/σ^2 can be obtained using (17.3.11). The values $msT = 86504.9394$, $msE = 6781.8727$, $v = 3$, $c = r = 11$, and $n = 33$ are obtained from Example 17.3.4. From the table in Appendix A.6, we have

$$F_{2,30,.05} = 3.32 \quad \text{and} \quad F_{2,30,.95} = (F_{30,2,.05})^{-1} = (19.5)^{-1} = 0.0513.$$

Therefore, the confidence interval (17.3.11) becomes

$$\frac{1}{11}\left(\frac{86504.9394}{(6781.8727)(3.32)} - 1\right) \le \frac{\sigma_T^2}{\sigma^2} \le \frac{1}{11}\left(\frac{86504.9394}{(6781.8727)(0.0513)} - 1\right),$$

that is,

$$\sigma_T^2/\sigma^2 \in (0.258, 22.513).$$

This interval is too wide to be of much practical use, since it says that with 95% confidence, σ_T^2 could be 4 times smaller or as much as 22 times bigger than σ^2. However, the result does agree with the test of the null hypothesis $H_0^{\gamma T}$ in Example 17.3.4, since the interval includes the value $\sigma_T^2/\sigma^2 = 1.0$. □

As can be seen from Example 17.3.5, a confidence interval for σ_T^2/σ^2 can be very wide. Not only do we need sufficient numbers of observations on each treatment in the experiment in order to keep a confidence interval narrow, but we also need a sufficiently large selection of treatments to represent the population. In Example 17.3.5, there were only $v = 3$ treatments to represent an entire population of ice cream flavors, and this has contributed to the lack of precision in the experiment. Calculation of sample sizes will be discussed in Section 17.4.

Confidence intervals for σ_T^2 There are various methods of obtaining approximate $100(1 - \alpha)\%$ confidence intervals for σ_T^2. The only method that we shall give here is one that is useful when σ_T^2 is not close to zero and that can be easily adapted when we have more complicated models.

First, remember that an unbiased estimator for σ_T^2 was obtained in equation (17.3.5), page 599, as

$$U = c^{-1}(MST - MSE), \tag{17.3.13}$$

where $c = (n^2 - \Sigma r_i^2)/(n(v - 1))$, and $c = r$ when the sample sizes are equal. If we can determine the distribution of U, then we can easily find a confidence interval for σ_T^2. We know that for the random-effects one-way model, $SST/(c\sigma_T^2 + \sigma^2) \sim \chi_{v-1}^2$ and $SSE/\sigma^2 \sim \chi_{n-v}^2$ and that SST and SSE are independent. The exact distribution of U is therefore based on the difference of two chi-squared distributions each multiplied by a constant of unknown value, and this is not a standard tabulated distribution. However, it can be shown that a reasonable approximation to the true distribution of U/σ_T^2 is a chi-squared distribution divided by its degrees of freedom x, where x is estimated by

$$x = \frac{(msT - msE)^2}{msT^2/(v - 1) + msE^2/(n - v)}. \tag{17.3.14}$$

In other words, the distribution of $xU/E[U]$ is approximately χ_x^2. This result is related to the Satterthwaite approximation that we used in Section 5.6.3, page 116 (Scheffé, 1959, Section 7.5, gives the general result). Using this approximation, we can write down the approximate probability statement

$$P\left(\chi_{x,1-\alpha/2}^2 \leq \frac{xU}{\sigma_T^2} \leq \chi_{x,\alpha/2}^2\right) \approx 1 - \alpha.$$

If we rearrange the left-hand inequality, we obtain

$$\sigma_T^2 \leq \frac{xU}{\chi_{x,1-\alpha/2}^2},$$

and if we rearrange the right-hand inequality, we obtain

$$\frac{xU}{\chi_{x,\alpha/2}^2} \leq \sigma_T^2.$$

Consequently, we obtain an approximate $100(1-\alpha)\%$ confidence interval for σ_T^2 as

$$\frac{xu}{\chi_{x,\alpha/2}^2} \leq \sigma_T^2 \leq \frac{xu}{\chi_{x,1-\alpha/2}^2}, \qquad (17.3.15)$$

where u is the observed value of U; that is,

$$u = c^{-1}(msT - msE). \qquad (17.3.16)$$

Example 17.3.4   Ice cream experiment, continued

Suppose we require a 90% confidence interval for the variance of the melting times of the population of ice creams in the ice cream experiment of Examples 17.3.3 and 17.3.4, pages 600 and 603. Using the information in those examples, we obtain the unbiased estimate (17.3.16) of σ_T^2 as $u = 7247.5526$ seconds^2. The degrees of freedom x of the approximate distribution of U are calculated using (17.3.14), that is,

$$x = \frac{(86504.9394 - 6781.8727)^2}{(86504.9394)^2/2 - (6781.8727)^2/30} \approx 1.7.$$

From Table A.5 we can guess at the approximate values of $\chi_{x,.05}^2$ and $\chi_{x,.95}^2$ as

$$\chi_{1.7,.05}^2 \approx 5.3 \quad \text{and} \quad \chi_{1.7,.95}^2 \approx 0.07.$$

So a 90% confidence interval for σ_T^2 is roughly

$$\sigma_T^2 \in \left(\frac{(1.7)(7247.5515)}{5.3}, \frac{(1.7)(7247.55)}{0.07}\right)$$
$$= (2{,}324.69, \ 176{,}011.97).$$

Taking square roots and converting to minutes, we obtain the approximate 90% confidence interval for the standard deviation of melting times as

$$\sigma_T \in (0.8, \ 7.0) \text{ minutes.}$$

Again, this interval is too wide for practical use, due to the small number of flavors examined from the population. □

17.4 Sample Sizes for an Experiment with One Random Effect

For the fixed-effects one-way analysis of variance model, we looked at two different ways of determining sample sizes. The first method (Section 3.6) was based on the required power of the hypothesis test for detecting whether two treatment effects differ by more than a chosen quantity Δ. The second method (Section 4.5) was based on the required length of confidence intervals for one or more treatment contrasts.

For the random-effects one-way model, we need to determine both the number v of levels of the treatment factor to be observed in the experiment and the number r of observations to be taken on each of these levels. A glance at the formulae (17.3.11) and (17.3.15) shows that a calculation of v and r based on the lengths of confidence intervals will not be straightforward. Both formulae depend on the values of msT and msE, which are unknown prior to the experiment. However, consideration of the variances of the estimators used to develop the confidence intervals helps us determine an appropriate balance between "more treatments" and "more replication."

Consider first the confidence interval for σ_T^2 given in (17.3.15). The confidence interval should be tight if the variance of the unbiased estimator U is small. Assuming equal sample sizes, $U = r^{-1}(MST - MSE)$ has variance

$$\text{Var}(U) = \left(\frac{2n^2}{v^2}\right)\left(\frac{(n\sigma_T^2/v + \sigma^2)^2}{v - 1} + \frac{\sigma^4}{n - v}\right) \quad (17.4.17)$$

for $n > v$. (This follows because MST and MSE are independent, $SST/(r\sigma_T^2 + \sigma^2) \sim \chi^2(v-1)$, $SSE/\sigma^2 \sim \chi^2(n-v)$, and the variance of a chi-squared random variable is twice its degrees of freedom.) We want this variance to be as small as possible. Suppose that the total number of observations $n = rv$ is fixed by budget considerations. We can see from (17.4.17) that if σ_T^2 is much larger than σ^2, the first term in the right-hand set of parentheses governs the size of Var(U), and we require v as large as possible. Even if σ^2 is much larger than σ_T^2, the variance of U is still made small by choosing v quite large. Taking this to the extreme, if we take $n = v$ and $r = 1$, then our estimator would be $U = MST$, for which Var(U) $= 2\sigma_T^4/(v - 1)$ is as small as possible.

We find a somewhat different requirement resulting from a confidence interval for σ_T^2/σ^2. The mean of an F-distribution with $v-1$ and $n-v$ degrees of freedom is $(n-v)/(n-v-2)$. So, from (17.3.10), page 604, with $c = r$, an unbiased estimator of σ_T^2/σ^2 is given by

$$U = \frac{1}{r}\left[\frac{(n-v-2)}{(n-v)}\frac{MST}{MSE} - 1\right],$$

and a narrow confidence interval should be obtained if we choose v and r to make $\text{Var}(U)$ small. The variance of an F-distribution with m and p degrees of freedom is

$$\frac{2p^2(m+p-2)}{m(p-2)^2(p-4)}.$$

It follows from (17.3.10), the definition of U, and $m = v-1$ and $p = n-v$ that when the sample sizes are all equal to r,

$$\text{Var}(U) = \left(r\frac{\sigma_T^2}{\sigma^2}+1\right)^2 \frac{1}{r^2}\left(\frac{2(n-v)^2(n-3)(n-v-2)^2}{(v-1)(n-v-2)^2(n-v-4)(n-v)^2}\right)$$

$$= \left(\frac{\sigma_T^2}{\sigma^2}+\frac{1}{r}\right)^2 \left(\frac{2(n-3)}{(v-1)(n-v-4)}\right),$$

So, if the number of observations n is fixed and if we expect that $\sigma_T^2 \geq \sigma^2$, then the squared term $(\sigma_T^2/\sigma^2)^2$ from the first set of parentheses will be more important for determining the size of the variance, and we need to minimize its coefficient, which is $2(n-3)/((v-1)(n-v-4))$. This requires that $v = (n-3)/2$, which suggests use of $v = n/2$ and $r = 2$. On the other hand, in the more unusual case when σ_T^2 is expected to be much smaller than σ^2, then the squared term $(1/r)^2$ from the first set of parentheses will be more important for determining the size of the variance, and we need the minimum value of

$$\frac{1}{r^2}\left(\frac{2(n-3)}{(v-1)(n-v-4)}\right) = \frac{2v^2(n-3)}{n^2(v-1)(n-v-4)},$$

and this occurs when v is as small as possible.

We can get a feel for how many observations $n = rv$ are needed in total if we examine the power of the hypothesis test for testing $H_0^{\gamma T} : \sigma_T^2 \leq \gamma\sigma^2$ against the alternative hypothesis $H_A^{\gamma T} : \sigma_T^2 > \gamma\sigma^2$ (for a chosen $\gamma \geq 0$). The decision rule was given in (17.3.9), page 602, as

$$\text{reject } H_0^{\gamma T} \text{ if } \frac{msT}{msE} > (c\gamma+1)F_{v-1,n-v,\alpha} = k, \text{ say}. \tag{17.4.18}$$

What is the probability of rejecting $H_0^{\gamma T}$ if the true value of σ_T^2/σ^2 is Δ? In other words, what is the probability that $MST/MSE > k$, when σ_T^2/σ^2 is equal to Δ? This is the power of the test at the value Δ. We can calculate the power from the knowledge that

$$\frac{MST}{MSE(c\sigma_T^2/\sigma^2+1)} \sim F_{v-1,n-v}$$

(see 17.3.12, page 604). If σ_T^2/σ^2 is equal to Δ, then

$$P\left(\frac{MST}{MSE} > k\right) = P\left(\frac{MST}{MSE(c\Delta+1)} > \frac{k}{c\Delta+1}\right).$$

Suppose we stipulate that the power must be π when $\sigma_T^2/\sigma^2 = \Delta$. Then, we must have that

$$\frac{k}{c\Delta+1} = F_{v-1,n-v,\pi}.$$

Remembering from (17.4.18) that $k = (c\gamma + 1)F_{v-1,n-v,\alpha}$, and that $(n-v) = v(r-1)$ and $c = r$ for equal sample sizes, we obtain the equality

$$\frac{F_{v-1,v(r-1),\alpha}}{F_{v-1,v(r-1),\pi}} = \frac{r\Delta + 1}{r\gamma + 1}.$$

So, we need to select γ and α for testing $H_0^{\gamma T}$ together with Δ and π. Then we can try to determine v and r by trial and error as illustrated in Example 17.4. Since $F_{v-1,v(r-1),\pi} = (F_{v(r-1),v-1,1-\pi})^{-1}$, we try to find values of v and r such that

$$(F_{v-1,v(r-1),\alpha})(F_{v(r-1),v-1,1-\pi}) \leq \frac{r\Delta + 1}{r\gamma + 1}. \tag{17.4.19}$$

Example 17.4.1 Ice cream experiment, continued

In Example 17.3.4, page 603, we were unable to reject the hypothesis $H_0^{\gamma T} : \sigma_T^2 \leq \sigma^2$ in favor of the hypothesis $H_0^{\gamma T} : \sigma_T^2 > \sigma^2$ at a significance level of $\alpha = 0.05$. Suppose we wish to repeat this experiment, still with $\gamma = 1.0$ and a Type I error probability of $\alpha = 0.05$. Suppose further that we would like to reject the hypothesis with high probability (say $\pi = 0.95$) if the true value of σ_T^2/σ^2 is as high as $\Delta = 2.0$. How many ice cream flavors should we look at and how many observations should we take on each?

From (17.4.19), we need to find v and r such that

$$(F_{v-1,v(r-1),.05})(F_{v(r-1),v-1,.05}) \leq \frac{23}{12} = 1.92.$$

For the moment set $r = 11$, which is the value used by the experimenter in the ice cream experiment. Then we have

v	$F_{v-1,10v,.05}$	$F_{10v,v-1,.05}$	Product	Action
4	2.84	8.59	24.40 > 1.92	Increase v
100	1.26	1.32	1.66 < 1.92	Decrease v
80	1.30	1.34	1.74 < 1.92	Decrease v
60	1.43	1.38	1.97 ≈ 1.92	Stop

So, v around 60 should be reasonable, requiring 660 observations in total. Now, let us double the value of r to 22. Then we have

v	$F_{v-1,21v,.05}$	$F_{21v,v-1,.05}$	Product	Action
50	1.38	1.46	2.01	Stop

So, the combination $v = 50$, $r = 22$ provides adequate power, but this requires $n = 1100$ observations. Let us now reduce r to 3 then

v	$F_{v-1,2v,.05}$	$F_{2v,v-1,.05}$	Product	Action
60	1.47	1.43	2.10	Increase v
80	1.38	1.40	1.93	Stop

So v in the region of 80 would be fine, requiring $n = 240$ observations. In Exercise 2, the reader is asked to determine whether the use of $r = 2$ would require a smaller total number of observations. To obtain a markedly smaller number of observations, we would need to relax our requirement of such a high power to detect $\sigma_T^2/\sigma^2 = 2$. □

17.5 Checking Assumptions on the Model

The simplest way to check the assumptions on the one-way random-effects model is to use residual plots in much the same way as for a fixed-effects one-way model. We need to check that the error assumptions are valid, that is,

$$\epsilon_{it} \sim N(0, \sigma^2), \quad t = 1, \ldots, r_i,$$

for each treatment factor level i ($i = 1, \ldots, v$), and also that the assumptions on the random effect T_i are valid, that is,

$$T_i \sim N(0, \sigma_T^2), \quad i = 1, \ldots, v,$$

and that all random variables are mutually independent.

Checking the error assumptions is straightforward, since we proceed in exactly the same way as for the fixed-effects one-way model. We replace T_i in the model, temporarily, by the fixed effect τ_i. Then the residuals are defined as usual as

$$\widehat{e}_{it} = y_{it} - \widehat{y}_{it} = y_{it} - \overline{y}_{i.}.$$

These are then standardized to obtain the standardized residuals z_{it} with standard deviation 1.0. We plot the standardized residuals versus treatment-factor levels, versus $\widehat{y}_{it}$, versus order and versus normal scores, as in Chapter 5, to check for outliers, independence, constant variance, and normality. Non-independence between the ϵ_{it}'s and the T_i's is not easy to detect, but unequal variances of the ϵ_{it}'s indicates one form of the problem.

The normality assumption on the random effect T_i can be checked when the sample sizes are equal, unless v is too small. The treatment averages $\overline{Y}_{i.}$ should have a $N(\mu, \sigma_T^2 + \sigma^2/r)$ distribution. So, if we plot the observed averages $\overline{y}_{i.}$ against their corresponding normal scores, we should roughly obtain a straight line that cuts the vertical axis at about μ and that has slope about $(\sigma_T^2 + \sigma^2/r)^{1/2}$. It is important to check the normality assumption, since the analysis for random-effects models is not robust to nonnormality of the random effects.

We can also use this plot to check for outliers among the observed treatments. In an experiment such as the ice cream experiment, where only $v = 3$ levels of the treatment factor were observed, there is not enough data to be able to examine the distribution of the T_i's in any detail. In Section 17.10 we will illustrate the assumption-checking procedures using the SAS computer software and the data from the clean wool experiment that was described in Section 17.2, page 595.

17.6 Two or More Random Effects

17.6.1 Models and Examples

In the ice cream experiment of Example 17.3.3, page 600, we modeled the ice cream effect as a random effect, since we were interested in the variability of the melting rates of varieties of a large population of all possible ice creams with similar ingredients. If the experimenter had also been interested in whether or not the container affects the melting time, then she might have randomly selected a number b of containers from the population of all possible containers. If one ice cream melts faster than another ice cream in one container, then it might be safe to assume that it melts faster, and by the same amount, in another container. In other words, the assumption of no ice cream×container interaction might be reasonable. In this case a random two-way main-effects model (with no interaction) would be a possible model; that is,

$$Y_{ijt} = \mu + A_i + B_j + \epsilon_{ijt}, \qquad (17.6.20)$$

$$A_i \sim N(0, \sigma_A^2), \quad B_j \sim N(0, \sigma_B^2), \quad \epsilon_{ijt} \sim N(0, \sigma^2),$$

A_i's, B_j's and ϵ_{ijt}'s are all mutually independent

$$t = 1, \ldots, r_{ij}, \quad i = 1, \ldots, a, \quad j = 1, \ldots, b.$$

where A_i is the effect of the ith ice cream randomly selected from the population of ice creams whose effects on melting times follow a normal distribution with variance σ_A^2 for each container, and where B_j is the effect of the jth container randomly selected from the population of containers whose effects on the melting times follow a normal distribution with variance σ_B^2 for each ice cream. The number of observations r_{ij} to be taken on the (ij)th ice cream–container combination needs to be determined. Normally, we would select the r_{ij}'s to be equal if possible.

Alternatively, it may be expected that a slightly thicker container would show a greater difference in melting times of ice creams than would a thinner container. In other words, an interaction may be expected. In this case, we would add to model (17.6.20) a random effect representing the interaction, as shown in the random-effects two-way complete model (17.6.21):

$$Y_{ijt} = \mu + A_i + B_j + (AB)_{ij} + \epsilon_{ijt} \qquad (17.6.21)$$

$$A_i \sim N(0, \sigma_A^2), \quad B_j \sim N(0, \sigma_B^2)$$

$$(AB)_{ij} \sim N(0, \sigma_{AB}^2), \quad \epsilon_{ijt} \sim N(0, \sigma^2)$$

A_i's, B_j's, $(AB)_{ij}$'s and ϵ_{ijt}'s are mutually independent

$$t = 1, \ldots, r_{ij}, \quad i = 1, \ldots, a, \quad j = 1, \ldots, b.$$

If σ_{AB}^2 is positive, then there are *AB effects* present—namely, main effects and interactions for the factors A and B. If σ_A^2 or σ_B^2 is positive, then the corresponding main effects are present.

Example 17.6.1 Ammunition experiment

W. A. Thompson, Jr. and J. R. Moore in the 1963 volume of *Technometrics* describe an experiment concerning the muzzle velocity characteristics of ammunition for a field artillery weapon. They describe the ammunition as follows:

> Propelling charges and projectiles for this type of weapon are manufactured and stored separately in a such a way that any charge might be employed by the user to propel any projectile. ... Both projectiles and charges are grouped into lots at the time of manufacture, each lot consisting of a large number of individual units assembled during a short period of time using essentially uniform components. Thus, it is hoped

Table 17.5 Data for the ammunition experiment

		Charge Lot			
		1	2	3	4
	1	63	56	69	78
		78	58	63	79
	2	71	60	64	65
Projectile		70	65	68	77
Lot	3	72	58	69	63
		55	55	71	72
	4	70	60	66	73
		64	71	68	79

Source: Thompson, W. A. Jr. and Moore, J. R. (1963).
Copyright © 1963 American Statistical Association.
Reprinted with permission.

that the round to round dispersion [variability] in velocity will be reduced by using charges and projectiles from within lots.

The experiment involved the examination of a random sample of four charge lots (factor A with levels 1, 2, 3, 4) selected at random from a large population of charge lots, and four projectile lots (factor B with levels 1, 2, 3, 4) selected at random from a large population of projectile lots. A weapon surveillance test was conducted using one weapon under uniform ballistic conditions. The muzzle velocities were measured to the nearest foot per second. These are shown in Table 17.5, except that a constant has been added to each recorded velocity.

Since the lots involved in the experiment were randomly selected from large populations, a random-effects two-way complete model (17.6.21) was used in the analysis. □

In an experiment with more than two random treatment factors, variables representing all of the main effects of the factors and some or all of their interactions would be included in the model in the obvious way. For example, an experiment with five random treatment factors A, B, C, D, G, in which interactions AB, AC, BC, CD, and ABC were thought to be nonnegligible, would be modeled as follows:

$$Y_{ijklmt} = \mu + A_i + B_j + C_k + D_l + G_m$$
$$+ (AB)_{ij} + (AC)_{ik} + (BC)_{jk} + (CD)_{kl} + (ABC)_{ijk} + \epsilon_{ijklmt},$$

$A_i \sim N(0, \sigma_A^2)$, $B_j \sim N(0, \sigma_B^2)$, $C_k \sim N(0, \sigma_C^2)$, $D_l \sim N(0, \sigma_D^2)$, $G_m \sim N(0, \sigma_G^2)$,
$(AB)_{ij} \sim N(0, \sigma_{AB}^2)$, $(AC)_{ik} \sim N(0, \sigma_{AC}^2)$, $(BC)_{jk} \sim N(0, \sigma_{BC}^2)$, $(CD)_{kl} \sim N(0, \sigma_{CD}^2)$,
$(ABC)_{ijk} \sim N(0, \sigma_{ABC}^2)$, $\epsilon_{ijklmt} \sim N(0, \sigma^2)$,

all random variables on the right-hand side
of the model are mutually independent,

$t = 1, \ldots, r_{ijklm}$, $i = 1, \ldots, a$, $j = 1, \ldots, b$,
$k = 1, \ldots, c$, $l = 1, \ldots, d$, $m = 1, \ldots, g$.

As for the fixed-effects models, if a high-order interaction is included in the model, then so are all of its "subinteractions" and constituent main effects; that is, if $(ABC)_{ijk}$ is in the model, so are $(AB)_{ij}$, $(AC)_{ik}$, $(BC)_{jk}$, A_i, B_j, and C_k.

17.6.2 Checking Model Assumptions

We may check the error assumptions by replacing temporarily all of the random effects by fixed effects, calculating the standardized residuals, and examining the residual plots in the usual way. Checking the assumptions of each random effect is not easy, since in a two-way or higher-way model there are generally few levels of each treatment factor observed, and the cell averages are not independent. Consequently, we will omit the model checks for the random-effect assumptions, and hope that any severe problems will show up through the analysis of the residuals.

17.6.3 Estimation of σ^2

In Section 17.3.2 we found that for the one-way random-effects model, an unbiased estimate of σ^2 was given by msE, where msE was calculated exactly as for the fixed-effects one-way model. Perhaps this should not be surprising, since msE measures the variability in the data that is not accounted for by those sources of variation that were ignored in the experiment. An unbiased estimate for σ^2 in *any* random-effects model can be obtained from its fixed-effects model counterpart.

Example 17.6.2 Unbiased estimate of σ^2

We will show that an unbiased estimate of σ^2 in the random-effects two-way complete model is $msE = ssE/(n-v)$, where

$$ssE = \left[\sum_i \sum_j \sum_t y_{ijt}^2 - \sum_i \sum_j r_{ij} \overline{y}_{ij.}^2 \right],$$

as in (6.4.17) for the fixed-effects two-way complete model. First, note that

$$E[Y_{ijt}] = \mu \quad \text{and} \quad \text{Var}(Y_{ijt}) = \sigma_A^2 + \sigma_B^2 + \sigma_{AB}^2 + \sigma^2$$

for the random-effects two-way complete model. Also,

$$\overline{Y}_{ij.} = \mu + A_i + B_j + (AB)_{ij} + \sum_t \epsilon_{ijt}/r_{ij}.$$

So,

$$E[\overline{Y}_{ij.}] = \mu \quad \text{and} \quad \text{Var}(\overline{Y}_{ij.}) = \sigma_A^2 + \sigma_B^2 + \sigma_{AB}^2 + \sigma^2/r_{ij}.$$

Thus, the expected value of the random variable SSE is

$$E[SSE] = E\left[\sum_i \sum_j \sum_t Y_{ijt}^2 - \sum_i \sum_j r_{ij} \overline{Y}_{ij.}^2\right]$$

$$= \sum_{i=1}^{a} \sum_{j=1}^{b} \sum_{t=1}^{r_{ij}} (\text{Var}(Y_{ijt}) + E[Y_{ijt}]^2)$$

$$- \sum_{i=1}^{a} \sum_{j=1}^{b} r_{ij}(\text{Var}(\overline{Y}_{ij.}) + E[\overline{Y}_{ij.}]^2)$$

$$= \left(\sum_{i=1}^{a}\sum_{j=1}^{b} r_{ij}\right)\sigma^2 - \sum_{i=1}^{a}\sum_{j=1}^{b}\sigma^2 = (n-v)\sigma^2,$$

where $n = \Sigma\Sigma r_{ij}$ and $v = ab$. Consequently,

$$E[MSE] = E[SSE/(n-v)] = \sigma^2. \qquad \square$$

17.6.4 Estimation of Variance Components

In Section 17.3.3, page 598, we found that for the random-effects one-way model,

$$E[MST] = c\sigma_T^2 + \sigma^2, \quad \text{where } c = \frac{n^2 - \sum r_i^2}{n(v-1)},$$

where MST is the mean square for treatments from the fixed-effects one-way model, and where $c = r$ if all the sample sizes are equal. From this, we were able to find an unbiased estimator for σ_T^2, namely $(MST - MSE)/c$.

For more complicated models, we will also be able to find unbiased estimators for the variance components using the fixed-effects mean squares, but each estimator must be calculated individually.

Example 17.6.3 Unbiased estimate of σ_B^2

Suppose an experiment involves three random treatment factors A, B, and D having a, b, and d levels, respectively, and suppose r observations are taken on each of the $v = abd$ combinations. If the only interactions that are expected to be nonnegligible are AB and BD, then, the model is

$$Y_{ijk} = \mu + A_i + B_j + D_k + (AB)_{ij} + (BD)_{jk} + \epsilon_{ijkt},$$

$$t = 1, \ldots, r, \quad i = 1, \ldots, a, \quad j = 1, \ldots, b, \quad k = 1, \ldots, d$$

with the usual assumptions about the distributions of the random treatment effects and error variables.

17.6 Two or More Random Effects

Suppose we want an unbiased estimator for σ_B^2. We start by investigating $E[MSB]$, where $MSB = SSB/(b-1)$. Using rule 4, page 202, the sum of squares for B is

$$ssB = adr \sum_{j=1}^{b} \overline{y}_{.j..}^2 - abdr\overline{y}_{....}^2 .$$

Now,

$$E[\overline{Y}_{.j..}] = E[\overline{Y}_{....}] = \mu$$

and

$$\text{Var}[\overline{Y}_{.j..}] = \frac{\sigma_A^2}{a} + \sigma_B^2 + \frac{\sigma_D^2}{d} + \frac{\sigma_{AB}^2}{a} + \frac{\sigma_{BD}^2}{d} + \frac{\sigma^2}{adr},$$

$$\text{Var}[\overline{Y}_{....}] = \frac{\sigma_A^2}{a} + \frac{\sigma_B^2}{b} + \frac{\sigma_D^2}{d} + \frac{\sigma_{AB}^2}{ab} + \frac{\sigma_{BD}^2}{bd} + \frac{\sigma^2}{abdr}.$$

Consequently,

$$E[SSB] = adr \sum_j \left[\text{Var}(\overline{Y}_{.j..}) + E[\overline{Y}_{.j..}]^2 \right] - abdr \left[\text{Var}(\overline{Y}_{....}) + E[\overline{Y}_{....}]^2 \right]$$

$$= adr(b-1)\sigma_B^2 + dr(b-1)\sigma_{AB}^2 + ar(b-1)\sigma_{BD}^2 + (b-1)\sigma^2.$$

So,

$$E[MSB] = adr\sigma_B^2 + dr\sigma_{AB}^2 + ar\sigma_{BD}^2 + \sigma^2. \tag{17.6.22}$$

Thus, if we wish to find an unbiased estimator for σ_B^2, we must find unbiased estimators also for σ_{AB}^2 and σ_{BD}^2. The logical place to look for these is at $E[MS(AB)]$ and $E[MS(BD)]$. We have

$$ss(AB) = dr \sum_{i=1}^{a} \sum_{j=1}^{b} \overline{y}_{ij..}^2 - bdr \sum_{i=1}^{a} \overline{y}_{i...}^2 - adr \sum_{j=1}^{b} \overline{y}_{.j..}^2 + abdr\overline{y}_{....}^2 ,$$

$$ss(BD) = ar \sum_{j=1}^{b} \sum_{k=1}^{d} \overline{y}_{.jk.}^2 - adr \sum_{j=1}^{b} \overline{y}_{.j..}^2 - abr \sum_{k=1}^{d} \overline{y}_{..k.}^2 + abdr\overline{y}_{....}^2 ,$$

and

$$E[\overline{Y}_{ij..}] = E[\overline{Y}_{i...}] = E[\overline{Y}_{.j..}] = E[\overline{Y}_{.jk.}] = E[\overline{Y}_{..k.}] = [\overline{Y}_{....}] = \mu,$$

$$\text{Var}[\overline{Y}_{ij..}] = \sigma_A^2 + \sigma_B^2 + \frac{\sigma_D^2}{d} + \sigma_{AB}^2 + \frac{\sigma_{BD}^2}{d} + \frac{\sigma^2}{dr},$$

$$\text{Var}[\overline{Y}_{i...}] = \sigma_A^2 + \frac{\sigma_B^2}{b} + \frac{\sigma_D^2}{d} + \frac{\sigma_{AB}^2}{b} + \frac{\sigma_{BD}^2}{bd} + \frac{\sigma^2}{bdr},$$

$$\text{Var}[\overline{Y}_{.jk.}] = \frac{\sigma_A^2}{a} + \sigma_B^2 + \sigma_D^2 + \frac{\sigma_{AB}^2}{a} + \sigma_{BD}^2 + \frac{\sigma^2}{ar},$$

$$\text{Var}[\overline{Y}_{..k.}] = \frac{\sigma_A^2}{a} + \frac{\sigma_B^2}{b} + \sigma_D^2 + \frac{\sigma_{AB}^2}{ab} + \frac{\sigma_{BD}^2}{b} + \frac{\sigma^2}{abr},$$

as well as

$$E[SS(AB)] = \left(dr\sum_i\sum_j \overline{Y}_{ij..}^2 - bdr\sum_i \overline{Y}_{i...}^2\right) - (E[SSB])$$

$$= adr(b-1)\sigma_B^2 + adr(b-1)\sigma_{AB}^2 + ar(b-1)\sigma_{BD}^2 + a(b-1)\sigma^2$$
$$- adr(b-1)\sigma_B^2 - dr(b-1)\sigma_{AB}^2 - ar(b-1)\sigma_{BD}^2 - (b-1)\sigma^2$$
$$= dr(a-1)(b-1)\sigma_{AB}^2 + (a-1)(b-1)\sigma^2.$$

So,

$$E[MS(AB)] = dr\sigma_{AB}^2 + \sigma^2.$$

Similarly,

$$E[MS(BD)] = ar\sigma_{BD}^2 + \sigma^2.$$

Thus, an unbiased estimator for σ_B^2 is

$$U = (MSB - MS(AB) - MS(BD) + MSE)/(adr),$$

and an unbiased estimate for σ_B^2 is therefore

$$u = (msB + ms(AB) + ms(BD) + msE)/(adr). \qquad \square$$

Calculation of expected mean squares is quite time-consuming, as was seen in Example 17.6.4. However, when sample sizes are all equal, we can exploit the pattern that emerges in studying such examples. All of the variance components that are involved in $E[MSB]$ in (17.6.22) are those whose random effects include the same subscript as for B in the model. Specifically, B has subscript j in the model, and a j also occurs as subscript in $(AB)_{ij}, (BD)_{jk}$, and ϵ_{ijkl}. The constant in front of each variance component is the number of observations taken on each combination of subscripts; that is, there are adr observations on each of the b levels of B, there are dr observations on each of the ab levels of AB, and so on.

A similar pattern can be seen for $E[MS(AB)]$ and $E[MS(BD)]$. This gives us a general rule when sample sizes are equal (which we add to the 16 rules in Chapter 7):

17. To obtain the expected mean square for a main effect or interaction in a random-effects model, first note the subscripts on the term representing that effect in the model. Write down a variance component σ^2 for the effect of interest, for the error, and for every interaction whose term in the model includes the noted set of subscripts. Multiply each variance component except σ^2 by the number of observations taken on each level or combination of levels of the corresponding main effect or interaction. Add up the terms.

17.6.5 Confidence Intervals for Variance Components

In the previous subsection a rule was given for calculating the expected mean square corresponding to each term in the model, when the sample sizes are equal. For unequal sample sizes, the mean squares and expected mean squares are best calculated by a computer program.

From the list of expected mean squares, we can find an unbiased estimator for any given variance component, say σ_*^2. Again, this was illustrated in Example 17.6.4. The estimator can always be a linear combination of mean squares, which, in general, we can write as $U = \Sigma k_i (MS)_i$, where k_i is the constant in front of the ith mean square in the linear combination. Then, an approximation to the distribution of xU/σ_*^2 is a chi-squared distribution with x degrees of freedom, where

$$x = \frac{(\Sigma k_i (ms)_i)^2}{\Sigma k_i^2 (ms)_i^2 / x_i} \qquad (17.6.23)$$

and where x_i is the number of degrees of freedom corresponding to the ith mean square and $(ms)_i$ is the observed value of the ith mean square in the linear combination. An example of this formula was given in Section 17.3.5, page 605, for the one-way model. A more complicated example is given below.

Example 17.6.4 Calculation of degrees of freedom

We continue Example 17.6.4, which involved a random-effects model with five random effects A_i, B_j, and D_k (corresponding to main effects of factors A, B, and D), and $(AB)_{ij}$ and $(BD)_{jk}$ (corresponding to interactions AB and BD). An unbiased estimator for σ_B^2 was shown to be

$$U = \Sigma k_i (MS)_i = MSB/(adr) - MS(AB)/(adr) - MS(BD)/(adr) + MSE/(adr).$$

An approximation to the distribution of xU/σ_B^2 is a χ_x^2 distribution, where x is given by (17.6.23), that is,

$$x = \frac{[(msB - ms(AB) - ms(BD) + msE)/(adr)]^2}{\frac{msB^2}{(adr)^2(b-1)} + \frac{ms(AB)^2}{(adr)^2(a-1)(b-1)} + \frac{ms(BD)^2}{(adr)^2(b-1)(d-1)} + \frac{msE^2}{(adr)^2 df}}$$

$$= \frac{[msB - ms(AB) - ms(BD) + msE]^2}{\frac{msB^2}{(b-1)} + \frac{ms(AB)^2}{(a-1)(b-1)} + \frac{ms(BD)^2}{(b-1)(d-1)} + \frac{msE^2}{df}},$$

where df is the number of degrees of freedom for error, which can be obtained, as usual, by subtraction. In this example, df is equal to

$$\begin{aligned} df &= (abdr - 1) - (a-1) - (b-1) - (d-1) \\ &\quad - (a-1)(b-1) - (b-1)(d-1) \\ &= ab(dr-1) - b(d-1) + 1. \end{aligned}$$
□

Once we know an approximate distribution for a variance component estimator, we can easily write down a probability statement and convert it to a confidence interval. Suppose that $U = \Sigma k_i (MS)_i$ is an unbiased estimator for σ_*^2 and that xU/σ_*^2 has approximately a χ_x^2 distribution; then

$$P\left(\chi_{x,1-\alpha/2}^2 \leq xU/\sigma_*^2 \leq \chi_{x,\alpha/2}^2\right) \approx 1 - \alpha.$$

Then, if we observe the value of U to be $u = \Sigma k_i (ms)_i$, by manipulating the two inequalities in the probability statement we can obtain the following approximate $100(1-\alpha)\%$ confidence interval:

$$\frac{xu}{\chi^2_{x,\alpha/2}} \le \sigma^2_* \le \frac{xu}{\chi^2_{x,1-\alpha/2}}, \qquad (17.6.24)$$

where x is calculated as in (17.6.23). If the estimate u is negative or the calculated degrees of freedom x is extremely small, then this approximate confidence interval procedure should not be used.

Example 17.6.5 Ammunition experiment, continued

The ammunition experiment was described in Example 17.6.1, page 611, and the data were given in Table 17.5. A random-effects two-way complete model (17.6.21) was used. The mean squares for this model are calculated in exactly the same way as for the fixed-effects two-way complete model, and these are shown in the analysis of variance table, Table 17.6. Also listed in the table are the expected mean squares calculated as in rule 17, page 616.

For example, to calculate the expected mean square for A, we note that the subscript for the term A_i in the model is i, and also that i is included among the subscripts of the terms $(AB)_{ij}$ and ϵ_{ijt}. This means that the expected mean square must include the three variance components

$$\sigma_A^2, \ \sigma_{AB}^2, \ \text{and} \ \sigma^2.$$

The constant in front of σ_A^2 is 8, the number of observations on each charge lot, whereas the constant in front of σ_{AB}^2 is 2, the number of observations on each combination of charge lot and projectile lot.

The expected mean square for AB, $E[MS(AB)] = 2\sigma_{AB}^2 + \sigma^2$, contains only two terms, since only the two terms $(AB)_{ij}$ and ϵ_{ijt} in the model contain both i and j as subscripts. An unbiased estimator for σ_A^2 is given by $U = (MSA - MS(AB))/8$. Also, xU/σ_A^2 has approximately a χ_x^2 distribution, where x is calculated as in (17.6.23). Thus, an unbiased estimate of σ_A^2 from this experiment is

$$u = (msA - ms(AB))/8 = (223.04 - 28.63)/8 = 24.30,$$

Table 17.6 Two-way analysis of variance table for the ammunition experiment

Source of Variation	Degrees of Freedom	Sum of Squares	Mean Square	Expected Mean Square
Charge (A)	3	669.12	223.04	$8\sigma_A^2 + 2\sigma_{AB}^2 + \sigma^2$
Projectile (B)	3	92.12	30.71	$8\sigma_B^2 + 2\sigma_{AB}^2 + \sigma^2$
Interaction (AB)	9	257.63	28.63	$2\sigma_{AB}^2 + \sigma^2$
Error	16	516.00	32.25	σ^2
Total	31	1534.87		

17.6 Two or More Random Effects

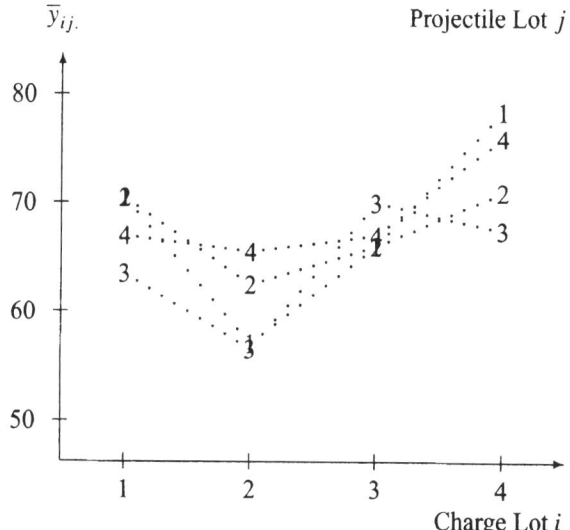

Figure 17.2
Plot of average velocity against charge lot i by projectile lot j for the ammunition experiment

and the number of degrees of freedom of the associated chi-squared distribution is

$$x = \frac{24.30^2}{\frac{223.04^2}{(8^2)(3)} + \frac{28.63^2}{(8^2)(9)}} = 2.27.$$

Therefore, an approximate 90% confidence interval (17.6.24) for σ_A^2, the variance of velocities arising from the population of charge lots, is

$$\frac{(2.27)(24.3)}{\chi^2_{2.27, 0.05}} \leq \sigma_A^2 \leq \frac{(2.27)(24.3)}{\chi^2_{2.27, 0.95}},$$

and since $\chi^2_{2.27, 0.05} \approx 6.5$ and $\chi^2_{2.27, 0.05} \approx 0.17$, the approximate 90% confidence interval, in units of (feet per second)^2, is

$$8.49 \leq \sigma_A^2 \leq 324.48.$$

An approximate 90% confidence interval for the standard deviation of the velocity (in feet per second), obtained by taking square roots, is

$$2.91 \leq \sigma_A \leq 18.01.$$

Before leaving this example, we note that an unbiased estimate for σ_{AB}^2 calculated in this way is actually negative, since

$$u = (ms(AB) - msE)/2 = -1.81,$$

and the calculation for x, the number of degrees of freedom of the associated χ^2 distribution, is

$$x = \frac{(-1.81)^2}{\frac{28.63^2}{(2^2)(9)} + \frac{32.25^2}{(2^2)(16)}} = 0.084.$$

Thus, we are not able to say anything sensible about the variance of the interaction, other than that it appears to be very small. The interaction plot in Figure 17.2 for the lots included in the experiment supports this conclusion.

17.6.6 Hypothesis Tests for Variance Components

In order to focus the discussion, we will use the random-effects model of Example 17.6.4 page 617; that is,

$$Y_{ijkt} = \mu + A_i + B_j + D_k + (AB)_{ij} + (BD)_{jk} + \epsilon_{ijkt},$$

$$t = 1, \ldots, r; \ i = 1, \ldots, a; \ j = 1, \ldots, b; \ k = 1, \ldots, d,$$

together with the usual assumptions about the distributions of the random variables. Some of the expected mean squares for this model were calculated in Example 17.6.4 and are listed, together with the remaining mean squares, in Table 17.7.

Testing the hypothesis $H_0^{AB} : \{\sigma_{AB}^2 = 0\}$ against its alternative hypothesis $H_A^{AB} : \{\sigma_{AB}^2 > 0\}$ is straightforward, since the corresponding expected mean square looks very similar to the situation that we had in the one-way model. If H_0^{AB} is true, then the numerator of the ratio $ms(AB)/msE$ is expected to be σ^2, the same as the denominator. Otherwise, the numerator is expected to be larger. Consequently, the decision rule is

$$\text{reject } H_0^{AB} \text{ if } \frac{ms(AB)}{msE} > F_{(a-1)(b-1),df,\alpha}$$

as usual, where the number of error degrees of freedom is

$$df = ab(dr - 1) - b(d - 1) + 1.$$

We could modify this test as in (17.3.9), page 602, so that the decision rule for testing $H_0^{\gamma AB} : \{\sigma_{AB}^2 \leq \gamma\sigma^2\}$ against $H_A^{\gamma AB} : \{\sigma_{AB}^2 > \gamma\sigma^2\}$ is

$$\text{reject } H_0^{\gamma AB} \text{ if } \frac{ms(AB)}{msE} > (1 + dr\gamma)F_{(a-1)(b-1),df,\alpha}.$$

We have similar tests for H_0^{BD} against H_A^{BD}, and $H_0^{\gamma BD}$ against $H_A^{\gamma BD}$.

Table 17.7 Expected mean squares and degrees of freedom for a random-effects three-way model with two interactions

Effect	Degrees of freedom	Expected mean square
A	$a - 1$	$bdr\sigma_A^2 + dr\sigma_{AB}^2 + \sigma^2$
B	$b - 1$	$adr\sigma_B^2 + dr\sigma_{AB}^2 + ar\sigma_{BD}^2 + \sigma^2$
D	$d - 1$	$abr\sigma_D^2 + ar\sigma_{BD}^2 + \sigma^2$
AB	$(a-1)(b-1)$	$dr\sigma_{AB}^2 + \sigma^2$
BD	$(b-1)(d-1)$	$ar\sigma_{BD}^2 + \sigma^2$
Error	df	σ^2

17.6 Two or More Random Effects

Testing $H_0^A : \sigma_A^2 = 0$ against $H_A^A : \sigma_A^2 > 0$ is more complicated. Until now, we have used the same test statistics as we used in the fixed-effects case. But if we try to use msA/msE to test H_0^A, we have a problem. If H_0^A is true, so that $\sigma_A^2 = 0$, the expected value of the numerator is $E[MSA] = d r \sigma_{AB}^2 + \sigma^2$, while that of the denominator is $E[msE] = \sigma^2$. This suggests two things:

(i) we should use $ms(AB)$ as the denominator, not msE, and

(ii) we should question whether it makes sense to test H_0^A if the interaction AB is significant.

The second point is, of course, exactly the same point that arose in the fixed-effects model, and the answer is usually "no, it makes no sense." Consequently, we generally test a main effect only when that factor is not involved in any significant interactions. Nevertheless, we shall still use the interaction mean square as the denominator in case an incorrect decision was made regarding the interaction. Consequently, the decision rule for testing H_0^A against H_A^A is

reject H_0^A if $MSA/MS(AB) > F_{a-1,(a-1)(b-1),\alpha}$.

Notice that the second set of degrees of freedom for the F-distribution is the degrees of freedom corresponding to the denominator of the ratio. The test for $H_0^D : \{\sigma_D^2 = 0\}$ is similar.

Obtaining a suitable denominator for testing the null hypothesis $H_0^B : \{\sigma_B^2 = 0\}$ versus the alternative hypothesis $H_0^B : \{\sigma_B^2 > 0\}$ is harder again. If H_0^B is true, so that $\sigma_B^2 = 0$, then the expected value of MSB is

$$E[MSB] = d r \sigma_{AB}^2 + a r \sigma_{BD}^2 + \sigma^2.$$

We would generally want to test this hypothesis only if we believed that the interactions AB and BD were both negligible. Yet to be on the safe side, we would like a denominator with the same expected value. It can be verified that

$$E[U] = E[MS(AB) + MS(BD) - MSE] = d r \sigma_{AB}^2 + a r \sigma_{BD}^2 + \sigma^2.$$

As in Section 17.3.5, $xU/(d r \sigma_{AB}^2 + a r \sigma_{BD}^2 + \sigma^2)$ has approximately a chi-squared distribution with degrees of freedom x calculated as in (17.6.23), page 617; that is,

$$x = \frac{[ms(AB) + ms(BD) - msE]^2}{\frac{[ms(AB)]^2}{(a-1)(b-1)} + \frac{[ms(BD)]^2}{(b-1)(d-1)} + \frac{[msE]^2}{df}}.$$

Therefore, if H_0 is true, msB/U has approximately an $F_{b-1,x}$ distribution. So to test H_0^B against H_A^B, the decision rule is

reject H_0^B if $\dfrac{msB}{ms(AB) + ms(BD) - msE} > F_{(b-1),x,\alpha}$.

Example 17.6.6 Ammunition experiment, continued

An unbiased estimate for the variance of the muzzle velocities due to the population of charge lots (factor A) was calculated to be $u = 24.3$ (feet per second)^2 in Example 17.6.5 for the ammunition experiment. A question is whether this value could be due to random error or whether the variance is really sizable; that is, we wish to test the hypothesis $H_0^A : \{\sigma_A^2 = 0\}$ against the alternative hypothesis $H_A^A : \{\sigma_A^2 > 0\}$. The interaction variability was found to be very small in Example 17.6.5, so the main-effect hypothesis makes sense. The expected mean squares for A and AB are listed in Table 17.6 as

$$E[MSA] = 8\sigma_A^2 + 2\sigma_{AB}^2 + \sigma^2$$

and

$$E[MS(AB)] = 2\sigma_{AB}^2 + \sigma^2,$$

with 3 and 9 corresponding degrees of freedom, respectively. The decision rule, therefore, is

$$\text{reject } H_0^A \text{ if } \frac{msA}{ms(AB)} > F_{3,9,\alpha}.$$

If we select a Type I error probability of $\alpha = 0.05$, then $F_{3,9,0.05} = 3.86$. Since $msA/ms(AB) = 223.04/28.63 = 7.79$, we can conclude that $\sigma_A^2 > 0$. □

17.6.7 Sample Sizes

If we test main effects and interactions only when the higher-order interactions involving those factors are negligible, then we can adapt (17.4.19) by changing the degrees of freedom to match those in the decision rule being used.

17.7 Mixed Models

Models that contain both random and fixed treatment effects are called *mixed models*. The analysis of random effects proceeds in exactly the same way as described in the previous sections. All that is needed is a way to write down the expected mean squares. The fixed effects can be analyzed as in Chapters 3–7, except that, here, too, we may need to replace the mean square for error by a different appropriate mean square. We show how to calculate the expected mean squares for a mixed model in Section 17.7.1.

An interaction between two or more factors any of which has random effects will be regarded as a random effect, since the combination of levels observed in the experiment depends upon the random selection of levels of those factors that have random effects.

17.7.1 Expected Mean Squares and Hypothesis Tests

Expected mean squares can be obtained for a mixed model when the sample sizes are equal by modifying rule 17 on page 616. We start by writing out the expected mean squares as

17.7 Mixed Models

though all the factors were random. We then collect all of the fixed effects and list them together as one "quadratic form." The quadratic form is a function of fixed-effect parameters such as $\alpha_i^* = \alpha_i + \overline{(\alpha\beta)}_{i.}$ (see Example 17.7.1) that typically feature in fixed-effects models.

As an example, consider a model containing the main effects of factors A, B, and D and the interactions AB and BD. Suppose that factors A and B are fixed, so that all of their levels of interest are observed in the experiment, and factor D is random, so that its levels form a large population of which only a random selection are observed in the experiment. Then interaction AB is a fixed effect, but interaction BD is a random effect.

We use α_i to represent the effect of the ith level of A, β_j to represent the effect of the jth level of B, and $(\alpha\beta)_{ij}$ to represent their interaction. The effect of the kth randomly selected level of D is represented by the random variable D_k, and the effect of the interaction between the jth specifically selected level of B and the kth randomly selected level of D is denoted by the random variable $(\beta D)_{jk}$. The model is then as follows:

$$Y_{ijt} = \mu + \alpha_i + \beta_j + D_k + (\alpha\beta)_{ij} + (\beta D)_{jk} + \epsilon_{ijkt}, \quad (17.7.25)$$

$$D_k \sim N(0, \sigma_D^2), \quad (\beta D)_{jk} \sim N(0, \sigma_{BD}^2), \quad \epsilon_{ijkt} \sim N(0, \sigma^2),$$

D_k's, $(\beta D)_{jk}$'s and, ϵ_{ijkt}'s are all mutually independent,

$t = 1, \ldots, r, \quad i = 1, \ldots, a, \quad j = 1, \ldots, b, \quad k = 1, \ldots, d.$

The expected mean squares for the corresponding random-effects model were calculated in Example 17.6.4 and are reproduced in the second column of Table 17.8. The expected mean squares for the above mixed model are given in the third column of Table 17.8 and are obtained by collecting the terms in the expected mean squares corresponding to the fixed effects into one quadratic form.

The expected mean squares can all be verified by direct calculation. We illustrate the calculation for B in the following example. The term $Q(B, AB)$ in $E[MSB]$ corresponds to a quadratic (i.e., squared) function of $\beta_j^* = \beta_j + \overline{(\alpha\beta)}_{.j}$, a quantity that we are used to dealing with in fixed-effects models.

Example 17.7.1 Calculation of expected mean squares

Consider an experiment with two fixed treatment factors A and B, and one random treatment factor D, and suppose that (17.7.25) is thought to be a reasonable model. Using rule 4 of

Table 17.8 Expected mean squares for a three-way mixed model

Effect	For random model	For mixed model
A	$bdr\sigma_A^2 + dr\sigma_{AB}^2 + \sigma^2$	$Q(A, AB) + \sigma^2$
B	$adr\sigma_B^2 + dr\sigma_{AB}^2 + ar\sigma_{BD}^2 + \sigma^2$	$Q(B, AB) + ar\sigma_{BD}^2 + \sigma^2$
D	$abr\sigma_D^2 + ar\sigma_{BD}^2 + \sigma^2$	$abr\sigma_D^2 + ar\sigma_{BD}^2 + \sigma^2$
AB	$dr\sigma_{AB}^2 + \sigma^2$	$Q(AB) + \sigma^2$
BD	$ar\sigma_{BD}^2 + \sigma^2$	$ar\sigma_{BD}^2 + \sigma^2$
Error	σ^2	σ^2

Chapter 7, the fixed-effect sum of squares for B is

$$SSB = adr \sum_{j=1}^{b} \overline{Y}_{.j..}^2 - abdr\overline{Y}_{....}^2.$$

Now,

$$\overline{Y}_{.j..} = \sum_{i=1}^{a}\sum_{k=1}^{d}\sum_{t=1}^{r} Y_{ijkt}/adr$$
$$= \mu + \overline{\alpha}_{.} + \beta_j + \overline{D}_{.} + \overline{(\alpha\beta)}_{.j} + \overline{(\beta D)}_{j.} + \overline{\epsilon}_{.j..}.$$

So,

$$E[\overline{Y}_{.j..}] = \mu + \overline{\alpha}_{.} + \beta_j + \overline{(\alpha\beta)}_{.j} \quad \text{and} \quad \text{Var}(\overline{Y}_{.j..}) = \frac{\sigma_D^2}{d} + \frac{\sigma_{BD}^2}{d} + \frac{\sigma^2}{adr}.$$

Similarly,

$$E[\overline{Y}_{....}] = \mu + \overline{\alpha}_{.} + \overline{\beta}_{.} + \overline{(\alpha\beta)}_{..} \quad \text{and} \quad \text{Var}(\overline{Y}_{....}) = \frac{\sigma_D^2}{d} + \frac{\sigma_{BD}^2}{bd} + \frac{\sigma^2}{abdr}.$$

Using the facts that $MSB = SSB/(b-1)$ and $E[X^2] = \text{Var}(X) + (E[X])^2$, we obtain

$$E[MSB] = \frac{adr}{(b-1)} \sum_{j=1}^{b} \left(\frac{\sigma_D^2}{d} + \frac{\sigma_{BD}^2}{d} + \frac{\sigma^2}{adr} + [\mu + \overline{\alpha}_{.} + \beta_j + \overline{(\alpha\beta)}_{.j}]^2 \right)$$
$$- \frac{abdr}{(b-1)} \left(\frac{\sigma_D^2}{d} + \frac{\sigma_{BD}^2}{bd} + \frac{\sigma^2}{abdr} + [\mu + \overline{\alpha}_{.} + \overline{\beta}_{.} + \overline{(\alpha\beta)}_{..}]^2 \right)$$
$$= ar\sigma_{BD}^2 + \sigma^2 + Q(B, AB),$$

where

$$Q(B, AB) = \frac{adr}{(b-1)} \sum_j \left[(\beta_j + \overline{(\alpha\beta)}_{.j}) - (\overline{\beta}_{.} + \overline{(\alpha\beta)}_{..}) \right]^2. \qquad \square$$

Notice that in Example 17.7.1, the quadratic form $Q(B, AB)$ is equal to zero when all the $\beta_j^* = \beta_j + \overline{(\alpha\beta)}_{.j}$ are equal. We can make use of this fact when looking for an appropriate denominator for the test ratio for testing $H_0^B : \{\beta_j + \overline{(\alpha\beta)}_{.j}$ are all equal$\}$. If this hypothesis is true, then

$$E[MSB] = ar\sigma_{BD}^2 + \sigma^2.$$

Consequently, a sensible denominator would be $MS(BD)$, which has the same expected value (see Table 17.8). Thus, the decision rule for testing H_0^B against the alternative hypothesis that the β_j^* are not all equal is

$$\text{reject } H_0^B \text{ if } \frac{msB}{ms(BD)} > F_{(b-1),(b-1)(d-1),\alpha}.$$

From Table 17.8 we can construct tests for the other relevant hypotheses in a similar manner. For example, to test the hypothesis

$$H_0^{AB} : \{(\alpha\beta)_{ij} - \overline{(\alpha\beta)}_{i.} - \overline{(\alpha\beta)}_{.j} + \overline{(\alpha\beta)}_{..} = 0, \text{ for all } i, j\}$$

against the alternative hypothesis that the interaction contrasts are not all zero, the decision rule is

$$\text{reject } H_0^{AB} \text{ if } \frac{ms(AB)}{msE} > F_{(a-1)(b-1),df,\alpha},$$

where df is the number of error degrees of freedom.

To test the hypothesis $H_0^D : \{\sigma_D^2 = 0\}$ against the alternative hypothesis $H_A^D : \{\sigma_D^2 > 0\}$, the decision rule is

$$\text{reject } H_0^D \text{ if } \frac{msD}{ms(BD)} > F_{d-1,(b-1)(d-1),\alpha}.$$

The test ratios are summarized in Table 17.9. Generally, we would not test a main-effect or interaction hypothesis unless all higher-order interactions involving these same factors were believed to be negligible. For some mixed models, as for random-effects models, the appropriate denominator for the test statistic may not be listed among the expected mean squares for the factors in the model. In this case, it would be necessary to calculate it, and the corresponding degrees of freedom, using (17.6.23), page 617.

17.7.2 Confidence Intervals in Mixed Models

Confidence intervals for fixed effects For fixed effects in a mixed model with equal sample sizes, we may use all of the rules of Section 7.3, page 201, exactly as if there were no random effects in the model, *except that we replace msE by the same mean square that was identified for hypothesis testing*—namely, used in the denominator of the test ratio—and the error degrees of freedom are also replaced. The necessity of doing this replacement is highlighted in Example 17.7.2. Apart from this, we may use the Bonferroni, Scheffé, Tukey, Dunnett, and Hsu methods of multiple comparisons in the usual way. When the sample sizes are unequal, we would obtain the least squares estimates and expected mean squares from a computer package.

Example 17.7.2

Consider an experiment with two fixed treatment factors and one random treatment factor, for which the following model is thought to be reasonable (this is the same model that has

Table 17.9 Test ratios for a three-way mixed model

Effect	E[MS]	Ratio
A	$Q(A, AB) + \sigma^2$	msA/msE
B	$Q(B, AB) + a r\sigma_{BD}^2 + \sigma^2$	$msB/ms(BD)$
D	$ab r\sigma_D^2 + a r\sigma_{BD}^2 + \sigma^2$	$msD/ms(BD)$
AB	$Q(AB) + \sigma^2$	$ms(AB)/msE$
BD	$a r\sigma_{BD}^2 + \sigma^2$	$ms(BD)/msE$

been discussed throughout this subsection):

$$Y_{ijkt} = \mu + \alpha_i + \beta_j + D_k + (\alpha\beta)_{ij} + (\beta D)_{jk} + \epsilon_{ijkt}.$$

The fixed part of the model is

$$\mu + \alpha_i + \beta_j + (\alpha\beta)_{ij},$$

which looks exactly like one of the two-way analysis of variance models that was studied in Chapter 6.

Suppose we need confidence intervals for pairwise comparisons in the levels of A and of B. Then, as usual, the least squares estimates for pairwise differences are

$$\widehat{\alpha}_i^* - \widehat{\alpha}_p^* = \left(\alpha_i + \widehat{(\alpha\beta)}_{i.}\right) - \left(\alpha - \widehat{(\alpha\beta)}_{p.}\right) = \overline{y}_{i...} - \overline{y}_{p...}$$

and

$$\widehat{\beta}_j^* - \widehat{\beta}_u^* = \left(\beta_j + \widehat{(\alpha\beta)}_{.j}\right) - \left(\beta_u + \widehat{(\alpha\beta)}_{.u}\right) = \overline{y}_{.j..} - \overline{y}_{.u..}.$$

Tables 17.8 (page 623) and 17.9 (page 625) suggest that msE should be used in the formulae for confidence intervals for $\alpha_i^* - \alpha_p^*$, as usual, but that $ms(BD)$ should be used in place of msE in the formulae for confidence intervals for $\beta_j^* - \beta_u^*$. All confidence intervals are of the form

(least squares estimate) $\pm (w) \times$ (standard error).

The standard error is the square root of the estimated variance of the least squares estimator. Now,

$$\text{Var}(Y_{ijkt}) = \sigma_D^2 + \sigma_{BD}^2 + \sigma^2 \quad \text{and} \quad \text{Var}(\overline{Y}_{.j..}) = \frac{\sigma^2}{d} + \frac{\sigma_{BD}^2}{d} + \frac{\sigma^2}{adr}.$$

The Y_{ijkt}'s are not independent. Observations on the same level of D are correlated. If two observations are taken on the same levels of B and D, we have

$$\text{Cov}(Y_{ijkt}, Y_{pjks}) = \sigma_D^2 + \sigma_{BD}^2.$$

If two observations are taken on the same level of D, but different levels of B, then

$$\text{Cov}(Y_{ijkt}, Y_{puks}) = \sigma_D^2.$$

All other pairs of response variables are independent. Consequently,

$$\text{Cov}(\overline{Y}_{.j..}, \overline{Y}_{.u..}) = \frac{1}{a^2 d^2 r^2} \left[\sum_{i=1}^{a}\sum_{p=1}^{a}\sum_{k=1}^{d}\sum_{t=1}^{r}\sum_{s=1}^{r} \text{Cov}(Y_{ijkt}, Y_{puks})\right]$$

$$= \frac{1}{a^2 d^2 r^2} [a^2 dr^2 \sigma_D^2]$$

$$= \frac{\sigma_D^2}{d},$$

and

$$\mathrm{Var}(\overline{Y}_{.j..} - \overline{Y}_{.u..}) = \mathrm{Var}(\overline{Y}_{.j..}) + \mathrm{Var}(\overline{Y}_{.u..}) - 2\mathrm{Cov}(\overline{Y}_{.j..}, \overline{Y}_{.u..})$$
$$= 2\left(\frac{\sigma_D^2}{d} + \frac{\sigma_{BD}^2}{d} + \frac{\sigma^2}{adr}\right) - 2\left(\frac{\sigma_D^2}{d}\right)$$
$$= \frac{2}{adr}\left(ar\sigma_{BD}^2 + \sigma^2\right),$$

which is of the form $(\Sigma c_i^2/(adr))(ar\sigma_{BD}^2 + \sigma^2)$. Thus, we need to estimate $(ar\sigma_{BD}^2 + \sigma^2)$ rather than σ^2, and an unbiased estimate is given by $ms(BD)$. So, the standard error for $\hat{\beta}_j^* - \hat{\beta}_u^* = \overline{Y}_{.j..} - \overline{Y}_{.u..}$ is $((2/(adr))\,ms(BD))^{1/2}$ with corresponding degrees of freedom $(b-1)(d-1)$. □

In some models, the necessary mean square will not be listed in the expected mean squares table, and (17.6.23), page 617, will need to be used to find an approximate mean square and degrees of freedom.

Confidence intervals for variance components In obtaining confidence intervals for variance components, only the random part of the model is used, or, equivalently, only the mean squares corresponding to random effects. Consequently, the formulae of Section 17.6.5 are used exactly as described for random-effects models.

17.8 Rules for Analysis of Random and Mixed Models

Rules 1–7 of Section 7.3, page 201, are valid for calculating degrees of freedom, sums of squares, and mean squares in random-effects and mixed models as well as in fixed-effects models. In addition, rules 8–16 are valid for analyzing fixed effects, except that σ^2 and msE may need to be replaced. Rules 17–22 below summarize the results of this chapter. Rule 17 is an expanded version of rule 17 on page 616.

17.8.1 Rules—Equal Sample Sizes

17. To obtain the expected mean square for a particular main effect or interaction, first make a note of the subscripts on the term representing that particular effect in the model. Write down variance components for the effect of interest, for the error, and for every interaction whose term in the model includes the noted set of subscripts. Gather up all variance components corresponding to fixed effects into one quadratic form Q. Multiply any remaining variance component except σ^2 by the number of observations taken on each level or combination of levels of the corresponding effect (main effect or interaction). Add up the terms.

18. To obtain the denominator of the test statistic for testing the null hypothesis that a main effect or interaction effect is zero, write down the expected mean square for the effect of interest (see rule 17). Cross out the term that would be zero if the null hypothesis were

true. The denominator of the test statistic is the mean square, or linear combination of mean squares, u, whose expected value is equal to the remaining expression.

19. For a random effect, let $U = \Sigma k_i MS_i$ be the mean square or linear combination of mean squares whose expected value is equal to the variance component corresponding to the random effect. An exact or approximate $100(1-\alpha)\%$ confidence interval for this variance component is

$$\left(\frac{xu}{\chi^2_{x,\alpha/2}}, \frac{xu}{\chi^2_{x,1-\alpha/2}}\right),$$

where

$$x = \frac{[\Sigma k_i (ms_i)]^2}{\Sigma [k_i (ms_i)]^2 / x_i}$$

and where u is the observed value of U, ms_i is the observed value of MS_i, and x_i is the number of degrees of freedom corresponding to ms_i.

20. For a fixed effect, confidence intervals are obtained as in rule 14, page 204, except that msE is replaced by the denominator u from rule 18, and the number of error degrees of freedom is replaced by x in rule 19.

21. For a fixed effect, the decision rule for testing the hypothesis that the effect is zero is the same as that in rule 8, page 203, for fixed-effects models, except that msE is replaced by the denominator u from rule 18, and the number of error degrees of freedom is replaced by x in rule 19.

22. For a random effect, the decision rule for testing the hypothesis H_0 that the corresponding variance component is zero against the alternative hypothesis that it is not zero is

$$\text{reject } H_0 \text{ if } \frac{ms}{u} > F_{v,x,\alpha},$$

where ms is the mean square for the effect of interest and v the corresponding degrees of freedom, u is the observed value of the denominator as in rule 18, and x is the corresponding degrees of freedom calculated as in rule 19.

17.8.2 Controversy (Optional)

Before proceeding, we should mention that some other textbooks may present slightly different tables of expected mean squares. For example, the expected mean square for D in Table 17.8, which we have calculated as

$$E[MSD] = abr\sigma_D^2 + ar\sigma_{BD}^2 + \sigma^2,$$

may in other texts be listed as

$$E[MSD] = abr\sigma_D^2 + \sigma^2.$$

17.8 Rules for Analysis of Random and Mixed Models

This alternative listing occurs when constraints are placed on the model parameters involving fixed factors, and it suggests use of the denominator msE rather than $ms(BD)$ in testing $H_0^D : \{\sigma_D^2 = 0\}$ against $H_A^D : \{\sigma_D^2 > 0\}$. A number of articles in the statistical literature have been written advocating one denominator rather than the other, and there still appears to be no consensus.

If we follow the line of reasoning that we have followed to this point, that normally we will examine main effects only when there is no interaction, then some of the controversy disappears. If σ_{BD}^2 is really zero, then $E[MSD] = abr\sigma_D^2 + \sigma^2$ in both cases. Of course, due to variability of the data and uncertainty about whether or not σ_{BD}^2 is really zero (or close to it), we still have to make the choice in practice. We have recommended using $ms(BD)$ as the denominator if the objective is to test $H_0^D : \sigma_D^2 = 0$. However, if interest is really in testing

$$H_0^{D+BD} : \{\sigma_D^2 + b^{-1}\sigma_{BD}^2 = 0\},$$

or equivalently

$$H_0^{D+BD} : \{\sigma_D^2 = \sigma_{BD}^2 = 0\},$$

then we would use msE as the denominator.

The controversy originally arose from the formulation of the model. In our example, the model was given in (17.7.25), page 623, and the controversy surrounds the random effect $(\beta D)_{jk}$. We have modeled this as a normally distributed random variable. Some authors add to the model the restriction $\Sigma_j(\beta D)_{jk} = 0$, and this leads to the canceling of the term in σ_{BD}^2 when the expected mean square of D is calculated.

Hocking (1996, page 569) shows that under this restriction, the hypothesis H_0^D is actually our hypothesis H_0^{D+BD}. An explanation for this is as follows. If constraints are placed on the parameters, then the $(\beta D)_{jk}$ effects truly represent interaction effects, and σ_{BD}^2 measures precisely variability in BD-interaction effects. However, if no constraints are placed on the parameters, then σ_{BD}^2 being positive implies the presence of main effects of B and D as well as the presence of BD-interaction effects. In other words, the parameters $(\beta D)_{jk}$ represent "BD effects" in model (17.7.25), though we have referred to them as BD-interaction effects. Thus, under our model (17.7.25), the hypothesis $H_0^{D+BD} : \{\sigma_D^2 = \sigma_{BD}^2 = 0\}$ is that there are no main effects of D (or BD interactions). Also, there are no BD interactions if $\sigma_{BD}^2 = 0$, and there are no main effects of B (or BD interactions) if $\beta_1 = \beta_2 = \cdots = \beta_b$ and $\sigma_{BD}^2 = 0$. From this viewpoint, the hypothesis $H_0^D : \sigma_D^2 = 0$ is that there are no main effects of D if σ_{BD}^2 is believed to be zero; otherwise, it is the hypothesis that main effects of D are no less negligible than BD interactions.

Since there are problems inherent in placing restrictions on the model parameters, we prefer not to do so, and we prefer to use the set of expected mean squares in Table 17.7. If the parameters in the model are properly interpreted, then there is no controversy, and the appropriate test is determined by what is most sensible for the experiment at hand.

17.9 Block Designs and Random Blocking Factors

In certain types of experiments, it is extremely common for the levels of a blocking factor to be randomly selected. For example, in medical, psychological, educational, or pharmaceutical experiments, blocks frequently represent subjects that have been selected at random from a large population of similar subjects. In agricultural experiments, blocks may represent different fields selected from a large variable population of fields. In industrial experiments, different machine operators may represent different levels of the blocking factor and may be similar to a random sample from a large population of possible operators. Raw material may be delivered to the factory in batches, a random selection of which are used as blocks in the experiment.

Since we are not interested in the blocking factor itself, its designation as random rather than fixed will affect the analysis only if the model includes a block × treatment interaction. For example, suppose that factor D in Table 17.8 represents a random blocking factor, and that A and B are two fixed treatment factors. The analysis of factor A, which has no interaction with D, is unaffected by the designation of D as a random effect. However, the analysis of factor B, which interacts with blocks, *is* affected, since msE in hypothesis tests and confidence intervals for contrasts in the levels of B will be replaced by $ms(BD)$.

Example 17.9.1 Temperature experiment

The temperature experiment was run by M. Bowe, J. Cooper, J. Donato, S. Giust, and H. Schieman in 1994 to compare the times required for three different digital thermometers (factor A at $a = 3$ levels) to register body temperature at two different sites—in the mouth and under the arm—(factor B at $b = 2$ levels). Thus, there were six treatment combinations. Four subjects were selected at random from the American statistics graduate students at The Ohio State University, and each treatment combination was measured once for each subject. The experiment was designed as a randomized complete block design, with subjects representing blocks. The recorded times are shown in Table 17.10.

The four subjects used in the experiment are not themselves of interest. Of more interest is how the thermometers react on average over a large population of subjects. The population of American statistics graduate students at the university is large, but not infinite. However, the four subjects used in the experiment are, hopefully, representative of all possible American graduate students, and it is reasonable to model the subject (block) effect as a random effect.

Since subjects vary in body heat, it is possible that factor B (site) might interact with subject. It is also possible that different thermometers might act differently at the two different

Table 17.10 Data (in seconds) for the temperature experiment

Subject	\multicolumn{6}{c}{Treatment Combination}					
	11	12	21	22	31	32
1	62.16	61.53	154.42	310.46	95.98	225.65
2	65.63	63.70	132.30	284.64	98.50	241.63
3	63.12	61.34	105.52	315.61	110.05	364.07
4	61.51	61.54	94.88	294.16	107.93	304.58

17.9 Block Designs and Random Blocking Factors

Table 17.11 Analysis of variance table for the mixed model temperature experiment

Source of Variation	Degrees of Freedom	Mean Square	p-value	Expected Mean Square
Subject (block)	3	570.04	–	–
Thermometer (A)	2	52879.34	0.0001	$Q(A, AB) + \sigma^2$
Site (B)	1	86029.60	0.0035	$Q(B, AB) + 3\sigma_{SB}^2 + \sigma^2$
Therm*Site (AB)	2	21897.23	0.0001	$Q(AB) + \sigma^2$
Subject*Site (SB)	3	1210.67	0.2625	$3\sigma_{SB}^2 + \sigma^2$
Error	12	3802.57		σ^2
Total	23			

sites. Consequently the following model might be reasonable for this experiment.

$$Y_{hij} = \mu + S_h + \alpha_i + \beta_j + (\alpha\beta)_{ij} + (S\beta)_{hj} + \epsilon_{hij},$$
$$h = 1, 2, 3, 4, \quad i = 1, 2, 3, \quad j = 1, 2,$$
$$S_h \sim N(0, \sigma_S^2), \quad (S\beta)_{hj} \sim N(0, \sigma_{SB}^2), \quad \epsilon_{hij} \sim N(0, \sigma^2),$$
$$S_h\text{'s, } (S\beta)_{hj}\text{'s and } \epsilon_{hij}\text{'s are all mutually independent,}$$

where all random variables on the right-hand side of the model are mutually independent, and where S_h represents the effect of the hth randomly selected subject (block), α_i represents the effect of the ith specifically selected thermometer, and β_j represents the effect of the ith specifically selected site. This model is similar to mixed model (17.7.25) with S_h replacing D_k. Consequently, the expected mean squares will be similar to those in Table 17.8, page 623. The analysis of variance table is shown in Table 17.11.

We start by testing the two interaction hypotheses. To test the hypothesis $H_0^{SB} : \{\sigma_{SB}^2 = 0\}$, that the subject by site interaction variance is negligible, against the alternative hypothesis that it is not negligible, using a significance level of 0.01 (so that the overall significance level will be at most 0.05), we

$$\text{reject } H_0^{SB} \text{ if } \frac{ms(SB)}{msE} > F_{3,12,0.01} = 5.95.$$

Since $ms(SB)/msE = 1.51$, there is not sufficient evidence to conclude that the interaction variance is greater than zero (equivalently, the p-value is greater than 0.01). Before we can examine the site main effect, however, we also need to look at the thermometer by site interaction.

To test the hypothesis

$$H_0^{AB} : \{(\alpha\beta)_{ij} - (\alpha\beta)_{ip} - (\alpha\beta)_{uj} + (\alpha\beta)_{up}, \text{ for all } i, j, u, p\}$$

against the alternative hypothesis that the interaction is not negligible, we

$$\text{reject } H_0^{AB} \text{ if } \frac{ms(AB)}{msE} > F_{2,12,0.01} = 6.93.$$

Since $ms(AB)/msE = 27.28$, we reject H_0^{AB} and conclude that there is a thermometer × site interaction. Thus, it is unlikely that the thermometer and site main effects are of interest.

However, for illustration purposes, we ask whether the *average* time taken for these three digital thermometers to register is the same whether used in the mouth or under the arm. Thus, we will test the hypothesis

$$H_0^B : \{\beta_1 + \overline{(\alpha\beta)}_{.1} = \beta_2 + \overline{(\alpha\beta)}_{.2}\}.$$

To test this hypothesis at significance level 0.01, we

$$\text{reject } H_0^B \text{ if } \frac{msB}{ms(SB)} > F_{1,3,0.01} = 29.5.$$

Since $msB/ms(SB) = 71.06$, we reject H_0^B and conclude that it does make a difference in registering temperature (on average for these three thermometers) as to whether the thermometer is used in the mouth or under the arm. This conclusion is made on average over the three thermometers and over the whole population of similar graduate students.

17.10 Using SAS Software

17.10.1 Checking Assumptions on the Model

Using the data of Table 17.1, page 596, for the clean wool experiment, we illustrate some methods of checking model assumptions for a random-effects one-way model. The experimenters took observations on $r = 11$ cores of wool from each of $v = 7$ randomly selected wool bales.

We let the random variable T_i represent the true clean content of the ith randomly selected bale of wool from the shipment, and let $Y_{it} = T_i + \epsilon_{it}$ represent the observed clean content of the tth core (observation) from the ith bale, where the error variable ϵ_{it} includes the deviation from the true average clean content of the tth core from the ith bale, the measurement error, environmental conditions, etc.

First, we check the error assumptions by calculating and plotting the standardized residuals obtained as though the bale effects were fixed. The standardized residuals are calculated in the usual way and plotted against the levels of the treatment factor and the predicted values (see Section 5.8, page 122). The latter plot, obtained from the statement

```
PLOT Z*PRED=BALE / VPOS=15 HPOS=50 VREF=0;
```

is shown in Figure 17.3. The plotted symbols indicate the bales from which the residuals arose.

Although eleven residuals are masked by other residuals, the most noticeable feature is that bale 1 gives rise to one very large standardized residual (an outlier). This means one of several things: Perhaps the data value is in error, so that this value is an outlier, or perhaps bale 1 is extremely more variable than the other bales in the population, or perhaps the error variables are not normally distributed. Let us suppose that we could go back to the original experimenters and that indeed, something unusual happened at this point during the time at which the observations were taken. If so, we could exclude this value. The new residual plot is shown in Figure 17.4.

17.10 Using SAS Software

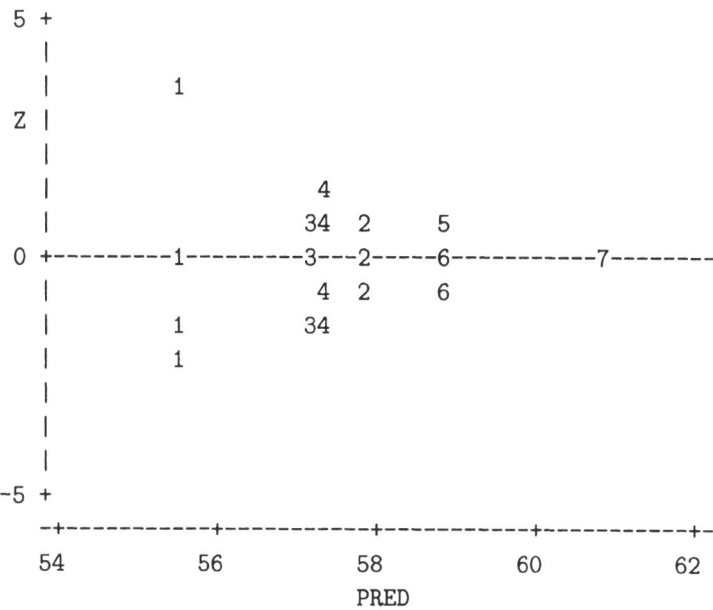

Figure 17.3
Residuals versus predicted values for the clean wool experiment

All standardized residuals now lie within the expected range for normally distributed errors. The plot gives us quite a lot of information about our sample of bales and possibly about the shipment of bales from which they were drawn. First, the average clean content of bale 1 is around 53, considerably below the others. This was the bale that had the supposed outlier. One might suspect that this bale either did not come from the same shipment or was contaminated at some point before being measured. On the other hand, the shipment may contain a number of "rogue bales," and this ought to be investigated. At the other end of the range, we see that bale 7 had the highest clean content and was least variable. Perhaps this is not too surprising, since a bale with 100% clean content would probably show no variability in the measurements taken on it. Thus, one might suspect that our model that includes normally distributed errors is not ideal for this situation. However, the plot of standardized residuals against normal scores does not show any anomalies (figure not shown).

In a one-way random-effects model, we can check the assumption that the treatment effects have a normal distribution by making a normal probability plot of the standardized treatment averages $\overline{Y}_{i.}$ against their normal scores. (This cannot be done for models with more than one random effect, since the treatment averages are not independent.) For the clean wool experiment, the normal probability plot is obtained by means of the statements in Table 17.12, and the resulting plot is shown in Figure 17.5. If the normality assumption for the population of bales is satisfied, the standardized bale averages—namely, the stan-

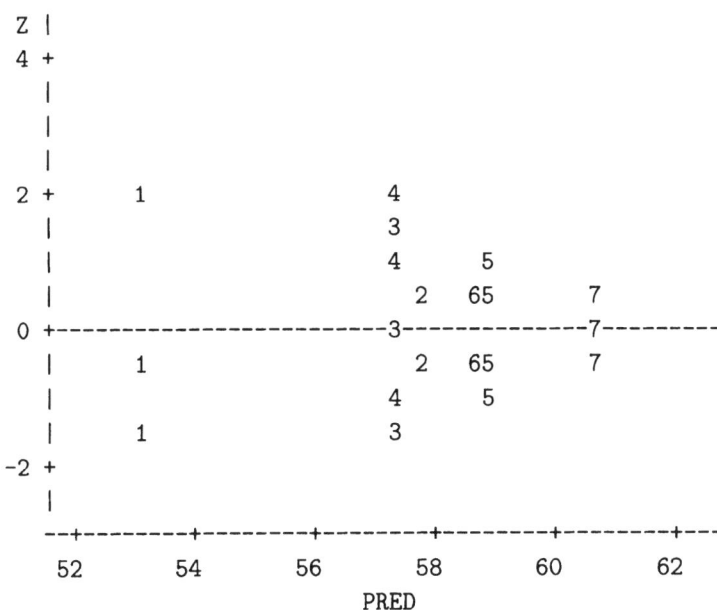

Figure 17.4
Residuals versus predicted values for the clean wool experiment, excluding the outlier.

Table 17.12 SAS program to plot treatment averages against their normal scores

```
PROC GLM;
  CLASS BALE;
  MODEL CONTENT = BALE;
  OUTPUT OUT=STATS P=PRED;
PROC STANDARD STD=1.0 MEAN=0.0;
  VAR PRED;
PROC RANK NORMAL=BLOM;
  VAR PRED;
  RANKS NSCORE;
PROC PLOT;
  PLOT PRED*NSCORE=BALE / VPOS=15 HPOS=40 HREF=0 VREF=0;
```

dardized predicted values—should roughly lie along a line (with slope 1.0) through (0, 0). In Figure 17.5, we see that this is roughly the case.

In summary, the random-effects one-way model with the standard distribution assumptions does not fit these data too well, since variances apparently are not constant or there is an outlier. Nevertheless, we have established that the population of bales in this shipment is extremely variable. Selected bales 1 and 7 appear to be somewhat different from the other

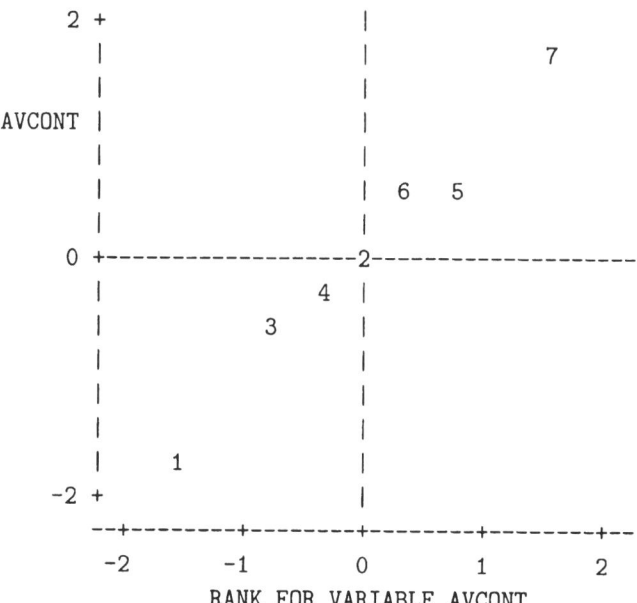

Figure 17.5
Normal probability plot of the standardized treatment averages for the clean wool experiment

five selected bales. Perhaps the shipment is made up of dissimilar subpopulations (perhaps from different sources). This should be checked, since it may give a clue as to how to improve the wool clean content in the future.

17.10.2 Estimation and Hypothesis Testing

Analysis of variance tables for random-effects and mixed models are obtained in exactly the same way as for fixed-effects models. The additional expected mean squares column can be obtained very easily by inserting a RANDOM statement immediately after the model statement. All random main effects and random interactions should be listed in the RANDOM statement, as shown, for example, in Table 17.13 for the temperature experiment of Example 17.9, page 630. The denominators, calculated as explained throughout this chapter, can be obtained by adding the option TEST to the RANDOM statement, as shown in the following example. The actual denominators are printed out as well as the p-values.

The output is shown in Table 17.14. The first few lines reproduce the expected mean squares that were calculated by hand in Table 17.7, page 620. The remainder of the output gives the TYPE III sums of squares, but instead of calculating the usual test ratios with msE as the denominator, the TEST option on the RANDOM statement has caused the denominator ms(SITE*SUBJ) to be used where appropriate.

Estimating contrasts for the fixed effects can be done as usual using the ESTIMATE statement. Confidence intervals can be calculated by hand and the mean squared error

Table 17.13 SAS program for the temperature experiment

```
DATA TEMPR;
  INPUT THERM SITE SUBJ TIME;
  LINES;
  1 1 1  62.16
  1 2 1  61.53
  : : :   :
  3 2 4 304.58
;
* Note that the option TEST gives correct denominators;
PROC PRINT;
PROC GLM;
  CLASSES THERM SITE SUBJ;
  MODEL TIME = SUBJ THERM SITE THERM*SITE SUBJ*SITE;
  RANDOM SUBJ SUBJ*SITE / TEST;
  OUTPUT OUT=RESIDS PREDICTED=PTIME RESIDUAL=Z;
  CONTRAST 'SITE1-SITE2' SITE 1 -1 / E=SUBJ*SITE;
```

replaced by the denominator used in the test procedures. For testing individual contrasts, the CONTRAST statement can be used and the required denominator can be specified. For example, the statement

CONTRAST 'SITE1-SITE2' SITE 1 -1 / E=SUBJ*SITE;

will use the subj × site interaction mean square as the variance estimate for comparing sites, rather than the error mean square. □

Other SAS procedures The SAS package includes an alternative procedure PROC VAR-COMP, explicitly designed to cope with random and mixed models. The call statement that generates the same set of information as in Table 17.14 is

```
PROC VARCOMP METHOD=TYPE1;
  CLASSES THERM SITE SUBJ;
  MODEL TIME = THERM SITE THERM*SITE SUBJ SUBJ*SITE / FIXED=3;
```

where FIXED=3 indicates that the first three terms of the model are fixed effects. Estimates of the variance components are also calculated by this procedure. As can be seen from the METHOD=TYPE1 statement, the type I sums of squares are used, so the estimates will agree with those obtained from the PROC GLM Type III sums of squares only when the sample sizes are equal.

Throughout this chapter we have discussed the estimation of random effects using the fixed-effects analysis of variance as a guide to finding unbiased estimators. There are other more sophisticated statistical procedures for estimating variance components that prevent the estimates from ever being negative. The procedure PROC VARCOMP, mentioned above, can access some of these methods (via the METHOD statement in the first line). There is also another procedure called PROC MIXED, which we will not discuss but which is designed for estimation and testing in mixed-effect models, including unbalanced designs.

Table 17.14 SAS analysis of variance for the temperature experiment

```
                        The SAS System
                 General Linear Models Procedure

Source        Type III Expected Mean Square
SUBJ          Var(Error) + 3 Var(SITE*SUBJ) + 6 Var(SUBJ)
THERM         Var(Error) + Q(THERM,THERM*SITE)
SITE          Var(Error) + 3 Var(SITE*SUBJ) + Q(SITE,THERM*SITE)
THERM*SITE    Var(Error) + Q(THERM*SITE)
SITE*SUBJ     Var(Error) + 3 Var(SITE*SUBJ)

                 General Linear Models Procedure
          Tests of Hypotheses for Mixed Model Analysis of Variance
Dependent Variable: TIME

Source: SUBJ
Error: MS(SITE*SUBJ)
                       Denominator   Denominator
   DF    Type III MS       DF            MS       F Value   Pr > F
    3    570.04094861       3       1210.6723042   0.4708   0.7240

Source: THERM *
Error: MS(Error)
                       Denominator   Denominator
   DF    Type III MS       DF            MS       F Value   Pr > F
    2    52879.34315       12        802.56832222 65.8877   0.0001
* - This test assumes one or more other fixed effects are zero.

Source: SITE *
Error: MS(SITE*SUBJ)
                       Denominator   Denominator
   DF    Type III MS       DF            MS       F Value   Pr > F
    1    86029.597838       3       1210.6723042  71.0594   0.0035
* - This test assumes one or more other fixed effects are zero.

Source: THERM*SITE
Error: MS(Error)
                       Denominator   Denominator
   DF    Type III MS       DF            MS       F Value   Pr > F
    2    21897.23105       12        802.56832222 27.2839   0.0001

Source: SITE*SUBJ
Error: MS(Error)
                       Denominator   Denominator
   DF    Type III MS       DF            MS       F Value   Pr > F
    3    1210.6723042      12        802.56832222  1.5085   0.2625
```

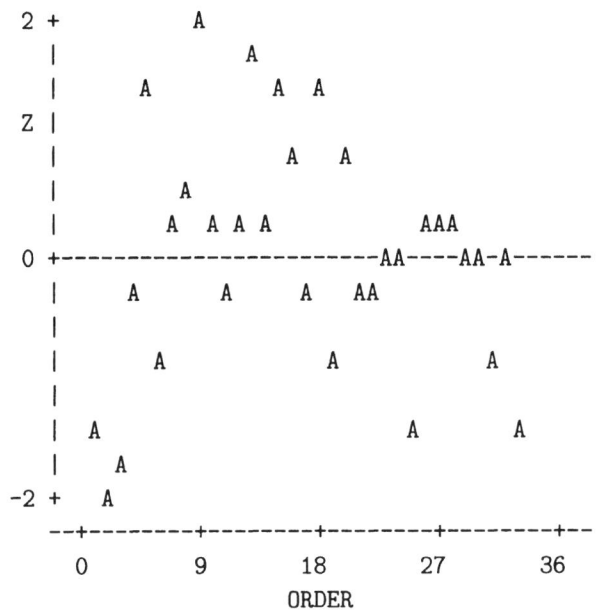

Figure 17.6
Plot of the standardized residuals against order of observation for the ice cream experiment

Covariates Before leaving this section, we will examine a more complicated model. The plot of standardized residuals against order of observation for the ice cream experiment (Example 17.3.3, page 600) is shown in Figure 17.6. This plot suggests that there may be a quadratic time trend in the data.

We define two extra variables X and X2 in the DATA statement as follows,

```
DATA ICE;
  INPUT FLAVOR MELTTIME ORDER;
  X=ORDER-16.5;
  X2=X*X;
  LINES;
  1 924 1
  : : :
```

and we add these variables to the model statement, so the code for PROC GLM becomes

```
PROC GLM;
  CLASS FLAVOR;
  MODEL MELTTIME = X X2 FLAVOR;
  RANDOM FLAVOR / TEST;
```

The variable X is just the same as ORDER, except that we have subtracted the average order 16.5. This helps to reduce computational problems in the model fitting. The Type III sums of squares and the expected mean squares are shown in Table 17.15. We see that the quadratic effect of time order is quite substantial and that from the list of expected mean squares, our

Table 17.15 SAS analysis of variance for the ice cream experiment

```
                      The SAS System
                General Linear Models Procedure
Dependent Variable: MELTTIME
                            Sum of         Mean
Source              DF      Squares     Square   F Value   Pr > F
Model                4    250538.12    62634.53    13.93   0.0001
Error               28    125927.94     4497.43
Corrected Total     32    376466.06


Source              DF   Type III SS  Mean Square  F Value   Pr > F
X                    1      6923.44      6923.44     1.54   0.2250
X2                   1     66292.64     66292.64    14.74   0.0006
FLAVOR               2    188394.28     94197.14    20.94   0.0001


Source       Type III Expected Mean Square
X            Var(Error) + Q(X)
X2           Var(Error) + Q(X2)
FLAVOR       Var(Error) + 9.6478 Var(FLAVOR)
```

estimate of the variance of melting times due to flavor (var(FLAVOR)) must be calculated as

$$\hat{\sigma}_T^2 = \frac{94179.139 - 4497.426}{9.6478} = 9359.62 \text{ seconds}^2$$

or $\hat{\sigma}_T = 96.75$ seconds, which is a little larger than the estimate of $\hat{\sigma}_T = 85.13$ seconds that we obtained in Example 17.3.5, page 606. Examination of the residuals in the new model shows that the error assumptions are fairly well satisfied. In Exercise 8, the reader is asked to recalculate the confidence intervals for σ_T^2 and σ_T^2/σ^2 using the new model.

Exercises

1. **Alcohol experiment**

    Solutions of alcohol are used for calibrating Breathalyzers. The data in Table 17.16 show the alcohol concentrations of samples of alcohol solutions taken from six bottles of alcohol solution randomly selected from a large batch. Concentrations are determined by gas chromatography.

    (a) Check the assumptions on the random-effects one-way model for these data.

    (b) Calculate a 95% upper confidence bound for the error variance.

    (c) Calculate a 95% confidence interval for the variance of the alcohol concentrations in the population of bottles in this large batch.

Table 17.16 Data for the alcohol experiment

Bottle	Concentration (mg/ml)			
1	1.4357	1.4348	1.4336	1.4309
2	1.4244	1.4232	1.4213	1.4256
3	1.4153	1.4137	1.4176	1.4164
4	1.4331	1.4325	1.4312	1.4297
5	1.4252	1.4261	1.4293	1.4272
6	1.4179	1.4217	1.4191	1.4204

(d) Test the hypothesis that the variance of the alcohol concentrations is at most five times the error variance versus the alternative hypothesis that it is not. Use a significance level of $\alpha = 0.05$.

2. **Ice cream experiment, continued**

As in Example 17.4.1, page 609, suppose the ice cream experiment is to be repeated with $\gamma = 1.0$ and with a Type I error probability of $\alpha = .05$. Suppose that we would like to reject the null hypothesis $H_0^{\gamma T} : \{\sigma_T^2 \leq \sigma^2\}$ with probability $\pi = 0.95$ if the true value of σ_T^2/σ^2 is greater than $\Delta = 2.0$. How many ice cream flavors should be included in the experiment if $r = 2$ observations are to be taken on each? How many observations are needed? Is this an improvement over the result in the example for $r = 3$? (Note: $F_{150,150,.05} = 1.309$, $F_{160,160,.05} = 1.298$, $F_{170,170,.05} = 1.288$, $F_{180,180,.05} = 1.279$).

3. Consider the following random-effects model:

$$Y_{ijkmt} = \mu + A_i + B_j + C_k + D_m$$
$$+ (AB)_{ij} + (BC)_{jk} + (BD)_{jm} + \epsilon_{ijkmt},$$
$$i = 1, \ldots, a, \quad j = 1, \ldots, b, \quad k = 1, \ldots, c,$$
$$m = 1, \ldots, d, \quad t = 1, \ldots, r,$$
$$A_i \sim N(0, \sigma_A^2), \quad B_j \sim N(0, \sigma_B^2), \quad C_k \sim N(0, \sigma_C^2),$$
$$D_m \sim N(0, \sigma_D^2), \quad (AB)_{ij} \sim N(0, \sigma_{AB}^2), \quad (BC)_{jk} \sim N(0, \sigma_{BC}^2),$$
$$(BD)_{jm} \sim N(0, \sigma_{BD}^2), \quad \epsilon_{ijkmt} \sim N(0, \sigma^2),$$

where all random variables on the right hand side of the model are mutually independent.

(a) Write out the expected mean squares for all main effects and interactions in the model.

(b) How would you test the null hypothesis $H_0^A : \{\sigma_A^2 = 0\}$ against the alternative hypothesis $H_A^A : \{\sigma_A^2 > 0\}$?

(c) How would you test the null hypothesis $H_0^B : \{\sigma_B^2 = 0\}$ against the alternative hypothesis $H_A^B : \{\sigma_B^2 > 0\}$?

(d) Give formulae for unbiased estimates of σ_{BD}^2 and σ_B^2.

(e) Give formulae for individual 95% confidence intervals for σ_{BD}^2 and σ_B^2. What is the overall confidence level?

Table 17.17 Treatments and percentage change in height for the buttermilk biscuit experiment

Block	Position					
	1	2	3	4	5	6
1	2 (150.0)	1 (188.2)	2 (177.8)	3 (166.7)	3 (187.5)	1 (182.4)
2	1 (183.3)	2 (183.3)	2 (183.3)	3 (176.5)	1 (160.0)	3 (187.5)
3	1 (178.9)	3 (182.4)	2 (193.8)	3 (176.5)	2 (188.9)	1 (188.9)
4	2 (177.8)	1 (145.5)	3 (155.0)	1 (173.7)	3 (200.0)	2 (187.5)
5	1 (205.6)	3 (188.2)	3 (142.9)	2 (161.9)	2 (177.8)	1 (159.1)

4. **Buttermilk biscuit experiment** (Stacie Taylor, 1995)

   The buttermilk biscuit experiment was run by Stacie Taylor in 1995 to find out which brands of refrigerated buttermilk biscuit give rise to the fluffiest biscuits. Three brands were examined (factor A, 3 levels, fixed effect), all of which had claims to be light, fluffy, or flaky in their advertising campaigns. The biscuits were baked on a baking tray for 7 minutes in the center of an oven set to 425°F. Since only six biscuits could be baked at a time, the experiment was run as a general complete block design with blocks of size $k = 6$.

   (a) Use a mixed model with interaction to represent the data, where the random effect represents the block (run of the oven) and the fixed effect represents the biscuit brand. Write out the model including all of the assumptions.

   (b) The data collected by the experimenter are shown in Table 17.17. As far as possible, check the assumptions on the model for these data.

   (c) Write out the expected mean squares for all terms in the model.

   (d) Draw a block×brand interaction plot for those blocks observed in the experiment.

   (e) Test the hypothesis that the variance in height of the biscuits due the population of block×brand interactions is negligible against the alternative hypothesis that it is not negligible. Interpret your conclusions in terms of the plot in part (d).

   (f) Calculate a set of 95% simultaneous confidence intervals for the pairwise comparisons between the brands. State your conclusions.

5. **Candle experiment, continued**

   The candle experiment was described in Example 6, page 326. The experiment was run to determine whether different colored candles (factor A, four fixed levels, red, white, blue, yellow, coded 1, 2, 3, 4) burn at different speeds. The design used was a general complete block design, with each of the four experimenters representing one block. Analyze the experiment as though the experimenters represent a random sample from a large population of people who might use these candles in practice. Use a two-way mixed model with interaction. The data are reproduced in Table 17.18.

Table 17.18 Data for the candle experiment (seconds)

Person	Red		White		Color Blue		Yellow	
1	989	1032	1044	979	1011	951	974	998
	1077	1019	987	1031	928	1022	1033	1041
2	899	912	847	880	899	800	886	859
	911	943	879	830	820	812	901	907
3	898	840	840	952	909	790	950	992
	955	1005	961	915	871	905	920	890
4	993	957	987	960	864	925	949	973
	1005	982	920	1001	824	790	978	938

Table 17.19 Distances (in yards) traveled by balls in the golf experiment

Golfer	Brand	Distance					
		1	2	3	4	5	6
1	1	209	204	179	230	233	245
	2	188	211	242	222	187	233
	3	219	204	247	215	197	161
2	1	240	207	192	190	226	188
	2	216	195	240	215	219	238
	3	195	221	205	192	183	230

number b of golfers to be determined. The experiment was to be run as a general complete block design with fixed treatment effects and random golfer effects. Since the golfer is aware of which brand of ball he or she is hitting, there may well be a golfer×brand interaction. However, the differences between brands averaged over the interaction is important here.

A small pilot experiment was conducted. There were only two golfers, and each hit $s = 6$ balls of each brand in a random order. Mis-hits were ignored. The distances that the balls traveled were recorded in yards and are shown in Table 17.19.

(a) Use the pilot experiment data to calculate a 95% upper bound for the error variance σ^2.

(b) Use the pilot experiment data to calculate a 95% confidence interval for σ^2_{GB}, the variance of the population of interactions between golfers and brands.

(c) The experimenter wanted the main experiment to be able to calculate a set of simultaneous 95% confidence intervals for the pairwise differences in the brands, and he wanted the widths of these intervals to be at most 20 yards. Assuming that the maximum block size would be about $k = 18$ as in the pilot experiment, how many randomly selected golfers would be needed?

7. Consider the following mixed model:

$$Y_{ijkmt} = \mu + \alpha_i + B_j + C_k + \delta_m + (\alpha B)_{ij} + (\alpha\delta)_{im}$$
$$+ (B\delta)_{jm} + (C\delta)_{km} + (\alpha B\delta)_{ijm} + \epsilon_{ijkmt},$$
$$i = 1, \ldots, a, \quad j = 1, \ldots, b, \quad k = 1, \ldots, c,$$
$$m = 1, \ldots, d, \quad t = 1, \ldots, r,$$
$$B_j \sim N(0, \sigma_B^2), \quad C_k \sim N(0, \sigma_C^2), \quad (\alpha B)_{ij} \sim N(0, \sigma_{AB}^2),$$
$$(B\delta)_{jm} \sim N(0, \sigma_{BD}^2), \quad (C\delta)_{km} \sim N(0, \sigma_{CD}^2),$$
$$(\alpha B\delta)_{ijm} \sim N(0, \sigma_{ABD}^2), \quad \epsilon_{ijkmt} \sim N(0, \sigma^2),$$

where α_i and δ_m are fixed effects, all other effects are random effects, and all random variables on the right-hand side of the model are mutually independent.

(a) Write out the expected mean squares for all main effects and interactions in the model.

(b) How would you test the hypothesis $H_0 : \{\delta_m + \overline{(\alpha\delta)}_{.m}$ all equal$\}$ against the alternative hypothesis that these parameters are not all equal?

(c) Give a formula for an unbiased estimate of σ_B^2.

(d) Give a formula for a 95% confidence interval for σ_B^2.

8. **Ice cream experiment, continued**

The ice cream experiment was described in Example 17.3.1, page 600, and was analyzed in Examples 17.3.2–17.3.5 and 17.4.1. In Section 17.10, a new model was suggested that involved a quadratic time trend.

(a) What could account for a quadratic time trend?

(b) Investigate the assumptions on the models with and without the quadratic time trend.

(c) Redo the analyses of Examples 17.3.2–17.3.5 and 17.4.1 for the new model and compare your answers with the original model.

(d) Which model do you prefer and why?

18 Nested Models

18.1 Introduction
18.2 Examples and Models
18.3 Analysis of Nested Fixed Effects
18.4 Analysis of Nested Random Effects
18.5 Using SAS Software
Exercises

18.1 Introduction

A factor is said to be *nested* within a second factor if each of its levels is observed in conjunction with just one level of the second factor. An example can be obtained from the clean wool experiment that was discussed in the last chapter. There, the objective of the experiment was to examine the variability of the "clean content" among bales of wool in a large shipment. Several bales were selected for examination, and several cores were taken from each bale and measured. Each core was taken from only one bale, so the cores (levels of the first factor) are observed in conjunction with only one bale (level of the second factor). In the above language, the cores are *nested within the bales*. In the original experiment, there was only one observation taken on each core. The variability of the different cores could not, therefore, be distinguished from measurement error, and their effects were not included explicitly in the model. Had there been more than one observation per core, we could have included in the model separate effects due to bales, cores nested within bales, and experimental error.

In this chapter we discuss how to recognize nested factors, how to formulate the associated models, and how to analyze the effects in these models. Many of the analysis techniques are similar to those in the previous chapter.

In the next section we discuss some examples of hypothetical experiments involving nested effects, and possible models to represent the data. In Section 18.3, we find the estimable contrasts for fixed-effects nested models and develop tests of hypotheses and confidence intervals for these. The more usual setting where the nested effects are random effects is discussed in Section 18.4 and, where possible, we borrow the formulae from the fixed effects setting as we did in Chapter 17. The rules of Chapters 7 and 17 for finding degrees of freedom, sums of squares and expected mean squares and variance components are then extended to encompass nested models. The analysis of nested models using the SAS computer package is discussed in Section 18.5.

18.2 Examples and Models

Nested factors are usually, but not always, random effects, and they are usually, but not always, blocking factors. In the following examples, we give a selection of different situations involving random effects and suggest some reasonable models to represent the data.

Example 18.2.1 Machine head experiment

Hicks (1956) describes a simple experiment to study the differences in the strain readings (the response) of four different heads on each of five different machines. The heads on each machine were supposedly all doing the same job and should have given rise to similar (nonvariable) readings.

Since each head was observed on only one machine, the heads were "nested within machines," giving twenty heads in total. Four observations were taken on each head. The usual two-way analysis of variance model is not appropriate here, since it would read

$$Y_{ijt} = \mu + \alpha_i + \beta_j + (\alpha\beta)_{ij} + \epsilon_{ijt},$$
$$\epsilon_{ijt} \sim N(0, \sigma^2),$$
ϵ_{ijt}'s are mutually independent,
$t = 1, \ldots, 4; \quad i = 1, \ldots, 5; \quad j = 1, \ldots, 4,$

where, α_i is the effect of the ith machine, β_j is the effect of the jth head, and $(\alpha\beta)_{ij}$ is the extra effect of observing the ith machine and jth head together. This suggests that every head is observed on every machine, which was not the case. Instead, we need a notation that will clearly indicate the nested nature of the factors. One popular notation, which we shall adopt here, is to replace $\beta_j + (\alpha\beta)_{ij}$ by $\beta_{j(i)}$, where the parentheses indicate that we are looking at the head that happens to be numbered as the jth head on the ith machine. The *two-way nested model* is then

$$Y_{ijt} = \mu + \alpha_i + \beta_{j(i)} + \epsilon_{ijt}, \tag{18.2.1}$$
$$\epsilon_{ijt} \sim N(0, \sigma^2),$$
ϵ_{ijt}'s are mutually independent,
$t = 1, \ldots, 4; \quad i = 1, \ldots, 5; \quad j = 1, \ldots, 4.$

18.2 Examples and Models

We note in passing that the response Y_{ijt} could also be written as a nested effect $Y_{t(ij)}$, since this represents the tth observation that is specific to the (ij)th machine head. However, since this representation is not crucial to the analysis, we will continue to use the notation Y_{ijt} that we have used so far throughout the book.

One final consideration is whether the machine effects and head effects should be fixed or random. Let us first suppose that the five machines are the only machines of this type in the factory and that they are not due for replacement. The experimenter would then be interested in these five machines specifically, and their effects on the response would be modeled as fixed effects. Let us alternatively suppose that machine heads wear out and are continually being replaced. The experimenter would then be interested in the population of heads from which the particular twenty in the experiment were drawn. Consequently, the nested head effect would be modeled as a random effect. The model would be written as

$$Y_{ijt} = \mu + \alpha_i + B_{j(i)} + \epsilon_{ijt}, \qquad (18.2.2)$$
$$\epsilon_{ijt} \sim N(0, \sigma^2), \quad B_{j(i)} \sim N(0, \sigma^2_{B(A)}),$$
$$\epsilon_{ijt}\text{'s and } B_{j(i)}\text{'s are all mutually independent,}$$
$$t = 1, \ldots, 4; \quad i = 1, \ldots, 5; \quad j = 1, \ldots, 4,$$

where α_i is the effect of the ith machine, and $\sigma^2_{B(A)}$ is the variance of responses from the population of machine heads that could be fitted on these five machines. Notice that all random variables on the right-hand side of the model are assumed to be mutually independent. □

In the previous example there were two treatment factors, one of whose levels were nested within those of the other. In the following experiment, there are two blocking factors, which are nested one within the other.

Example 18.2.2 Efficiency experiment

An experiment was run in 1997 by Carina Dalton, Greg Krzys, Scott O'Dee, and Brad Welch to examine the assertion that "a person works more efficiently when there is no one looking over his or her shoulder." Twelve subjects were recruited for the experiment, and three of these were assigned to each of the four experimenters. Each subject was asked to complete a simple task—crossing through every occurrence of the letter "e" on a page of prose. There were two levels of the treatment factor. Level 1 required the assigned experimenter to look over the subject's shoulder while the task was being completed, and level 2 required the experimenter to be elsewhere in the room absorbed in a book. The response was the time taken to complete the task. Each subject was assigned both treatments, but in a randomized order.

The blocking factor in this experiment was subject. However, the subjects each worked with only one experimenter, and so the subject effects were nested within the experimenter effects.

The subjects were graduate students at The Ohio State University. Although they were not selected according to the rules of a simple random sample, let us suppose that they were a reasonable representation of that population. Let us also suppose that the variation among the techniques of the experimenters, who were also graduate students, was representative of a population of student experimenters. It might also be reasonable to assume that some subjects may be more perturbed than others about an experimenter watching them complete the task. In this case, we might wish to include a subject–treatment interaction in the model. However, there is only one observation per subject per treatment, so the subject–treatment interaction could not be distinguished from the random error.

A second possible model would be to include an experimenter–treatment interaction instead of a subject–treatment interaction. Such an interaction might occur if the actions of the four experimenters were not all identical. In this case the model would be

$$Y_{hqi} = \mu + E_h + S_{q(h)} + \alpha_i + (\alpha E)_{hi} + \epsilon_{hqi},$$
$$\epsilon_{hqi} \sim N(0, \sigma^2), \quad S_{q(h)} \sim N(0, \sigma^2_{S(E)}), \quad (\alpha E)_{hi} \sim N(0, \sigma^2_{EA}),$$
$$\epsilon_{hqi}\text{'s, } E_h\text{'s, } S_{q(h)}\text{'s and } (\alpha E)_{hi}\text{'s are all mutually independent,}$$
$$h = 1, \ldots, 4; \quad i = 1, 2, 3; \quad i = 1, 2.$$

where E_h is the effect of the hth randomly selected experimenter, $S_{q(h)}$ is the effect of the qth randomly selected subject assigned to the hth experimenter, α_i is the effect of the ith treatment, and $(\alpha E)_{hi}$ is the random effect representing the interaction between the hth experimenter and the ith treatment.

Lastly, we may also wish to include a time-order effect in the model, since the subjects may have been able to complete the task faster on the second occasion just due to familiarity. So we could add the extra term γx_{hqi}, where x_{hqi} is 1 or 2 according to whether the (hq)th subject is assigned treatment i on the first or second occasion. □

18.3 Analysis of Nested Fixed Effects

18.3.1 Least Squares Estimates

Consider first the simplest possible fixed-effects nested model—the two-way nested model (18.2.1) that was suggested for the machine head experiment of Example 18.2; that is,

$$Y_{ijt} = \mu + \alpha_i + \beta_{j(i)} + \epsilon_{ijt},$$
$$\epsilon_{ijt} \sim N(0, \sigma^2),$$
$$\epsilon_{ijt}\text{'s are mutually independent,}$$
$$t = 1, \ldots, r_{ij}; \quad i = 1, \ldots, a; \quad j = 1, \ldots, b.$$

The error assumptions are examined in the same way as in Chapter 5 for the one-way analysis of variance model. In any model, the estimable contrasts are functions of the expected values of the response variables (see, for example, Section 3.4.1, page 37). In the present model, $E[Y_{ijt}]$ is equal to

$$E[Y_{ijt}] = \mu + \alpha_i + \beta_{j(i)}.$$

18.3 Analysis of Nested Fixed Effects

If we take an average over the subscripts t and j, we find that a comparison of the levels of A averaged over the levels of B is estimable; that is, we can estimate pairwise comparisons such as

$$[\alpha_i + \overline{\beta}_{.(i)}] - [\alpha_s + \overline{\beta}_{.(s)}],$$

and we can estimate general contrasts such as

$$\sum_{i=1}^{a} c_i [\alpha_i + \overline{\beta}_{.(i)}], \quad \text{with} \quad \sum_{i=1}^{a} c_i = 0.$$

We can also compare the effects of those levels of B that were observed in conjunction with the *same* level of A; that is,

$$[\alpha_i + \beta_{j(i)}] - [\alpha_i + \beta_{u(i)}] = \beta_{j(i)} - \beta_{u(i)},$$

or, in general,

$$\sum_{j=1}^{b} d_j \beta_{j(i)}, \quad \text{with} \quad \sum_{j=1}^{b} d_j = 0, \quad \text{for any given } i.$$

To obtain the least squares estimators of estimable contrasts, we use the method of least squares to find parameter estimates that minimize the sum of squared errors

$$\sum_{i=1}^{a} \sum_{j=1}^{b} \sum_{t=1}^{r_{ij}} e_{ijt}^2 = \sum_{i=1}^{a} \sum_{j=1}^{b} \sum_{t=1}^{r_{ij}} \left(y_{ijt} - \mu - \alpha_i - \beta_{j(i)}\right)^2.$$

Readers with a knowledge of calculus may verify (see Exercise 7) that the least squares estimate of $\mu + \alpha_i + \beta_{j(i)}$ is $\overline{y}_{ij.}$. Consequently, the least squares estimator of

$$\sum_{i=1}^{a} c_i [\alpha_i + \overline{\beta}_{.(i)}] \quad \text{is} \quad \sum_{i=1}^{a} c_i \overline{Y}_{i..}$$

with $\Sigma c_i = 0$. The corresponding variance is $\Sigma c_i^2 \sigma^2 / r_{i.}$. Similarly, the least squares estimator of

$$\sum_{j=1}^{b} d_j \beta_{j(i)} \quad \text{is} \quad \sum_{j=1}^{b} d_j \overline{Y}_{ij.} \quad \text{for any } i$$

with $\Sigma d_j = 0$. The corresponding variance is $\Sigma d_j^2 \sigma^2 / r_{ij}$.

All of these formulae can easily be adapted to the case where B has a different number of levels for each level of A by replacing b by b_i.

18.3.2 Estimation of σ^2

The error sum of squares is

$$ssE = \sum_{i=1}^{a} \sum_{j=1}^{b} \sum_{t=1}^{r_{ij}} \left(y_{ijt} - \hat{\mu} - \hat{\alpha}_i - \hat{\beta}_{j(i)}\right)^2$$

$$= \sum_{i=1}^{a}\sum_{j=1}^{b}\sum_{t=1}^{r_{ij}} (y_{ijt} - \bar{y}_{ij.})^2 \qquad (18.3.3)$$

$$= \sum_{i=1}^{a}\sum_{j=1}^{b}\sum_{t=1}^{r_{ij}} y_{ijt}^2 - \sum_{i=1}^{a}\sum_{j=1}^{b} r_{ij}\bar{y}_{ij.}^2 . \qquad (18.3.4)$$

A comparison with the formulae in Section 6.4 shows that everything that we have written so far about the fixed-effects two-way nested model could have been deduced from the fixed-effects two-way complete model after replacing $\beta_j + (\alpha\beta)_{ij}$ by $\beta_{j(i)}$. Therefore, we may also deduce that the error mean square, $msE = ssE/(n-v)$, gives an unbiased estimate for σ^2, and the corresponding random variable MSE has a chi-squared distribution with $n-v$ degrees of freedom (where $n = r$ and $v = ab$).

18.3.3 Confidence Intervals

We may obtain a $100(1-\alpha)\%$ confidence bound for σ^2 from the information in the previous subsection; that is,

$$\sigma^2 \leq \frac{ssE}{\chi^2_{n-v,1-\alpha}} .$$

The derivation of the bound was explained in Section 3.4.10.

Confidence intervals for $\Sigma c_i(\alpha_i + \bar{\beta}_{.(i)})$ and for $\Sigma d_j \beta_{j(i)}$ may be obtained using the relevant methods from Chapter 4 together with the formulae

$$\Sigma c_i \bar{Y}_{i..} \pm w \sqrt{\sum_i \left(\frac{c_i^2}{r_i}\right) msE}$$

and

$$\Sigma d_j \bar{Y}_{ij.} \pm w \sqrt{\sum_j \left(\frac{d_j^2}{r_{ij}}\right) msE} .$$

18.3.4 Hypothesis Testing

We may obtain a test of the null hypothesis that the levels of B have the same effect on the response within every given level of A, that is,

$$H_0^{B(A)} : \{\beta_{1(i)} = \beta_{2(i)} = \ldots = \beta_{b(i)}, \text{ for every } i = 1, \ldots, a\},$$

against the alternative hypothesis $H_A^{B(A)} : \{H_0^{B(A)} \text{ is not true}\}$ by comparing the sum of squares for error (18.3.3) in the fixed-effects two-way nested model with the sum of squares for error in the reduced (one-way) model. The reduced model is

$$Y_{ijt} = \mu^* + \alpha_i + \epsilon_{ijt} ,$$

18.3 Analysis of Nested Fixed Effects

and the error sum of squares is given by (3.4.4), page 42, with an extra subscript; that is,

$$ssE_0 = \sum_{i=1}^{a}\sum_{j=1}^{b}\sum_{t=1}^{r_{ij}}(y_{ijt} - \bar{y}_{i..})^2.$$

The numerator of the test statistic is then

$$msB(A) = \frac{ssB(A)}{a(b-1)},$$

where the number of degrees of freedom for $B(A)$ is obtained as the difference between the error degrees of freedom in the reduced and full models; that is,

$$(n-a) - (n-v) = v - a = ab - a = a(b-1),$$

and where

$$\begin{aligned}ssB(A) &= ssE_0 - ssE\\
&= \sum_i\sum_j\sum_t(y_{ijt} - \bar{y}_{i..})^2 - \sum_i\sum_j\sum_t(y_{ijt} - \bar{y}_{ij.})^2\\
&= \sum_i\sum_j\sum_t y_{ijt}^2 - \sum_i r_{i.}\bar{y}_{i..}^2 - \sum_i\sum_j\sum_t y_{ijt}^2 + \sum_i\sum_j r_{ij}\bar{y}_{ij.}^2\\
&= \sum_i\sum_j r_{ij}\bar{y}_{ij.}^2 - \sum_i r_{i.}\bar{y}_{i..}^2 \quad\quad\quad\quad\quad\quad\quad\quad\quad (18.3.5)\\
&= \sum_i\sum_j r_{ij}(\bar{y}_{ij.} - \bar{y}_{i..})^2. \quad\quad\quad\quad\quad\quad\quad (18.3.6)\end{aligned}$$

The decision rule for testing $H_0^{B(A)}$ versus $H_A^{B(A)}$ at significance level α is

$$\text{reject } H_0^{B(A)} \text{ if } \frac{ssB(A)/a(b-1)}{ssE/(n-ab)} > F_{a(b-1),n-ab,\alpha}.$$

Similarly, the decision rule for testing

$$H_0^A : \{\alpha_i + \overline{\beta}_{.(i)} \text{ all equal}\}$$

against the alternative hypothesis $H_A^A : \{H_0^A \text{ is false}\}$ is

$$\text{reject } H_0^A \text{ if } \frac{ssA/(a-1)}{ssE/(n-ab)} > F_{a-1,n-ab,\alpha},$$

where

$$ssA = \sum_i r_{i.}(\bar{y}_{i..} - \bar{y}_{...})^2 = \sum_i r_{i.}\bar{y}_{i..}^2 - n\bar{y}_{...}^2.$$

Notice that $ssB(A)$ in the two-way nested model is equal to $ssB + ssAB$ in the two-way complete model. Also, the degrees of freedom for $B(A)$ in the nested model can be obtained as the sum of the degrees of freedom for B and AB in the complete model; that is,

$$(b-1) + (b-1)(a-1) = a(b-1).$$

This link between the nested model and the corresponding complete model means that when the sample sizes are equal, we can obtain all the formulae we need from the rules in

Chapter 7. This remains true for more complicated models also. For example, if we take the nested model

$$Y_{ijkt} = \mu + \alpha_i + \beta_{j(i)} + \gamma_{k(ij)} + \epsilon_{ijkt},$$

we have the following equivalences with the terms of the three-way complete model:

$$\beta_{j(i)} = \beta_j + (\alpha\beta)_{ij},$$
$$\gamma_{k(ij)} = \gamma_k + (\alpha\gamma)_{ik} + (\beta\gamma)_{jk} + (\alpha\beta\gamma)_{ijk};$$

so, for example, the sum of squares for $C(AB)$ is

$$ssC(AB) = ssC + ssAC + ssBC + ssABC$$
$$= \sum_i \sum_j \sum_k r_{ijk}\bar{y}^2_{ijk.} - \sum_i \sum_j r_{ij.}\bar{y}^2_{ij..}$$
$$= \sum_i \sum_j \sum_k r_{ijk}(\bar{y}_{ijk.} - \bar{y}_{ij..})^2,$$

with degrees of freedom

$$(c-1) + (a-1)(c-1) + (b-1)(c-1) + (a-1)(b-1)(c-1) = ab(c-1).$$

As with the crossed model, the degrees of freedom for $C(AB)$ give a clue to the subscripts needed in the formula for the sum of squares for $C(AB)$; that is, the degrees of freedom $ab(c-1) = abc - ab$ suggest that the sum of squares for $C(AB)$ must contain the terms $\bar{y}_{ijk.}$ and $\bar{y}_{ij..}$, the latter with a minus sign.

To obtain the degrees of freedom corresponding to any effect, we notice that the degrees of freedom for A are the same as in the crossed model; that is, $(a-1)$. The degrees of freedom for $B(A)$ are $(b-1) + (a-1)(b-1) = a(b-1)$, and those for $C(AB)$ are $ab(c-1)$. Thus we see a pattern. The number of degrees of freedom is the product of the numbers of levels corresponding to the factors in parentheses and one less than the numbers of levels corresponding to the factors not in parentheses. We may now modify rules 1 and 2 in Section 7.3 for equal sample sizes listed below. We also include rules 3 and 4 here for easy reference, although these remain the same.

1. Write down the name of the main effect or interaction of interest and the corresponding number of levels and subscripts. Include parentheses to denote nesting of factors.

2. The number of degrees of freedom ν for any effect is the product of the numbers of levels corresponding to the factors in parentheses and one less than the numbers of levels corresponding to the factors not in parentheses.

3. Multiply out the number of degrees of freedom and replace each letter with the corresponding subscripts.

4. The sum of squares for testing the hypothesis that a main effect or an interaction is negligible is obtained as follows. Use each group of subscripts in rule 3 as the subscripts of a term $\bar{y}$, averaging over all subscripts not present and keeping the same signs. Put the resulting estimate in parentheses, square it and sum over all possible subscripts. To

18.3 Analysis of Nested Fixed Effects

expand the parentheses, square each term in the parentheses, keep the same signs, and sum over all possible subscripts.

The other rules remain the same. In particular, confidence intervals for $\sum_{i=1}^{a} c_i \left(\alpha_i + \overline{\beta}_{.(i)}\right)$ and for $\sum_{j=1}^{b} d_j \beta_{j(i)}$ may be calculated using the usual multiple-comparison techniques of Chapter 4.

Example 18.3.1 Plastic experiment

Consider the following hypothetical experiment in which a manufacturer of molded plastic wishes to replace a standard ingredient by a cheaper alternative. The two ingredients form the two levels of the treatment factor to be studied. The manufacturing company has factories in three different parts of the country, and since different climates may affect the product differently, the experiment is to take place in each of the three locations. Within each factory, two operators oversee two machines each. The experiment will be run during the usual downtime of the machines.

A possible model for the experiment is

$$Y_{ijkut} = \mu + \alpha_i + \beta_{j(i)} + \gamma_{k(ij)} + \tau_u + (\tau\gamma)_{uk(ij)} + \epsilon_{ijkut},$$

$$\epsilon_{ijkut} \sim N(0, \sigma^2),$$

ϵ_{ijkut}'s are mutually independent,

$$t = 1, \ldots, r; \quad i = 1, 2, 3; \quad j = 1, 2; \quad k = 1, 2;$$

where α_i is the effect of the ith location, $\beta_{j(i)}$ is the effect of the jth operator at the ith location, $\gamma_{k(ij)}$ is the effect of the kth machine that is looked after by the jth operator at the ith location, τ_u is the effect of the uth treatment, $(\tau\gamma)_{uk(ij)}$ is the interaction effect between the uth treatment and (ijk)th machine, Y_{ijkut} is the tth response (strength measurement) on the uth treatment and (ijk)th machine, and ϵ_{ijkut} is the corresponding random error, assumed to have a normal distribution with mean 0 and variance σ^2. We also assume that the error variables are mutually independent.

Table 18.1 Degrees of freedom and sums of squares

Effect	Degrees of Freedom	Sum of Squares
A	$a - 1 = 2$	$bcdr\Sigma_i \overline{y}_{i...}^2 - abcdr\overline{y}_{....}^2$
B(A)	$a(b-1) = 3$	$cdr\Sigma_i \Sigma_j \overline{y}_{ij..}^2 - bcdr\Sigma_i \overline{y}_{i...}^2$
C(AB)	$ab(c-1) = 6$	$dr\Sigma_i \Sigma_j \Sigma_k \overline{y}_{ijk.}^2 - cdr\Sigma_i \Sigma_j \overline{y}_{ij..}^2$
Trt	$d - 1 = 1$	$abcr\Sigma_u \overline{y}_{...u.}^2 - abcdr\overline{y}_{....}^2$
Trt×C(AB)	$ab(c-1)(d-1) = 6$	$r\Sigma_i \Sigma_j \Sigma_k \Sigma_u \overline{y}_{ijku.}^2 - dr\Sigma_i \Sigma_j \Sigma_k \overline{y}_{ijk..}^2$ $- cr\Sigma_i \Sigma_j \Sigma_u \overline{y}_{ij.u.}^2 + cdr\Sigma_i \Sigma_j \overline{y}_{ij..}^2$
Error	$24r - 19$, by subtraction	Obtain by subtraction
Total	$n - 1 = 24r - 1$	$\Sigma_i \Sigma_j \Sigma_k \Sigma_u \Sigma_t y_{ijkut}^2 - abcdr\overline{y}_{....}^2$

The degrees of freedom and sums of squares for each effect are obtained from rules 1–listed above this example and are shown in Table 18.1. Using the formula for confidence intervals in Section 18.3.3, we may obtain a confidence interval for $\Sigma_u h_u \tau_u$ as

$$\sum_u h_u \bar{y}_{...u.} \pm w \sqrt{\sum_u \frac{h_u^2}{12r} msE} \,.$$

18.4 Analysis of Nested Random Effects

18.4.1 Expected Mean Squares

In Chapter 17 we found that we could modify many of the formulae arising from the fixed-effect crossed models to obtain confidence intervals and hypothesis tests for variance components in the corresponding random-effects models. To find the denominators for the hypothesis tests and to find the estimates for variance components, all we need to do is to calculate the expected values of the mean squares arising from the corresponding fixed-effect models. For equal sample sizes, expected mean squares can be obtained using rule 17 of Section 17.8.1, page 627, exactly as for the random and mixed-effects crossed models. The rules 18–22 for calculating test ratios and confidence intervals also follow exactly as for the crossed models. We will illustrate these via the model that was suggested for the machine head experiment in Example 18.2. A suggested model was

$$Y_{ijt} = \mu + \alpha_i + B_{j(i)} + \epsilon_{ijt},$$
$$\epsilon_{ijt} \sim N(0, \sigma^2), \quad B_{j(i)} \sim N(0, \sigma^2_{B(A)}),$$

ϵ_{ijt}'s and $B_{j(i)}$'s are all mutually independent,

$$t = 1, \ldots, 4; \quad i = 1, \ldots, 5; \quad j = 1, \ldots, 4,$$

where α_i is the effect of the ith machine, $B_{j(i)}$ is the effect of the jth randomly selected head on the ith machine, and $\sigma^2_{B(A)}$ is the variance of responses from the population of machine heads that could be fitted on these five machines. The error assumptions are examined in the same way as in Chapter 5 for the one-way analysis of variance model. More sophisticated techniques of checking other assumptions on a mixed model with nested effects are discussed by Beckman, Nachtsheim, and Cook (1987).

The degrees of freedom and sums of squares for the fixed-effects two-way nested model were calculated in Section 18.3. These are listed, for equal sample sizes, in Table 18.2.

We first verify that the fixed-effects mean square for error also provides an unbiased estimate for σ^2 in the mixed effects two-way nested model. From Section 18.3.2, we know that $MSE = SSE/(ab(r-1))$, where

$$SSE = \sum_{i=1}^{a} \sum_{j=1}^{b} \sum_{t=1}^{r} Y_{ijt}^2 - r \sum_{i=1}^{a} \sum_{j=1}^{b} \bar{Y}_{ij.}^2 \,.$$

For the mixed-effects two-way nested model (18.2.2), we have

$$E[Y_{ijt}] = E[\bar{Y}_{ij.}] = E[\bar{Y}_{i..}] = \mu + \alpha_i \quad \text{and} \quad E[\bar{Y}_{...}] = \mu + \bar{\alpha}_. \,.$$

18.4 Analysis of Nested Random Effects

Table 18.2 Analysis of variance table for a mixed-effects two-way nested model

Source of Variation	Deg. of Freedom	Sum of Squares	Mean Square	Expected Mean Square
A	$(a-1)$	ssA	msA	$Q(\alpha_i) + r\sigma^2_{B(A)} + \sigma^2$
B(A)	$a(b-1)$	ssB(A)	msB(A)	$r\sigma^2_{B(A)} + \sigma^2$
Error	$ab(r-1)$	ssE	msE	σ^2
Total	$abr-1$	sstot		

Formulae for Equal Sample Sizes	
$ssA = br\Sigma_i \bar{y}^2_{i..} - abr\bar{y}^2_{...}$	$ssB(A) = r\Sigma_i\Sigma_j \bar{y}^2_{ij.} - br\Sigma_i \bar{y}^2_{i..}$
$ssE = \Sigma_i\Sigma_j\Sigma_t y^2_{ijt} - r\Sigma_i\Sigma_j \bar{y}^2_{ij.}$	$sstot = \Sigma_i\Sigma_j\Sigma_t y^2_{ijt} - abr\bar{y}^2_{...}$

Also,
$$\text{Var}(Y_{ijt}) = \sigma^2_{B(A)} + \sigma^2$$

and
$$\text{Var}(\bar{Y}_{ij.}) = \text{Var}\left(\mu + \alpha_i + B_{j(i)} + \frac{1}{r}\sum_{t=1}^{r}\epsilon_{ijt}\right) = \sigma^2_{B(A)} + \frac{\sigma^2}{r}.$$

So,
$$E[SSE] = \left[abr(\sigma^2_{B(A)} + \sigma^2) + br\sum_i(\mu+\alpha_i)^2\right]$$
$$- \left[abr\left(\sigma^2_{B(A)} + \frac{\sigma^2}{r}\right) + br\sum_i(\mu+\alpha_i)^2\right]$$
$$= ab(r-1)\sigma^2.$$

So, $E[MSE] = \sigma^2$ as required. We also have that
$$\text{Var}(\bar{Y}_{i..}) = \text{Var}\left(\mu + \alpha_i + \frac{1}{b}\sum_{j=1}^{b}B_{j(i)} + \frac{1}{br}\sum_{j=1}^{b}\sum_{t=1}^{r}\epsilon_{ijt}\right)$$
$$= \frac{\sigma^2_{B(A)}}{b} + \frac{\sigma^2}{br}.$$

Similarly,
$$\text{Var}(\bar{Y}_{...}) = \frac{\sigma^2_{B(A)}}{ab} + \frac{\sigma^2}{abr}.$$

Consequently, the expected value of the sum of squares for A is
$$E[SSA] = E\left[br\sum_{i=1}^{a}\bar{Y}^2_{i..} - abr\bar{Y}^2_{...}\right]$$

$$= \left[abr\left(\frac{\sigma_{B(A)}^2}{b} + \frac{\sigma^2}{br}\right) + br\sum_i(\mu+\alpha_i)^2\right]$$

$$- \left[abr\left(\frac{\sigma_{B(A)}^2}{ab} + \frac{\sigma^2}{abr}\right) + abr\sum_i(\mu+\overline{\alpha}_.)^2\right]$$

$$= r(a-1)\sigma_{B(A)}^2 + (a-1)\sigma^2 + br\sum_i(\alpha_i - \overline{\alpha}_.)^2.$$

Then, since $MSA = SSA/(a-1)$, we have

$$E[MSA] = \frac{br}{a-1}\sum_i(\alpha_i - \overline{\alpha}_.)^2 + r\sigma_{B(A)}^2 + \sigma^2$$

$$= Q(\alpha_i) + r\sigma_{B(A)}^2 + \sigma^2.$$

Similarly, the expected value of the sum of squares for B nested within A is

$$E[SSB(A)] = E\left[r\sum_{i=1}^{a}\sum_{j=1}^{b}\overline{Y}_{ij.}^2 - br\sum_{i=1}^{1}\overline{Y}_{i..}^2\right]$$

$$= \left[abr\left(\sigma_{B(A)}^2 + \frac{\sigma^2}{r}\right) + br\sum_i(\mu+\alpha_i)^2\right]$$

$$- \left[abr\left(\frac{\sigma_{B(A)}^2}{b} + \frac{\sigma^2}{br}\right) + br\sum_i(\mu+\alpha_i)^2\right]$$

$$= ar(b-1)\sigma_{B(A)}^2 + a(b-1)\sigma^2,$$

and, since $MSB(A) = SSB(A)/(a(b-1))$, we have

$$E[MSB(A)] = r\sigma_{B(A)}^2 + \sigma^2.$$

These expected mean squares are listed in the fourth column of Table 18.2, and we may verify that they can all be obtained from rule 17 of Chapter 17. This rule, which applies also to more complicated mixed-effects nested models, says

17. To obtain the expected mean square for a particular main effect or interaction, first make a note of the subscripts on the term representing that particular effect in the model. Write down variance components for the effect of interest, for the error, and for every interaction whose term in the model includes the noted set of subscripts. Gather up all variance components corresponding to fixed effects into one quadratic form Q. Multiply any remaining variance component except σ^2 by the number of observations taken on each level or combination of levels of the corresponding effect (main effect or interaction). Add up the terms.

18.4.2 Estimation of Variance Components

The rules for obtaining confidence intervals for fixed effects or variance components also remain the same as those in Chapter 17 for non-nested models. Thus, we may obtain a confidence interval for a variance component in a mixed-effects nested model as follows:

18.4 Analysis of Nested Random Effects

19. For a random effect, let $U = \Sigma k_i MS_i$ be the mean square or linear combination of mean squares whose expected value is equal to the variance component corresponding to the random effect. An exact or approximate $100(1-\alpha)\%$ confidence interval for this variance component is

$$\left(\frac{xu}{\chi^2_{x,\alpha/2}}, \frac{xu}{\chi^2_{x,1-\alpha/2}} \right),$$

where

$$x = \frac{[\Sigma k_i(ms_i)]^2}{\Sigma [k_i(ms_i)]^2/x_i},$$

and where u is the observed value of U, ms_i is the observed value of MS_i, and x_i is the number of degrees of freedom corresponding to ms_i.

For example, for the mixed-effects two-way nested model, we may estimate the variability of the response due to the effect of B within A as

$$u = \frac{msB(A) - msE}{r}.$$

Then, using rule 19, we can obtain a $100(1-\alpha)\%$ confidence interval for $\sigma^2_{B(A)}$ as

$$\left(\frac{xu}{\chi^2_{x,\alpha/2}}, \frac{xu}{\chi^2_{x,1-\alpha/2}} \right),$$

where

$$x = \frac{(msB(A) - msE)^2}{\frac{msB(A)^2}{a(b-1)} + \frac{msE^2}{ab(r-1)}}.$$

18.4.3 Hypothesis Testing

Hypothesis testing rules are also obtained from the rules in Chapter 17:

18. To obtain the denominator of the test statistic for testing the null hypothesis that a main effect or interaction effect is zero, write down the expected mean square for the effect of interest (see rule 17). Cross out the term that would be zero if the null hypothesis were true. The denominator of the test statistic is the mean square, or linear combination of mean squares, u, whose expected value is equal to the remaining expression.

21. For a fixed effect, the decision rule for testing the hypothesis that the effect is zero is the same as that in rule 8, page 203, for fixed-effects models except that msE is replaced by the denominator u from rule 18 and the number of error degrees of freedom is replaced by x in rule 19.

22. For a random effect, the decision rule for testing the hypothesis H_0 that the corresponding variance component is zero against the alternative hypothesis that it is not zero

is

$$\text{reject } H_0 \text{ if } \frac{ms}{u} > F_{v,x,\alpha},$$

where ms is the mean square for the effect of interest and v the corresponding degree of freedom, u is the observed value of the denominator as in rule 18, and x is the corresponding degrees of freedom calculated as in rule 19.

For example, using the information in the expected mean squares column of Table 18.2 for the mixed-effects two-way nested model, the decision rule for testing the null hypothesis $H_0^{B(A)}: \{\sigma_{B(A)}^2 = 0\}$ of no variability in the effect of B within each level of A against the alternative hypothesis $H_A^{B(A)}: \{\sigma_{B(A)}^2 > 0\}$ is

$$\text{reject } H_0^{B(A)} \text{ if } \frac{msB(A)}{msE} > F_{a(b-1),ab(r-1),\alpha}, \tag{18.4.7}$$

at chosen significance level α.

To test the hypothesis $H_0^A: \{\alpha_1 = \alpha_2 = \cdots = \alpha_a\}$ that the machine effects are the same averaged over their four heads, the decision rule at significance level α is

$$\text{reject } H_0^A \text{ if } \frac{msA}{msB(A)} > F_{a-1,a(b-1),\alpha}. \tag{18.4.8}$$

18.4.4 Some Examples

Example 18.4.1 Machine head experiment, continued

The data for the machine head experiment are listed in Table 18.3, and the analysis of variance table is shown in Table 18.4. We see that the p-value for testing the hypothesis of no machine differences is 0.67, and we would conclude no difference in the effect on strain readings of the five machines. The test of the null hypothesis that the variance $\sigma_{B(A)}^2$ of the population of possible heads fitted to the machines is zero has p-value 0.065. Only if our choice of significance level is greater than this value would we conclude nonzero variability among the heads.

Table 18.3 Data for the machine head experiment

Mach.	Head 1				Head 2				Head 3				Head 4			
1	6	2	0	8	13	3	9	8	1	10	0	6	7	4	7	9
2	10	9	7	12	2	1	1	10	4	1	7	9	0	3	4	1
3	0	0	5	5	10	11	6	7	8	5	0	7	7	2	5	4
4	11	0	6	4	5	10	8	3	1	8	9	4	0	8	6	5
5	1	4	7	9	6	7	0	3	3	0	2	2	3	7	4	0

Source: Hicks, C. R. (1956). Copyright © 1956 American Society for Quality. Reprinted with permission.

18.4 Analysis of Nested Random Effects

Table 18.4 Analysis of variance table for the machine head experiment

Effect	d.f.	Sum of Squares	Mean Square	Expected Mean Square	Ratio
Machine	4	45.075	11.2688	$Q(\alpha_i) + 4\sigma^2_{B(A)} + \sigma^2$	0.5975
Head(mach.)	15	282.875	18.8583	$4\sigma^2_{B(A)} + \sigma^2$	1.7625
Error	60	642.000	10.7000	σ^2	
Total	79	969.950			

An unbiased estimate of $\sigma^2_{B(A)}$ is given by

$$\frac{msB(A) - msE}{r} = \frac{18.8583 - 10.7000}{4} = 2.0396,$$

and since,

$$x = \frac{(2.0396)^2}{\frac{(18.8583/4)^2}{15} + \frac{(10.70/4)^2}{60}} = 2.598,$$

a 90% confidence interval for $\sigma^2_{B(A)}$ is given by

$$\left(\frac{(2.598)(2.0396)}{\chi^2_{2.598,.05}}, \frac{(2.598)(2.0396)}{\chi^2_{2.598,.95}} \right) \approx \left(\frac{5.299}{6.90}, \frac{5.299}{0.22} \right) = (0.77, 24.09)$$

measured in squared units of strain. □

Example 18.4.2 Soil experiment

Consider an experiment to compare analyses of soil samples with four treatment factors A, B, C, and D, where

- A is "method of analysis" and involves $a = 2$ specifically selected methods.
- B is "laboratory" and involves $b = 4$ specifically selected labs.
- C is "operator conducting the analysis" and there are $c = 3$ randomly selected operators in each lab.
- D is "location from which soil was taken" and involves $d = 3$ randomly selected locations.

The model is

$$Y_{ijkut} = \mu + \alpha_i + \beta_j + (\alpha\beta)_{ij} + C_{k(j)} + (\alpha C)_{ik(j)} + D_u$$
$$+ (\alpha D)_{iu} + (\beta D)_{ju} + (\alpha\beta D)_{iju} + \epsilon_{ijkut},$$

$C_{k(j)} \sim N(0, \sigma^2_{C(B)})$; $(\alpha C)_{ik(j)} \sim N(0, \sigma^2_{AC(B)})$; $D_u \sim N(0, \sigma^2_D)$; $(\alpha D)_{iu} \sim N(0, \sigma^2_{AD})$

$(\beta D)_{ju} \sim N(0, \sigma^2_{BD})$; $(\alpha\beta D)_{iju} \sim N(0, \sigma^2_{ABD})$; $\epsilon_{ijkut} \sim N(0, \sigma^2)$

$i = 1, 2$; $j = 1, 2, 3, 4$; $k = 1, 2, 3$; $u = 1, 2, 3$; $t = 1, 2$;

where α_i is the effect of the ith method of analysis, β_j is the effect of the jth laboratory and $(\alpha\beta)_{ij}$ is the effect of their interaction; $C_{k(j)}$ is the effect of the kth randomly selected operator in the kth laboratory and $(\alpha C)_{ik(j)}$ is the operator×analysis method interaction. D_u is the effect of the uth randomly selected location from which the soil was selected and $(\alpha D)_{iu}$, $(\beta D)_{ju}$ and $(\alpha\beta D)_{iju}$ are respectively the interactions of the uth soil location with the ith method of analysis and with the jth laboratory, and the three-factor interaction of the uth soil location, ith method of analysis, and jth laboratory. Two observations are taken on each soil sample via each method of analysis by each operator.

The degrees of freedom, sums of squares, and expected mean squares for this model are obtained using rules 17–21 in Sections 18.4.1–18.4.3 and are shown in Table 18.5.

The decision rule for testing the null hypothesis $H_0^{ABD} : \{\sigma_{ABD}^2 = 0\}$ against the alternative hypothesis $H_A^{ABD} : \{\sigma_{ABD}^2 > 0\}$ is given by

$$\text{reject } H_0^{ABD} \text{ if } \frac{msABD}{msE} > F_{6,104,\alpha}.$$

If this hypothesis is not rejected, we may wish to examine the AB, AD, and BD interactions. The decision rule for testing the null hypothesis $H_0^{BD} : \{\sigma_{BD}^2 = 0\}$ against the alternative hypothesis $H_A^{BD} : \{\sigma_{BD}^2 > 0\}$ is given by

$$\text{reject } H_0^{BD} \text{ if } \frac{msBD}{msABD} > F_{6,6,\alpha}.$$

The test for the AD interaction is similar. To obtain a suitable denominator for testing

$$H_0^{AB} : \{(\alpha\beta)_{ij} - \overline{(\alpha\beta)}_{i.} - \overline{(\alpha\beta)}_{.j} + \overline{(\alpha\beta)}_{..} = 0, \quad \text{for all } i, j\}$$

against the alternative hypothesis that the interaction is not zero, we need the denominator of the test statistic to be an unbiased estimator for

$$6\sigma_{AC(B)}^2 + 6\sigma_{ABD}^2 + \sigma^2.$$

Such an estimator is

$$U = MS(AC(B)) + MS(ABD) - MSE.$$

This has approximately a χ_x^2 distribution with

$$x = \frac{[MS(AC(B)) + MS(ABD) - MSE]^2}{\frac{MSAC(B)^2}{8} + \frac{MS(ABD)^2}{6} + \frac{MSE^2}{104}}.$$

Thus the decision rule for testing H_0^{AB} against H_A^{AB} is

$$\text{reject } H_0^{AB} \text{ if } \frac{msAB}{U} > F_{3,x,\alpha}.$$

An unbiased estimate of σ_{BD}^2 is

$$U = \frac{msBD - msABD}{12}.$$

18.4 Analysis of Nested Random Effects

Table 18.5 Degrees of freedom, sums of squares, and expected mean squares for the soil experiment

Effect	Degrees of Freedom	Expected Mean Square
A	$a - 1 = 1$	$Q(\alpha, \alpha\beta) + 6\sigma^2_{AC(B)} + 24\sigma^2_{AD}$ $+ 6\sigma^2_{ABD} + \sigma^2$
B	$b - 1 = 3$	$Q(\beta, \alpha\beta) + 12\sigma^2_{C(B)} + 6\sigma^2_{AC(B)}$ $+ 12\sigma^2_{BD} + 6\sigma^2_{ABD} + \sigma^2$
AB	$(a-1)(b-1) = 3$	$Q(\alpha\beta) + 6\sigma^2_{AC(B)} + 6\sigma^2_{ABD} + \sigma^2$
C(B)	$b(c-1) = 8$	$12\sigma^2_{C(B)} + 6\sigma^2_{AC(B)} + \sigma^2$
AC(B)	$(a-1)b(c-1) = 8$	$6\sigma^2_{AC(B)} + \sigma^2$
D	$d-1 = 2$	$48\sigma^2_D + 24\sigma^2_{AD} + 12\sigma^2_{BD}$ $+ 6\sigma^2_{ABD} + \sigma^2$
AD	$(a-1)(d-1) = 2$	$24\sigma^2_{AD} + 6\sigma^2_{ABD} + \sigma^2$
BD	$(b-1)(d-1) = 6$	$12\sigma^2_{BD} + 6\sigma^2_{ABD} + \sigma^2$
ABD	$(a-1)(b-1)(d-1) = 6$	$6\sigma^2_{ABD} + \sigma^2$
Error	subtraction $= 104$	σ^2
Total	$n - 1 = 143$	

Formulae

$ssA = 72\Sigma_i \bar{y}^2_{i...} - 144\bar{y}^2_{....}$

$ssB = 36\Sigma_j \bar{y}^2_{.j..} - 144\bar{y}^2_{....}$

$ssAB = 18\Sigma_i\Sigma_j \bar{y}^2_{ij..} - 72\Sigma_i \bar{y}^2_{i...} - 36\Sigma_j \bar{y}^2_{.j..} + 144\bar{y}^2_{....}$

$ssC(B) = 12\Sigma_j\Sigma_k \bar{y}^2_{.jk.} - 36\Sigma_j \bar{y}^2_{.j..}$

$ssAC(B) = 6\Sigma_i\Sigma_j\Sigma_k \bar{y}^2_{ijk.} - 18\Sigma_i\Sigma_j \bar{y}^2_{ij..} - 12\Sigma_j\Sigma_k \bar{y}^2_{.jk.} + 36\Sigma_j \bar{y}^2_{.j..}$

$ssD = 48\Sigma_u \bar{y}^2_{...u.} - 144\bar{y}^2_{....}$

$ssAD = 24\Sigma_i\Sigma_u \bar{y}^2_{i..u.} - 72\Sigma_j \bar{y}^2_{.j..} - 48\Sigma_u \bar{y}^2_{...u.} + 144\bar{y}^2_{....}$

$ssBD = 12\Sigma_j\Sigma_u \bar{y}^2_{.j.u.} - 36\Sigma_j \bar{y}^2_{.j..} - 48\Sigma_u \bar{y}^2_{...u.} + 144\bar{y}^2_{....}$

$ssABD = 6\Sigma_i\Sigma_j\Sigma_u \bar{y}^2_{ij.u.} - 18\Sigma_i\Sigma_j \bar{y}^2_{ij..} - 24\Sigma_i\Sigma_u \bar{y}^2_{i..u.} - 12\Sigma_j\Sigma_u \bar{y}^2_{.j.u.}$
$\qquad + 72\Sigma_i \bar{y}^2_{i...} + 36\Sigma_j \bar{y}^2_{.j..} + 48\Sigma_u \bar{y}^2_{...u.} - 144\bar{y}^2_{....}$

$sstot = \Sigma_i\Sigma_j\Sigma_k\Sigma_u\Sigma_t y^2_{ijkut} - 144\bar{y}^2_{....}$

ssE is obtained by subtraction

This has approximately a χ^2_x distribution, where

$$x = \frac{u^2}{\frac{(msBD/12)^2}{6} + \frac{(msABD/12)^2}{6}},$$

and an approximate 95% confidence interval for σ^2_{BD} is

$$\left(u/\chi^2_{x,\alpha/2},\ u/\chi^2_{x,1-\alpha/2}\right).$$

Any one of the main effects A, B, or $C(B)$ can be investigated if the interactions involving the corresponding factor are all negligible. The relevant formulae can be obtained along the same lines as those described above.

18.5 Using SAS Software

The SAS procedure PROC GLM can handle nested effects when they are described in the MODEL statement using notation of the form $B(A)$. The RANDOM statement is used to obtain expected mean squares. The procedure PROC VARCOMP, which was described briefly in Chapter 17, can also be used. We will illustrate these procedures via the experiment in Section 18.5.1.

18.5.1 Voltage Experiment

An experiment was described by David Desmond in the 1954 issue of *Applied Statistics* on reducing the variability of voltage regulators fitted to motor cars. The voltage regulator was required to operate within a range of 15.8 to 16.4 volts. When the experiment took place records showed that about 18% of regulators required readjustment during inspection, and sometimes this figure rose to 50%. Despite the inspection procedure, some of the regulators reaching customers were still outside the specification limits, and complaints from customers were considered to be excessive.

The experiment was run in order to measure the variability in the regulator setting operation. Table 18.6 shows the measurements taken on 64 voltage regulators at each of four testing stations. The 64 regulators were selected at random from several different setting stations, and we have reproduced the data for six of these. Since the regulators are selected at random, we model their effects as random effects nested within setting station. For purposes of illustration, we consider the four testing stations and six setting stations as the only stations of interest and model them as fixed effects. In the original article, these were modeled as random effects.

The effect of testing station is crossed with the effect of setting station and with regulator. A model to describe the data can be written as

$$Y_{ijk} = \mu + \alpha_i + \beta_j + C_{k(j)} + \epsilon_{ijk},$$
$$\epsilon_{ijk} \sim N(0, \sigma^2), \quad C_{k(j)} \sim N(0, \sigma^2_{C(B)}),$$

ϵ_{ijk}'s and $C_{k(j)}$'s are all mutually independent

$$i = 1, \ldots, 4; \quad j = 1, \ldots, 6; \quad k = 1, \ldots, r_j,$$

where α_i is the effect of the ith testing station, β_j is the effect of the jth setting station, and $C_{k(j)}$ is the effect of the kth randomly selected regulator from the jth setting station.

There is no reason to suspect that the testing stations would differ in their comparative results for different regulators, so there is no reason to expect a regulator×testing station interaction. Since there is only one observation per regulator–testing station combination, we would not be able to distinguish such an interaction from experimental error. A SAS program for analyzing this model is shown in Table 18.7.

18.5 Using SAS Software

Table 18.6 Voltages for the voltage experiment

Set. Sta.	Regulator	Testing Station 1	2	3	4	Set. Sta.	Regulator	Testing Station 1	2	3	4
1	1	16.5	16.5	16.6	16.6	4	1	16.1	16.0	16.0	16.2
	2	15.8	16.7	16.2	16.3		2	16.5	16.1	16.5	16.7
	3	16.2	16.5	15.8	16.1		3	16.2	17.0	16.4	16.7
	4	16.3	16.5	16.3	16.6		4	15.8	16.1	16.2	16.2
	5	16.2	16.1	16.3	16.5		5	16.2	16.1	16.4	16.2
	6	16.9	17.0	17.0	17.0		6	16.0	16.2	16.2	16.1
	7	16.0	16.2	16.0	16.0		7	16.0	16.0	16.1	16.0
	8	16.0	16.0	16.1	16.0						
2	1	16.0	16.1	16.0	16.1	5	1	15.5	15.6	15.4	15.8
	2	*15.4*	16.4	16.8	16.7		2	15.8	16.2	16.0	16.2
	3	16.1	16.4	16.3	16.3		3	16.2	*15.4*	16.1	16.3
	4	15.9	16.1	16.0	16.0		4	16.2	16.2	16.0	16.1
							5	16.1	16.2	16.3	16.2
							6	16.1	16.1	16.0	16.1
3	1	16.0	16.0	15.9	16.3	6	1	15.5	15.5	15.3	15.6
	2	15.8	16.0	16.3	16.0		2	16.0	15.6	15.7	16.2
	3	15.7	16.2	15.3	15.8		3	16.0	16.4	16.2	16.2
	4	16.2	16.4	16.4	16.6		4	15.8	16.5	16.2	16.2
	5	16.0	16.1	16.0	15.9		5	15.9	16.1	15.9	16.0
	6	16.1	16.1	16.1	16.1		6	15.9	16.1	15.8	15.7
	7	16.1	16.0	16.1	16.0		7	16.0	16.4	16.0	16.0
							8	16.1	16.2	16.2	16.1

Source: Desmond, D. J. (1954). Copyright © 1956 Blackwell Publishers. Reprinted with permission.

A plot of the standardized residuals (not shown) highlights two rather large outliers. The two outlying observations are those highlighted in italics in Table 18.6, and we notice that they are from different regulators and different testing stations. If these outliers are removed, the output shown in Table 18.8 is obtained. The TEST option produces the correct denominators for the tests of $H_0^A : \{\alpha_1 = \alpha_2 = \alpha_3 = \alpha_4\}$, $H_0^B : \{\beta_1 = \beta_2 = \cdots = \beta_6\}$ and $H_0^{C(B)} : \{\sigma_{C(B)}^2 = 0\}$. If we select an overall significance level of $\alpha = 0.06$ for the three tests and do each test at level $\alpha^* = 0.02$, we see that there is a significant difference between testing stations, but not between setting stations. Also, the variance of the regulators within setting stations appears to be significantly different from zero.

Unbiased estimates of σ^2 and $\sigma_{C(B)}^2$ can be obtained from the listed expected mean squares as $msE = 0.0268$ and

$$u = \frac{msC(B) - msE}{3.9499} = \frac{0.2405 - 0.0268}{3.9499} = 0.0541,$$

respectively. Thus the variability of the regulator strain readings is estimated to be about twice as large as the experimental error.

Table 18.7 SAS program to analyze a mixed-effects nested model

```
DATA VLT;
  INPUT SETTING REGUL TESTING VOLTG;
  LINES;
  1 1 1 16.5
  1 1 2 16.5
  : : :  :
  6 8 4 16.1
;
* Plot standardized residuals versus predicted values for all data;
PROC GLM;
  CLASSES TESTING SETTING REGUL;
  MODEL VOLTG = TESTING SETTING REGUL(SETTING);
  RANDOM REGUL(SETTING);
  OUTPUT OUT=RESIDS PREDICTED=PRED RESIDUAL=Z;
PROC STANDARD STD=1.0;
   VAR Z;
PROC PLOT;
   PLOT Z*PRED=SETTING Z*PRED=REGUL Z*PRED=TESTING
      / VREF=0 VPOS=19 HPOS=50;
;
* Analysis without two outliers;
DATA VLT; SET VLT;
  IF SETTING=2 AND REGUL=2 AND TESTING=1 THEN DELETE;
  IF SETTING=5 AND REGUL=3 AND TESTING=2 THEN DELETE;
;
PROC GLM;
  CLASSES TESTING SETTING REGUL;
  MODEL VOLTG = TESTING SETTING REGUL(SETTING);
  RANDOM REGUL(SETTING);
  OUTPUT OUT=RESIDS PREDICTED=PREDS RESIDUAL=Z;
  LSMEANS TESTING SETTING / PDIFF=ALL CL ADJUST=TUKEY ALPHA=0.05;
;
DATA; SET VLT;
PROC VARCOMP METHOD=TYPE1;
  CLASSES TESTING SETTING REGUL;
  MODEL VOLTG = TESTING SETTING REGUL(SETTING) / FIXED=2;
;
**************** Proc Nested can only be used when each factor
**************** is nested within the previously listed factor,
**************** which is not the case here;
DATA;  SET VLT;
PROC NESTED;
  CLASSES TESTING SETTING REGUL;
```

18.5 Using SAS Software

Table 18.8 SAS output for the voltage experiment

```
                        The SAS System
                  General Linear Models Procedure
Dependent Variable: VOLTG
                           Sum of          Mean
Source              DF     Squares        Square    F Value   Pr > F
Model               42    11.64681161    0.27730504   10.34    0.0001
Error              115     3.08287193    0.02680758
Corrected Total    157    14.72968354

                  General Linear Models Procedure
Source             Type III Expected Mean Square
TESTING            Var(Error) + Q(TESTING)
SETTING            Var(Error) + 3.9059 Var(REGUL(SETTING)) + Q(SETTING)
REGUL(SETTING)     Var(Error) + 3.9499 Var(REGUL(SETTING))

                  General Linear Models Procedure
          Tests of Hypotheses for Mixed Model Analysis of Variance
Dependent Variable: VOLTG

Source: TESTING
Error: MS(Error)
                       Denominator   Denominator
  DF   Type III MS         DF            MS       F Value   Pr > F
   3   0.194598245         115       0.026807582   7.2591   0.0002

Source: SETTING
Error: 0.9889*MS(REGUL(SETTING)) + 0.0111*MS(Error)

                       Denominator   Denominator
  DF   Type III MS         DF            MS       F Value   Pr > F
   5   0.5659272096       34.09     0.2381015368   2.3768   0.0593

Source: REGUL(SETTING)
Error: MS(Error)
                       Denominator   Denominator
  DF   Type III MS         DF            MS       F Value   Pr > F
  34   0.240481856         115       0.026807582   8.9707   0.0001
```

A 90% confidence interval for $\sigma^2_{C(B)}/\sigma^2$ can be obtained by adapting the formula (17.3.11) as follows:

$$\frac{1}{c}\left[\frac{msC(B)}{msE\ F_{v_1,v_2,\alpha/2}} - 1\right] \leq \frac{\sigma^2_{C(B)}}{\sigma^2} \leq \frac{1}{c}\left[\frac{msC(B)}{msE\ F_{v-1,n-v,1-\alpha/2}} - 1\right]$$

$$= \frac{1}{c}\left[\frac{0.2405}{0.0268\ F_{34,115,0.1}} - 1\right] \leq \frac{\sigma^2_{C(B)}}{\sigma^2} \leq \frac{1}{c}\left[\frac{0.2405}{0.0268\ F_{34,115,0.9}} - 1\right].$$

Table 18.9 Output from PROC VARCOMP for the voltage experiment

```
                        The SAS System
                 Variance Components Estimation Procedure
Dependent Variable: VOLTG

Source              DF      Type I SS        Type I MS
TESTING              3      0.66029252       0.22009751
SETTING              5      2.81013599       0.56202720
REGUL(SETTING)      34      8.17638310       0.24048186
Error              115      3.08287193       0.02680758
Corrected Total    157     14.72968354

                 Variance Components Estimation Procedure
Dependent Variable: VOLTG
Source                    Expected Mean Square
TESTING           Var(Error) + 0.0127 Var(REGUL(SETTING))
                                   + Q(TESTING,SETTING)
SETTING           Var(Error) + 3.9407 Var(REGUL(SETTING)) + Q(SETTING)
REGUL(SETTING)    Var(Error) + 3.9499 Var(REGUL(SETTING))
Error             Var(Error)

                 Variance Components Estimation Procedure
Dependent Variable: VOLTG

Variance Component           Estimate
Var(REGUL(SETTING))          0.05409609
Var(Error)                   0.02680758
```

Since $E[MSC(B)] = \sigma^2 + 3.9499\sigma^2_{C(B)}$, the value of $c =$ is 3.9499. So, using $F_{34,115,0.1} \approx 2.0$ and $F_{34,115,0.9} = (F_{115,34,0.1})^{-1} \approx 1.8^{-1} = 0.5556$, the confidence interval becomes

$$0.883 \leq \frac{\sigma^2_{C(B)}}{\sigma^2} \leq 3.836.$$

The general conclusion of the experiment was that the differences between the four testing stations were of little practical importance. However, we note that the residual plots still indicate one or two large residuals, especially from testing station 2, so perhaps testing station 2 should have been examined a little more closely.

Most of the variability in the regulators appeared to be due to the inherent measurement error, and the experimenters concluded that it was not possible to set the regulators within the desired tolerance limits. A quality control scheme to ensure that the current quality did not deteriorate was put in place.

We note in passing that the effect of the outliers on the analysis was actually very small. If the two original outliers had been included in the analysis, the estimates $\hat{\sigma}^2 = 0.027$ and $\hat{\sigma}^2_{C(B)} = 0.054$ would have changed to 0.039 and 0.046, respectively. There would also be little change in the p-values of the hypothesis tests. There is some benefit in retaining the entire data set, since the coefficient of $\sigma^2_{C(B)}$ in the expected mean squares is then 4.0, as stated by the rules on page 616.

The model can also be analyzed using the SAS procedure PROC VARCOMP as in Chapter 17. The input statements are shown in Table 18.7. The output from PROC VARCOMP is shown in Table 18.9. This output is based on Type I sums of squares, and since the removal of the two outliers results in unequal sample sizes, the expected mean squares and the estimated variance components differ slightly from those obtained above.

When all effects are nested one within another, the procedure PROC NESTED can be used. These statements are also shown in Table 18.7. However, for our current experiment, PROC NESTED is not suitable, since setting station is not nested within testing station.

Exercises

1. **Viscosity experiment**

   An experiment was described by Johnson and Leone (1977, page 744) to determine the viscosity of a polymeric material. The material was divided into two samples. The two samples were each divided into ten "aliquots." After preparation of these aliquots, they were divided into two subaliquots and a further step in the preparation made. Finally, each subaliquot was divided into two parts and the final step of the preparation made. The viscosity determinations are listed in Table 18.10.

   (a) Write down a model for the viscosity determinations allowing for variability in the samples, aliquots, subaliquots and parts.

   (b) Examine the error assumptions on your model.

   (c) Estimate the variances of all the random effects in the model.

   (d) Give a set of confidence intervals for the variances of all the random effects in the model at overall significance level 90%. At which step of the preparation is most of the variability introduced?

2. **Sleep experiment**

   Sleeping patterns can be classified according to periods of "deep sleep" and of "REM sleep" (rapid eye movement). An experiment is done to see how sleeping tablets and amount of daily activity affect the proportion of REM sleep. Three types of sleeping tablets are to be tested, coded 1, 2, 3 (where type 3 is a placebo).

   Twelve subjects are selected at random from a large population and are assigned at random to the levels of A, four to each level. Each subject is assigned an activity level for the day, and the proportion of REM sleep is monitored during that night. The four activity levels are:

   B1 = read quietly all day,   B2 = walk 10 miles during the day,
   B3 = spend the day shopping,   B4 = play video games all day.

Table 18.10 Viscosity determinations for the viscosity experiment

Sample	Aliquot	Subaliquot 1		Subaliquot 2	
		Part 1	Part 2	Part 1	Part 2
1	1	59.8	59.4	58.2	63.5
	2	66.6	63.9	61.8	62.0
	3	64.9	68.8	66.3	63.5
	4	62.7	62.2	62.9	62.8
	5	59.5	61.0	54.6	61.5
	6	69.0	69.0	60.6	61.8
	7	64.5	66.8	60.2	57.4
	8	61.6	56.6	64.5	62.3
	9	64.5	61.3	72.7	72.4
	10	65.2	63.9	60.8	61.2
2	1	59.8	61.2	60.0	65.0
	2	65.0	65.8	64.5	64.5
	3	65.0	65.2	65.5	63.5
	4	62.5	61.9	60.9	61.5
	5	59.8	60.9	56.0	57.2
	6	68.8	69.0	62.5	62.0
	7	65.2	65.6	61.0	59.3
	8	59.6	58.5	62.3	61.5
	9	61.0	64.0	73.0	71.7
	10	65.0	64.0	62.0	63.0

Source: Johnson, N. L. and Leone, F. C. (1977). Copyright © 1977 Johnson and Leone. Reprinted with permission.

The experiment continues for four days, so that each subject is observed at each activity level in a random order. The model is assumed to be

$$Y_{hijt} = \mu + S_{h(i)} + \alpha_i + \beta_j + (\alpha\beta)_{ij} + \epsilon_{hijt},$$

where α_i is the effect of the ith sleeping tablet, β_j is the effect of the jth activity level, $(\alpha\beta)_{ij}$ is the effect of their interaction, $S_{h(i)}$ is the effect of the hth random subject assigned to the ith sleeping tablet, and $S_{h(i)} \sim N(0, \sigma^2_{S(A)})$ and $\epsilon_{hijt} \sim N(0, \sigma^2)$ and all $S_{h(i)}$ and ϵ_{hijt} are independent.

(a) Write down the degrees of freedom and expected mean squares for the analysis of variance table.

(b) Explain how to test the null hypothesis $H_0^A : \{\alpha_1 = \alpha_2 = \alpha_3\}$ against the alternative hypothesis that at least two of the α_i differ.

(c) Explain how to test the null hypothesis $H_0^{S(A)} : \{\sigma^2_{S(A)} = 0\}$ against the alternative hypothesis $H_A^{S(A)} : \{\sigma^2_{S(A)} > 0\}$.

(d) Suppose that the null hypothesis

$$H_0^{AB} : \{(\alpha\beta)_{ij} - \overline{(\alpha\beta)}_{i.} - \overline{(\alpha\beta)}_{.j} + \overline{(\alpha\beta)}_{..}, \text{ for all } i, j\}$$

appears to be correct. Which contrasts would be of particular interest to the experimenter? Why? Give formulas that would provide an overall 95% set of confidence intervals for your chosen contrasts. Give reasons for your choice of formula(s).

(e) If the experimenter thought that a day effect would be important, how would you modify the design of the experiment and the model?

3. Consider the model

$$Y_{ijkl} = \mu + \alpha_i + B_{j(i)} + C_{k(ji)} + \delta_l + (\alpha\delta)_{il} + (B\delta)_{lj(i)} + \epsilon_{ijkl},$$

(a) Calculate the expected mean squares for all effects in the model.

(b) Which ratio would you use to test $H_0 : \{\delta_l + \overline{(\alpha\delta)}_{\cdot l}$ all equal$\}$?

(c) Which ratio would you use to test $H_0 : \sigma_A^2 = 0$?

4. **Alloy experiment**

An experiment described by Johnson and Leone (1977, page 758) was performed by a company to investigate the effects of various factors on the "yield strength" of a particular titanium alloy. The factors investigated were:

A: vendors (4 fixed levels representing suppliers of raw material).

C: bar size (2 fixed levels representing standard sizes of bars of raw material).

B: batch (3 randomly selected levels nested within each combination of levels of A and C).

D: product type (2 fixed levels representing different types of finished product—forgedown and finished-forge blades).

Three observations were taken on each treatment combination. A reasonable model was thought to be

$$\begin{aligned} Y_{ijklt} &= \mu + \alpha_i + \gamma_j + (\alpha\gamma)_{ij} + B_{k(ij)} + \delta_l \\ &\quad + (\alpha\delta)_{il} + (\gamma\delta)_{jl} + (B\delta)_{kl(ij)} + \epsilon_{ijklt}, \end{aligned}$$

$\epsilon_{ijklt} \sim N(0, \sigma^2)$, $B_{k(ij)} \sim N(0, \sigma_{B(AC)}^2)$, $(B\delta)_{kl(ij)} \sim N(0, \sigma_{BD(AC)}^2)$,

$i = 1, 2, 3, 4;\ j = 1, 2;\ k = 1, 2, 3;\ l = 1, 2;\ t = 1, 2, 3,$

where α_i, γ_j, and δ_l represent the effects of the ith vendor, jth bar size, and lth product type, respectively, and $B_{k(ij)}$ represents the effect of the kth randomly selected batch of the jth bar size made with bar stock from the ith vendor, and random variables on the right-hand side of the model are assumed to be mutually independent.

(a) Write down the degrees of freedom and expected mean squares column of the analysis of variance table.

(b) Give a formula for an approximate 95% confidence interval for $\sigma_{B(AC)}^2$.

(c) How would you test the hypothesis

$H_0 :$ {no differences in yield strength of the titanium alloy

can be attributed to the two test specimens}

against the alternative hypothesis $H_A : \{\ H_0$ is false$\}$?

5. **Alloy experiment, continued**

   Suppose that factors C and D are to be investigated further in a followup experiment. Suppose that two new factors P and Q ("heat setting during processing" and "cooling method") are also to be investigated at two levels each. A followup experiment is required with the four factors C, D, P, and Q at two levels each (a 2^4 experiment). Only sixteen observations will be taken, four for each vendor. It is known that the interactions CP, CQ, PQ, CPQ, and $CDPQ$ are likely to be negligible. Also, there was information gained from the previous parts to Exercise 4 to suggest that all interactions of treatment factors with vendor can be assumed negligible.

   (a) Divide the 16 treatment combinations into four blocks of size four (one block for each vendor). Show your design explicitly, and indicate what should be randomized.

   (b) Write down a suitable model and the degrees of freedom column for the analysis of variance table for your design in part (a).

   (c) Before your design in part (a) is run, the management announces that in future, only one vendor will be used by the company. Also, your budget is cut, so that you can take only 8 observations. Thus, you need to design a $\frac{1}{2}$–fraction of a 2^4 experiment. In reviewing the list of negligible interactions above, you discover that two have been omitted. Interactions DP and CDQ are also known to be negligible. Choose a design and list the treatment combinations explicitly. (Hint: Try $I = CPQ$. State the aliasing scheme and a suitable model. Will there be any problems in interpreting the results of this experiment?

6. **Operator experiment**

   An experiment to identify the causes of variability in readings of a spectrometer was described in Exercise 10 of Chapter 7, page 236. The same authors (J. Inman, J. Ledolter, R. Lenth, and L. Niemi, *Journal of Quality Technology*, 1992) also described a study to determine how much of the variation in measured manganese concentration in steel was due to operator variation.

   Ten steel samples were sliced form a steel billet. Each operator was asked to measure the manganese content of each sample twice. The measurements taken by any one operator were done in a random order on a single day. There were four operators, who were regarded as representative of a large population of potential operators.

   (a) Write down a model for this experiment. Indicate clearly which effects are fixed, random, crossed, and nested.

   (b) Write down the degrees of freedom, the sums of squares, and the expected mean squares for each of the sources of variation in your model.

   (c) The authors analyzed this experiment using a gamma distribution to model the distribution of the error terms. Using the data in Table 6, investigate whether or not the normal distribution could be used (it may be necessary to take a transformation).

   (d) If the normal distribution can be used as a reasonable approximation to the error distribution, then analyze the experiment. In particular, obtain estimates of the variances of the random effects and identify the major sources of variation.

Table 18.11 Manganese concentrations (percentages) for the operator experiment

Sample	Operator 1		Operator 2		Operator 3		Operator 4	
1	0.63	0.60	0.62	0.62	0.60	0.60	0.59	0.61
2	0.64	0.63	0.63	0.64	0.67	0.65	0.62	0.64
3	0.60	0.58	0.60	0.61	0.60	0.60	0.58	0.60
4	0.75	0.74	0.74	0.74	0.74	0.73	0.73	0.76
5	0.71	0.68	0.69	0.70	0.69	0.67	0.68	0.71
6	0.65	0.63	0.62	0.65	0.63	0.64	0.62	0.64
7	0.67	0.64	0.66	0.67	0.65	0.65	0.64	0.66
8	0.65	0.63	0.65	0.64	0.62	0.62	0.60	0.62
9	0.68	0.66	0.67	0.68	0.67	0.67	0.65	0.68
10	0.67	0.64	0.66	0.66	0.65	0.64	0.64	0.66

Source: Inman, J., Ledolter, J., Lenth, R. V. and Niemi, L. (1992). Copyright © 1997 American Society for Quality. Reprinted with Permission.

7. For the two-way nested fixed-effects model (18.2.1) on page 646, show that the least squares estimator of $\mu + \alpha_i + \beta_{j(i)}$ is given by $\overline{Y}_{ij..}$.
   [Hint: Differentiate the sum of squared errors with respect to μ, α_i ($i = 1, \ldots, a$), and $\beta_{j(i)}$ ($j = 1, \ldots, b; i = 1, \ldots, a$), in turn. Set the resulting three sets of normal equations equal to zero. Show that the third set of equations adds to the first equation, and that the ith portion of the third set of equations adds to the ith equation in the second set. Thus, the first and second sets of equations are redundant, and $a + 1$ extra equations must be added to the set.]

8. **Erythrocite experiment**

   The trout experiment reported by J. S. Gutsell (*Biometrics*, 1951) was described in Exercise 15 of Chapter 3. As part of the same experiment, the erythrocite counts in the blood of brown trout were measured.

   Fish were put at random into eight troughs of water. Two troughs were assigned to each of the four levels of the treatment factor "sulfamerazine" (0, 5, 10, 15 grams per 100 pounds of fish added to the diet per day). After 42 days, five fish were selected at random from each trough and divided between two counting chambers. The erythrocite counts were then measured. The observations reported in Table 18.12, when multiplied by 5000, give the number of erythrocites per cubic millimeter of blood.
   A possible model for these data is

   $$Y_{ijklt} = \mu + \alpha_i + B_{j(i)} + C_{k(ij)} + \epsilon_{ijklt},$$
   $$\epsilon_{ijklt} \sim N(0, \sigma^2), \quad B_{j(i)} \sim N(0, \sigma_B^2), \quad C_{k(ij)} \sim N(0, \sigma_{C(B)}^2),$$
   $$t = 1, 2, 3, 4, 5; \quad i = 1, 2, 3, 4; \quad j = 1, 2; \quad k = 1, 2;$$

   where α_i is the effect of the ith level of sulfamerazine in the diet, $B_{j(i)}$ is the effect of the jth randomly selected trough assigned to the ith level of sulfamerazine, and C_k is the effect of the kth randomly selected fish from the (ij)th trough, and random variables on the right hand-side of the model are assumed to be mutually independent.

Table 18.12 Erythrocite counts from brown trout for the erythrocite experiment

	0 gm sulph.				5 gm sulph.			
Fish	Trough 1		Trough 2		Trough 1		Trough 2	
1	213	230	166	157	296	319	310	309
2	253	231	206	185	278	258	241	270
3	195	164	245	250	345	307	272	311
4	193	203	213	181	322	372	254	237
5	191	195	198	169	248	274	266	275

	10 gm sulph.				15 gm sulph.			
Fish	Trough 1		Trough 2		Trough 1		Trough 2	
1	339	322	196	232	278	212	287	280
2	282	285	205	186	275	311	221	243
3	236	262	252	274	186	158	331	309
4	252	209	245	216	301	281	231	244
5	263	296	249	260	223	246	292	295

Source: Gutsell, J. S. (1951). Copyright © 1951 International Biometric Society. Reprinted with permission.

(a) Since the data are counts, examine the assumption of normally distributed errors and equal variances. If the assumptions are not approximately satisfied, is there a transformation that can be used to correct the problem?

(b) Write out the degrees of freedom and the expected mean squares for each term in the model.

(c) Test the hypothesis that sulfamerazine has no effect on the erythrocite counts. Examine the linear and quadratic trends.

(d) If the test in part (c) is rejected, calculate a 95% set of confidence intervals for pairwise comparisons in the effects of the sulfamerazine levels.

9. **Aerosol experiment**

An experiment was described by R. Beckman, C. Nachtsheim, and R. D. Cook in the 1987 issue of *Technometrics* to illustrate the use of diagnostic procedures in the mixed model. They used part of the data set that was originally collected by H. Kershner, H. Ettinger, J. DeField, and R. Beckman (1984) at Los Alamos National Laboratories. Here, we use a subset of the data presented by Beckman, Nachtsheim, and Cook.

The goal of the study was to determine whether the current standard aerosol (level 1 of factor A) that was used for testing respirator filters could be replaced by an alternative aerosol (level 2). The experimenters also wanted to investigate the variability of the filters.

Consider using the following model:

$$Y_{ijkt} = \mu + \alpha_i + \beta_j + C_{k(j)} + (\alpha C)_{ik(j)} + \epsilon_{ijkt},$$
$$\epsilon_{ijkt} \sim N(0, \sigma^2), \quad C_{k(j)} \sim N(0, \sigma^2_{C(B)}),$$
ϵ_{ijkt}'s and $C_{k(j)}$'s are all mutually independent,

Table 18.13 Percent penetration for the aerosol experiment

Manufacturer	Filter	Aerosol 1			Aerosol 2		
1	1	0.750	0.770	0.840	1.120	1.100	1.120
	2	0.082	0.076	0.077	0.150	0.120	0.120
2	1	0.600	0.680	0.870	0.910	0.830	0.950
	2	1.000	1.800	2.700	2.170	1.520	1.580

Source: Beckman, R. J., Nachtsheim, C. J., and Cook, R. D. (1987). Copyright © 1987 American Statistical Association. Reprinted with Permission.

Source: Kershner, H. F., Ettinger, H. J., DeField, J. D., and Beckman, R. J. (1984).

$$i = 1, 2; \quad j = 1, 2; \quad k = 1, 2; \quad t = 1, 2, 3,$$

where α_i is the effect of the ith aerosol, β_j is the effect of the jth manufacturer, and $C_{k(j)}$ is the effect of the kth randomly selected filter from the jth manufacturer, and $(\alpha C)_{ik(j)}$ is the interaction effect of this filter with the ith aerosol. The tth response, Y_{ijkt}, obtained from the kth filter from the jth manufacturer and ith aerosol, is the percent penetration.

Analyze the data shown in Table 18.13 and state your conclusions.

19 Split-Plot Designs

19.1 Introduction
19.2 Designs and Models
19.3 Analysis of a Split-Plot Design with Complete Blocks
19.4 Split-Split-Plot Designs
19.5 Split-Plot Confounding
19.6 Using SAS Software
Exercises

19.1 Introduction

Split-plot designs are needed when the levels of some treatment factors are more difficult to change during the experiment than those of others. The designs have a nested blocking structure. In a block design, the experimental units are nested within the blocks, and a separate random assignment of units to treatments is made within each block. In a split-plot design, the experimental units are called *split plots*, and are nested within *whole plots*, which themselves may or may not be nested within blocks.

The split plots within each whole plot are assigned at random to the levels of one or more of the treatment factors. The levels of other treatment factors are assigned to whole plots and remain constant for all split plots within a whole plot. Typically, these will be the factors whose levels are difficult to change, and the effects of their levels will be less precisely compared than those assigned to the split plots.

In Section 19.2 we show an example of an experiment designed as a split-plot design, together with a typical model for this type of design. The analysis of split-plot designs is discussed in Section 19.3 and illustrated via a second experiment. Designs with an extra level of nesting (split-split-plot designs) are briefly described in Section 19.4, and the issue of confounding treatment contrasts is introduced in Section 19.5. In Section 19.6, the use of SAS for analyzing split-plot designs is illustrated.

19.2 Designs and Models

When a factorial experiment is run as a completely randomized design or a randomized complete block design, the levels of all the factors generally have to be changed frequently during the course of the experiment. For example, in Block I of the design in Table 13.13, page 443, we see that as the experiment progressed on day 1, the level of the first factor had to be changed from 1 to 0 to 1 to 0 to 1, and the level of the second factor had to be changed from 0 to 1 to 0 to 1 to 0. The levels of the third and fourth factors also had to be changed four times. In most experiments this is no particular problem, but sometimes the level of one of the factors is *not* particularly easy to change.

An experiment is described by Munro in his 1986 University of Southampton dissertation on the effect of lighting conditions (factor A) and the speed of a rotating drum (factor B) on a subject's ability to focus on the center of the drum. In this experiment, it was easy to change the speed of rotation by the turn of a dial. The lighting conditions, however, took time to set up, and Munro wished to change these as seldom as possible. He therefore asked each subject to view all the rotation speeds (in a randomized order) under one set of lighting conditions during one session and return for a second session with different lighting conditions at a later date. Part of a possible design is shown in Table 19.1.

A whole plot is defined by a session for a particular subject. A split plot is defined by a time slot nested in a particular session for a particular subject. The two whole plots (sessions) within each block are assigned at random to the levels of one factor (A), and the four split plots (time slots) within each whole plot are assigned at random to the levels of the other factor (B).

If we look at the design in Table 19.1 and ignore factor B, we see that the levels of A are assigned according to a randomized complete block design, where the s subjects play the role of blocks, the 2 whole plots per block play the role of experimental units, and the 2 levels of A are assigned at random to the 2 whole plots within each block. Assuming

Table 19.1 Part of a split-plot design for the rotating drum experiment

Block (Subject)	Whole plot 1 (Session 1)		Whole plot 2 (Session 2)	
	Level of A (Lighting)	Levels of B (Speed)	Level of A (Lighting)	Levels of B (Speed)
I	0	0 3 1 2	1	1 0 2 3
II	0	1 0 2 3	1	2 1 3 0
III	1	2 1 3 0	0	3 2 0 1
⋮	⋮	⋮ ⋮ ⋮ ⋮	⋮	⋮ ⋮ ⋮ ⋮
s	0	0 1 3 2	1	2 3 1 0

19.2 Designs and Models

no block×A interaction, the difference in the two levels of A could be analyzed like any randomized complete block design, using the whole-plot totals as the observations.

If we now look at the levels of B, they have also been assigned according to a randomized complete block design, but this time, the whole plots play the role of the blocks, and the four split plots nested within each whole plot are assigned to the four levels of B.

The analysis of the split-plot design is divided into two parts, reflecting this nested blocking system, each part with its own error. Analysis of the main effect of A involves comparisons of responses from split plots in different whole plots, whereas analysis of the main effect of B and the AB interaction involve comparisons of responses from split plots within the same whole plots.

In general, split plots within a whole plot will be more similar than split plots in different whole plots. Consequently, within-whole-plot comparisons will generally be more precise than between-whole-plot comparisons. So, in the rotating drum experiment, the main effect of B and the AB interaction will very likely be more precisely estimated than the main effect of A. In general, if the levels of all factors are easy to change, split-plot designs are recommended only when there is considerably less interest in one or more of the treatment factors.

Ignoring the effects of the treatment factors for the moment, the response could be modeled as

$$Y_{hpq} = \mu + \theta_h + \epsilon^W_{p(h)} + \epsilon^S_{q(hp)},$$

where θ_h is the effect of the hth block, $\epsilon^W_{p(h)}$ is the effect of the pth whole plot nested within the hth block, and $\epsilon^S_{q(hp)}$ is the effect of the qth split plot nested within the pth whole plot in the hth block. We model the whole-plot and split-plot effects, and possibly the block effects, as random effects that are independent and normally distributed with mean 0 and variances σ^2_W, σ^2_S, and σ^2_θ, respectively.

Now, suppose that the levels of factor A are assigned to the whole plots, and in the hth block, the pth whole plot receives the uth assignment of the ith level of A ($i = 1, \ldots, a$; $u = 1, \ldots, \ell$). Also, suppose that the levels of factor B are assigned to the split plots, and in the (hp)th whole plot, the qth split plot receives the tth assignment of the jth level of B ($j = 1, \ldots, b; t = 1, \ldots, m$). Then the model includes the effects of A, B and AB, and p is replaced by iu and q is replaced by jt, as follows:

$$\begin{aligned} Y_{hiujt} &= \mu + \theta_h + \alpha_i + \epsilon^W_{iu(h)} \\ &\quad + \beta_j + (\alpha\beta)_{ij} + \epsilon^S_{jt(hiu)}, \end{aligned} \tag{19.2.1}$$

$$\epsilon^W_{iu(h)} \sim N(0, \sigma^2_W), \quad \epsilon^S_{jt(hiu)} \sim N(0, \sigma^2_S),$$

$\epsilon^W_{iu(h)}$'s and $\epsilon^S_{jt(hiu)}$'s are all mutually independent,

$h = 1, \ldots, s; \; i = 1, \ldots, a; \; u = 1, \ldots, \ell; \; j = 1, \ldots, b; \; t = 1, \ldots, m$,

where θ_h is the effect of the hth block, α_i is the effect of the ith level of factor A measured on the whole plots, the random variables $\epsilon^W_{iu(h)}$ represent the random effects of the whole plots, β_j is the effect of the jth level of B measured on the split plots, $(\alpha\beta)_{ij}$ is the interaction effect of A at level i and B at level j, and the random variables $\epsilon^S_{jt(hiu)}$ represent the random effects of the split plots.

The model (19.2.1) has been written on two lines to emphasize the two different parts of the design. In the design of Table 19.1, each level of A appears exactly once per block and each level of B appears exactly once per whole plot, so we may drop the subscripts u and t, and the model for this design becomes

$$Y_{hij} = \mu + \theta_h + \alpha_i + \epsilon^W_{i(h)}$$
$$+ \beta_j + (\alpha\beta)_{ij} + \epsilon^S_{j(hi)}, \qquad (19.2.2)$$
$$\epsilon^W_{i(h)} \sim N(0, \sigma^2_W), \quad \epsilon^S_{j(hi)} \sim N(0, \sigma^2_S),$$
$\epsilon^W_{i(h)}$'s and $\epsilon^S_{j(hi)}$'s are all mutually independent,
$h = 1, \ldots, s; \ i = 1, \ldots, a; \ j = 1, \ldots, b.$

In some experiments the whole plot is the largest unit, which is equivalent to there being only one block ($s = 1$). In this case, model (19.2.1) becomes simpler, since the block effect θ_h and all subscripts h are omitted:

$$Y_{iujt} = \mu + \alpha_i + \epsilon^W_{iu}$$
$$+ \beta_j + (\alpha\beta)_{ij} + \epsilon^S_{jt(iu)}, \qquad (19.2.3)$$
$$\epsilon^W_{iu} \sim N(0, \sigma^2_W), \quad \epsilon^S_{jt(iu)} \sim N(0, \sigma^2_S),$$
ϵ^W_{iu}'s and $\epsilon^S_{jt(iu)}$'s are all mutually independent,
$i = 1, \ldots, a; \ u = 1, \ldots, \ell; \ j = 1, \ldots, b; \ t = 1, \ldots, m.$

For unequal sample sizes, the ranges of the subscripts would be modified in models (19.2.1) and (19.2.3).

19.3 Analysis of a Split-Plot Design with Complete Blocks

In this section we consider only the case of equal sample sizes and randomized complete block designs for each of the treatment factors. There are then s blocks, each of which is divided into a whole plots, and each of these is subdivided into b split plots, giving a total of sab observations. Model (19.2.2) is used, and the degrees of freedom and sums of squares are calculated according to the rules in Chapter 18, as shown in the following two subsections. The analysis of variance is outlined in Table 19.2.

19.3.1 Split-Plot Analysis

Consider first the *split-plot analysis*, which is that part of the analysis (shown in the bottom half of the analysis of variance table, Table 19.2) that is based on the observations arising from the split plots within whole plots. There are $sab - 1$ total degrees of freedom, and the

19.3 Analysis of a Split-Plot Design with Complete Blocks

Table 19.2 Outline analysis of variance table for the rotating drum split-plot design

Source of Variation	Degrees of Freedom	Sum of Squares	Mean Square	Ratio
Block (Subjects)	$s-1$	$ss\theta$	–	–
A (Lighting)	$a-1$	ssA	msA	msA/msE_W
Whole-plot error	$(s-1)(a-1)$	ssE_W	msE_W	
Whole-plot total	$sa-1$	ssW		
B (Speed)	$b-1$	ssB	msB	msB/msE_S
AB	$(a-1)(b-1)$	$ss(AB)$	$ms(AB)$	$ms(AB)/msE_S$
Split-plot error	$a(b-1)(s-1)$	ssE_S	msE_S	
Total	$abs-1$	$sstot$		

Formulae

$ss\theta = \Sigma_h y_{h..}^2/(ab) - y_{...}^2/(sab)$ $\qquad ssW = \Sigma_h \Sigma_i y_{hi.}^2/b - y_{...}^2/(sab)$

$ssA = \Sigma_i y_{.i.}^2/(sb) - y_{...}^2/(sab)$ $\qquad ssB = \Sigma_j y_{..j}^2/(sa) - y_{...}^2/(sab)$

$ssE_W = ssW - ss\theta - ssA$ $\qquad ss(AB) = \Sigma_i \Sigma_j y_{.ij}^2/s - \Sigma_i y_{.i.}^2/(sb)$

$sstot = \Sigma_h \Sigma_i \Sigma_j y_{hij}^2 - y_{...}^2/(sab)$ $\qquad\qquad - \Sigma_j y_{..j}^2/(sa) + y_{...}^2/(sab)$

$\qquad\qquad\qquad\qquad\qquad\qquad ssE_S = sstot - ssW - ssB - ss(AB)$

total sum of squares is

$$sstot = \sum_h \sum_i \sum_j y_{hij}^2 - sab\bar{y}_{...}^2 = \sum_h \sum_i \sum_j y_{hij}^2 - y_{...}^2/(sab). \tag{19.3.4}$$

The b levels of factor B are assigned to the split plots within each whole plot according to a randomized complete block design. The sa whole plots are playing the role of sa blocks, so there are $sa-1$ whole-plot degrees of freedom, and the whole-plot-total sum of squares is

$$ssW = b \sum_h \sum_i \bar{y}_{hi.}^2 - sab\bar{y}_{...}^2 = \sum_h \sum_i y_{hi.}^2/b - y_{...}^2/(sab). \tag{19.3.5}$$

Due to the fact that all levels of B are observed in every whole plot as in a randomized complete block design, the sum of squares for B needs no adjustment for whole plots, and is given by

$$ssB = sa \sum_j \bar{y}_{..j}^2 - sab\bar{y}_{...}^2 = \sum_j y_{..j}^2/(sa) - y_{...}^2/(sab) \tag{19.3.6}$$

corresponding to $b-1$ degrees of freedom. The interaction between the factors A and B is also calculated as part of the split-plot analysis. Again, due to the complete block structure of both the whole-plot design and the split-plot design, the interaction sum of squares needs no adjustment for blocks. The number of interaction degrees of freedom is $(a-1)(b-1) = ab - a - b + 1$, and the sum of squares is

$$ss(AB) = \sum_i \sum_j y_{.ij}^2/s - \sum_i y_{.i.}^2/(sb) - \sum_j y_{..j}^2/(sa) + y_{...}^2/(sab). \tag{19.3.7}$$

Since there are b split plots nested within the sa whole plots, there are, in total, $sa(b-1)$ split-plot degrees of freedom. Of these, $b-1$ are used to measure the main effect of B, and $(a-1)(b-1)$ are used to measure the AB interaction, leaving

$$sa(b-1) - (b-1) - (a-1)(b-1) = a(s-1)(b-1)$$

degrees of freedom for error. Equivalently, this can be obtained by subtraction of the whole plot, B, and AB degrees of freedom from the total

$$(sab-1) - (sa-1) - (b-1) - (a-1)(b-1) = a(s-1)(b-1).$$

The split-plot error sum of squares can also be calculated by subtraction:

$$ssE_S = sstot - ssW - ssB - ss(AB). \tag{19.3.8}$$

The split-plot error mean square $msE_S = ssE_S/[a(s-1)(b-1)]$ is used as the error estimate in testing hypotheses and calculating confidence intervals for contrasts in B and AB. Notice that we cannot compare the levels of factor A on the split plots, since within each whole plot the level of A is held constant. The A contrasts are, in fact, confounded with whole plots.

The sums of squares (19.3.4)–(19.3.8) and their associated degrees of freedom are summarized in the bottom half of the analysis of variance table shown in Table 19.2.

19.3.2 Whole-Plot Analysis

We now move on to the *whole-plot analysis*, which is the part of the analysis based on comparisons of whole-plot totals. The levels of A are assigned to the whole plots within blocks according to a randomized complete block design, and so the sum of squares for A needs no block adjustment. There are $a-1$ degrees of freedom for A, so the sum of squares is given by

$$ssA = sb \sum_i \bar{y}_{.i.}^2 - sab\bar{y}_{...}^2 = \sum_i y_{.i.}^2/(sb) - y_{...}^2/(sab). \tag{19.3.9}$$

There are $s-1$ degrees of freedom for blocks, giving a block sum of squares of

$$ss\theta = ab \sum_h \bar{y}_{h..}^2 - sab\bar{y}_{...}^2 = \sum_h y_{h..}^2/(ab) - y_{...}^2/(sab). \tag{19.3.10}$$

There are a whole plots nested within each of the s blocks, so there are, in total, $s(a-1)$ whole-plot degrees of freedom. Of these, $a-1$ are used to measure the effects of A leaving $(s-1)(a-1)$ degrees of freedom for whole-plot error. Equivalently, this can be obtained by the subtraction of the block and A degrees of freedom from the whole-plot total degrees of freedom

$$(sa-1) - (s-1) - (a-1) = (s-1)(a-1).$$

Similarly, the whole-plot error sum of squares, which is used in testing hypotheses and calculating confidence intervals for contrasts in factor A, is obtained by subtraction:

$$ss(E_W) = ssW - ss\theta - ssA. \tag{19.3.11}$$

The sums of squares (19.3.9)–(19.3.11) and their corresponding degrees of freedom are summarized in the top half of Table 19.2 (page 679).

19.3.3 Contrasts Within and Between Whole Plots

The formulae for the least squares estimates of the main effect and interaction treatment contrasts are similar to those given by the rules of Section 7.3 for fixed effects, since no block adjustments are needed. Thus

$$\sum_i c_i \hat{\alpha}_i^* = \sum_i c_i \bar{y}_{i..} \,, \tag{19.3.12}$$

$$\sum_j d_j \hat{\beta}_j^* = \sum_j d_j \bar{y}_{.j.} \,,$$

$$\sum_i \sum_j k_{ij} \widehat{(\alpha\beta)}_{ij} = \sum_i \sum_j k_{ij} \bar{y}_{.ij} \,,$$

where $\sum_i c_i = 0$, $\sum_j d_j = 0$, and $\sum_i k_{ij} = \sum_j k_{ij} = 0$. The corresponding estimated variances reflect whether the contrasts are measured in terms of whole-plot differences (as for contrasts in the levels of A) or split-plot (within whole-plot) differences (as for contrasts in B or AB). The former use the whole-plot error mean square, and the latter use the split-plot error mean square as follows.

$$\widehat{\text{Var}}\left(\sum_i c_i \hat{\alpha}_i^*\right) = \sum_i \frac{c_i^2}{sb} \, msE_W \,, \tag{19.3.13}$$

$$\widehat{\text{Var}}\left(\sum_j d_j \hat{\beta}_j^*\right) = \sum_j \frac{d_j^2}{sa} \, msE_S \,,$$

$$\widehat{\text{Var}}\left(\sum_i \sum_j k_{ij} \widehat{(\alpha\beta)}_{ij}\right) = \sum_i \sum_j \frac{k_{ij}^2}{s} \, msE_S \,.$$

For main effect and interaction contrasts, the methods of multiple comparison of Bonferroni, Scheffé, Tukey, Dunnett, and Hsu can be used as usual (incorporating either the whole-plot or split-plot error mean square as above). Inferences for other contrasts in the treatment effects $\tau_{ij} = \alpha_i + \beta_j + (\alpha\beta)_{ij}$, such as all pairwise comparisons, are more complicated and are not discussed here.

19.3.4 A Real Experiment—Oats Experiment

An experiment on the yield of three varieties of oats (factor A) and four different levels of manure (factor B) was described by F. Yates in his 1935 paper *Complex Experiments*. The experimental area was divided into $s = 6$ blocks. Each of these was then subdivided into $a = 3$ whole plots. The varieties of oat were sown on the whole plots according to a randomized complete block design (so that every variety appeared in every block exactly once). Each whole plot was then divided into $b = 4$ split plots, and the levels of manure were applied to the split plots according to a randomized complete block design (so that every level of B appeared in every whole plot exactly once). The design, after randomization, is shown in Table 19.3, together with the yields in quarter pounds. Model (19.2.2) was used.

Table 19.3 Split-plot design and yields (in quarter lb) for the oats experiment

Block	Level of A	Level of B (yield)		Block	Level of A	Level of B (yield)	
I	2	3 (156)	2 (118)	II	2	2 (109)	3 (99)
		1 (140)	0 (105)			0 (63)	1 (70)
	0	0 (111)	1 (130)		1	0 (80)	2 (94)
		3 (174)	2 (157)			3 (126)	1 (82)
	1	0 (117)	1 (114)		0	1 (90)	2 (100)
		2 (161)	3 (141)			3 (116)	0 (62)
III	2	2 (104)	0 (70)	IV	1	3 (96)	0 (60)
		1 (89)	3 (117)			2 (89)	1 (102)
	0	3 (122)	0 (74)		0	2 (112)	3 (86)
		1 (89)	2 (81)			0 (68)	1 (64)
	1	1 (103)	0 (64)		2	2 (132)	3 (124)
		2 (132)	3 (133)			1 (129)	0 (89)
V	1	1 (108)	2 (126)	VI	0	2 (118)	0 (53)
		3 (149)	0 (70)			3 (113)	1 (74)
	2	3 (144)	1 (124)		1	3 (104)	2 (86)
		2 (121)	0 (96)			0 (89)	1 (82)
	0	0 (61)	3 (100)		2	0 (97)	1 (99)
		1 (91)	2 (97)			2 (119)	3 (121)

Source: Yates, F. (1935). Copyright © 1935 Blackwell Publishers. Reprinted with permission. (Reprinted in *Experimental Design* (1970), Charles Griffin and Company, Ltd., London. Copyright 1970 Edward Arnold/Hodder & Stoughton Educational. Reprinted with permission.)

Analysis of variance—oats experiment Using the formulae (19.3.4)–(19.3.11), we obtain the sums of squares shown in Table 19.4. To test, at significance level $\alpha = 0.01$, the hypothesis H_0^{AB} that the interaction between variety and manure level is negligible against the alternative hypothesis that the interaction is not negligible, we reject H_0^{AB} if

$$\frac{ms(AB)}{msE_S} = \frac{53.63}{177.08} = 0.30 > F_{6,45,0.01} .$$

Since $F_{6,45,0.01} \approx 3.2$, we do not reject H_0^{AB}, and we conclude that the interaction is negligible.

The hypothesis H_0^B of no difference in yield due to the different levels of manure (averaged over variety) is also tested using the split-plot error mean square as the denominator. We reject H_0^B in favor of the alternative hypothesis, that the manure levels do affect yield

19.3 Analysis of a Split-Plot Design with Complete Blocks

Table 19.4 Analysis of variance for the oats split-plot experiment

Source of Variation	Degrees of Freedom	Sum of Square	Mean Square	Ratio	p-value
Blocks	5	15875.28	3175.06	–	
A (variety)	2	1786.36	893.18	1.49	0.2724
Whole-plot error	10	6013.31	601.33		
Whole-plot total	17	23674.94	1392.64		
B (manure)	3	20020.50	6673.50	37.69	0.0001
AB	6	321.75	53.63	0.30	0.9322
Split-plot error	45	7968.75	177.08		
Total	71	51985.94			

of oats, if

$$\frac{msB}{msE_S} = \frac{6673.50}{177.08} = 37.69 > F_{3,45,0.01} \, .$$

Since $F_{3,45,0.01} \approx 4.3$, we conclude that these four manure levels have different effects on the yield of the oat varieties tested.

Factor A is measured on the whole plots, so the whole-plot error is used as the denominator of the test statistic. We reject the hypothesis H_0^A of no difference in the average yields of the different varieties averaged over the manure levels if

$$\frac{msA}{msE_W} = \frac{893.18}{601.33} = 1.49 > F_{2,10,0.01} \, .$$

Since $F_{2,10,0.01} = 7.56$, there is no evidence to conclude a difference in average yields of the three varieties of oats.

The same conclusions could be reached from the p-values in Table 19.4.

Multiple comparisons—oats experiment Suppose that level 0 of A was the currently used variety and that level 0 of B was the usual level of manure, and suppose that two-sided treatment-versus-control intervals had been required for both A and B at an overall level of 98% (that is, 99% for each set of Dunnett intervals).

Since both the split-plot and whole-plot designs are randomized complete block designs, the least squares estimates of the treatment contrasts are given by the formula in (19.3.12), so

$$\hat{\alpha}_0^* - \hat{\alpha}_1^* = \overline{y}_{.0.} - \overline{y}_{.1.} = -6.875,$$
$$\hat{\alpha}_0^* - \hat{\alpha}_2^* = \overline{y}_{.0.} - \overline{y}_{.2.} = -12.167,$$

$$\hat{\beta}_0^* - \hat{\beta}_1^* = \overline{y}_{..0} - \overline{y}_{..1} = -19.500,$$
$$\hat{\beta}_0^* - \hat{\beta}_2^* = \overline{y}_{..0} - \overline{y}_{..2} = -34.833,$$
$$\hat{\beta}_0^* - \hat{\beta}_3^* = \overline{y}_{..0} - \overline{y}_{..3} = -44.000,$$

where $\overline{y}_{.i.}$ is an average over the $b = 4$ split plots within the $s = 6$ whole plots (one per block) on which level i of A is measured. Similarly, $\overline{y}_{..j}$ is an average over the $sa = 18$ split plots (one per whole plot) on which level j of B is measured. The confidence intervals are obtained from (19.3.12) and (19.3.13) as follows:

$$\alpha_0 - \alpha_1 \in \left((\bar{y}_{.0.} - \bar{y}_{.1.}) \pm t^{(0.5)}_{2,10,0.01} \sqrt{\frac{2}{24} msE_W} \right)$$

$$= \left(-6.875 \pm 3.53 \sqrt{\frac{2}{24}(601.33)} \right)$$

$$= (-6.875 \pm 24.99) = (-31.87, 18.12),$$

$$\alpha_0 - \alpha_2 = (-37.16, 18.12),$$

$$\beta_0 - \beta_1 \in \left((\bar{y}_{..0} - \bar{y}_{..1}) \pm t^{(0.5)}_{3,45,0.01} \sqrt{\frac{2}{18} msE_S} \right)$$

$$= \left(-19.5 \pm 3.09 \sqrt{\frac{2}{18}(177.03)} \right)$$

$$= (-19.5 \pm 13.70) = (-33.20, -5.79),$$

$$\beta_0 - \beta_2 \in (-48.54, -21.13),$$

$$\beta_0 - \beta_3 \in (-57.70, -30.30).$$

It is clear that the treatment-versus-control comparisons for the factor B manure levels are made more precisely ($msd = 13.70$) than those for the factor A oat varieties ($msd = 24.99$). This is primarily due to the much smaller error variance estimate, $msE_S < msE_W$, which reflects the fact that split plots within a whole plot are generally more similar than split plots in different whole plots. There are also more degrees of freedom associated with the split-plot error than with the whole-plot error, which also helps to reduce the minimum significant difference. Comparisons for factor B are more precise, despite the fact that the means $\bar{y}_{.i.}$ for factor A involve more observations.

19.4 Split-Split-Plot Designs

In the split-plot designs illustrated in Section 19.2, the factor A contrasts were confounded with the whole-plot contrasts, so that the main effect of A was assessed against the whole-plot variability, while the main effect of B and the AB interaction were assessed against the split-plot variability. It is possible to extend this idea, and to divide the split plots into split split plots on which are assigned the levels of a third factor.

For example, in the drum rotation experiment described in Section 19.2, the experimenter used a third factor C, the direction of rotation of the drum. A possible design for the experiment would be to ask each subject to be present at two sessions (whole plots) with a different lighting condition (A) at each session. In the first half of a session (split plot), set the direction of rotation (C), and run through each speed (B) in a random order (split split plots), changing the direction of rotation in the second half of the session. The design would then appear as in Table 19.5.

19.4 Split-Split-Plot Designs

Table 19.5 Part of a split-split-plot design for the rotating drum experiment

			Split-plot 1 First half session		Split-plot 2 Second half session	
Block (Subject)	Whole- plot (Session)	Level of A (Light)	Level of C (Direction)	Levels of B (Speed)	Level of C (Direction)	Levels of B (Speed)
I	1	0	1	0 3 1 2	0	2 1 3 0
	2	1	0	1 0 2 3	1	3 2 0 1
II	1	1	1	1 0 2 3	0	0 3 1 2
	2	0	0	2 1 3 0	1	3 2 0 1
:	:	:	:	: : : :	:	: : : :

The model and analysis of variance table would have three parts, one for the whole plots nested within blocks together with the factor A effect, one for the split plots nested within whole plots together with the factor C effect and the AC interaction, and one for the split split plots nested within split plots together with the factor B effect and the other interactions, as shown in model (19.4.14):

$$Y_{hijk} = \mu + \theta_h + \alpha_i + \epsilon^W_{i(h)} \qquad (19.4.14)$$
$$+ \gamma_j + (\alpha\gamma)_{ij} + \epsilon^S_{j(hi)}$$
$$+ \beta_k + (\alpha\beta)_{ik} + (\gamma\beta)_{jk} + (\alpha\gamma\beta)_{ijk} + \epsilon^{SS}_{k(hij)}.$$

The analysis of variance table, shown in Table 19.6, has three sections, reflecting the three parts of the model.

Table 19.6 Analysis of variance for a split-split-plot design

Source of Variation	Degrees of Freedom	Mean Square	Ratio
Blocks (subjects)	$s-1$	$ms\theta$	
A (lighting)	$a-1$	msA	msA/msE_W
Whole-plot error	$(s-1)(a-1)$	msE_W	
Whole-plot total	$sa-1$	msW	
C (direction)	$c-1$	msC	msC/msE_S
AC	$(a-1)(c-1)$	$ms(AC)$	$ms(AC)/msE_S$
Split-plot error	$a(s-1)(c-1)$	msE_S	
Split-plot total	$sac-1$	msE_S	
B	$b-1$	msB	msB/msE_{SS}
AB	$(a-1)(b-1)$	$ms(AB)$	$ms(AB)/msE_{SS}$
CB	$(c-1)(b-1)$	$ms(CB)$	$ms(CB)/msE_{SS}$
ACB	$(a-1)(c-1)(b-1)$	$ms(ACB)$	$ms(ACB)/msE_{SS}$
Split-split-plot error	$ac(s-1)(b-1)$	msE_{SS}	
Total	$sacb-1$	$mstot$	

Table 19.7 A split-plot confounded 2^4 experiment in 8 whole plots of size 4

Whole plot	A	Levels of B, C, D on the split plots	Whole plot	A	Levels of B, C, D on the split plots
I	0	000 011 101 110	II	1	001 010 100 111
III	0	001 010 100 111	IV	1	000 011 101 110
V	0	000 011 101 110	VI	1	001 010 100 111
VII	0	001 010 100 111	VIII	1	000 011 101 110

Table 19.8 Outline analysis of variance table for a split-plot confounded 2^4 experiment in 8 whole plots of size 4

Source of Variation	Degrees of Freedom	Mean Square	Ratio
A	1	msA	msA/msE_W
BCD	1	ms(BCD)	$ms(BCD)/msE_W$
ABCD	1	ms(ABCD)	$ms(ABCD)/msE_W$
Whole-plot error	4	msE_W	
Whole-plot total	7	msW	
B	1	msB	msB/msE_S
C	1	msC	msC/msE_S
D	1	msD	msD/msE_S
BC	1	ms(BC)	$ms(BC)/msE_S$
BD	1	ms(BD)	$ms(BD)/msE_S$
CD	1	ms(CD)	$ms(CD)/msE_S$
AB	1	ms(AB)	$ms(AB)/msE_S$
AC	1	ms(AC)	$ms(AC)/msE_S$
AD	1	ms(AD)	$ms(AD)/msE_S$
ABC	1	ms(ABC)	$msABC/msE_S$
ABD	1	ms(ABD)	$ms(ABD)/msE_S$
ACD	1	ms(ACD)	$ms(ACD)/msE_S$
Split-plot Error	12	msE_S	
Total	31		

19.5 Split-Plot Confounding

If there are a number of different factors involved in a split-plot design, the size of the whole plots may not be large enough to allow a randomized complete block design to be used for the split-plot factors. We can obtain smaller blocks by confounding one or more interaction contrasts as we did for the single-replicate designs in Chapter 13. For example, suppose a two-replicate 2^4 experiment is to be conducted for treatment factors A, B, C, and D, and for practical reasons, the observations are to be divided into eight whole plots of size four. Suppose that the levels of A are sufficiently difficult to change that it is decided to change the level only after each whole plot of four observations is taken. The A contrasts are confounded with the whole plots, since each whole plot is assigned only one level of A.

Now, only four of the eight combinations of factors B, C, and D can be taken in any one whole plot. Thus, ignoring factor A, a design with $b = 8$ whole plots of size $k = 4$ and $v = 8$

treatment labels is required. An incomplete block design, such as a cyclic design, would be a possible choice. However, since a split-plot design is a complex design to analyze, it is better to select a repeated single-replicate design, so that we know exactly what is confounded with the whole plots. If we choose to confound BCD with the whole plots, as well as A, then $ABCD$ will also be confounded. The single-replicate design that confounds A, BCD, and $ABCD$ is obtained from the equations

$$a_1 = 0 \text{ or } 1 \bmod 2,$$
$$a_2 + a_3 + a_4 = 0 \text{ or } 1 \bmod 2.$$

If we repeat this single-replicate design twice, we obtain the split-plot plan shown in Table 19.7. Before this plan can be used, the eight whole plots would need to be randomly ordered, and the four split plots within each whole plot would need to be randomly ordered. An outline analysis of variance table is shown in Table 19.8.

19.6 Using SAS Software

In this section we illustrate the use of the SAS software in analyzing a split-plot design. Care needs to be taken with the PROC GLM statements in order to obtain the two separate parts of the analysis of variable table. We illustrate two different methods of obtaining the information needed to construct an analysis of variance table similar to the one that was presented for the oats experiment in Example 19.3.4, page 681.

Method 1—Complete blocks When the whole-plot and split-plot designs are randomized complete block designs, we can make use of the fact that the Type I sums of squares are the same as the Type III sums of squares. In the first call of PROC GLM in Table 19.9, the sources of variation are entered into the model in the same order as in model (19.2.2). The whole-plot error term is represented by WP(BLOCK), which is the nested effect of whole plots within blocks. No term is needed to represent the split-plot error term $\epsilon^S_{j(hi)}$, since this plays the role of the usual error variable, and the corresponding sum of squares is automatically calculated by SAS.

The option E1 in the model statement asks for the expected mean squares to be calculated as though the model were being constructed sequentially (that is, Type I expected mean squares). Inclusion of the statement

RANDOM BLOCK WP(BLOCK) / TEST;

ensures that the correct denominators are used for all of the hypothesis tests.

The output is shown in Table 19.10. The Type I sums of squares are listed, but the p-values are not correct for all tests, since the split-plot error mean square is used throughout. The expected mean squares are listed, and we can verify that the error estimate for A differs from those of B and AB. The correct hypothesis tests are listed in the bottom half of the output.

Since both the whole-plot design and the split-plot design are randomized complete block designs, the MEANS option can be used to obtain standard multiple comparison procedures.

Table 19.9 SAS input statements for the oats split-plot experiment

```
******** to input the data;
DATA OAT;
  BLOCK WP A B  Y;
  LINES;
  1   1   2 3 156
  1   1   2 2 118
  1   1   2 1 140
  1   1   2 0 105
  1   2   0 0 111
  :   :   : :  :
  6   3   2 3 121
  ;
PROC PRINT;
*** analysis of variance; * method 1;
PROC GLM;
  CLASSES BLOCK A B WP;
  MODEL Y = BLOCK A WP(BLOCK) B A*B / E1;
  RANDOM BLOCK WP(BLOCK) / TEST;
  MEANS   A / DUNNETT('0')   ALPHA=0.01   CLDIFF   E=WP(BLOCK);
  MEANS   B / DUNNETT('0')   ALPHA=0.01   CLDIFF;
  ;
*** analysis of variance; * method 2;
DATA; SET OAT;
PROC GLM;
  CLASSES  BLOCK A B;
  MODEL  Y  =  BLOCK   A   BLOCK*A   B   A*B;
  RANDOM BLOCK A*BLOCK/TEST;
  MEANS   A / DUNNETT('0')   ALPHA=0.01   CLDIFF   E=BLOCK*A;
  MEANS   B / DUNNETT('0')   ALPHA=0.01   CLDIFF;
```

For example, Dunnett's procedure for comparing each level of A with control level 0 and each level of B with control level 0 is obtained via the statements shown in Table 19.9. In the MEANS procedure, SAS will use the split-plot term as the error mean square unless told to do otherwise, which is correct for the B contrasts but not for the A contrasts. To obtain the whole-plot error mean square for the A contrasts, we include the option E=WP(BLOCK). The output is shown in Table 19.11.

Method 2—Complete or incomplete blocks The second method makes use of the fact that the whole-plot error sum of squares uses the same degrees of freedom as the interactions between the block factor and the whole-plot treatment factors (if they have been deemed negligible and omitted from the model). The sources of variation are entered into the model in the same order as in model (19.2.2), but with $\epsilon_{i(h)}^{W}$ in model (19.2.2) replaced by the negligible block×whole-plot-treatment interactions.

Table 19.10 Output for the oats split-plot experiment (method 1)

```
                         The SAS System
                   General Linear Models Procedure
Dependent Variable: Y
                             Sum of         Mean
Source              DF      Squares       Square   F Value   Pr > F
Model               26    44017.194     1692.969      9.56   0.0001
Error               45     7968.750      177.083
Corrected Total     71    51985.944

Source              DF    Type I SS  Mean Square   F Value   Pr > F
BLOCK                5    15875.278     3175.056     17.93   0.0001
A                    2     1786.361      893.181      5.04   0.0106
WP(BLOCK)           10     6013.306      601.331      3.40   0.0023
B                    3    20020.500     6673.500     37.69   0.0001
A*B                  6      321.750       53.625      0.30   0.9322

Source      Type I Expected Mean Square
BLOCK       Var(Error) + 4 Var(WP(BLOCK)) + 12 Var(BLOCK)
A           Var(Error) + 4 Var(WP(BLOCK)) + Q(A,A*B)
WP(BLOCK)   Var(Error) + 4 Var(WP(BLOCK))
B           Var(Error) + Q(B,A*B)
A*B         Var(Error) + Q(A*B)

                   General Linear Models Procedure
          Tests of Hypotheses for Mixed Model Analysis of Variance
Dependent Variable: Y

Source: A *
Error: MS(WP(BLOCK))
               Denominator   Denominator
  DF   Type I MS        DF            MS   F Value   Pr > F
   2   893.18055556     10   601.33055556   1.4853   0.2724
* - This test assumes one or more other fixed effects are zero.

Source: B *
Error: MS(Error)
               Denominator   Denominator
  DF   Type I MS        DF            MS   F Value   Pr > F
   3      6673.5         45   177.08333333  37.6856   0.0001
* - This test assumes one or more other fixed effects are zero.

Source: A*B
Error: MS(Error)
               Denominator   Denominator
  DF   Type I MS        DF            MS   F Value   Pr > F
   6      53.625         45   177.08333333   0.3028   0.9322
```

Table 19.11 Multiple comparisons for the oats split-plot experiment

```
                         The SAS System
                  General Linear Models Procedure

                  Dunnett's T tests for variable: Y
NOTE: This tests controls the type I experimentwise error for
      comparisons of all treatments against a control.

      Alpha= 0.01   Confidence= 0.99   df= 10   MSE= 601.3306
           Critical Value of Dunnett's T= 3.531
           Minimum Significant Difference= 24.998
Comparisons significant at the 0.01 level are indicated by '***'.

                       Simultaneous              Simultaneous
                          Lower      Difference     Upper
              A         Confidence    Between    Confidence
          Comparison      Limit        Means        Limit
           2   - 0       -12.831      12.167       37.165
           1   - 0       -18.123       6.875       31.873

      Alpha= 0.01   Confidence= 0.99   df= 45   MSE= 177.0833
           Critical Value of Dunnett's T= 3.071
           Minimum Significant Difference= 13.62
Comparisons significant at the 0.01 level are indicated by '***'.

                       Simultaneous              Simultaneous
                          Lower      Difference     Upper
              B         Confidence    Between    Confidence
          Comparison      Limit        Means        Limit
           3   - 0        30.380      44.000       57.620   ***
           2   - 0        21.213      34.833       48.454   ***
           1   - 0         5.880      19.500       33.120   ***
```

The whole-plot error degrees of freedom and whole-plot error sum of squares are obtained as the sums of the degrees of freedom and Type I sums of squares for the negligible block × whole-plot-treatment interactions. All other sums of squares are obtained from the Type III sums of squares. For the oats experiment there is only one such interaction, namely BLOCK*A. The SAS output is not shown, but is identical to that in Table 19.10 with WP(BLOCK) replaced by BLOCK*A, and the Type I expected mean squares replaced by the Type III expected mean squares. For complete blocks, Dunnett's method of multiple comparisons proceeds as in Method 1 except that the whole-plot error sum of squares may be a sum of more than one interaction sum of squares and if so, the calculations must be completed by hand.

Table 19.12 Fluid (milliliters) in pleural cavity for the drug experiment

Block I Whole Plot	Dose B	Time C	Drug A							
			1	2	3	4	5	6	7	8
1	1	1	5.7	8.6	6.9	6.6	6.7	7.4	5.7	6.7
2	1	2	8.4	9.6	9.3	11.1	12.5	8.7	9.3	9.5
3	2	1	5.1	7.2	6.8	6.4	6.6	8.7	6.7	7.0
4	2	2	7.3	8.7	7.9	6.9	8.9	9.5	8.3	11.3
Block II Whole Plot	Dose B	Time C	Drug A							
			1	2	3	4	5	6	7	8
5	1	1	5.8	6.8	7.0	8.5	7.8	7.3	6.4	8.5
6	1	2	9.1	10.8	6.9	12.2	9.9	10.4	10.6	10.5
7	2	1	5.4	7.9	8.0	6.4	8.4	7.1	6.4	7.2
8	2	2	5.3	10.4	8.2	8.1	10.9	9.8	8.4	14.6

Source: Wooding, W. M. (1973). Copyright © 1997 American Society for Quality. Reprinted with Permission.

Exercises

1. **Drug experiment**

   An experiment designed as a split-plot design was described by W. M. Wooding in the *Journal of Quality Technology* in 1973. The experiment concerned the evaluation of eight drugs (factor A at $a = 8$ levels) for the treatment of arthritis. A second factor was the dose of the drug (factor B at $b = 2$ levels), and the third factor was the length of time (factor C at $c = 2$ levels) that a measurement was taken after injection by a substance known to cause an inflammatory reaction. The experimental units used in the study were $n = 64$ rats. The response was the amount of fluid (in milliliters) measured in the pleural cavity of an animal after having been administered a particular treatment combination.

   In many pharmacological studies, time of day has an effect on the response due to changing laboratory conditions, etc. Consequently, the experiment was divided into blocks, whole plots and split plots. The blocks were of size 32, each set of 32 observations being measured on a single day. Each treatment combination was measured once per day. Each day was then subdivided into 4 whole plots of size 8, where the eight measurements within a whole plot were taken fairly close together in time.

   Since the effect of the drug (A) was of primary importance, and since the effects of B and C were of interest only in the form of an interaction with A, the main effects of B and C and the BC interaction were confounded with the whole plots. The data are shown in Table 19.12, and the experimenter used the logarithms of the data in his analysis. Notice that the design for A on the split plots is a randomized block design, and the design for the BC combinations on the whole plots is also a randomized block design.

   (a) Write out a model for this experiment.

(b) Calculate an analysis of variance table using the logarithms of the data. Distinguish between the effects measured on the whole plots and those measured on the split plots. Identify the whole-plot error and split-plot error.

(c) Test any hypotheses of interest and state your conclusions clearly.

(d) Examine interaction plots of any important interactions. Calculate a set of 95% confidence intervals for the differences between pairs of drugs. State your conclusions.

2. **Fishing line experiment**

The fishing line experiment was run by C. Reynolds, B. Grunden, and K. Taylor in 1997 in order to compare the strengths of two brands of fishing line exposed to two different levels of stress. Two different reels of fishing line were purchased for each of the two brands, and sections of line were cut from each reel. Thus the reels were automatically assigned to the levels of factor A (Brand), and constituted the four whole plots. There were no blocks in this experiment. The split plots constituted sixteen sections of line, four cut from each of the four reels (that is, 16 split plots in total, 4 per whole plot). The split plots were randomly assigned to two different stress levels (stressed, "S", nonstressed, "N") so that each stress level was assigned two split plots per whole plot. The stress was induced by hanging a brick from the assigned section of line for 1 hours. Although this did not precisely mimic the stress induced during fishing, it was still expected to give some information about the strength of the lines under stress.

The strength test was accomplished by hanging a bucket on the end of the line, which was suspended from a beam. The bucket was gradually filled with water through a small hole in the lid until the line broke. The data are the resulting weights of water to the nearest 0.01 lb and are shown in Table 19.13.

(a) Write down a model for this experiment.

(b) Construct an analysis of variance table and test the hypotheses that you think are of interest. State your conclusions.

(c) If you were to repeat this experiment, suggest ways in which you would try to improve it.

3. **Cigarette experiment**

The cigarette experiment was run by J. Edwards, H. Hwang, S. Jamison, J. Kindelberger, and J. Steinbugl in 1996 in order to determine the factors that affect the length of time that a cigarette will burn. There were three factors of interest:

"Tar" (factor A) at two levels, "regular" and "ultra-light,"

Table 19.13 Strength of line for the fishing line experiment

Whole plot (Reel)	A (Brand)	Level of B (weight)			
1	1	N (6.70)	S (6.40)	S (7.20)	N (7.00)
2	2	S (8.10)	S (8.90)	N (8.00)	N (6.10)
3	2	S (8.00)	S (8.00)	N (8.75)	N (8.50)
4	1	N (8.50)	S (9.50)	N (9.70)	S (9.40)

Table 19.14 Burning times for the cigarette experiment

Whole plot (Time)	A (Tar)	Levels of BC (Burning times in seconds)
1	1	22 (301) 11 (326) 23 (260) 13 (290) 12 (312) 21 (292)
2	2	11 (329) 12 (331) 13 (285) 21 (306) 22 (258) 23 (276)
3	2	22 (290) 11 (380) 12 (335) 13 (309) 23 (243) 21 (334)
4	2	11 (321) 21 (337) 23 (275) 12 (316) 13 (307) 22 (250)
5	2	22 (308) 11 (345) 21 (307) 23 (288) 13 (321) 12 (330)
6	1	11 (344) 23 (283) 21 (281) 22 (261) 13 (307) 12 (292)
7	1	21 (274) 13 (310) 12 (304) 22 (279) 23 (277) 11 (330)
8	1	13 (302) 12 (325) 22 (301) 11 (338) 23 (270) 21 (297)
9	2	12 (323) 13 (334) 23 (265) 11 (326) 22 (269) 21 (297)
10	1	23 (309) 13 (314) 22 (259) 11 (344) 21 (310) 12 (322)

"Brand" (factor B) at two levels, "name brand" and "generic brand" (coded 1 and 2),

"Age" (factor C) at three levels, "fresh," "24 hour air exposure," "48 hour air exposure."

The cigarettes were to be burned in whole plots of size six. This was to help with the difficulty of recording burning times and to help keep the amount of smoke in the room at a reasonable level. There were ten whole plots, and these were assigned at random to the tar levels so that each tar level was assigned five whole plots.

The six split plots (time slots) in each whole plot were assigned at random to the six brand/age treatment combinations. Marks were made across the seam of each cigarette at a given distance apart. Each cigarette was lit at the beginning of its allotted time slot, and the time taken to burn between the two marks was recorded. The data are shown in Table 19.14.

(a) Write down a model for this experiment.

(b) Construct an analysis of variance table and test the hypotheses that you think are of interest. State your conclusions.

(c) Use Tukey's method to examine the pairwise differences in the effects on burning time of the six BC treatment combinations (averaged over tar levels).

(d) Examine the linear and quadratic trends of burning time due to different ages, for each brand separately.

(e) State your conclusions about the experiment, including your choice of overall significance levels and overall confidence levels.

4. **Injection molding experiment, continued**

The injection molding experiment, introduced in Exercise 10 of Chapter 15, was run in order to examine the effect of six factors on the shrinkage of a part produced by injection molding. The six factors were injection velocity (factor A), cooling time (factor B), barrel zone temperature (factor C), mold temperature (factor D), hold pressure (factor E), and back pressure (factor F). Each factor had two levels coded 0 and 1. The treatment combinations, which are shown in Table 15.52, page 538, were not completely randomized. The levels of factor D were time-consuming to change, so for the first four observations D was held at its low level and the combinations of the other factors were randomly ordered. For the last four observations, D was held at its high level and the combinations of the other factors were randomly ordered. Thus we can think of this design as having two whole plots assigned to the two levels of D, and having four split plots nested within each whole plot assigned at random to combinations of levels of A, B, C, E, and F.

(a) Sketch an outline analysis of variance table for the average response of the length data for this experiment, explaining what can and cannot be estimated.

(b) Is it possible to analyze this experiment in a sensible way? If so, present your results.

A Tables

Table	Contents	Page
A.1	Random numbers	696
A.2	Orthogonal polynomial trend contrast coefficients	702
A.3	Standard normal distribution	703
A.4	Student's t-distribution	704
A.5	Chi-squared distribution	705
A.6	F-distribution	706
A.7	Power of the F-test	712
A.8	Studentized range distribution (Tukey's method)	718
A.9	Multivariate t-distribution maximum (Dunnett's one-sided method; Hsu's method)	720
A.10	Multivariate t-distribution absolute maximum (Dunnett's two-sided method)	722
A.11	Critical coefficients for the Voss–Wang method	724

Table A.1 Random numbers* (section 1)

261801	405393	795629	984038	937262	246802
455002	022803	942084	282366	862592	208908
401740	342531	110624	556808	929467	890422
588997	239885	861372	174674	802976	282967
878442	246174	934034	742133	788160	605104
158548	322915	195521	983592	287197	830518
419143	997169	938007	082445	423073	519993
468385	248119	799893	063443	821373	704116
352182	301533	866569	449214	257435	604647
881577	417401	914692	217629	342337	450742
761539	294275	904508	663363	198103	529062
648011	303897	815210	185000	035892	480611
596382	818454	382630	113271	042611	207228
821102	165656	622192	468490	255562	823977
472026	149443	564800	889812	196473	653488
681894	132802	471119	153007	617730	685888
133778	808504	097707	851472	057572	050069
830760	701419	536410	808673	426637	399767
366782	135519	726629	862649	075541	397422
321998	149556	680438	422543	848364	306968
642738	281911	765212	135961	427055	902029
066663	851516	403807	602870	105581	909955
056204	979791	258619	779554	427618	575144
033241	108783	671985	058186	077758	054072
403903	074870	873724	674234	619433	307203
991027	677059	586841	917044	894829	379409
836509	522824	988627	383546	584577	618169
722169	041552	050665	023694	081899	589699
854044	734187	033408	962651	546079	509095
425201	043159	215327	508295	931671	964890
915378	082157	907379	037189	732182	432281
934288	871662	971697	666687	561764	676994
113504	564494	700751	839657	275097	926682
961588	702512	550995	312095	033745	730169
940589	655496	568514	252098	670421	378339
957832	004510	407427	178467	096529	896791

* Random numbers were generated using the SAS statements "retain seed 1613126064;" and "rn=floor(10*ranuni(seed));".

Table A.1 Random numbers (section 2)

772357	993916	529577	678243	561863	847933
432529	970061	557500	798091	179138	481792
307505	525070	258159	554655	575046	678559
466350	487907	299729	248259	480128	346994
617303	468101	318577	923429	038402	757219
593585	243209	814395	862822	746040	791440
001642	339004	160458	600772	662567	368585
149273	042555	774276	484976	819195	397235
650084	170717	096777	646438	049121	435982
291696	199860	255968	977965	704832	152207
380204	676820	172541	838781	875496	866057
601789	268945	637114	763853	972534	321807
034840	046770	352998	091564	836769	679665
021222	269557	705784	227686	602822	502110
657269	645233	803339	576422	510363	650292
496756	485525	641125	398402	351104	116450
837444	628196	558438	029761	119038	624675
488075	739348	349563	070527	096075	150099
704658	615314	955986	559666	475521	954801
102178	988809	827881	636004	439186	939951
693252	078895	586050	545691	461858	159346
252611	441081	786315	281005	010070	205751
280273	117608	080447	202151	108901	325238
453702	029011	657141	620700	604050	013203
327097	317434	109367	377563	036065	203413
786141	861409	263375	518852	087913	861715
194910	643115	993964	022892	046775	061654
875988	516714	699665	303752	334663	145398
370225	083754	123844	056596	275830	997716
060627	822010	449146	256578	444509	074713
536122	366557	689934	795977	120966	293430
413614	370683	263343	215245	320757	951902
722209	875974	091346	411939	116215	304007
661019	124710	939043	865547	430354	627979
631837	027762	468015	102214	231263	950303
751081	600240	089651	253141	586782	632880

Table A.1 Random numbers (section 3)

059546	019867	891131	219608	173193	734618
176484	135942	054815	297442	998872	645538
052799	283293	624836	780633	998025	994382
490320	135340	917771	861186	092763	140293
978406	693078	035305	548903	669275	913423
503664	229298	561181	108615	391336	315399
978589	171347	250219	115588	116702	928926
308761	190680	538063	250996	045474	183835
393619	650606	523668	256623	023861	001087
894218	530435	255864	410171	294324	298222
087945	008260	939499	504474	432769	457539
864409	902698	565732	934650	688277	927769
492768	775393	750111	405919	025468	318104
843672	633027	110801	875047	295640	159752
961195	008740	286149	342458	354060	395475
153985	621691	205500	784869	590596	216745
532725	257818	410867	430907	397363	823657
066251	237385	491172	852331	749317	870908
849872	417854	375970	970009	258362	689175
605204	959747	687531	185277	704122	455741
139687	548165	188942	191439	020545	008885
665434	415021	014415	363450	845716	338980
747965	348796	602064	736557	173122	088897
070671	978664	679967	551926	584739	875137
367207	150356	985292	225591	614974	546526
133364	130583	171077	591404	458326	989837
750462	837858	752783	034800	224890	303605
661831	942203	640775	553730	172198	613256
571491	742343	546985	291036	129295	092863
733891	915752	993268	161851	581437	310503
860024	372853	655098	860970	835755	851476
693411	850277	659850	450238	891661	833743
496459	267113	803320	951153	296721	724280
041583	582606	563804	291119	571246	779088
820676	609521	563473	685897	392666	640840
235342	259362	833616	965799	738929	309229

Table A.1 Random numbers (section 4)

016142	555203	337411	794826	713966	709974
978894	911393	618079	204585	836997	998908
678441	751010	977003	501509	309106	120588
620001	359523	273064	751323	180015	270379
948285	975435	696866	279117	039232	576484
927964	446542	181604	237955	204176	239554
051454	087543	736477	327127	875598	152154
293224	763344	831295	678062	350044	697498
921180	741707	429758	879961	077184	796016
198788	997136	659392	029750	290365	343074
568833	057583	981327	556771	948875	122583
530095	245050	419676	273606	426779	587147
916245	783959	537085	612017	322930	588669
298122	300842	211852	293561	364121	122979
734366	970620	945898	471468	699734	435229
456614	268986	327941	303186	728708	507817
254953	280996	188769	520142	922840	785908
724665	374545	680830	659533	902847	232275
097212	307241	239959	688514	751627	247604
672881	324642	750191	371545	448195	277834
281900	769322	174336	545677	112313	933411
615009	687396	574876	813695	737331	089523
722940	863951	870968	179200	803442	700204
263346	922254	075810	083638	427701	558054
970212	185873	338039	716699	873664	140895
474212	828410	407079	839120	428749	942084
180898	123880	750659	550722	340545	552695
189104	806512	971589	794609	977397	597013
281079	947929	814252	381829	963815	052642
101405	749903	786162	715052	078825	599255
483562	696164	660208	304095	640899	681489
616631	612340	689906	434929	864908	485266
695033	292608	472863	401903	962354	609183
924279	180637	627669	740082	626024	669425
153076	294647	891462	034556	163029	112465
604409	335384	672094	246758	202371	516723

Table A.1 Random numbers (section 5)

178620	984332	234149	062802	203306	409996
857451	696719	245360	219945	554014	514398
363613	427959	500852	392452	867274	977098
934682	279698	307888	039229	143251	417387
355775	341604	728003	783298	021655	069731
665699	212463	461254	662458	822056	868018
490532	385220	245113	034522	661336	735326
177715	678060	444410	984458	394908	203395
247042	584562	285965	331489	536429	150486
247037	647390	958757	889429	320390	733635
037684	319850	319453	708172	600706	116880
163936	682211	276583	695090	141138	775518
936059	110744	230502	313261	037153	962437
723498	531743	968071	809740	728333	510996
260338	879915	375157	296812	522971	126148
224336	392250	500589	802202	461732	596494
983601	286360	962976	318965	952288	371597
950090	003498	944526	890667	177182	557799
517339	436412	783049	641166	652511	767894
192027	132666	760861	620786	271242	132728
503888	846019	856424	233882	942325	511328
796439	106427	930753	519367	261714	969157
354978	019946	325140	390753	674999	265593
881853	487399	093950	592023	463922	642370
834658	788932	959353	528824	971467	890240
429335	115816	387007	698632	332875	685767
502244	063360	540982	294989	737673	576578
230094	859807	408288	251055	307629	402631
108970	046489	583823	027768	600042	788227
985730	336414	654852	476272	011614	049166
797389	447994	431552	294961	293555	180306
967876	584097	346493	644527	708534	540623
488214	736136	666810	990017	550198	332353
594677	673425	481734	749014	380123	200747
216666	893249	173621	966579	934953	101352
829921	075112	978105	427576	707853	535755

Table A.1 Random numbers (section 6)

797468	484252	023853	192278	937112	102503
869406	106834	234802	302737	222508	069132
084703	456756	742761	693385	713106	704798
953014	558897	076970	286515	902394	008147
068334	355648	814540	713770	474093	866359
291019	695796	692442	099917	354198	854868
829045	784428	553248	413099	160541	531210
347351	078917	067538	591284	925325	240708
961281	639335	808637	097914	370752	132493
237408	602057	784998	569469	812566	754613
949564	679405	303591	345128	074802	919895
611262	590822	844714	210655	727420	947239
382111	838410	235673	240015	226367	794195
017136	100540	728145	640255	869932	716829
832572	001164	362585	877812	958998	300684
559820	106016	954619	384380	539268	309304
567125	265846	256533	543082	341655	423495
135117	247924	017309	392054	724319	938577
491976	446939	602958	446781	684067	223566
031030	690179	893112	500077	640800	683948
821529	204137	910310	756711	114949	579204
889160	294936	024169	683546	518509	569679
089569	665239	349069	158437	093839	659940
713873	707840	086598	612305	446369	210679
436460	508406	967487	781618	226710	547386
613674	173160	898727	687279	638647	536756
366699	338577	843373	151717	430829	944266
828010	825546	202510	695671	505800	680682
340755	280452	630101	428665	695421	280009
358767	365611	845098	195830	471533	344402
495054	142723	141577	566053	231325	044934
495165	641454	169446	498379	947861	680617
326075	085211	962907	515048	074505	256236
897734	334967	269056	297336	570543	835922
039744	115260	301407	795887	731089	488919
279055	728952	717435	600372	561076	481948

Table A.2 Coefficients c_i for orthogonal polynomial trend contrasts

$v = 3$

Trend	c_1	c_2	c_3
Linear	−1	0	1
Quadratic	1	−2	1

$v = 4$

Trend	c_1	c_2	c_3	c_4
Linear	−3	−1	1	3
Quadratic	1	−1	−1	1
Cubic	−1	3	−3	1

$v = 5$

Trend	c_1	c_2	c_3	c_4	c_5
Linear	−2	−1	0	1	2
Quadratic	2	−1	−2	−1	2
Cubic	−1	2	0	−2	1
Quartic	1	−4	6	−4	1

$v = 6$

Trend	c_1	c_2	c_3	c_4	c_5	c_6
Linear	−5	−3	−1	1	3	5
Quadratic	5	−1	−4	−4	−1	5
Cubic	−5	7	4	−4	−7	5
Quartic	1	−3	2	2	−3	1
Quintic	−1	5	−10	10	−5	1

$v = 7$

Trend	c_1	c_2	c_3	c_4	c_5	c_6	c_7
Linear	−3	−2	−1	0	1	2	3
Quadratic	5	0	−3	−4	−3	0	5
Cubic	−1	1	1	0	−1	−1	1
Quartic	3	−7	1	6	1	−7	3
Quintic	−1	4	−5	0	5	−4	1
Sextic	1	−6	15	−20	15	−6	1

Table A.3 Standard normal distribution:* Upper α critical coefficients, z_α, and upper-tail probabilities, $\alpha = P(Z > z_\alpha)$

α	0.10	0.05	0.025	0.01	0.005	0.0025	0.001	0.0005	0.00025	0.0001
z_α	1.282	1.645	1.960	2.326	2.576	2.807	3.090	3.291	3.481	3.719

z_α	0.00	0.01	0.02	0.03	0.04	0.05	0.06	0.07	0.08	0.09
0.0	0.5000	0.4960	0.4920	0.4880	0.4840	0.4801	0.4761	0.4721	0.4681	0.4641
0.1	0.4602	0.4562	0.4522	0.4483	0.4443	0.4404	0.4364	0.4325	0.4286	0.4247
0.2	0.4207	0.4168	0.4129	0.4090	0.4052	0.4013	0.3974	0.3936	0.3897	0.3859
0.3	0.3821	0.3783	0.3745	0.3707	0.3669	0.3632	0.3594	0.3557	0.3520	0.3483
0.4	0.3446	0.3409	0.3372	0.3336	0.3300	0.3264	0.3228	0.3192	0.3156	0.3121
0.5	0.3085	0.3050	0.3015	0.2981	0.2946	0.2912	0.2877	0.2843	0.2810	0.2776
0.6	0.2743	0.2709	0.2676	0.2643	0.2611	0.2578	0.2546	0.2514	0.2483	0.2451
0.7	0.2420	0.2389	0.2358	0.2327	0.2296	0.2266	0.2236	0.2206	0.2177	0.2148
0.8	0.2119	0.2090	0.2061	0.2033	0.2005	0.1977	0.1949	0.1922	0.1894	0.1867
0.9	0.1841	0.1814	0.1788	0.1762	0.1736	0.1711	0.1685	0.1660	0.1635	0.1611
1.0	0.1587	0.1562	0.1539	0.1515	0.1492	0.1469	0.1446	0.1423	0.1401	0.1379
1.1	0.1357	0.1335	0.1314	0.1292	0.1271	0.1251	0.1230	0.1210	0.1190	0.1170
1.2	0.1151	0.1131	0.1112	0.1093	0.1075	0.1056	0.1038	0.1020	0.1003	0.0985
1.3	0.0968	0.0951	0.0934	0.0918	0.0901	0.0885	0.0869	0.0853	0.0838	0.0823
1.4	0.0808	0.0793	0.0778	0.0764	0.0749	0.0735	0.0721	0.0708	0.0694	0.0681
1.5	0.0668	0.0655	0.0643	0.0630	0.0618	0.0606	0.0594	0.0582	0.0571	0.0559
1.6	0.0548	0.0537	0.0526	0.0516	0.0505	0.0495	0.0485	0.0475	0.0465	0.0455
1.7	0.0446	0.0436	0.0427	0.0418	0.0409	0.0401	0.0392	0.0384	0.0375	0.0367
1.8	0.0359	0.0351	0.0344	0.0336	0.0329	0.0322	0.0314	0.0307	0.0301	0.0294
1.9	0.0287	0.0281	0.0274	0.0268	0.0262	0.0256	0.0250	0.0244	0.0239	0.0233
2.0	0.0228	0.0222	0.0217	0.0212	0.0207	0.0202	0.0197	0.0192	0.0188	0.0183
2.1	0.0179	0.0174	0.0170	0.0166	0.0162	0.0158	0.0154	0.0150	0.0146	0.0143
2.2	0.0139	0.0136	0.0132	0.0129	0.0125	0.0122	0.0119	0.0116	0.0113	0.0110
2.3	0.0107	0.0104	0.0102	0.0099	0.0096	0.0094	0.0091	0.0089	0.0087	0.0084
2.4	0.0082	0.0080	0.0078	0.0075	0.0073	0.0071	0.0069	0.0068	0.0066	0.0064
2.5	0.0062	0.0060	0.0059	0.0057	0.0055	0.0054	0.0052	0.0051	0.0049	0.0048
2.6	0.0047	0.0045	0.0044	0.0043	0.0041	0.0040	0.0039	0.0038	0.0037	0.0036
2.7	0.0035	0.0034	0.0033	0.0032	0.0031	0.0030	0.0029	0.0028	0.0027	0.0026
2.8	0.0026	0.0025	0.0024	0.0023	0.0023	0.0022	0.0021	0.0021	0.0020	0.0019
2.9	0.0019	0.0018	0.0018	0.0017	0.0016	0.0016	0.0015	0.0015	0.0014	0.0014
3.0	0.0013	0.0013	0.0013	0.0012	0.0012	0.0011	0.0011	0.0011	0.0010	0.0010
3.1	0.0010	0.0009	0.0009	0.0009	0.0008	0.0008	0.0008	0.0008	0.0007	0.0007
3.2	0.0007	0.0007	0.0006	0.0006	0.0006	0.0006	0.0006	0.0005	0.0005	0.0005

* Values were generated using the SAS statements "z_alpha = probit(1-alpha);" and "alpha = 1 - probnorm(z_alpha);".

Table A.4 Student's t-distribution:* Upper α critical coefficients, $t_{df,\alpha}$, where $\alpha = P(t_{df} > t_{df,\alpha})$

df	\multicolumn{9}{c}{α}
	0.10	0.05	0.025	0.01	0.005	0.0025	0.001	0.0005	0.0001
1	3.078	6.314	12.71	31.82	63.66	127.3	318.3	636.6	3183
2	1.886	2.920	4.303	6.965	9.925	14.09	22.33	31.60	70.70
3	1.638	2.353	3.182	4.541	5.841	7.453	10.21	12.92	22.20
4	1.533	2.132	2.776	3.747	4.604	5.598	7.173	8.610	13.03
5	1.476	2.015	2.571	3.365	4.032	4.773	5.893	6.869	9.678
6	1.440	1.943	2.447	3.143	3.707	4.317	5.208	5.959	8.025
7	1.415	1.895	2.365	2.998	3.499	4.029	4.785	5.408	7.063
8	1.397	1.860	2.306	2.896	3.355	3.833	4.501	5.041	6.442
9	1.383	1.833	2.262	2.821	3.250	3.690	4.297	4.781	6.010
10	1.372	1.812	2.228	2.764	3.169	3.581	4.144	4.587	5.694
11	1.363	1.796	2.201	2.718	3.106	3.497	4.025	4.437	5.453
12	1.356	1.782	2.179	2.681	3.055	3.428	3.930	4.318	5.263
13	1.350	1.771	2.160	2.650	3.012	3.372	3.852	4.221	5.111
14	1.345	1.761	2.145	2.624	2.977	3.326	3.787	4.140	4.985
15	1.341	1.753	2.131	2.602	2.947	3.286	3.733	4.073	4.880
16	1.337	1.746	2.120	2.583	2.921	3.252	3.686	4.015	4.791
17	1.333	1.740	2.110	2.567	2.898	3.222	3.646	3.965	4.714
18	1.330	1.734	2.101	2.552	2.878	3.197	3.610	3.922	4.648
19	1.328	1.729	2.093	2.539	2.861	3.174	3.579	3.883	4.590
20	1.325	1.725	2.086	2.528	2.845	3.153	3.552	3.850	4.539
21	1.323	1.721	2.080	2.518	2.831	3.135	3.527	3.819	4.493
22	1.321	1.717	2.074	2.508	2.819	3.119	3.505	3.792	4.452
23	1.319	1.714	2.069	2.500	2.807	3.104	3.485	3.768	4.415
24	1.318	1.711	2.064	2.492	2.797	3.091	3.467	3.745	4.382
25	1.316	1.708	2.060	2.485	2.787	3.078	3.450	3.725	4.352
26	1.315	1.706	2.056	2.479	2.779	3.067	3.435	3.707	4.324
27	1.314	1.703	2.052	2.473	2.771	3.057	3.421	3.690	4.299
28	1.313	1.701	2.048	2.467	2.763	3.047	3.408	3.674	4.275
29	1.311	1.699	2.045	2.462	2.756	3.038	3.396	3.659	4.254
30	1.310	1.697	2.042	2.457	2.750	3.030	3.385	3.646	4.234
35	1.306	1.690	2.030	2.438	2.724	2.996	3.340	3.591	4.153
40	1.303	1.684	2.021	2.423	2.704	2.971	3.307	3.551	4.094
45	1.301	1.679	2.014	2.412	2.690	2.952	3.281	3.520	4.049
50	1.299	1.676	2.009	2.403	2.678	2.937	3.261	3.496	4.014
55	1.297	1.673	2.004	2.396	2.668	2.925	3.245	3.476	3.986
60	1.296	1.671	2.000	2.390	2.660	2.915	3.232	3.460	3.962
70	1.294	1.667	1.994	2.381	2.648	2.899	3.211	3.435	3.926
80	1.292	1.664	1.990	2.374	2.639	2.887	3.195	3.416	3.899
90	1.291	1.662	1.987	2.368	2.632	2.878	3.183	3.402	3.878
100	1.290	1.660	1.984	2.364	2.626	2.871	3.174	3.390	3.862
110	1.289	1.659	1.982	2.361	2.621	2.865	3.166	3.381	3.848
120	1.289	1.658	1.980	2.358	2.617	2.860	3.160	3.373	3.837
∞	1.282	1.645	1.960	2.326	2.576	2.807	3.090	3.291	3.719

* Values $t_{df,\alpha}$ were generated using the SAS statements "t = tinv(1-alpha,df);" for $df < \infty$ and "t = probit(1-alpha)" for $df = \infty$.

Table A.5 Chi-squared distribution:* Upper α critical coefficients, $\chi^2_{df,\alpha}$, where $\alpha = P(\chi^2_{df} > \chi^2_{df,\alpha})$

df	α								
	0.999	0.99	0.975	0.95	0.90	0.10	0.05	0.025	0.01	0.001
1	0.000	0.000	0.001	0.004	0.016	2.706	3.841	5.024	6.635	10.83
2	0.002	0.020	0.051	0.103	0.211	4.605	5.991	7.378	9.210	13.82
3	0.024	0.115	0.216	0.352	0.584	6.251	7.815	9.348	11.34	16.27
4	0.091	0.297	0.484	0.711	1.064	7.779	9.488	11.14	13.28	18.47
5	0.210	0.554	0.831	1.145	1.610	9.236	11.07	12.83	15.09	20.52
6	0.381	0.872	1.237	1.635	2.204	10.64	12.59	14.45	16.81	22.46
7	0.598	1.239	1.690	2.167	2.833	12.02	14.07	16.01	18.48	24.32
8	0.857	1.646	2.180	2.733	3.490	13.36	15.51	17.53	20.09	26.12
9	1.152	2.088	2.700	3.325	4.168	14.68	16.92	19.02	21.67	27.88
10	1.479	2.558	3.247	3.940	4.865	15.99	18.31	20.48	23.21	29.59
11	1.834	3.053	3.816	4.575	5.578	17.28	19.68	21.92	24.72	31.26
12	2.214	3.571	4.404	5.226	6.304	18.55	21.03	23.34	26.22	32.91
13	2.617	4.107	5.009	5.892	7.042	19.81	22.36	24.74	27.69	34.53
14	3.041	4.660	5.629	6.571	7.790	21.06	23.68	26.12	29.14	36.12
15	3.483	5.229	6.262	7.261	8.547	22.31	25.00	27.49	30.58	37.70
16	3.942	5.812	6.908	7.962	9.312	23.54	26.30	28.85	32.00	39.25
17	4.416	6.408	7.564	8.672	10.09	24.77	27.59	30.19	33.41	40.79
18	4.905	7.015	8.231	9.390	10.86	25.99	28.87	31.53	34.81	42.31
19	5.407	7.633	8.907	10.12	11.65	27.20	30.14	32.85	36.19	43.82
20	5.921	8.260	9.591	10.85	12.44	28.41	31.41	34.17	37.57	45.31
21	6.447	8.897	10.28	11.59	13.24	29.62	32.67	35.48	38.93	46.80
22	6.983	9.542	10.98	12.34	14.04	30.81	33.92	36.78	40.29	48.27
23	7.529	10.20	11.69	13.09	14.85	32.01	35.17	38.08	41.64	49.73
24	8.085	10.86	12.40	13.85	15.66	33.20	36.42	39.36	42.98	51.18
25	8.649	11.52	13.12	14.61	16.47	34.38	37.65	40.65	44.31	52.62
26	9.222	12.20	13.84	15.38	17.29	35.56	38.89	41.92	45.64	54.05
27	9.803	12.88	14.57	16.15	18.11	36.74	40.11	43.19	46.96	55.48
28	10.39	13.56	15.31	16.93	18.94	37.92	41.34	44.46	48.28	56.89
29	10.99	14.26	16.05	17.71	19.77	39.09	42.56	45.72	49.59	58.30
30	11.59	14.95	16.79	18.49	20.60	40.26	43.77	46.98	50.89	59.70
35	14.69	18.51	20.57	22.47	24.80	46.06	49.80	53.20	57.34	66.62
40	17.92	22.16	24.43	26.51	29.05	51.81	55.76	59.34	63.69	73.40
45	21.25	25.90	28.37	30.61	33.35	57.51	61.66	65.41	69.96	80.08
50	24.67	29.71	32.36	34.76	37.69	63.17	67.50	71.42	76.15	86.66
55	28.17	33.57	36.40	38.96	42.06	68.80	73.31	77.38	82.29	93.17
60	31.74	37.48	40.48	43.19	46.46	74.40	79.08	83.30	88.38	99.61
70	39.04	45.44	48.76	51.74	55.33	85.53	90.53	95.02	100.4	112.3
80	46.52	53.54	57.15	60.39	64.28	96.58	101.9	106.6	112.3	124.8
90	54.16	61.75	65.65	69.13	73.29	107.6	113.1	118.1	124.1	137.2
100	61.92	70.06	74.22	77.93	82.36	118.5	124.3	129.6	135.8	149.4
120	77.76	86.92	91.57	95.70	100.6	140.2	146.6	152.2	159.0	173.6

* Values were generated using the SAS statement "chi2 = cinv(1-alpha,df);".

Table A.6 F-distribution:* Upper α critical coefficients, $F_{\nu_1,\nu_2,\alpha}$, where $\alpha = P(F_{\nu_1,\nu_2} > F_{\nu_1,\nu_2,\alpha})$

		ν_1								
ν_2	α	1	2	3	4	5	6	7	8	9
1	0.100	39.9	49.5	53.6	55.8	57.2	58.2	58.9	59.4	59.9
	0.050	161	200	216	225	230	234	237	239	241
	0.010	4052	5000	5403	5625	5764	5859	5928	5981	6022
2	0.100	8.53	9.00	9.16	9.24	9.29	9.33	9.35	9.37	9.38
	0.050	18.5	19.0	19.2	19.3	19.3	19.3	19.4	19.4	19.4
	0.010	98.5	99.0	99.2	99.3	99.3	99.3	99.4	99.4	99.4
	0.001	999	999	999	999	999	999	999	999	999
3	0.100	5.54	5.46	5.39	5.34	5.31	5.28	5.27	5.25	5.24
	0.050	10.1	9.55	9.28	9.12	9.01	8.94	8.89	8.85	8.81
	0.010	34.1	30.8	29.5	28.7	28.2	27.9	27.7	27.5	27.4
	0.001	167	149	141	137	135	133	132	131	130
4	0.100	4.54	4.32	4.19	4.11	4.05	4.01	3.98	3.95	3.94
	0.050	7.71	6.94	6.59	6.39	6.26	6.16	6.09	6.04	6.00
	0.010	21.2	18.0	16.7	16.0	15.5	15.2	15.0	14.8	14.7
	0.001	74.1	61.3	56.2	53.4	51.7	50.5	49.7	49.0	48.5
5	0.100	4.06	3.78	3.62	3.52	3.45	3.40	3.37	3.34	3.32
	0.050	6.61	5.79	5.41	5.19	5.05	4.95	4.88	4.82	4.77
	0.010	16.3	13.3	12.1	11.4	11.0	10.7	10.5	10.3	10.2
	0.001	47.2	37.1	33.2	31.1	29.8	28.8	28.2	27.7	27.2
6	0.100	3.78	3.46	3.29	3.18	3.11	3.05	3.01	2.98	2.96
	0.050	5.99	5.14	4.76	4.53	4.39	4.28	4.21	4.15	4.10
	0.010	13.8	10.9	9.78	9.15	8.75	8.47	8.26	8.10	7.98
	0.001	35.5	27.0	23.7	21.9	20.8	20.0	19.5	19.0	18.7
7	0.100	3.59	3.26	3.07	2.96	2.88	2.83	2.78	2.75	2.72
	0.050	5.59	4.74	4.35	4.12	3.97	3.87	3.79	3.73	3.68
	0.010	12.3	9.55	8.45	7.85	7.46	7.19	6.99	6.84	6.72
	0.001	29.3	21.7	18.8	17.2	16.2	15.5	15.0	14.6	14.3
8	0.100	3.46	3.11	2.92	2.81	2.73	2.67	2.62	2.59	2.56
	0.050	5.32	4.46	4.07	3.84	3.69	3.58	3.50	3.44	3.39
	0.010	11.3	8.65	7.59	7.01	6.63	6.37	6.18	6.03	5.91
	0.001	25.4	18.5	15.8	14.4	13.5	12.9	12.4	12.1	11.8
9	0.100	3.36	3.01	2.81	2.69	2.61	2.55	2.51	2.47	2.44
	0.050	5.12	4.26	3.86	3.63	3.48	3.37	3.29	3.23	3.18
	0.010	10.6	8.02	6.99	6.42	6.06	5.80	5.61	5.47	5.35
	0.001	22.9	16.4	13.9	12.6	11.7	11.1	10.7	10.4	10.1
10	0.100	3.29	2.92	2.73	2.61	2.52	2.46	2.41	2.38	2.35
	0.050	4.96	4.10	3.71	3.48	3.33	3.22	3.14	3.07	3.02
	0.010	10.0	7.56	6.55	5.99	5.64	5.39	5.20	5.06	4.94
	0.001	21.0	14.9	12.6	11.3	10.5	9.93	9.52	9.20	8.96

* Values $F_{\nu_1,\nu_2,\alpha}$ were generated using the SAS statement "f = round(finv(1-alpha,df1,df2),0.01);".

Table A.6 (continued) F-distribution: Upper α critical coefficients, $F_{\nu_1,\nu_2,\alpha}$

ν_2	α	\multicolumn{9}{c}{ν_1}
		1	2	3	4	5	6	7	8	9
11	0.100	3.23	2.86	2.66	2.54	2.45	2.39	2.34	2.30	2.27
	0.050	4.84	3.98	3.59	3.36	3.20	3.09	3.01	2.95	2.90
	0.010	9.65	7.21	6.22	5.67	5.32	5.07	4.89	4.74	4.63
	0.001	19.7	13.8	11.6	10.4	9.58	9.05	8.66	8.35	8.12
12	0.100	3.18	2.81	2.61	2.48	2.39	2.33	2.28	2.24	2.21
	0.050	4.75	3.89	3.49	3.26	3.11	3.00	2.91	2.85	2.80
	0.010	9.33	6.93	5.95	5.41	5.06	4.82	4.64	4.50	4.39
	0.001	18.6	13.0	10.8	9.63	8.89	8.38	8.00	7.71	7.48
13	0.100	3.14	2.76	2.56	2.43	2.35	2.28	2.23	2.20	2.16
	0.050	4.67	3.81	3.41	3.18	3.03	2.92	2.83	2.77	2.71
	0.010	9.07	6.70	5.74	5.21	4.86	4.62	4.44	4.30	4.19
	0.001	17.8	12.3	10.2	9.07	8.35	7.86	7.49	7.21	6.98
14	0.100	3.10	2.73	2.52	2.39	2.31	2.24	2.19	2.15	2.12
	0.050	4.60	3.74	3.34	3.11	2.96	2.85	2.76	2.70	2.65
	0.010	8.86	6.51	5.56	5.04	4.69	4.46	4.28	4.14	4.03
	0.001	17.1	11.8	9.73	8.62	7.92	7.44	7.08	6.80	6.58
15	0.100	3.07	2.70	2.49	2.36	2.27	2.21	2.16	2.12	2.09
	0.050	4.54	3.68	3.29	3.06	2.90	2.79	2.71	2.64	2.59
	0.010	8.68	6.36	5.42	4.89	4.56	4.32	4.14	4.00	3.89
	0.001	16.6	11.3	9.34	8.25	7.57	7.09	6.74	6.47	6.26
16	0.100	3.05	2.67	2.46	2.33	2.24	2.18	2.13	2.09	2.06
	0.050	4.49	3.63	3.24	3.01	2.85	2.74	2.66	2.59	2.54
	0.010	8.53	6.23	5.29	4.77	4.44	4.20	4.03	3.89	3.78
	0.001	16.1	11.0	9.01	7.94	7.27	6.80	6.46	6.19	5.98
17	0.100	3.03	2.64	2.44	2.31	2.22	2.15	2.10	2.06	2.03
	0.050	4.45	3.59	3.20	2.96	2.81	2.70	2.61	2.55	2.49
	0.010	8.40	6.11	5.18	4.67	4.34	4.10	3.93	3.79	3.68
	0.001	15.7	10.7	8.73	7.68	7.02	6.56	6.22	5.96	5.75
18	0.100	3.01	2.62	2.42	2.29	2.20	2.13	2.08	2.04	2.00
	0.050	4.41	3.55	3.16	2.93	2.77	2.66	2.58	2.51	2.46
	0.010	8.29	6.01	5.09	4.58	4.25	4.01	3.84	3.71	3.60
	0.001	15.4	10.4	8.49	7.46	6.81	6.35	6.02	5.76	5.56
19	0.100	2.99	2.61	2.40	2.27	2.18	2.11	2.06	2.02	1.98
	0.050	4.38	3.52	3.13	2.90	2.74	2.63	2.54	2.48	2.42
	0.010	8.18	5.93	5.01	4.50	4.17	3.94	3.77	3.63	3.52
	0.001	15.1	10.2	8.28	7.27	6.62	6.18	5.85	5.59	5.39
20	0.100	2.97	2.59	2.38	2.25	2.16	2.09	2.04	2.00	1.96
	0.050	4.35	3.49	3.10	2.87	2.71	2.60	2.51	2.45	2.39
	0.010	8.10	5.85	4.94	4.43	4.10	3.87	3.70	3.56	3.46
	0.001	14.8	9.95	8.10	7.10	6.46	6.02	5.69	5.44	5.24

Table A.6 (continued) F-distribution: Upper α critical coefficients, $F_{v_1,v_2,\alpha}$

v_2	α	\multicolumn{9}{c}{v_1}								
		1	2	3	4	5	6	7	8	9
22	0.100	2.95	2.56	2.35	2.22	2.13	2.06	2.01	1.97	1.93
	0.050	4.30	3.44	3.05	2.82	2.66	2.55	2.46	2.40	2.34
	0.010	7.95	5.72	4.82	4.31	3.99	3.76	3.59	3.45	3.35
	0.001	14.4	9.61	7.80	6.81	6.19	5.76	5.44	5.19	4.99
25	0.100	2.92	2.53	2.32	2.18	2.09	2.02	1.97	1.93	1.89
	0.050	4.24	3.39	2.99	2.76	2.60	2.49	2.40	2.34	2.28
	0.010	7.77	5.57	4.68	4.18	3.85	3.63	3.46	3.32	3.22
	0.001	13.9	9.22	7.45	6.49	5.89	5.46	5.15	4.91	4.71
30	0.100	2.88	2.49	2.28	2.14	2.05	1.98	1.93	1.88	1.85
	0.050	4.17	3.32	2.92	2.69	2.53	2.42	2.33	2.27	2.21
	0.010	7.56	5.39	4.51	4.02	3.70	3.47	3.30	3.17	3.07
	0.001	13.3	8.77	7.05	6.12	5.53	5.12	4.82	4.58	4.39
35	0.100	2.85	2.46	2.25	2.11	2.02	1.95	1.90	1.85	1.82
	0.050	4.12	3.27	2.87	2.64	2.49	2.37	2.29	2.22	2.16
	0.010	7.42	5.27	4.40	3.91	3.59	3.37	3.20	3.07	2.96
	0.001	12.9	8.47	6.79	5.88	5.30	4.89	4.59	4.36	4.18
40	0.100	2.84	2.44	2.23	2.09	2.00	1.93	1.87	1.83	1.79
	0.050	4.08	3.23	2.84	2.61	2.45	2.34	2.25	2.18	2.12
	0.010	7.31	5.18	4.31	3.83	3.51	3.29	3.12	2.99	2.89
	0.001	12.6	8.25	6.59	5.70	5.13	4.73	4.44	4.21	4.02
60	0.100	2.79	2.39	2.18	2.04	1.95	1.87	1.82	1.77	1.74
	0.050	4.00	3.15	2.76	2.53	2.37	2.25	2.17	2.10	2.04
	0.010	7.08	4.98	4.13	3.65	3.34	3.12	2.95	2.82	2.72
	0.001	12.0	7.77	6.17	5.31	4.76	4.37	4.09	3.86	3.69
80	0.100	2.77	2.37	2.15	2.02	1.92	1.85	1.79	1.75	1.71
	0.050	3.96	3.11	2.72	2.49	2.33	2.21	2.13	2.06	2.00
	0.010	6.96	4.88	4.04	3.56	3.26	3.04	2.87	2.74	2.64
	0.001	11.7	7.54	5.97	5.12	4.58	4.20	3.92	3.70	3.53
100	0.100	2.76	2.36	2.14	2.00	1.91	1.83	1.78	1.73	1.69
	0.050	3.94	3.09	2.70	2.46	2.31	2.19	2.10	2.03	1.97
	0.010	6.90	4.82	3.98	3.51	3.21	2.99	2.82	2.69	2.59
	0.001	11.5	7.41	5.86	5.02	4.48	4.11	3.83	3.61	3.44
120	0.100	2.75	2.35	2.13	1.99	1.90	1.82	1.77	1.72	1.68
	0.050	3.92	3.07	2.68	2.45	2.29	2.18	2.09	2.02	1.96
	0.010	6.85	4.79	3.95	3.48	3.17	2.96	2.79	2.66	2.56
	0.001	11.4	7.32	5.78	4.95	4.42	4.04	3.77	3.55	3.38
1000	0.100	2.71	2.31	2.09	1.95	1.85	1.78	1.72	1.68	1.64
	0.050	3.85	3.00	2.61	2.38	2.22	2.11	2.02	1.95	1.89
	0.010	6.66	4.63	3.80	3.34	3.04	2.82	2.66	2.53	2.43
	0.001	10.9	6.96	5.46	4.65	4.14	3.78	3.51	3.30	3.13

Table A.6 (continued) F-distribution: Upper α critical coefficients, $F_{v_1,v_2,\alpha}$

v_2	α	\multicolumn{10}{c}{v_1}									
		10	12	15	20	25	30	40	60	120	1000
1	0.100	60.2	60.7	61.2	61.7	62.1	62.3	62.5	62.8	63.1	63.3
	0.050	242	244	246	248	249	250	251	252	253	254
	0.010	6056	6106	6157	6209	6240	6261	6287	6313	6339	6363
2	0.100	9.39	9.41	9.42	9.44	9.45	9.46	9.47	9.47	9.48	9.49
	0.050	19.4	19.4	19.4	19.5	19.5	19.5	19.5	19.5	19.5	19.5
	0.010	99.4	99.4	99.4	99.5	99.5	99.5	99.5	99.5	99.5	99.5
	0.001	999	999	999	999	999	999	999	999	999	1000
3	0.100	5.23	5.22	5.20	5.18	5.17	5.17	5.16	5.15	5.14	5.13
	0.050	8.79	8.74	8.70	8.66	8.63	8.62	8.59	8.57	8.55	8.53
	0.010	27.2	27.1	26.9	26.7	26.6	26.5	26.4	26.3	26.2	26.1
	0.001	129	128	127	126	126	125	125	124	124	124
4	0.100	3.92	3.90	3.87	3.84	3.83	3.82	3.80	3.79	3.78	3.76
	0.050	5.96	5.91	5.86	5.80	5.77	5.75	5.72	5.69	5.66	5.63
	0.010	14.6	14.4	14.2	14.0	13.9	13.8	13.8	13.7	13.6	13.5
	0.001	48.1	47.4	46.8	46.1	45.7	45.4	45.1	44.8	44.4	44.1
5	0.100	3.30	3.27	3.24	3.21	3.19	3.17	3.16	3.14	3.12	3.11
	0.050	4.74	4.68	4.62	4.56	4.52	4.50	4.46	4.43	4.40	4.37
	0.010	10.1	9.89	9.72	9.55	9.45	9.38	9.29	9.20	9.11	9.03
	0.001	26.9	26.4	25.9	25.4	25.1	24.9	24.6	24.3	24.1	23.8
6	0.100	2.94	2.90	2.87	2.84	2.81	2.80	2.78	2.76	2.74	2.72
	0.050	4.06	4.00	3.94	3.87	3.83	3.81	3.77	3.74	3.70	3.67
	0.010	7.87	7.72	7.56	7.40	7.30	7.23	7.14	7.06	6.97	6.89
	0.001	18.4	18.0	17.6	17.1	16.9	16.7	16.4	16.2	16.0	15.8
7	0.100	2.70	2.67	2.63	2.59	2.57	2.56	2.54	2.51	2.49	2.47
	0.050	3.64	3.57	3.51	3.44	3.40	3.38	3.34	3.30	3.27	3.23
	0.010	6.62	6.47	6.31	6.16	6.06	5.99	5.91	5.82	5.74	5.66
	0.001	14.1	13.7	13.3	12.9	12.7	12.5	12.3	12.1	11.9	11.7
8	0.100	2.54	2.50	2.46	2.42	2.40	2.38	2.36	2.34	2.32	2.30
	0.050	3.35	3.28	3.22	3.15	3.11	3.08	3.04	3.01	2.97	2.93
	0.010	5.81	5.67	5.52	5.36	5.26	5.20	5.12	5.03	4.95	4.87
	0.001	11.5	11.2	10.8	10.5	10.3	10.1	9.92	9.73	9.53	9.36
9	0.100	2.42	2.38	2.34	2.30	2.27	2.25	2.23	2.21	2.18	2.16
	0.050	3.14	3.07	3.01	2.94	2.89	2.86	2.83	2.79	2.75	2.71
	0.010	5.26	5.11	4.96	4.81	4.71	4.65	4.57	4.48	4.40	4.32
	0.001	9.89	9.57	9.24	8.90	8.69	8.55	8.37	8.19	8.00	7.84
10	0.100	2.32	2.28	2.24	2.20	2.17	2.16	2.13	2.11	2.08	2.06
	0.050	2.98	2.91	2.85	2.77	2.73	2.70	2.66	2.62	2.58	2.54
	0.010	4.85	4.71	4.56	4.41	4.31	4.25	4.17	4.08	4.00	3.92
	0.001	8.75	8.45	8.13	7.80	7.60	7.47	7.30	7.12	6.94	6.78

Table A.6 (continued) F-distribution: Upper α critical coefficients, $F_{\nu_1,\nu_2,\alpha}$

ν_2	α	ν_1 10	12	15	20	25	30	40	60	120	1000
11	0.100	2.25	2.21	2.17	2.12	2.10	2.08	2.05	2.03	2.00	1.98
	0.050	2.85	2.79	2.72	2.65	2.60	2.57	2.53	2.49	2.45	2.41
	0.010	4.54	4.40	4.25	4.10	4.01	3.94	3.86	3.78	3.69	3.61
	0.001	7.92	7.63	7.32	7.01	6.81	6.68	6.52	6.35	6.18	6.02
12	0.100	2.19	2.15	2.10	2.06	2.03	2.01	1.99	1.96	1.93	1.91
	0.050	2.75	2.69	2.62	2.54	2.50	2.47	2.43	2.38	2.34	2.30
	0.010	4.30	4.16	4.01	3.86	3.76	3.70	3.62	3.54	3.45	3.37
	0.001	7.29	7.00	6.71	6.40	6.22	6.09	5.93	5.76	5.59	5.44
13	0.100	2.14	2.10	2.05	2.01	1.98	1.96	1.93	1.90	1.88	1.85
	0.050	2.67	2.60	2.53	2.46	2.41	2.38	2.34	2.30	2.25	2.21
	0.010	4.10	3.96	3.82	3.66	3.57	3.51	3.43	3.34	3.25	3.18
	0.001	6.80	6.52	6.23	5.93	5.75	5.63	5.47	5.30	5.14	4.99
14	0.100	2.10	2.05	2.01	1.96	1.93	1.91	1.89	1.86	1.83	1.80
	0.050	2.60	2.53	2.46	2.39	2.34	2.31	2.27	2.22	2.18	2.14
	0.010	3.94	3.80	3.66	3.51	3.41	3.35	3.27	3.18	3.09	3.02
	0.001	6.40	6.13	5.85	5.56	5.38	5.25	5.10	4.94	4.77	4.62
15	0.100	2.06	2.02	1.97	1.92	1.89	1.87	1.85	1.82	1.79	1.76
	0.050	2.54	2.48	2.40	2.33	2.28	2.25	2.20	2.16	2.11	2.07
	0.010	3.80	3.67	3.52	3.37	3.28	3.21	3.13	3.05	2.96	2.88
	0.001	6.08	5.81	5.54	5.25	5.07	4.95	4.80	4.64	4.47	4.33
16	0.100	2.03	1.99	1.94	1.89	1.86	1.84	1.81	1.78	1.75	1.72
	0.050	2.49	2.42	2.35	2.28	2.23	2.19	2.15	2.11	2.06	2.02
	0.010	3.69	3.55	3.41	3.26	3.16	3.10	3.02	2.93	2.84	2.76
	0.001	5.81	5.55	5.27	4.99	4.82	4.70	4.54	4.39	4.23	4.08
17	0.100	2.00	1.96	1.91	1.86	1.83	1.81	1.78	1.75	1.72	1.69
	0.050	2.45	2.38	2.31	2.23	2.18	2.15	2.10	2.06	2.01	1.97
	0.010	3.59	3.46	3.31	3.16	3.07	3.00	2.92	2.83	2.75	2.66
	0.001	5.58	5.32	5.05	4.78	4.60	4.48	4.33	4.18	4.02	3.87
18	0.100	1.98	1.93	1.89	1.84	1.80	1.78	1.75	1.72	1.69	1.66
	0.050	2.41	2.34	2.27	2.19	2.14	2.11	2.06	2.02	1.97	1.92
	0.010	3.51	3.37	3.23	3.08	2.98	2.92	2.84	2.75	2.66	2.58
	0.001	5.39	5.13	4.87	4.59	4.42	4.30	4.15	4.00	3.84	3.69
19	0.100	1.96	1.91	1.86	1.81	1.78	1.76	1.73	1.70	1.67	1.64
	0.050	2.38	2.31	2.23	2.16	2.11	2.07	2.03	1.98	1.93	1.88
	0.010	3.43	3.30	3.15	3.00	2.91	2.84	2.76	2.67	2.58	2.50
	0.001	5.22	4.97	4.70	4.43	4.26	4.14	3.99	3.84	3.68	3.53
20	0.100	1.94	1.89	1.84	1.79	1.76	1.74	1.71	1.68	1.64	1.61
	0.050	2.35	2.28	2.20	2.12	2.07	2.04	1.99	1.95	1.90	1.85
	0.010	3.37	3.23	3.09	2.94	2.84	2.78	2.69	2.61	2.52	2.43
	0.001	5.08	4.82	4.56	4.29	4.12	4.00	3.86	3.70	3.54	3.40

Table A.6 (continued) F-distribution: Upper α critical coefficients, $F_{\nu_1,\nu_2,\alpha}$

ν_2	α	\multicolumn{10}{c}{ν_1}									
		10	12	15	20	25	30	40	60	120	1000
22	0.100	1.90	1.86	1.81	1.76	1.73	1.70	1.67	1.64	1.60	1.57
	0.050	2.30	2.23	2.15	2.07	2.02	1.98	1.94	1.89	1.84	1.79
	0.010	3.26	3.12	2.98	2.83	2.73	2.67	2.58	2.50	2.40	2.32
	0.001	4.83	4.58	4.33	4.06	3.89	3.78	3.63	3.48	3.32	3.17
25	0.100	1.87	1.82	1.77	1.72	1.68	1.66	1.63	1.59	1.56	1.52
	0.050	2.24	2.16	2.09	2.01	1.96	1.92	1.87	1.82	1.77	1.72
	0.010	3.13	2.99	2.85	2.70	2.60	2.54	2.45	2.36	2.27	2.18
	0.001	4.56	4.31	4.06	3.79	3.63	3.52	3.37	3.22	3.06	2.91
30	0.100	1.82	1.77	1.72	1.67	1.63	1.61	1.57	1.54	1.50	1.46
	0.050	2.16	2.09	2.01	1.93	1.88	1.84	1.79	1.74	1.68	1.63
	0.010	2.98	2.84	2.70	2.55	2.45	2.39	2.30	2.21	2.11	2.02
	0.001	4.24	4.00	3.75	3.49	3.33	3.22	3.07	2.92	2.76	2.61
35	0.100	1.79	1.74	1.69	1.63	1.60	1.57	1.53	1.50	1.46	1.42
	0.050	2.11	2.04	1.96	1.88	1.82	1.79	1.74	1.68	1.62	1.57
	0.010	2.88	2.74	2.60	2.44	2.35	2.28	2.19	2.10	2.00	1.90
	0.001	4.03	3.79	3.55	3.29	3.13	3.02	2.87	2.72	2.56	2.40
40	0.100	1.76	1.71	1.66	1.61	1.57	1.54	1.51	1.47	1.42	1.38
	0.050	2.08	2.00	1.92	1.84	1.78	1.74	1.69	1.64	1.58	1.52
	0.010	2.80	2.66	2.52	2.37	2.27	2.20	2.11	2.02	1.92	1.82
	0.001	3.87	3.64	3.40	3.14	2.98	2.87	2.73	2.57	2.41	2.25
60	0.100	1.71	1.66	1.60	1.54	1.50	1.48	1.44	1.40	1.35	1.30
	0.050	1.99	1.92	1.84	1.75	1.69	1.65	1.59	1.53	1.47	1.40
	0.010	2.63	2.50	2.35	2.20	2.10	2.03	1.94	1.84	1.73	1.62
	0.001	3.54	3.32	3.08	2.83	2.67	2.55	2.41	2.25	2.08	1.92
80	0.100	1.68	1.63	1.57	1.51	1.47	1.44	1.40	1.36	1.31	1.25
	0.050	1.95	1.88	1.79	1.70	1.64	1.60	1.54	1.48	1.41	1.34
	0.010	2.55	2.42	2.27	2.12	2.01	1.94	1.85	1.75	1.63	1.51
	0.001	3.39	3.16	2.93	2.68	2.52	2.41	2.26	2.10	1.92	1.75
100	0.100	1.66	1.61	1.56	1.49	1.45	1.42	1.38	1.34	1.28	1.22
	0.050	1.93	1.85	1.77	1.68	1.62	1.57	1.52	1.45	1.38	1.30
	0.010	2.50	2.37	2.22	2.07	1.97	1.89	1.80	1.69	1.57	1.45
	0.001	3.30	3.07	2.84	2.59	2.43	2.32	2.17	2.01	1.83	1.64
120	0.100	1.65	1.60	1.55	1.48	1.44	1.41	1.37	1.32	1.26	1.20
	0.050	1.91	1.83	1.75	1.66	1.60	1.55	1.50	1.43	1.35	1.27
	0.010	2.47	2.34	2.19	2.03	1.93	1.86	1.76	1.66	1.53	1.40
	0.001	3.24	3.02	2.78	2.53	2.37	2.26	2.11	1.95	1.77	1.57
1000	0.100	1.61	1.55	1.49	1.43	1.38	1.35	1.30	1.25	1.18	1.08
	0.050	1.84	1.76	1.68	1.58	1.52	1.47	1.41	1.33	1.24	1.11
	0.010	2.34	2.20	2.06	1.90	1.79	1.72	1.61	1.50	1.35	1.16
	0.001	2.99	2.77	2.54	2.30	2.14	2.02	1.87	1.69	1.49	1.22

Table A.7 Power of the F-test: $\pi(\phi) = P(F_{v_1,v_2,\phi} > F_{v_1,v_2,\alpha})$

$v_1 = 1, \quad \alpha = 0.05$

v_2	\multicolumn{10}{c}{ϕ}
	1.00	1.33	1.67	2.00	2.33	2.67	3.00	3.33	3.67	4.00	4.33
5	0.21	0.33	0.48	0.62	0.75	0.85	0.92	0.96	0.98	0.99	1.00
6	0.22	0.36	0.51	0.66	0.78	0.88	0.94	0.97	0.99	1.00	1.00
7	0.23	0.37	0.53	0.68	0.81	0.90	0.95	0.98	0.99	1.00	1.00
8	0.24	0.38	0.54	0.70	0.82	0.91	0.96	0.98	0.99	1.00	1.00
9	0.24	0.39	0.56	0.71	0.83	0.92	0.96	0.99	1.00	1.00	1.00
10	0.25	0.40	0.57	0.72	0.84	0.92	0.97	0.99	1.00	1.00	1.00
12	0.26	0.41	0.58	0.74	0.86	0.93	0.97	0.99	1.00	1.00	1.00
15	0.26	0.42	0.60	0.75	0.87	0.94	0.98	0.99	1.00	1.00	1.00
20	0.27	0.43	0.61	0.77	0.88	0.95	0.98	0.99	1.00	1.00	1.00
30	0.28	0.45	0.63	0.78	0.89	0.95	0.98	1.00	1.00	1.00	1.00
60	0.29	0.46	0.64	0.79	0.90	0.96	0.99	1.00	1.00	1.00	1.00
1000	0.29	0.47	0.65	0.81	0.91	0.96	0.99	1.00	1.00	1.00	1.00

$v_1 = 1, \quad \alpha = 0.01$

v_2	\multicolumn{10}{c}{ϕ}
	1.67	2.00	2.33	2.67	3.00	3.33	3.67	4.00	4.33	4.67	5.00
5	0.18	0.27	0.38	0.50	0.61	0.72	0.80	0.87	0.92	0.95	0.97
6	0.21	0.31	0.44	0.57	0.69	0.79	0.87	0.92	0.96	0.98	0.99
7	0.23	0.35	0.48	0.62	0.74	0.84	0.91	0.95	0.98	0.99	1.00
8	0.25	0.38	0.52	0.66	0.78	0.87	0.93	0.97	0.99	0.99	1.00
9	0.26	0.40	0.55	0.69	0.81	0.89	0.95	0.98	0.99	1.00	1.00
10	0.28	0.42	0.57	0.71	0.83	0.91	0.96	0.98	0.99	1.00	1.00
12	0.30	0.45	0.61	0.75	0.86	0.93	0.97	0.99	1.00	1.00	1.00
15	0.32	0.48	0.64	0.78	0.88	0.94	0.98	0.99	1.00	1.00	1.00
20	0.34	0.51	0.67	0.81	0.90	0.96	0.98	1.00	1.00	1.00	1.00
30	0.36	0.54	0.71	0.84	0.92	0.97	0.99	1.00	1.00	1.00	1.00
60	0.39	0.57	0.74	0.86	0.94	0.98	0.99	1.00	1.00	1.00	1.00
1000	0.41	0.60	0.76	0.88	0.95	0.98	1.00	1.00	1.00	1.00	1.00

* Power was computed using the SAS statements "nc=v*phi**2;", "Falpha=finv(1-alpha,nu1,nu2);", "power = 1 - probf(Falpha,nu1,nu2,nc);", and "power = round(power,.01);" for values of the parameters "v", "phi", "alpha", "nu1=v-1" and "nu2". "Falpha" is the upper-α critical value of the F-distribution with "nu1" and "nu2" degrees of freedom, and "nc" is the noncentrality parameter for the corresponding noncentral F-distribution.

Table A.7 (continued) Power of the F-test: $\pi(\phi) = P(F_{v_1,v_2,\phi} > F_{v_1,v_2,\alpha})$

$v_1 = 2$, $\alpha = 0.05$

v_2	\multicolumn{11}{c}{ϕ}
	1.00	1.33	1.67	2.00	2.33	2.67	3.00	3.33	3.67	4.00	4.33
5	0.20	0.32	0.46	0.61	0.75	0.85	0.92	0.96	0.98	0.99	1.00
6	0.21	0.34	0.50	0.66	0.79	0.89	0.95	0.98	0.99	1.00	1.00
7	0.22	0.37	0.53	0.70	0.83	0.91	0.96	0.99	1.00	1.00	1.00
8	0.23	0.38	0.56	0.72	0.85	0.93	0.97	0.99	1.00	1.00	1.00
9	0.24	0.40	0.58	0.74	0.87	0.94	0.98	0.99	1.00	1.00	1.00
10	0.25	0.41	0.59	0.76	0.88	0.95	0.98	1.00	1.00	1.00	1.00
12	0.26	0.43	0.62	0.78	0.90	0.96	0.99	1.00	1.00	1.00	1.00
15	0.27	0.45	0.64	0.81	0.91	0.97	0.99	1.00	1.00	1.00	1.00
20	0.28	0.47	0.67	0.83	0.93	0.98	0.99	1.00	1.00	1.00	1.00
30	0.29	0.49	0.69	0.85	0.94	0.98	1.00	1.00	1.00	1.00	1.00
60	0.31	0.51	0.71	0.87	0.95	0.99	1.00	1.00	1.00	1.00	1.00
1000	0.32	0.53	0.73	0.88	0.96	0.99	1.00	1.00	1.00	1.00	1.00

$v_1 = 2$, $\alpha = 0.01$

v_2	\multicolumn{11}{c}{ϕ}
	1.67	2.00	2.33	2.67	3.00	3.33	3.67	4.00	4.33	4.67	5.00
5	0.16	0.25	0.36	0.48	0.60	0.70	0.80	0.87	0.92	0.95	0.97
6	0.20	0.30	0.43	0.57	0.69	0.80	0.88	0.93	0.96	0.98	0.99
7	0.22	0.35	0.50	0.64	0.76	0.86	0.92	0.96	0.98	0.99	1.00
8	0.25	0.39	0.54	0.69	0.81	0.90	0.95	0.98	0.99	1.00	1.00
9	0.27	0.42	0.58	0.73	0.85	0.92	0.97	0.99	1.00	1.00	1.00
10	0.29	0.45	0.62	0.76	0.87	0.94	0.98	0.99	1.00	1.00	1.00
12	0.32	0.49	0.67	0.81	0.91	0.96	0.99	1.00	1.00	1.00	1.00
15	0.35	0.54	0.71	0.85	0.93	0.98	0.99	1.00	1.00	1.00	1.00
20	0.39	0.58	0.76	0.88	0.96	0.99	1.00	1.00	1.00	1.00	1.00
30	0.43	0.63	0.80	0.91	0.97	0.99	1.00	1.00	1.00	1.00	1.00
60	0.47	0.68	0.84	0.94	0.98	1.00	1.00	1.00	1.00	1.00	1.00
1000	0.51	0.72	0.87	0.96	0.99	1.00	1.00	1.00	1.00	1.00	1.00

Table A.7 (continued) Power of the F-test: $\pi(\phi) = P(F_{v_1,v_2,\phi} > F_{v_1,v_2,\alpha})$

$v_1 = 3, \quad \alpha = 0.05$

v_2	\multicolumn{10}{c}{ϕ}									
	1.00	1.33	1.67	2.00	2.33	2.67	3.00	3.33	3.67	4.00	4.33
5	0.19	0.31	0.46	0.61	0.75	0.86	0.93	0.97	0.99	0.99	1.00
6	0.21	0.35	0.51	0.67	0.81	0.90	0.96	0.98	0.99	1.00	1.00
7	0.22	0.37	0.55	0.72	0.85	0.93	0.97	0.99	1.00	1.00	1.00
8	0.24	0.40	0.58	0.75	0.87	0.95	0.98	0.99	1.00	1.00	1.00
9	0.25	0.41	0.60	0.77	0.89	0.96	0.99	1.00	1.00	1.00	1.00
10	0.25	0.43	0.63	0.79	0.91	0.97	0.99	1.00	1.00	1.00	1.00
12	0.27	0.45	0.66	0.82	0.93	0.98	0.99	1.00	1.00	1.00	1.00
15	0.28	0.48	0.69	0.85	0.94	0.98	1.00	1.00	1.00	1.00	1.00
20	0.30	0.51	0.72	0.87	0.96	0.99	1.00	1.00	1.00	1.00	1.00
30	0.32	0.54	0.75	0.90	0.97	0.99	1.00	1.00	1.00	1.00	1.00
60	0.34	0.57	0.78	0.92	0.98	1.00	1.00	1.00	1.00	1.00	1.00
1000	0.36	0.60	0.81	0.93	0.98	1.00	1.00	1.00	1.00	1.00	1.00

$v_1 = 3, \quad \alpha = 0.01$

v_2	\multicolumn{10}{c}{ϕ}									
	1.67	2.00	2.33	2.67	3.00	3.33	3.67	4.00	4.33	4.67	5.00
5	0.16	0.25	0.36	0.48	0.60	0.71	0.80	0.87	0.92	0.95	0.98
6	0.20	0.31	0.44	0.58	0.71	0.81	0.89	0.94	0.97	0.99	0.99
7	0.23	0.36	0.52	0.66	0.79	0.88	0.94	0.97	0.99	1.00	1.00
8	0.26	0.41	0.57	0.72	0.84	0.92	0.96	0.99	1.00	1.00	1.00
9	0.29	0.45	0.62	0.77	0.88	0.95	0.98	0.99	1.00	1.00	1.00
10	0.31	0.48	0.66	0.81	0.91	0.96	0.99	1.00	1.00	1.00	1.00
12	0.35	0.54	0.72	0.86	0.94	0.98	0.99	1.00	1.00	1.00	1.00
15	0.39	0.59	0.77	0.90	0.96	0.99	1.00	1.00	1.00	1.00	1.00
20	0.44	0.65	0.83	0.93	0.98	1.00	1.00	1.00	1.00	1.00	1.00
30	0.49	0.71	0.87	0.96	0.99	1.00	1.00	1.00	1.00	1.00	1.00
60	0.55	0.77	0.91	0.97	0.99	1.00	1.00	1.00	1.00	1.00	1.00
1000	0.60	0.82	0.94	0.99	1.00	1.00	1.00	1.00	1.00	1.00	1.00

Table A.7 (continued) Power of the F-test: $\pi(\phi) = P(F_{v_1,v_2,\phi} > F_{v_1,v_2,\alpha})$

$v_1 = 4$, $\alpha = 0.05$

v_2	\multicolumn{10}{c}{ϕ}									
	1.00	1.33	1.67	2.00	2.33	2.67	3.00	3.33	3.67	4.00	4.33
5	0.19	0.31	0.46	0.62	0.76	0.86	0.93	0.97	0.99	1.00	1.00
6	0.21	0.35	0.52	0.69	0.82	0.91	0.96	0.99	1.00	1.00	1.00
7	0.23	0.38	0.56	0.73	0.86	0.94	0.98	0.99	1.00	1.00	1.00
8	0.24	0.41	0.60	0.77	0.89	0.96	0.99	1.00	1.00	1.00	1.00
9	0.25	0.43	0.63	0.80	0.91	0.97	0.99	1.00	1.00	1.00	1.00
10	0.26	0.45	0.65	0.82	0.93	0.98	0.99	1.00	1.00	1.00	1.00
12	0.28	0.48	0.69	0.85	0.95	0.98	1.00	1.00	1.00	1.00	1.00
15	0.30	0.51	0.73	0.88	0.96	0.99	1.00	1.00	1.00	1.00	1.00
20	0.32	0.54	0.76	0.91	0.97	0.99	1.00	1.00	1.00	1.00	1.00
30	0.34	0.58	0.80	0.93	0.98	1.00	1.00	1.00	1.00	1.00	1.00
60	0.37	0.62	0.83	0.95	0.99	1.00	1.00	1.00	1.00	1.00	1.00
1000	0.39	0.65	0.86	0.96	0.99	1.00	1.00	1.00	1.00	1.00	1.00

$v_1 = 4$, $\alpha = 0.01$

v_2	\multicolumn{10}{c}{ϕ}									
	1.67	2.00	2.33	2.67	3.00	3.33	3.67	4.00	4.33	4.67	5.00
5	0.16	0.25	0.36	0.48	0.60	0.71	0.80	0.88	0.92	0.96	0.98
6	0.20	0.32	0.45	0.59	0.72	0.83	0.90	0.95	0.97	0.99	1.00
7	0.24	0.38	0.53	0.68	0.81	0.89	0.95	0.98	0.99	1.00	1.00
8	0.27	0.43	0.60	0.75	0.86	0.94	0.97	0.99	1.00	1.00	1.00
9	0.30	0.47	0.65	0.80	0.90	0.96	0.99	1.00	1.00	1.00	1.00
10	0.33	0.51	0.70	0.84	0.93	0.97	0.99	1.00	1.00	1.00	1.00
12	0.38	0.58	0.76	0.89	0.96	0.99	1.00	1.00	1.00	1.00	1.00
15	0.43	0.64	0.82	0.93	0.98	0.99	1.00	1.00	1.00	1.00	1.00
20	0.49	0.71	0.87	0.96	0.99	1.00	1.00	1.00	1.00	1.00	1.00
30	0.55	0.78	0.92	0.98	1.00	1.00	1.00	1.00	1.00	1.00	1.00
60	0.62	0.83	0.95	0.99	1.00	1.00	1.00	1.00	1.00	1.00	1.00
1000	0.69	0.88	0.97	1.00	1.00	1.00	1.00	1.00	1.00	1.00	1.00

Table A.7 (continued) Power of the F-test: $\pi(\phi) = P(F_{v_1,v_2,\phi} > F_{v_1,v_2,\alpha})$

$v_1 = 5, \quad \alpha = 0.05$

v_2	\multicolumn{10}{c}{ϕ}									
	1.00	1.33	1.67	2.00	2.33	2.67	3.00	3.33	3.67	4.00	4.33
5	0.19	0.31	0.47	0.62	0.76	0.87	0.93	0.97	0.99	1.00	1.00
6	0.21	0.35	0.53	0.70	0.83	0.92	0.97	0.99	1.00	1.00	1.00
7	0.23	0.39	0.58	0.75	0.88	0.95	0.98	0.99	1.00	1.00	1.00
8	0.24	0.42	0.62	0.79	0.90	0.97	0.99	1.00	1.00	1.00	1.00
9	0.26	0.44	0.65	0.82	0.93	0.98	0.99	1.00	1.00	1.00	1.00
10	0.27	0.46	0.67	0.84	0.94	0.98	1.00	1.00	1.00	1.00	1.00
12	0.29	0.50	0.72	0.88	0.96	0.99	1.00	1.00	1.00	1.00	1.00
15	0.31	0.54	0.76	0.90	0.97	0.99	1.00	1.00	1.00	1.00	1.00
20	0.34	0.58	0.80	0.93	0.98	1.00	1.00	1.00	1.00	1.00	1.00
30	0.36	0.62	0.84	0.95	0.99	1.00	1.00	1.00	1.00	1.00	1.00
60	0.40	0.66	0.87	0.97	1.00	1.00	1.00	1.00	1.00	1.00	1.00
1000	0.43	0.71	0.90	0.98	1.00	1.00	1.00	1.00	1.00	1.00	1.00

$v_1 = 5, \quad \alpha = 0.01$

v_2	\multicolumn{10}{c}{ϕ}									
	1.67	2.00	2.33	2.67	3.00	3.33	3.67	4.00	4.33	4.67	5.00
5	0.16	0.25	0.36	0.48	0.61	0.72	0.81	0.88	0.93	0.96	0.98
6	0.20	0.32	0.46	0.60	0.73	0.84	0.91	0.95	0.98	0.99	1.00
7	0.24	0.39	0.55	0.70	0.82	0.91	0.96	0.98	0.99	1.00	1.00
8	0.28	0.44	0.62	0.77	0.88	0.95	0.98	0.99	1.00	1.00	1.00
9	0.31	0.49	0.68	0.82	0.92	0.97	0.99	1.00	1.00	1.00	1.00
10	0.35	0.54	0.72	0.86	0.94	0.98	0.99	1.00	1.00	1.00	1.00
12	0.40	0.61	0.79	0.91	0.97	0.99	1.00	1.00	1.00	1.00	1.00
15	0.46	0.68	0.85	0.95	0.99	1.00	1.00	1.00	1.00	1.00	1.00
20	0.53	0.76	0.91	0.97	0.99	1.00	1.00	1.00	1.00	1.00	1.00
30	0.60	0.82	0.95	0.99	1.00	1.00	1.00	1.00	1.00	1.00	1.00
60	0.68	0.88	0.97	1.00	1.00	1.00	1.00	1.00	1.00	1.00	1.00
1000	0.76	0.93	0.99	1.00	1.00	1.00	1.00	1.00	1.00	1.00	1.00

Table A.7 (continued) Power of the F-test: $\pi(\phi) = P(F_{v_1,v_2,\phi} > F_{v_1,v_2,\alpha})$

$v_1 = 6, \quad \alpha = 0.05$

v_2	\multicolumn{10}{c}{ϕ}
	1.00	1.33	1.67	2.00	2.33	2.67	3.00	3.33	3.67	4.00	4.33
5	0.19	0.31	0.47	0.63	0.77	0.87	0.94	0.97	0.99	1.00	1.00
6	0.21	0.36	0.53	0.70	0.84	0.92	0.97	0.99	1.00	1.00	1.00
7	0.23	0.39	0.59	0.76	0.88	0.95	0.98	1.00	1.00	1.00	1.00
8	0.25	0.43	0.63	0.80	0.91	0.97	0.99	1.00	1.00	1.00	1.00
9	0.26	0.45	0.66	0.83	0.94	0.98	1.00	1.00	1.00	1.00	1.00
10	0.28	0.48	0.69	0.86	0.95	0.99	1.00	1.00	1.00	1.00	1.00
12	0.30	0.52	0.74	0.89	0.97	0.99	1.00	1.00	1.00	1.00	1.00
15	0.32	0.56	0.78	0.92	0.98	1.00	1.00	1.00	1.00	1.00	1.00
20	0.35	0.60	0.82	0.95	0.99	1.00	1.00	1.00	1.00	1.00	1.00
30	0.39	0.65	0.87	0.97	0.99	1.00	1.00	1.00	1.00	1.00	1.00
60	0.42	0.70	0.90	0.98	1.00	1.00	1.00	1.00	1.00	1.00	1.00
1000	0.47	0.75	0.93	0.99	1.00	1.00	1.00	1.00	1.00	1.00	1.00

$v_1 = 6, \quad \alpha = 0.01$

v_2	\multicolumn{10}{c}{ϕ}
	1.67	2.00	2.33	2.67	3.00	3.33	3.67	4.00	4.33	4.67	5.00
5	0.16	0.25	0.36	0.49	0.61	0.72	0.81	0.88	0.93	0.96	0.98
6	0.20	0.33	0.47	0.61	0.74	0.84	0.91	0.96	0.98	0.99	1.00
7	0.25	0.39	0.56	0.71	0.83	0.91	0.96	0.98	0.99	1.00	1.00
8	0.29	0.46	0.63	0.79	0.89	0.95	0.98	0.99	1.00	1.00	1.00
9	0.33	0.51	0.70	0.84	0.93	0.97	0.99	1.00	1.00	1.00	1.00
10	0.36	0.56	0.74	0.88	0.95	0.98	1.00	1.00	1.00	1.00	1.00
12	0.42	0.64	0.82	0.93	0.98	0.99	1.00	1.00	1.00	1.00	1.00
15	0.49	0.71	0.88	0.96	0.99	1.00	1.00	1.00	1.00	1.00	1.00
20	0.57	0.79	0.93	0.98	1.00	1.00	1.00	1.00	1.00	1.00	1.00
30	0.65	0.86	0.96	0.99	1.00	1.00	1.00	1.00	1.00	1.00	1.00
60	0.74	0.92	0.98	1.00	1.00	1.00	1.00	1.00	1.00	1.00	1.00
1000	0.81	0.96	0.99	1.00	1.00	1.00	1.00	1.00	1.00	1.00	1.00

Table A.8 Tukey's method:* Upper α critical values, $q_{v,df,\alpha}$, of the Studentized range distribution

df	α	\multicolumn{13}{c}{v}
		2	3	4	5	6	7	8	9	10	12	14	16	18	20
2	0.01	14.0	19.0	22.3	24.7	26.6	28.2	29.5	30.7	31.7	33.4	34.8	36.0	37.0	37.9
	0.05	6.08	8.33	9.80	10.9	11.7	12.4	13.0	13.5	14.0	14.7	15.4	15.9	16.4	16.8
	0.10	4.13	5.73	6.77	7.54	8.14	8.63	9.05	9.41	9.72	10.3	10.7	11.1	11.4	11.7
3	0.01	8.26	10.6	12.2	13.3	14.2	15.0	15.6	16.2	16.7	17.5	18.2	18.8	19.3	19.8
	0.05	4.50	5.91	6.82	7.50	8.04	8.48	8.85	9.18	9.46	9.95	10.3	10.7	11.0	11.2
	0.10	3.33	4.47	5.20	5.74	6.16	6.51	6.81	7.06	7.29	7.67	7.98	8.25	8.48	8.68
4	0.01	6.51	8.12	9.17	9.96	10.6	11.1	11.5	11.9	12.3	12.8	13.3	13.7	14.1	14.4
	0.05	3.93	5.04	5.76	6.29	6.71	7.05	7.35	7.60	7.83	8.21	8.52	8.79	9.03	9.23
	0.10	3.01	3.98	4.59	5.03	5.39	5.68	5.93	6.14	6.33	6.65	6.91	7.13	7.33	7.50
5	0.01	5.70	6.98	7.81	8.42	8.91	9.32	9.67	9.97	10.2	10.7	11.1	11.4	11.7	11.9
	0.05	3.64	4.60	5.22	5.67	6.03	6.33	6.58	6.80	6.99	7.32	7.60	7.83	8.03	8.21
	0.10	2.85	3.72	4.26	4.66	4.98	5.24	5.46	5.65	5.82	6.10	6.34	6.54	6.71	6.86
6	0.01	5.24	6.33	7.03	7.56	7.97	8.32	8.61	8.87	9.10	9.48	9.81	10.1	10.3	10.5
	0.05	3.46	4.34	4.90	5.30	5.63	5.90	6.12	6.32	6.49	6.79	7.03	7.24	7.43	7.59
	0.10	2.75	3.56	4.07	4.44	4.73	4.97	5.17	5.34	5.50	5.76	5.98	6.16	6.32	6.47
7	0.01	4.95	5.92	6.55	7.02	7.39	7.70	7.98	8.21	8.43	8.80	9.11	9.38	9.62	9.84
	0.05	3.34	4.16	4.68	5.06	5.36	5.61	5.81	5.99	6.15	6.42	6.65	6.84	7.00	7.15
	0.10	2.68	3.45	3.93	4.28	4.55	4.78	4.97	5.14	5.28	5.53	5.74	5.91	6.06	6.20
8	0.01	4.74	5.64	6.21	6.63	6.97	7.24	7.48	7.69	7.88	8.20	8.46	8.70	8.90	9.08
	0.05	3.26	4.04	4.53	4.89	5.17	5.40	5.60	5.77	5.92	6.17	6.39	6.57	6.72	6.86
	0.10	2.63	3.37	3.83	4.17	4.43	4.65	4.83	4.99	5.13	5.36	5.56	5.73	5.87	6.00
9	0.01	4.60	5.43	5.96	6.35	6.66	6.92	7.14	7.33	7.50	7.79	8.03	8.23	8.41	8.57
	0.05	3.20	3.95	4.41	4.76	5.02	5.24	5.43	5.59	5.74	5.98	6.19	6.36	6.51	6.64
	0.10	2.59	3.32	3.76	4.08	4.34	4.54	4.72	4.87	5.01	5.23	5.42	5.58	5.72	5.85
10	0.01	4.48	5.27	5.77	6.14	6.43	6.67	6.88	7.05	7.21	7.48	7.71	7.90	8.07	8.22
	0.05	3.15	3.88	4.33	4.65	4.91	5.12	5.30	5.46	5.60	5.83	6.03	6.19	6.34	6.47
	0.10	2.56	3.27	3.70	4.02	4.26	4.47	4.64	4.78	4.91	5.13	5.32	5.47	5.61	5.73
11	0.01	4.39	5.15	5.62	5.97	6.25	6.48	6.67	6.84	6.99	7.25	7.46	7.64	7.80	7.94
	0.05	3.11	3.82	4.26	4.57	4.82	5.03	5.20	5.35	5.49	5.71	5.90	6.06	6.20	6.33
	0.10	2.54	3.23	3.66	3.96	4.20	4.40	4.57	4.71	4.84	5.05	5.23	5.38	5.51	5.63

* Values $q_{v,df,\alpha}$ were generated using the SAS statement "qT = probmc('range',..,prob,df,v);", where "prob" $= 1 - \alpha$, and "df=." for $df = \infty$.

Table A.8 (continued) Tukey's method: Upper α critical coefficients, $q_{v,df,\alpha}$ of the Studentized range distribution

df	α	2	3	4	5	6	7	8	9	10	12	14	16	18	20
12	0.01	4.32	5.05	5.50	5.84	6.10	6.32	6.51	6.68	6.82	7.07	7.28	7.46	7.62	7.76
	0.05	3.08	3.77	4.20	4.51	4.75	4.95	5.12	5.26	5.39	5.61	5.80	5.95	6.08	6.20
	0.10	2.52	3.20	3.62	3.92	4.16	4.35	4.51	4.65	4.78	4.99	5.16	5.31	5.44	5.55
14	0.01	4.21	4.89	5.32	5.63	5.88	6.09	6.26	6.41	6.55	6.77	6.97	7.13	7.27	7.40
	0.05	3.03	3.70	4.11	4.41	4.64	4.83	4.99	5.13	5.25	5.46	5.64	5.78	5.91	6.03
	0.10	2.49	3.16	3.56	3.85	4.08	4.27	4.42	4.56	4.68	4.88	5.05	5.19	5.32	5.43
16	0.01	4.13	4.79	5.19	5.49	5.72	5.92	6.08	6.22	6.35	6.56	6.74	6.90	7.03	7.15
	0.05	3.00	3.65	4.05	4.33	4.56	4.74	4.90	5.03	5.15	5.35	5.52	5.66	5.79	5.90
	0.10	2.47	3.12	3.52	3.80	4.03	4.21	4.36	4.49	4.61	4.80	4.97	5.11	5.23	5.33
18	0.01	4.07	4.70	5.09	5.38	5.60	5.79	5.94	6.08	6.20	6.41	6.58	6.73	6.85	6.97
	0.05	2.97	3.61	4.00	4.28	4.49	4.67	4.82	4.96	5.07	5.27	5.43	5.57	5.69	5.79
	0.10	2.45	3.10	3.49	3.77	3.98	4.16	4.31	4.44	4.55	4.75	4.90	5.04	5.16	5.26
20	0.01	4.02	4.64	5.02	5.29	5.51	5.69	5.84	5.97	6.09	6.28	6.45	6.59	6.71	6.82
	0.05	2.95	3.58	3.96	4.23	4.45	4.62	4.77	4.90	5.01	5.20	5.36	5.49	5.61	5.71
	0.10	2.44	3.08	3.46	3.74	3.95	4.12	4.27	4.40	4.51	4.70	4.85	4.99	5.10	5.20
24	0.01	3.96	4.55	4.91	5.17	5.37	5.54	5.68	5.81	5.92	6.11	6.26	6.39	6.51	6.61
	0.05	2.92	3.53	3.90	4.17	4.37	4.54	4.68	4.81	4.92	5.10	5.25	5.38	5.49	5.59
	0.10	2.42	3.05	3.42	3.69	3.90	4.07	4.21	4.34	4.44	4.63	4.78	4.91	5.02	5.12
30	0.01	3.89	4.45	4.80	5.05	5.24	5.40	5.54	5.65	5.76	5.93	6.08	6.20	6.31	6.41
	0.05	2.89	3.49	3.85	4.10	4.30	4.46	4.60	4.72	4.82	5.00	5.15	5.27	5.38	5.47
	0.10	2.40	3.02	3.39	3.65	3.85	4.02	4.16	4.28	4.38	4.56	4.71	4.83	4.94	5.03
40	0.01	3.82	4.37	4.70	4.93	5.11	5.26	5.39	5.50	5.60	5.76	5.90	6.02	6.12	6.21
	0.05	2.86	3.44	3.79	4.04	4.23	4.39	4.52	4.63	4.73	4.90	5.04	5.16	5.27	5.36
	0.10	2.38	2.99	3.35	3.60	3.80	3.96	4.10	4.21	4.32	4.49	4.63	4.75	4.86	4.95
60	0.01	3.76	4.28	4.59	4.82	4.99	5.13	5.25	5.36	5.45	5.60	5.73	5.84	5.93	6.01
	0.05	2.83	3.40	3.74	3.98	4.16	4.31	4.44	4.55	4.65	4.81	4.94	5.06	5.15	5.24
	0.10	2.36	2.96	3.31	3.56	3.75	3.91	4.04	4.16	4.25	4.42	4.56	4.67	4.78	4.86
120	0.01	3.70	4.20	4.50	4.71	4.87	5.01	5.12	5.21	5.30	5.44	5.56	5.66	5.75	5.83
	0.05	2.80	3.36	3.68	3.92	4.10	4.24	4.36	4.47	4.56	4.71	4.84	4.95	5.04	5.13
	0.10	2.34	2.93	3.28	3.52	3.71	3.86	3.99	4.10	4.19	4.35	4.48	4.60	4.69	4.78
∞	0.01	3.64	4.12	4.40	4.60	4.76	4.88	4.99	5.08	5.16	5.29	5.40	5.49	5.57	5.65
	0.05	2.77	3.31	3.63	3.86	4.03	4.17	4.29	4.39	4.47	4.62	4.74	4.85	4.93	5.01
	0.10	2.33	2.90	3.24	3.48	3.66	3.81	3.93	4.04	4.13	4.28	4.41	4.52	4.61	4.69

Table A.9 Dunnett's one-sided method; Hsu's method:* Upper α critical coefficients
$w_{D1} = t_{v-1,df,\alpha}^{(0.5)}$

		\multicolumn{13}{c}{$v - 1$}
df	α	2	3	4	5	6	7	8	9	10	12	14	16	18	20
2	0.01	8.88	10.0	10.9	11.5	12.0	12.5	12.8	13.2	13.5	14.0	14.4	14.7	15.1	15.3
	0.05	3.80	4.34	4.71	5.00	5.24	5.43	5.60	5.75	5.88	6.11	6.29	6.45	6.59	6.72
	0.10	2.54	2.92	3.19	3.40	3.57	3.71	3.83	3.94	4.03	4.19	4.32	4.44	4.54	4.62
3	0.01	5.48	6.04	6.44	6.74	6.99	7.20	7.38	7.53	7.67	7.91	8.11	8.28	8.43	8.56
	0.05	2.94	3.28	3.52	3.70	3.85	3.97	4.08	4.17	4.25	4.39	4.51	4.61	4.70	4.78
	0.10	2.13	2.41	2.61	2.75	2.87	2.97	3.06	3.13	3.20	3.31	3.41	3.49	3.56	3.62
4	0.01	4.41	4.80	5.07	5.28	5.45	5.59	5.71	5.82	5.92	6.08	6.22	6.34	6.44	6.53
	0.05	2.61	2.88	3.08	3.22	3.34	3.44	3.52	3.59	3.66	3.77	3.86	3.94	4.01	4.07
	0.10	1.96	2.20	2.37	2.50	2.60	2.68	2.75	2.82	2.87	2.97	3.05	3.11	3.17	3.22
5	0.01	3.90	4.21	4.43	4.60	4.73	4.85	4.94	5.03	5.11	5.24	5.34	5.44	5.52	5.59
	0.05	2.44	2.68	2.85	2.98	3.08	3.16	3.24	3.30	3.36	3.45	3.53	3.60	3.66	3.71
	0.10	1.87	2.09	2.24	2.36	2.45	2.53	2.59	2.65	2.70	2.78	2.86	2.92	2.97	3.02
6	0.01	3.61	3.88	4.06	4.21	4.32	4.42	4.51	4.58	4.64	4.76	4.85	4.93	5.00	5.06
	0.05	2.34	2.56	2.71	2.83	2.92	3.00	3.06	3.12	3.17	3.26	3.33	3.40	3.45	3.50
	0.10	1.82	2.02	2.17	2.27	2.36	2.43	2.49	2.54	2.59	2.67	2.74	2.79	2.84	2.89
7	0.01	3.42	3.66	3.83	3.96	4.06	4.15	4.22	4.29	4.35	4.45	4.53	4.60	4.67	4.72
	0.05	2.27	2.48	2.62	2.73	2.82	2.89	2.95	3.00	3.05	3.13	3.20	3.26	3.31	3.36
	0.10	1.78	1.98	2.11	2.22	2.30	2.37	2.42	2.47	2.52	2.59	2.66	2.71	2.76	2.80
8	0.01	3.29	3.51	3.66	3.78	3.88	3.96	4.03	4.09	4.14	4.23	4.31	4.38	4.43	4.49
	0.05	2.22	2.42	2.55	2.66	2.74	2.81	2.87	2.92	2.96	3.04	3.11	3.16	3.21	3.25
	0.10	1.75	1.94	2.08	2.17	2.25	2.32	2.38	2.42	2.47	2.54	2.60	2.65	2.70	2.74
9	0.01	3.19	3.40	3.54	3.66	3.75	3.82	3.89	3.94	3.99	4.08	4.15	4.21	4.26	4.31
	0.05	2.18	2.37	2.50	2.60	2.68	2.75	2.81	2.86	2.90	2.97	3.04	3.09	3.14	3.18
	0.10	1.73	1.92	2.05	2.14	2.22	2.28	2.34	2.39	2.43	2.50	2.56	2.61	2.65	2.69
10	0.01	3.11	3.31	3.45	3.56	3.64	3.72	3.78	3.83	3.88	3.96	4.03	4.08	4.14	4.18
	0.05	2.15	2.34	2.47	2.56	2.64	2.70	2.76	2.81	2.85	2.92	2.98	3.03	3.08	3.12
	0.10	1.71	1.90	2.02	2.12	2.19	2.26	2.31	2.35	2.40	2.46	2.52	2.57	2.61	2.65
11	0.01	3.06	3.25	3.38	3.48	3.56	3.63	3.69	3.74	3.79	3.86	3.93	3.99	4.03	4.08
	0.05	2.13	2.31	2.43	2.53	2.60	2.67	2.72	2.77	2.81	2.88	2.94	2.99	3.03	3.07
	0.10	1.70	1.88	2.01	2.10	2.17	2.23	2.29	2.33	2.37	2.44	2.49	2.54	2.58	2.62

* Values $w_{D1} = t_{v-1,df,\alpha}^{(0.5)}$ were generated using the SAS statement "wD1 = probmc('dunnett1',,,prob,df,vm1);", where "prob" $= 1 - \alpha$, "vm1" $= v - 1$, $t_{v-1,df,\alpha}^{(0.5)}$ is the upper α critical value for the maximum of a $(v - 1)$-variate t-distribution with common correlation $\rho = 0.5$ and degrees of freedom df, and "df=." for $df = \infty$.

Table A.9 (continued) Dunnett's one-sided method, Hsu's method: Upper α critical coefficients, $w_{D1} = t_{v-1,df,\alpha}^{(0.5)}$

df	α	\multicolumn{13}{c}{$v-1$}
		2	3	4	5	6	7	8	9	10	12	14	16	18	20
12	0.01	3.01	3.19	3.32	3.42	3.50	3.56	3.62	3.67	3.71	3.79	3.85	3.91	3.95	3.99
	0.05	2.11	2.29	2.41	2.50	2.58	2.64	2.69	2.74	2.78	2.84	2.90	2.95	2.99	3.03
	0.10	1.69	1.87	1.99	2.08	2.16	2.22	2.27	2.31	2.35	2.42	2.47	2.52	2.56	2.60
14	0.01	2.94	3.11	3.23	3.33	3.40	3.46	3.51	3.56	3.60	3.67	3.73	3.78	3.83	3.87
	0.05	2.08	2.25	2.37	2.46	2.53	2.59	2.64	2.69	2.73	2.79	2.85	2.89	2.93	2.97
	0.10	1.67	1.85	1.97	2.06	2.13	2.19	2.24	2.28	2.32	2.38	2.44	2.48	2.52	2.56
16	0.01	2.88	3.05	3.17	3.26	3.33	3.39	3.44	3.48	3.52	3.59	3.65	3.70	3.74	3.77
	0.05	2.06	2.23	2.34	2.43	2.50	2.56	2.61	2.65	2.69	2.75	2.81	2.85	2.89	2.93
	0.10	1.66	1.83	1.95	2.04	2.11	2.17	2.22	2.26	2.30	2.36	2.41	2.46	2.50	2.53
18	0.01	2.84	3.01	3.12	3.21	3.27	3.33	3.38	3.42	3.46	3.53	3.58	3.63	3.67	3.71
	0.05	2.04	2.21	2.32	2.41	2.48	2.53	2.58	2.62	2.66	2.72	2.78	2.82	2.86	2.89
	0.10	1.65	1.82	1.94	2.02	2.09	2.15	2.20	2.24	2.28	2.34	2.39	2.44	2.48	2.51
20	0.01	2.81	2.97	3.08	3.17	3.23	3.29	3.34	3.38	3.42	3.48	3.53	3.58	3.62	3.65
	0.05	2.03	2.19	2.30	2.39	2.46	2.51	2.56	2.60	2.64	2.70	2.75	2.80	2.83	2.87
	0.10	1.64	1.81	1.93	2.01	2.08	2.14	2.19	2.23	2.26	2.33	2.38	2.42	2.46	2.49
24	0.01	2.77	2.92	3.03	3.11	3.17	3.22	3.27	3.31	3.35	3.41	3.46	3.50	3.54	3.57
	0.05	2.01	2.17	2.28	2.36	2.43	2.48	2.53	2.57	2.60	2.66	2.72	2.76	2.80	2.83
	0.10	1.63	1.80	1.91	2.00	2.06	2.12	2.17	2.21	2.24	2.30	2.35	2.40	2.43	2.47
30	0.01	2.72	2.87	2.97	3.05	3.11	3.16	3.21	3.25	3.28	3.34	3.39	3.43	3.46	3.50
	0.05	1.99	2.15	2.25	2.34	2.40	2.45	2.50	2.54	2.57	2.63	2.68	2.72	2.76	2.79
	0.10	1.62	1.79	1.90	1.98	2.05	2.10	2.15	2.19	2.22	2.28	2.33	2.37	2.41	2.44
40	0.01	2.68	2.82	2.92	2.99	3.05	3.10	3.14	3.18	3.21	3.27	3.32	3.36	3.39	3.42
	0.05	1.97	2.13	2.23	2.31	2.37	2.42	2.47	2.51	2.54	2.60	2.65	2.69	2.72	2.75
	0.10	1.61	1.77	1.88	1.96	2.03	2.08	2.13	2.17	2.20	2.26	2.31	2.35	2.39	2.42
60	0.01	2.64	2.78	2.87	2.94	3.00	3.04	3.08	3.12	3.15	3.20	3.25	3.29	3.32	3.35
	0.05	1.95	2.10	2.21	2.28	2.34	2.40	2.44	2.48	2.51	2.57	2.61	2.65	2.69	2.72
	0.10	1.60	1.76	1.87	1.95	2.01	2.06	2.11	2.15	2.18	2.24	2.29	2.33	2.36	2.39
120	0.01	2.60	2.73	2.82	2.89	2.94	2.99	3.03	3.06	3.09	3.14	3.18	3.22	3.25	3.28
	0.05	1.93	2.08	2.18	2.26	2.32	2.37	2.41	2.45	2.48	2.53	2.58	2.62	2.65	2.68
	0.10	1.59	1.75	1.85	1.93	1.99	2.05	2.09	2.13	2.16	2.22	2.27	2.31	2.34	2.37
∞	0.01	2.56	2.69	2.77	2.84	2.89	2.93	2.97	3.00	3.03	3.08	3.12	3.15	3.18	3.21
	0.05	1.92	2.06	2.16	2.23	2.29	2.34	2.38	2.42	2.45	2.50	2.55	2.58	2.62	2.64
	0.10	1.58	1.73	1.84	1.92	1.98	2.03	2.07	2.11	2.14	2.20	2.24	2.28	2.32	2.35

Table A.10 Dunnett's two-sided method:* Upper α critical coefficients $w_{D2} = |t|_{v-1,df,\alpha}^{(0.5)}$

df	α	$v-1$ = 2	3	4	5	6	7	8	9	10	12	14	16	18	20
2	0.01	12.4	13.8	14.8	15.6	16.2	16.7	17.1	17.5	17.8	18.4	18.8	19.2	19.6	19.9
	0.05	5.42	6.06	6.51	6.85	7.12	7.35	7.54	7.71	7.85	8.10	8.31	8.49	8.64	8.77
	0.10	3.72	4.18	4.50	4.74	4.93	5.09	5.23	5.34	5.45	5.62	5.77	5.89	6.00	6.09
3	0.01	6.97	7.64	8.10	8.46	8.74	8.98	9.19	9.37	9.52	9.79	10.0	10.2	10.4	10.5
	0.05	3.87	4.26	4.54	4.75	4.92	5.06	5.18	5.28	5.37	5.53	5.66	5.77	5.87	5.95
	0.10	2.91	3.23	3.45	3.62	3.75	3.87	3.96	4.04	4.12	4.24	4.34	4.43	4.51	4.58
4	0.01	5.36	5.81	6.12	6.36	6.55	6.72	6.85	6.98	7.08	7.27	7.42	7.55	7.66	7.77
	0.05	3.31	3.62	3.83	3.99	4.13	4.23	4.33	4.41	4.48	4.60	4.71	4.79	4.87	4.94
	0.10	2.60	2.86	3.05	3.18	3.30	3.39	3.47	3.54	3.60	3.70	3.79	3.86	3.92	3.98
5	0.01	4.63	4.97	5.22	5.41	5.56	5.68	5.79	5.89	5.97	6.11	6.24	6.34	6.43	6.51
	0.05	3.03	3.29	3.48	3.62	3.73	3.82	3.90	3.97	4.03	4.14	4.23	4.30	4.37	4.42
	0.10	2.43	2.67	2.83	2.96	3.05	3.14	3.21	3.27	3.32	3.41	3.49	3.56	3.61	3.66
6	0.01	4.21	4.51	4.71	4.87	5.00	5.10	5.20	5.28	5.35	5.47	5.57	5.66	5.74	5.80
	0.05	2.86	3.10	3.26	3.39	3.49	3.57	3.64	3.71	3.76	3.86	3.94	4.00	4.06	4.11
	0.10	2.33	2.55	2.70	2.81	2.91	2.98	3.05	3.10	3.15	3.24	3.31	3.37	3.42	3.47
7	0.01	3.95	4.21	4.39	4.53	4.64	4.74	4.82	4.89	4.96	5.07	5.16	5.24	5.31	5.37
	0.05	2.75	2.97	3.12	3.24	3.33	3.41	3.48	3.53	3.58	3.67	3.75	3.81	3.86	3.91
	0.10	2.26	2.47	2.61	2.72	2.81	2.88	2.94	2.99	3.04	3.12	3.18	3.24	3.29	3.33
8	0.01	3.77	4.00	4.17	4.29	4.40	4.48	4.56	4.62	4.68	4.78	4.86	4.93	5.00	5.05
	0.05	2.67	2.88	3.02	3.13	3.22	3.29	3.35	3.41	3.46	3.54	3.61	3.67	3.72	3.76
	0.10	2.22	2.41	2.55	2.65	2.73	2.80	2.86	2.91	2.96	3.03	3.10	3.15	3.20	3.24
9	0.01	3.63	3.85	4.01	4.12	4.22	4.30	4.37	4.43	4.48	4.57	4.65	4.71	4.77	4.82
	0.05	2.61	2.81	2.95	3.05	3.14	3.20	3.26	3.32	3.36	3.44	3.51	3.56	3.61	3.65
	0.10	2.18	2.37	2.50	2.60	2.68	2.74	2.80	2.85	2.89	2.97	3.03	3.08	3.13	3.17
10	0.01	3.53	3.74	3.88	3.99	4.08	4.16	4.22	4.28	4.33	4.42	4.49	4.55	4.60	4.65
	0.05	2.57	2.76	2.89	2.99	3.07	3.14	3.19	3.24	3.29	3.36	3.43	3.48	3.53	3.57
	0.10	2.15	2.34	2.46	2.56	2.64	2.70	2.75	2.80	2.84	2.92	2.98	3.03	3.07	3.11
11	0.01	3.45	3.65	3.79	3.89	3.98	4.05	4.11	4.16	4.21	4.29	4.36	4.42	4.47	4.52
	0.05	2.53	2.72	2.84	2.94	3.02	3.08	3.14	3.19	3.23	3.30	3.36	3.42	3.46	3.50
	0.10	2.13	2.31	2.43	2.53	2.60	2.66	2.72	2.76	2.80	2.87	2.93	2.98	3.03	3.06

* Values $w_{D1} = t_{v-1,df,\alpha}^{(0.5)}$ generated using the SAS statement "wD2 = probmc('dunnett2',.,prob, df,vm1);", where "prob" = $1 - \alpha$, "vm1" = $v - 1$, $|t|_{v-1,df,\alpha}^{(0.5)}$ is the upper α critical value for the maximum absolute value of a $(v-1)$-variate t-distribution with common correlation $\rho = 0.5$ and degrees of freedom df, and "df=." for $df = \infty$.

Table A.10 (continued) Dunnett's two-sided method: Upper α critical coefficients, $w_{D2} = |t|^{(0.5)}_{v-1,df,\alpha}$

df	α	\multicolumn{13}{c}{$v-1$}
		2	3	4	5	6	7	8	9	10	12	14	16	18	20
12	0.01	3.39	3.58	3.71	3.81	3.89	3.96	4.02	4.07	4.12	4.19	4.26	4.32	4.37	4.41
	0.05	2.50	2.68	2.81	2.90	2.98	3.04	3.09	3.14	3.18	3.25	3.31	3.36	3.41	3.45
	0.10	2.11	2.29	2.41	2.50	2.57	2.64	2.69	2.73	2.77	2.84	2.90	2.95	2.99	3.03
14	0.01	3.29	3.47	3.59	3.69	3.76	3.83	3.88	3.93	3.97	4.05	4.11	4.16	4.20	4.25
	0.05	2.46	2.63	2.75	2.84	2.91	2.97	3.02	3.07	3.11	3.18	3.23	3.28	3.32	3.36
	0.10	2.08	2.25	2.37	2.46	2.53	2.59	2.64	2.68	2.72	2.79	2.84	2.89	2.93	2.97
16	0.01	3.22	3.39	3.51	3.60	3.67	3.73	3.78	3.83	3.87	3.94	4.00	4.05	4.09	4.13
	0.05	2.42	2.59	2.71	2.80	2.87	2.92	2.97	3.02	3.06	3.12	3.18	3.22	3.26	3.30
	0.10	2.06	2.23	2.34	2.43	2.50	2.56	2.61	2.65	2.69	2.75	2.80	2.85	2.89	2.93
18	0.01	3.17	3.33	3.44	3.53	3.60	3.66	3.71	3.75	3.79	3.86	3.91	3.96	4.00	4.04
	0.05	2.40	2.56	2.68	2.76	2.83	2.89	2.94	2.98	3.01	3.08	3.13	3.18	3.22	3.25
	0.10	2.04	2.21	2.32	2.41	2.47	2.53	2.58	2.62	2.66	2.72	2.77	2.82	2.86	2.89
20	0.01	3.13	3.29	3.40	3.48	3.55	3.60	3.65	3.69	3.73	3.80	3.85	3.90	3.94	3.97
	0.05	2.38	2.54	2.65	2.73	2.80	2.86	2.90	2.95	2.98	3.05	3.10	3.14	3.18	3.22
	0.10	2.03	2.19	2.30	2.39	2.46	2.51	2.56	2.60	2.64	2.70	2.75	2.79	2.83	2.87
24	0.01	3.07	3.22	3.32	3.40	3.47	3.52	3.57	3.61	3.64	3.70	3.76	3.80	3.84	3.87
	0.05	2.35	2.51	2.61	2.70	2.76	2.81	2.86	2.90	2.94	3.00	3.05	3.09	3.13	3.16
	0.10	2.01	2.17	2.28	2.36	2.43	2.48	2.53	2.57	2.60	2.66	2.71	2.76	2.79	2.83
30	0.01	3.01	3.15	3.25	3.33	3.39	3.44	3.49	3.52	3.56	3.62	3.66	3.71	3.74	3.77
	0.05	2.32	2.47	2.58	2.66	2.72	2.77	2.82	2.86	2.89	2.95	3.00	3.04	3.08	3.11
	0.10	1.99	2.15	2.25	2.33	2.40	2.45	2.50	2.54	2.57	2.63	2.68	2.72	2.76	2.79
40	0.01	2.95	3.09	3.19	3.26	3.32	3.37	3.41	3.44	3.48	3.53	3.58	3.62	3.65	3.68
	0.05	2.29	2.44	2.54	2.62	2.68	2.73	2.77	2.81	2.85	2.90	2.95	2.99	3.02	3.06
	0.10	1.97	2.13	2.23	2.31	2.37	2.42	2.47	2.51	2.54	2.60	2.65	2.69	2.72	2.75
60	0.01	2.90	3.03	3.12	3.19	3.25	3.29	3.33	3.37	3.40	3.45	3.49	3.53	3.56	3.59
	0.05	2.27	2.41	2.51	2.58	2.64	2.69	2.73	2.77	2.80	2.86	2.90	2.94	2.97	3.00
	0.10	1.95	2.10	2.21	2.28	2.34	2.40	2.44	2.48	2.51	2.57	2.61	2.65	2.69	2.72
120	0.01	2.85	2.97	3.06	3.12	3.18	3.22	3.26	3.29	3.32	3.37	3.41	3.45	3.48	3.50
	0.05	2.24	2.38	2.47	2.55	2.60	2.65	2.69	2.73	2.76	2.81	2.86	2.89	2.93	2.95
	0.10	1.93	2.08	2.18	2.26	2.32	2.37	2.41	2.45	2.48	2.53	2.58	2.62	2.65	2.68
∞	0.01	2.79	2.91	3.00	3.06	3.11	3.15	3.19	3.22	3.25	3.29	3.33	3.37	3.40	3.42
	0.05	2.21	2.35	2.44	2.51	2.57	2.61	2.65	2.69	2.72	2.77	2.81	2.85	2.88	2.91
	0.10	1.92	2.06	2.16	2.23	2.29	2.34	2.38	2.42	2.45	2.50	2.55	2.58	2.62	2.64

Table A.11 Voss–Wang method: Upper α critical coefficients $w_V = v_{m,d,\alpha}$ for m orthogonal contrasts and d sums of squares pooled into each quasi mean squared error.

even m	d	α 0.10	0.05	0.01	0.001	odd m	d	α 0.10	0.05	0.01	0.001
2	1	12.4	25.0	126.	1384.	3	2	5.31	7.61	17.0	54.4
4	2	9.08	13.1	29.9	105.	5	3	6.35	8.27	14.7	32.0
6	3	8.57	11.2	19.9	50.0	7	4	6.82	8.56	13.8	26.8
8	4	8.37	10.5	16.6	33.2	9	5	7.18	8.73	13.3	24.7
10	5	8.39	10.3	16.0	29.5	11	6	7.42	8.89	13.0	21.4
12	6	8.37	10.1	14.7	24.1	13	7	7.64	8.95	12.4	20.4
14	7	8.43	9.89	14.0	22.5	15	8	7.76	9.04	12.4	19.2
16	8	8.45	9.88	13.5	21.8	17	9	7.85	9.10	12.4	19.1
18	9	8.51	9.77	13.5	19.9	19	10	7.98	9.10	12.2	17.8
20	10	8.50	9.70	12.9	19.0	21	11	8.03	9.17	12.1	16.8
22	11	8.53	9.78	12.9	18.2	23	12	8.12	9.19	11.9	17.2
24	12	8.55	9.76	12.7	18.6	25	13	8.15	9.28	11.9	17.2
26	13	8.60	9.80	12.5	18.2	27	14	8.22	9.33	11.8	16.5
28	14	8.61	9.78	12.6	18.3	29	15	8.27	9.36	11.8	16.7
30	15	8.63	9.78	12.4	17.5	31	16	8.32	9.38	11.8	16.3
32	16	8.70	9.76	12.4	16.9	33	17	8.35	9.38	11.8	15.9
34	17	8.66	9.70	12.2	16.5	35	18	8.38	9.35	11.7	15.8
36	18	8.69	9.71	12.1	16.3	37	19	8.41	9.40	11.6	15.6
38	19	8.70	9.67	12.0	16.5	39	20	8.43	9.37	11.6	16.0
40	20	8.71	9.69	12.0	16.5	41	21	8.45	9.36	11.5	15.5
42	21	8.72	9.64	12.0	15.8	43	22	8.48	9.39	11.6	15.5
44	22	8.73	9.68	12.0	15.7	45	23	8.52	9.40	11.6	15.2
46	23	8.76	9.69	11.9	15.7	47	24	8.53	9.42	11.5	15.1
48	24	8.75	9.65	11.8	15.8	49	25	8.55	9.44	11.5	15.1
50	25	8.76	9.68	11.8	15.7	51	26	8.57	9.42	11.5	15.0
52	26	8.78	9.66	11.8	15.6	53	27	8.60	9.45	11.5	15.0
54	27	8.80	9.69	11.8	15.5	55	28	8.61	9.46	11.4	15.0
56	28	8.82	9.70	11.8	15.7	57	29	8.64	9.51	11.5	15.1
58	29	8.81	9.71	11.7	15.3	59	30	8.63	9.49	11.4	15.0
60	30	8.83	9.70	11.7	15.3	61	31	8.66	9.49	11.4	15.1
62	31	8.84	9.68	11.7	15.9	63	32	8.67	9.50	11.4	15.4

Bibliography

Anderson, V. L. and McLean, R. A. (1974). *Design of Experiments: A Realistic Approach*. Marcel Dekker, Inc., New York.

Bainbridge, J. R. (1951). Factorial experiments in pilot plant studies. *Industrial and Engineering Chemistry* **43**, 1300–1306.

Barnabas, I. J., Dean, J. R., Tomlinson, W. R., and Owen, S. P. (1995). Experimental design approach for the extraction of polycyclic aromatic hydrocarbons form soil using supercritical carbon dioxide. *Analytical Chemistry* **67**, 2064–2069.

Barnett, M. K. and Mead, F. C. Jr. (1956). A 2^4 factorial experiment in four blocks of eight: a study in radioactive decontamination, *Applied Statistics* **5**, 122–131.

Baten, W. D. (1956). An analysis of variance applied to screw machines. *Industrial Quality Control* **12**, 8–9.

Beckman, R. J., Nachtsheim, C. J., and Cook, R. D. (1987). Diagnostics for mixed-model analysis of variance, *Technometrics* **29**, 413–426.

Bigham, R. (1987). Frame Torque Optimization via Taguchi Methods, *Fifth Symposium on Taguchi Methods, American Supplier Institute, Inc.*, ASI Press, Dearborn, Michigan, 439–455.

Blom, G. (1958). *Statistical Estimates and Transformed Beta Variables*. John Wiley and Sons, New York.

Box, G. E. P. and Behnken, D. W. (1960). Some new three level designs for the study of quantitative variables. *Technometrics* **2**, 455–475.

Box, G. E. P. and Cox, D. R. (1964). An analysis of transformations, *Journal of the Royal Statistical Society*, B **26**, 211–243.

Box, G. E. P. and Hunter, J. S. (1957). Multi-factor experimental designs for exploring response surfaces. *Annals of Mathematical Statistics* **28**, 195–241.

Box, G. E. P. and Wilson, K. B. (1951). On the experimental attainment of optimum conditions. *Journal of the Royal Statistical Society, Series B* **13**, 1–45.

Brickell, J. and Knox, K. (1992). Designed experiments case history—environmental engineering operation of an activated sludge system. In *Understanding Industrial Designed Experiments*, Schmidt, S. R. and Launsby, R. G. Third Edition, 8.187–8.194. Air Academy Press, Colorado.

Bullough, R. C. and Melby, C. L. (1993). Effect of inpatient versus outpatient measurement protocol on resting metabolic rate and respiratory exchange ratio. *Annals of Nutrition and Metabolism* **37**, 24–32.

Cameron, J. M. (1951). The use of components of variance in preparing schedules for sampling of baled wool, *Biometrics* **7**, 83–96.

Clatworthy, W. H. (1973). Tables of two-associate-class partially balanced designs, *National Bureau of Standards, Applied Mathematics Series* **63**.

Cochran, W. G. and Cox, G. M. (1957). *Experimental Designs*, Second Edition, John Wiley and Sons, New York.

Corlett, E. N. and Gregory, G. (1960). The consistency of setting of a machine tool handwheel. *Applied Statistics* **9**, 92–102.

Coward, K. H. and Kassner, E. W. (1941). A comparison between interlitter and intralitter variation in rats with respect to the healing of rachitic bones by Vitamin D. *Biochemical Journal* **35**, 979–982.

Cunningham, A. and O'Connor, N. (1968). Consumer reaction to retail price and display changes. *British Journal of Marketing* **2**, 147–151.

Cuq, B., Aymard, C., Cuq, J.-L., and Guilbert, S. (1995). Edible packaging films based on fish myofibrillar proteins: formulation and functional purposes. *Journal of Food Science* **60**, 1369–1373.

Daniel, C. (1976). *Applications of Statistics to Industrial Experimentation*, John Wiley and Sons, New York.

Das, M. N. and Kulkarni, G. A. (1966). Incomplete block designs for bio-assays. *Biometrics* **22**, 706–729.

Davies, O. L., editor (1963). *Design and Analysis of Industrial Experiments*. Second Edition, Oliver and Boyd, London.

Dean, A. M. and Draper, N. R. (1998). Saturated main-effect designs for factorial experiments. *Statistics and Computing*, in press.

Desmond, D. J. (1954). Quality control on the setting of voltage regulators. *Applied Statistics* **3**, 65–73.

Draper, N. R. (1982). Center points in second-order response surface designs. *Technometrics* **24**, 127–133.

Draper, N. R. and Smith, H. (1981). *Applied Regression Analysis*, Second Edition, John Wiley and Sons, New York.

Dunnett, C. W. (1955). A multiple comparisons procedure for comparing several treatments with a control. *Journal of the American Statistical Association* **50**, 1096–1121.

Dunnett, C. W. (1980). Pairwise multiple comparisons in the unequal variance case. *Journal of the American Statistical Association* **75**, 796–800.

Durbin, J. and Watson, G. S. (1951). Testing for serial correlation in least squares regression II. *Biometrika* **38**, 159–178.

Eibl, S., Kess, U., and Pukelsheim, F. (1992). Achieving a target value for a manufacturing process: a case study. *Journal of Quality Technology* **24**, 22–26.

Ertas, A., Carper, H. J., and Blackstone, W. R., "Development of a test machine and method for galling studies," *Experimental Mechanics*, Vol. 32, No. 4, December 1992, pp. 340–347, published by the Society for Experimental Mechanics.

Feuell, H. J. and Wagg, R. E. (1949). Statistical methods in detergency investigations. Research **2**, 334–337.

Fisher, R. A. and Yates, F. (1973). *Statistical Tables for Biological, Agricultural and Medical Research*, Oliver and Boyd, Edinburgh.

Graybill, F.A. (1976). *Theory and Application of the Linear Model*. Duxbury Press.

Gutsell, J. S. (1951). The effect of sulfamerazine on the erythrocyte and hemoglobin content of trout blood, *Biometrics* **7**, 171–179.

Hare, L. B. (1988). In the soup: a case study to identify contributors to filling variability. *Journal of Quality Technology* **20**, 36–43.

Hayter, A. J. (1984). A proof of the conjecture that the Tukey-Kramer multiple comparisons procedure is conservative. *Annals of Statistics* **12**, 61–75.

Hicks, C. R. (1956). Fundamentals of Analysis of Variance, Part III—Nested designs in analysis of variance. *Industrial Quality Control* **13**, part 4, 13–16.

Hicks, C. R. (1965). The analysis of covariance. *Industrial Quality Control* **22**, 282–286.

Hicks, C. R. (1993). *Fundamental Concepts in the Design of Experiments*, Fourth Edition, Oxford University Press Inc., New York.

Hochberg, Y. and Tamhane, A. C. (1987). *Multiple Comparison Procedures*. John Wiley and Sons, New York.

Hocking, R. R. (1996). *Methods and Applications of Linear Models; regression and the analysis of variance*. John Wiley and Sons, New York.

Hoerl, A. E. (1988). The 3-7 Phenomena. *Royal Statistical Society News and Notes*, January issue.

Hollander, M. and Wolfe, D. A. (1973). *Nonparametric Statistical Methods*. John Wiley and Sons, New York.

Hsu, J. C. (1984). Ranking and selection and multiple comparisons with the best. In *Design of Experiments: Ranking and Selection (Essays in Honor of Robert E. Bechhofer)*. Editors: T. J. Santner and A. C. Tamhane). 23–33, Marcel Dekker, New York.

Hsu, J. C. (1996). *Multiple Comparisons: Theory and Methods*. Chapman and Hall, London.

Inman, J., Ledolter, J., Lenth, R. V. and Niemi, L. (1992). Two case studies involving an optical emission spectrometer. *Journal of Quality Technology* **24**, 27–36.

Jeffers, J. N. R. (1987). Acid rain and tree roots: an analysis. In *The Statistical Consultant in Action*. Editors: D. J. Hand and B. S. Everitt. Cambridge University Press, New York.

John, J. A. (1987). *Cyclic Designs*, Chapman and Hall, London.

John, J. A. and Turner, G. (1977). Some new group divisible designs. *Journal of Statistical Planning and Inference* **1**, 103–107.

John, J. A., Wolock, F. W., and David, H. A. (1972). Cyclic Designs, *National Bureau of Standards, Applied Mathematics Series*, **62**.

John, P. W. M. (1961). An application of a balanced incomplete design, *Technometrics* **3**, 51–54.

John, P. W. M. (1971). *Statistical Design and Analysis of Experiments*, Macmillan, New York.

John, P. W. M. (1980). *Incomplete Block Designs*, Lecture Notes in Statistics, Volume I, Marcel Dekker, New York.

Johnson, N. L. and Leone, F. C. (1977). *Statistics and Experimental Design in Engineering and the Physical Sciences*, Volume II, Second Edition. John Wiley and Sons, New York.

Kackar, R. N. and Shoemaker, A. C. (1986). Robust design: a cost-effective method for improving manufacturing processes. *AT&T Technical Journal* **65**, Issue 2, 39–50.

Kempthorne, O. (1976). *Design and Analysis of Experiments*, John Wiley and Sons, New York.

Kempthorne, O. (1977). Why Randomize? *Journal of Statistical Planning and Inference* 1-25.

Kershner, H. F., Ettinger, H. J., DeField, J. D., and Beckman, R. J. (1984). A Comparative Study of HEPA Filter Efficiencies When Challenged With Thermal and Air-Jet Generated Di-2-Ethylhexyl Sebecate, Di-2-Ethylhexyl Phthalate and Sodium Chloride. *Laboratory Report* LA-9985-MS, Los Alamos National Laboratory, Los Alamos, NM.

Khuri, A. I. and Cornell, J. A. (1987). *Response Surfaces: Designs and Analyses*, Marcel Dekker, New York.

Kuehl, R. O. (1994). *Statistical Principles of Research Design and Analysis*. Duxbury Press, Belmont, California.

Lewis, S. M. and Dean, A. M. (1980). Factorial Experiments in Resolvable Generalized Cyclic Designs. *Bulletin in Applied Statistics* **7**, 159–167.

Lewis, S. M., Hodgson, B. A., New, R. E., and Sexton, C. J. (1989). The application of Taguchi methods at the design analysis stage. *Proceedings of the Institute of Mechanical Engineering. International Conference on Engineering Design* **1**, 283–294.

Lorenz, R. C., Hsu, J. C., and Tuovinen, O. H. (1982). Performance variability, ranking and selection analysis of membrane filters for enumerating coliform bacteria in river water. *Journal of the American Water Works Association* **74**, 429–437.

Moore, M. A. and Epps H. H., (1992). Accelerated weathering of marine fabrics. *Journal of Testing and Evaluation* **20**, 139–143.

Munro, K. J. (1986). Investigation of optokinetic nystagmus under different visual conditions. M.Sc. dissertation, Department of Audiology, University of Southampton.

Neter, J., Kutner, M. H., Nachtsheim, C. J., and Wasserman, W. (1996). *Applied Linear Statistical Models*, Fourth ed., Richard D. Irwin, Inc., Chicago.

Peake, R.E. (1953). Planning an experiment in a cotton spinning mill. *Applied Statistics* **2**, 184–192.

Peiser, A. M. (1943). Asymptoptic formulas for significance levels of a certain distributions. *Annals of Mathematical Statistics* **14**, 56–62. (Correction 1949, *Annals of Mathematical Statistics* **20**, 128–129).

Plackett, R. L. and Burman, J. P. (1946). The design of optimum multifactorial experiments. *Biometrika* **33**, 305–325.

Poon, G. K. K. (1995) Sequential experimental study and optimisation of an acid copper pattern plating process. *Circuit World* **22**, 7–9 and 13.

Pugh, C. (1953). The evaluation of detergent performance in domestic dishwashing, *Applied Statistics* **2**, 172–179.

Ratkowski, D. A., Evans, M. A., and Alldredge, J. R. (1993). Cross-over experiments. Design, analysis and application. Marcel Dekker, New York, Statistics Textbooks and Monographs, 135.

Salvadori, M. (1980). *Why Buildings Stand Up*, Norton, New York.

SAS/STAT User's Guide (1990), Volume 1, Version 6, Fourth Edition, SAS Institute Inc., Cary, NC.

Satterthwaite, F. E. (1946). An approximate distribution of estimates of variance components, *Biometrics* **2**, 110–114.

Scheffé, H. (1959). *The Analysis of Variance*. John Wiley and Sons, New York.

Schilling, E. G. (1973). A systematic approach to the analysis of means. Part II. Analysis of contrasts. *Journal of Quality Technology* **5**, 147–155.

Schmidt, S. R. and Launsby, R. G. (1992). Experimental design on an injection molded plastic part. In *Understanding Industrial Designed Experiments*, Third Edition. Air Academy Press, Colorado.

Shahani, A. K. (1970). A saturated experiment in sequential determination of operating conditions. *The Statistician* **19**, 403–408.

Smith, H. F. (1936). The problem of comparing the results of two experiments with unequal errors, *Journal of the Council of Scientific and Industrial Research*, **9**, 211–212.

Smith, H. Jr. (1969). The analysis of data from a designed experiment. *Journal of Quality Technology* **1**, 259–263.

Sosada, M. (1993). Optimal conditions for fractionation of rapeseed lecithin with alcohols. *Journal of the American Oil Chemists' Society* **70**, 405–410.

Thompson, W. A. Jr. and Moore, J. R. (1963). Non-negative estimates of variance components. *Technometrics* **5**, 441–449.

Tuck, M. G., Lewis, S. M., and Cottrell, J. I. L. (1993). Response surface methodology and Taguchi: a quality improvement study from the milling industry. *Applied Statistics* **42**, 671–681.

Tukey, J. W. (1949). One degree of freedom for non-additivity. *Biometrics* **5**, 232–242.

Tukey, J. W. (1953). The problem of multiple comparisons. *Dittoed manuscript of 396 pages, Department of Statistics*, Princeton University.

Vance, F. P. (1962). Optimization study of lube oil treatment by process 'X'. *Proceedings of the Symposium on Application of Statistics and Computers to Fuel Lubricant Research Problems*. Office of Chief Ordinance, U.S. Army, May 1962.

Voss, D. T. and Wang, W. (1997). Exact simultaneous confidence intervals in the analysis of orthogonal saturated designs. *Wright State University Technical Report*, 1997.01.

Welch, B. L. (1938). The significance of the difference between two means when the population variances are unequal, *Biometrika* **29**, 350–362.

Willke, D. (1962). A method of analysis of mixed level factorial experiments. *Applied Statistics* **11**, 184–195.

Wood, S. R. and Hartvigsen, D. E. (1964). Statistical design and analysis of qualification test program for a small rocket engine. *Industrial Quality Control* **20**, 14–18.

Wooding, W. M. (1973). The split-plot design. *Journal of Quality Technology* **5**, 16–33.

Wu, S. M. (1964). Analysis of rail steel bar welds by two-level factorial design. *Welding Journal Research Supplement* **43**, 179s–183s. (Reprinted University of Wisconsin Engineering Experiment Station, Reprint 684.)

Yates, F. (1935). Complex experiments. *Supplement to the Journal of the Royal Statistical Society* **2**, 181–247. (Reprinted in *Experimental Design* (1970), Charles Griffin and Company, Ltd., London.

Index of Authors

Alldredge, J. R., 390
Anderson, V. L., 185
Aymard, C., 588

Bainbridge, J. R., 459
Barnabas, I. J., 573
Barnett, M. K., 442, 455
Baten, W. D., 240
Beckman, R. J., 654, 672
Behnken, D. W., 576
Bigham, R., 218
Blackstone, W. R., 237
Blom, G., 119
Box, G. E. P., 189, 547, 562, 570, 572, 573, 576
Brickell, J., 492
Bullough, R. C., 302
Burman, J. P., 512

Cameron, J. M., 595
Carper, H. J., 237
Clatworthy, W. H., 346
Cochran, W. G., 345, 381, 402, 426
Cook, R. D., 654, 672
Corlett, E. N., 531
Cornell, J. A., 570
Cottrell, J. I. L., 503, 571, 573
Coward, K. H., 383
Cox, D. R., 189
Cox, G. M., 345, 381, 402, 426
Cunningham, A., 417
Cuq, B., 588
Cuq, J.-L., 588

Daniel, C., 212
Das, M. N., 383

David, H. A., 347
Davies, O. L., 456, 467
Dean, A. M., 29, 513
Dean, J. R., 573
DeField, J. D., 672
Desmond, D. J., 662
Draper, N. R., 71, 513, 570
Dunnett, C. W., 87, 117
Durbin, J., 111

Eibl, S., 533, 552, 586
Epps, H. H., 232
Ertas, A., 237
Ettinger, H. J., 672
Evans, M. A., 390

Feuell, H. J., 238
Fisher, R. A., 345

Graybill, F. A., 74
Gregory, G., 531
Guilbert, S., 588
Gutsell, J. S., 64, 671

Hare, L. B., 487
Hartvigsen, D. E., 235
Hayter, A. J., 86
Hicks, C. R., 292, 446, 646
Hochberg, Y., 88
Hocking, R. R., 629
Hodgson, B. A., 516
Hoerl, A. E., 4
Hollander, M., 121
Hsu, J. C., 76, 81, 89, 92
Hunter, J. S., 570, 572, 573

Inman, J., 236, 670

Jeffers, J. N. R., 49
John, J. A., 346, 347
John, P. W. M., 345, 346, 355, 497
Johnson, N. L., 454, 667, 669

Kackar, R. N., 509
Kassner, E. W., 383
Kempthorne, O., 4, 437
Kershner, H. F., 672
Kess, U., 533, 552, 586
Khuri, A. I., 570
Knox, K., 492
Kuehl, R. O., 379
Kulkarni, G. A., 383
Kutner, M. H., 111, 112

Launsby, R. G., 492
Ledolter, J., 236, 670
Lenth, R. V., 236, 670
Leone, F. C., 454, 667, 669
Lewis, S. M., 29, 503, 516, 571, 573
Lorenz, R. C., 76, 81

McLean, R. A., 185
Mead, F. C. Jr., 442, 455
Melby, C. L., 302
Moore, J. R., 611
Moore, M. A., 232
Munro, K. J., 676

Nachtsheim, C. J., 111, 112, 654, 672
Neter, J., 111, 112
New, R. E., 516
Niemi, L., 236, 670

O'Connor, N., 417
Owen, S. P., 573

Peake, R. E., 14
Peiser, A. M., 81
Plackett, R. L., 512
Poon, G. K. K., 558, 579

Pugh, C., 32
Pukelsheim, F., 533, 552, 586

Ratkowsky, D. A., 390

Salvadori, M., 13
Satterthwaite, F. E., 117, 606
Scheffé, H., 83, 606
Schilling, E. G., 238
Schmidt, S. R., 492
Sexton, C. J., 516
Shahani, A. K., 494
Shoemaker, A. C., 509
Smith, H., 71
Smith, H. F., 117
Smith, H. Jr., 130
Sosada, M., 587

Tamhane, A. C., 88
Thompson, W. A. Jr., 611
Tomlinson, W. R., 573
Tuck, M. G., 503, 571, 573
Tukey, J. W., 85, 172
Tuovinen, O. H., 76, 81
Turner, G., 346

Vance, F. P., 497
Voss, D. T., 216

Wagg, R. E., 238
Wang, W., 216
Wasserman, W., 111, 112
Watson, G. S., 111
Welch, B. L., 117
Wilkie, D., 173
Wilson, K. B., 547, 562
Wolfe, D. A., 121
Wolock, F. W., 347
Wood, S. R., 235
Wooding, W. M., 691
Wu, S. M., 227

Yates, F., 345, 457, 480, 681

Index of Experiments

Absorbancy, paper towel, 293
acid copper pattern plating, 558, 563, 565, 568, 570, 579
aerosol, 672
air freshener, 417, 418
air rifle, 385
air velocity, 173, 191
alchohol, 639
algorithm, 332
alloy, metal, 341
alloy, titanium, 669, 670
ammunition, 611, 618, 622

Balloon, 62, 110, 285, 287
banana, 318
battery, 26, 44, 48, 49, 69, 71, 86, 94, 99, 100, 108, 114, 120, 143, 147, 170, 185
bean-soaking, 92, 262, 268, 275
beef, 381
bicycle, 131, 273
bicycle, exercise, 393
biscuit, 331
bleach, 151, 186
bread-baking, 299
buttermilk biscuit, 641

Cake-baking, 29
candle, 326, 641
catalyst, 130, 294
catalytic reaction, 459
cigarette, 692
clean wool, 595
coil, 446, 452

colorfastness, 315, 333
cotton-spinning, 14, 97, 325

Dairy cow, 402, 404, 409
decontamination (alpha), 442
decontamination (beta), 455, 529
dessert, 131
detergent, 355
drill advance, 212, 214, 216
drug, 691
dye, 467, 476, 479, 530

Efficiency, 647
erythrocite, 671
evaporation, 234
exam paper, 335
exercise, 336
exercise bicycle, 393

Field, 426, 454, 458
film viscosity, 588
filter, 76, 81, 84, 90
fishing line, 692
flour, 503, 531, 536
flour production, 571, 573, 589
fractionation, 587, 588

Galling, 237
gasoline, 325
golf ball, 641

Handwheel, 531
heart–lung pump, 40, 62, 72, 74, 75, 254, 276

Ice cream, 600, 603, 604, 606, 609, 640, 643
inclinometer, 516, 524, 526
injection molding, 537, 693
ink, 188
insole cushion, 334
isomer, 379

Laser printer, 534
length perception, 329
light bulb, 300, 311, 313
load-carrying, 330

Machine head, 646, 658
mangold, 437, 455, 529
margarine, 129
memory, 187, 188
metal alloy, 341
microwave popcorn, 205, 232
mung bean, 122

Nail varnish, 159, 163, 164, 167, 190

Oats, 681
operator, 670

PAH recovery, 573, 582
paint, 533, 552, 554, 555, 560, 586
paint followup, 586
paper towel absorbancy, 293
paper towel strength, 234, 239, 242
peas, 457
pedestrian light, 63, 98, 100, 127
penicillin, 456
perception, length, 329
perception, quantity, 418
plasma, 362, 382
plastic, 653
popcorn–microwave, 205, 232
popcorn–robust, 239
projectile, 454

Quantity perception, 418

Rail weld, 227
reaction time, 98, 129, 148, 150, 157, 328
refinery, 497, 540
resin impurity, 590
resin moisture, 591
respiratory exchange ratio, 325
resting metabolic rate, 325
rocket, 235

Saltwater, 327
sludge, 492
soap, 22, 53, 64, 89, 99, 129
soil, 659
soup, 487
spaghetti sauce, 133
spectrometer, 236, 538
steel bar, 240
step, 370, 381
sugar beet, 480, 530
survival, 189
systolic blood pressure, 274

Temperature, 630
titanium alloy, 669, 670
torque optimization, 218, 222
trout, 64, 99, 100, 106, 113, 117, 274

Video game, 416, 417
viscosity, 667
viscosity, film, 588
vitamin D, 383
voltage, 662

Wafer, 509
washing power, 238
water boiling, 190
weathering, 232, 233
weight lifting, 184
weld strength, 185, 186
welding, 494
wildflower, 132

Zinc plating, 292

Index of Subjects

Additive model, 139
adjusted sum of squares, 350, 406
adjustment factor, 220
aliasing scheme, 486
analysis of covariance, 278, 282, 288, 638
   adjusted means, 281
   assumption checking, 279
   confidence intervals, 286
   least squares estimators, 281
   models, 278
analysis of variance
   balanced incomplete block designs, 355, 373
   confounded factorial experiments, 424, 452, 477
   crossed treatment factors, 156, 167, 175, 203, 209
   fixed effects, 44, 47, 58
   fractional factorial experiments, 523, 526, 528
   incomplete block designs, 350, 376
   mixed effects, 628, 635
   nested factors, 650, 654, 657, 658, 662
   one-way, 44, 47, 58, 602
   polynomial regression, 256, 268
   random effects, 602, 635
   randomized complete block designs, 302, 321
   response surface methods, 554, 564, 579, 584
   row–column designs, 401, 406, 411
   split-plot designs, 678, 685, 687, 688
   two-way, 156, 167, 175
assumption checking, 104, 140
   constant error variance, 111
   independent errors, 109
   lack of fit, 247, 249, 273
   model fit, 107
   normality of errors, 119
   outliers, 107
   random-effects model, 610, 613
asymmetric factorial, 422, 501
axial points, 562, 572

Balanced incomplete block designs, 343
   adjusted plots, 357
   analysis of variance, 355, 373
   assumption checking, 348
   existence, 345
   factorial experiments, 369, 449
   least squares estimators, 353, 374
   multiple comparisons, 354, 374
   orthogonal contrasts, 355
   randomization, 344
   sample sizes, 368
bias, 3
block designs, 18, 296
block–treatment model, 348, 423
blocking factors, 3, 10, 19, 20, 296, 387, 582
Blom's normal scores, 119
Bonferroni method, 79, 80
Box–Behnken designs, 576

Canonical analysis, 585
carryover effects, 390
cell-means model, 138, 194
center points, 551

centered regressors, 259
central composite designs, 562
checklist, 7
coefficient list, 143, 200
coefficient of determination, 257
complete block designs, 295, 298
    see randomized complete block designs
complete model, 139
completely randomized designs, 18, 33, 135
    see crossed treatment factors
    see fractional factorial experiments
confidence band, 253
confidence bounds, 43, 74
confidence intervals, 73, 205, 215, 258, 271, 286, 617
confidence region, 83
confirmatory experiment, 440
confounded factorial experiments, 421, 461, 462
    analysis of variance, 424, 452, 477
    asymmetric factorial, 472
    complete confounding, 442, 450
    confounding with contrasts, 424, 463
    confounding with equations, 433, 464
    four-level factors, 471
    partial confounding, 441, 450, 464
    plans, 441, 470
    pseudofactors, 471, 472
    three-level factors, 462, 464
    two-level factors, 421
connected design, 343
connectivity graph, 342
contrasts, 37, 68
    difference of averages, 70
    interaction, 141, 199
    least squares estimator, 203
    main-effect, 199
    normalized, 68
    pairwise comparison, 69
    preplanned, 79
    simple, 142
    standard error, 68
    sum of squares, 76, 204
    three-factor interaction, 199
    treatment versus control, 70
    trend, 71, 144, 261
    two-factor interaction, 199

variance, 204
control factors, 217
control of noise variability, 217, 515
control treatment, 70
covariates, 10, 278, 296, 638
critical coefficient, 80
critical value, 585
crossed array, 473
crossed blocking factors, 19, 387
crossed treatment factors, 135, 193
    analysis of variance, 152, 156, 165, 167 175, 203, 209, 212
    assumption checking, 140, 181
    cell-means model, 138, 194
    complete model, 139
    confidence intervals, 141, 149, 205
    hypothesis tests, 204
    interaction plots, 136, 195, 209
    least squares estimators, 146, 158, 161, 162, 203
    main-effects model, 139, 158, 195
    multiple comparisons, 141, 149, 163, 177, 205
    randomization, 135
    residual plots, 181
    rules for estimation and testing, 202
    sample sizes, 168
    single replicate experiment, 169, 182, 211
    three-way complete model, 194
crossover experiments, 390
cyclic designs, 346, 393
cyclic Latin squares, 390
cyclic Youden designs, 392

Data adjusted for block effects, 357
data snooping, 79
data transformation, 113
defining contrast, 484
defining relation, 484
degrees of freedom, 202
design array, 221
design factors, 217
design-by-noise interactions, 218
difference-of-averages contrast, 70
disconnected designs, 342, 346
Dunnett method, 80, 87

Index of Subjects

bf Effect sparsity, 169, 211, 215
eigenvalues, 585
eigenvectors, 585
empty cells, 227
error assumptions, 36, 104
   constant variance, 111
   independence, 109
   normality, 119
error sum of squares, 42, 203
error variable, 35
estimable functions, 37, 202, 343
experimental plan, 340
experimental units, 9
experimentwise error rate, 78, 149

Factorial experiments, 9, 193, 317, 369,
    410, 421
   asymmetric, 422, 501
   confounding, 421, 461
   fractions, 483
   single replicate, 422
   symmetric, 422
   three-level factors, 462
   two-level factors, 421
factorial points, 551, 572
factorial structure, 369
first associates, 345, 361
Fisher's inequality, 345
fixed effects, 12, 593
fixed-effects models, 36, 260
   analysis of variance, 44, 47, 58
   assumption checking, 103
   confidence intervals, 73, 75, 94
   hypothesis tests, 75, 76, 94
   least squares estimates, 39
   multiple comparisons, 78, 80, 83, 85,
     88, 90, 96
   normal equations, 38
   randomization, 34, 57
   residual plots, 105
   sample sizes, 49, 52, 92
   see crossed treatment factors
fractional factorial experiments, 483, 487
   analysis of variance, 523, 526, 528
   asymmetric fractions, 501, 502
   blocking, 503 528
   orthogonal arrays, 506, 513, 514
   pseudofactors, 501

   saturated designs, 512
   symmetric fractions, 483, 487
   Taguchi experiments, 515, 524
   three-level factors, 496
   two-level factors, 484, 506

Gauss–Markov Theorem, 40
group divisible designs, 345
   see incomplete block designs

Hsu method, 80, 89
hypothesis testing rules, 202, 204

Incomplete block designs, 298, 339, 341
   adjusted plots, 357, 377
   analysis of variance, 350, 376
   assumption checking, 348
   balanced incomplete block designs, 343
   block sizes, 340
   block–treatment model, 348
   cyclic designs, 346
   factorial experiments, 369, 421, 461
   group divisible designs, 345
   least squares estimators, 349, 352, 361
   multiple comparisons, 349, 376
   randomization, 341
   sample sizes, 368
independent contrasts, 433
initial block, 347
interaction, 136, 138
   contrasts, 141, 199
   design-by-noise, 218
   line graph, 198
   plots, 136, 195, 209, 528
   three-factor, 195
   two-factor, 136

Lack of fit, 247, 249, 555, 564, 582
Latin square designs, 389
   analysis of variance, 401
   assumption checking, 409
   least squares estimators, 397, 402, 404
   multiple comparisons, 404
   randomization, 389
   row–column–treatment model, 395
   sample sizes, 405
least squares estimators, 39, 146, 158, 161,
    162, 248, 259, 260, 281, 305

line graph, 198
linear effect, 550
linear model, 36
local experiment, 549
loss of information, 442

Main effects, 138
main-effect contrasts, 149, 199
main-effects model, 139, 158, 195
mean square, 203
method of least squares, 38
minimum significant difference, 80
mixed arrays, 218, 515
mixed models, 12, 594, 622, 648
   analysis of variance, 628, 635
   confidence intervals, 625, 627, 628, 635
   controversy, 628
   expected mean squares, 622, 627, 635
   hypothesis tests, 622, 628, 635
   least squares estimators, 635
   multiple comparisons, 625
model building, 168
multiple comparisons, 78, 149, 163, 177, 205, 287, 312, 313
   Bonferroni method, 79, 80
   combination of methods, 91
   Dunnett method, 80, 87
   Hsu method, 80, 89
   Scheffé method, 79, 83
   Tukey method, 79, 85

Nested blocking factors, 19, 20, 675
nested factors, 645
   analysis of variance, 650, 654, 657, 658, 662
   assumption checking, 648
   confidence intervals, 650, 653, 657
   estimation and testing, 652, 654
   expected mean squares, 654, 656, 662
   fixed-effects model, 648
   least squares estimators, 649
   mixed effects, 656
   two-way nested models, 647
noise array, 221
noise factors, 10, 218, 296
normal equations, 38, 248, 260, 280
normal probability plots
   contrast estimates, 213, 224
   residuals, 119
normal scores, 119
normalized contrasts, 68
nuisance factors, 8, 296

Observational study, 11
one source of variation, 33
   see analysis of covariance
   see analysis of variance
   see fixed-effects models
   see random-effects models
orthogonal arrays, 425, 483, 506, 513, 514
orthogonal blocking, 571, 572
orthogonal central composite designs, 570
orthogonal contrasts, 169, 355, 424
orthogonal first-order designs, 551
orthogonal main-effect plans, 512
orthogonal polynomials, 261
orthogonal second-order designs, 570
outliers, 107
overall confidence level, 78
overall significance level, 78
overfit, 246

P-value, 48
pairwise comparisons, 69
parameter design, 217
partial confounding, 441, 446, 450, 464, 477
partially balanced incomplete block designs, 345
path of steepest ascent, 549, 559
pilot experiment, 11
Plackett–Burman designs, 512
polynomial regression, 243
   analysis of variance, 256, 268
   assumption checking, 247
   confidence intervals, 252, 253, 258, 271
   hypothesis testing, 271
   lack of fit, 247, 249, 273
   least squares estimators, 246, 248, 259, 260
   model, 245, 550
   orthogonal polynomials, 261
   prediction intervals, 253, 271
   quadratic regression, 245, 260
   simple linear regression, 245, 251, 258
power, 51

Index of Subjects

prediction intervals, 253, 271
preplanned contrasts, 79
product arrays, 221, 515
pseudofactors, 471, 501
pure error, 249, 555

Quadratic regression, 245, 260
quasi mean squared error, 216

Random blocking factors, 630
random effects, 12, 593, 596, 610, 654
random numbers, 5
random-effects models, 593, 596, 610, 611
   analysis of covariance, 638
   analysis of variance, 602, 628, 635
   assumption checking, 610, 613, 632
   confidence intervals, 603, 604, 606, 617, 628
   expected mean squares, 616, 627, 635
   hypothesis tests, 620, 628, 635
   intermediate random-effects model, 613
   least squares estimates, 39
   normal equations, 38
   randomization, 34, 57
   sample sizes, 607, 622
   variance components, 597, 599, 613, 616
randomization, 3, 34, 57, 299, 341, 344
randomized complete block designs, 18, 295, 298, 299, 676
   analysis of variance, 302, 310, 321
   assessment of blocking, 302
   assumption checking, 316
   block-treatment models, 301, 309, 317
   factorial experiments, 317
   least squares estimators, 305
   multiple comparisons, 305, 312, 313
   randomization, 299
   sample sizes, 301, 315
reduced model, 45
replication, 2
residual effects, 390
residual plots, 104, 105, 247
   see assumption checking
residuals, 42, 104
resolution, 487
response surface methods, 244, 547
   analysis of variance, 554, 564, 579, 584

   analysis with blocking factors, 582
   assumption checking, 553
   Box–Behnken designs, 576
   canonical analysis, 585
   central composite designs, 562
   first-order designs, 551
   first-order model, 550
   lack of fit, 564, 582
   orthogonal blocking, 571, 572
   orthogonal designs, 570
   path of steepest ascent, 559
   rotatable designs, 569, 570
   second-order designs, 561
   second-order model, 557, 582
robust design, 10, 217, 218, 226
   randomization, 221
rotatable central composite designs, 570
rotatable second-order designs, 569
row–column designs, 19, 388
   adjusted plots, 415
   analysis of variance, 411
   factorial experiments, 410
   least squares estimators, 415
   row–column–treatment model, 395
   see Latin square designs
   see Youden designs
rules for estimation and testing, 202, 628, 653, 657, 658
run order, 288

Saddle point, 562
sample correlation, 257
sample sizes, 49, 92, 168, 301, 315, 368
SAS, 5
   analysis of covariance, 288
   analysis of variance, 58
   assumption checking, 122
   confidence intervals, 94
   data input, 57
   hypothesis tests, 94
   means, 60, 125
   multiple comparisons, 96
   nested effects, 662
   plots, 59, 122
   random effects, 635
   regression, 268
   transforming data, 126
Satterthwaite's approximation, 117, 151, 165

saturated designs, 512
Scheffé method, 79, 83
screening experiments, 485
second associates, 345, 361
second-order designs, 561, 569, 570
separability of factorial effects, 197
sequential sums of squares, 176, 270
several crossed treatment factors
   see crossed treatment factors
significance level, 47
simple contrasts, 142
simple linear regression, 245, 258
simple pairwise differences, 142
simultaneous confidence intervals, 78
   see multiple comparisons
simultaneous hypothesis tests, 78
single replicate experiments, 169, 182, 211, 213, 215, 422
split-plot designs, 21, 675
   analysis of variance, 678, 687, 688
   confounding, 686
   expected mean squares, 687
   least squares estimators, 681
   models, 677, 678
   multiple comparisons, 681, 688, 690
   randomization, 675
   split-plot analysis, 678
   whole-plot analysis, 680
split-split-plot designs, 684
standard error, 68
star points, 562
stationary point, 562
Studentized range distribution, 85
sum of squares, 42, 46, 47, 76, 250, 350, 406, 555
   Type I, 175, 270
   Type III, 175
symmetric factorial experiments, 422

T-distribution approximation, 81
Taguchi, 218, 515
three-way complete model, 194

total sum of squares, 47, 202
transformation, 113
treatment combinations, 9
treatment contrasts, 37, 68, 286, 424
treatment factors, 8
treatment sum of squares, 46
treatment-versus-control contrast, 70
treatments adjusted for blocks, 350, 452
trend contrasts, 71, 144, 261
Tukey method, 79, 85
Tukey's test for additivity, 172
two crossed treatment factors
   see crossed treatment factors
two-factor interaction, 136, 199
two-way analysis of variance, 156
   see crossed treatment factors, 156
Type I sums of squares, 175, 270
Type III sums of squares, 175

Unequal variances, 151, 165

Variance components, 597
   see random effects models
Voss–Wang method, 216, 224

Washout periods, 390
whole plots, 675
whole-plot analysis, 680
words, 487

Youden designs, 392
   analysis of variance, 401, 406
   assumption checking, 409
   confidence intervals, 407
   least squares estimators, 398, 406
   multiple comparisons, 407
   randomization, 392
   replication, 392
   row–column–treatment model, 395
   sample sizes, 407
Youden square, 391

Springer Texts in Statistics *(continued from page ii)*

Madansky: Prescriptions for Working Statisticians
McPherson: Applying and Interpreting Statistics: A Comprehensive Guide, Second Edition
Mueller: Basic Principles of Structural Equation Modeling: An Introduction to LISREL and EQS
Nguyen and Rogers: Fundamentals of Mathematical Statistics: Volume I: Probability for Statistics
Nguyen and Rogers: Fundamentals of Mathematical Statistics: Volume II: Statistical Inference
Noether: Introduction to Statistics: The Nonparametric Way
Nolan and Speed: Stat Labs: Mathematical Statistics Through Applications
Peters: Counting for Something: Statistical Principles and Personalities
Pfeiffer: Probability for Applications
Pitman: Probability
Rawlings, Pantula and Dickey: Applied Regression Analysis
Robert: The Bayesian Choice: From Decision-Theoretic Foundations to Computational Implementation, Second Edition
Robert and Casella: Monte Carlo Statistical Methods
Rose and Smith: Mathematical Statistics with *Mathematica*
Santner and Duffy: The Statistical Analysis of Discrete Data
Saville and Wood: Statistical Methods: The Geometric Approach
Sen and Srivastava: Regression Analysis: Theory, Methods, and Applications
Shao: Mathematical Statistics, Second Edition
Shorack: Probability for Statisticians
Shumway and Stoffer: Time Series Analysis and Its Applications
Simonoff: Analyzing Categorical Data
Terrell: Mathematical Statistics: A Unified Introduction
Timm: Applied Multivariate Analysis
Toutenburg: Statistical Analysis of Designed Experiments, Second Edition
Whittle: Probability via Expectation, Fourth Edition
Zacks: Introduction to Reliability Analysis: Probability Models and Statistical Methods

Made in the USA
San Bernardino, CA
15 August 2016